Phytophthora

Phytophthora

Symposium of the British Mycological Society, the British Society for Plant Pathology and the Society of Irish Plant Pathologists held at Trinity College, Dublin September 1989

Edited by

J. A. Lucas, R. C. Shattock, D. S. Shaw & Louise R. Cooke

Published for the British Mycological Society by
CAMBRIDGE UNIVERSITY PRESS
Cambridge
New York Port Chester Melbourne Sydney

CAMBRIDGE UNIVERSITY PRESS
Cambridge, New York, Melbourne, Madrid, Cape Town,
Singapore, São Paulo, Delhi, Tokyo, Mexico City

Cambridge University Press
The Edinburgh Building, Cambridge CB2 8RU, UK

Published in the United States of America by Cambridge University Press, New York

www.cambridge.org
Information on this title: www.cambridge.org/9780521189767

First published 1991
First paperback edition 2011

A catalogue record for this publication is available from the British Library

ISBN 978-0-521-40080-0 Hardback
ISBN 978-0-521-18976-7 Paperback

Additional resources for this publication at www.cambridge.org/9780521189767

Contents

Contributors

S. A. Archer *Department of Biology, Imperial College of Science, Technology and Medicine, London SW7 2BB, UK*

G. Bahnweg *Abt. für Molekulare Genetik der Gesellsschaft für Strahlen- und Umweltforschung mbH München in Göttingen, Grisebachstrasse 6, D-3400 Göttingen, Germany*

A. Bourke *143 Ballymun Road, Glasnevin, Dublin 9, Ireland*

C. M. Brasier *Forest Research Station, Alice Holt Lodge, Farnham, Surrey GU10 4LH UK*

K. W. Buck, *Department of Biology, Imperial College of Science, Technology and Medicine, London SW7 2BB, UK*

Stefan T. Buczacki *'Dodds', Clifford Chambers, Stratford upon Avon, Warwickshire CV37 8HX UK*

D. A. Carter *Department of Biology, Imperial College of Science, Technology and Medicine, London SW7 2BB, UK*

M. D. Coffey *Department of Plant Pathology, University of California, Riverside, California 92521-0122 USA*

Louise R. Cooke *Plant Pathology Research Division, Department of Agriculture for Northern Ireland, Newforge Lane, Belfast BT9 5PX UK*

L. C. Davidse *Department of Phytopathology, Wageningen Agricultural University, POB 8025/6700 EE Wageningen, The Netherlands [Present address: Royal Sluis, Koninklijke Zaaizaadbedrijven Gebroeders Sluis B.V., Postbox 22, 1600 AA Enkhuizen, The Netherlands]*

M. A. Doster *Department of Plant Pathology, Cornell University, 334 Plant Science Building, Ithaca, New York 14853-5908 USA*

L. J. Dowley *The Agricultural Institute, Oak Park Research Centre, Carlow, Ireland*

J. M. Duncan *Scottish Crop Research Institute, Invergowrie, Dundee DD2 5DA UK*

J. Duniec *Plant Cell Biology Group, Research School of Biological Sciences, Australian National University, GPO Box 475, Canberra ACT 2601 Australia*

A. R. Finlay *Department of Mycology and Plant Pathology, Faculty of Agriculture and Food Science, The Queen's University of Belfast, Newforge Lane, Belfast BT9 5PX UK*

H. Förster *Department of Plant Pathology, University of California, Riverside, California 92521-0122 USA*

J. Friend *Department of Plant Biology and Genetics, University of Hull, Hull HU6 7RX UK*

W. E. Fry *Department of Plant Pathology, Cornell University, 334 Plant Science Building, Ithaca, New York 14853-5908 USA*

U. Gisi *Sandoz AG, Agrobiologie, IHR Zeichen, CH-4108 Witterswil Switzerland*

S. B. Goodwin *Department of Plant Pathology, Cornell University, 334 Plant Science Building, Ithaca, New York 14853-5908 USA*

F. Gubler *Plant Cell Biology Group, Research School of Biological Sciences, Australian National University, GPO Box 475, Canberra ACT 2601 Australia*

E. M. Hansen *Department of Botany and Plant Pathology, Oregon State University, Corvallis, Oregon 97331-2902 USA*

A. R. Hardham *Plant Cell Biology Group, Research School of Biological Sciences, Australian National University, GPO Box 475, Canberra ACT 2601 Australia*

H. R. Hohl *Institut für Pflanzenbiologie, Zollikerstrasse 107, CH-8008 Zürich Switzerland*

A. B. K. Jespers *Department of Phytopathology, Wageningen Agricultural University, POB 8025/6700 EE Wageningen The Netherlands*

P. Karlovsky *Abt. für Molekulare Genetik der Gesellsschaft für Strahlen- und Unweltforschung mbH München in Göttingen, Grisebachstrasse 6, D-3400 Göttingen, Germany*

H. W. Kehoe *The Agricultural Institute, Oak Park Research Centre, Carlow, Ireland*

D. M. Kennedy *Scottish Crop Research Institute, Invergowrie, Dundee DD2 5DA UK*

J. R. Kinghorn *Plant Molecular Genetics Unit, University of St. Andrews, Sir Harold Mitchell Building, Greenside Place, St. Andrews, Fife KY16 9AL UK*

W. Knogge *Max Planck Institut für Züchtungsforschung, Abteilung Biochemie, Carl-von-Linné-Weg 10, D-5000 Koln 30, Germany*

D. N. Kuhn *Department of Biological Sciences, Florida International University, University Park, Miami, Florida 33199 USA*

R. G. Linderman *USDA Agricultural Research Service, Horticultural Crops Research Laboratory, 3420 N.W. Orchard Avenue, Corvallis, Oregon 97330 USA*

J. A. Lucas *Department of Botany, School of Biological Sciences, University of Nottingham, University Park, Nottingham NG7 2RD UK*

B. C. Mantel *Department of Phytopathology, Wageningen Agricultural University, POB 8025/6700 EE Wageningen, The Netherlands*

A. R. McCracken *Department of Mycology and Plant Pathology, Faculty of Agriculture and Food Science, Queen's University of Belfast, Newforge Lane, Belfast BT9 5PX UK*

R. P. Moon *Plant Molecular Genetics Unit, University of St. Andrews, Sir Harold Mitchell Building, Greenside Place, St. Andrews, Fife KY16 9AL UK*

J. S. Niederhauser *2474 N. Camino Valle Verde, Tucson, Arizona 85715 USA*

E. O'Sullivan *The Agricultural Institute, Oak Park Research Centre, Carlow, Ireland*

P. Oudemans *Department of Plant Pathology, University of California, Riverside, California 92521-0122 USA*

J. E. Parker *Max Planck Institut für Züchtungsforschung, Abteilung Biochemie, Carl-von-Linné-Weg 10, D-5000 Koln 30, Germany*

H. H. Prell *Abt. für Molekulare Genetik der Gesellsschaft für Strahlen- und Umweltforschung mbH München in Göttingen, Grisebachstrasse 6, D-3400 Göttingen, Germany*

O. K. Ribeiro *Microbiotica International Inc., 10744 NE Manitou Beach Drive, Bainbridge Island, Washington 98110 USA*

D. Scheel *Max Planck Institut für Züchtungsforschung, Abteilung Biochemie, Carl-von-Linné-Weg 10, D-5000 Koln 30, Germany*

P. H. Scott *Scottish Crop Research Institute, Invergowrie, Dundee DD2 5DA UK*

R. C. Shattock *School of Biological Sciences, University College of North Wales, Bangor, Gwynedd LL57 2UW UK*

D. S. Shaw *School of Biological Sciences, University College of North Wales, Bangor, Gwynedd LL57 2UW UK*

L. J. Spielman *Department of Plant Pathology, Cornell University, 334 Plant Science Building, Ithaca, New York 14853 USA*

C. D. Therrien *Departments of Plant Pathology, Pennsylvania State University, University Park, Pennsylvania 16802 USA*

P. W. Tooley *USDA, ARS, Fort Detrick, Building 1301, Frederick, Maryland 21701 USA*

S. E. Unkles *Plant Molecular Genetics Unit, University of St. Andrews, Sir Harold Mitchell Building, Greenside Place, St. Andrews, Fife KY16 9AL UK*

G. C. M. van den Berg-Velthuis *Department of Phytopathology, Wageningen Agricultural University, POB 8025/6700 EE Wageningen, The Netherlands*

Preface

The fungal genus *Phytophthora* achieved notoriety due primarily to an historical event. In 1845-6 the late blight pathogen devastated potato crops throughout northwestern Europe. The blight epidemic, and the ensuing tragic famine in Ireland, stirred a debate about the nature of plant disease which, for the first time, put plant pathology on the scientific agenda. One crucial figure in the debate was the mycologist M. J. Berkeley who argued, correctly, that the causal agent was a fungus, now known as *Phytophthora infestans*, which spread from plant to plant. In subsequent years many other destructive plant diseases were found to be due to related *Phytophthora* species. Despite the pioneering work of Berkeley, and a century of research, this group of fungi still causes major losses in crops as diverse as potato, soybean, alfalfa, peppers, citrus and avocado, as well as in natural plant communities.

1989 was the centenary of Berkeley's death and to mark the occasion The British Mycological Society, together with the British Society for Plant Pathology and the Society of Irish Plant Pathologists, held an international symposium on the theme *Phytophthora* at Trinity College Dublin; the venue reflected the unique significance of one of these pathogens in Irish history. More than 200 delegates from 22 different countries attended the meeting, a testimony to the continuing worldwide importance of diseases caused by *Phytophthora* species.

While the original impetus for the symposium was commemorative, the timing was appropriate for other reasons. Like many areas of experimental biology, the study of plant pathogenic fungi has been recently revolutionised by the advent of powerful techniques for analyzing molecular events and probing genome structure. The application of such new technology to questions of taxonomy, pathogenic variation, and mechanisms of host-pathogen interaction, has opened up fresh perspectives on old problems. At the same time there has been accelerating progress in the genetics of *Phytophthora* and other Oomycete fungi, long considered an intractable area.

The symposium programme was designed to reflect the diversity of research on *Phytophthora* species and diseases, but with a strong emphasis on molecular and genetic approaches. Following a biographical paper on Berkeley and accounts of the origin and evolution of the potato late blight problem, the main sessions of the meeting concerned progress in host pathogen interaction, mechanisms of variation and speciation, classical and molecular genetics, and current strategies for the control of diseases caused by airborne and soilborne *Phytophthora* species. This book arises from the programme, and follows the same broad plan, but is not simply the conference proceedings. Contributors have been encouraged not only to review progress in particular areas, but also to include the latest research results and to look ahead to future possibilities. Some speculation was invited, especially in rapidly advancing areas such as DNA polymorphisms, parasexuality, and the development of transformation systems for *Phytophthora*. The concluding chapters on control cover progress in disease forecasting, fungicides, cultivar resistance, and the developing interest in biological and integrated approaches to disease management. The coverage is inevitably not comprehensive, but taken together the contributions provide a wide-ranging and up to date account of current areas of research activity. Some unresolved points remain, particularly in the nomenclature of certain taxa, as new information arising from molecular studies has yet to be incorporated into a formal taxonomy. For example, *P. nicotianae*, formerly divided into varieties *nicotianae* and *parasitica*, may be referred to as such or simply as *P. parasitica*. Current studies (e.g. Chapters 11 and 12) suggest that no subdivision of this species is justified, and thus *P. nicotianae* is likely to become the accepted species name. Further progress in molecular systematics should shortly resolve such questions. We hope that the contents communicate some of the enthusiasm, sense of novelty, and occasional controversy provoked by the symposium itself, and point the way towards the next decade of research on *Phytophthora*.

Many people have helped, both in organizing and running the original meeting, and in the subsequent production of the book. Una Lee and Mike Cooke were involved in initial programme plans. The heavy burden of local organization was borne by Kevin Clancy and Paul Dowding. Trinity College Dublin deserves our thanks for providing a memorable venue, and the City and Mayor of Dublin for a lively Civic Reception. Several organizations and companies generously provided sponsorship towards the costs of supporting speakers, including Bayer UK Ltd., Du Pont (UK) Ltd., ICI Agrochemicals, Schering Agrochemicals, and Irish Potato Marketing Ltd. Great energy, enthusiasm and technical expertise have been contributed by David Moore, the production editor. Thanks are also due to Alan Crowden at Cambridge University Press, and Peter Ayres, the BMS Symposium Series Editor, who was instrumental in launching the original concept for a Berkeley Centenary Meeting based on *Phytophthora*.

J. A. Lucas, R. C. Shattock, D. S. Shaw & Louise R. Cooke

The Reverend Miles Joseph Berkeley (1803-1889)

Chapter 1

The Reverend Miles Berkeley

Stefan T. Buczacki

This symposium is a joint celebration: of a man and a fungus. Understandably, most of the deliberations and discussions are accorded to the fungus but present day mycologists and plant pathologists sometimes too swiftly neglect how the foundations of the modern sciences were laid. It is all too easy to forget the difficulties that faced the founding practitioners of our disciplines. Thus, it is wholly appropriate that the proceedings should be prefaced with an account that will serve to remind us of the magnitude of the debt that we owe to those who went before.

Whilst the main theme of the symposium is *Phytophthora*, its timing is the centenary of a death, and an opportunity therefore to celebrate the life of one of the most remarkable figures among many such in a remarkable century. The eighty six years of the Reverend Miles Joseph Berkeley's life all but matched the span of the nineteenth century, itself a time within the memory of the grandparents of some of us, yet an age away in many of its attitudes and characteristics. It is Berkeley and his time that I shall examine; and I shall do so with reference to his published outpourings on botanical science but especially by reference to his unpublished correspondence. I believe that I have read every surviving letter written by or to Berkeley, a total of many thousands (Table 1.1), and they paint a graphic and intriguing picture of him and the scientific and cultural world within which he lived.

I must first declare my personal interest. I first became aware of Berkeley (or at least of those celebrated initials MJB) at a very early age as my developing interest in mushrooms and toadstools revealed how frequently Berkeley's name appeared after those of very many of the fungi that were collected. Later, I was for many years a professional plant pathologist and Berkeley's name appeared consistently near to the top of almost every plant disease bibliography.

Let us therefore cast back our thoughts to the early years of the nineteenth century and to rural Northamptonshire, a time when homes were without electricity, when water came from the well, and when transport was by horse, trap or stagecoach. Communication and correspondence were slow and the community, be it in a rural village or a town

Table 1.1 Selected sources of information: a brief summary intended to provide starting points for further information without being complete or definitive

Location	Material
Academy of Natural Sciences, Philadelphia	E. A. Rau Collection
Bath Reference Library	Letters of Naturalists to Leonard Jenyns
Clemson University	H. W. Ravenal Collection
Linnean Society, London	Miscellaneous manuscripts
Lund University	Agardh correspondence
Musée de Science Naturelle, Paris	Miscellaneous correspondence
Natural History Museum London	Berkeley correspondence
	Broome correspondence
Prague Museum	Corda correspondence
Royal Botanic Gardens, Kew	J. G. Baker letters
	Bentham correspondence
	M. C. Cooke letters
	J. G. Duthie letters
	English letters
	East African letters
	J. S. Henslow letters
	J. D. Hooker letters
	Indian letters
	Lindley letters
	W. Mitten letters
	Thistleton Dyer letters
Royal Society London	Sowerby letters
	Miscellaneous correspondence
University of Geneva	Duby and De Candolle correspondence
University of North Carolina, Chapel Hill	Botany Department Papers
	M. A. Curtis Collection
University of Oxford, Department of Botany	Druce correspondence
	MSS Sherard 411
University of Uppsala	Fries correspondence

In this table I have listed the most important primary sources of information on M. J. Berkeley. Background and general biographical information is readily obtainable from the standard scientific biographies. Much family information is contained in the Parish Records of Oundle, Benefield and Sibbertoft, Northamptonshire and Margate, Kent. Other useful sources of local information are Rugby School records and the archives of the *Stamford Mercury*.

street, was highly self-contained. The butcher, the baker and the all-important candlestick maker were nearby.

The Berkeley family is an ancient one; the famous Irish philosopher and Anglican Bishop Berkeley was probably a member of one branch and thus perhaps its best known son. Their most celebrated seat was at Berkeley Castle in Gloucestershire but in 1650 one of the family, Maurice Berkeley, moved to Northamptonshire where he settled, apparently as land agent to one of the local nobility. The family was, in modern terms, middle class and in keeping with the middle class traditions of the time, they were good breeders, but short on inspiration for names. Charles, Miles, Maurice, Rowland and Joseph among the men, and Charlotte, Cecilia, Mary and Elizabeth among the women crop up repeatedly in archives, so it is sometimes extremely difficult to trace particular individuals. Although I have been unable to piece together an entire family tree, I am fairly certain that in one generation, two brothers actually had the same christian name. This was not unknown in the eighteenth and nineteenth centuries, but normally occurred when one child had died in infancy and the name was resurrected for a later sibling. In the Berkeley instance both seem to have survived and it is tempting to suggest that the occurrence arose because the family was so large that the parents forgot one of their existing children.

Our Maurice Berkeley and his wife Susanna had seven children. Their eldest son, also called Maurice, and his wife Mary produced, rather sparingly, only four offspring. Their third son was named Miles and when he married Ann, they set about restoring the family tradition of fecundity and spawned six. Their eldest son Charles was born in 1752 and became a land agent, as some of his family seem to have done before him. He and his wife Charlotte Elizabeth did particularly well and produced eight children. For our purposes the most important event noted amongst the archives occurred on 4 April 1803 when the Parish registers for Oundle record the baptism (at the age of three days) of Miles Joseph, their second son and fourth child. Baptisms were normal at such a tender age because of the high rate of infant mortality in the nineteenth century.

From his mother's side young MJB (as I shall now refer to him) inherited much artistic talent. Charlotte was the sister of Paul Sandby Munn, the distinguished water colour painter of the Norwich school and intimate friend of John Sell Cotman. Their father was described as 'a landscape painter and carriage decorator' and the very evident abilities of MJB and his own offspring clearly spring from this pedigree. The Munns lived in Greenwich and this locality provided MJB in later years with a valuable London base.

MJB's own birthplace in Northamptonshire is not known with precision. It is listed on his tombstone and in biographies as Biggin Hall but this is merely the name of the house for which his father worked as land agent and it is my belief that he was actually born in a smaller property on the estate, still known as Berkeley House.

So, young Miles grew up in a middle class home, surrounded by his siblings (three brothers and four sisters) and with a handful of servants – probably a cook, nurse and butler. He attended the local grammar school in Oundle and then, in midsummer 1817, at the age of 14, went up to Rugby school. Unfortunately, no records survive of MJB's schooldays although letters written to his mother clearly existed immediately after his death for they are referred to in family correspondence. MJB made few lasting friendships at Rugby, although among his contemporaries were Francis Wedgwood, youngest son of Josiah Wedgwood of Etruria, and also a young man named Andrew Bloxham, who came from a family with long Rugby associations. Later, Bloxham became a cleric in Berkeley's neighbouring county of Leicestershire, and developed an amateur interest in mycology; some years ago I acquired Bloxham's copy of MJB's *Outlines of British Fungology*. It is also fascinating to realise that MJB was at Rugby with one, William Webb Ellis who, a few years later and 'with a fine disregard for the rules of football as played at that time', picked up a ball and ran with it, so inventing the game of rugby football.

From Rugby, Miles went on to Christ's College, Cambridge as a scholar. I am sure that his interest in natural history had been well stirred during his childhood (he later hankered for 'that wooded county of Northamptonshire') but of course, it was impossible then to study botany as an undergraduate. Thomas Martyn held the botanical chair but had not delivered a lecture since 1796 and would not allow anyone else to do so. So Miles studied for the mathematics tripos and graduated in 1825 as fifth senior optime (in other words, fifth in the second class, below the wranglers) more or less an Upper Second Class Honours degree in modern parlance.

Although there was no family precedent for the motivation, MJB decided to study for holy orders and was ordained in 1825. He obtained a local curacy in the parish of Thornhaugh before going to Margate as curate in 1829, clearly building on his mother's family's connection with Kent. Indeed, in later years, several of the family including his widowed mother, returned to the Greenwich-Blackheath area of North Kent. In Margate, there are still several fine trees in and around the church gardens that seem to date from MJB's time there, and could perhaps have been planted at his instigation. There is also, in the garden of the old curate's house, a small summerhouse known as Berkeley's Box.

At this time, MJB's natural history interests seem to have been in algology and to some extent in conchology; his first scientific paper, published in 1828, was on molluscs. He had made visits, whilst a student, to stay in the West of Scotland during the summers with his Cambridge friend The Reverend Richard Lowe, a man who remained close to MJB until his untimely death whilst collecting marine plants in Madeira in 1874. In Scotland they met and stayed with Captain Dugald Carmichael, a Scottish soldier, traveller and noted naturalist, who was undoubtedly a great influence on MJB. In the first recorded reference to MJB as a naturalist Carmichael wrote to Sir William Hooker: 'I had two Cambridge students with me some time ago who passed the last summer at Oban. A Mr Berkeley and a Mr Lowe. They are keen botanists and conchologists'.

In 1828, after Carmichael's death, MJB himself wrote to Hooker and so began a correspondence that continued with Hooker, father and then son, throughout his life. By 1831/2, the interest in agaric fungi had begun and for once in his life, MJB was pushing. Although it has often been said that Hooker asked him to help with the fungi in his forthcoming *Flora*, it is sometimes necessary to ask to be asked. In 1832, therefore, he wrote to Hooker 'It would give me a great pleasure if I could be of service to you in taking a share of the labour of the British Flora off your hands'. Hooker agreed with some relief (mycologists were somewhat thin on the ground in the eighteen thirties) but MJB had clearly bitten off rather more than he could chew. A few months later, on his return to Northamptonshire, he was in trouble and wrote again to Sir William 'I am very worried about *Clavaria* of which I am profoundly ignorant. Like *Rosa* and *Rubus*, it has many intermediate species'. He would have given more time to the project but had been appointed Perpetual Curate of the twin Northamptonshire parishes of Apethorpe with Woodnewton, with a residence nearby at King's Cliffe and 'must rise early'.

Before Miles left Kent, he achieved one more success – his marriage on 28 January 1833 to Cecilia Emma Campbell of Blackheath, a wife who was to become his devoted companion and to live up admirably to the great Berkeley tradition of procreation. She bore him Emeric, Richard, Cecilia, Miles, Charlotte, Grace, Ruth, Margaret, Isabella, Paul, Charles, Julian, Anna, Rose, Mary and Frances; sixteen children, of whom thirteen survived to adulthood, a remarkable achievement for the time.

On his return to Northamptonshire, the interest in algae remained – his *Gleanings of British Algae* was published in 1833 – but fungi were taking over, and I am sure that the English botanical fraternity was delighted to find someone who would take on this burden. MJB took such opportunity as he could to correspond with and meet other mycologists; he met Agardh, for instance, at Cambridge in 1833 and visited France in 1843 after which he was very uncharitable about the pettiness of the French scientific community. But travel wasn't particularly easy; there was a coach from London to King's Cliffe every Tuesday, Thursday and Saturday – *The Oundle* – from the George and Blue Boar in Holborn.

It is easy, nonetheless, in tracing MJB's scientific advancement to forget that he was by profession a cleric with all of the duties that this entailed; and that family life was no easy matter. Frequently the correspondence is edged in black as the Berkeleys mourned a close relative or friend. Indeed, in 1841, the entire family escaped narrowly from a fire in the house and MJB was laid up with burns after rescuing the nurse. Nor was MJB a man of means. Writing paper was costly and there were periods when the letters were written on both sides of extremely thin paper, so rendering them exceptionally difficult to read. When times were really hard, the page was turned through 90° and written over again. MJB actually ran a small school for a time to help supplement his income and charged the pupils £50 per year, excluding books.

After visiting a sick parishioner, MJB wrote to Hooker 'I contracted smallpox and thought I would die', so 'had made arrangements for my wife to return your books and specimens'. But in the event he recovered and whilst convalescing, sat in bed with his herbarium, checking the specimens which by then were flooding in from far and wide. To help his recovery he went to stay with friends in Margate for three weeks to convalesce. He wrote to Hooker that this would enable him to complete agarics and boleti but *Thelephora* remained to do and part of *Polyporus* but 'I am still worried about *Clavaria*'.

Then came the first child and for once a mystifying departure from the old family christian names with the strangely named Emeric after whom *Emericella* and several other fungi were subsequently named. Emeric later became a Major General in the Indian Army and was clearly a man of talent. After only a few months, he became so fluent in Hindustani as to be an official army interpreter and in his spare time became an authority on orchids.

In 1834, MJB began to study the 'microscopic' structure of fungi, working, of course, with doublets (hand lens) but he obtained a micrometer to measure spores and finer structure. Sir William Hooker was still Professor in Glasgow at this time – he went to Kew as Director in 1841 – and this made communication harder. The cost of sending specimens was proving crippling; parcels of fungi cost more than 'two portmanteaus which were sent from Edinburgh by Highflier coach for only £1-0-8', and, of course, if the specimens had been incorrectly preserved, they would rot *en route*, something that became a major problem when specimens were sent to him from the tropics. Then finally, the work on *British Fungi* was complete, but on seeing the first copy, MJB was mortified for on the title page he was described as FLS, which he wasn't. He immediately applied to the Linnean Society in order to put the matter right.

In 1837, as so often throughout his life, MJB was slow and too cautious at coming forward and not for the first time missed the boat of mycological immortality. He wrote 'I have made some curious discoveries with respect to the pileate and clavate fungi...the sporidia are not contained within asci but are naked seated on the top of little papillae'. In passing, of course, MJB had discovered the structure of basidia although sadly, and unknown to him, Ascherson and Leveille had published the same findings a few months earlier. MJB had in fact been very close to this discovery in 1836 but poor microscopes and the English climate didn't help. One of the most telling remarks about the gulf that exists between his time and ours is contained in a letter written later, for on 3 December 1844, he wrote 'The weather is dreadfully dark for the microscope'. Working of necessity early in the morning by candlelight before parish and other duties, made mycology a very seasonal pursuit. And the volume of work to be done increased constantly as specimens were now coming from Australia, Van Diemen's Land (Tasmania) and from many other parts of the world too. Every botanical collector from Britain, wherever he journeyed, sent back his fungal material to Berkeley. His reputation grew by the month and there is a fine story of MJB later sitting down to dinner with Sir Joseph Hooker and Darwin (what a dinner party that must have been) where a dish of pickles was served – MJB was the only one present able to identify all of the ingredients.

Yet still there was no sign of interest in microfungi – understandably because microfungi were so little understood and because MJB still had no compound microscope. However, by 1840 he had begun to make some preliminary observations on *Cladosporium*, *Helminthosporium* and *Alternaria* – honey dew moulds growing on orange trees in conservatories. Study in other directions continued nonetheless. In 1842, MJB reported the connection between *Boletus* and some subterranean fungi – in this case a stipitate *Hymenangium* as he described his finding. And about the same time as this interest in things underground developed he received a letter on the subject from a Mr Christopher Broome of Bath. So began a long and close association, especially on the subject of subterranean fungi, and the beginning of the most famous joint attribution in all of mycology – Berkeley and Broome. Their correspondence is tedious *in extremis* and I must vote Broome one of the most boring men in the history of mycology. Never a quip is made in all those thousands of letters. Whilst Berkeley didn't seem to have much of a sense of humour, at least he permitted himself a few acid comments such as his criticism of the state of Mordecai Cooke's fingernails which 'disgusted everyone at Glasgow'.

But, as always, mycology was to the fore and MJB's priorities always centred on things fungal, no matter what else the distraction. Writing to Broome in 1848, he said 'I have not written to you for some days for we have been in great trouble on account of the death of our infant of inflammation of the lungs. All was done that skill could do but it died in a few hours. I send you another *Hydnangium*...'

In 1844, MJB moved house within King's Cliffe and transferred to the building that was later to become the village school. It was to be his home for the next 35 years and when I visited it in 1976, there was still, in an attic room, some Victorian wallpaper, probably put there by MJB. It was to this house that the foreign mycological dignitaries of the world came to pay their respects when visiting Britain. In 1853 'M. de Candolle arrived with his son last night and left this morning'. Communications were improving too, so making it easier for visitors to reach remote Northamptonshire; on 3 June 1845, MJB could write 'our railroad is now open to Oundle and we are only 5 hours distant from London'. It became easier for the professional colleagues to exchange gifts as well as pleasantries; a note to Hooker read 'I am sending you a pamphlet of Montagne's also a couple of geese'.

Around this time, MJB was in active correspondence with several of those who were to become players in the great *Phytophthora* controversy that was about to burst onto the scientific stage. Notwithstanding his generally poor view of the French scientific establishment, he wrote very favourably of Montagne, with whom he had stayed in Paris; and also in kindly and glowing terms of Decaisne, despite being on the opposite side to him in the ensuing potato blight debate. He was already in correspondence with Lindley in February 1844 regarding what was obviously a smut fungus and so the scene was set; the personalities were familiar to each other and reputations were ready to be made or broken.

I do not intend to enter into the detail of the debate itself for that is admirably covered elsewhere in this volume (Bourke, chapter 2), but it is inescapable that had Berkeley put his case more forcefully and come down less equivocally for a fungal cause for potato blight, the matter might have been resolved sooner and Berkeley's standing in the matter might have been much higher still. As it was, he discussed the affair very little in his private correspondence and even to Lindley himself, was cautious. Writing to him on 7 September 1845, he said 'I have little doubt that half the debates on the subject arise from parties having different diseases in mind'. But he confided in Broome that he was certain that he had grasped the situation and on 15 September, he wrote to him: 'The mould is decidedly the same with what causes the potato murrain amongst us'. This view was reinforced; witness MJB to Broome on 26 September: 'I see the mould regularly coming out of the mouths of the stomata and nowhere else' and then on 14 October, 'The mould which attacks the potato tubers is strange to say precisely the same species which attacks the leaves'. On 5 October, he wrote to Lindley 'I suspect there is not only the potato disease in Ireland but an extension rot into the bargain'. 'An extension rot' MJB wrote, leaving no doubt that he really did know exactly what was happening. He was also well in tune with the importance of the American observations. On 30 October, he wrote 'Dr Morren tells me that Teschemacker published an account at Boston in 1844 of a disease similar to that of the present year and ascribed it to Botrytis'. If only he had said so more emphatically and earlier.

But potato blight was not the only thing to occupy MJB's time and 1846 was an eventful year for him in several ways. He acquired a new microscope whilst in Paris (and was later, in 1854 and again in 1868, presented with larger ones still by Hooker). His perception of microscopy and the microscopical method was considerable and many of his words are relevant today. Writing to Broome of the lichenologist Leighton, he said: 'Measuring spores is exactly like estimating in astronomical science. Truth can only be arrived at by means of many experiments. The eye is not always equally accurate in its estimation and sporidia vary immensely. Measuring single spores is useless. All my numbers are means and I doubt not will be of very great use in the end'. MJB was himself no mean contributor to early lichenology and in his *Introduction to Cryptogamic Botany* enterprisingly grouped them together with fungi.

The mid-nineteenth century was a time when it was hard not to express a political opinion, and even MJB found time from his mycological utterances to express support for his chosen party. An English country parson so firmly behind Disraeli at first sight seems anachronistic but as MJB himself said in 1852 it seemed that 'we are on the way to becoming a republic', although his priorities remained in place. Witness the remarks made to Broome in 1854 'I have no faith in Prussia and less in Lord Aberdeen. My melons and cucumbers have died off this year very unpleasantly'. Yet, he continued, 'I have implicit confidence in Lord Raglan'.

There also began in 1846 another long and fruitful collaboration, which was to have political overtones, with the Reverend Moses Ashley Curtis of North Carolina which continued, with a break for the American Civil War, until the latter's death in 1872. Accounts have been published elsewhere on the Berkeley-Curtis correspondence but it is pertinent to place on record once again the debt that American mycology owes to MJB who took the untutored Curtis by the metaphorical hand and so provided fledgling American mycology with a worthy successor to Schweinitz.

In 1861, Joseph Hooker urged MJB to apply for the chair of Botany at Cambridge to succeed Henslow – it was Henslow's wish that he should do so. After much enquiry, Berkeley was persuaded that the University Senate would favour Babington, so in the event he did not apply. Then in 1867, he was asked, and put under considerable pressure, to apply for the Oxford Chair in succession to Daubeny. He was enthusiastic, but apparently it was decided that a clergyman was not eligible for this post, so once again the academic heights for which he would have been so admirably suited were denied him.

He returned again, quite literally, to more parochial things and in 1868, his Bishop offered him the post of Vicar of Sibbertoft. This was a parish some few miles distant from King's Cliffe and he decided to move – 'the country is extremely pretty,' he wrote, 'a capital school, the Church just restored, a very good house just built, a nice garden with greenhouse, a well conditioned Parish with a population just exceeding 400, excellent neighbours and two miles off, Mr Thompson who married Dr Lindley's daughter...'

But a long and active life was beginning to take its toll. Gout was an ever present and worsening problem and MJB took to more relaxing pursuits – 'By way of amusement, I have been learning Welsh', he wrote in 1861, and continued to read deeply, as he had done throughout his life. 'I think Darwin's *Orchids* is the most wonderful book ever published' he said in 1863. By 1876 and with failing health, he decided to make provision for perpetuity and presented his herbarium to Kew. Gradually, the specimens were catalogued and transferred; more than 6000 species had been sent by 1878. Eventually, the old man faded and MJB died on 30 July 1889, to be buried alongside his beloved wife close by the porch in the churchyard at Sibbertoft.

Berkeley was never a seeker after fame, although mycologically he certainly found it; nor indeed a seeker after fortune, which equally certainly eluded him. He was awarded a Civil List pension of £100 a year in 1867 for his services to mycology, but the family were not left well provided for. His stipend at King's Cliffe was £403 per year but with a gradually decreasing deduction of £70 per year until the debt incurred by his predecessor for rebuilding the vicarage was paid off. His furniture and books fetched about £1000; Wheldon bought the library for £250 – £100 down, £100 in three months and £50 in 6 months. For the scientific correspondence, the British Museum paid ten guineas. A few hundreds more were realised by the sale of some Munn pictures.

He led a life as full as a nineteenth century cleric could possibly have lived. His statistics speak for themselves – a very large family to care for, a parish, a school and the burden of examining for several bodies. For a long time, he gave up two days a week to go to London on Royal Horticultural Society business. His collection of fungi numbered some 10,000 species with 5,000 of them types. He wrote several hundred papers, several books and, of course, all of those letters.

It is indeed fitting that The British Mycological Society should mark the death of this remarkable man by holding a meeting at this time. M. C. Cooke called him the father of English mycology – with sixteen children, that he should also find time to sire our discipline too was a quite remarkable achievement. As far as I can determine, despite his large family, no direct descendants survive. But mycology certainly does. MJB died before the Society was founded although undoubtedly he would have approved of it. And I am sure that he would have approved too that this little celebration has chosen as its theme perhaps the most practical of the many contributions that he made. That he might have played his part in sparing future generations the misery of potato blight and famine would surely have brought a faint smile to that wonderfully distinguished face.

Chapter 2

Potato blight in Europe in 1845: the scientific controversy

Austin Bourke

Two years before the first great European epidemic of potato blight (*Phytophthora infestans* (Mont.) de Bary) the disease had broken out in the U.S.A. in the hinterland of the East coast ports. Fig. 2.1 illustrates the rapid spread of the disease in 1843–5. Americans thus had the earliest opportunity to speculate on the cause of the outbreak. In a pamphlet published in Boston, Bosson (1846) attributed the disease to an exhaustion of the vital forces of the potato, similar to the deterioration which he observed in the royal families of Europe, where successive generations of inbreeding had culminated in 'a race of dwarfs and idiots'. Not everyone subscribed to this sturdy republican diagnosis. Bosson reports that J. E. Teschemacker of Boston had suggested that fungi were responsible for the disease, a theory which, in Bosson's view, had 'the merit of novelty, if little else'. Indeed, the idea of a fungal cause of the disease was summarily dismissed in the U.S.A. as an amusing aberration. Teschemacker had submitted an account of his views to the New York State Agricultural Society in December 1844, but his paper was completely ignored; even the author failed to keep a copy. In the following year, when the agricultural section of the annual report of the U.S. Commissioner of Patents reproduced European publications which initially seemed to favour a fungal explanation of the disease, Teschemacker, now anxious to establish priority, had to confess that his evidence had disappeared 'amongst things lost or forgotten' (*Report for 1845*, p. 514).

Ailments of the European potato crop prior to 1845

By 1845 over two and a half centuries had elapsed since the Spaniards first brought the potato to Europe from South America. Although the potato had always been subject to direct damage by extreme weather conditions – sustained drought, excessive rainfall, severe frost – it was not until the 1760s that a specific major disease of the crop was recognised. This was the 'curl', later to be identified as a virus disease, but attributed at the time to a gradual 'degeneration' of the seed tuber +, a diagnosis which provided a ready-made explanation of later diseases, as each new ailment was seen as the consequence of progressive degeneration. The same cause was

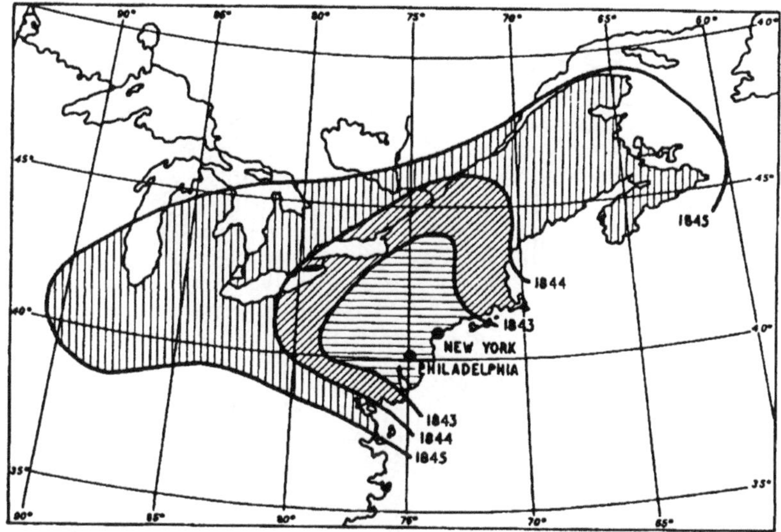

Fig. 2.1. Approximate extent of potato blight attacks in the USA and Canada between 1843 and 1845 (based on Stevens (1933) and Bourke (1969)).

blamed for the 'taint', a dry rot of the potato tuber caused by *Fusarium* spp. which aroused concern in Britain and Ireland in the early 1830s and later affected continental Europe. It reached its greatest intensity in Germany in 1841, and inspired the Bavarian government to commission a world-famous botanist, ethnologist and explorer, C. F. P. von Martius, to investigate and report on the problems besetting the potato. This he did in a classic study (von Martius, 1842) in which he made it clear that he was satisfied that dry rot was caused by a parasitic fungus. No one of von Martius' scientific and social standing had previously advanced so unorthodox an opinion, which, to some, smacked of heresy. In those pre-Pasteur days, long before the germ theory of disease began to find favour, the accepted wisdom was that spores, bacteria and germs were not a cause but a consequence of disease; they appeared by spontaneous generation after the organism had fallen ill. A few idiosyncratic thinkers took the contrary view, but the organized ranks of science and, in particular, the learned societies, as always staunch supporters of orthodox opinion, stood firm in their distrust of novelty. Thus von Martius's report created something of a sensation. It was welcomed by progressive thinkers in many countries; Berkeley, for instance, translated it into English and had much of it published in the *Gardener's Chronicle*. It would shortly encourage some other 'fungalists' to declare themselves.

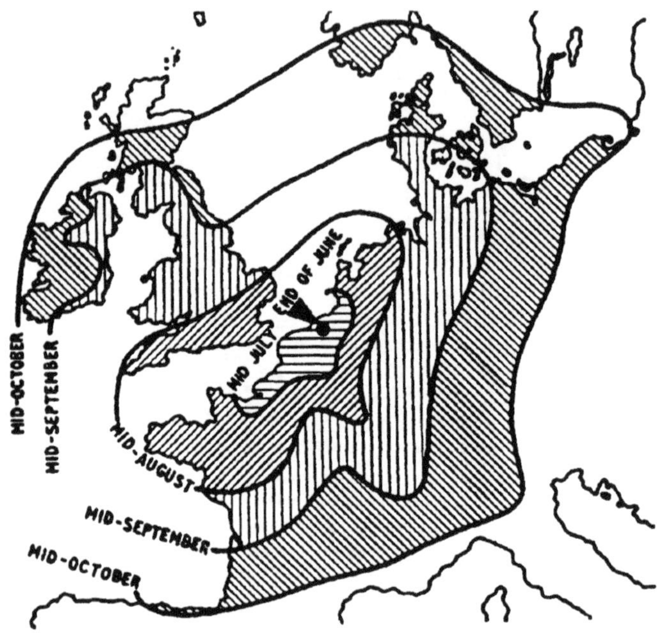

Fig. 2.2. Approximate dates of first reports of potato blight in Europe in 1845 (after Bourke, 1964).

The dry rot outbreak in continental Europe in the 1840s may well have led to the accidental introduction of an even more serious disease. In a move to improve the health of the crop, the provincial government of West Flanders voted funds for the importation of fresh potato varieties from North and South America, and these imports were the subject of field trials in 1843 to 1845. It was in West Flanders that potato blight was first reported in 1845.

First theories as to the cause of blight

The broad timetable of the European blight epidemic of 1845 is illustrated in Fig. 2.2. The outbreak first came to attention in Belgium in the last week of June and speculations on its cause were being put forward in the Low Countries as early as July. The disease did not appear in the environs of Paris until mid-August; specimens of affected foliage were first exhibited to a learned society there on 20th August. Simultaneously with publication of the news of the arrival of the epidemic in southern England in the *Gardener's Chronicle* of 23rd August, the editor, Professor John Lindley, committed himself to the view that the cold and cheerless summer of 1845 was to blame. The weather theory, which had already been advanced by several Belgian commentators, was widely adopted in England thanks to the prestige and influence of Lindley; thus the Royal Agricultural Society awarded all three of its prizes for essays on the potato disease to protagonists of a meteorological cause.

While the majority of the earliest commentators held the weather culpable, some argued that blight was merely a further downward step in the progressive deterioration of the potato; others that it was due to insects or worms. A small group opted for a poisonous 'miasma ' borne on the air, but differing as they did as to its origin, published individual pamphlets in support of industrial pollution (Peeters, 1855), volcanic exhalations (Parkin, 1846, 1847), gases from the recently introduced sulphur matches (Zuppinger, 1847), or 'some aerial taint originating in outer space' (Bain, 1848).

A remarkable feature in retrospect is that the true explanation was advanced at the very earliest stages in a series of newspaper articles written independently by a number of Belgian commentators, so for a short time, until massive opposition overwhelmed it, it looked as if the truth might prevail. Writing in *L'Organe des Flanders* of 31st July and 1st August, the Abbé Edouard van den Hecke, who was Grand Vicar to the bishop of Versailles and a keen amateur mycologist, reported on his examination of diseased potato leaves under high magnification. He discovered a fungus of the Botrytis family and was particularly struck by the unusual mechanism by which the spores were projected to a distance. He believed the fungus to be the true cause of the disease but modestly sought confirmation by others. Meanwhile he advised the early removal and destruction of affected potato foliage.

Professor Martin Martens of the University of Louvain, in a contribution to the *Journal de Bruxelles* of 14 August, also gave the apparent cause of the disease as a fungus which attacked the leaves, especially on their lower faces. He too advised the removal of diseased tops. He repeated his opinion in a letter dated 10th September, although he admitted there were difficulties. However, in a paper read to the Royal Academy of Brussels some months later, when the fungal theory was under severe attack, he prudently distanced himself from a cause which he confessed he had 'to a certain degree adopted' in his earlier writings.

On the 19th August the *Journal de Liège* published a letter, written on 14th August, from Anne-Marie Libert, a cultured lady of leisure and self-taught mycologist of international repute. She gave a clear and detailed description of the fungus responsible for the disease, but mistakenly believed it to be identical with one, *Botrytis farinacea*, described some years earlier by Fries. She suggested, in view of the devastation which it was causing, that the specific name should be changed to *vastatrix*.

At a meeting of the Agronomic Society of Thorout held on 18th August and reported in the *Nouvelliste des Flanders* of 25th August, Dr Rene Van

Oye said that the one true determining cause of the potato disease was a fungus which, reproducing itself with astonishing rapidity and profuseness, had infected all the potato fields, and that the disease was clearly contagious. In these early commentaries there was thus a broad measure of agreement among some prominent Belgian savants, who put forward their opinions with restraint. The tone became more strident in the next stage of the debate.

The protagonists

The subsequent phase of the controversy on the role of the fungus in the potato disease can be better understood against the background of the careers of three of the principal protagonists.

Two Belgian orphans, Charles Morren and Joseph Decaisne grew up as close neighbours and schoolmates in Brussels; both embarked on scientific careers which soon diverged. After earning a doctorate in natural philosophy and mathematical science in 1829, Morren went to Paris on a travelling studentship, only to be recalled when Belgium became independent to assume the professorship of physics at Gand in place of his departing Dutch predecessor. His subsequent career was one of unbroken success. While at Gand he widened his scope by taking a medical degree, although he never practiced as a doctor. Finding the climate of Gand uncongenial, he transferred to the professorship of botany at Liège. In 1842 a special new professorship of botany, agriculture and economic forestry was created for him. This enabled him to offer public courses popularising agricultural science, designed particularly for extensive landowners. Over all these years Morren's output of technical papers was prodigious and often controversial. He sat on numerous commissions, many of them governmental, served on university examination boards, and edited several scholarly periodicals to which he was often the main contributor. He prided himself on revising his university courses each year to take account of the latest developments. Inevitably, there were those, particularly in academic circles, who looked on Morren's high-flying career with less than complete enthusiasm, and who would not be broken-hearted should he 'come a cropper' in one or other of his manifold activities.

Meanwhile, Morren's schoolmate Decaisne had been pursuing a less spectacular scientific career. By 1845 he was employed as assistant naturalist at the Museum of Natural History in Paris. This post, as Semal, Joly & Lamy (1983) stress, was more important than its title might suggest, but it clearly represented a lower level of attainment than Morren's academic honours.

Also in Paris, working as a freelance mycologist, was Jean Françaís Camille Montagne who had served in the French armed forces, first as a boy sailor and later as a military doctor. Retired from the army in 1831 at the age of 47, he devoted the rest of his life to mycology. He was driven by two ambitions: firstly, to fill a gap in French expertise which forced naturalist explorers to send their botanical specimens to Sweden, Germany or England for identification and description, and, secondly, to earn for himself election to membership of the French Academy of Sciences. He is credited with working on his speciality for ten hours per day over twenty years, and with introducing, describing and depicting nearly 2,000 specimens. For all his industry, he gathered a mere seven votes when he first tried for membership of the Academy in 1837. Clearly he needed a spectacular addition to his mycological credits to attract the support of the scientific electorate.

Entry of the 'big guns'

On 18th August Morren released his first broadside on the subject of the potato disease in the form of a letter to the newspaper *L'Independence Belge* in which it was given page one billing in the issue of 20th August. It was written with all of Morren's accustomed assurance and in his most trenchant style. He dismissed out of hand all earlier speculations with the exception of those of the two pro-fungal exponents (van den Hecke and Martens) which had first been published. The true cause of potato blight, he maintained, was a fungus of the Botrytis kind, similar to that which von Martius associated with dry rot of the potato. It was just one further example of the series of parasitic plants such as caused ergot of rye, rust of wheat and smut of oats and maize. He promised that in a paper he would submit to the Royal Academy of Science of Brussels, he would depict all phases of the development and reproduction of the fatal mould. Meanwhile he was publishing the present simplified account to counter the propaganda of those whose blind hatred of what they called the theory of agriculture led them to mock scientific knowledge which alone could enlighten practice and put an end to empiricism.

The article concluded with ten items of practical advice on how to counter the disease, beginning with a repetition of van den Hecke's suggestion of removing and burning all diseased foliage. Affected tubers should also be destroyed and maximum sanitation practiced. A mixture of copper sulphate, lime, salt and water should be used to dip potato seed and to disinfect the soil. Autumnal sowing was also suggested.

Morren made further contributions to the same newspaper on 2, 4 and 9 September. He combined and edited the content of all these articles in a major brochure (Morren, 1845a) which was published in Brussels on 22nd September in both French and Flemish. In view of French interest in the work, a revised version of the brochure (Morren, 1845b) was published in Paris on 1st October. In the middle of all this activity, Morren

(1845c) managed to complete his promised scientific article on 24th September and read it to the Brussels Academy on 4th October.

Of all these publications, the one which created the greatest impact was the initial article of 18th August. It was distributed, translated and reprinted throughout Western Europe with remarkable speed and was reproduced even in North America (*Monthly Journal of Agriculture, New York*, 1845, p. 387). More significantly, it appeared as early as 21 August in the official French publication *Journal des Débats* and was there seen by Montagne who was himself hard at work on specimens of diseased potato foliage which he had received three days earlier from a grower in the outskirts of Paris. Montagne had written a letter to Berkeley, which the latter received on 26th August, enclosing specimens and recounting the description of the fungus which he planned to publish in Paris very shortly. In this letter he expressed some uncertainty as to the basic cause of the disease, but he recognised that it was closely connected with the fungus.

On the day he received Montagne's letter, Berkeley wrote to the *Gardener's Chronicle* (issue of 30th August) saying that he had so far failed to locate the disease in Northamptonshire and that, in the absence of direct evidence, he was inclined to go along with Lindley's view in the previous issue that the continued wet was the primary cause of the malady. However, by the date of the following issue (6th September), the disease had made a dramatic appearance around Berkeley's home at King's Cliffe, and he wrote that having now seen the disease *in situ* he had no doubt that it was caused by the fungus *Botrytis infestans* Mont., a title which Montagne had registered a few days earlier and which Berkeley was the first to use.

Counter-attack

In his second letter to *L'Independence* (2nd September), Morren commented on the controversy which his earlier letter had aroused. The discussion, he said, had taken a rancorous turn in some journals. He himself would refrain from responding to insults, or malicious insinuations, or to comments inspired by evil passion. Quoting Linnaeus, he intended, he said, to devote the years which remained to him to useful work, and to waste no time in replying to his enemies. Errors in natural history could not long be defended; truth could not long be concealed. It was to time that he would entrust his reputation.

One of the earlier critics was Dieudonné (1845) who, in a report commissioned by the public health authority of Brussels, had denounced Morren's thesis which, he said, would hang for ever like a sword of Damocles over the heads of potato growers, if common sense, observation and long experience did not enable them to judge it aright. In order that the fungus might not become the terror of the farmers, this unfortunate and depressing thesis must be exposed. Even more barbed criticism came from some who had earlier rushed to clamber on to what they had mistakenly thought to be Morren's bandwagon but who now found it advisable, in view of the violent reaction in many influential quarters, to declare themselves as 'born-again' antifungalists. Typical of these was Boset (1846) who admitted that he had at first been deluded by Morren's thesis which he later recognised to have been modelled on 'the vagaries of von Martius'. In a bitter personal diatribe he condemned Morren as an armchair scientist who rushed into print in pursuit of precedence and publicity, a man notorious for his anxiety to add his torch of wisdom to the communal fire, a controversialist too eager to break a lance in all kinds of scientific tournaments.

Morren's vow of silence was not proof against such taunts and he was stung to reply in kind. Thus, for example: 'not a fortnight ago some one wrote in all solemnity: "The sun will soon dispose of M. Morren's fungi and make short work of all these microscopic fabrications." Today it is my turn to answer: the only microscopic factor involved is the range of vision of some people. If the sun, which has spared both the fungus and the disease, has disposed of anything it is the inanity of the hare-brained' (Morren, 1845a).

Heavily beleaguered as he was in Belgium, Morren was heartened by news of support from abroad, firstly the declaration of the 'learned M. Berkeley' (Morren, 1845b) and secondly the avowal of Dr Montagne who, 'like all true savants', had recognised that the disease was caused by a fungus (*L'Independence Belge*, 4th September). It was true that the doubts which Montagne had earlier expressed in his letter to Berkeley appear to have melted away in the blazing conviction of Morren's first article which he had read in the *Journal des Débats* of 21st August. Certainly there was a tone of utter conviction in the presentation of Montagne to the Société Philomatique of Paris on 30th August in which he laid claim to priority in describing and naming the fungus (Montagne, 1845). Indeed he praised Morren's article in which, he said, the deleterious effects of the parasite had been perfectly depicted, and the cause and nature of the malady fully explained. Had Morren gone on to name and describe the microscopic fungus which was responsible for all the damage, there would have been nothing left to add. As it was, Montagne proceeded to propose the name, 'only too well deserved', of *Botrytis infestans* and to add a description in Latin of the fungus.

No sooner had Montagne finished his presentation than two members of his audience rose to speak in turn. The first was Decaisne, one-time schoolmate of Morren, and the other was Duchartre, editor of the *Revue botanique*. Their purpose was not to congratulate, but to denounce the concept of a fungal parasite; indeed this was merely the first engagement in a running battle, directed primarily against Morren, which they kept up for months. It culminated in Decaisne's pamphlet (Decaisne, 1846) which was designed to be a counter blast to Morren's tracts (Roze, 1898). In the concluding paragraph, Decaisne argued that Morren 's thesis, which had done so much to spread panic among the population, was based on nothing more than observational error, and that, no matter what tortuous arguments he advanced, Morren would soon find himself alone if he persisted with his hypothesis.

Montagne, seeing that he risked losing favour with French Academicians who, with one exception (A. Payen), were strongly opposed to the fungal theory, tried at once to retreat into a neutral corner, and later took up a mildly antifungal stand. At the *Société Philomatique* meeting he expressed to his critics his regrets that the terms in which he had spoken could be interpreted as support for Morren's thesis, and protested that his sole purpose had been to describe and register the fungus, irrespective of whether it was the cause or consequence of the disease.

Montagne wrote forthwith to Berkeley describing the controversy into which he had unwittingly been drawn, and recalling the doubts which he had expressed in a previous letter, and to which he was now reverting. Berkeley, in his kindly way, replied that he could understand his friend's uncertainty; he acknowledged the difficulties of the case and himself remained open to conviction. Never given to dogmatic claims, he continued, in his 1846 paper, to state his opinion with a moderation unusual at the time: 'After an attentive consideration of the progress of the disease and of almost everything of value that has been written on the subject, and after duly weighing the peculiar difficulties with which it is attended, I must candidly confess, that, with a becoming share of philosophical doubt where such authorities are ranged upon the opposite side, I believe the fungal theory to be the correct one'.

Such convoluted restraint was wide open to misinterpretation by a foreigner; certainly, Montagne understood from Berkeley's letter that the latter had modified his original views. He hastened to convey the happy tidings to Duchartre who gave prominence in the *Revue Botanique* (Vol. 1, 1845, p. 177) to news of the defection of the 'talented English mycologist', who was now disposed to seek the prime cause of the disease elsewhere than in the fungus, and to the claim that, in Duchartre's view, Morren now stood in absolute isolation. Curiously, Berkeley (1846) made almost the same comment: the fungal theory 'is now peculiar almost to Dr Morren, unless M. Payen is to be reckoned also as its advocate'. In fact, A. Payen, like Berkeley himself, was a firm believer but no missionary; he was satisfied to eschew controversy and wait for time to prove him right. Active propagandists for the fungal theory, now in low repute, were hard to find, and there were few new adherents. A remarkable exception was the Scot, David Moore, curator of the Dublin Botanic Gardens, who had at first subscribed to the weather theory but who, as a result of his own painstaking experiments, became a public convert to the fungal hypothesis at a time when it was being generally derided (*Gardener's Chronicle*, 1846, 531; Nelson, 1983).

And afterwards ...

The fungal theory

By the end of 1845 the vocal adherents of a fungal cause of the potato disease had dwindled to a beleaguered handful. At the 1846 meeting of the British Association in Southampton, the one point of virtual unanimity about the potato disease was that a fungus was not responsible. 'That the fungi were the cause had now been disproved by the best chemists' said Dr Lankaster and Dr Solly agreed that the fungal theory 'had lost ground, latterly, very materially' (*Plough*, 1846, 516-519). In Britain it was recorded that 'the notion that the fungi which have been detected are the cause of the disease has been abandoned by most of the writers on the subject, and they are all but proved, in most persons' estimation, to be the effect: amongst practical men it has been, all along, treated as a chimera' (*Annals of Horticulture*, 1847, 227). It was the same in France: the argument that fungi caused the disease was 'almost completely abandoned' (*Journal d'Agriculture Pratique*, Paris, 1845-6, p. 236). The idea was derided even in contemporary popular ballads (McKay, 1961):

Some cock their glasses up to their eye,
And mushrooms in the cells descry,
But we, my Lords, have looked as well,
And think such notions all a sell;
Decaisne in France, in Germany Kützing,
Have sought the rot all manner of roots in,
And proved that those have looked with a loose eye
Who said 'twas caused by fungi and fuci
(Sure, never since the days of Plato
Was there such a row about a rotten potato.)

Ten years after blight had first ravaged Europe, Berkeley was still writing with circumspection in the *Cyclopedia of Agriculture* (Morton, 1855): 'There are two theories which, in the present position of the subject, seem to us at all tenable the fungal theory, and that which ascribes the disease to a vicious mode of cultivation pursued through many years. It is not our intention to decide between them, and although we are certainly disposed to lay the greater stress upon the former, which is gaining ground and gives full conviction to many minds, we acknowledge that there are great difficulties about it, and such as to make it far from becoming to speak dogmatically'.

Berkeley was correct in saying that the pendulum of scientific opinion was slowly swinging in his direction, although the more diehard opponents such as Baron de Croeser were calling for the final abandonment of the theory as late as 1860 (Croeser, 1860). In the next year, de Bary (1861, 1863) published the first of the two famous papers in which the life cycle of the potato blight fungus was described, and its responsibility for the disease irrefutably established.

The name of the fungus

It was soon realised that the blight fungus did not fit well into the *Botrytis* genus, and it was reclassified as *Peronospora* by Unger in 1847 and Caspary in 1852. Much later, in 1876, de Bary renamed it *Phytophthora*.

The specific epithet *infestans* survived, but only after fierce controversy. Its connotation of something aggressive, hostile and dangerous remained an embarrassing reminder to Montagne of his earlier acceptance of the parasitic role of the fungus. Indeed in a letter to Montagne reported by Duchartre (1845), Demazières suggested that *fallax* be used in place of *infestans* as a permanent reminder of how an innocuous fungus had led so many scientists astray. In a move towards an acceptable compromise Harting (1846) suggested the neutral *solani* but gained no support. There was also a school of thought that would assign the credit for naming the fungus to Miss Libert and use her adjective *vastatrix* (Morren, 1845c). Berkeley (1846) supported the attribution to Montagne, and although the argument was revived much later by Puttemans (1938), Montagne's name and adjective have survived.

The principals

Berkeley had always remained on the outer fringe of the blight controversy, partly because of his courteous and pacific nature, partly because his scientific paper (Berkeley, 1846) was published at a time when acceptance of the fungal theory was at its lowest ebb and earned little immediate recognition. In his prize essay on the potato disease Graham (1847) referred dismissively to Berkeley's 'elaborate memoir', in which blame was assigned 'to a species of botrytis which, singularly enough, had never been observed before'.

Outside Britain, the erroneous reports that Berkeley had abandoned the fungal theory persisted for years, despite a correction by Berkeley in a footnote to his 1846 paper, and a withdrawal and apology by Duchartre (*Revue Botanique*, 1845-6, p. 562). The rumour was repeated as fact by Roze (1898) in his *History of the Potato*. Prunet (1902) takes the story further in claiming that Berkeley, in time, took his place among the advocates of antiparasitism. Jones, Giddings & Lutman (1912) brought the legend to its finest flowering in suggesting that Berkeley had lent his authority to supporting 'all the ignorance and superstition of the time in opposition to Morren'. Even in England official recognition and honours came belatedly to Berkeley, but the passage of time (he lived to 86 years) brought his genius into clearer focus.

As for Decaisne, who had attacked Morren's thesis so doggedly, there is a certain irony in the fact that he was, in the year 1863, joint editor of the French journal in which was published the second part of de Bary's justification of the views of Decaisne's old antagonist. But Morren alas, did not live long enough to savour the triumph of his views. He became seriously ill in 1854 at the age of 47, and died four years later without resuming his labours. The Belgian *Biographie Nationale* attributed his early death to an excessive work load over almost 30 years. The hyperactivity which marked his whole life suggests some inner imbalance which prevented him from earning the recognition that his brilliant intelligence merited, and aroused the fierce opposition to his views which, without much doubt, contributed to his early death.

On a happier note, Montagne gradually re-established himself in good favour with the French scientific establishment, and was elected virtually unanimously to the Academy of Sciences in 1852. He died in 1866 at the age of 82.

References

Bain, D. (1848). *Observations upon the potatoe disease of 1845 and 1846.* Edinburgh.

Berkeley, M. J. (1846). Observations, botanical and physiological, on the potato murrain. *Journal of the Horticultural Society of London*, **1**, 9-34.

Boset, C. J. (1846). *Plantation des pommes de terre en 1846, d'après la théorie démontrée de leur maladie*. Liège.

Bosson, C. P. (1846). *Observations on the potatoe, and a remedy for the potatoe plague*. Boston.

Bourke, P. M. A. (1964). Emergence of potato blight, 1843-46. *Nature*, **203**, 805-808.

Bourke, P. M. A. (1969). Potato late blight in Canada in 1844-45. *Canadian Plant Disease Survey*, **49**, 29-31.

Croeser, E. (1860). *Études sur l'histoire et la culture de la pomme de terre*. Bruges.

de Bary, A. (1861). *Die gegenwärtig herrschende Kartoffelkrankheit, ihre Ursache und ihre Verhütung*. Felix: Leipzig.

de Bary, A. (1863). Recherches sur le développement de quelques champignons parasites, etc. *Annals des Sciences Naturelles*, 4e sér. Bot., **20**, 1-148.

Decaisne, M. J. (1846). *Histoire de la maladie des pommes de terre*. Dusacq: Paris.

Dieudonné, M. M. (1845). *Rapport fait au conseil central de salubrité publique de Bruxelles sur la maladie des pommes de terre*. Brussels.

Duchartre, P. (1845). Sur la maladie des pommes de terre. *Revue Botanique*, Paris, **1**, 223-229.

Graham, F. J. (1847). The potato disease. *Journal of the Royal Agricultural Society of England*, **7**, 357-391.

Harting, P. (1846). *Recherches sur la nature et les causes de la maladie des pommes de terre en 1845*. Amsterdam.

Jones, L. R., Giddings, N. J. & Lutman, B. J. (1912). Investigations on the potato fungus, *Phytophthora infestans*. *Bureau of Plant Industry, Bulletin No. 245*. US Department of Agriculture: Washington.

Martius, C. F. P. von (1842). *Die Kartoffel Epidemie der letzten Jahre, oder die Stockfäule und Räude der Kartoffeln*. Munich.

McKay, R. (1961). *An Anthology of the Potato*. Figgis: Dublin.

Montagne, C. (1845). Observations sur la maladie des pommes de terre. *L'Institut, Bulletin de la Société Philomatique de Paris*, **13**, 312-314.

Morren, C. (1845a). *Instructions populaires sur les moyens de combattre et de détruire la maladie actuelle (gangrène humide) des pommes de terre ...* Perichon: Brussels.

Morren, C. (1845b). *Nouvelles instructions populaires sur les moyens de combattre et de détruire la maladie actuelle (gangrène humide) des pommes de terre ...* Roret: Paris.

Morren, C. (1845c). Notice sur le *Botrytis* devastateur ou le champignon des pommes de terre. *Annales de la Société Royale d'Agriculture et de Botanique de Gand*, **1**, 287-292.

Morton, J. C. (1855). *A Cyclopaedia of Agriculture, Practical and Scientific*. vol 2, pp. 667-694. London.

Nelson, E. C. (1983). David Moore, Miles J. Berkeley and scientific studies of potato blight in Ireland, 1845-1847. *Archives of Natural History*, **11**, 249-261.

Parkin, J. (1846). *The cause of blight and pestilence in the animal kingdom*. London.

Parkin, J. (1847). *The prevention and treatment of disease in the potato and other crops*. London.

Peeters, L. (1855). *Guerison radicale de la maladie des pommes de terre et d'autres végétaux*. Namur.

Prunet, A. (1902). Les mildiou de la pomme de terre. *Revue de Viticulture*, Paris, **17**, 663-666; **18**, 97-104.

Puttemans, A. (1938). Reinvindicação visando a denominação scientifica da doenca da batateira. *Rodriguésia*, **2**, 341-350.

Roze, E. (1898). *Histoire de la pomme de terre*. Rothschild: Paris.

Semal, J., Joly, P. & Lamy, D. (1983). L'épidémie de 'maladie des pommes de terre' causée en Europe en 1845 par le *Phytophthora infestans* (Mont.) de Bary: les faits et les auteurs. *Annales de Gembloux*, **89**, 79-99.

Stevens, N. E. (1933). Phytopathology – the dark ages in plant pathology in America: 1830-1870. *Journal Of the Washington Academy of Sciences*, **23**, 435-446.

Zuppinger, F. (1847). *Die glücklich entdeckte Ursache der Kartoffelkrankheit*. Zürich.

Chapter 3

Phytophthora infestans: the Mexican connection

John S. Niederhauser

The Mexican national potato programme was created in 1947 to improve potato production and productivity in Mexico, and is an integral part of the National Institute of Agricultural Research of the Mexican Ministry of Agriculture. The potato late blight project is a priority activity of this national programme. The Mexican research reported here on potato late blight and its control has been conducted in the fields and laboratories of the government experiment stations, by Mexican scientists, and in collaboration with Mexican potato farmers. These research results on potato late blight in Mexico soon attracted international attention, and in the ensuing years a close co-operation was established with many potato scientists and institutions all over the world. Mexico proved to be an unique location to study this potato disease of world-wide importance.

This paper will address three principal topics:

- the origin of *Phytophthora infestans*, beginning with a review of the early assumption that this pathogen came from South America, and concluding with what I consider convincing evidence of its origin in Central Mexico;
- some specific proposals for future research on potato late blight and its cause, *P. infestans*. These proposals are selected to emphasize the special role that Mexico might play in future international research co-operation; and
- a new strategy for international co-operation, designed to encourage a broader and more efficient collaboration in the implementation of these research proposals.

Origin of *P. infestans*

Potato late blight was first reported in Europe and North America in the 1840's (Bourke, chapter 2), nearly 300 years after the cultivated potato was introduced into Europe from South America. Where had it come from? Why did it suddenly appear only after the potato had become an important food crop?

South American origin of P. infestans

Speculation on the origin of *P. infestans* began soon after the appearance of potato late blight in Europe in 1845. The initial, widely-accepted theory was that *P. infestans* must have come from the South American home of the cultivated potato. And 'evidence' for this theory was soon forthcoming. Even so eminent an authority as Anton de Bary (1861), quoting from Munter's *Krankheiten der Kartoffeln*, said that as early as 1571 the Jesuit priest, Joseph Acosta, had observed in Bolivia 'that after damp, cold weather the tubers were often destroyed in the ground through gangrene or mildew'. And Boussingault (1845) communicated to the French Academy of Sciences a paper by another Acosta (an army colonel in Colombia) about the same time, in which he stated that on the high table lands of Bogota the potatoes 'on low ground were everywhere destroyed in wet years'. These statements are as true today as they were in 1571. However, we have since learned that tuber rots, in the cold wet weather at harvest time in the high Andean region, may be due to a wide range of pathogens and physiological disorders associated with water-logged soils, and that even in modern times, *P. infestans* has rarely played a major role.

Assuming, as almost all observers did, that the fungus *P. infestans* was native on the cultivated or wild potatoes of the Andean region, the reasonable conclusion was that the introduction of the fungus into Europe from its Andean habitat would have occurred not long after the potato was brought to Europe. For nearly a century many dedicated botanists and historians attempted to establish why it took *P. infestans* so long to escape from its so-called 'Andean home' and attack the European potato crop. However, the persistent curiosity of these investigators only led to a growing frustration, due to the simple fact that they could not uncover any evidence of the presence of *P. infestans* in its alleged ancestral Andean home – it just wasn't there. Indeed observations and reports gathered since 1950 suggest that *P. infestans* appeared only recently on potatoes in South America, becoming widely established only during the 20th century. The evidence on the first confirmed reports of *P. infestans* in South America is considered below.

Peru

Coastal region Abbott (1929) was the first to report the presence of potato late blight in Peru. He observed that occasional spraying with fungicides kept the disease in check on the coast. However, two years later (Abbott, 1931) he noted that potato blight was not nearly as widespread as he had supposed, and caused damage only along the coast, particularly in the vicinity of Lima. This is the same area where introductions of seed potatoes from Europe and the U.S.A. had been made. The first severe blight year recorded was 1947. Bazan-Segura (1952) states that 'before 1947, the blight disease passed completely unnoticed by farmers, making the use of fungicides unnecessary ... [since then] late blight has appeared annually with destructive effect, obliging farmers to use intensive spray schedules to save their crop'. By 1952, late blight had become the limiting phytopathological factor in potato production on the coast of Peru (Niederhauser, 1954a)

Montañas (intermediate elevations) First reports of late blight in the Arequipa region were in 1928. Although susceptible potato varieties had been grown in a few fields in the area for many years, only a small amount of late blight was noted. Since then, the disease gradually became more prevalent, and by 1949 late blight had become a severe and annual problem. Though some growers began successful spray programmes, the total acreage planted in Arequipa dropped sharply because of late blight (Niederhauser, 1954a).

Sierra (high elevations) Abbott (1931) indicated that all of the damage previously attributed to late blight in the Andean region was due to frost. At these higher altitudes potato late blight was first noted in 1949, and since that time became widely distributed, causing serious losses in some areas and in some seasons (Bazan-Segura, 1952). However, although 1953 was an unusually rainy season in the Sierra, there was surprisingly little late blight, even on susceptible cultivars grown for 4 to 5 months with little or no protective spraying. While an occasional foliar blight lesion could be found, damage was minimal. The relatively light foliage attack was probably due to low temperatures (day and night) during the growing season inhibiting rapid spread of the pathogen (Niederhauser, 1954a).

The traditional methods of potato cultivation practiced in the Andes have evolved through many centuries of keen observation and practical experience by subsistence farmers. It is curious that these potato growers should persist in the belief that the foliage blight is a quite different disease problem from tuber blight. The latter was often confused with other problems, including frost damage. It would appear that these very knowledgeable potato growers had only a brief experience with *P. infestans*. Indeed, in the central Andean region the common name given to late blight is 'hielo', the same name given to frost damage, and the potato agricultural specialist must remind the farmers of the difference (Bazan-Segura, 1952).

Argentina

Late blight was first reported in Argentina in 1887 (Rieder, 1887). Reports for the following 70 years indicate sporadic attacks in certain potato growing regions, often separated by intervals of 5 to 15 years (Calderoni, 1960).

Brazil

Potel (1900) observed potato late blight in Brazil in 1898, although the attack was sporadic and not considered to be a serious annual problem.

However, some 50 years later I observed severe losses due to late blight in unsprayed potato fields. By that time, most potato fields in Southern Brazil were protected by fungicides.

Chile

The presence of late blight in the Chilean potato crop was verified for the first time in 1949 in Mallarauco, a valley southwest of Santiago (Montaldo, 1953). This was found not to be an isolated case, and it was soon established that *P. infestans*, although previously unnoticed, was present in other potato growing areas of the country. Later evidence indicated that a severe outbreak of late blight occurred in some of the smaller islands of Chiloé in 1949 (Bourke, 1956).

It seems likely that late blight was *not* present in 1937, when a careful search by Chilean and American plant pathologists (in connection with a seed certification scheme) failed to find the slightest evidence of *P. infestans* in the field in Chiloé, although optimum weather conditions for late blight attack prevailed.

The best theory advanced so far is that the blight fungus came to Chile in infected tubers from Argentina, perhaps following the severe late blight epiphytotic there in 1940-1941. During the war years, with European sources of seed cut off, there was a sharp increase in potato trade between Argentina and Chile. The pathogen was supposedly introduced at this time from Argentina into Chiloé, from where it subsequently spread in seed tubers sent to the rest of the potato growing areas of Chile. With susceptible cultivars being grown in Chile in a favourable climate for late blight, the discovery of *P. infestans* in 1949 prompted the use of protective sprays, and by 1955 these fungicide applications had become essential.

Bolivia

The first report of late blight in Bolivia was in 1943 near Cochabamba (Anon., 1943). Potatoes grown below 3000 metres were attacked often in subsequent years, but the late blight disease was of little importance or concern at higher altitudes.

Venezuela

In Aragua, Venezuela, potatoes have traditionally been grown at all altitudes during the 'dry' season, but during the rainy period their cultivation has been primarily at lower altitudes. However, by the 1940's these 'temporal' plantings at lower altitudes were on a reduced scale, due to late blight caused by *P. infestans* (Ciccarone, 1949).

Summary of evidence concerning a South American origin

It seems that *Phytophthora infestans* was not known to exist in South America until very late in the 19th century, when it was first reported in

Table 3.1. Date of first recorded appearance of *Phytophthora infestans* in South America

Country	Year	Authority
Argentina	1887	Rieder (1887)
Brazil	1898	Potel (1900)
Peru	1929	Abbott (1929)
Bolivia	1943	Anonymous (1943)
Chile	1949	Montaldo (1953)

Argentina and Brazil. Several decades later it was found to be present in several Andean countries, including Peru (1929), Bolivia (1943), and Chile (1949). Curiously enough, those South American countries that had regularly imported seed potatoes from Europe during the latter part of the 19th century (Argentina and Brazil), were the first to report potato late blight. Those countries where the cultivated potato originated (Peru, Bolivia and Chile), discovered *P. infestans* in their potato fields some 40 to 50 years later. This circumstantial evidence argues against a South American origin of *P. infestans* (Table 3.1).

Mexican origin of **P. infestans**

The occurrence in Mexico of a native population of wild potato species resistant to P. infestans

Reddick (1943) summed up his search for late blight resistance with the statement '...no one has brought out of South America a potato, either seed or tuber, wild or cultivated, that has any appreciable resistance to the disease'. The eminent Russian geneticist, botanist and explorer, N. I. Vavilov (1935) declared that the 'investigations of S. M. Bukasov have clearly shown that all of the Mexican group of varieties of wild potatoes are distinct from the South American forms in their immunity from *Phytophthora*'. Granted that the Mexican wild potato species might have a unique contribution to make as a source of blight resistance, what has this got to do with the origin of *P. infestans*?

According to a theory first advanced by Vavilov (1935), the existence in a region of a group of native plant species resistant to a common pathogen, suggests that this region is also the place of origin of that pathogen. He postulated that the existing resistant species are the survivors of a selection process in which the more susceptible clones and species have been eliminated. Thus, for thousands of years in Central

Mexico the fungus and its native plant hosts must have undergone an annual biological struggle for survival, in which the wild potato population developed sufficient resistance to survive comfortably, while the pathogen was evolving into a heterogeneous population capable of maintaining itself on the host potato plants from one season to the next. This biological balance is apparently what has emerged in Central Mexico and illustrates the 'Vavilov paradox'.

Of the nearly 40 tuber-bearing *Solanum* species occurring in Mexico, nearly all of those which are found in the Central Plateau (Mesa Central) with a clearly defined rainy season, show moderate to high levels of resistance to *P. infestans*. In the drier coastal and northern regions of Mexico more susceptible species have prospered, apparently because their survival did not depend on a resistance to annual late blight epiphytotics (Hawkes, 1950).

It is curious to note that certain species occur in contiguous areas of both climatic zones. Within the same species, clones from the drier zones are often susceptible, while their counterparts from the nearby higher rainfall areas are uniformly resistant (e.g. *S. cardiophyllum*). Taken together, these observations provide firm evidence that this region was the place of origin of *P. infestans*.

Occurrence of the sexual stage of P. infestans in Central Mexico

The sexual stage of *P. infestans* remained a mystery for over 100 years after this fungus had been identified as the cause of potato late blight. There had been infrequent reports of oospores of *P. infestans* in Europe and the United States, sometimes in infected potato tubers or leaves, as well as in pure cultures of the fungus on artificial media (Clinton, 1911; Pethybridge & Murphy, 1913; Lohnis, 1922: de Bruyn, 1923; Eriksson, 1923). However, there has never been an adequate explanation for these sporadic appearances of the sexual stage, and often the reported observations could not be repeated experimentally.

After collecting resistant potato species in Mexico, and observing *P. infestans* there, Reddick was one of the first to postulate that Central Mexico might indeed be the place of origin of *P. infestans* (Reddick & Crosier, 1933). In lectures given at Cornell University, he even went so far as to speculate that one day the elusive oospore stage might be found in Mexico in nature. This astute hypothesis was substantiated nearly 20 years later; in the early 1950's, the sexual oospore of *P. infestans* was discovered to be prevalent in nature in Central Mexico (Niederhauser, 1956; Gallegly & Galindo, 1957).

Shortly thereafter, two compatibility groups (mating types) of *P. infestans*, A1 and A2, were found to be present in Mexico. To form the sexual oospore two isolates, one from each of these groups, must pair. At that time the A1 compatibility group was found to be world-wide in its distribution, while the A2 compatibility group was found only in Mexico. Thus the sexual oospore occurred only there (Niederhauser, 1956; Gallegly & Galindo, 1958). Most *Phytophthora* specialists consider this to be conclusive proof of the Mexican origin of *P. infestans*.

The range of distribution of the A2 compatibility group in Mexico is apparently limited to the same ecological zone in which the population of resistant wild potato species also prevails. Field isolations of the fungus made in Mexico from 1950 to 1970 failed to reveal the presence of the A2 compatibility group in the drier areas of northern Mexico, although the A1 compatibility group has been found there. Likewise, cultures of the A2 compatibility group have not been found further south than the Oaxaca-Chiapas area near the Isthmus of Tehuantepec in southern Mexico (Fig. 3.1).

It is now postulated that when *P. infestans* was taken from Mexico in the early 19th century and introduced into the potato-growing regions of Europe and North America, only cultures of the A1 compatibility group made the trip. This suggests a single, or very few, such introductions of the fungus into Europe and the United States. The comparatively low level of potato commerce between Mexico and the rest of the world might account for this. The discovery, in the early 1950's, of the widespread occurrence of the sexual stage of *P. infestans* in nature in Mexico, is perhaps the single most exciting event in the history of this fungus since Berkeley identified (and de Bary later proved) it as the causal pathogen of potato late blight.

Variability in isolates of P. infestans *in Mexico*

An enormous range of variability has been found in the population of isolates of *P. infestans* in Central Mexico. This variation includes characteristics such as types of growth on artificial media, pathogenicity, tolerance to chemicals, and homothallism. Such a variable population of strains is typical of a biological species in its native habitat.

Gene-for-gene relationship between the blight-resistant population of Mexican wild potato species and P. infestans

According to Vavilov (1935), the struggle for co-existence between a local population of host plants and an indigenous pathogen, would result in the creation of a complex genetical relationship between host and pathogen. Thanks to the research of Black, Reddick, Mastenbroek and

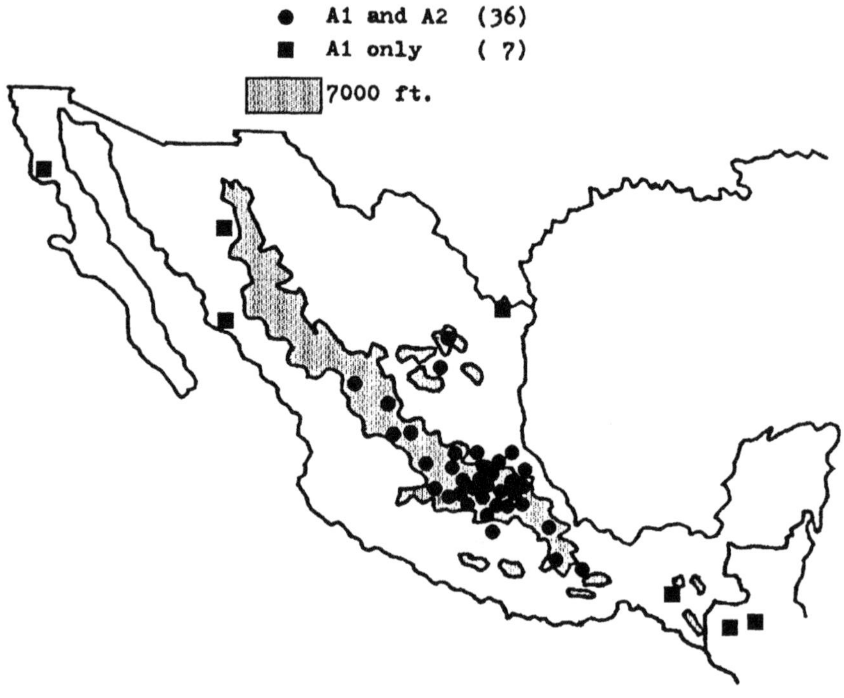

Fig. 3.1. Locations for collections of *Phytophthora infestans*, 1950 to 1970, and mating types found.

---

others, a gene-for-gene system has been proposed for *P. infestans* and the wild potato species in Mexico.

Each new resistance gene identified in a wild potato species was found to have its counterpart in a pathogenic race of the pathogen. Today more than 15 such dominant R-genes have been identified (primarily from *S. demissum*, with a few from other wild species such as *S. stoloniferum*) and each such 'major' R-gene has its virulence counterpart in *P. infestans*. The number of potential combinations in such a complex of genes is extremely large. But to date, *P. infestans* has demonstrated its ability to attack any combination of R-genes that the plant breeder has been able to concentrate in a potato seedling or selection. The fact that all of these R-genes are found in the wild potatoes only in Central Mexico together with all of their virulence counterparts in *P. infestans*, is additional evidence that this is the region of origin of the pathogen, as well as of the resistant potato species.

It may be pertinent here to mention that in forty years of field testing for late blight resistance in Central Mexico, no tuber-bearing *Solanum* species, wild or cultivated, has shown immunity to *P. infestans*. When exposed in the field to late blight, every potato species or selection has been observed to have at least a few late blight lesions from which the pathogen can be isolated. The severity of late blight on each species is determined by its level of horizontal (non-specific) resistance.

Proposals for international collaboration with the Mexican connection

The organizing committee of this *Phytophthora 89* Symposium, in extending the invitation to participate in this programme, indicated that the aim of this international meeting is not only to commemorate a centenary, but to point the way for *Phytophthora* research over the next decade.

In response to this suggestion, I would like to present several proposals for international co-operation in research on potato late blight and *P. infestans*. These proposals were selected to emphasize the special role that Mexico and its unique natural resources for late blight research might play in their implementation during the next decade. Most of these proposals are not new and have been recommended on several occasions during the past 30 to 40 years. Why is this new co-operative international strategy so critical and urgent today?

Firstly, in an era of increasingly sophisticated technologies and methods of communication the opportunity to improve the efficiency of research is greatly enhanced by closer international co-operation. This becomes even more important as budget restrictions in many countries become a limiting factor, and valuable research projects may be reduced or even abandoned.

Secondly, world food production must continue to rise during the next few decades. However, the rate of increase in potato production in many countries of the developing world is levelling off after nearly four decades of marked improvement. New irrigated land is not available. If more potatoes are to be grown, they must be grown under rainfall. In most developing countries the fungicides needed for late blight control under these conditions, are either too expensive or not available. Thus, if potato production is going to continue to expand in countries of the developing world, it must be with cultivars with higher levels of durable horizontal resistance to *P. infestans*. International co-operation will be vital to the success of such a programme.

Thirdly, further increases in food production must be obtained by rising productivity, rather than by expanding the area planted. Potato growers throughout the world, but particularly low-income subsistence farmers, are more willing to invest in improved production technologies (such as better seed, fertilizer, etc.) if there is security against sudden and catastrophic loss due to late blight.

Fourthly, more chemicals are applied to the potato than to any other food plant. Increasing numbers of concerned citizens throughout the world are searching for ways to reduce the use of agricultural chemicals, help preserve the environment and improve the quality of the foods we eat. A higher level of durable late blight resistance in potatoes, wherever they are grown, would be a definite contribution toward a solution to this serious problem of deterioration of the environment.

Some progress has been made towards these goals but the basic problems, and lack of knowledge in important research areas, persist. Today, more than at any other time during the past 150 years, we are better armed and prepared to establish international co-operative projects designed to reduce or solve the problem of potato late blight. Such an initiative would be a lasting and fitting tribute to great pioneers such as the Reverend Berkeley, Anton de Bary and others. The following are proposed areas for international research cooperation.

Nature and inheritance of a durable horizontal resistance

We have a wealth of knowledge about the vertical (specific) resistance in potatoes to *P. infestans*. Late blight research, particularly in Japan, on the hypersensitivity reaction expressed by R-genes, and the role of phytoalexins, has provided valuable information on such resistance. But, in sharp contrast, we have comparatively little information about the basic biochemical factors contributing to a durable horizontal resistance. Is our lack of basic knowledge and wider utilization of horizontal resistance due to neglect? Or are we pursuing an unattainable goal? Large (1940), in his book *The Advance of the Fungi*, concluded with the memorable statement: 'Those who believed in genes, postulated the existence of an eternal quality, R, which they could take from wild plants, build into the genetical constitution of cultivated ones, and so make them disease-resistant forever. Those who thought, not in terms of mathematical abstractions, but of the green flux of ever-changing nature, saw little hope of such permanency, and no end to man's labours in defending the crops upon which he depended for his life'. This eloquent statement was made at a time when the frustration of potato breeders all over the world was at a height, when their annual expectation of an easily attained immunity to *P.*

infestans, due to a few R-genes, was shattered repeatedly as the resourceful pathogen formed new races. However, this statement should not be taken as a final judgment that it is futile to search for effective levels of horizontal resistance, but as a warning that the search for any such durable resistance must be based on continued testing in a shifting environment, with constant exposure to a versatile and changing pathogen.

Recognition of the world-wide need for a durable, horizontal resistance to potato late blight has been increasing over the past 40 years, and some important progress has been made during this time, particularly in Mexico. More than 20 Mexican blight resistant potato varieties have been named and released for cultivation in Mexico and are replacing older susceptible cultivars. In many other countries in the developing world (e.g. Costa Rica, Rwanda and the Philippines), new varieties of Mexican origin have made a significant contribution to national potato production. Throughout the world, most potato breeding programmes interested in resistance to *P. infestans* are using Mexican germplasm as the source of resistance but are still searching for efficient and reliable methods to screen for horizontal resistance to late blight, and to measure its durability.

Factors contributing to horizontal resistance

Several general factors are characteristic of, or contribute to, horizontal resistance to *P. infestans* (Niederhauser, 1954b): (i) longer incubation period; (ii) slow growth of the pathogen in invaded tissue; (iii) reduced sporulation by the pathogen. These are little more than descriptions of certain phenomena exhibited by potato plants with horizontal resistance. We know little about the basic biochemistry, physiology, and inheritance of these factors.

If we are to make significant progress in the conquest of potato late blight in the near future, we must intensify our efforts to analyze and utilize a durable horizontal resistance. Specifically, it is recommended that a co-ordinated international research effort be initiated to determine:

- what basic chemical and physiological factors contribute to horizontal resistance;
- how such factors are genetically controlled;
- how to improve screening techniques for the physiological and chemical mechanisms involved in horizontal resistance;
- what levels of horizontal resistance are needed to minimize dependence on fungicides in integrated disease control programmes.

This proposal is made while recognizing that excellent research projects addressing some of these questions have already been initiated at various institutions in the world, and that such research can be conducted anywhere that interest, facilities, and resources are available. But I suggest it would be more efficient and cost-effective if a co-ordinated international research effort were launched, and if more of the resultant information and data could be checked and proven in the field in the home of *P. infestans*, in Central Mexico. The initiation and implementation of such an international co-operative research project is a formidable task. Are we equal to it?

Identification of new sources of horizontal resistance to potato late blight

In breeding programmes for resistance to potato late blight, a limited number of Mexican wild potato species, primarily *Solanum demissum*, have been utilized as sources of resistance. Restricted by the genetic barriers of incompatibility, the breeder has not been able to tap a number of wild potato species in Mexico that have exhibited high levels of horizontal resistance in the field (*S. bulbocastanum*, *S. cardiophyllum*, *S. pinnatisectum*, etc.). Blight resistance in these other wild potato species may be distinct from that now being used (*S. demissum*, *S. stoloniferum*, *S. verrucosum*, etc.), and they are potentially valuable sources of a new resistance to potato late blight.

Although late blight resistance is not common in South American potato species or cultivars, several of these genotypes have made a useful contribution to the resistance shown in a number of blight-resistant selections and varieties. *Solanum andigenum* (the parent species of the cultivated potato) and *Solanum phureja* (a cultivated diploid potato species) have apparently contributed to the level of resistance obtained in a number of selected new varieties (e.g. Monserrate and Atzimba).

Ochoa (1954, 1981), in Peru, has indicated that several Peruvian wild potato species (*S. chiquidenum*, *S. piurae*, *S. irosinum*) have resistance to *P. infestans*. But to date no one has evaluated this resistance or determined if it is useful for potato breeding programmes. Likewise, *S. andreanum* was described by Hawkes (1950) as having late blight resistance in southern Colombia and Ecuador, but to this day we know nothing more about this potential source of blight resistance.

Finally, we have not yet begun to tap even more exotic sources of resistance, such as the many non-tuberiferous species in the genus *Solanum* that are highly resistant or immune to *P. infestans*. Could resistance from these sources be used to create new potato cultivars? Today new, sophisticated technologies, such as genetic engineering and somatic fusion, are available to overcome the incompatibility barriers and mobilize new sources of resistance. It is time they were put to work.

Evaluation of durability of horizontal resistance

An evaluation of the durability, or stability, of horizontal resistance takes time. The word 'durability' implies that we measure how many years the horizontal resistance will last, and the significance of the test and results is in direct relation to the number of years the evaluation was conducted. There has been no way to speed up this evaluation.

The Toluca Valley in Central Mexico has gained an international reputation for the severe annual field trial for horizontal resistance to *P. infestans*. For a period of 21 years, from 1951 until 1972, two strategies for testing the durability of horizontal resistance were established. In the first, called the 'sprayed–unsprayed trial', two plots of each selection (a total of 350 to 400 from the national breeding programme for late blight resistance) were planted each year. In one plot the plants were sprayed with fungicide and late blight was controlled. In the other, the plants were unprotected, and exposed to the severe field attack by *P. infestans*. The yield in the unsprayed plot was weighed and expressed as a percentage of the total yield in the sprayed plot. Since late blight was the only important foliar disease controlled by the fungicide, the difference in yield between the sprayed and the unsprayed plots was considered a practical measure of late blight resistance in that selection. Though there was an 'erosion' of horizontal resistance observed in a few selections, most of them retained fairly stable levels of horizontal resistance over a period of 20 years.

In the second strategy, called the 'germplasm bank trial', a row of 25 plants of each selection maintained by the breeding programme was planted annually in one comprehensive unsprayed experiment. This large collection of selections (over 1800) represented a spectrum of horizontal resistance from 'highly resistant' to 'susceptible' clones. The only requirement of the 'susceptible' clones was that enough tubers were produced each year to maintain the clone. The susceptible clones were maintained year after year so as to ensure the prevalence in the field of a broad spectrum of races of *P. infestans*. In this trial the selections were rated annually for their degree of horizontal resistance. Each successive year they were planted in the order of their resistance, beginning with the most resistant and ending with the most susceptible. Thus, if a new race of *P. infestans* appeared and attacked a previously 'resistant' clone that was mistakenly thought to have horizontal resistance, that clone would appear out of place in the 'resistant' sector of the spectrum. Similarly, if certain races of the pathogen were lacking during a given season, some clones previously classified as susceptible might escape severe infection that year, and appear to be resistant and out of place in the 'susceptible' end of the spectrum.

During the 21 years that these trials were conducted, less than 5% of the selections had to be reclassified after the first two years of exposure in the field. This stability and consistency of performance is surprising and encouraging. Apparently, we are dealing with a stable, durable resistance to *P. infestans* in these field trials in the Toluca Valley, at least for the conditions under which they were annually exposed.

Utilizing the 'Mexican connection' in an international co-operative programme, two new strategies are proposed below to screen more efficiently for horizontal resistance to late blight in potato breeding programmes.

Screening of segregating resistant seedling populations in the field in Mexico: Potato breeders might find it very useful to test, in the field in Mexico, a sample of each of their resistant seedling selections at successive stages in their breeding programme. Such an additional screening trial would not substitute for the established national breeding and selecting procedures which would continue to be conducted to meet national priorities. However, the additional and severe screening for horizontal resistance in Mexico might ensure success in obtaining high levels of a durable resistance far more quickly and reliably. By testing in an environment where R-genes provide little or no protection from late blight attack, a more realistic appraisal of the level of horizontal resistance is possible.

International standard potato late blight resistance trial: It would be valuable to establish a standard field trial, not only in the Toluca Valley, but in five or six other locations in the world (Asia, Africa, Europe, North and South America) selected for the severity of their annual late blight outbreaks. If a standard world set of about 25 selections with promising levels of horizontal resistance were exposed simultaneously to late blight attack in each of these several climates, we might soon reach the long-desired goal of a high level of horizontal resistance in selections of agronomic value for many countries.

Seed-tuber samples of all selections tested in the international standard trial would be disease-free and observe all national quarantine regulations. Selections which failed to perform satisfactorily in one or more locations could be replaced annually by more promising clones. We have the interesting example of the Mexican potato cultivar Atzimba. In field trials in Mexico, it still maintains the medium level of durable horizontal resistance that it showed as a seedling selection in 1957. Yet today, as the principal potato variety planted in Costa Rica, it must be given fungicide protection against late blight. Does the horizontal resistance selected in Mexico perform differently in warmer climates or shorter photoperiods? We do not know.

Range of late blight resistance in Mexican wild potato species

Clones of wild potato species in Mexico may demonstrate a wide range of resistance to *P. infestans*. This is true even for *Solanum demissum*, the species most widely used in potato breeding programmes. It is obviously important to the potato breeder that he utilizes a clone of *S. demissum* with a high level of horizontal resistance, rather than a less promising source of resistance. These high levels of horizontal resistance in a species like *S. demissum* can be distinguished at the moment only by an exposure to a severe late blight attack in the field in Central Mexico– the home of this wild potato species.

As we expand our international co-operation in search of a practical, long-term solution to the late blight problem, a field trial should be conducted annually in Central Mexico to determine the most promising sources of resistant germplasm, even within the species widely considered to be uniformly resistant. It might also be useful to conduct such a field trial not only in Central Mexico, but in several locations in the world under other environments favouring severe late blight epidemics. Further, we should examine the levels and frequency of horizontal resistance in segregating seedling populations of these Mexican wild potato species to determine their usefulness as genetic sources of horizontal resistance.

Asexual variation in **P. infestans**

The opportunity for asexual variation in a coenocytic mycelium is immense. However, little is known about the specific mechanisms involved although their effects are evident when a 'new' race of the pathogen attacks a previously 'resistant' R-gene potato selection. Reddick & Mills (1934) noted this build up of virulence in *P. infestans* and called it 'adaptation' of the pathogen to previously-resistant cultivars. More recently, the widespread appearance of strains resistant to fungicides (Chapters 22 and 23) has caused alarm in those who rely on these chemical control measures.

In the Mexican home of *P. infestans*, a wide range of asexual variability has been repeatedly observed in cultural characteristics, pathogenicity, and even sexuality. An outstanding example of this variability was discovered by Castro (1968) while working with the phenomenon of homothallism in *P. infestans* (Fig. 3.2). Single zoospore cultures derived from homothallic isolates were found to segregate into A1, A2 and homothallic strains. Single zoospore cultures from A2 isolates segregated into A2 and homothallic strains, while those from A1 isolates were uniformly of the A1 strain. This pattern of asexual segregation continued through four generations of zoospore propagation.

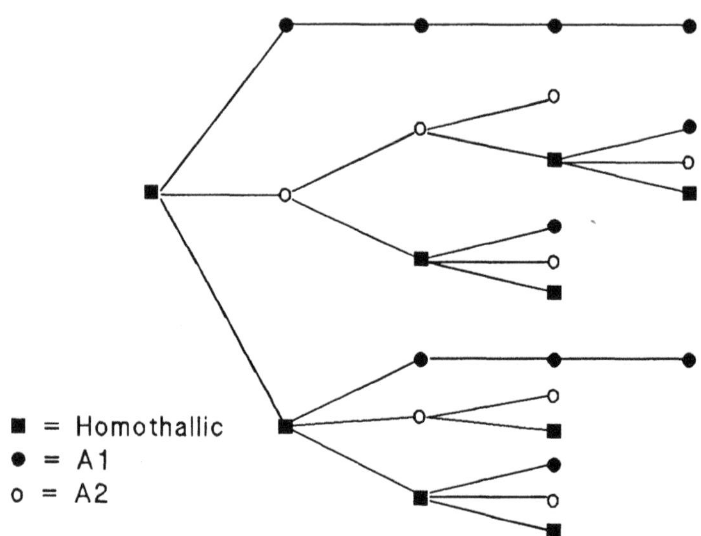

Fig. 3.2. Zoospore segregation in *Phytophthora infestans*. Single zoospore cultures derived from homothallic isolates segregate into A1, A2 and homothallic strains. Single zoospore cultures from A2 isolates segregate into A2 and homothallic strains, while those from A1 isolates were uniformly of the A1 strain. This pattern of asexual segregation continued through four generations of zoospore propagation (after Castro, 1968).

This example of asexual variation illustrates the need for a better understanding of the mechanisms involved if we are to proceed towards a solution of the potato late blight problem. Such an understanding can be attained sooner and more efficiently by closer international collaboration comparing the capacity for asexual variation in the native population of *P. infestans* in Central Mexico with that in the populations of the pathogen in other parts of the world.

Sexual stage of **P. infestans**

A thorough study of the distribution and co-existence of the A2 compatibility group and resistant wild potato species in Mexico still needs to be completed. It is important that this study be conducted soon, because of the recent spread of commercial potato cultivation in Mexico, and the corresponding expansion of the zone in which A2 is found. It is of vital

Table 3.2. Proportion of A2 isolates in different countries 1981-88 (after Hohl, unpublished)

Country	% of A2 isolates in samples of *P. infestans*
Brazil	98
Egypt	100
Israel	80-90
Japan	61
Mexico	50
Netherlands	12
Poland	1-2
Sweden	1-2
UK	10-15
West Germany	33

importance to know the effect of the arrival of the A2 strain of *P. infestans* in a new region upon the virulence phenotypes currently prevalent in that region. This is not only of academic interest, but of interest to all involved in the production of potato crops. Equally, the need to study distribution of A2 in regions of the world where it has been discovered during the past decade is clear so that the routes for spread of *P. infestans* can be established (Chapter 15). For example, there is at least circumstantial evidence that the A2 compatibility group was taken from Mexico to Europe in 1976, when some 25 000 tons of potatoes were exported. Some of the potatoes in this shipment were from regions of Mexico where the A2 strains of *P. infestans* are prevalent. This supposed introduction of these A2 strains into Europe might be substantiated using molecular markers. Another intriguing question is whether A2 will remain a rarity when introduced into new regions or increase to the 50:50 ratio found in Central Mexico (Chapter 16). Recent surveys, summarized in Table 3.2, have detected A2 strains in many countries. Further information of this sort, and especially information about its change with time, is essential.

Genetic recombination in P. infestans

As the A2 strain and the sexual stage of *P. infestans* become more common in the potato growing areas of the world, more complex populations of the pathogen can be expected to arise through genetic recombination. We have limited data indicating the range and potential

Table 3.3. Genetic recombination in *Phytophthora infestans* (from Niederhauser, 1961)

Races used in crosses		Races identified in progeny
A1	A2	
1.2.3.4	0	0; 1; 1.4; 2.4; 1.2.3.4
0	1.2.3.4	1; 1.3.4; 1.2.3.4
1	1.4	0; 1; 2; 1.4

importance of this variation through sexual crossing in *P. infestans*. An indication of what might be expected comes from experimental crosses shown in Table 3.3 (Niederhauser, 1961). The reciprocal crosses between race 0 and race 1.2.3.4 revealed the expected progeny of parental strains and various recombinant strains. Of special interest was the unexpected appearance of a race 2 progeny from a cross between race 1 and race 1,4. More information on the inheritance of virulence is now available and is summarized in Chapter 15. The variability of the fungus in Central Mexico promises a wide genetic reservoir for the expansion of such studies.

Late blight etiology and the sexual stage of **P. infestans**

The arrival of the A2 compatibility group in a new region indicates that the sexual oospore may soon become of frequent occurrence in local populations of the pathogen. This is of interest to the potato breeder because of the potential for genetic recombination and the increasing complexity of pathogenic races of *P. infestans*. But of perhaps even greater importance, is the effect of the natural occurrence of the oospore in the field on the etiology of potato late blight. Will the oospore become, as it is in Mexico, a source of inoculum in the early days of the growing season, when rainfall and cool weather could provoke oospore germination and infection of the young potato plants as they emerge? The annual course of development of the disease could be radically changed. These potential etiological changes should be anticipated and Central Mexico is an ideal location to study methods for control of such a source of inoculum. It is uncertain how many years the oospores might remain viable in the soil. Oospores will survive up to 2 years in the soil in Mexico. How will other climates affect oospore viability in soil?

International co-operative potato late blight project (ICPLBP)

Several specific suggestions have been made above for expanding international co-operation to solve some of the world-wide problems in the control of potato light blight. If we are to implement any of these suggested proposals there is an obvious and urgent need to improve the channels for international co-operation in potato late blight research projects of mutual interest to the participants. The proposal to establish an international co-operative potato late blight project was first made in 1981 at the International Symposium on *Phytophthora*, held at the University of California in Riverside. Progress has been slow, but in May 1989 representatives from Mexico, Poland, and the United States met in Poland, and formed a Technical Committee to take the first steps to establish this new International Co-operative Potato Late Blight Project (ICPLBP). This Technical Committee will hold its second meeting here in Dublin this week. Not only will they be discussing further co-operation among scientists from these three countries, but the committee will explore the possibility of expanding this collaborative programme to include interested scientists from other countries. The ICPLBP will have its official base in Mexico, and be administered by an Executive Committee composed of scientists internationally recognized for their interest in potato late blight research. Research projects sponsored by institutions included in the ICPLBP could operate as independent units within the ICPLBP or could collaborate with other institutions with similar goals, in a co-operative effort made possible through the ICPLBP.

World interest in potato late blight was awakened by a tragedy in the 1840s – the Irish Famine. Today, potato late blight remains a major problem in potato production all over the world. This symposium provides an auspicious opportunity to promote the establishment of a new project in international cooperation. Let us proceed into the next decade, and into the next millennium, in a coordinated world-wide effort to conquer this infamous plant disease that has threatened man's food supply for so many years.

References

Abbott, E. V. (1929). Diseases of economic plants in Peru. *Phytopathology* **19**, 645.

Abbott, E. V. (1931). Further notes on plant diseases in Peru. *Phytopathology* **21**, 645-656.

Anonymous (1943). In *Revista de Agricultura*, Cochabamba, Bolivia.

Bazan-Segura, C. (1952). El 'hielo' o 'la rancha' de la papa en el Peru. *Centro Nacional de Investigaciones y Experimentacion Agricola Boletin 45*.

Bourke, P. M. A. (1956). *Report to the Government of Chile on the control of potato blight*, pp. 1-112, (mimeographed).

Boussingault, J. B. (1845). Sur la maladie des pommes de terre dans la Nouvelle Grenade. *Comptes rendu Hebdomadaire des séances de l'Académie des Sciences*, **21**, 1114-1115.

Calderoni, A. (1960). *Phytophthora infestans* en el sudeste de la Provincia de Buenos Aires. Instituto de Investigaciones Agricolas (IDIA) Reunion **2**, 28-32.

Castro, J. (1968). Observaciones sobre la variacion asexual en *Phytophthora infestans* (Mont.) de Bary. Tesis Profesional, Escuela Nacional de Agricultura, Chapingo, Mexico, pp. 1-37.

Ciccarone, R. (1949). Osservazioni sulle epifizie di *Phytophthora infestans* (Mont.) de Bary nel Valle de Aragua, Venezuela. *Revista Agricola Subtropical* **43**, 22-28.

Clinton, N. P. (1911). Oospores of potato blight. *Science* **33**, 744-747.

de Bary, A. (1861). *Die Gegenwartig herrschende Kartoffelkrankheit ihre Ursache und ihre Verhuting*. Leipzig.

de Bruyn, H. L. G. (1923). The oospores of *Phytophthora infestans* (Mont.) de Bary. *Report of the International Conference of Phytopathology and Economic Entomology, Holland*, 30-31.

Eriksson, J. (1923). Das Mykoplasma-stadium von *Phytophthora infestans*. *Report of the International Conference of Phytopathology and Economic Entomology, Holland*, 33-34.

Gallegly, M. E. & Galindo, J. (1957). The sexual stage of *Phytophthora infestans* in Mexico. *Phytopathology* **47**, 13 (abstract).

Gallegly, M. E. & Galindo, J. (1958). Mating types and oospores of *Phytophthora infestans* in nature in Mexico. *Phytopathology* **48**, 274-277.

Hawkes, J. G. (1950). Algunas observaciones sobre la papa del Ecuador. *Flora (Quinto)* **7** (nos. 17-20), 93-96.

Large, E. C. (1940). *The Advance of the Fungi*. Henry Holt & Co.: New York.

Lohnis, M. P. (1922). *Onderzock over Phytophthora infestans (Mont.) de Bary. op de aardappelplant*. H. Veenman: Wageningen, Netherlands.

Montaldo, A. (1953). El cultivo de variedades de papas resistentes al tizon. *Deptartamento Investigaciones Agricolas Boletin Tecnico, Ministerio de Agricultura: Santiago*.

Niederhauser, J. S. (1954a). Observations on late blight and other potato diseases in Peru. *Plant Disease Reporter* **38**, 81-82.

Niederhauser, J. S. (1954b). Late Blight in Mexico. *American Potato Journal* **31**: 233-237.

Niederhauser, J. S. (1956). The blight, the blighter, and the blighted. *Transactions of the New York Academy of Sciences, Ser.* II **19**, 55-63.

Niederhauser, J. S. (1961). Genetic studies of *Phytophthora infestans* and *Solanum* species in relation to late blight resistance in the potato. In *Recent Advances in Botany*, pp. 491-497. University of Toronto Press: Canada.

Ochoa, C. (1954). Northern Peru, a possible new source of potatoes resistant to *Phytophthora infestans*. *Phytopathology* **44**, 500.

Ochoa, C. (1981). *Solanum irosinum*, new Peruvian tuber-bearing *Solanum* species resistant to *Phytophthora infestans*. *American Potato Journal* **58**, 131-133.

Pethybridge, G. H. & Murphy, P. A. (1913). On pure culture of *Phytophthora infestans* de Bary and the development of oospores. *Science Proceedings of the Royal Dublin Society* **13**, 566-588.

Potel, H. (1900). As molestias cryptogamicas da batata ingieza (s.t.) a seu tratamento. *Bol. Agric. (Sao Paulo)* **1**, 45.

Reddick, D. (1943). Development of blight-immune varieties. *American Potato Journal* **20**, 118-126.

Reddick, D. & Crosier, W. (1933). Biological specialization in *Phytophthora infestans*. *American Potato Journal* **10**, 131-134.

Reddick, D. & Mills, W. (1938). Building up virulence in *Phytophthora infestans*. *American Potato Journal* **15**, 29-34.

Rieder, R. (1887). La enfermedad de la papa. *Anales Santa Catalina (Buenos Aires)* **1**, 256-258.

Vavilov, N. I. (1935). *The origin, variation, immunity and breeding of cultivated plants*. Chronica Botanica, 1-366.

Chapter 4

Host-pathogen interactions: current questions

John Friend

A large proportion of the published work on host-pathogen interactions in *Phytophthora* has concerned a single species on a single host, namely *P. infestans* on the potato. My own interests are centred on biochemical studies of this particular interaction.

Although there have been innumerable papers published in the past 25 years or so, many questions still need to be answered to obtain a complete understanding. The first point to emphasize is that much of the published biochemical work has used the tuber, and often tuber slices or discs, as the experimental material. This, of course, introduces an artefact since the first stage of these investigations, namely cutting the tuber, triggers a whole series of wound reactions. The initial wound reactions, which commence within seconds of cutting the tissue, are the hydrolysis of the complex lipids (many of them in membranes) and starch of the tuber. These reactions are followed, possibly as long as 24 hours after cutting, by biosynthetic reactions leading to the synthesis of suberin, lignin and new complex lipids (and new membranes). These wound reactions are so rapid that they could easily mask the biochemical reactions occurring specifically in response to the pathogen. Even if we can differentiate between the responses to wounding and to the pathogen, how valid are tuber reactions to a study of the disease *in vivo*? *P. infestans* is primarily a leaf pathogen and whole tuber infection is only a late stage of the disease.

Investigations of the mechanism of penetration of tuber cell walls have shown that galactanase enzymes produced by the pathogen appear to be of far greater importance than the polygalacturonase enzymes (Keenan & Friend, 1989). One useful 'spin-off' from these studies has been that we have been able to use these two types of enzyme to investigate, in some detail, the structure of potato pectin. Nevertheless, we still need to determine the nature of the phenolic cross-links in potato pectin; in particular, we need to know whether these phenolic cross-links have any role in protecting the host cell wall against the degradative enzymes of the pathogen. Fortunately, chemists are developing new methods for determination of phenols in plant cell walls and these new chemical approaches should be tried on potatoes.

The biochemistry of resistance is another area where we have some knowledge, but each piece of extra information seems to raise additional questions. The results of the experiments on the interaction between *P. infestans* and potato which formed the basis of the phytoalexin hypothesis were first published in 1940 by Müller, but the structure of the first potato phytoalexin, rishitin, was not published until 1968. A great deal is now known about the structures of the rishitin series of sesquiterpenoid phytoalexins and their biosynthetic and metabolic pathways. However, we know little about the role of rishitin in the process of resistance. It is not clear whether this compound accumulates locally in sufficiently high concentrations to cause fungal death in all R-gene resistant tubers. Some tubers do not accumulate any rishitin; others only react to produce rishitin after a long period of post-harvest storage. The rishitin series of phytoalexins are not produced in leaves; we do not know whether there are other types of phytoalexin present in leaves. What role in the resistance process is played by *p*-hydroxybenzoic, vanillic and salicylic acids, produced in the incompatible interaction of cell suspension cultures from cv. Orion containing the R1 gene and race 4 of *P. infestans*? What is the role of 'lignification' and of callose encasements of fungal hyphae, both of which occur in incompatible interactions between avirulent races of the fungus and in tuber tissue of resistant cultivars? The really important question is whether the accumulation of sesquiterpenoid phytoalexins, hydroxybenzoic acids, lignin-like compounds or callose are important in causing fungal death which follows hypersensitive death of the penetrated and some adjacent host cells. Since *P. infestans* is biotrophic in its early stages of growth on potato tissue, the biotrophy could be inhibited by the process of hypersensitive cell death. It is therefore possible that the accumulation of the antifungal chemicals may be a secondary defence against a subsequent attack, rather than a defence against the initial challenge by the avirulent race of the fungus. We must remember Müller's initial experiments (Müller & Börger, 1940) in which half tubers were inoculated with an avirulent and a virulent race 24 hours after each other. The growth of the second, virulent, inoculum was inhibited in those regions where there had been the first inoculation by the avirulent race. In other words there was presumably sufficient phytoalexin present after 24 hours to inhibit either spore germination or growth of the second inoculum. This experimental result gives little indication of the mechanism by which the avirulent race itself was inhibited.

Turning to the published work on elicitors for an explanation of specificity, we find ourselves in a quandary. So far, only race-non-specific elicitors have been isolated from fungal mycelium, spore germination fluids and culture filtrates (Keenan, *et al.*, 1986; Preisig & Kuć, 1987). The isolated elicitors are of two types. Carbohydrate-containing elicitors all seem to have a peptide component and are probably glycoproteins. These elicitors are inhibited by β-1,3-glucans or mycolaminarins isolated from the fungus. Lipid-containing elicitors have either arachidonic or eicosapentaenoic acids in their fatty acid moieties. There is some published evidence that the mycolaminarins have some race-specificity, since compounds isolated from virulent races appear to be more effective suppressors of phytoalexin accumulation than similar compounds isolated from avirulent races of the fungus (Doke, Garas & Kuć, 1979). It has accordingly been proposed that race-specificity in the *P. infestans*-potato interaction resides in the mycolaminarins. Unfortunately, this hypothesis conflicts with the genetic evidence that avirulence is dominant. Moreover, the mycolaminarins enhance the phytoalexin eliciting activity of the lipid-containing elicitors. It must be remembered that both the carbohydrate- and lipid-containing elicitor preparations have been made from *in vitro* fungal cultures; it is possible, therefore, that neither of them is important in the interaction between the fungus and the plant. If so, we should probably be looking for race-specific elicitors which are produced *in planta* and which interact with specific host receptors. A specific interaction of this type would probably occur at the host-pathogen interface; exploration of the chemistry of these interfaces might yield interesting information.

Another current question is the possible role of superoxide anion as a trigger for the hypersensitive response. Doke has published a series of papers which show the production of superoxide in tuber discs of the Japanese cultivar, Rishiri, challenged either with an avirulent race of *P. infestans* or with an elicitor preparation derived from hyphal walls. In addition, he has evidence that a membrane-bound NADPH-oxidase system which can generate superoxide is induced prior to the appearance of the hypersensitive reaction (Doke, 1983). However, there is one published paper (Moreau & Osman, 1989) and unpublished reports from two laboratories in the United States (J. Kuć; R. Bostock) and two in the United Kingdom (J. Ellis & J. Friend, Hull; K. Beckman & D. Ingram, Cambridge) suggesting that Doke's experiments cannot be repeated. Nevertheless, it should be reiterated that whereas all Doke's experiments were carried out with cv. Rishiri potatoes those carried out by others used cv. Kennebec. Doke has offered, in a personal communication to Beckman, Ingram and myself, an explanation of why it will only be possible to demonstrate superoxide production in certain cultivars. His explanation is related to the number of hypersensitively reacting cells at cut surfaces of tuber discs; there are far more in Rishiri than in Kennebec where some of the reacting cells are well below the surface layer. Since the half-life of superoxide anion is very short (0·25 s), it will have dismutated by the time it diffuses through the tuber disc to reach the external solution.

Moreover, there is indirect evidence from our own experiments (Keenan *et al.*, 1986) that oxygen free radicals such as superoxide and hydroxyl could be involved in the hypersensitive reaction of Kennebec tuber discs treated with either elicitors or sporangia of an avirulent race of the fungus. Compounds which either act as general free radical scavengers or which specifically scavenge either superoxide or hydroxyl all inhibited phytoalexin accumulation and hypersensitive browning. From the results of these indirect experiments, I am prepared to accept that superoxide is involved in the early stages of the elicitation of the hypersensitive response. However, I would argue the case much more forcibly if there were more evidence for the direct demonstration of superoxide.

In summary, the fundamental question which needs to be answered is: how do the fungus and the plant recognize each other? Following on from this, subsidiary questions concern how this recognition leads to either compatibility and on to disease or to incompatibility and on to resistance. Finally, I would suggest that we have insufficient knowledge at this stage to ask the questions in the correct manner, let alone to answer them correctly.

References

Doke, N. (1983). Generation of superoxide anion by potato tuber protoplasts during the hypersensitive response to hyphal wall components of *Phytophthora infestans* and specific inhibition of the reaction by suppressors of hypersensitivity. *Physiological Plant Pathology* **23**, 359-367.

Doke, N., Garas, N. A. & Kuć, J. (1979). Partial characterization and aspects of the mode of action of a hypersensitivity-inhibiting factor (HIF) isolated from *Phytophthora infestans*. *Physiological Plant Pathology* **15**, 127-140.

Keenan, P. J. & Friend, J. (1989). The degradation of potato cell walls by pathogens. In *Cell Separation in Plants*, (ed. D. J. Osborne & M. B. Jackson), pp. 179-187. Springer-Verlag: Berlin.

Keenan, P. J., Ellis, J. S., Rathmell, W. G. & Friend, J. (1986). Carbohydrate and lipid-containing elicitors from *Phytophthora infestans*. Do they have a common mechanism of action? In *Biology and Molecular Biology of Plant-Pathogen Interactions*, (ed. J. Bailey), pp. 185-189. Springer-Verlag: Berlin.

Moreau, R. A. & Osman, S. F. (1989). The properties of reducing agents released by treatment of *Solanum tuberosum* with elicitors from *Phytophthora infestans*. *Physiological and Molecular Plant Pathology* **35**, 1-10.

Müller, K. O. & Börger, H. (1940). Experimentelle Untersuchungen über die *Phytophthora*-Resistenz der Kartoffel. *Arbeiten aus der biologischen Bundesanstalt für Land- und Forstwirtschaft (Berlin)* **23**, 189-231.

Preisig, C. L. & Kuć, J. (1987). Phytoalexins, elicitors, enhancers, suppressors and other considerations in the regulation of R-gene resistance to *Phytophthora infestans* in potato. In *Molecular Determinants of Plant Diseases*, (ed. S. Nishimura, C. A. Vance & N. Doke), pp. 203-221. Japan Science Press: Tokyo & Springer-Verlag: Berlin.

Chapter 5

Ultrastructural and immunological studies of zoospores of *Phytophthora*

A. R. Hardham, F. Gubler & J. Duniec

During the years since the first International *Phytophthora* conference in 1981, our understanding of the ultrastructure and cell biology of pathogenic fungi, including species of *Phytophthora*, has been greatly augmented by the use of two techniques previously only rarely applied to studies of fungi. One is freeze-substitution, the other immunocytochemistry. Application of freeze-substitution (Howard & O'Donnell, 1987) has followed the first report of excellent ultrastructural preservation of hyphae of filamentous fungi using this technique (Howard & Aist, 1979). Its use has revealed new details of the structure of apical organelles and of the organisation of cytoskeletal components in hyphal tips and their role in tip growth and differentiation (e.g. Heath *et al.*, 1985; McKerracher & Heath, 1985; Hoch, Tucker & Staples, 1987; Bourett, Hoch & Staples, 1987). The first use of freeze-substitution to study cells of *Phytophthora* has appeared recently (Cho & Fuller, 1989a & b). Although immunological techniques were used as early as 1961 in attempts to obtain sera diagnostic for particular fungal pathogens (Buxton, Culbreth & Esposito, 1961), it has only recently been demonstrated that immunological probes can help determine the functions of selected fungal components during plant infection. Immunofluorescence microscopy has been employed extensively to study the distribution of microtubules and actin microfilaments in fungal cells (e.g. Hoch & Staples, 1985; Hoch, Bourett & Staples, 1986) including *Phytophthora* (Hardham, 1987a). Monoclonal antibodies raised against a variety of fungal antigens have also been used to study non-cytoskeletal components in infective cells, zoospores, and cysts of *Phytophthora*. These latter studies are the main focus of this chapter.

For many species of *Phytophthora*, motile zoospores are instrumental in initiating the infection of host plants. Under suitable conditions, large numbers of zoospores are produced by the fungus. The zoospores may be moved passively over long distances, and once near a potential host are chemotactically attracted to the plant surface where they encyst. Encystment involves loss of motility through the detachment of the flagella, a change in cell shape from ovoid to spherical (Fig. 5.1), secretion of adhesive material which bonds the spores to the plant surface, and the formation of a cellulosic cell wall. About 20 to 30 min later, the cysts germinate and the hyphae grow towards and into the plant tissues.

Molecules on the surface of *Phytophthora* zoospores and cysts must play many important roles in the infection process. They are likely to function in the recognition events that occur during chemotaxis, docking, the induction of encystment and chemotropic growth of the hyphae. They will also facilitate adhesion of the fungal cells to the host surface (Chapter 6) and protect the cyst via mucilaginous or wall components. In addition, these and other molecules at the fungal cell surface are prime candidates for recognition by the host of the invading pathogen.

This chapter will review what is now known about the surface properties of *Phytophthora* zoospores and will examine the variety of storage vesicles whose exocytosis changes the properties of the surface during early infection. Both stored and surface components have been studied with lectins and with monoclonal antibodies raised against *P. cinnamomi*. Monoclonal antibodies have been especially valuable, allowing a completely new approach to research on the infection of plants by *P. cinnamomi*. In addition, varying levels of cross-reactivity make them ideal tools for comparative studies of other *Phytophthora* species.

Surface properties of *Phytophthora* zoospores

Unlike most plant and fungal cells, zoospores of fungi lack a microfibrillar cell wall and their outer surface is that of the plasma membrane; they are therefore unable to osmoregulate by building up turgor pressure. Water flows into the zoospores down its concentration gradient and must be continually pumped out through the action of a water expulsion vacuole. These features mean that the zoospores are easily damaged during fixation. Even after careful adjustment of fixative osmolarity (Hardham, 1985), in material prepared conventionally using chemical fixation, the zoospore surface is irregular, with dilation of peripheral vesicles apparently causing the plasma membrane to bulge (e.g. Figs. 5.4, 5.11 to 5.14; Bimpong & Hickman, 1975; Pinto da Silva & Nogueira, 1977; Sing & Bartnicki-Garcia, 1975a; Hemmes, 1983). By contrast, the elegant freeze-substitution work of Cho & Fuller (1989a) demonstrates that the surface topography of *Phytophthora* zoospores is normally quite smooth.

Phytophthora and *Saprolegnia* zoospores possess a carbohydrate cell coat analogous to the glycocalyx of animal cells. The outer leaflet of the plasma membrane bilayer is thicker than the inner (Fig. 5.2) and carbohydrates on the outer surface of the plasma membrane can be visualised cytochemically with the periodic acid-Schiff stain (Fig. 5.3; Lehnen & Powell, 1988) or with ruthenium red (Fig. 5.4; Hardham, 1989). They can also be demonstrated with lectins (Figs. 5.5, 5.6), proteins which bind specifically to certain sugar residues.

The lectin concanavalin A (ConA), specific for mannosyl and/or glucosyl residues, binds to most plant and animal plasma membranes (e.g. Burgess & Linstead, 1976; Cuatrecasas, 1963), and to the surface of zoospores of *P. boehmeriae* (Fig. 5.5), *P. cinnamomi* (Bacic, Williams & Clarke, 1985; Hardham, 1985; Hardham & Suzaki, 1990), *P. heveae*, *P. infestans* (Garas & Kuć, 1981), *P. meadii* (Fig. 5.6), *P. palmivora* (Sing & Bartnicki-Garcia, 1975b), *P. vignae* and *Saprolegnia ferax* (Lehnen & Powell, 1989). There is also indirect evidence that potato lectin binds to zoospores of *P. infestans*, indicating the presence of N-acetyl-glucosamine residues (Garas & Kuć, 1981). Binding of other lectins to the zoospore surface has not yet been demonstrated.

ConA binding sites are evenly distributed over the surface of the zoospores, including the two flagella. However, immunological probes raised against *P. cinnamomi* reveal that the surface is not homogeneous (Hardham, Suzaki & Perkin, 1986). In zoospores of all *Phytophthora* species examined (see Table 5.1) there is evidence for four molecular domains on the surface. Each flagellum constitutes a separate domain, the

---

Figs. 5.1 to 5.10 (facing page). The surface of *Phytophthora* zoospores. 1, zoospores of *Phytophthora* are elliptical in outline with a groove (g) running along the length of the ventral surface. Two flagella (small arrows) emerge from the centre of the groove. A water expulsion vacuole (large arrows) lies at the anterior end of the zoospore. Cysts (c) are spherical, lack flagella and are surrounded by a cell wall (cells shown are of *P. cinnamomi*; bar = 10 μm); 2, the outer leaflet of the plasma membrane and the non-cytoplasmic leaflet of the peripheral cisternae are thicker than the inner and cytoplasmic leaflets, respectively, of these membranes (*P. cinnamomi*, post-osmication in 0·025% OsO_4 (bar = 0·1 μm); 3, the plasma membrane and peripheral cisternae react with the periodic acid-Schiff-thiocarbohydrazide-silver proteinate cytochemical tests for carbohydrate (Edgar & Pickett-Heaps, 1982) (bar = 0·1 μm); 4, carbohydrates on the surface of the zoospore plasma membrane are stained by ruthenium red (*P. cinnamomi*, bar = 3 μm); 5 & 6, concanavalin A-FITC binds to the entire surface of *P. boehmeriae* (Fig. 5.5) and *P. meadii* (Fig. 5.6) zoospores (bar = 5 μm); 7, monoclonal antibody Zt-1 binds to the surface of the anterior flagellum of *P. cinnamomi* zoospores [in this and all other immunofluorescence micrographs, monoclonal antibody binding sites are revealed by secondary incubation in sheep-antimouse immunoglobulin coupled to FITC (SAM-FITC)] (bar = 5 μm); 8, monoclonal antibody Zf-1 binds to the surface of both flagella of zoospores of all species of *Phytophthora* as illustrated here by *P. meadii* (bar = 5 μm); 9, monoclonal antibody Zw-1 binds to the surface of the plasma membrane associated with the water expulsion vacuole in zoospores of all species of *Phytophthora* (*P. heveae*, bar = 5 μm); 10, monoclonal antibody Zp-1 binds to the entire surface of zoospores of *P. cinnamomi* isolate 6BR (bar = 10 μm).

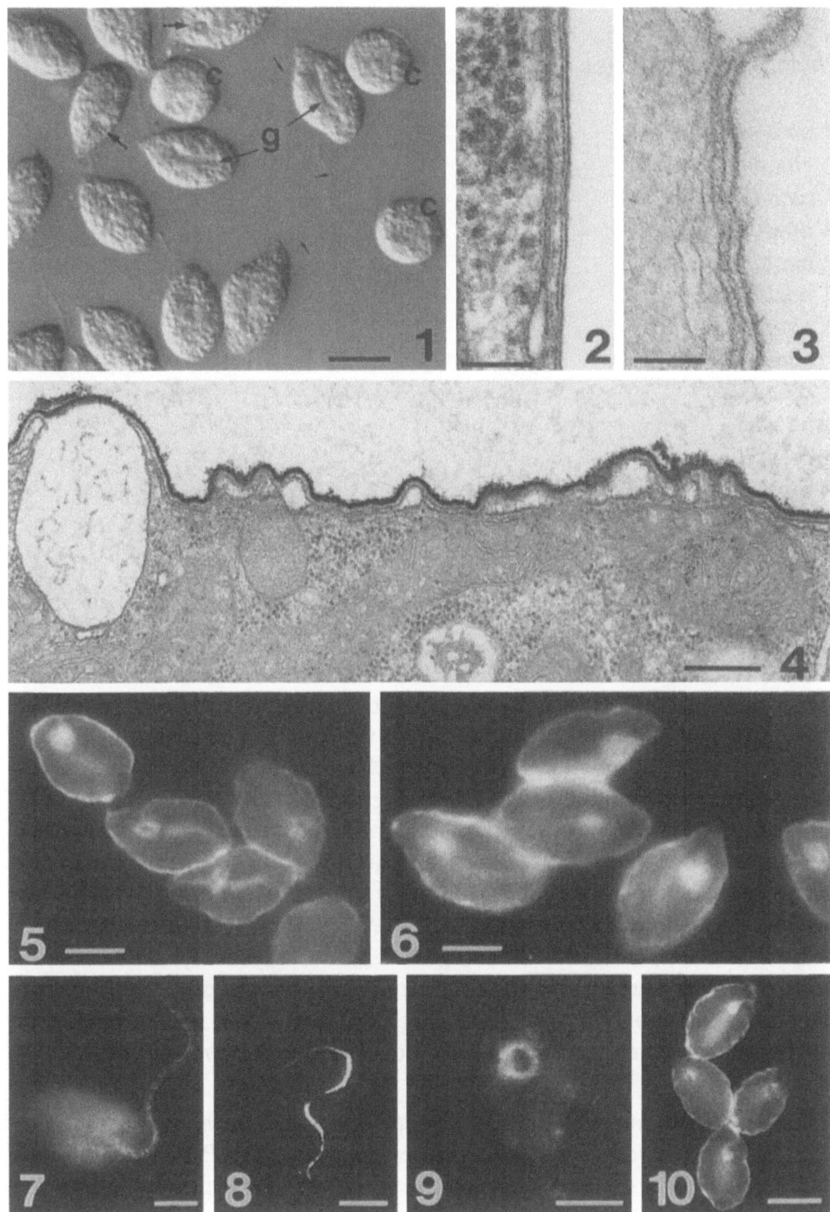
c
c
g
c
1
2
3
4
5
6
7
8
9
10

Table 5.1 Reactions of monoclonal antibodies raised against *P. cinnamomi* 6BR with other *Phytophthora* species

Phytophthora species	Antibody				
	Zp-1	Zt-1	Zg-1	Zf-1	Zw-1
P. cinnamomi 6BR	+	+	+	+	+
P. cinnamomi A278	+	+	+	+	+
P. cinnamomi A2420	-	+	+	+	+
P. boehmeriae	-	-	-	+	+
P. cactorum	-	-	-	+	+
P. castanae	-	-	-	+	+
P. citricola	-	-	-	+	+
P. citrophthora	-	-	-	+	+
P. drechsleri	-	-	-	+	+
P. heveae	-	-	-	+	+
P. meadii	-	-	-	+	+
P. megasperma	-	-	-	+	+
P. nicotianae var. *nicotianae*	-	-	-	+	+
P. palmivora	-	-	-	+	+
P. nicotianae var. *parasitica*	-	-	-	+	+
P. vignae	-	-	-	+	+

membrane associated with the water expulsion vacuole is another and the remaining cell body plasma membrane is the fourth. The antibodies which reveal these domains bind to both flagella only (Zf-1; Fig. 5.8) and the water expulsion vacuole only (Zw-1; Fig. 5.9). These antibodies bind to all *Phytophthora* species examined, indicating that these four domains are of common occurrence (Table 5.1). The antibodies which bind to the entire zoospore surface (Fig. 5.10) have a similar distribution to ConA but bind to only two out of 17 isolates of *P. cinnamomi* tested. It remains to be determined whether these antibodies bind to glycocalyx determinants or to another surface component. Epitopes confined to the anterior flagellum and labelled with Zt-1 (Fig. 5.7) and Zg-1, are restricted to isolates of *P. cinnamomi* and do not occur in other *Phytophthora* species (Table 5.1).

There is evidence that the components localised on the surface of both flagella are involved in the induction of encystment in *P. cinnamomi* and *P. drechsleri* (Hardham & Suzaki, 1986). Other surface components are likely to be chemoreceptors but as yet we have no idea as to their identity or distribution. As for the glycocalyx, even in animal cells it has proved difficult to demonstrate its function unambiguously. Likely roles include cell-cell and cell-matrix recognition but there is at this time only indirect evidence for this in some systems (Alberts *et al.*, 1989). The antigens confined to the water expulsion vacuole may play a role in water expulsion or in endocytosis, which has been shown to be restricted to this region in *Aphanomyces* zoospores (Cerenius, Rennie & Fowke, 1988).

The zoospore cortex

The cytoplasm of *Phytophthora* zoospores is highly structured (see Hemmes, 1983). The water expulsion vacuole lies at the anterior end of the cell (Fig. 5.1). The nucleus and fingerprint vesicles dominate the central region, while most mitochondrial profiles, and a variety of vesicles are confined to the cell cortex (Hemmes, 1983; Hardham, 1987b). Polarity exists even within the cell cortex. Vesicles and mitochondrial profiles are absent from the cortex lining the lower half of the groove, and the vesicle complement in the ventral (the surface containing the groove) cortex is different from that on the dorsal surface of the cell (Hardham, 1987b).

Underneath the entire zoospore plasma membrane, except for that within the groove or associated with the water expulsion vacuole, a system of flat, disc-like vesicles lies close to the cytoplasmic face of the plasma membrane (Figs. 5.2 to 5.4, 5.12 to 5.15). These vesicles are generally called peripheral cisternae (see Hardham, 1987b). In chemically fixed material they are usually dilated at their edges and adjacent cisternae are closely appressed. Freeze-substituted cells, however, lack cisternal dilation and adjacent cisternae are from 50 to 500 nm apart (Cho & Fuller, 1989a & b). The membrane of the peripheral cisternae is morphologically and cytochemically similar to the plasma membrane (Grove & Bracker, 1978; Hardham, 1987b). It is asymmetrical (Fig. 5.2), and carbohydrate moieties are present on its non-cytoplasmic face (Fig. 5.3). The membrane of strands of ER which lie below the peripheral cisternae are thinner than either the plasma membrane or cisternal membrane and do not react with carbohydrate stains.

Ultrastructural observations of *Phytophthora* zoospores show that, apart from the peripheral cisternae, there are vesicles of two size classes in the peripheral cytoplasm (Table 5.2; Beakes, 1987); the larger are very sensitive to fixative osmolarity, displaying a tendency to extend perpendicularly to the cell surface and to become elliptical as a result (Figs. 5.4, 5.11, 5.13). In *P. cinnamomi* and *P. megasperma* f. sp. *glycinea* they are close to $0\cdot6\,\mu$m in diameter in the plane parallel to the cell surface (Table 5.2). In most other species, they are between $0\cdot4$ to $0\cdot5\,\mu$m in diameter, while in *P. nicotianae* var. *parasitica* they are only $0\cdot3\,\mu$m in diameter. Their contents appear granular after OsFeCN fixation (Figs. 5.11 to 5.14), flocculent after conventional post-osmication (Fig. 5.4), and homogeneous and moderately electron-dense in non-osmicated material embedded in Lowicryl K4M (Figs. 5.21, 5.33, 5.34; Gubler & Hardham, 1988). Often, in osmicated cells, a shell of more densely packed material lines the inside of the vesicle membrane (Fig. 5.13). In Lowicryl K4M, this shell appears electron-lucent (Figs. 5.33, 5.34; Gubler & Hardham, 1988). Because of their size and distribution they are often called large peripheral vesicles (but see Hemmes, 1983, and Hardham, 1987b).

Until recently, only one type of vesicle in the smaller size class has been recognised. Because some gave a positive reaction with the diaminobenzidine test, they have been considered to be microbodies (Philippi, Parish & Hohl, 1975; Powell, Lehnen & Bortnick, 1984; Powell & Bracker, 1986). Detailed examination of *P. cinnamomi* zoospores showed that there are vesicles with two predominant structures in this size class (Hardham, 1987b). Some are filled with moderately electron-dense material and contain plate-like inclusions which often lie parallel to the adjacent surface of the vesicle. These vesicles usually have angular profiles and occur mainly on the ventral surface. In other vesicles, visible contents occupy only a small part of the organelle. These vesicles are usually circular in cross-section and occur mainly on the dorsal surface. Intermingling of the two types occurs near the edges of the ventral surface.

Examination of zoospores of other *Phytophthora* species reveals that two types of small peripheral vesicle can be distinguished in some species. As in *P. cinnamomi*, there is a tendency for vesicles on the ventral surface to be slightly larger than those on the dorsal surface (Table 5.2). *P. cactorum* (Figs. 5.11, 5.12), *P. castanae*, *P. citrophthora* (Fig. 5.13) and *P. heveae* contain two types of small vesicles which are very similar in morphology to those in *P. cinnamomi*. In *P. citricola*, *P. palmivora* (Fig.

Figs. 5.11 to 5.15 (facing page). Peripheral vesicles in zoospores of *Phytophthora*. 11, transverse section through the anterior end of a zoospore of *P. cactorum*. Large (large arrows) and small peripheral vesicles are present. Small peripheral vesicles containing plate-like inclusion (arrowheads) occur above the groove on the ventral surface. Small peripheral vesicles whose contents only partially fill the vesicle (small arrows) occur predominantly on the dorsal surface (bar = 0·5 μm); 12, the four types of peripheral vesicles in a zoospore of *P. cactorum*. Small arrowheads: peripheral cisternae; large arrows: large peripheral vesicle; small arrows: small peripheral vesicle predominating on the dorsal surface; large arrowheads: small peripheral vesicle predominating on the ventral surface (bar = 0·25 μm); 13, the four types of peripheral vesicles in a zoospore of *P. citrophthora*. Labelling of micrograph as in Fig. 5.12 (bar = 0·5 μm); 14, peripheral vesicles in a zoospore of *P. palmivora*. In this species, the dorsal surface small peripheral vesicles have a homogeneous appearance. Labelling of micrograph as in Fig.12 (bar = 0·5 μm); 15, peripheral cisternae (small arrowheads) and small peripheral vesicles in zoospores of *P. vignae*. In this species it is difficult to distinguish two different types of small peripheral vesicle. Those present contain a shell of rod-like fibres near the vesicle membrane (bar = 0·25 μm).

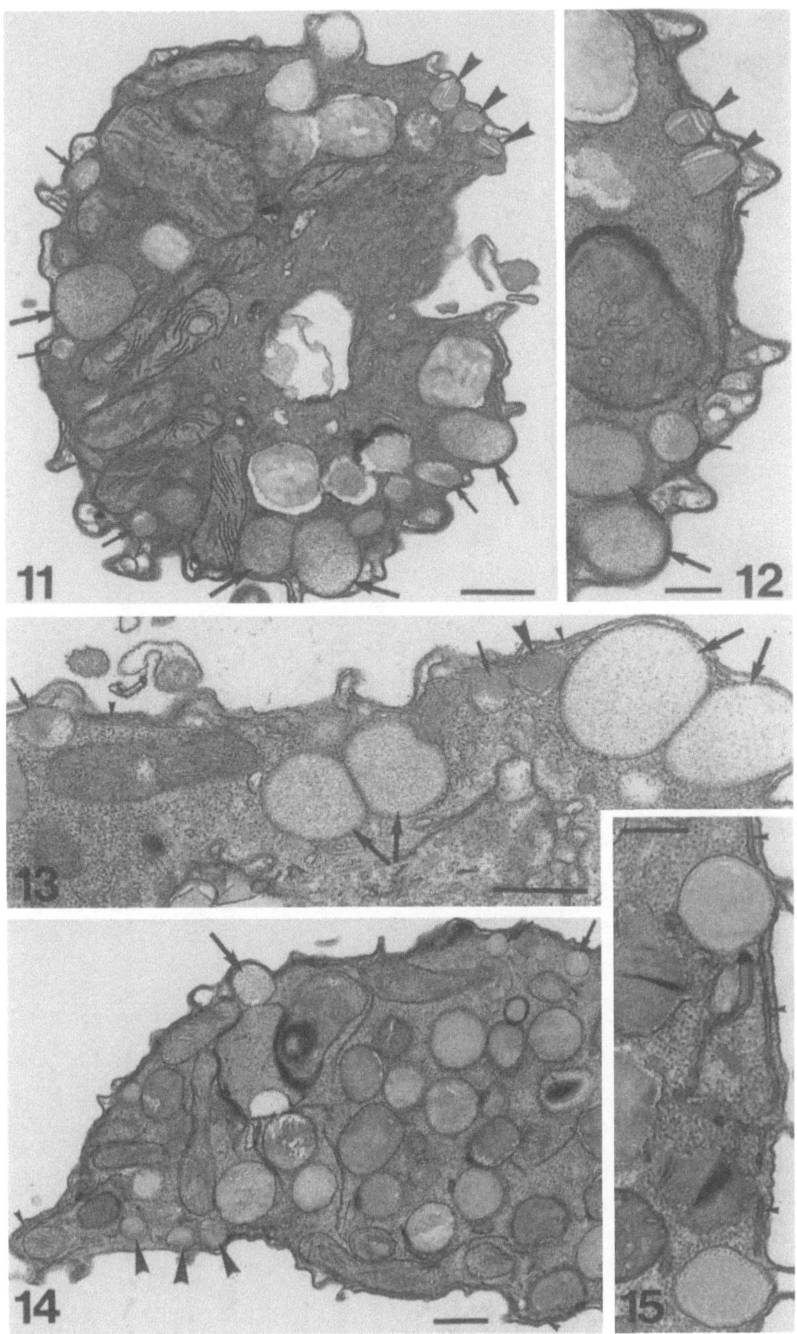
11
12
13
14
15

Table 5.2. Dimensions of peripheral vesicles in *Phytophthora* zoospores

Phytophthora species	Lpv[a] diameter	Vsv[b] length	Vsv[b] width	Dsv[c] length	Dsv[c] width	Spv[d] length	Spv[d] width
P. cactorum	0·49±0·13(66)[e]	0·26±0·06	0·17±0·03(30)	0·22±0·06	0·18±0·04(39)		
P. castanae	0·38±0·07(23)	0·30±0·06	0·23±0·03(12)	0·24±0·05	0·23±0·05(7)		
P. cinnamomi	0·59±0·13(121)	0·32±0·05	0·31±0·05(39)	0·35±0·06	0·33±0·06(47)		
P. citricola	0·49±0·11(48)	0·21±0·03	0·15±0·03(9)	0·19±0·01	0·16±0·03(4)		
P. citrophthora	0·52±0·11(47)	0·28±0·06	0·20±0·03(26)	0·26±0·06	0·22±0·04(20)		
P. drechsleri	0·53±0·12(56)					0·32±0·07	0·26±0·05(50)
P. heveae	0·36±0·05(24)	0·30±0·09	0·20±0·03(12)	0·22±0·09	0·19±0·06(8)		
P. meadii	0·50±0·06(5)						
P. megasperma[f]	0·57±0·10(23)					0·41±0·08	0·37±0·06(32)
P. palmivora	0·42±0·11(75)	0·29±0·06	0·24±0·05(45)	0·23±0·04	0·22±0·04(40)		
P. nicotianae[g]	0·28±0·05(32)	0·26±0·06	0·21±0·03(25)	0·21±0·02	0·21±0·02(2)		
P. vignae	0·40±0·08(10)					0·31±0·07	0·27±0·06(47)
P. palmivora[h]	0·38±0·04(17)						

[a]Lpv: Large peripheral vesicle diameter in plane parallel to cell surface.

[b]Small peripheral vesicles predominating on the ventral surface.

[c]Small peripheral vesicles predominating on the dorsal surface.

[d]Small peripheral vesicles in species in which two types cannot be distinguished by morphological criteria alone.

[e]Numbers in parentheses indicate the number of vesicles measured.

[f]*P. megasperma* f. sp. *glycinea*.

[g]*P. nicotianae* var. *parasitica*

[h]From micrographs of freeze-substituted material in Cho & Fuller, 1989a & b.

5.14) and *P. nicotianae* var. *parasitica* the ventral surface vesicles are similar to those in *P. cinnamomi* but contents of the dorsal vesicles are homogeneous, and electron-lucent areas are absent. In *P. drechsleri*, *P. megasperma* f. sp *glycinea* and *P. vignae* (Fig. 5.15) it is not possible clearly to identify two types of small peripheral vesicle. Even in species in which two types are evident, the occurrence of vesicles whose structure appears to be intermediate between these two extremes makes it difficult from morphological criteria alone to determine if the two types are different or simply variations in ultrastructural preservation of the same vesicle. However, recent immunolabelling has revealed that there are two distinct types of small peripheral vesicle in *Phytophthora* zoospores, as described in the next section.

Immunolabelling of peripheral vesicles in *Phytophthora* zoospores

Monoclonal antibodies, Cpa-2, Cpw-1 (Hardham *et al*., 1986), Lpv-1 (Gubler & Hardham, 1988), and Vsv-1 (Hardham & Gubler, unpublished) were raised against components in spores of *P. cinnamomi* and are specific for each of the four types of peripheral vesicles in zoospores of this species.

Well-preserved zoospores are not labelled by any of these antibodies when pre-embedding methods are used. Low concentrations of glutaraldehyde, however, do not preserve the large peripheral vesicles reliably and after fixation in the 0·2% glutaraldehyde/4% paraformaldehyde combination useful for fluorescence microscopy (Hardham, 1985), most zoospores have at least some large peripheral vesicles which have dilated and ruptured through the plasma membrane. The contents of these vesicles are then accessible to Lpv-1 and appear as plaques of fluorescence on the zoospores (Fig. 5.16). The small peripheral vesicles are more consistently preserved intact even with low glutaraldehyde concentrations, but after fixation in paraformaldehyde alone, the plasma membrane and vesicle membranes are fragmented and the contents of both large and small peripheral vesicles are able to be labelled by pre-embedding methods. Lpv-1 labels over 100 disc-shaped organelles (Fig. 5.17). Cpa-2 labelling yields small spots of fluorescence over most of the zoospore except in the central regions of the ventral surface (Fig. 5.18). Vsv-1 labelling appears as small spots of fluorescence that predominate on the ventral surface, often being concentrated along the ridges lining the groove (Fig. 5.19). Pre-embedding labelling of zoospores with Cpw-1 has not been observed.

Post-embedding labelling using secondary antibodies conjugated to colloidal gold particles reveals details of the binding sites of the monoclonal antibodies at the ultrastructural level. Cpw-1 labels the peripheral cisternae, albeit at very low levels (Fig. 5.20). Lpv-1 labels the contents of the large peripheral vesicles (Fig. 5.21). Cpa-2 labels the contents of the small peripheral vesicles that predominate on the dorsal surface (Fig. 5.22). Vsv-1 labels the contents of the small peripheral vesicles that predominate on the ventral surface (Fig. 5.23). Double-labelling using gold particles of different sizes shows that the peripheral vesicles are not labelled by more than one antibody (e.g. Gubler & Hardham, 1988).

Immunofluorescence screening of other *Phytophthora* species reveals epitopes recognized by Lpv-1, Cpa-2 and Vsv-1 (Figs. 5.24, 5.25; Table 5.3). Lpv-1 gives a positive reaction with *P. heveae*, *P. nicotianae* var. *nicotianae* and *P. nicotianae* var. *parasitica*. Cpa-2 reacts with *P. cactorum*, *P. castanae* and *P. megasperma* f. sp. *glycinea*. Vsv-1 reacts with all *Phytophthora* species. In some cases post-embedding labelling at the ultrastructural level has also been obtained. Lpv-1 binds to the contents of large peripheral vesicles in *P. nicotianae* var. *nicotianae* and *P. nicotianae var. parasitica* (Fig. 5.27). Vsv-1 labels the small peripheral vesicles along the ventral surface in *P. heveae*, *P. megasperma* f. sp. *glycinea* (Fig. 5.26) and *P. palmivora*.

Figs. 5.16 to 5.23 (facing page). Immunolabelling of peripheral vesicles in zoospores of *P. cinnamomi*. 16 & 17, zoospores labelled with monoclonal antibody Lpv-1. Zoospores fixed in 0·2% glutaraldehyde/4% paraformaldehyde contain a small number of plaques of fluorescent material (Fig. 5.16); those fixed in 4% paraformaldehyde alone contain over 100 fluorescent circular structures (bar = 10 μm); 18, zoospores stained with monoclonal antibody Cpa-2. Brightly fluorescent spots are seen all over the dorsal surface (d) but only along the edges of the ventral surface (v); g = groove (bar = 10 μm); 19, monoclonal antibody Vsv-1 stains small components that occur predominantly along the sides of the groove (arrows) on the ventral surface. The dorsal surface (d) contains few structures that react with this antibody (bar = 10 μm); 20, post-embedding labelling of a zoospore embedded in Lowicryl K4M and incubated with monoclonal antibody Cpw-1 followed by SAM coupled to 10 nm gold particles (SAM-Au_{10}). The density of labelling with this antibody is low but its specificity shows it to associate with the peripheral cisternae in the zoospores (bar = 0·2 μm); 21, section of a zoospore embedded in Lowicryl K4M and labelled with Lpv-1 then SAM-Au_{10}. Only the large peripheral vesicles are labelled (bar = 0·2 μm); 22, section of a zoospore embedded in Lowicryl K4M and labelled with Cpa-2 then SAM-Au_{10}. Only the small peripheral vesicles that predominate on the dorsal surface are labelled [arrow = large peripheral vesicle] (bar = 0·2 μm); 23, ventral surface of a zoospore embedded in Lowicryl K4M and labelled with Vsv-1 then SAM-Au_{10}. Labelling only occurs on small peripheral vesicles which contain plate-like inclusions and which predominate on the ventral surface. The groove lies to the left of the area shown (bar = 0·2 μm).

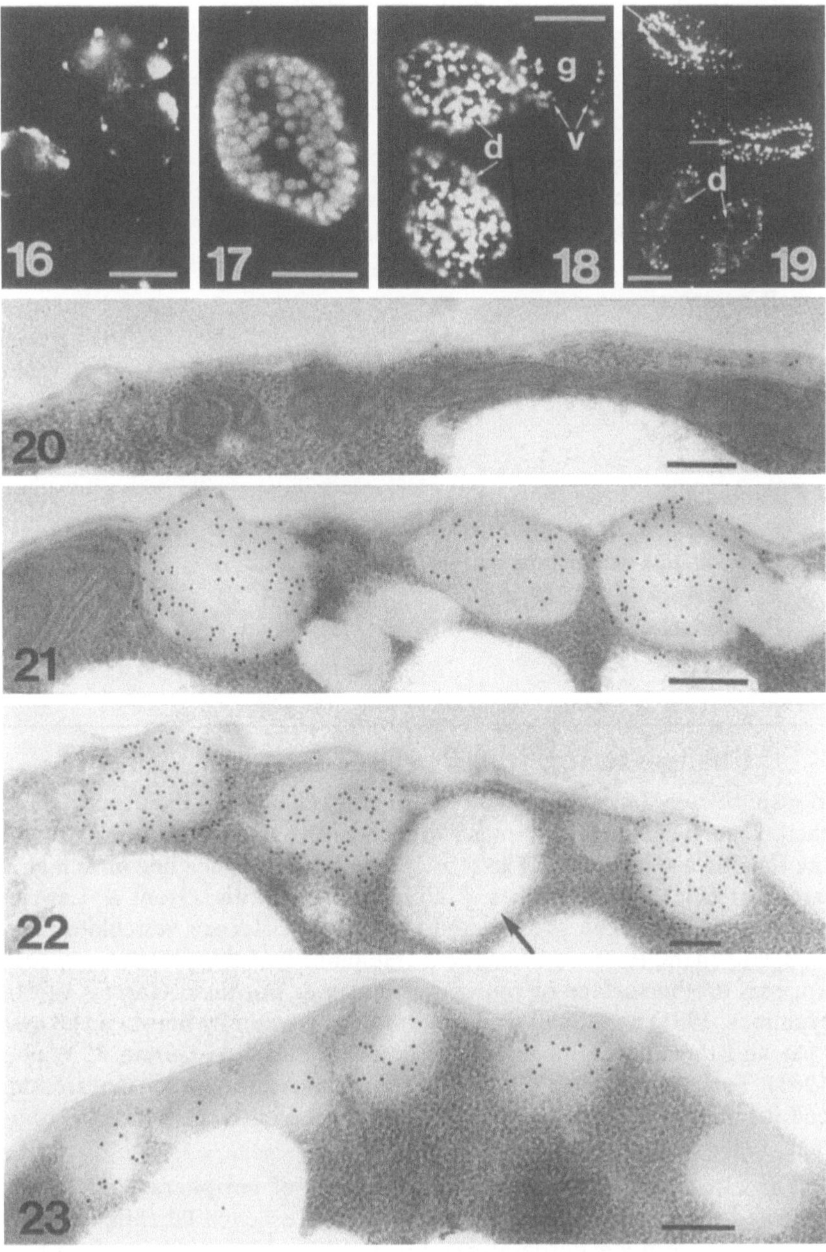
16
17
18
g
d
v
19
d
20
21
22
23

Table 5.3. Reactions of monoclonal antibodies raised against *P. cinnamomi* 6BR with other *Phytophthora* species

Phytophthora species	Antibody			
	Cpa-2	Lpv-1	Cpw-1	Vsv-1
P. cinnamomi 6BR	+	+	+	+
P. cinnamomi A278	+	+	+	+
P. cinnamomi A2420	+	+	+	+
P. boehmeriae	-	-	+	+
P. cactorum	+	-	+	+
P. castanae	+	-	+	+
P. citricola	-	-	+	+
P. citrophthora	-	-	+	+
P. drechsleri	-	-	+	+
P. heveae	-	+	+	+
P. meadii	-	-	+	+
P. megasperma f.sp. *glycinea*	+	-	+	+
P. nicotianae var. *nicotianae*	-	+	+	+
P. palmivora	-	-	+	+
P. nicotianae var. *parasitica*	-	+	+	+
P. vignae	-	-	+	+

Changes in surface properties during encystment

Zoospore encystment brings about dramatic changes to the properties of the cell surface. Surface area is reduced as the zoospores round up and the flagella are detached. The topography of the surface becomes highly irregular (Fig. 5.28; Hemmes & Hohl, 1971; Paktitis, Grant & Lawrie, 1986). Just below the surface, the peripheral cisternae vesiculate (Fig. 5.28; Hardham, 1989) and disappear (Hemmes & Hohl, 1971). A cyst coat appears on the surface of young cysts (Sing & Bartnicki-Garcia, 1975b; Hemmes, 1983) and a cell wall is formed subsequently between the cyst coat and the plasma membrane (Fig. 5.29; Bartnicki-Garcia & Wang, 1983). Ten minutes after the onset of encystment, the wall is mature, the cell turgid and few peripheral vesicles are seen in the cell cortex (Hemmes, 1983; Hardham, 1989).

The fate of the contents of the four types of peripheral vesicle was followed by pre-embedding and immunofluorescence labelling of cells fixed in glutaraldehyde/paraformaldehyde at various stages of encystment. Cpw-1 staining of the cyst surface becomes detectable about 2 min

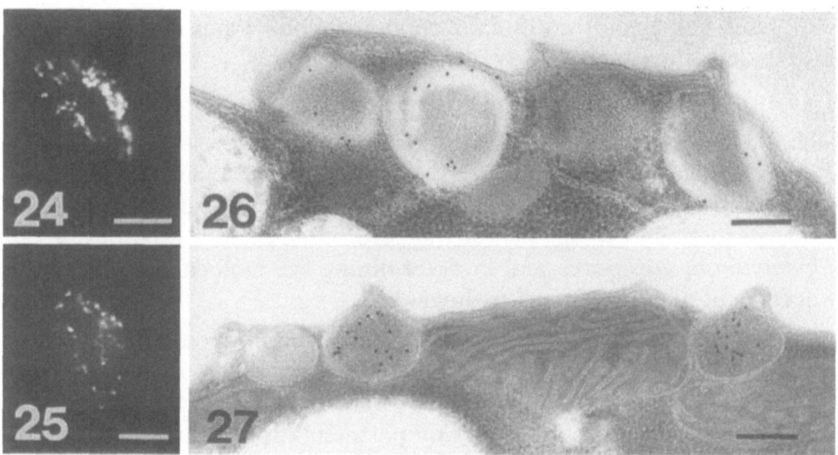

Figs. 5.24 to 5.27. Immunolabelling of peripheral vesicles in *Phytophthora* zoospores. 24 & 25, Vsv-1 labelling of zoospores of *P. castanae* (Fig. 5.24) and *P. boehmeriae* (Fig. 5.25). Components along the sides of the groove on the ventral surface are fluorescent (bar = 5 μm); 26, Vsv-1 (then SAM-Au_{10}) binds to the contents of vesicles that predominate along the ventral surface in zoospores of *P. megasperma* f. sp. *glycinea* (bar = 0·2 μm); 27, Lpv-1 (then SAM-Au_{10}) labels large peripheral vesicles in zoospores of *P. nicotianae* var. *parasitica* (bar = 0·2 μm).

---

after commencement of encystment. Labelling increases as encystment proceeds and gives a well-defined staining of the entire cyst periphery (Fig. 5.30; Hardham *et al.*, 1986). This antibody reacts with components on the surface of all species of *Phytophthora* tested (Table 5.3). Labelling of cysts by Lpv-1 was not observed at any stage of encystment. Material which bound Cpa-2 or Vsv-1 appeared at about 1 min. The distribution of Cpa-2 was at first patchy but later the antibody surrounded most of the cyst surface (Hardham *et al.*, 1986). In most cysts, Vsv-1 binding was concentrated on one side of the cell (Fig. 5.31).

Ultrastructural details of the changes in surface properties were determined by post-embedding, immunogold labelling. Cpw-1 labels the vesiculated peripheral cisternae at early stages of encystment (Fig. 5.33) and binds to the cyst wall once it appears at the cell surface (Fig. 5.34). Lpv-1 staining reveals that the large peripheral vesicles do not undergo exocytosis but instead, about 5 to 10 mins after the induction of encystment, they move away from the plasma membrane and become dispersed throughout the cytoplasm (Gubler & Hardham, 1988). This behaviour is also observed in *P. nicotianae* var. *nicotianae* and *P. nicotianae* var.

parasitica (Fig. 5.35). Labelling with Cpa-2 (Fig. 5.36; Gubler & Hardham, 1988) and Vsv-1 confirms that the contents of both types of small peripheral vesicle are secreted early in encystment.

Concluding remarks

The studies described here have begun to characterise the surface of *Phytophthora* zoospores and the changes that occur in the cell cortex and at the cell surface during encystment. They demonstrate the potential value of immunocytology in studying structure/function relationships in *Phytophthora* zoospores and in determining the molecular basis of the infection of plants by fungal pathogens.

We now know that some molecules are confined to certain regions of the zoospore plasma membrane while carbohydrates containing glucosyl/mannosyl residues occur over the entire surface of zoospores of a number of *Phytophthora* species. Compartmentation of surface molecules is likely to be important for zoospore function and the infection process; observations of the involvement of a flagellum-specific molecule in the triggering of encystment (Hardham & Suzaki, 1986) are consistent with this idea.

Figs. 5.28 to 5.36 (facing page). Ultrastructure and immunolabelling of *Phytophthora* spores during encystment. 28, zoospores of *P. cinnamomi* 1 min after induction of encystment. Many of the peripheral cisternae have moved away from the plasma membrane and have vesiculated. The vesicles lie in an area of the cytoplasm containing very few ribosomes. The contours of the cell surface are irregular (bar = 0· 25 μm); 29, surface of a cyst of *P. cinnamomi* 10 min after induction of encystment. A microfibrillar cell wall surrounds the cyst (bar = 0· 1 μm); 30, cysts of *P. heveae* are labelled by Cpw-1 (bar = 10 μm); 31 & 32, cyst of *P. cinnamomi* labelled with Vsv-1 (SAM-FITC) (Fig. 5.31) and soybean agglutinin (SBA)-rhodamine which is specific for N-acetyl-D-galactosamine (Fig. 5.32). SBA stains the material which reacts with Cpa-2. Vsv-1 binds to material localised on one side of the cyst. SBA binds to all parts of the cyst surface except for the region labelled by Vsv-1 (bar = 10 μm); 33, during encystment Cpw-1 (SAM-Au_{10}) labels the vesicles derived from the peripheral cisternae (bar = 0· 2 μm); 34, Cpw-1 (SAM-Au_{10}) binds to the cell wall of a young *P. cinnamomi* cyst embedded in Lowicryl K4M. The wall is electron-lucent in this non-osmicated material (bar = 0· 2 μm); 35, a cyst of *P. nicotianae* var. *parasitica* embedded in Lowicryl K4M labelled with Lpv-1 then SAM-Au_{10}. This antibody labels the large peripheral vesicles which are still present. Some have moved away from the cell periphery and are distributed throughout the cell (bar = 0· 25 μm); 36, young cyst of *P. cinnamomi* embedded in Lowicryl K4M and labelled with Cpa-2 then SAM-Au_{10}. The material contained in the small peripheral vesicles that predominate on the dorsal surface has been secreted and coats the surface of the cell (bar = 0·2 μm).

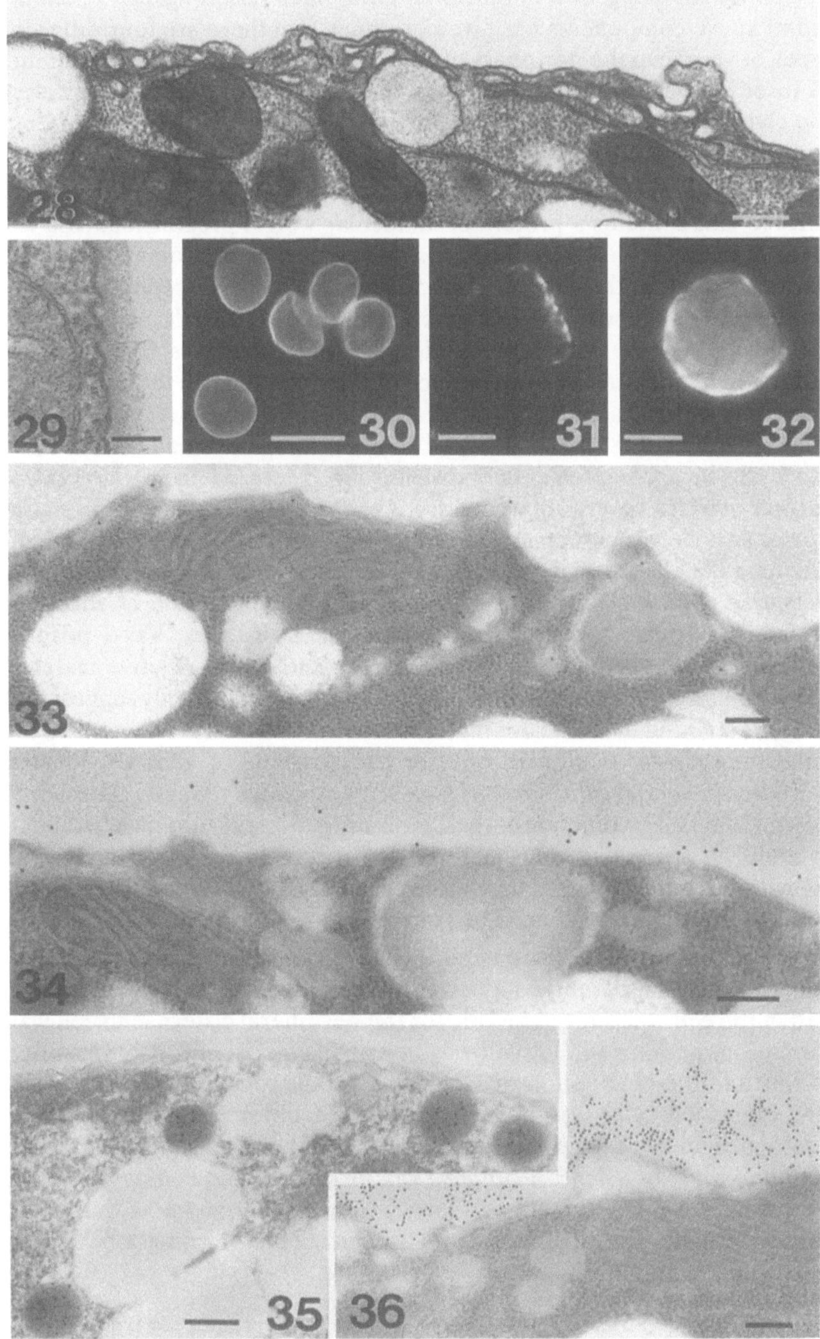
28
29
30
31
32
33
34
35
36

Immunolabelling with monoclonal antibodies raised against *P. cinnamomi* spore components has given evidence that there are four different types of vesicle in the periphery of *Phytophthora* zoospores. The contents of three of these are secreted during encystment and consequently alter the characteristics of the cell surface.

There are two immunologically distinct types of vesicles about $0 \cdot 2 \times 0 \cdot 3\,\mu m$ in diameter (Table 5.2). One type predominates on the ventral surface and contains a protein of molecular weight greater than 200 kD (Hardham & Gubler, unpublished). The other type predominates on the dorsal surface and contains glycoproteins rich in N-acetyl-D-galactosamine and of molecular weight greater than 300 kD (Gubler & Hardham, 1988). Vesicles of both types undergo rapid exocytosis following the induction of encystment. Appearance of their contents on the cell surface coincides with the acquisition of adhesiveness by the cysts (Gubler, Hardham & Duniec, 1989). The material from each type of small vesicle coats the cysts in a complementary distribution. Material from the ventral surface vesicles covers about one third of the cyst surface. That from the dorsal surface vesicles coats the rest of the cell (Figs. 5.31, 5.32). *In situ* labelling of cells that have encysted on the surface of roots of *Eucalyptus sieberi* or *Allium cepa* reveals a polarity in the orientation of zoospore attachment to the root surface. In about 90% of cases, Vsv-1-positive material lies between the cyst and the root and Cpa-2-positive material faces away from the root surface. These observations strongly suggest that Vsv-1 is staining the adhesive material which mediates attachment to the adjacent surface. A similar role for the contents of ventrally-located K-bodies in *Saprolegnia ferax* has also been proposed recently (Lehnen & Powell, 1989). The function of the Cpa-2 antigen is still unknown, although it could be a mucilage-like material that protects the cysts during the transition from osmoregulation via operation of the water expulsion vacuole to the generation of cell turgor through the formation of a cell wall.

The occurrence of common epitopes in the zoospore peripheral cisternae, in vesicles apparently derived from the cisternae during encystment, and in the cell wall suggests a role for the peripheral cisternae in cell wall formation, perhaps in the synthesis of matrix components. Confirmation of this and the elucidation of molecular details remain for future work. The high molecular weight glycoproteins (Gubler & Hardham, 1988) contained in the large peripheral vesicles labelled by Lpv-1 in *P. cinnamomi*, *P. heveae*, *P. nicotianae* var. *nicotianae* and *P. nicotianae* var. *parasitica* are not secreted. We have evidence that suggests that these vesicles provide a store of protein for use during cyst germination (Gubler & Hardham, unpublished).

Immunological studies of other *Phytophthora* species described here have begun to reveal the generality of the observations that have been made in *P. cinnamomi*. There is little doubt that future research will expand the use of immunological probes to identify specific cell components and determine their distribution and function. Such research should result in a much better understanding of the molecular basis of the infection of plants by *Phytophthora* and other fungal pathogens.

Acknowledgements We are grateful to Mr D. Abbott, Drs J. Heather, J. Hinch, J. Ralton, J. Walker, W. Thompson, N. Keen and I. Pascoe for supplying fungal cultures, and to the Australian Government for support through the National Research Fellowship Scheme.

References

Alberts, B., Bray, D., Lewis, J., Raff, M., Roberts, K. & Watson, J. D. (1989). *Molecular Biology of the Cell*. Garland Publishing Inc.: New York.

Bacic, A., Williams, M. L. & Clarke, A. E. (1985). Studies on the cell surface of zoospores and cysts of the fungus *Phytophthora cinnamomi*: nature of the surface saccharides as determined by quantitative lectin binding studies. *Journal of Histochemistry & Cytochemistry* **33**, 384-388.

Bartnicki-Garcia, S. & Wang, M. C. (1983). Biochemical aspects of morphogenesis in *Phytophthora*. In *Phytophthora. Its Biology, Taxonomy, Ecology, and Pathology*, ed. D. C. Erwin, S. Bartnicki-Garcia & P. H. Tsao, pp. 121-137. The American Phytopathological Society: St. Paul, Minnesota.

Beakes, G. (1987). Oomycete phylogeny: ultrastructural perspectives. In *Evolutionary Biology of the Fungi*, ed. A. D. M. Rayner, C. M. Brasier & D. Moore, pp. 405-421. Cambridge University Press.

Bimpong, C. E. & Hickman, C. J. (1975). Ultrastructural and cytochemical studies of zoospores, cysts, and germinating cysts of *Phytophthora palmivora*. *Canadian Journal of Botany* **53**, 1310-1327.

Bourett, T., Hoch, H. C. & Staples, R. C. (1987). Association of the microtubule cytoskeleton with the thigmotropic signal for appressorium formation in *Uromyces*. *Mycologia* **79**, 540-545.

Burgess, J. & Linstead, P. J. (1976). Ultrastructural studies of the binding of concanavalin A to the plasmalemma of higher plant protoplasts. *Planta* **130**, 73-79.

Buxton, E. W., Culbreth, W. & Esposito, R. G. (1961). Serological separation of forms and physiologic races of pathogenic *Fusarium oxysporum*. *Phytopathology* **51**, 575 (abstract).

Cerenius, L., Rennie, P. & Fowke, L. C. (1988). Endocytosis of cationized ferritin by zoospores of the fungus *Aphanomyces euteiches*. *Protoplasma* **144**, 119-124.

Cho, C. W. & Fuller, M. S. (1989a). Ultrastructural organization of freeze-substituted zoospores of *Phytophthora palmivora*. *Canadian Journal of Botany* **67**, 1493-1499.

Cho, C. W. & Fuller, M. S. (1989b). Observations of the water expulsion vacuole of *Phytophthora palmivora*. *Protoplasma* **149**, 47-55.

Cuatrecasas, P. (1963). Interaction of wheat germ agglutinin and concanavalin A with isolated fat cells. *Biochemistry* **12**, 1312-1323.

Edgar, L. A. & Pickett-Heaps, J. D. (1982). Ultrastructural localization of polysaccharides in the motile diatom *Navicula cuspidata*. *Protoplasma* **113**, 10-22.

Garas, N. A. & Kuć, J. (1981). Potato lectin lyses zoospores of *Phytophthora infestans* and precipitates elicitors of terpenoid accumulation produced by the fungus. *Physiological Plant Pathology* **18**, 227-237.

Grove, S. N. & Bracker, C. E. (1978). Protoplasmic changes during zoospore encystment and cyst germination in *Pythium aphanidermatum*. *Experimental Mycology* **2**, 51-98.

Gubler, F. & Hardham, A. R. (1988). Secretion of adhesive material during encystment of *Phytophthora cinnamomi* zoospores, characterized by immunogold labelling with monoclonal antibodies to components of peripheral vesicles. *Journal of Cell Science* **90**, 225-235.

Gubler, F., Hardham, A. R. & Duniec, J. (1989). Characterising adhesiveness of *Phytophthora cinnamomi* zoospores during encystment. *Protoplasma* **149**, 24-30.

Hardham, A. R. (1985). Studies on the cell surface of zoospores and cysts of the fungus *Phytophthora cinnamomi*: the influence of fixation on patterns of lectin binding. *Journal of Histochemistry & Cytochemistry* **33**, 110-118.

Hardham, A. R. (1987a). Microtubules and the flagellar apparatus in zoospores and cysts of the fungus *Phytophthora cinnamomi*. *Protoplasma* **137**, 109-124.

Hardham, A. R. (1987b). Ultrastructure and serial section reconstruction of zoospores of the fungus *Phytophthora cinnamomi*. *Experimental Mycology* **11**, 297-306.

Hardham, A. R. (1989). Lectin and antibody labelling of surface components of spores of *Phytophthora cinnamomi*. *Australian Journal of Plant Physiology* **16**, 19-32.

Hardham, A. R. & Suzaki, E. (1986). Encystment of zoospores of the fungus, *Phytophthora cinnamomi*, is induced by specific lectin and monoclonal antibody binding to the cell surface. *Protoplasma* **133**, 165-173.

Hardham, A. R. & Suzaki, E. (1990). Glycoconjugates on the surface of spores of the pathogenic fungus *Phytophthora cinnamomi* studied using fluorescence and electron microscopy and flow cytometry. *Canadian Journal of Microbiology* **36**, 183-192.

Hardham, A. R., Suzaki, E., & Perkin, J. L. (1986). Monoclonal antibodies to isolate-, species- and genus-specific components on the surface of zoospores and cysts of the fungus *Phytophthora cinnamomi*. *Canadian Journal of Botany* **64**, 311-321.

Heath, I. B., Rethoret, K., Arsenault, A. L. & Ottensmeyer, F. P. (1985). Improved preservation of the form and contents of wall vesicles and the golgi apparatus in freeze substituted hyphae of *Saprolegnia*. *Protoplasma* **128**, 81-93.

Hemmes, D. E. (1983). Cytology of *Phytophthora*. In *Phytophthora: Its Biology, Taxonomy, Ecology, and Pathology*, (ed. D. C. Erwin, S. Bartnicki-Garcia & P. H. Tsao), pp. 9-40. The American Phytopathological Society: St. Paul, Minnesota.

Hemmes, D. E. & Hohl, H.R. (1971). Ultrastructural aspects of encystation and cyst-germination in *Phytophthora parasitica*. *Journal of Cell Science* **9**, 175-191.

Hoch, H. C., Bourett, T. M. & Staples, R. C. (1986). Inhibition of cell differentiation in *Uromyces* with D_2O and taxol. *European Journal of Cell Biology* **41**, 290-297.

Hoch, H. C. & Staples, R. C. (1985). The microtubule cytoskeleton in hyphae of *Uromyces phaseoli* germlings: its relationship to the region of nucleation and to the f-actin cytoskeleton. *Protoplasma* **124**, 112-122.

Hoch, H. C., Tucker, B. E. & Staples, R. C. (1987). An intact microtubule cytoskeleton is necessary for mediation of the signal for cell differentiation in *Uromyces*. *European Journal of Cell Biology* **45**, 209-218.

Howard, R. J. & Aist, J. R. (1979). Hyphal tip ultrastructure of the fungus *Fusarium*: improved preservation by freeze-substitution. *Journal of Ultrastructural Research* **66**, 224-234.

Howard, R. J. & O'Donnell, K. L. (1987). Freeze substitution of fungi for cytological analysis. *Experimental Mycology* **11**, 250-269.

Lehnen, L. P., Jr. & Powell, M. J. (1988). Cytochemical localization of carbohydrates in zoospores of *Saprolegnia ferax*. *Mycologia* **80**, 423-432.

Lehnen, L. P., Jr. & Powell, M. (1989). The role of kinetosome-associated organelles in the attachment of encysting secondary zoospores of *Saprolegnia ferax* to substrates. *Protoplasma* **149**, 163-174.

McKerracher, L. & Heath, I. (1985). Microtubules around migrating nuclei in conventionally-fixed and freeze-substituted cells. *Protoplasma* **125**, 162-172.

Paktitis, S., Grant, B. & Lawrie, A. (1986). Surface changes in *Phytophthora palmivora* zoospores following induced differentiation. *Protoplasma* **135**, 119-129.

Philippi, M. L., Parish, R. W. & Hohl, H. R. (1975). Histochemical and biochemical evidence for the presence of microbodies in *Phytophthora palmivora*. *Archives of Microbiology* **103**, 127-132.

Pinto Da Silva, P. & Nogueira, M. L. (1977). Membrane fusion during secretion. A hypothesis based on electron microscope observations of *Phytophthora palmivora* zoospores during encystment. *Journal of Cell Biology* **73**, 161-181.

Powell, M. J. & Bracker, C. E. (1986). Distribution of diaminobenzidine reaction products in zoospores of *Phytophthora palmivora*. *Mycologia* **78**, 892-900.

Powell, M. J., Lehnen, L. P., Jr. & Bortnick, R. N. (1984). Microbody-like organelles as taxonomic markers among Oomycetes. *BioSystems* **18**, 321-334.

Sing, V. O. & Bartnicki-Garcia, S. (1975a). Adhesion of *Phytophthora palmivora* zoospores: electron microscopy of cell attachment and cyst wall fibril formation. *Journal of Cell Science* **18**, 123-132.

Sing, V. O. & Bartnicki-Garcia, S. (1975b). Adhesion of *Phytophthora palmivora* zoospores: detection and ultrastructural visualization of concanavalin A receptor sites appearing during encystment. *Journal of Cell Science* **19**, 11-20.

Chapter 6

Surface-related host-pathogen interactions in *Phytophthora*

H. R. Hohl

Interactions involving cell-cell adhesion are important in animal development and in bacterial mutualism and parasitism. However, investigation of contact phenomena in fungal-fungal or fungal-plant interactions has only recently begun (Nicholson, 1984; Ralton, Howlett & Clarke, 1986; Ralton, Smart & Clarke, 1987). Adhesion plays a role in morphogenesis of rhizomorphs and fruiting bodies, self-fusion (Ainsworth & Rayner, 1986; Aylmore & Todd, 1986), pathogenesis (Beckett & Woods, 1987; Samson, Evans & Latgé, 1988), mycoparasitism (Barak *et al*., 1985), mycorrhizal associations (Bonfante-Fasolo *et al*., 1987; Massicotte, Ackerley & Peterson, 1987a & b), nematode trapping fungi, microbial fouling and other circumstances where establishment of colonies on dry or wet (Hyde *et al*., 1986), organic or inorganic substrata is of importance, including biotechnology (Amory & Rouxhet, 1988). Adhesion may involve the surfaces of spores (Young & Kauss, 1984; Hamer, Howard & Chumley, 1988) or cysts (Sing & Bartnicki-Garcia, 1975a & b; Gubler & Hardham, 1988; Gubler, Hardham & Duniec, 1989; Chapter 5), hyphae or germ tubes (Hyde, Jones & Moss, 1986; Hohl & Balsiger, 1986a; Beckett & Porter, 1988), or specialised structures such as appressoria (Nicholson, 1984; Gold & Mendgen, 1984; Epstein *et al*., 1985, 1987; Mims & Richardson, 1989), hyphopodia (Onyile, Edwards & Gessner, 1982), or other attachment organs (Jarvis & Berry, 1986).

For attachment to the outer surface of the host, fungi often but not always produce some adhesive material (Nicholson, 1984). The latter probably increases the chances of infection by preventing the fungal propagules from becoming dislodged by wind or water. Infection structures become even more difficult to remove after appressoria or other structures of attachment have been produced.

The heightened interest in adhesion is due to the possibility that apart from the purely mechanical aspects of stickiness, contact between host and parasite may lead to more refined interactions between surface molecules such as recognition and the induction of pathogenesis-related events (Keen, 1982; Ralton *et al*., 1986, 1987). The idea that critical steps for recognition occur at the wall surface or at the plasma membrane has spurred a new wave of activity in this field. However, even though specificity of the interaction is observed during mating in yeast (Kervin *et al.*, 1987; Lipke, Terrance & Wu, 1987; Sijmons *et al.*, 1987; Terrance *et al.*, 1987) or during recognition of target organs such as stomatal openings (Epstein *et al.*, 1985), too little is known to draw any general conclusions. In particular it is unclear whether contact at the level of the host cell wall, the plasma membrane, or both, is of importance.

Perhaps the most intensive studies on fungal adhesion have been made with the human pathogen *Candida albicans* (Douglas, 1987). For plant pathogens, Hoch & Staples (1987) and Staples & Hoch (1987) summarized the first molecular approaches to unravel the structural and chemical parameters of appressorium initiation and development in rusts and anthracnose fungi (Hoch, Staples & Bourett, 1987a; Hoch *et al.*, 1987b). Tactic (thigmotropic), chemical and temperature triggers are involved which perhaps act through receptors in the extracellular matrix and surface of the pathogen. From there the signals might be transmitted by second messengers such as calcium ions and cAMP to act on protein synthesis, nuclear division and the cytoskeleton. The latter has been shown by Bourett, Hoch & Staples (1987) to be part of the thigmotropic signal for appressorium formation in *Uromyces*.

A fibrillar, brush-like border on the surface of *Uromyces appendiculatus* correlates with the biotrophic phase of fungal growth and could possibly represent material involved in adhesion and recognition (Welter, Muller & Mendgen, 1988; Mendgen *et al.*, 1988). Differentiation of the fungal surface during the infection process has been demonstrated previously by the use of enzymes and carbohydrate-specific lectins (Kapooria & Mendgen, 1985; Mendgen, Lange & Bretschneider, 1985).

Cytochemical and biochemical methods are being employed to understand the mechanisms of adhesion. One approach aims at demonstrating lectin-ligand type interactions and involves the characterization of surface accessible sugar residues and lectins. In this context, studies are also being aimed at isolating fungal surface components from spores (Latgé *et al.*, 1986) or hyphae (Teepe, Boettge & Woestemeyer, 1988; Herrero, Sanz & Sentandreu, 1987).

Data of Epstein *et al.* (1987) suggest that in *Uromyces appendiculatus* extracellular protein is required for germling adhesion to a substratum and that adhesion is required for germ tube orientation and contact-stimulated differentiation. That surface topography as well as surface chemistry might lead to specific fungal responses has been beautifully illustrated by Hoch *et al.* (1987b) for germ tubes of *Uromyces*. There is good evidence that glycoproteins of fungal origin are involved in adhesion of encysting zoospores of *Phytophthora cinnamomi* to root surfaces (Gubler & Hardham, 1988).

Studies of surface interactions between filamentous fungi and host plants have concentrated on fungal cysts or spores deposited and germinating on leaf surfaces (cuticles), root hairs or non-biological material such as glass or plastic. The first contact of the pathogen is with the cell wall or cuticle. Interactions between parasite and host wall have been described, e.g. release of elicitors from the host wall by pathogen-derived enzymes. Contact phenomena within the host tissue, or between the parasite and the plasma membrane of the host, might, however, be of equal importance, as the outcome of the interaction between the two partners often occurs only after the host epidermis has been penetrated. Fungal elicitors present on the surface of the pathogen or released by it appear to have receptors on the plasma membrane (Yoshikawa, Keen & Wang, 1983; Schmidt & Ebel, 1987).

One obvious reason for the lack of information in this area is the technical difficulty encountered in studying contact phenomena after the epidermal outer cell wall has been breached. At this stage the pathogen penetrates further into the host tissue and makes contact with host cells at unpredictable sites. It is also very hard to study separately *in situ* the interaction of the pathogen with the host wall and the plasma membrane. However, some observations have been made. Using plasmolysis techniques Nozue, Tomiyama & Doke (1979) demonstrated tight adherence of host plasma membranes (extrahaustorial membranes) to haustoria of *Phytophthora infestans* in potato, and Coffey (1983) has made similar observations with *Albugo candida* on cabbage.

In *Phytophthora*, recent attempts to uncover the mechanisms and role of adhesion centre on two aspects. Adhesion of developing cysts has been discussed in Chapter 5; here I will concentrate on our own work on adhesion of germ tubes and hyphae of *Phytophthora* to walls and plasma membranes (protoplasts) of host cells. The cytology of interactions will be covered, and the possible involvement of lectin-ligand type linkages in these events will be explored; such interactions are likely to participate in recognition of plant roots by *Phytophthora* (Hinch & Clarke, 1980) and by *Pythium aphanidermatum* (Longman & Callow, 1987).

Adhesion of *Phytophthora* to protoplasts and walls of the host.

To study the adhesion of fungal walls to the host plasma membrane and cell wall, a test was developed based on the attachment of protoplasts or isolated single mesophyll cells to a pregrown network of hyphae (Hohl & Balsiger, 1986a). Adhesion of soybean protoplasts to *P. megasperma* f. sp. *glycinea* with time is shown in Fig. 6.1. Following an initial lag phase of

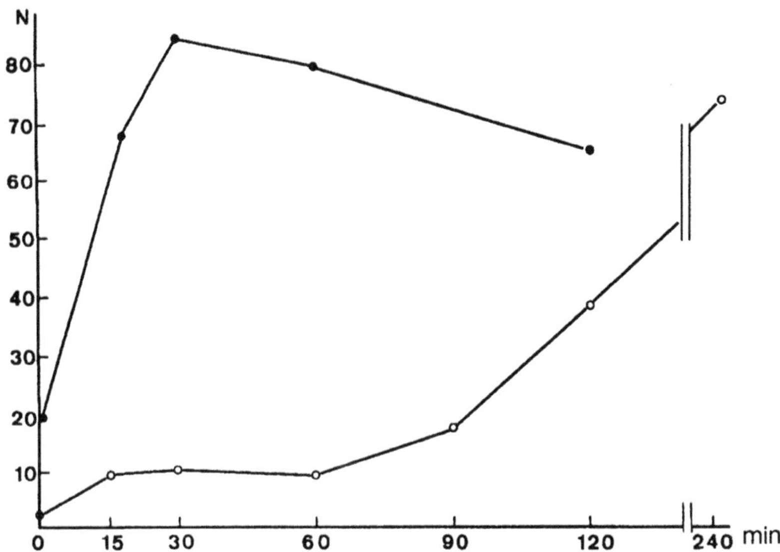

Fig. 6.1. Kinetics of adhesion of plant protoplasts to hyphae: soybean mesophyll and *Phytophthora megasperma* f. sp. *glycinea* (open circles), potato mesophyll and *P. infestans* (closed circles). N = number of protoplasts mm^{-2} of mycelial mat.

about 90 min adhesion of protoplasts increased and reached levels of 70 to 100 protoplasts mm^{-2} after 4 h. For protoplasts of potato, adhesion to hyphae of *P. infestans* was faster and there was apparently no lag phase (Fig. 6.1). Adhesion of walled mesophyll cells to a hyphal network of *P. megasperma* f.sp. *glycinea* also revealed no clear-cut lag phase.

The lag phase observed during adhesion of soybean protoplasts to the soybean pathogen suggested that adhesive material was only produced following contact and its formation was probably triggered by the contact. Several lines of evidence (Table 6.1) support this idea and also demonstrate that the adhesive material is produced by the fungus and not by the host. Addition of the protein synthesis inhibitors actinomycin D and cycloheximide prevented adhesion, as did killing of the hyphal network with paraformaldehyde. However, fixing the host protoplasts with the same chemical did not interfere with adhesion, suggesting that the fungus plays the active part in this process. Several enzymes also inhibited adhesion, implying the participation of some protein or glycoprotein in this process. Of several sugars tested, deoxyglucose, N-acetyl-glucosamine and methyl-mannoside inhibited adhesion when added to the test system while galactose, N-acetyl-galactosamine and rhamnose did not.

Table 6.1. Effect of various compounds or treatments on adhesion between soybean protoplasts and hyphae of *P. megasperma* f. sp. *glycinea*

Compound/treatment	Resulting adhesion
none (control)	+
cycloheximide	–
actinomycin D	–
protoplasts fixed with paraformaldehyde	+
fungal hyphae fixed with paraformaldehyde	–
snail digestive juice	–
proteinase K	–
cellulase	–
Ca^{2+} ions 0-10 mM	+
galactose	+
N-acetyl-galactosamine	+
rhamnose	+
2-deoxyglucose	–
N-acetyl-glucosamine	–
methyl-mannoside	–

+ = adhesion occurred, – = no adhesion

This last result led us to suspect that a lectin-ligand type of interaction might be involved in adhesion of the pathogen to the host plasma membrane. Interestingly, the results obtained from protoplasts and walled cells were not identical (Table 6.2). Thus, depending on the situation, several types of interaction involving different molecules appear to be involved in adhesion of *Phytophthora* to various surfaces.

We did not observe differences in the adhesion of protoplasts from susceptible and resistant cultivars. Thus, adhesion *per se* is certainly not a specific response. Nozue, Tomiyama & Doke (1979) also described non-specific adherence of host plasma membranes to infecting hyphae of compatible and incompatible races of *P. infestans*. However, the possibility should not be discarded that the tight adherence monitored with our test system is possibly accompanied, or perhaps preceded by mechanically weaker specific and hapten-reversible interactions; as is evident in bacterial adhesion to plant surfaces (Dazzo & Gardiol, 1984) and from our experiment with hapten sugars. Therefore, an analysis of sugar haptens and lectins on the surfaces of host and pathogen was undertaken.

Table 6.2. Agglutination by lectins of soybean protoplasts and walled cells

lectin	sugar hapten	agglutination of protoplasts	agglutination of cells
Concanavalin A	glucose, mannose	+	+
lentil agglutinin	glucose, mannose	+	+
garden pea agglutinin	glucose, mannose	–	–
castor bean agglutinin I	galactose, N-acetylgalactosamine	+	–
peanut agglutinin	galactose–N-acetylgalactos-amine	+	–
soybean agglutinin VI	N-acetylgalactosamine	+	–
red kidney bean agglutinin	galactose–N-acetylglucos-amine–mannose	–	–
wheat germ agglutinin	$(N\text{-acetylglucosamine})_2$ [=chitobiose]	–	–
gorse agglutinin I	fucose	–	–

Agglutinations were inhibited by the corresponding hapten sugar(s)

Lectins and sugar ligands present on the surface of *Phytophthora* and the host plasma membrane and wall

A series of lectins was used to detect accessible sugar ligands on the surfaces of pathogen and host. Agglutination tests and FITC-labelled lectins were used in these studies. To detect lectins, adhesion of fungal germlings and host cells to beads coated with specific sugars was studied.

Glycosyl residues on host plasma membranes

Five of the 11 lectins tested agglutinated soybean protoplasts, pointing to the presence of mannosyl/glucosyl and galactosyl/NAC-galactosyl residues on the cell surface (Table 6.3). Agglutination was noticeable as early as 15 min and was maximal after 120 min. Potato and barley showed a somewhat different pattern (Hohl & Balsiger, 1986b).

A quantitative comparison of agglutination revealed that the soybean VI agglutinin (SBA) had the highest and the lentil agglutinin (LCA) the lowest activity (Table 6.4). There were slight differences between the soybean cv. Harosoy and Williams but none between the near-isogenic lines Harosoy and Harosoy 63, and Williams and Williams 79, respectively.

Table 6.3. Agglutination by lectins (after 90 min) of soybean, potato and barley protoplasts

lectin	sugar hapten	agglutination of soybean	potato	barley
Concanavalin A	glucose, mannose	+++	++	—
lentil agglutinin	glucose, mannose	+++	—	—
garden pea agglutinin	glucose, mannose	—	—	++++
castor bean agglutinin I	galactose, N-acetylgalactosamine	++++	++++	++++
peanut agglutinin	galactose–N-acetylgalactosamine	++++	++++	++++
soybean agglutinin VI	N-acetylgalactosamine	++++	—	++++
red kidney bean agglutinin	galactose–N-acetylglucosamine–mannose	—	—	—
wheat germ agglutinin	(N-acetylglucosamine)$_2$ [= chitobiose]	—	—	—
poke weed agglutinin	(N-acetylglucosamine)$_3$	—	—	—
winged pea agglutinin	fucose	—	—	—
gorse agglutinin I	fucose	—	+	++++

Sugar residues on the parasite

Cysts and germlings were not agglutinated by ConA, wheat germ agglutinin (WGA), peanut agglutinin (PNA) or soybean agglutinin VI (SBA), but by using FITC-labelled molecules lectin binding could be demonstrated for ConA only. FITC-labelled ConA adhered strongly to germ tubes and hyphae of all four species tested (*P. megasperma* f. sp. *glycinea*, *P. cactorum*, *P. palmivora* and *P. infestans*). The tip exhibited stronger fluorescence than the main body of the germ tube, as has been shown before (Keen & Legrand, 1980; Hermanns & Ziegler, 1984; Sing & Bartnicki-Garcia, 1975b; Galun *et al.*, 1976; Barkai-Golan, Mirelman & Sharon, 1978). Furthermore, Keen & Legrand (1980) found only glucose and mannose in the glycoproteins isolated and purified from cell walls of *P. megasperma* f. sp. *glycinea*.

Bacic, Williams & Clarke (1985) were able to demonstrate binding of ConA and SBA in about equal amounts to cysts of *P. cinnamomi* using ^{125}I-iodinated lectins. This is rather surprising since, as the authors state, there are no reports of galactosyl or N-acetyl-D-galactosaminosyl residues in the walls of *Phytophthora* species. However, an electron microscopic study using ferritin-labelled lectins (Hardham, 1985) confirmed

Table 6.4. Minimum concentration of lectins capable of agglutinating protoplasts of soybean (cvs. Williams W; Williams W79; Harosoy H; Harosoy H63) and of barley

Lectin	MW (kDa)	Sugar hapten	Soybean cultivars W	W79	H	H63	Barley
LCA	49	mannose	500	500	125	125	na
ConA	102	mannose	125	125	250	250	na
SBA VI	110	N-acetyl-galactosamine	7·8	7·8	3·9	3·9	3·9
PNA	120	galactose	31·2	31·2	62·5	62·5	125
RCA I	60	galactose	31·2	31·2	31·2	31·2	7·8
UEA I	170	fucose	na	na	na	na	125

Entries show the minimum quantity of lectin (μg ml^{-1}) able to agglutinate the protoplasts; na = no agglutination. ConA = Concanavalin A, LCA = lentil agglutinin, RCA I = castor bean agglutinin I, PNA = peanut agglutinin, SBA VI = soybean agglutinin VI, UEA I = gorse agglutinin I.

the observations of Bacic *et al.* (1985). The discrepancy between these results and those of others (see above) has not yet been resolved.

Using gold-labelled lectin we were able to confirm, at the electron microscope level, the adherence of ConA to cysts and germ tubes of *Phytophthora palmivora*. With this technique we also obtained positive evidence for WGA binding to the 'fluffy' outer coat but not the actual wall of cysts (Guhl, 1983).

It might be noted that WGA was the only lectin that inhibited growth of these four *Phytophthora* species, and that only *Phytophthora megasperma* f. sp. *glycinea* was also inhibited by garden pea agglutinin (PSA) and, less so, by SBA (Gibson *et al.*, 1982; Hohl & Balsiger, 1986b). Inhibition of vegetative growth of *Phytophthora* has been demonstrated before (Mirelman *et al.*, 1975), and various lectins are capable of disturbing and inhibiting fungal growth (Barkai-Golan, Mirelman & Sharon., 1978; Brambl & Gade, 1985). All of these results on the growth inhibitory action of lectins should, however, be re-evaluated in view of the finding by Schlumbaum *et al.* (1986) that among commercial preparations of chitin-binding lectins, only those containing chitinase activity inhibited fungal growth.

Lectins on host plasma membrane

Protoplasts of Harosoy and Harosoy 63 were agglutinated by the glucosylated Yariv reagent (Hohl & Balsiger, 1988) indicating the presence

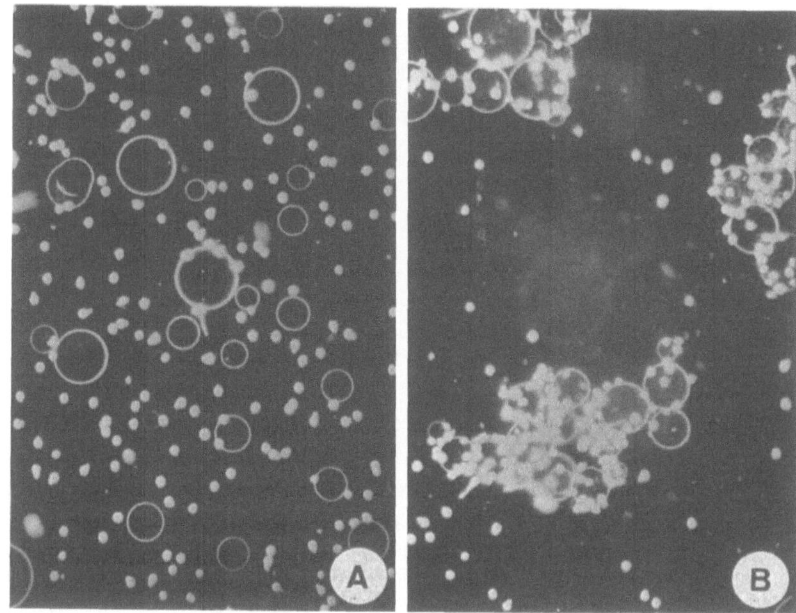

Fig. 6.2. Agglutination test. Mixed incubation of germinating cysts of *Phytophthora megasperma* f. sp. *glycinea* and glycosyl-coated agarose beads. In A the beads are coated with glucose residues (no agglutination), in B with galactosamine residues (strong agglutination).

of a β-glucose receptor on the host plasma membrane. The Yariv reagent is a synthetic, trivalent sugar ligand which binds to arabino-galactans called β-lectins (Clarke *et al*., 1978). No agglutination occurred with the galactosylated Yariv reagent. However, there is evidence for galactosyl receptors on soybean protoplast surfaces as the galactose-specific soybean lectin has been localized there with immunohistological methods (Ho *et al*., 1986).

Lectins on parasite surfaces

Germinated cysts of all three species tested agglutinated with agarose beads coated with galactosamine residues, indicating the presence of the corresponding receptors on the surface of the pathogen (Fig. 6.2). *P. megasperma* f. sp. *glycinea P. infestans*, but not isolates of *P. palmivora*, agglutinated in the presence of beads coated with fucose residues (Hohl & Balsiger, 1988). Thus, the fucose receptor is not present on all *Phytophthora* cysts. The other ligands tested, namely glucose, N-acetyl-glucosamine, mannose, methyl-α-D-mannopyranoside, lactose, and N-acetyl-galactosamine did not lead to attachment to or agglutination of the pathogen with the beads.

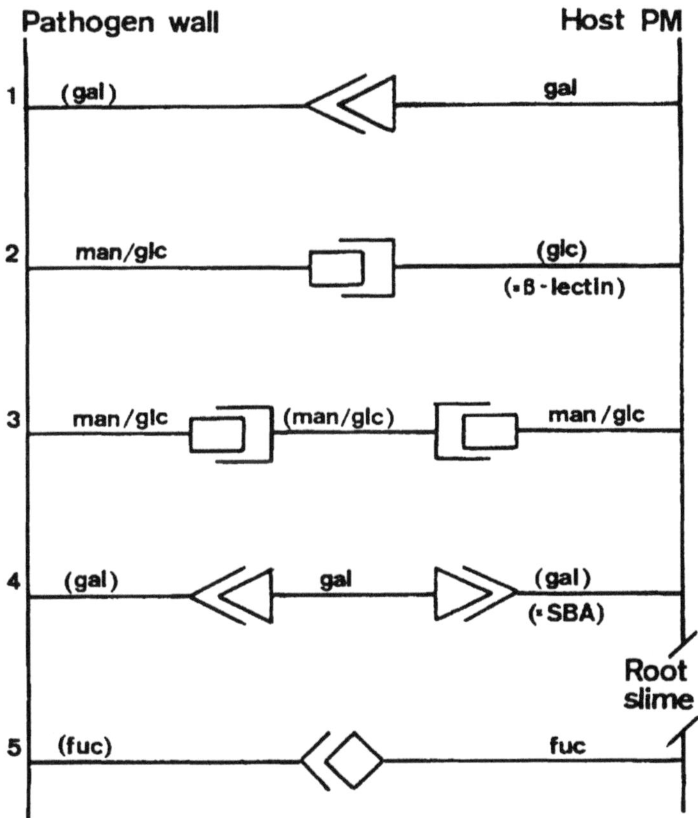

Fig. 6.3. Potential linkages between glycosyl receptors (lectins, with sugar specificity shown in parentheses) and terminal glycosyl residues demonstrated on the outer surface of germinated cysts of *Phytophthora megasperma* f. sp. *glycinea* (pathogen) and the surface of protoplasts (plasma membrane) of its soybean host. Based on experimental evidence, example 2 is the linkage most likely involved in adhesion. The presence of central, bridging linkers shown in examples 3 and 4, respectively, has not been demonstrated experimentally. A potential link between the fucosyl-specific lectin of the pathogen and putative fucosyl residues of the root slime (5) is drawn in analogy to the situation described by Hinch & Clarke (1980) for corn.

Glycosyl receptors on fungal surfaces have been described in a few other cases. A fungal lectin and its apparent receptor on a nematode surface is described by Borrebaeck, Mattiasson & Nordbring-Hertz (1985). The presence of lectins binding to glucose and N-acetyl-glucosamine on the surface of *Conidiobolus obscurus* spores was demonstrated but no corresponding sugar ligands were found on the surface of the insect host (Latgé, Monsigny & Prévost, 1988).

Summary: lectins and ligands on host and parasite

Fig. 6.3 and Table 6.5 summarize the glycosyl receptors and the ligands shown to be accessible on the surfaces of germinated cysts and hyphae of *P. megasperma* f. sp. *glycinea* and of soybean protoplasts. Several types of linkages between the two partners may be imagined. Some of the possibilities are indicated. The galactose/N-acetyl-galactosamine ligand described for *P. cinnamomi* by Bacic *et al.* (1985) and by Hardham (1985) is not included because we were not able to demonstrate its presence on the surface of *P. megasperma* f. sp. *glycinea*.

While the lectins and their respective ligands of linkage types 1 and 2 in Fig. 6.3 have been shown to be present, combinations 3 and 4 each require a component which has not yet been demonstrated experimentally. Combination 3 would require a soluble, non-attached glucose/mannose-specific lectin present at the interface which could possibly form a link between the glucose/mannose residues present on the two surfaces. Combination 4 necessitates the existence of a carbohydrate component with more than one terminal galactose residue which could act as a linker between the soybean lectin SBA and the galactose-specific receptor on the fungal surface.

Sugars of the glucose/mannose type have been shown to inhibit adhesion between soybean protoplasts and growing germ tubes of *P. megasperma* f. sp. *glycinea* (Table 6.1). These sugars potentially inhibit formation of the bonds shown diagrammatically in combinations 2 and 3 of Fig. 6.3. Since combination 3 contains an unproven element, a soluble glucose/mannose specific lectin, combination 2 represents the type of linkage most likely to be involved in adhesion, as it best fits the results obtained so far. However, it should be emphasized that we do not yet know whether adhesion is essential for infection or whether any of the potential bonds between sugar ligands and receptors also participate in recognition, i.e. in the triggering of a reaction in either host or parasite. Furthermore, working with lectins may lead to ambiguous results since hapten inhibition assays using simple sugars cannot always be correlated with the structure of the binding sites on the cell or cell wall surface (Osawa & Tsuji, 1987). Generally, lectins do not bind exclusively to individual sugar ligands but rather to specific sequences of several sugar moieties.

The identification of specific glycoproteins involved in adhesion and/or recognition will probably not be possible using lectins as probes. We have found that many glycoproteins on gels from host and pathogen surfaces are stained with a variety of lectins. Possibly monoclonal antibodies to specific epitopes will help to resolve these problems. However, so far our monoclonal antibodies to surfaces of host and parasite also exhibit ex-

Table 6.5. Summary of lectins and sugar ligands exposed on the surfaces of soybean (cv. Harosoy and Harosoy 63) protoplasts and of germinated cysts of *Phytophthora megasperma* f. sp. *glycinea* race 1

	lectin specific for	sugar ligand(s)[a]
protoplasts	β-D-glucose[b]	glucose/mannose
	galactose, N-acetylgalactosamine[c]	galactose, N-acetylgalactosamine
germinated cysts	galactosamine	—
	fucose	—
	—	glucose/mannose

(from Hohl & Balsiger, 1988). [a]from Hohl & Balsiger (1986b); [b]a β-lectin identified with the Yariv reagent; [c]soybean agglutinin (SBA) demonstrated by Ho *et al.* (1986).

tended cross-reactivity, indicating that some carbohydrate-epitope is highly antigenic and common to many glycoconjugates, as observed by others (Ayers & Wycoff, 1987). Our efforts to raise antibodies against deglycosylated surface glycoproteins may help to alleviate this difficulty.

Induction of contact-induced, surface-related reactions by host and parasite

Reactions of the parasite

Germlings and young hyphae of *Phytophthora* may change their direction of growth upon making contact with the plasma membrane. They also produce adhesive material with which they attach to biological (wall, cuticle, plasma membrane) and non-biological (glass, plastic) surfaces. As outlined before, there is evidence that in *P. megasperma* f. sp. *glycinea* the adhesive material is formed upon contact with a surface, i.e. is induced. We still do not know whether different surfaces induce production of the same or different adhesive substances but there is good evidence that at least in *Phytophthora* (Gubler & Hardham, 1988; Gubler *et al.*, 1989), *Colletotrichum* (Lapp & Skoropod, 1978) and *Uromyces* (Epstein *et al.*, 1987) the material involved is composed of glycoproteins.

Reaction of the host protoplast

Physical contact between fungus and host plasma membrane also induced a local reaction by the host cell which can be visualised. For these experiments (Odermatt *et al.*, 1988) protoplasts were embedded in agarose together with germinating cysts of *Phytophthora*, and the encounters between protoplasts and germ tubes were monitored. Germ tubes or hyphae either penetrated the protoplasts (without, however, forming appressoria), grew along the protoplast surface, or continued their growth with no directional change. Contact between protoplast and pathogen frequently triggered production of finely fibrillar material at the sites of contact (Fig. 6.4a) which stains with the fluorescing dye 'diethanol' (Fig. 6.4b), indicative of polysaccharides. Since adhesion normally occurred in the absence of this material it obviously is not essential for attachment but probably represents a response of the host cell to contact with the pathogen.

The fibrillar material produced by the protoplast is reminiscent of collar or papilla formation as observed at penetration sites during infection of intact host tissue (Fig. 6.4c & d). These wall-like appositions to the host cell wall represent a defence and/or wound reaction known to produce β-1,3- and β-1,4-linked glucans (Aist, 1976; Hächler & Hohl, 1982; Hinch & Clarke, 1982). They have been well-studied in soybeans infected by *Phytophthora megasperma* f. sp. *glycinea*. Although termed wall-appositions this material apparently forms in the absence of any cell wall. The process is probably closely related to early wall regeneration by soybean (Klein, Montezinos & Delmer, 1981) and other protoplasts (Burgess, 1983) known to produce similar glucan polymers.

Since the enzyme glucan synthase is located in the plasma membrane (Kauss, 1987), it is a likely target of the fungal triggering mechanism. It has been shown that elicitors of phytoalexin production from fungal pathogens act at the plasma membrane (Pelissier *et al*., 1986; Schmidt & Ebel, 1987). Our results suggest that this triggering mechanism requires physical contact between protoplast and pathogen since no induction of fibrils by non-contacting hyphae has been observed. However, this does not preclude that other, possibly soluble, materials may also induce a similar reaction, as shown for the toxin victorin which elicits similar fibril formation in oat protoplasts (Walton & Earle, 1985).

Fig. 6.4. (opposite) A, B: Soybean (cv. Williams) protoplast and *Phytophthora megasperma* f. sp. *glycinea*, early phase of penetration with a hyphal branch entering a protoplast. In B the specimen is stained with the fluorescing dye 'diethanol' and viewed under UV illumination. Note the fluorescent material formed at the point of penetration. Scale bar = 20 μm. C: Potato (cv. Eba) leaf epidermis infected by *P. infestans*; a single cell has reacted hypersensitively and fluorescing material (papilla) has formed at the point of penetration (top right) and at some distance (lower right corner). Scale bar = 10 μm. D: Potato (cv. Eba) tuber tissue infected by *P. infestans*; a strong papilla has formed at the site of penetration, visualized under UV illumination after staining with aniline blue, a stain indicative for polysaccharides such as callose. Scale bar = 20 μm. A & B from Odermatt *et al.* (1988); microphotograph C taken by Dr R. Gees, and D by Dr H. Hächler.

Host-pathogen interactions in *Phytophthora*

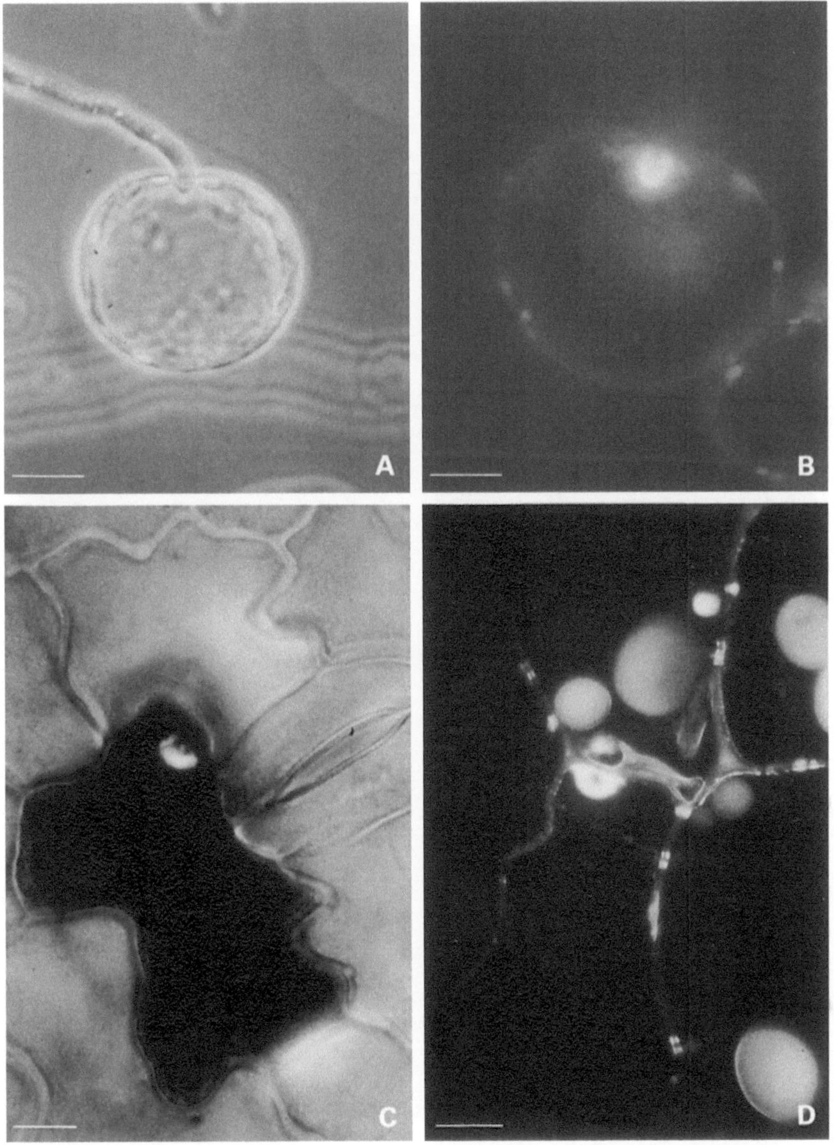

(*Facing p.* 82)

This plate is available for download in colour from www.cambridge.org/9780521189767

Conclusions

Adhesion is the most visible type of surface interaction between fungal plant pathogens and their hosts. In the case of diseases caused by *Phytophthora* and perhaps many others, the adhesive material is produced by the pathogen. This material appears to consist of proteins and/or glycoproteins, or in some instances polysaccharides, but little definite information is available. From our results with *Phytophthora*, it appears that different kinds of bonds are established between different kinds of surfaces. This implies that different types of molecule or parts of molecules may be involved in adhesion of the pathogen to, for example, the plasma membrane, the host cell wall or to a plastic petri dish. Good evidence has been obtained to implicate lectin-ligand type interactions in adhesion of *Phytophthora* to the host surfaces; surface lectins and exposed sugar ligands were detected and adhesion may be prevented by specific sugars. Physical contact triggers reactions in both the pathogen (production of adhesive material) and the host cell (production of callose or some related polysaccharide). Characteristically, these reactions manifest themselves locally where host and pathogen meet which emphasizes their contact-mediated origin. However, major questions remain. It has not been possible to show any degree of specificity in adhesion. Perhaps some specific interactions do occur but are hidden by concomitant strong and unspecific adhesion processes. Furthermore, it has not been established that adhesion is a prerequisite for infection, nor that lectin-ligand interactions are involved in recognition, i.e. in triggering a response in either host or parasite. A more detailed analysis of the molecules involved is certainly needed. Attempts to interfere with the natural infection process, for example by adding potential ligands and lectins (Margan, Jarvis & Friend, 1979; Furuichi, Tomiyama & Doke, 1980; Nozue, Tomiyama & Doke, 1980; Kogel *et al.*, 1985) and purified surface components, or the use of monospecific antibodies against surface components (Chapter 5), may help to define more clearly the role of surface contacts in fungus-plant interactions.

Acknowledgements Financial support for the research in our laboratory has been provided by Swiss National Science Foundation grant 3.240-0.85.

References

Ainsworth, A. M. & Rayner, A. D. M. (1986). Responses of living hyphae associated with self and non-self fusions in the basidiomycete *Phanerochaete velutina*. *Journal of General Microbiology* **132**, 191-201.

Aist, J. R. (1976). Papillae and related wound plugs of plant cells. *Annual Review of Phytopathology* **14**, 145-163.

Amory, D. E. & Rouxhet, P. G. (1988). Surface properties of *Saccharomyces cerevisiae* and *Saccharomyces carlsbergensis*: chemical composition, electrostatic charge and hydrophobicity. *Biochimica et Biophysica Acta* **938**, 61-70.

Aylmore, R. C. & Todd, N. K. (1986). Cytology of self fusion in hyphae of *Phanerochaete velutina*. *Journal of General Microbiology* **132**, 571-579.

Ayers, A. R. & Wycoff, K. L. (1987). Carbohydrate-specific monoclonal antibodies and immunosuppression in the study of plant disease resistance. In *Molecular Strategies for Crop Protection*, ed. Ch. J. Arntzen & C. Ryan, pp. 71-81. Alan R. Liss, Inc.: New York.

Bacic, A., Williams, M. L. & Clarke, A. E. (1985). Studies on the cell surface of zoospores and cysts of the fungus *Phytophthora cinnamomi*: nature of the surface saccharides as determined by quantitative lectin binding studies. *Journal of Histochemistry and Cytochemistry* **33**, 384-388.

Barak, R., Elad, Y., Mirelman, D. & Chet, I. (1985). Lectins: a possible basis for specific recognition in the interaction of *Trichoderma* and *Sclerotium rolfsii*. *Phytopathology* **75**, 458-462.

Barkai-Golan, R., Mirelman, D. & Sharon, N. (1978). Studies on growth inhibition by lectins of penicillia and aspergilli. *Archives of Microbiology* **116**, 119-124.

Beckett, A. & Porter, R. (1988). The use of complementary fractures and low-temperature scanning electron microscopy to study hyphal host cell surface adhesion between *Uromyces viciae-fabae* and *Vicia faba*. *Canadian Journal of Botany* **66**, 645-652.

Beckett, A. & Woods, A. M. (1987). The pattern of colony development and the formation of uredinium of *Uromyces viciae-fabae* on *Vicia faba*. *Canadian Journal of Botany* **65**, 1998-2006.

Bonfante-Fasolo, P., Perotto, S., Testa, B. & Faccio, A. (1987). Ultrastructural localization of cell surface sugar residues in ericoid mycorrhizal fungi by gold-labeled lectins. *Protoplasma* **139**, 25-35.

Borrebaeck, C. A. K., Mattiasson, B. & Nordbring-Hertz, B. (1985). A fungal lectin and its apparent receptors on a nematode surface. *FEMS Microbiology Letters* **22**, 35-39.

Bourett, T., Hoch, H. C. & Staples, R. C. (1987). Association of the microtubule cytoskeleton with the thigmotropic signal for appressorium formation in *Uromyces*. *Mycologia* **79**, 540-545.

Burgess, J. (1983). Wall regeneration around isolated protoplasts. *International Review of Cytology* (suppl.) **16**, 55-75.

Brambl, R. & Gade, W. (1985). Plant seed lectins disrupt growth of germinating fungal spores. *Physiologia Plantarum* **64**, 402-408.

Clarke, A. E., Gleeson, P. A., Jermyn, M. A. & Knox, R. B. (1978). Characterization and localization of β-lectins in lower and higher plants. *Australian Journal of Plant Physiology* **5**, 707-722.

Coffey, M. D. (1983). Cytochemical specialization at the haustorial interface of a biotrophic fungal parasite, *Albugo candida*. *Canadian Journal of Botany* **61**, 2004-2014.

Dazzo, F. B. & Gardiol, A. E. (1984). Host specificity in *Rhizobium* legume interactions. In *Genes Involved in Microbe Plant Interactions*, ed. D. P. S. Verna & T. Hohn, pp. 3-31. Springer Verlag: Wien.

Douglas, L. J. (1987). Adhesion of *Candida* species to epithelial surfaces. *CRC Critical Reviews in Microbiology* **15**, 27-45.

Epstein, L., Laccetti, L., Staples, R. C., Hoch, H. C. & Hoose, W. A. (1985). Extracellular proteins associated with induction of differentiation in bean rust uredospore germlings. *Phytopathology* **75**, 1073-1076.

Epstein, L., Laccetti, L. B., Staples, R. C. & Hoch, H. C. (1987). Cell-substratum adhesive protein involved in surface contact responses of the bean rust fungus. *Physiological & Molecular Plant Pathology* **30**, 373-388.

Furuichi, N., Tomiyama, K. & Doke, N. (1980). The role of potato lectin in the binding of germ tubes of *Phytophthora infestans* to potato cell membrane. *Physiological Plant Pathology* **16**, 249-256.

Galun, M., Braun, A., Frensdorff, A. & Galun, E. (1976). Hyphal walls of isolated lichen fungi. Autoradiographic localization of precursor incorporation and binding of fluorescein-conjugated lectins. *Archives of Microbiology* **108**, 9-16.

Gibson, D. M., Stack, S., Krell, K. & House, J. (1982). A comparison of soybean agglutinin in cultivars resistant and susceptible to *Phytophthora megasperma* var. *sojae* (race 1). *Plant Physiology* **70**, 560-566.

Gold, E. R. & Mendgen, K. (1984). Cytology of basidiospore germination, penetration and early colonization of *Phaseolus vulgaris* by *Uromyces appendiculatus* var. *appendiculatus*. *Canadian Journal of Botany* **62**, 1989-2002.

Gubler, F. & Hardham, A. R. (1988). Secretion of adhesive material during encystment of *Phytophthora cinnamomi* zoospores, characterized by immunogold labelling with monoclonal antibodies to components of peripheral vesicles. *Journal of Cell Science* **90**, 225-235.

Gubler, F., Hardham, A. R. & Duniec, J. (1989). Characterising adhesiveness of *Phytophthora cinnamomi* zoospores during encystment. *Protoplasma* **149**, 24-30.

Guhl, B. (1983). Cytochemischer Nachweis von Lektin-Bindungsstellen an gekeimten Zysten von *Phytophthora palmivora*. M.S. thesis, University of Zürich.

Hächler, H. & Hohl, H. R. (1982). Histochemistry of papillae in potato tuber tissue infected with *Phytophthora infestans*. *Botanica Helvetica* **92**, 23-31.

Hamer, J. E., Howard, R. J. & Chumley, F. G. (1988). A mechanism for surface attachment in spores of a plant pathogenic fungus. *Science* **239**, 288-290.

Hardham, A. R. (1985). Studies on the cell surface of zoospores and cysts of the fungus *Phytophthora cinnamomi*: the influence of fixation on patterns of lectin binding. *Journal of Histochemistry & Cytochemistry* **33**, 110-118.

Hermanns, R. & Ziegler, E. (1984). Localization of α-mannan in the hyphal wall of *Phytophthora megasperma* f. sp. *glycinea* and its possible relevance to the host-pathogen interaction of the fungus with soybeans (*Glycine max*). *Phytopathologische Zeitschrift* **109**, 363-366.

Herrero, E., Sanz, P. & Sentandreu, R. (1987). Cell wall proteins liberated by zymolase from several ascomycetous and imperfect yeasts. *Journal of General Microbiology* **133**, 2895-2903.

Hinch, J. M. & Clarke, A. E. (1980). Adhesion of fungal zoospores to root surfaces is mediated by carbohydrate determinants of the root slime. *Physiological Plant Pathology* **16**, 303-307.

Hinch, J. M. & Clarke, A. E. (1982). Callose formation in *Zea mays* as a response to infection with *Phytophthora cinnamomi*. *Physiological Plant Pathology* **21**, 113-124.

Ho, S. -C., Malek-Hedayat, S., Wang, J. L. & Schindler, M. (1986). Endogenous lectins from cultured soybean cells: isolation of a protein immunologically cross-reactive with seed soybean agglutinin and analysis of its role in binding of *Rhizobium japonicum*. *Journal of Cell Biology* **103**, 1043-1054.

Hoch, H.C. & Staples, R. C. (1987). Structural and chemical changes among the rust fungi during appressorium development. *Annual Review of Phytopathology* **25**, 231-247.

Hoch, H. C., Staples, R. C. & Bourett, T. (1987a). Chemically induced appressoria in *Uromyces appendiculatus* are formed aerially, apart from the substrate. *Mycologia* **78**, 418-424.

Hoch, H. C., Staples, R. C., Whitehead, B., Comeau, J. & Wolf, E. D. (1987b). Signaling for growth orientation and cell differentiation by surface topography in *Uromyces*. *Science* **235**, 1659-1662.

Hohl, H. R. & Balsiger, S. (1986a). A model system for the study of fungus host surface interactions: adhesion of *Phytophthora megasperma* to protoplasts and mesophyll cells of soybean. In *Recognition in Microbe-Plant Symbiosis*, NATO Advanced Research Workshop, ASI Series vol. H4, ed. B. Lugtenberg, pp. 259-272. Springer-Verlag: Berlin.

Hohl, H. R. & Balsiger, S. (1986b). Probing the surfaces of soybean protoplasts and of germ tubes of the soybean pathogen *Phytophthora megasperma* f.sp. *glycinea* with lectins. *Botanica Helvetica* **98**, 289-297.

Hohl, H. R. & Balsiger, S. (1988). Surface glycosyl receptors of *Phytophthora megasperma* f. sp. *glycinea* and its soybean host. *Botanica Helvetica* **98**, 271-277.

Hyde, H. D., Jones, E. B. G., Moss, S. T. (1986). Mycelial adhesion to surfaces. In *The Biology of Marine Fungi*, ed. S. T. Moss, pp. 331-340. Cambridge University Press.

Jarvis, W. R. & Berry, J. W. (1986). The attachment organ of *Botryotinia porri*. *Transactions of the British Mycological Society* **82**, 543-549.

Kapooria, R. G. & Mendgen, K. (1985). Infection structures and their surface changes during differentiation in *Uromyces fabae*. *Phytopathologische Zeitschrift* **113**, 317-323.

Kauss, H. (1987). Some aspects of calcium-dependent regulation in plant metabolism. *Annual Review of Plant Physiology* **38**, 47-72.

Keen, N. T. (1982). Specific recognition in gene-for-gene host-parasite systems. *Advances in Plant Pathology* **1**, 35-82.

Keen, N. T. & Legrand, M. (1980). Surface glycoproteins: evidence that they may function as the race specific phytoalexin elicitors of *Phytophthora megasperma* f.sp. *glycinea*. *Physiological Plant Pathology* **17**, 175-192.

Kervin, T., Heller, P., Wu, Y. -S. & Lipke, P. N. (1987). Identification of glycoprotein components of alpha-agglutinin, a cell adhesion protein from *Saccharomyces cerevisiae*. *Journal of Bacteriology* **169**, 475-482.

Klein, A. S., Montezinos, D. & Delmer, D. P. (1981). Cellulose and 1,3-glucan synthesis during the early stages of wall regeneration in soybean protoplasts. *Planta* **152**, 105-114.

Kogel, K. H., Schrenk, F., Sharon, N. & Reisener, H. J. (1985). Suppression of the hypersensitive response in wheat stem rust interaction by reagents with affinity for wheat plasma membrane conjugates. *Journal of Plant Physiology* **118**, 343-352.

Lapp, M. S. & Skoropod, W. P. (1978). Nature of adhesive material of *Colletotrichum graminicola* appressoria. *Transactions of the British Mycological Society* **70**, 221-223.

Latgé, J. -P., Cole, G. T., Horisberger, M. & Prévost, M. -C. (1986). Ultrastructure and chemical composition of the ballistospore wall of *Conidiobolus obscurus*. *Experimental Mycology* **10**, 99-113.

Latgé, J. -P., Monsigny, M. & Prévost, M. -C. (1988). Visualization of exocellular lectins in the entomopathogenic fungus *Conidiobolus obscurus*. *Journal of Histochemistry & Cytochemistry* **36**, 1419-1424.

Lipke, P., Terrance, K. & Wu, Y. -S. (1987). Interaction of α-agglutinin with *Saccharomyces cerevisiae* **a** cells. *Journal of Bacteriology* **169**, 483-488.

Longman, D. & Callow, J. A. (1987). Specific saccharide residues are involved in the recognition of plant root surfaces by zoospores of *Pythium aphanidermatum*. *Physiological & Molecular Plant Pathology* **30**, 139-150.

Margan, H., Jarvis, M. C. & Friend, J. (1979). Effect of methyl glycosides and oligosaccharides on cell death and browning of potato tuber discs induced by mycelial components of *Phytophthora infestans*. *Physiological Plant Pathology* **14**, 1-9.

Massicotte, H. B., Ackerley, C. A. & Peterson, R. L. (1987a). The root fungus interface as an indicator of symbiont interaction in ectomycorrhizae. *Canadian Journal of Forestry Research* **17**, 846-854.

Massicotte, H. B., Ackerley, C. A. & Peterson, R. L. (1987b). Localization of three sugar residues in the interface of ectomycorrhizae synthesized between *Alnus crispa* and *Alpova diplophloeus* as demonstrated by lectin binding. *Canadian Journal of Botany* **65**, 1127-1132.

Mendgen, K., Lange, M. & Bretschneider, K. (1985). Quantitative estimation of the surface carbohydrates in the infection structures of rust fungi with enzymes and lectins. *Archives of Microbiology* **140**, 307-311.

Mendgen, K., Schneider, A., Sterk, M. & Fink, W. (1988). The differentiation of infection structures as a result of recognition events between some biotrophic parasites and their hosts. *Journal of Phytopathology* **123**, 259-272.

Mims, C. W. & Richardson, E. A. (1989). Ultrastructure of appressorium development by basidiospore germlings of the rust fungus *Gymnosporangium juniperi-virginianae*. *Protoplasma* **148**, 111-119.

Mirelman, D., Galun, E., Sharon, N. & Lotan, R. (1975). Inhibition of fungal growth by wheat germ agglutinin. *Nature* **256**, 414-416.

Nicholson, R. L. (1984). Adhesion of fungi to plant cuticles. In *Infection Processes of Fungi*, ed. D. W. Roberts & J. R. Aist, pp. 74-89. Rockefeller Foundation: New York.

Nozue, M., Tomiyama, K. & Doke, N. (1979). Evidence for adherence of host plasmalemma to infection hyphae of both compatible and incompatible races of *Phytophthora infestans*. *Physiological Plant Pathology* **15**, 111-115.

Nozue, M., Tomiyama, K. & Doke, N. (1980). Effect of N,N'-diacetyl-D-chitobiose, the potato lectin hapten and other sugars on hypersensitive reaction of potato tuber cells infected by incompatible and compatible races of *Phytophthora infestans*. *Physiological Plant Pathology* **17**, 221-228.

Odermatt, M., Röthlisberger, A., Werner, Ch. & Hohl, H. R. (1988). Interactions between agarose-embedded plant protoplasts and germ tubes of *Phytophthora*. *Physiological and Molecular Plant Pathology* **33**, 209-220.

Onyile, A. B., Edwards, H. H. & Gessner, R.V. (1982). Adhesive material of the hyphopodia in *Buergenerula spartinae*. *Mycologia* **74**, 777-784.

Osawa, T. & Tsuji, T. (1987). Fractionation and structural assessment of oligosaccharides and glycopeptides by use of immobilized lectins. *Annual Review of Biochemistry* **56**, 21-42.

Pelissier, B., Thibaud, J. B., Grignon, C. & Esquerré-Tugayé, M. T. (1986). Cell surfaces in plant-microorganism interaction. VII. Elicitor preparations from two fungal pathogens depolarize plant membranes. *Plant Science* **46**, 103-109.

Ralton, J. E., Howlett, B. J. & Clarke, A. E. (1986). Receptors in host-pathogen interactions. In *Hormones, Receptors and Cellular Interactions in Plants*, ed. C. M. Chadwick & D. R. Garrod, pp. 281-318. Cambridge University Press.

Ralton, J. E., Smart, M. G. & Clarke, A. E. (1987). Recognition and infection processes in host-plant interactions. In *Plant-Microbe Interactions*, vol. 2, ed. T. Kosuge & E. W. Nester, pp. 217-252. Macmillan Publishing Company: New York.

Samson, R. A., Evans, H. C. & Latgé, J. -P. (1988). *Atlas of Entomopathogenic Fungi*. Springer-Verlag: Berlin & Heidelberg.

Schlumbaum, A., Mauch, F., Vögeli, U. & Boller, T. (1986). Plant chitinases are potent inhibitors of fungal growth. *Nature* **324**, 365-367.

Schmidt, W. E. & Ebel, J. (1987). Specific binding of a fungal glucan elicitor to membrane fractions from soybean *Glycine max*. *Proceedings of the National Academy of Sciences, USA* **84**, 4117-4121.

Sijmons, P. C., Nederbragt, A. J. A., Klis, F. M., Van den Ende, H. (1987). Isolation and composition of the constitutive agglutinins from haploid *Saccharomyces cerevisiae* cells. *Archives of Microbiology* **148**, 208-212.

Sing, V. O. & Bartnicki-Garcia, S. (1975a). Adhesion of *Phytophthora palmivora* zoospores: electron microscopy of cell attachment and cyst wall fibril formation. *Journal of Cell Science* **18**, 123-132.

Sing, V. O. & Bartnicki-Garcia, S. (1975b). Adhesion of *Phytophthora palmivora* zoospores: detection and ultrastructural visualization of Concanavalin A receptor sites appearing during encystment. *Journal of Cell Biology* **19**, 11-20.

Staples, R. C. & Hoch, H. C. (1987). Infection structures–form and function. *Experimental Mycology* **11**, 163-169.

Teepe, H., Boettge, J. -A. & Woestemeyer, J. (1988). Isolation and electrophoretic analysis of surface proteins of the zygomycete *Absidia glauca*. *FEBS Letters* **234**, 100-106.

Terrance, K., Heller, P., Wu, Y. -S. & Lipke, P. N. (1987). Identification of glycoprotein components of α-agglutinin, a cell adhesion protein from *Saccharomyces cerevisiae*. *Journal of Bacteriology* **169**, 475-482.

Walton, J. D. & Earle, E. D. (1985). Stimulation of extracellular polysaccharide synthesis in oat protoplasts by the host-specific phytotoxin victorin. *Planta* **165**, 407-415.

Welter, K., Müller, M. & Mendgen, K. (1988). The hyphae of *Uromyces appendiculatus* within the leaf tissue after high pressure freezing and freeze substitution. *Protoplasma* **147**, 91-99.

Yoshikawa, M., Keen, N. T. & Wang, M. -C. (1983). A receptor on soybean membranes for a fungal elicitor of phytoalexin accumulation. *Plant Physiology* **73**, 497-506.

Young, D. H. & Kauss, H. (1984). Adhesion of *Colletotrichum lindemuthianum* spores to *Phaseolus vulgaris* hypocotyls and to polystyrene. *Applied & Environmental Microbiology* **47**, 616-619.

Chapter 7

Molecular aspects of host-pathogen interactions in *Phytophthora*

Jane E. Parker, Wolfgang Knogge & Dierk Scheel

Plants respond to pathogen attack with the rapid activation of a complex pattern of defence reactions, which appear to be similar in many diverse plant species (Hahlbrock & Scheel, 1987). The same resistance responses are induced by different pathogens in non-host and host-incompatible interactions. Some of these reactions involve gene activation, such as phytoalexin synthesis, whereas others do not, for example callose formation. No matter what the mechanisms of stimulation are, they require recognition of the pathogen as non-self and consequent signal transduction to activate the different defence responses. Since most processes of a successful resistance reaction occur in only a few cells at the site of infection, the molecular events are difficult to study in the intact plant. Cultured cells of a number of plants synchronously respond to treatment with plant- or pathogen-derived elicitors in a way similar to the plant reaction to pathogen attack (DiCosmo & Misawa, 1985). These cultured cell systems have been used successfully to elucidate the biochemistry and molecular biology of defence responses (Ebel, 1986; Hahlbrock & Scheel, 1989) and we are now employing them to study the mechanisms of pathogen recognition and signal transduction.

We will describe our recent studies on the interaction of *Phytophthora infestans* with its host plant potato and the non-host resistance response of parsley to *Phytophthora megasperma* f. sp. *glycinea*. Soybean is included in some experiments as it is the host plant of *P. megasperma*. The three plant-fungal interactions are illustrated in Fig. 7.1.

Defence responses

Parsley leaves respond hypersensitively to inoculation with *P. megasperma* zoospores (Jahnen & Hahlbrock, 1988). Growth of fungal hyphae is restricted to small necrotic areas at the penetration sites. Cells surrounding this region excrete furanocoumarin phytoalexins into the infection droplets (Scheel *et al.*, 1986). Their accumulation results from a local and specific increase in mRNA and protein, and levels of enzymes involved in their synthesis (Fig. 7.2), such as phenylalanine ammonia-lyase (PAL), 4-coumarate:CoA ligase (4CL) and S-adenosyl-L-methionine:bergaptol

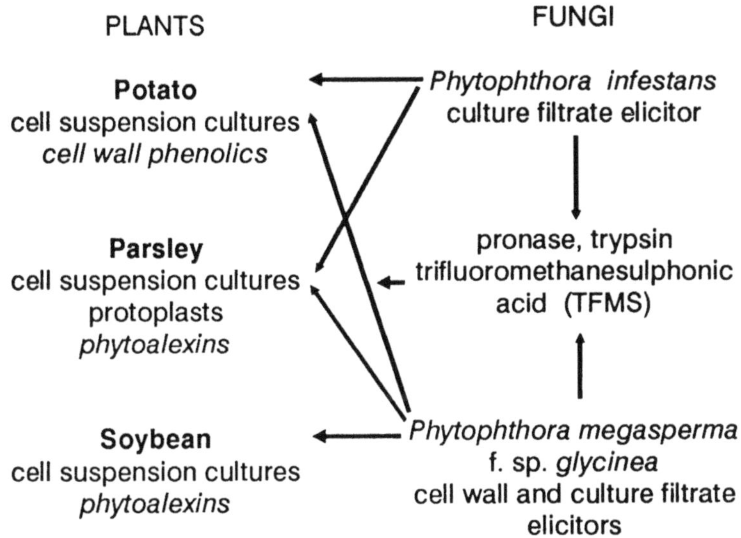

Fig. 7.1. Three plant-fungal interactions showing the plant tissues tested and fungal elicitor treatments.

O-methyltransferase (Jahnen & Hahlbrock, 1988; Schmelzer, Krüger-Lebus & Hahlbrock, 1989). Other mRNAs of unknown function are also induced (Somssich *et al.*, 1989). Suspension-cultured parsley cells and freshly prepared protoplasts respond to treatment with crude elicitor preparations from *P. megasperma* by excretion of the same furanocoumarin products into the suspension medium (Dangl *et al.*, 1987) and have therefore been extremely helpful tools for detailed analysis of the induction processes.

Potato leaves are the primary target of *P. infestans* and the relevant organs for successful late blight resistance (Thurston & Schultz, 1981). They respond to inoculation with avirulent races of *P. infestans* by undergoing hypersensitive cell death, limiting fungal invasion to the necrotic area (Cuypers & Hahlbrock, 1988). In incompatible interactions all cells neighbouring the fungal hyphae rapidly collapse, whereas cells of susceptible leaves appear not to detect the invading pathogen. In contrast to tuber tissue, leaves do not synthesize sesquiterpenoid phytoalexins such as rishitin and lubimin (Rohwer *et al.*, 1987), but respond to infection or crude elicitor treatment with rapid synthesis and incorporation into the cell wall of phenylpropanoid derivatives, a reaction which occurs with similar kinetics in cultured cells upon addition of elicitor (Scheel *et al.*,

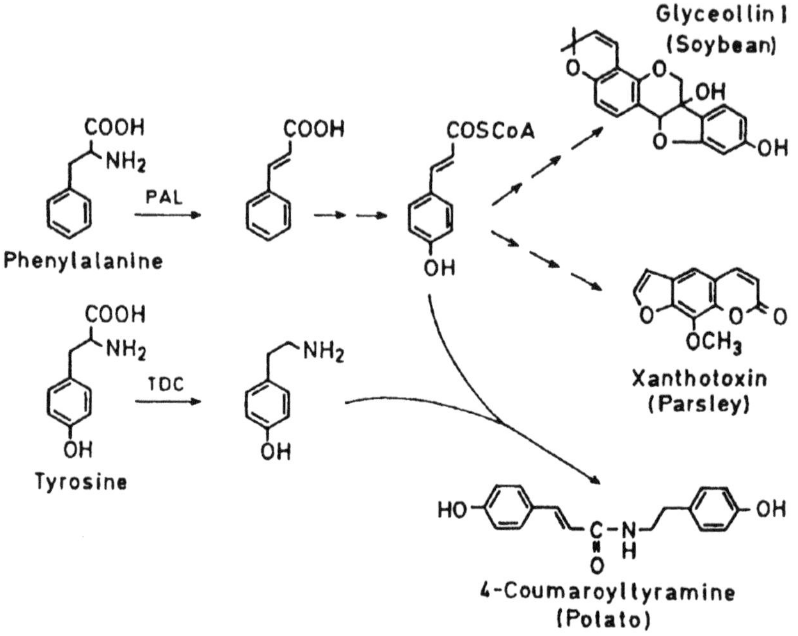

Fig. 7.2. Biosynthetic pathways of the plant defence compounds glyceollin I, xanthotoxin and 4-coumaroyltyramine. PAL = phenylalanine ammonia-lyase; TDC = tyrosine decarboxylase.

1989). The major components are 4-coumaroyltyramine and feruloyl-tyramine which both originate from phenylalanine and tyrosine (Fig. 7.2). The biosynthetic enzymes PAL and tyrosine decarboxylase (TDC) are stimulated by elicitor in a coordinated manner. Treatment of potato leaves with the PAL inhibitor L-2-aminooxy-3-phenylpropionic acid (AOPP) prior to infection results in a complete collapse of late blight resistance (Fig. 7.3), emphasizing the importance of phenylpropanoid derivatives in a successful defence response. The phenylpropanoid enzymes PAL and 4CL, as well as a pathogenesis-related protein of unknown function (PR1), are transcriptionally activated and the mRNAs of PAL and PR1 accumulate around infection sites, as shown by *in situ* hybridization studies (Fritzemeier *et al.*, 1987: Taylor *et al.*, 1990). A number of other unidentified proteins also accumulate rapidly upon infection (Fritzemeier *et al.*, 1987), whereas levels of chitinases and 1,3-β-glucanases increase more slowly (Kombrink, Schröder & Hahlbrock, 1988).

The race-specific resistance of soybean to *P. megasperma* involves hypersensitive cell death and local synthesis of callose and isoflavonoid phytoalexins, such as glyceollin (Fig. 7.2) (Bonhoff *et al.*, 1987; Ebel &

Fig. 7.3. Infection of potato leaves (cv. Datura, R1) with spores from *Phytophthora infestans* (races 1 and 4). The leaves were placed into water or a solution of 1 mM L-2-aminooxy-3-phenylpropionic acid (AOPP) 4 h prior to inoculation with 2×10^5 zoospores per leaf.

Grisebach, 1988; Hahn, Bonhoff & Grisebach, 1985). Again, phytoalexin synthesis, but not callose formation, appears to be regulated by specific gene activation. Inhibition of PAL activity in soybean roots by AOPP or R-(1-amino-2-phenylethyl)-phosphonic acid converts an incompatible interaction into a compatible one, indicating that phenylpropanoid products play a decisive role in this defence reaction (Moesta & Grisebach, 1982; Waldmüller & Grisebach, 1987).

Activation of these responses requires perception of appropriate signals by the plant cell. Suitable screening systems are needed in order to identify and purify both pathogen- and plant-derived signal molecules. We are using the onset of an indicative resistance reaction after elicitor addition to cultured cells to detect the recognition event. In parsley and soybean, phytoalexin synthesis is taken as a measure of elicitation, whereas in potato cells activities of PAL and TDC are determined.

Recognition

The molecular basis of non-self recognition in plants is not yet understood but is likely to involve a complex array of plant and pathogen molecules. Fungal wall-derived and extracellular molecules act as elicitors of defence reactions in many plants (Darvill & Albersheim, 1984; Ebel, 1986) and may therefore be excellent tools to study primary recognition events. Elicitor-active fractions isolated from different *Phytophthora* species have been identified in potato (Bostock, Kuć & Laine, 1981; Keenan, Bryan & Friend, 1985; Rohwer *et al.*, 1987), parsley (Kombrink & Hahlbrock, 1986), soybean (Ayers *et al.*, 1976a & b) and tobacco (Farmer & Helgeson, 1987). The best characterized of these are β-glucan fragments extracted from the cell walls of *P. megasperma*, which are potent elicitors of glyceollin accumulation in soybean cotyledons (Ayers *et al.*, 1976b & c). The smallest elicitor-active oligoglucan isolated from the cell walls by acid hydrolysis is a hepta-β-glucoside alditol (Sharp, Valent & Albersheim, 1984b). Only one out of eight stereoisomeric forms of the heptaglucoside examined possessed elicitor activity, demonstrating a high degree of structural specificity (Sharp et al., 1984 a & b). Saturable, high affinity binding sites were located on microsomal membranes of soybean cells (Schmidt & Ebel, 1987; Cosio *et al.*, 1988) and on the surface of isolated protoplasts (Cosio *et al.*, 1988) using an elicitor-active radiolabelled oligoglucan fraction from *P. megasperma* walls. Interestingly, there was a strong positive correlation between the ability of structurally related oligosaccharides to compete for binding of this glucan to membranes and their activity as elicitors of glyceollin in soybean cotyledons, suggesting a causal relationship between binding and elicitation.

Although crude cell wall preparations from *P. megasperma* are potent elicitors in soybean and parsley, pure glucan fractions isolated from this material are highly active in soybean but completely inactive in parsley (Parker, Hahlbrock & Scheel, 1988). In preliminary experiments the crude *P. megasperma* and *P. infestans* elicitors were treated with proteases or chemically deglycosylated using trifluoromethane sulphonic acid in order to determine which constituents are active in each plant system (see Fig. 7.1). Our results show that the cultured cells of the three plants recognize different components of *P. megasperma* and *P. infestans*. For example, proteins from *P. megasperma* elicit furanocoumarin accumulation in parsley cells and protoplasts, whereas carbohydrates are active in soybean cotyledons and cultured cells. Interestingly, in similar preparations from *P. infestans*, both parsley and potato respond to carbohydrate constituents (Scheel *et al.*, 1989). Therefore, there seems to be no consensus in the structural determinants for elicitor activity. It appears that parsley, potato and soybean not only perceive different signals from *P. megasperma* and *P. infestans*, but are also unable to respond to fungal components recognized by the other plant.

We have taken particular advantage of parsley cells for elicitor recognition studies since the excreted furanocoumarin phytoalexins are readily detected due to their autofluorescence under UV light. In addition,

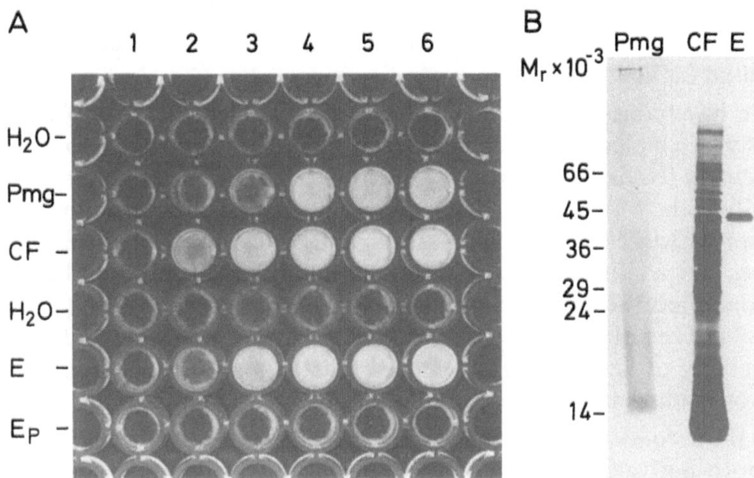

Fig. 7.4A. Parsley protoplast microassay. Increasing concentrations of elicitors in 1 to 8 μl were added to 5×10^4 protoplasts in 0·1 ml. Fluorescence was tested under UV light after 24 h incubation at 26°C. Pmg = *P. megasperma* crude wall elicitor (0·4 to 8 μg ml^{-1} protein); CF = *P. megasperma* crude culture filtrate elicitor (0·5 to 10 μg ml^{-1}); E = *P. megasperma* pure 42 kDa elicitor (5 to 100 ng ml^{-1}): Ep = *P. megasperma* pure 42 kDa elicitor (5 to 100 ng ml^{-1}) after prior treatment with pronase. All elicitor preparations were autoclaved.

Fig. 7.4B. SDS-polyacrylamide gel showing protein from the *P. megasperma* crude wall elicitor (Pmg, 2 μg), *P. megasperma* crude culture filtrate elicitor (CF, 2 μg) and *P. megasperma* pure 42 kDa elicitor (E, 0·1 μg) after silver staining.

protoplasts freshly prepared from parsley cell cultures are fully responsive to the crude *P. megasperma* wall elicitor (Dangl *et al.*, 1987). This points to the existence of recognition sites on the parsley plasma membrane. It is possible, however, that plant cell wall components are involved in a finer tuning or enhancement of fungal elicitor recognition in the whole cell environment.

We have used parsley protoplasts to test for active components of the crude *P. megasperma* cell wall and culture filtrate material. For this purpose we developed a microassay to measure the response of the protoplasts in a total volume of 0·1 ml (Fig. 7.4A). This enables rapid fluorimetric screening of active molecules during purification. A 42 kDa glycoprotein, which is a potent elicitor of phytoalexin synthesis in parsley protoplasts and cells, has been purified to homogeneity from the *P.*

megasperma culture filtrate (Fig. 7.4B). It induces the full complement of furanocoumarins in protoplasts at very low concentrations (2 pmol 10^{-6} protoplasts) and is therefore an ideal candidate for recognition and mode of action studies. Complementary proteinase and deglycosylation treatments have established that the protein, and not the N-linked sugar, is essential for activity.

We have considered three strategies for the identification of parsley cell surface molecules which interact with the pure *P. megasperma* glycoprotein. In the first approach, the elicitor is radiolabelled to high specific activity with ^{125}I (Bolton, 1986). Assuming the labelled protein retains elicitor activity, it can be used as a ligand to test parsley plasma membranes for specific, saturable binding. If high affinity, specific binding sites are identified, the elicitor can be immobilized on a gel as an affinity matrix for solubilized plasma membrane proteins. However, this depends on the availability of large amounts of pure elicitor and also on retention of elicitor binding activity after solubilization of the plant proteins.

In the second approach the elicitor protein can be covalently linked to a photo-activatable reagent in photo-affinity labelling experiments (Jacobs *et al.*, 1979; Ji & Ji, 1980). The derivatized elicitor is allowed to react with parsley plasma membranes in the dark. Upon illumination at selected wavelengths, the photo-activatable group becomes covalently attached to neighbouring molecules. The recent introduction of iodinatable, cleavable, photo-affinity reagents (Sorensen, Farber & Krystal, 1986) allows detection of putative receptor proteins on SDS-polyacrylamide gels after detachment of the ligand.

In the third, somewhat different strategy, it is possible to generate monoclonal antibodies (mabs) to plant plasma membrane epitopes (Brewin *et al.*, 1985; Norman *et al.*, 1986), some of which may recognize components of an elicitor-binding protein. The parsley protoplast microassay system lends itself to such an approach since mabs could be efficiently screened for interference or inhibition of binding or for positive binding in the absence of elicitor. Mabs were raised against gently fixed protoplasts and isolated plasma membrane vesicles. About 1000 hybridoma supernatants proved to be positive in ELISA studies using right-side-out plasma membrane vesicles as immobilized antigens. However, none of these antibodies altered the induction of furanocoumarin synthesis in parsley protoplasts by the crude *P. megasperma* elicitor. We assume that the presence of more than one species of elicitor in the crude preparation, together with the possible existence of different binding sites on the plasma membrane, hampered our approach. An enriched or pure elicitor appears to be a prerequisite for this kind of experiment which would then provide key information on external and functional domains of a putative binding site. Mabs have also proved to be invaluable tools for the localization of membrane-bound proteins in different plant tissues and plant species (Bradley *et al.*, 1988; Brewin *et al.*, 1985; Napier *et al.*, 1988). The availability of mabs recognizing the plant elicitor binding protein(s) also provides a means for their isolation by immunoaffinity chromatography as well as discrimination of membrane-spanning and extrinsic domains.

Signal transduction

At present very little is known about signal transduction in response to environmental and developmental stimuli in plants. Most recent studies, however, suggest that key elements of some but not all signalling cascades found in animals are present. There is substantial evidence for Ca^{2+} regulation of plant physiological processes (Kauss, 1987) and the existence of Ca^{2+}/calmodulin-dependent protein kinases is clearly established (Klucis & Polya, 1988; Poovaiah & Reddy, 1987). Phosphoinositides and enzymes involved in their metabolism have been identified in plant lipid fractions (Blowers & Trewavas, 1988; Irvine *et al.*, 1989; McMurray & Irvine, 1988; Wheeler & Boss, 1987). Limited experimental data also suggest that inositol-1, 4, 5-trisphosphate acts as a signalling molecule (Ettlinger & Lehle, 1988; Rincon & Boss, 1987). Evidence for signal activity of diacylglycerol is less clear (Morré *et al.*, 1989; Schäfer *et al.*, 1985). Interestingly, phosphoinositide metabolism does not appear to be involved in the response of soybean and parsley cells to the crude *P. megasperma* elicitor (Strasser *et al.*, 1986) but more detailed analyses should be performed to confirm this.

We are now conducting experiments to examine the transduction of elicitor signals from the surface of potato and parsley cells where they are perceived to the nucleus where specific defence-related genes are activated. Gene activation is measured by *in vitro* transcription in isolated nuclei (nuclear run-off transcription). This requires isolation of nuclei from elicitor-treated plant cells, elongation of pre-initiated RNA in the presence of radiolabelled nucleotides, extraction of the labelled transcripts and hybridization to cloned cDNAs of interest (Chappell & Hahlbrock, 1984). In parsley cell cultures the earliest transcriptional events after elicitor application are detected within 10 to 15 min in nuclear run-off transcription experiments (Somssich *et al.*, 1986; 1989). Indicative genes of the general phenylpropanoid pathway reach maximal transcription rates within 2 h of elicitor treatment (Somssich *et al.*, 1986). Second messengers are therefore transduced very rapidly to the nucleus and we would expect transmembrane and intracellular changes to occur within the first minutes of induction. Of course, we cannot rule out that the elicitor or an elicitor-recognition complex is internalized and transported to an intracellular location.

A monoclonal antibody was used to determine by radioimmunoassay (Colling *et al.*, 1988) the levels of cAMP in parsley cells and protoplasts after *P. megasperma* elicitor application. Amounts of cAMP remained constant within the first 2 h of treatment (Scheel *et al.*, 1989). Similar experiments have dismissed cAMP as a second messenger in the response of soybean to *P. megasperma* infection (Hahn & Grisebach, 1983). In contrast, convincing evidence has been obtained for the role of Ca^{2+} in elicitation of glyceollin synthesis in soybean cells by *P. megasperma* crude wall elicitor (Stäb & Ebel, 1987) and of rishitin and lubimin in potato tubers by fatty acid elicitors of *P. infestans* (Zook, Rush & Kuć, 1987). Our experiments also show that omission of Ca^{2+} from the medium of parsley and potato cells substantially reduces induction of their respective defence reactions (Scheel *et al.*, 1989). Therefore, it appears that Ca^{2+} stores in the cell wall (Kauss, 1987) are not sufficient or not available for induction. Elicitor treatment of suspension-cultured parsley cells results in rapid uptake of Ca^{2+}, alkalization of the culture medium and massive efflux of K^{+} all starting within 2 min of elicitor addition (Colling *et al.*, unpublished). We do not know if elicitor recognition directly triggers Ca^{2+} uptake, possibly through activated Ca^{2+} channels (Schröder & Hagiwara, 1989; Thuleau *et al.*, 1988), or if Ca^{2+} is taken up by the cell as an indirect response to depletion of mobilized internal stores. Nuclear run-off experiments with parsley protoplasts show that Ca^{2+} but not Mg^{2+} omission reduces the transcriptional rates of elicitor-responsive genes (Colling *et al.*, unpublished). Constitutive and UV-induced genes were not affected, showing that the requirement is specific to some degree and that protoplast viability is not diminished due to the lack of Ca^{2+} for the duration of the experiment. It will be of great interest to test for Ca^{2+} and other ion channels upon elicitation using the patch-clamp technique (Schröder & Hedrich, 1989) on single parsley protoplasts. The observed changes in membrane potential in response to various elicitor applications (Mayer & Ziegler, 1988; Pelissier *et al.*, 1986) are indicative of alterations in membrane-bound ATPase activity and consequent ion fluxes. However, care must be taken with crude elicitor preparations to distinguish events which are directly associated with particular defence reactions from non-specific effects. In this respect, the pure *P. megasperma* glycoprotein elicitor will be very useful in future experiments with parsley.

Recent *in vivo* ^{32}P-labelling experiments with parsley cells show that treatment with crude *P. megasperma* wall elicitor leads to a rapid, transient and reversible phosphorylation of several proteins in both cytoplasmic and membrane fractions (Dietrich, Mayer & Hahlbrock, 1990). One 45 kDa protein becomes phosphorylated within 1 min of elicitor treatment, and a 26 kDa nuclear protein within 10 min. The specific changes in phosphorylation are not observed in cells exposed to heat-shock or UV light. Also, phosphorylation is dependent on Ca^{2+} provision in the medium and on integrity of proteins in the crude elicitor.

Taken together, the results of the experiments described strongly implicate Ca^{2+}-mediated protein kinase activation and consequent phosphorylation cascades early in the response of parsley cells to elicitor.

Conclusions

We have used plant cell cultures in conjunction with fungal elicitors to examine recognition and early signalling events leading to assayable defence responses. Plant cells may possess plasma membrane binding sites for structurally distinct surface components of potential fungal pathogens. We envisage the presence of many different target sites for pathogen recognition on the plant cell surface funnelling the signal through a transduction chain, which in turn switches on multiple defence reactions. These signal transduction processes appear to involve Ca^{2+} and H^{+} uptake, K^{+} efflux and Ca^{2+}-dependent protein phosphorylation. Ultimately, the transcription of a typical set of defence-related genes is initiated. In compatible plant/pathogen and symbiotic interactions any step of this process may be specifically inhibited or altered, resulting in delay or failure of the resistance response.

Acknowledgements We thank Andrea Gierten, Magdalena Jung and Rosemarie Voigt for skilled technical assistance and Dr Klaus Hahlbrock for valuable discussions and steady support. The work in our laboratory was supported by an EMBO postdoctoral fellowship to JEP, a research grant to DS from the Deutsche Forschungsgemeinschaft (Sche 235/2-1) and by the Max-Planck-Gesellschaft.

References

Ayers, A. R., Ebel, J., Finelli, F., Berger, N. & Albersheim, P. (1976a). Host-pathogen interactions IX. Quantitative assays of elicitor activity and characterization of the elicitor present in the extracellular medium of cultures of *Phytophthora megasperma* var. *sojae*. *Plant Physiology* **57**, 751-759.

Ayers, A. R., Ebel, J., Valent, B. & Albersheim, P. (1976b). Host-pathogen interactions X. Fractionation and biochemical activity of an elicitor isolated from the mycelial walls of *Phytophthora megasperma* var. *sojae*. *Plant Physiology* **57**, 760-765.

Ayers, A. R., Valent, B., Ebel, J. & Albersheim, P. (1976c). Host-pathogen interactions XI. Composition and structure of wall-released elicitor fractions. *Plant Physiology* **57**, 766-774.

Blowers, D. P. & Trewavas, A. J. (1988). Phosphatidylinositol kinase activity of a plasma membrane-associated calcium-activated protein kinase from pea. *FEBS Letters* **238**, 87-89.

Bolton, A. E. (1986). Comparative methods for the radiolabelling of peptides. *Methods in Enzymology* **124**, 18-29.

Bonhoff, A., Rieth, B., Golecki, J. & Grisebach, H. (1987). Race cultivar-specific differences in callose deposition in soybean roots following infection with *Phytophthora megasperma* f. sp. *glycinea*. *Planta* **172**, 101-105.

Bostock, R. M., Kuć, J. A. & Laine, R. A. (1981). Eicosapentaenoic and arachidonic acids from *Phytophthora infestans* elicit fungitoxic sesquiterpenes in the potato. *Science* **212**, 67-69.

Bradley, D. J., Wood, E. A., Larkins, A. P., Galfre, G., Butcher, G. W. and Brewin, N. J. (1988). Isolation of monoclonal antibodies reacting with peribacteroid membranes and other components of pea root nodules containing *Rhizobium leguminosarum*. *Planta* **173**, 149-160.

Brewin, N. J., Robertson, J. G., Wood, E. A., Wells, B., Larkins, A. P., Galfre, G. & Butcher, G. W. (1985). Monoclonal antibodies to antigens in the peribacteroid membrane from *Rhizobium*-induced root nodules of pea cross-react with plasma membranes and golgi bodies. *EMBO Journal* **4**, 605-611.

Chappell, J. & Hahlbrock, K. (1984). Transcription of plant defence genes in response to UV light or fungal elicitor. *Nature* **311**, 76-78.

Colling, C., Gilles, R., Cramer, M., Nass, N., Moka, R. & Jaenicke, L. (1988). Measurement of 3':5'-cyclic AMP in biological samples using a specific monoclonal antibody. *Second Messengers and Phosphoproteins* **12**, 123-133.

Cosio, G., Pöpperl, H., Schmidt, W. E. & Ebel, J. (1988). High-affinity binding of fungal β-glucan fragments to soybean (*Glycine max* L.) microsomal fractions and protoplasts. *European Journal of Biochemistry* **175**, 309-315.

Cuypers, B. & Hahlbrock, H. (1988). Immunohistochemical studies of compatible and incompatible interactions of potato leaves with *Phytophthora infestans* and of the nonhost response to *Phytophthora megasperma*. *Canadian Journal of Botany* **66**, 700-705.

Dangl, J. L., Hauffe. K. D., Lipphardt, S., Hahlbrock, K. & Scheel, D. (1987). Parsley protoplasts retain differential responsiveness to u.v. light and fungal elicitor. *EMBO Journal* **6**, 2551-2556.

Darvill. A. G. & Albersheim, P. (1984). Phytoalexins and their elicitors. *Annual Review of Plant Physiology* **35**, 243-275.

DiCosmo, F. & Misawa, M. (1985). Eliciting secondary metabolism in plant cell cultures. *Trends in Biotechnology* **3**, 318-322.

Dietrich, A., Mayer, J. E. & Hahlbrock, K. (1990). Fungal elicitor triggers rapid, transient and specific protein phosphorylation in parsley cell suspension cultures. *Journal of Biological Chemistry* **265**, 6360-6368.

Ebel, J. (1986). Phytoalexin synthesis: the biochemical analysis of the induction process. *Annual Review of Phytopathology* **24**, 235-264.

Ebel, J. & Grisebach, H. (1988). Defense strategies of soybean against the fungus *Phytophthora megasperma* f. sp. *glycinea*: a molecular analysis. *Trends in Biochemical Science* **13**, 23-27.

Ettlinger, C. & Lehle, L. (1988). Auxin induces rapid changes in phosphatidylinositol metabolites. *Nature* **331**, 176-178.

Farmer, E. E. & Helgeson, J. P. (1987). An extracellular protein from *Phytophthora parasitica* var. *nicotianae* is associated with stress metabolite accumulation in tobacco callus. *Plant Physiology* **85**, 733-740.

Fritzemeier, K. -H., Cretin, C., Kombrink, E., Rohwer, F., Taylor, J., Scheel, D. & Hahlbrock, K. (1987). Transient induction of phenylalanine ammonia-lyase and 4-coumarate:CoA ligase mRNAs in potato leaves infected with virulent or avirulent races of *Phytophthora infestans*. *Plant Physiology* **85**, 34-41.

Hahlbrock. K. & Scheel, D. (1987). Biochemical responses of plants to pathogens. In *Innovative Approaches to Plant Disease Control*, ed. I. Chet, pp. 229-254. Wiley & Sons: New York.

Hahlbrock, K. & Scheel, D. (1989). Physiology and molecular biology of phenylpropanoid metabolism. *Annual Review of Plant Physiology and Plant Molecular Biology* **40**, 347-369.

Hahn, M. G., Bonhoff, A. & Grisebach, H. (1985). Quantitative localization of the phytoalexin glyceollin I in relation to fungal hyphae in soybean roots infected with *Phytophthora megasperma* f. sp. *glycinea*. *Plant Physiology* **77**, 591-601.

Hahn, M. G. & Grisebach, H, (1983). Cyclic AMP is not involved as a second messenger in the response of soybean to infection by *Phytophthora megasperma* f. sp. *glycinea*. *Zeitschrift für Naturforschung* **38c**, 578-582.

Irvine, R. F., Letcher, A. J., Lander, D. J., Drobak, B. K., Dawson, A. P. & Musgrave, A. (1989). Phosphatidylinositol(4,5)bisphosphate and phosphatidylinositol(4)phosphate in plant tissues. *Plant Physiology* **89**, 888-892.

Jacobs, S., Hazum, E., Shechter, Y. & Cuatrecasas, P. (1979). Insulin receptor: covalent labeling and identification of subunits. *Proceedings of the National Academy of Sciences, USA* **76**, 4918-4921.

Jahnen, W. & Hahlbrock, K. (1988) Cellular localization of nonhost resistance reactions of parsley (*Petroselinum crispum*) to fungal infection. *Planta* **173**, 197-204.

Ji, I. & Ji, T. H. (1980). Macromolecular photoaffinity labeling of the lutropin receptor on granulosa cells. *Proceedings of the National Academy of Sciences, USA* **77**, 7167-7170.

Kauss, H. (1987). Some aspects of calcium-dependent regulation in plant metabolism. *Annual Review of Plant Physiology* **38**, 47-72.

Keenan, P. J., Bryan, I. B. & Friend, J. (1985). The elicitation of the hypersensitive response of potato tuber tissue by a component of the culture filtrate of *Phytophthora infestans*. *Physiological Plant Pathology* **26**, 343-355.

Klucis, E. & Polya, G. M. (1988). Localization, solubilization and characterisation of plant membrane-associated calcium-dependent protein kinases. *Plant Physiology* **88**, 164-171.

Kombrink, E. & Hahlbrock, K. (1986). Responses of cultured parsley cells to elicitors from phytopathogenic fungi. *Plant Physiology* **81**, 216-221.

Kombrink, E., Schröder, M. & Hahlbrock, K. (1988). Several pathogenesis-related proteins in potato are 1,3-β-glucanases and chitinases. *Proceedings of the National Academy of Sciences, USA* **85**, 782-786.

Mayer, M. G. & Ziegler, E. (1988). An elicitor from *Phytophthora megasperma* f. sp. *glycinea* influences the membrane potential of soybean cotyledonary cells. *Physiological and Molecular Plant Pathology* **33**, 397-407.

McMurray, W. C. & Irvine, R. F. (1988). Phosphatidylinositol 4,5-bisphosphate phosphodiesterase in higher plants. *Biochemical Journal* **249**, 877-881.

Moesta, P. & Grisebach, H. (1982). L-2-Aminooxy-3-phenylpropionic acid inhibits phytoalexin accumulation in soybean with concomitant loss of resistance against *Phytophthora megasperma* f. sp. *glycinea*. *Physiological Plant Pathology* **21**, 65-70.

Morré, D. J., Pfaffmann, H., Drobes, B., Wilkinson, F. E. & Hartman, E. (1989). Diacylglycerol levels unchanged during auxin-stimulated growth of excised hypocotyl segments of soybean. *Plant Physiology* **90**, 275-279.

Napier, R. M., Venis, M. A., Bolton, M. A., Richardson, L. I. & Butcher, G. W. (1988). Preparation and characterisation of monoclonal and polyclonal antibodies to maize membrane auxin-binding protein. *Planta* **176**, 519-526.

Norman, P. M., Wingate, V. P. M., Fitter, M. S. & Lamb, C. J. (1986). Monoclonal antibodies to plant plasma-membrane antigens. *Planta* **167**, 452-459.

Parker, J. E., Hahlbrock, K. & Scheel, D. (1988). Different cell-wall components from *Phytophthora megasperma* f. sp. *glycinea* elicit phytoalexin production in soybean and parsley. *Planta* **176**, 75-82.

Pelissier, B., Thibaud, J. B., Grignon, C. & Esquerré-Tugayé, M. T. (1986). Cell surfaces in plant-microorganism interactions. VII. Elicitor preparations from two fungal pathogens depolarize plant membranes. *Plant Science* **46**, 103-109.

Poovaiah, B. W. & Reddy, A. S. N. (1987). Calcium messenger systems in plants. *CRC Critical Reviews of Plant Science* **6**, 47-103.

Rincon, M. & Boss, W. F. (1987). Myo-inositol trisphosphate mobilizes calcium from fusogenic carrot (*Daucus carota* L.) protoplasts. *Plant Physiology* **83**, 395-398.

Rohwer, F., Fritzemeier, K. -H., Scheel, D. & Hahlbrock, K. (1987). Biochemical reactions of different tissues of potato (*Solanum tuberosum*) to zoospores or elicitors from *Phytophthora infestans*. *Planta* **170**, 556-561.

Schäfer, A., Bygrave, F., Matzenauer, S. & Marmé, D. (1985). Identification of a calcium- and phospholipid-dependent protein kinase in plant tissue. *FEBS Letters* **18**, 25-28.

Scheel, D., Hauffe, K. D., Jahnen, W. & Hahlbrock, K. (1986). Stimulation of phytoalexin formation in fungus-infected plants and elicitor-treated cell cultures of parsley. In *Recognition in Microbe-Plant Symbiotic and Pathogenic Interactions*, ed. B. Lugtenberg, pp. 325-331. Springer-Verlag: Heidelberg.

Scheel, D., Colling, C., Keller, H., Parker, J., Schulte, W. & Hahlbrock, K. (1989). Studies on elicitor recognition and signal transduction in host and non-host plant/fungus pathogenic interactions. In *Signal Molecules in Plants and Plant-Microbe Interactions*, ed. B. J. J. Lugtenberg, pp. 211-218. Springer-Verlag: Berlin & Heidelberg.

Schmelzer, E., Krüger-Lebus, S. & Hahlbrock, K. (1989). Temporal and spatial patterns of gene expression around sites of attempted fungal infection in parsley leaves. *The Plant Cell* **1**, 993-1001.

Schmidt, W. E. & Ebel, J. (1987). Specific binding of a fungal glucan phytoalexin elicitor to membrane fractions from soybean *Glycine max*. *Proceedings of the National Academy of Sciences, USA* **84**, 4117-4121.

Schröder, J. I. & Hedrich, R. (1989). Involvement of ion channels and active transport in osmoregulation and signalling of higher plants. *Trends in Biochemical Sciences* **14**, 187-192.

Schröder, J. I. & Hagiwara, S. (1989). Cytosolic calcium regulates ion channels in the plasma membrane of *Vicia faba* guard cells. *Nature* **338**, 427-430.

Sharp, J. K., Albersheim, P., Ossowski, P., Pilotti, Å., Garegg, P. & Lindberg, B. (1984a). Comparison of the structures and elicitor activities of a synthetic and mycelial-wall-derived hexa(β-D-gluco-pyranosyl)-D-glucitol. *Journal of Biological Chemistry* **259**, 11341-11345.

Sharp, J. K., Valent, B. & Albersheim, P. (1984b). Purification and partial characterisation of a β-glucan fragment that elicits phytoalexin accumulation in soybean. *Journal of Biological Chemistry* **259**, 11312-11320.

Somssich, I. E., Bollman, J., Hahlbrock, K., Kombrink, E. & Schulz, W. (1989). Differential early activation of defense-related genes in elicitor-treated parsley cells. *Plant Molecular Biology* **12**, 227-234.

Somssich, I. E., Schmelzer, E., Bollman, J. & Hahlbrock, K. (1986). Rapid activation by fungal elicitor of genes encoding pathogenesis-related proteins in cultured parsley cells. *Proceedings of the National Academy of Sciences, USA* **83**, 2427-2430.

Sorensen, P., Farber, N. M. & Krystal, G. (1986). Identification of the interleukin-3 receptor using an iodinatable, cleavable, photoreactive cross-linking agent. *Journal of Biological Chemistry* **261**, 9094-9097.

Stäb, M. R. & Ebel, J. (1987). Effects of Ca^{2+} on phytoalexin induction by fungal elicitor in soybean cells. *Archives of Biochemistry and Biophysics* **257**, 416-423.

Strasser, H., Hoffmann, C., Grisebach, H. & Matern, U. (1986). Are polyphosphoinositides involved in signal transduction of elicitor-induced phytoalexin synthesis in cultured plant cells? *Zeitschrift für Naturforschung* **41c**, 717-724.

Taylor, J. L,, Fritzemeier, K. -H., Häuser, I., Kombrink, E., Rohwer, F., Schröder, M., Strittmatter, G. & Hahlbrock, K. (1990). Structural analysis and activation by fungal infection of a gene encoding a pathogenesis-related protein in potato. *Molecular Plant-Microbe Interactions* **3**, 72-77.

Thuleau, P., Graziana, A., Rossignol, M., Kauss, H., Auriol, P. & Ranjeva, R. (1988). Binding of the phytotoxin zinniol stimulates the entry of calcium into plant protoplasts. *Proceedings of the National Academy of Sciences, USA* **85**, 5932-5935.

Thurston, H. D. & Schultz. O. (1981). Late blight. In *Compendium of Potato Diseases*, ed. W. J. Hooker, pp. 40-42. American Phytopathological Society: St. Paul, Minnesota.

Waldmüller, T. & Grisebach, H. (1987). Effects of R-(1-amino-2-phenylethyl)-phosphonic acid on glyceollin accumulation and expression of resistance to *Phytophthora megasperma* f. sp. *glycinea* in soybean. *Planta* **172**, 424-430.

Wheeler, J. J. & Boss, W. F. (1987). Polyphosphoinositides are present in plasma membranes isolated from fusogenic carrot cells. *Plant Physiology* **85**, 389-392.

Zook, M. N., Rush, J. S. & Kuć, J. A. (1987). A role for Ca^{2+} in the elicitation of rishitin and lubimin accumulation in potato tuber tissue, *Plant Physiology* **84**, 520-525.

Chapter 8

Current questions in *Phytophthora* systematics: the role of the population approach

C. M. Brasier

'There is no such thing as typical or atypical, merely variability of living things' (Leonian, 1934).

The genus *Phytophthora* contains some of the world's most destructive plant pathogens, while collectively the Phytophthoras present a fascinating array of biological structures and mechanisms. Research into their biology and control requires a vigorous and healthy approach to their systematics. It must be outward and exploratory, yet rest upon secure foundations so that new information can be accommodated in a meaningful framework. Currently, *Phytophthora* systematics is undergoing an information revolution that belies the rather slower metamorphosis anticipated a few years ago (Gallegly, 1983a; Brasier, 1983). This chapter examines the historical and methodological basis of that revolution from the standpoint of a genetically oriented mycologist, with emphasis on the increasing importance of molecular characters, on the role of the population approach, and on the biological species concept.

The operational system

The present operational framework of *Phytophthora* systematics is that encapsulated in the tabular keys of Newhook, Waterhouse & Stamps (1978) and Stamps *et al.* (1990). It is based largely on morphological criteria, and recognises about fifty species, together with a small number of varieties and *formae speciales*. It derives from the early work of Rosenbaum (1917), Tucker (1931) and others; and more directly from the system of Waterhouse (1963) in which the described species are arranged in six groups distinguished mainly on the structure of the sporangial apex and on the occurrence of paragynous or amphigynous antheridia.

The Waterhouse system is familiar to most current *Phytophthora* workers. It has brought sense and cohesion to a rather loosely structured mass of information (Gallegly, 1983a), has provided a solid and workable basis for our identifications, and has fuelled academic progress by providing a framework against which to test alternatives. Waterhouse was nevertheless concerned that her key was 'put forward as a first effort, to be tested and moulded by use' (Waterhouse, 1963). The tabular key of Newhook *et al.* (1978) represents just such a remoulding and includes, for example, an important new emphasis on the sporangial pedicel, a previously under-utilised criterion in caducous species.

Many morphological taxa continue to be identified with a high level of confidence by the Waterhouse approach, particularly in the hands of the experienced. Its widespread use for over a quarter of a century is in itself a tribute to its success. This is not to say, however, that there have been no nomenclatural or conceptual problems. In fact, there has been a continuing debate on the utility of morphological criteria and the breadth of the species concept since at least the 1920s. Thus, Tucker (1931) in a scholarly study, emphasised the value of sporangial and gametangial characteristics as taxonomic criteria (together with temperature relations), reduced many early names to synonymy, and accepted seventeen taxa; while Leonian (1925, 1934) concerned at the morphological plasticity of Phytophthoras, strongly advocated use of physiological characters and, though incorporating twenty-two taxa in a key, also suggested that there were 'not more than two or three good species in the genus *Phytophthora*' (Leonian, 1934).

In recent decades the comparative utility of morphological and non-morphological criteria and the nature of the 'species' have been increasingly debated. A wide range of taxonomic issues were explored during the International Symposium on *Phytophthora* at Riverside, California in 1981 (Brasier, 1983; Erwin, 1983; Gallegly, 1983a & b; Waterhouse *et al.*, 1983). Similar topics have been covered elsewhere in reviews and research articles (e.g. Waterhouse, 1970; Al-Hedaithy &Tsao, 1979; Brasier & Griffin, 1979; Hansen *et al.*, 1986; Ho & Jong, 1987). Some authors have emphasized the problems inherent in applying a predominantly morphological taxonomy to *Phytophthora*. The issues raised include the following:

- The small number of morphological characters available in *Phytophthora*, which are based principally on the structure of sporangia and gametangia.
- The plasticity of the morphological structures themselves, e.g. variation in sporangial size in *P. nicotianae* (syn. *parasitica*) (Fig. 8.1) or in sporangial and oogonial morphology in *P. megakarya* (Fig. 8.2).
- Overlap between spore sizes in different taxa, as shown for example in the oogonial sizes in *P. megasperma* var. *megasperma* and *P. megasperma* var. *sojae* (Figs. 8.3 & 8.4).
- Variation in morphological structures under different cultural and environmental conditions; e.g. *P. capsici* forms rounded persistent

sporangia in darkness but elongated caducous sporangia in light (Hendrix, 1967; Brasier, 1969; Alizadeh & Tsao, 1985).

- The special techniques needed to obtain some morphological structures; e.g. the mycelial rinsings or other methods required to induce sporangial formation in *P. megasperma* (see Ribeiro, 1978); or use of polycarbonate membranes (Ko, 1978) or the *Trichoderma* method (Brasier, 1975) to induce 'pure' gametangia in otherwise self-sterile heterothallic Phytophthoras.
- Difficulties in working within the International Code of Botanical Nomenclature (cf. Brasier & Rayner, 1987). To this issue is sometimes added the rider that fungi are not plants, and therefore deserve independent nomenclatural treatment (e.g. Brasier, 1983).
- The role under the Code of the type concept in naming species, which has tended to narrow the basis of species descriptions.
- Improper adherence to ranks under the Code; e.g. varieties should show morphological differences, while *formae speciales* should not.
- The continuing use of informal classifications originally intended to be temporary, so undermining formal naming procedures under the Code, e.g. the continued use of '*P. palmivora* MF4' (= *P. capsici*, see Zentmyer *et al.*, 1981; Alizadeh & Tsao, 1985).
- The inadequacy of many species descriptions, both early and recent (cf. Waterhouse, 1963).
- The number of species assignations in the literature that are incorrect even within the context of contemporary knowledge. This devalues the literature base, particularly with regard to species host lists which, if compiled solely on the basis of literature searches, are likely to be inaccurate.

Waterhouse (1970) recognised the limitations of morphological criteria and the common problem of overlap of form and dimension when distinguishing taxa. Leonian (1934) had earlier emphasized the considerable plasticity of sporangia (Fig. 8.1) and of the mycelium, the interaction of such characters with cultural conditions and what he perceived as a danger in relying upon these characters in classification. He was also opposed to the role of the type species under the Botanical Code. This role was later restated by Waterhouse: 'the specific epithet is indissolubly tied to the type. A type, be it the original specimen, culture or description with figures, or all three, must be the basis for classification' (Waterhouse, 1970). Leonian strongly disapproved of this constraint, pointing out that the choice of type material was often subject to chance and might therefore be unrepresentative: 'The accident of a first discovery and description determines the typical strain, while a subsequent discovery

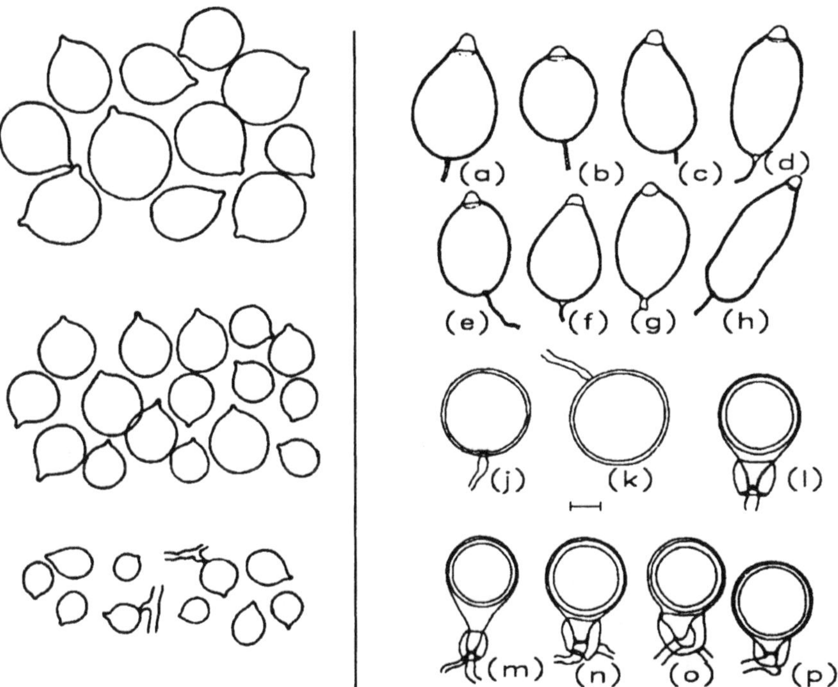

Figs. 8.1 & 8.2. Morphological variability in *Phytophthora*. Fig. 8.1 (left hand panel). Variation in size among sporangia of three dissociants of *Phytophthora parasitica* var. *rhei* (Leonian, 1934). Fig. 8.2 (right hand panel) Variation in shape of sporangia (a-h), Chlamydospores (j, k) and gametangia (l-p) in isolates of *Phytophthora megakarya* (Brasier & Griffin, 1979)(scale bar = 10 μm).

and observation of another strain which may sharply differ from the original, determines the atypical strain. Thus nothing more tangible than a mere state of mind is brought forth' (Leonian, 1934).

Associated with the above issues are the various nomenclatural problems that have been a focus of debate in recent decades. These were reviewed in some detail by Erwin (1983). They have tended to involve prominent taxa of major economic importance or worldwide distribution and include:

- The continuing question of the conspecificity of the morphologically similar *P. cryptogea* and *P. drechsleri* (e.g. Bumbieris, 1974; Ho & Jong, 1986). The latter species was erected by Tucker (1931) mainly on the basis of its tolerating a higher temperature, but was later morphologically distinguished by Waterhouse (1963).

- The much debated and rather complex history of the taxonomic structure of '*P. megasperma*' (Chapter 10).
- Disagreement over the status of the two varieties of *P. nicotianae*, var. *nicotianae* and var. *parasitica*, a problem considered by Erwin (1983) to have arisen in part from Tucker's (1931) use of 'variety' rather than '*forma specialis*' in discriminating the biotype of the fungus on tobacco (see also Tsao & Sisemore, 1978). Also the continuing confusion in the literature over the use of the specific epithet '*nicotianae*' versus '*parasitica*', the former being considered by Waterhouse (1963) to have priority.
- The now largely resolved issue of the occurrence of distinct morphospecies within '*P. palmivora*', including the separation from *P. palmivora sensu stricto* of *P. capsici* and *P. meadii*, and the erection of *P. megakarya* (e.g. Sansome, Brasier & Griffin., 1975; Brasier & Griffin, 1979; Kaosiri & Zentmyer, 1980; Dantanarayana, Peries & Liyanage, 1984; Tsao & Alizadeh, 1988).
- The status of the raspberry root rot *Phytophthora* (Chapter 9).

The role of the population approach

Implicit in many of Leonian's pithy statements is an objection to the necessarily simplistic and more static view of *Phytophthora* species embodied in dichotomous morphological keys. Thus: 'A given species of *Phytophthora* ... is not an absolute entity ... but a fluctuating group of organisms more or less loosely bound in a flexible orbit. It is unwise to regard any species as firmly fixed' (Leonian, 1934). Similar concerns can be seen in Gallegly's (1983a) appeal for the development of a 'more natural system of classification ... beginning with the characteristics of sexuality rather than with the characteristics of sporangial papillae'; and in the suggestion that there may be 'morphologically similar but physiologically distinct species in the genus', (Brasier, 1983). Indeed, since the 1970s, a number of *Phytophthora* workers, myself included, have called for a shift towards a more population oriented approach to *Phytophthora* systematics so that the limits of our taxa may be more accurately defined. This requires the estimation of variation in larger samples using discontinuous and continuous, molecular and behavioural characters; the analysis of groupings of arrays of major characters and the identification of more homogeneous genetic units within a wide spectrum of variation; and the recognition that the units themselves are highly dynamic and potentially changing entities (e.g. Brasier, 1983, 1987; Hansen, 1987).

This 'population approach' is essentially that advocated, if indirectly, by Leonian (1925, 1934). It was probably first applied by Shepherd & Pratt (1973) in their identification of two ecotypes within *P. drechsleri* in Australia; and later by Brasier & Griffin (1979) when defining morphological

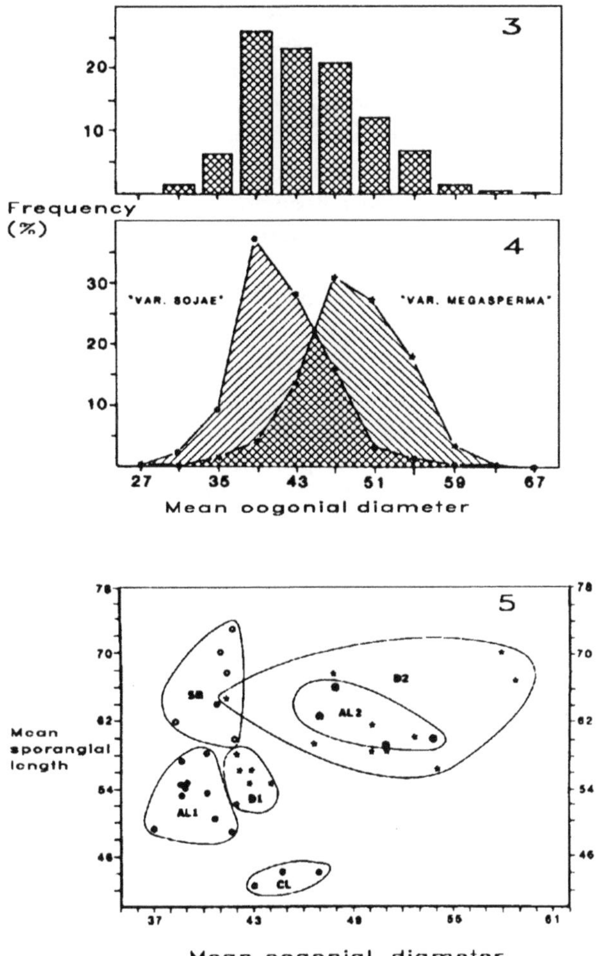

Figs. 8.3-8.5. Continuous variation in *Phytophthora*. Fig. 8.3 (top). The continuum of oogonial sizes in isolates of *P. megasperma* (Hamm & Hansen, 1982). Fig. 8.4 (middle). The same isolates arranged according to original varietal assignation, showing the considerable overlap in oogonial size between the two varieties (Hamm & Hansen, 1982). Fig. 8.5 (bottom). The separation of isolates of *P. megasperma* into six host associated groups on the basis of sporangial length and oogonial diameter: SB, soybean group; AL1, alfalfa group 1; AL2, alfalfa group 2; D1, Douglas fir group 1; D2, Douglas fir group 2; CL, clover group (Hansen & Hamm, 1983).

taxa within '*P. palmivora*' on cacao using a wide array of characters; and by Hansen & Hamm (1983, and see Chapter 10) in their identification of six host-associated groups within '*P. megasperma*' using two simple continuous variables, oogonial size and sporangial size (Fig. 8.5) plus other correlated characters. It is important to note that oogonial size alone (Figs. 8.3 & 8.4) would not have discriminated these groups.

The population approach can be extended via a genetical dimension to the breeding unit approach. Thus it could be further defined as the identification of more homogeneous genetic units within a wide spectrum of variation, combined with an assessment of gene flow potential within and between the units. This now approaches the sort of taxon definition favoured by evolutionary biologists, in which gene flow defines the boundaries of populations.

By emphasizing gene flow, the population approach requires an understanding not only of the current size, structure and modes of reproduction of population units, but also of the processes of speciation that have moulded our taxa. The ingredients of speciation can be theoretically defined as: the structure of the original population, natural selection, the genetic system, and reproductive isolation (cf. Burnett, 1975, 1983; Brasier, 1987). Our understanding of the role of these factors in the biology of Phytophthoras is often unsatisfactorily sketchy. In particular, and for obvious reasons, something akin to an original or ancestral population may not be extant, although the question of geographical origins is nevertheless being pursued, as in the case of *P. cinnamomi* (Crandall & Gravatt, 1967; and see Zentmyer, 1988). Molecular information (see below) may strengthen these perspectives.

Despite the difficulty of reconstructing the ancestral origins of modern taxa, it is relatively easy to envisage the role of natural selection in their evolution. In particular, a strong influence can be invoked for the host plant (including different parts of the plant such as fruits, shoots and roots), and for climatic and edaphic factors (Brasier, 1983). Indeed, the geographical spread of crop plants by man and man's development of crop monocultures may have had a special role, perhaps even resulting in rapid evolution of endemic or introduced Phytophthoras under intense host selection (cf. Burnett, 1975; Brasier, 1983, 1987; Hansen 1987). Examples of response to host selection may include the characteristic biotype of *P. palmivora* found on cocoa pods (Brasier, Griffin & Maddison, 1981), and the host-associated biological species within *P. megasperma* (Hansen *et al.*, 1986; Chapter 10).

Much also needs to be understood about the genetic system in *Phytophthora* (Shaw 1983, 1988), and its role in generating the variation on which selection acts. This includes the role of selfing versus outcrossing, and of homozygosity versus heterozygosity; of diploidy versus polyploidy; of linkage and of recombination frequency; of sexual versus asexual reproduction; of cytoplasmic inheritance and so on. It can also mean understanding the genetic control of characters used in the assessment of variation, from the structure of the sporangial apex to host range (Gallegly 1983a).

Finally, critical to a deeper understanding of our taxa is knowledge of the processes that restrict gene flow between population units, such as geographical isolation, habitat isolation, and pre- and post-zygotic isolating mechanisms (Brasier, 1983, 1987). Geographical and habitat isolation can be assumed to have played a major role in the radiation of Phytophthoras but little factual information is available despite emerging theories on the geographical origins of some species (Crandall & Gravatt, 1967; Zentmyer, 1988). More amenable to research but also poorly understood are isolating mechanisms acting at the level of somatic and sexual fusion. One can speculate that barriers to gene flow might exist at the vegetative incompatibility level, or be behavioural, chromosomal, at the level of gamete recognition, gametic abortion, zygotic abortion, or, in the event of zygote germination, be reflected in poor hybrid vigour (Brasier, 1983, 1987). In heterothallic Phytophthoras such mechanisms are beginning to be demonstrated in terms of the observed failure of hybrid oospore formation and of germling survival in interspecific crosses (Sansome *et al.*, 1975; Boccas & Zentmyer, 1976; Boccas, 1981; Erselius & Shaw, 1982). Breeding barriers in homothallic species are also beginning to be examined, for example by attempting protoplast fusion between biological species within *P. megasperma* (Chapter 16).

Instant species?

It has been argued that in the fungi, total reproductive isolation could come about almost instantly via mutations which prevent hyphal or gametangial fusion (Burnett, 1983), and furthermore, that speciation in response to intense selection, as in exposure to crop monoculture, might also be rapid (Brasier, 1983, 1987; Hansen, 1988). Such processes could lead to the appearance of morphologically similar sibling or biological species because taxonomically useful morphological differences would not have had time to accumulate. The biological species within '*P. megasperma*' (Hansen *et al.*, 1986, chapter 10) may be examples of such a phenomenon. More research on isolating mechanisms and gene flow potential is needed so that the significance of population units defined by other methods can be fully evaluated.

Role of molecular approaches

Molecular characters in *Phytophthora* systematics fall under the broad umbrella of what Leonian (1934) termed physiological, as opposed to

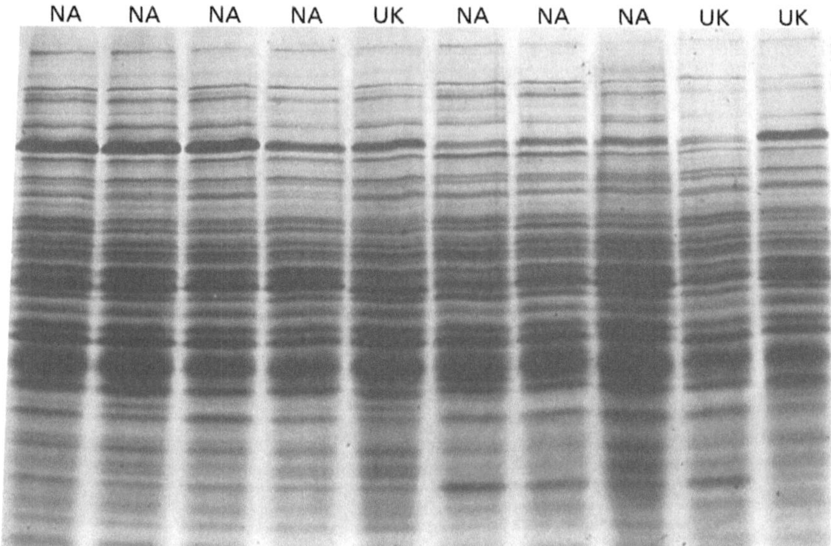

Fig. 8.6. Comparative homology of electrophoretic banding patterns of proteins among isolates of *Phytophthora gonapodyides* from Britain (UK) and North America (NA) (Brasier, Hamm & Hansen, unpublished).

morphological, criteria. In early fungal systematics morphological criteria predominated because of the necessary reliance on optical evidence. Recent technological advances have created a role for molecular methods in fungal systematics, challenging the orthodoxy of reliance on optical techniques. The molecular approach allows analysis of the very fundamentals of variation in our taxa. Leonian, as an early radical, would probably have revelled in the opportunity now presented. An interesting question is the extent to which molecular techniques can be harnessed to the more practical diagnostic functions of *Phytophthora* systematics. Another is the extent to which, together with the population approach, they may alter our systematic concepts. A third is the extent to which these various needs and approaches can be harmonised.

Early applications of molecular methods to *Phytophthora* systematics were those of Clare (1963), Clare & Zentmyer (1966) and Gill & Powell (1968) who compared patterns of buffer-soluble proteins; and Burrell *et al.* (1966) and Merz, Burrell & Gallegly (1969) who analyzed polyclonal antisera. There are now many more methods available, ranging from monoclonal antibody techniques to analysis of DNA polymorphism, and

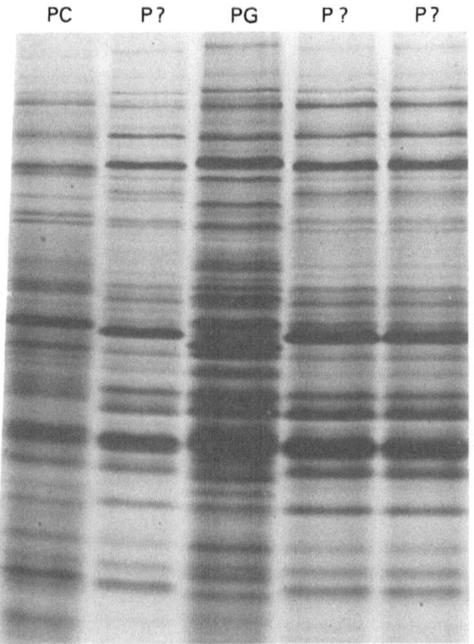

Fig. 8.7. The clear separation of three morphologically similar taxa by electrophoretic protein banding pattern: Pg, *P. gonapodyides*; Pc, *P. cryptogea*; P?, an unidentified sexually sterile taxon, possibly a species hybrid related to *P. cryptogea* (Brasier, Hamm & Hansen, unpublished).

their sensitivity and scope grows almost daily. Selected examples will now be examined to illustrate progress and to demonstrate additional points.

Protein electrophoresis

The early promise of protein electrophoresis has been more than confirmed. It can be a clear and effective systematic tool, as shown by its successful application to a variety of important nomenclatural problems, including support for the separation of morphospecies within '*P. palmivora*' on cocoa, (Kaosiri & Zentmyer, 1980; Erselius & Shaw, 1982); for the separation of *P. cactorum* from *P. pseudotsugae* sp. nov. (Hamm & Hansen, 1983); and for the separation of biological species within '*P. megasperma*' (Chapter 10). It has also provided further evidence indicating that *P. cryptogea* and *P. drechsleri* may be conspecific (Masago & Yoshikawa, unpublished; Brasier, Hamm & Hansen, 1989).

The general similarity in protein profiles of isolates of *P. cryptogea* to those of *P. drechsleri* is shown in Fig. 8.8. Discrimination of three morpho-

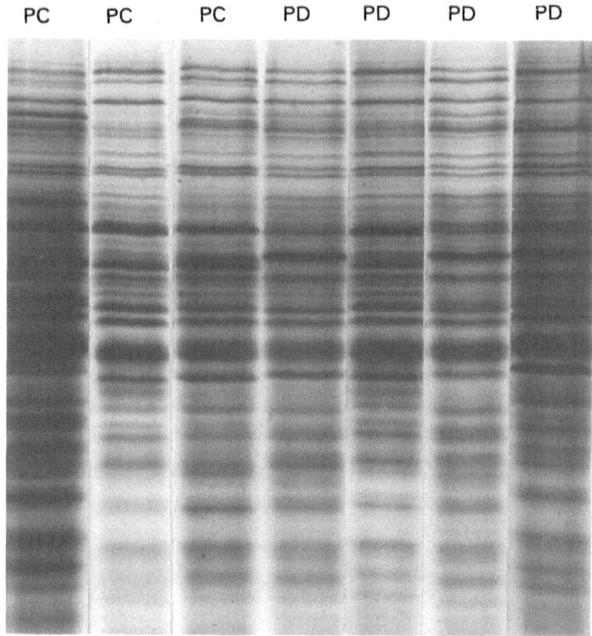

Fig. 8.8. Comparative homology of protein banding patterns among isolates previously attributed to *P. cryptogea* (Pc) and *P. drechsleri* (Pd)(Brasier, Hamm & Hansen, unpublished).

logically rather similar taxa is illustrated by the gel in Fig. 8.6. This and similar gels have been used in a reclassification of a '*P. drechsleri*-like' fungus from remote forested areas in Alaska and Oregon (Hansen *et al.*, 1988). This has now been assigned to the aquatic species *P. gonapodyides* (Brasier *et al.*, 1989). The comparative uniformity in banding patterns of *P. gonapodyides* isolates from both Europe and North America is shown in Fig. 8.6.

By applying a cluster analysis to protein patterns among isolates of six morphological species, Erselius & de Vallavieille (1984) obtained good discrimination of all six taxa, and found little intraspecific variation (Fig. 8.9). They did, however, find that two isolates identified as *P. citricola* on morphological grounds had a protein pattern similar to that of *P. citrophthora* isolates from the same habitat. In another study, de Vallavieille & Erselius (1984) detected two genetic groups within *P. citrophthora*, and also found citrus isolates of *P. nicotianae* with a *P. citrophthora* protein pattern. These observations raise intriguing questions about gene flow between morphological species.

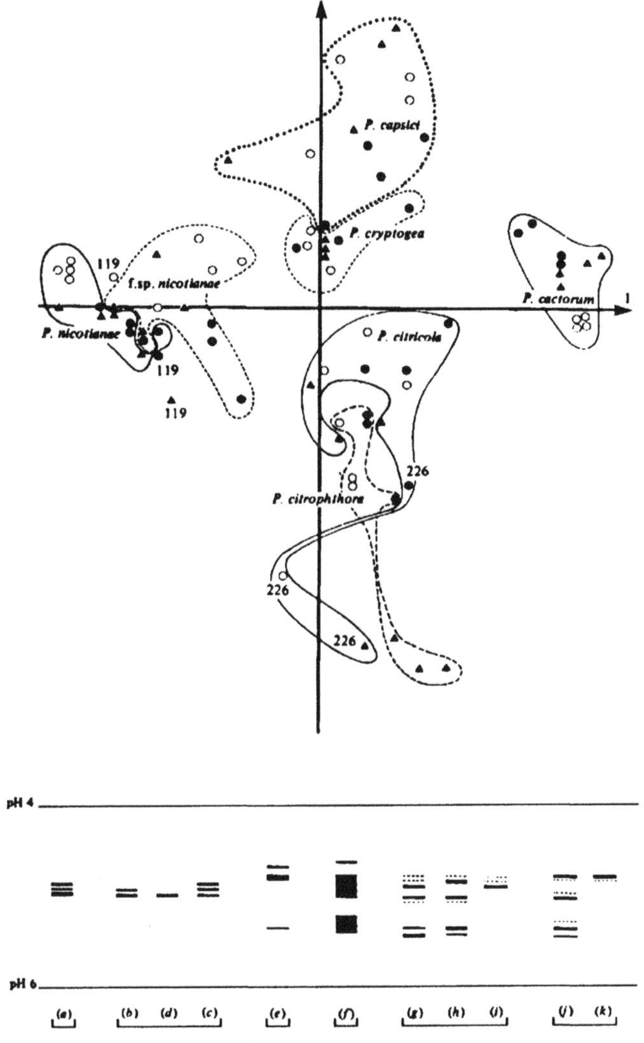

Figs. 8.9 & 8.10. Protein and isozyme electrophoresis of *Phytophthora* taxa (Erselius & de Vallavieille, 1984). Fig. 8.9 (top). The grouping of six morphological taxa on the basis of a cluster analysis of protein patterns. Fig. 8.10 (bottom). Acid phosphatase isozyme profiles of the same isolates, showing the heterogeneity of banding patterns: a, all *P. cactorum* isolates; b-d, *P. citricola* isolates; e, all *P. citrophthora* isolates; f, all *P. nicotianae* isolates; g-i, *P. capsici* isolates; j & k, *P. cryptogea* isolates.

Present evidence suggests that protein electrophoresis can give excellent discrimination at about the morphospecies level. Intraspecific variation seems generally to be small, and a high level of similarity may indicate conspecificity. Since the method requires only fairly simple apparatus, it appears to be under-utilised as a systematic tool for *Phytophthora*. Its value could be further enhanced by use of two-dimensional gels (Masago & Yoshikawa, unpublished).

Isozymes

Given a sufficient number of polymorphic enzymes, or the occurrence of unique or rare enzyme patterns, isozymes can be used to discriminate taxa. Until recently, however, their application to *Phytophthora* taxonomy has been surprisingly limited. Fig. 8.10 shows acid phosphatase patterns obtained by Erselius & de Vallavieille (1984) for the same isolates and taxa as in Fig. 8.9. Since more intrataxon variation is usually revealed than with total protein profiles (e.g. see *P. citricola* and *P. capsici*, Fig. 8.10) isozymes are better at discriminating complexity below the 'species' level, and have been used to measure intraspecific variation in *P. cinnamomi* and *P. infestans* (Old, Moran & Bell, 1984; Ko *et al.*, 1978; Tooley, Fry & Villarreal Gonzalez, 1985). Their potential for systematic and population studies is well illustrated by Oudemans & Coffey (Chapter 12) who, applying 15 enzymes to a wide range of *Phytophthora* isolates, have provided a wealth of information, including evidence of distinctive cacao and citrus groups within *P. citrophthora*, and evidence favouring the amalgamation of the two varieties in *P. nicotianae*. Similarly, Nygaard *et al.*, (1989) have produced isozyme evidence supporting the occurrence of biological species within '*P. megasperma*'.

Isozymes can also be used to assess allele frequencies, Hardy Weinberg equilibria and homozygosity and heterozygosity levels in population samples (Tooley *et al.*, 1985; Shattock, Tooley & Fry, 1986; Chapter 15) and to estimate inbreeding or outbreeding frequency. Their analysis will therefore contribute to our understanding of the dynamics and fine structure of *Phytophthora* populations, including information on the geographical spread of Phytophthoras from putative centres of origin (Old *et al.*, 1984; Ko *et al.*, 1978), and on host plant or habitat adaptation, all of which are important factors in microevolution of the genus.

DNA and RNA polymorphisms

Many questions of systematic relationship are likely to be resolved by analysis of DNA and RNA polymorphisms. Moreover, the method has the added appeal that a wide range of powerful genetic engineering techniques can be brought to bear in concert. The promises held in this approach are both academic and practical, and include:

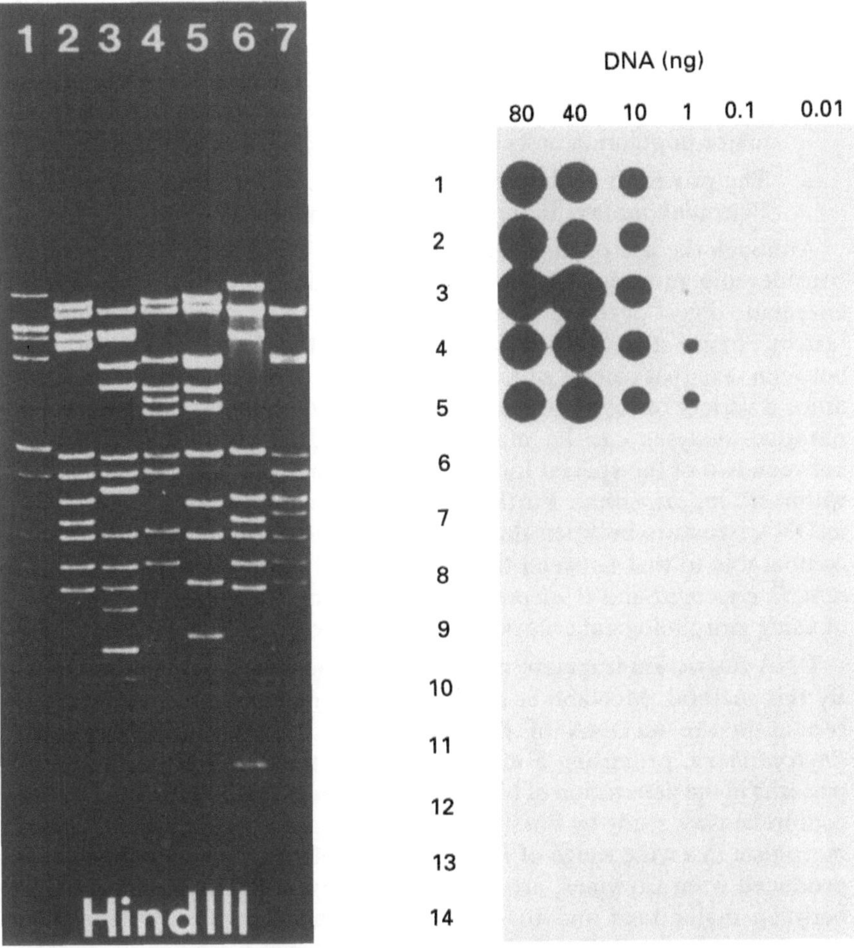

Figs. 8.11 & 8.12. DNA polymorphism in *Phytophthora*. Fig. 8.11 (left). Banding patterns of mitochondrial DNA of six *Phytophthora* taxa digested with the *Hin*dIII restriction enzyme. Lanes 1-4, *P. megasperma* soybean form, alfalfa form, Douglas fir form and apple-cherry form, respectively *sensu* Hansen *et al.* (1986); Lane 5, *P. nicotianae* var. *nicotianae*; Lane 6, *P. cryptogea*; Lane 7, *P. cactorum* (Förster *et al.*, 1988). Note the distinctive banding patterns of all seven taxa, including the four host-associated forms within '*P. megasperma*'. Fig. 8.12 (right). Dot hybridization of a cloned nuclear DNA fragment of *P. nicotianae* to fungal and plant total nuclear DNA. Lanes 1-5, various isolates of *P. nicotianae; Lanes 6-12, isolates of P. cactorum*, *P. cinnamomi*, *P. citricola*, *P. citrophthora*, *P. cryptogea*, *P. megasperma* and *P. erythroseptica* respectively; Lane 13, *Pythium ultimum*; Lane 14, tomato roots. Note that only the DNA of the five *P. nicotianae* isolates hybridizes (Goodwin, Kirkpatrick & Duniway, 1989).

- The identification of differences in chromosomal architecture of major phyletic significance.
- The ability to construct hierarchical phenograms that, when refined, should approximate to natural phylogenies ranging from major population units down to unique clones or genets.
- The provision of simple and effective tools for rapid diagnosis of *Phytophthora* taxa in pure culture and in host material.

Although the use of such techniques in *Phytophthora* is in its infancy, considerable progress has already been made. Figure 8.11 shows an enzymatic digest of the mitochondrial (mt) DNA of seven *Phytophthora* taxa by Förster *et al.* (1988, 1989). It illustrates the characteristic patterns between taxa that can be generated by a single DNA restriction enzyme. Since a variety of restriction enzymes are now available, extensive comparative analyses can be made. This particular study discriminated between two of the special forms, '*f. sp. glycinea*' and '*f. sp. medicaginis*', within '*P. megasperma*'. Furthermore the genetic distance in terms of mtDNA structure between these two morphologically similar forms was comparable to that between the three distinct morphospecies, *P. cactorum*, *P. cryptogea* and *P. nicotianae*, highlighting once more the problem of using morphological criteria in the separation of Phytophthoras.

DNA restriction fragment patterns can also be used to map the DNA. By this method, McNabb *et al.* (1987) have demonstrated an inverted repeat in the mtDNA of *Pythium* which is apparently absent in *Phytophthora*, promising a major generic distinction. The use of these patterns in the generation of hierarchical phenograms is illustrated by the comprehensive study by Förster & Coffey (chapter 11) of mtDNA polymorphism in a wide range of *Phytophthora* isolates. Such data, like those produced from isozymes, are beginning to reveal phyletic relationships between major taxa and to demonstrate much complexity below the morphospecies level. Thus, the study by Förster & Coffey indicates the existence of two population units within *P. megakarya* on cocoa in West Africa and provides further evidence of cacao and citrus forms within '*P. citrophthora*'. Mills, Förster & Coffey (1991) have demonstrated several major population units within *P. cryptogea* and *P. drechsleri* on the basis of both mtDNA and isozyme polymorphisms (Fig. 8.13) revealing much greater complexity than is shown by either protein polymorphism (Fig. 8.8) or by morphological characters. Similarly, on the basis of probes revealing polymorphisms in repetitive nuclear DNA Goodwin *et al.* (1990) have identified three groups within *P. citrophthora*, a walnut-cherry, a Californian citrus and an Australian citrus group. Panabieres *et al.* (1989) have shown polymorphisms in repetitive whole-cell DNA to have good potential for separating morphologically similar taxa. Like Förster &

Coffey (Chapter 11) and Oudemans & Coffey (Chapter 12) they also found a close homology between tobacco and non-tobacco isolates of *P. nicotianae* (*P. parasitica*), and further questioned the maintenance of two varieties in this species. Two isolates (only) of *P. cryptogea* and two of *P. drechsleri* showed distinct differences, but also shared more bands in common than other taxa.

The potential of DNA probes in rapid diagnosis was recently demonstrated by Goodwin *et al.* (1989). They selected DNA clones which clearly discriminated multiple copy nuclear DNA of *P. nicotianae* (*P. parasitica*) from that of seven other *Phytophthora* taxa, from *Pythium*, and from tomato DNA, when using a simple dot hybridisation technique (Fig. 8.12). Such methods hold the promise, along with highly sensitive monoclonal antibody techniques (Chapter 5), of rapid identification of Phytophthoras in soil and contaminated plant material. They may in some circumstances, obviate the need for tedious isolation procedures, opening the way for development of sensitive diagnostic kits, and for the identification of Phytophthoras at all hierarchical levels in both field and laboratory.

Karyotype analysis and chromosome electrophoresis

While differences in chromosome size and number can be used to define taxa from a morphological standpoint, they may also indicate restricted gene flow, and so have a role in defining taxa from a population standpoint. Light microscope studies of *Phytophthora* chromosomes (Sansome, 1987) have already influenced major nomenclatural questions in *Phytophthora*. The discovery of striking differences in chromosome size and number within '*P. palmivora*' (Sansome *et al.*, 1975) initiated a major taxonomic revision of the species (Brasier & Griffin, 1979; Kaosiri & Zentmyer, 1980). Distinct diploid and polyploid karyotypes were shown to be associated with the two 'varieties' in *P. megasperma* (Sansome & Brasier, 1974); and subsequently the demonstration of four karyotypes, KI-KIV ranging from n = about 10 to n = about 30 contributed to the identification of biological species in '*P. megasperma*' by Hansen *et al.*, (1986).

Pulse field gel electrophoresis of chromosomal DNA promises greatly to extend such information. For example, it is already being applied to the intriguing question of whether the occurrence of all four karyotypes KI-KIV in the 'BHR' biological species group in *P. megasperma* simply reflect duplication of homologous chromosomes, or whether differences in chromosome size and structure are involved, in which case additional sub-populations may exist within the BHR group itself (Chapter 10). One application of karyotype electrophoresis, therefore, lies in assessing the role of chromosome structure in reproductive isolation. When combined with other molecular methods, such as the use of rare cutting restriction enzymes and DNA hybridisation and sequencing techniques, it may reveal the fine details of comparative genome organisation, and so influence questions of relationship at many taxonomic levels.

New and not so new problems

New questions are likely to be raised by the increasing emphasis on population and molecular approaches to *Phytophthora* systematics, while other more classical problems will remain to interact with the new. Potential issues include the question of interspecies hybridisation, the classification of complex hierarchies, the species concept, the approach to naming of new taxa; and the molecular picture, still to emerge, of the *Phytophthora* evolutionary tree.

Species hybrids

The occurrence of species hybrids has scarcely been addressed in *Phytophthora* systematics. Their potential as a complicating factor needs to be recognised (Waterhouse, 1970). They are more likely to be detected by population and molecular approaches than by orthodox taxonomy, under which they might often be labelled as 'atypical' or 'unidentified'.

Owing to the operation of various genetic isolating mechanisms many hybridisation attempts in nature could be expected to fail. Indeed, in sexual pairings between the heterothallic *P. nicotianae* and *P. capsici*, and *P. megakarya* and *P. palmivora* in the laboratory hybrids have failed to develop, and only selfed (i.e. parental) offspring have survived (Sansome *et al.*, 1979; Boccas & Zentmyer, 1976; Boccas, 1981; Erselius & Shaw, 1982). However, Sansome, Brasier & Hamm (1991) have obtained cytological evidence that the heterothallic morphospecies *P. meadii* may have originated as a species hybrid. It occurs in both diploid and tetraploid forms and the diploids show gametangial abnormality and a high level of zygotic abortion, suggesting chromosome imbalance. *P. meadii* may therefore be a balanced allopolyploid. Also, Erselius & de Vallavieille (1984) and de Vallavieille & Erselius (1984) have identified isolates of *P. citricola* (Fig. 8.9) and *P. nicotianae* with protein profiles resembling *P. citrophthora*, raising the question of interspecific hybridisation. Perhaps significantly, the *Phytophthora* population from which the *P. citricola* and *P. nicotianae* isolates came was dominated by *P. citrophthora*. Again, in a recent study of *P. gonapodyides*, *P. cryptogea* and *P. drechsleri*, two small groups of isolates, one from the UK (Fig. 8.6) and one from California each showed a unique protein banding pattern, a unique colony growth pattern, and each was intrinsically sterile. These seem more likely to be local hybrids than rare major taxa (Brasier *et al.*, 1989). When Phytophthoras showing developmental abnormalities, full or partial sterility and unusual morphological or molecular properties are found, interspecies hybridisation may be indicated. Occasional spread of more viable hybrids

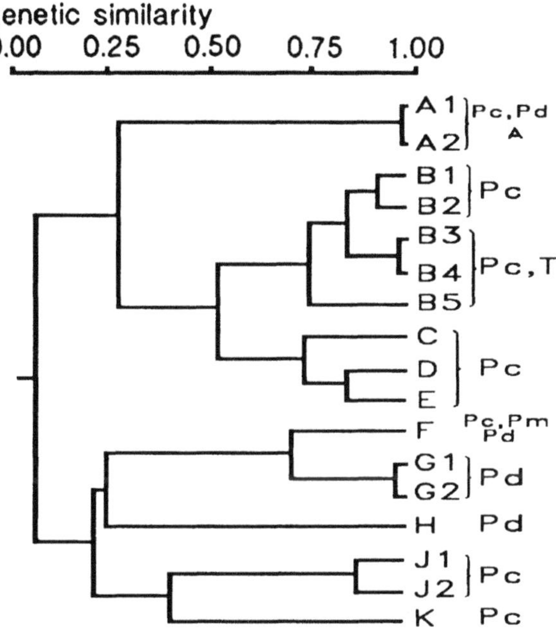

Fig. 8.13. A phenogram of molecular relationships between isolates previously attributed to *Phytophthora cryptogea* and *P. drechsleri*, *P. melonis* and *P. sinensis*, based on their mitochondrial DNA polymorphisms. Pc = predominantly isolates of *P. cryptogea*; Pd = predominantly isolates of *P. drechsleri*; Pc/Pd, includes both *P. cryptogea* and *P. drechsleri* isolates; Pd/Pm/Ps, includes isolates of *P. drechsleri*, *P. melonis*, and *P. sinensis*. A, includes an authentic isolate of *P. drechsleri*; T, includes the type isolate of *P. cryptogea*. Note the complex hierarchy revealed including the occurrence of several highly divergent genetic units (Mills *et al.*, 1991).

hybridisation may be indicated. Occasional spread of more viable hybrids could also account for some observed phenetic complexity, e.g. that recently demonstrated in molecular analyses of the *P. cryptogea*/*P. drechsleri* group (Mills *et al.*, 1991; and see Fig. 8.13).

Hierarchy, terminology, and the biological species concept

Molecularly derived phenograms are ultimately likely to produce far greater phylogenetic accuracy, especially at middle and lower orders of *Phytophthora* relationships, than that obtainable by weighted analysis of morphological criteria (e.g. Ho, 1982). However, given the complex 'trees' that these methods may often generate (Fig. 8.13), a challenging question is: where, among these multiple discontinuities, will lie the species, subspecies, variety or below as conceived under the present Code of Botanical Nomenclature? Also, as biological and genetical information alter, sometimes quite rapidly as with the various units within '*P. megasperma*' (Chapter 10). This mobility problem has already led to a call for a more flexible system of rankings in fungi (Brasier & Rayner, 1987). At the very least, better use may need to be made of the existing rankings than at present. Apart from the non-morphological *forma specialis*, of the six formal morphological ranks from species to subform only two, species and variety, are in common usage with *Phytophthora*. It is particularly notable that no subspecies have been recognised. Such a two-rank operational system is unlikely to reflect reality (again cf. Fig. 8.13). It may simply reflect the practical difficulty of constructing hierarchies using orthodox systematic criteria, and a resulting tendency to polarise around the species on the one hand and the variety on the other.

We are also dealing with several concepts of a species ranging from the solely morphological to the genetical. It could be argued that a species concept based on the breeding unit, the sibling or biological species, should begin to assume precedent. However it should be noted that, if defined on the basis of total reproductive isolation, a 'species' might occur at almost any hierarchical position in a phenogram, high or low, the latter particularly if recent or sudden reproductive isolation has occurred. This will clearly add to our problem of sustaining a hierarchical system of terms. Nevertheless, as information on gene flow between population units accumulates, it seems inevitable that the fungal species concept will tend to shift downwards, with increasing emphasis on the sibling or biological species (Burnett, 1983; Brasier, 1987; Brasier & Rayner, 1987; Hansen *et al*., 1986; Chapter 10), the original morphospecies sometimes being elevated to a status equivalent to a superspecies.

The increasing revelation of sibling species in fungi further highlights the problem of maintaining the current operational systematic system alongside the new information deriving from the population approach. The current system has historically relied on the morphological species. The population approach, while often supportive, may sometimes produce a profoundly different systematic picture as in the cases of '*P. megasperma*' and *P. cryptogea*/*P. drechsleri* (Fig. 8.13). Compared with the conventional approach it is also, at least initially, more experimental, longer term and in some respects more academic in that the objective itself may have the status of a research project, rather than an immediate communicative function such as the attachment of a label to a *Phytophthora* for use in a report or a publication on a plant disease. The population approach should not, however, be seen solely in academic terms since it represents the road to the much vaunted more natural system of classification. Indeed, with a disease organism like *Phytophthora*, it is the genetically and behaviourally distinct unit and not the morphological unit that the pathologist most needs to distinguish for communication purposes. It is also desirable that the labelling system used is one that most accurately reflects a unit's relationship to other units.

The naming question

If we are within grasp of a more natural system of classification, then, given the present state of flux and uncertainty in *Phytophthora* systematics, and the impending explosion of phyletic information, and considering the nomenclatural difficulties of the past, there may be a case for caution in the naming or renaming of taxa, particularly in the resolution of complex situations, until their status relative to other taxa in the same system has become clearer. This somewhat echoes an earlier view of Waterhouse (1963) that 'There is still much information needed concerning many of the species ... and it would be wise to defer the name changes until more is known'. The problem of a moratorium, however, is that, to the scientist(s) concerned, it frustrates the seemingly urgent imperative of attaching a label to an apparently valid and unique taxon. Moreover, stasis, however well meaning, may act as a brake on progress and, to remain healthy, systematics must be dynamic. A degree of systematic confusion must always be accepted, partly because new information takes time to be disseminated. Nevertheless there is also a strong case now, given the ease of modern communications, for a greater consensus of approach to naming on the part of different research groups than has existed previously. As more hierarchical complexity is revealed and as taxonomy becomes an increasingly computer based science, attempts at consensus may become a necessity.

Towards an evolutionary tree?

The gradual refinement of molecular phenograms should lead to something approaching a true phylogenetic tree for *Phytophthora* and related genera. An intriguing question is the extent to which the tree will be found either to support or to cut across traditional morphological groupings. For example, will some Phytophthoras with different sporangial apex types prove to be close relatives in the molecular phylogeny? What relationships will be found between the many heterothallic taxa in Waterhouse's (1963) Group II? Might more Pythiums prove to be Phytophthoras, as *P. undulata* recently has (Dick, 1989)? Might *Pythium* itself prove to be the progenitor of *Phytophthora* as is sometimes supposed, or a sibling genus, or a descendant? Or again, will the separation of the monotypic genus *Peronophythora* (Ko *et al.*, 1978) or of the new marine genus, *Halophytophthora* (Ho & Jong, 1990) be sustained? The prospects are clearly exciting.

Concluding comments

A considerable debt is owed to C.M. Tucker, Grace Waterhouse and others for the construction of the conceptual framework of *Phytophthora* systematics that has taken us to the onset of the present information revolution and will continue to serve us in a variety of ways. A fascinating new period of experimental systematics is now unfolding, and perhaps none would have enjoyed this more than L.H. Leonian, who was probably ahead of his time in his scientific thoughts and instincts, if perhaps a little hasty in his 'only two or three good species in *Phytophthora*' judgement. Rapid and accurate molecular techniques may soon begin to play an increasing role in routine diagnosis. Some major phylogenetic and nomenclatural questions seem likely to be resolved. At the same time, patterns of complexity may emerge that will considerably challenge our skills of interpretation. Our species concept may need to be revised before it is so stretched as to become meaningless. Indeed there is no point in generating patterns unless we can attach meaning to them. While absorbed by the excitement of the population and molecular approaches, therefore, we should ensure that the broader biological information on our taxa, their morphology, ecology, physiology and pathology, are kept abreast of genetical developments. Otherwise the fuller significance of our research efforts may sometimes be overlooked.

References

Al-Hedaithy, S. S. A. & Tsao, P. H. (1979). Sporangium pedicel length in *Phytophthora* species and the consideration of its uniformity in determining sporangium caducity. *Transactions of the British Mycological Society* **72**, 1-13.

Alizadeh, A. & Tsao, P. H. (1985). Effect of light on sporangium formation, morphology, ontogeny, and caducity of *Phytophthora capsici* and '*P. palmivora*' MF4 isolates from black pepper and other hosts. *Transactions of the British Mycological Society* **85**, 47-69.

Boccas, B. R. (1981). Interspecific crosses between closely related heterothallic *Phytophthora* species. *Phytopathology* **71**, 60-65.

Boccas, B. & Zentmyer, G. A. (1976). Genetic studies with interspecific crosses between *P. cinnamomi* and *P. palmivora*. *Phytopathology* **66**, 477-484.

Brasier, C. M. (1969). Formation of oospores *in vivo* by *Phytophthora palmivora*. *Transactions of the British Mycological Society* **52**, 273-279.

Brasier, C. M. (1975). Stimulation of sex organ formation in *Phytophthora* by antagonistic species of *Trichoderma*. I. The effect *in vitro*. *New Phytologist* **74**, 183-194.

Brasier, C. M. (1983). Problems and prospects in *Phytophthora* research. In *Phytophthora: Its Biology, Taxonomy, Ecology and Pathology*, (ed. D. C. Erwin, S. Bartnicki-Garcia & P. H. Tsao), pp. 353-364. American Phytopathological Society: St. Paul, Minnesota.

Brasier, C. M. (1987). The dynamics of fungal speciation. In *Evolutionary Biology of the Fungi*, (ed. A. D. M. Rayner, C. M. Brasier & D. Moore), pp. 231-260. Cambridge University Press.

Brasier, C. M. & Griffin, M. J. (1979). The taxonomy of '*Phytophthora palmivora*' on cocoa. *Transactions of the British Mycological Society* **72**, 111-143.

Brasier, C. M., Griffin, M. J. & Maddison, A. C. (1981). The cocoa black pod Phytophthoras. In *Epidemiology of Phytophthora on Cocoa in Nigeria*, (ed. P. H. Gregory & A. C. Maddison), pp. 18-30. Phytopathological Paper No. 25, Commonwealth Mycological Institute: Kew, U.K.

Brasier, C. M., Hamm, P. H. & Hansen, E. M. (1989). *Phytophthora* diseases. Status of *P. gonapodyides*, *P. drechsleri* and *P. cryptogea*. In *Report on Forest Research 1989*, pp. 49-50. HMSO: London.

Brasier, C. M. & Rayner, A. D. M. (1987). Whither terminology below the species level in the fungi? In *Evolutionary Biology of the Fungi*, (ed. A. D. M. Rayner, C. M. Brasier & D. Moore), pp. 379-388. Cambridge University Press.

Bumbieris, M. (1974). Characteristics of two *Phytophthora* species. *Australian Journal of Botany* **22**, 655-660.

Burnett, J. H. (1975). *Mycogenetics*. John Wiley & Sons: London and New York.

Burnett, J. H. (1983). Speciation in fungi. *Transactions of the British Mycological Society* **81**, 1-14.

Burrell, R. G., Clayton, C. W., Gallegly, M. E. & Lilly, V. G. (1966). Factors affecting the antigenicity of the mycelium of three species of *Phytophthora*. *Phytopathology* **56**, 422-426.

Clare, B. G. (1963). Starch gel electrophoresis of proteins as an aid in identifying fungi. *Nature* **200**, 803-804.

Clare, B. G. & Zentmyer, G.A. (1966). Starch-gel electrophoresis of proteins from species of *Phytophthora*. *Phytopathology* **56**, 1334-1335.

Crandall, B. S. & Gravatt, G. F. (1967). The distribution of *Phytophthora cinnamomi*. *Ceiba* **13**, 57-78.

Dantanarayana, D. M., Peries, O. S. & Liyanage, A. de S. (1984). Taxonomy of *Phytophthora* species isolated from rubber in Sri Lanka. *Transactions of the British Mycological Society* **82**, 113-126.

de Vallavieille, C. & Erselius, L. J. (1984). Variation in protein profiles of *Phytophthora*: survey of a composite population of three species on citrus. *Transactions of the British Mycological Society* **83**, 473-479.

Dick, M. W. (1989). *Phytophthora undulata* comb. nov. *Mycotaxon* **35**, 449-451.

Erselius, L. J. & Shaw, D. S. (1982). Protein and enzyme differences between *Phytophthora palmivora* and *P. megakarya*: evidence for self-fertilization in pairings of the two species. *Transactions of the British Mycological Society* **78**, 227-238.

Erselius, L. J. & de Vallavieille, C. (1984). Variation in protein profiles of *Phytophthora*: comparison of six species. *Transactions of the British Mycological Society* **83**, 463-472.

Erwin, D. C. (1983). Variability within and among species of *Phytophthora*. In *Phytophthora: Its Biology, Taxonomy, Ecology and Pathology*, (ed. D. C. Erwin, S. Bartnicki-Garcia & P. H. Tsao), pp. 149-165. American Phytopathological Society: St. Paul, Minnesota.

Förster, H., Kinsherf, T. G., Leong, S. A. & Maxwell, D. P. (1988). Estimation of relatedness between *Phytophthora* species by analysis of mitochondrial DNA. *Mycologia* **80**, 466-478.

Förster, H., Kinscherf, T. G., Leong, S. A. & Maxwell, D. P. (1989). Restriction fragment length polymorphisms of the mitochondrial DNA of *Phytophthora megasperma* isolated from soybean, alfalfa and fruit trees. *Canadian Journal of Botany* **67**, 529-537.

Gallegly, M. E. (1983a). New criteria for classifying *Phytophthora* and critique of existing approaches. In *Phytophthora: Its Biology, Taxonomy, Ecology and Pathology*, (ed. D. C. Erwin, S. Bartnicki-Garcia & P. H. Tsao), pp. 167-172. American Phytopathological Society: St. Paul, Minnesota.

Gallegly, M. E. (1983b). Summary of the open discussion session on taxonomy of *Phytophthora*. In *Phytophthora: Its Biology, Taxonomy, Ecology and Pathology*, (ed. D. C. Erwin, S. Bartnicki-Garcia & P. H. Tsao), pp. 173-174. American Phytopathological Society: St. Paul, Minnesota.

Gill, H. S. & Powell, D. (1968). The use of polyacrylamide gel disc electrophoresis in delimiting three species of *Phytophthora*. *Phytopathologische Zeitschrift* **63**, 23-29.

Goodwin, P. H., Kirkpatrick, B. C. & Duniway, J. M. (1989). Cloned DNA probes for identification of *Phytophthora parasitica*. *Phytopathology* **79**, 716-721.

Goodwin, P. H., Kirkpatrick, B. C. & Duniway, J. M. (1990). Identification of *Phytophthora citrophthora* with cloned DNA probes. *Applied and Environmental Microbiology* **56**, 669-674.

Hamm, P. B. & Hansen, E. M. (1982). Single-spore isolate variation: the effect on varietal designation in *Phytophthora megasperma*. *Canadian Journal of Botany* **60**, 2931-2938.

Hamm, P. B. & Hansen, E. M. (1983). *Phytophthora pseudotsugae*: a new species causing root-rot of Douglas Fir. *Canadian Journal of Botany* **61**, 2626-2631.

Hansen, E. M. (1987). Speciation in *Phytophthora*: evidence from the *Phytophthora megasperma* complex. In *Evolutionary Biology of the Fungi*, (ed. A. D. M. Rayner, C. M. Brasier & D. Moore), pp 325-338. Cambridge University Press.

Hansen, E. M. (1988). Speciation in plant pathogenic fungi. *Canadian Journal of Plant Pathology* **9**, 403-410.

Hansen, E. M. & Hamm, P. B. (1983). Morphological differentiation of host-specialized groups of *Phytophthora megasperma*. *Phytopathology* **73**, 129-134.

Hansen, E. M., Brasier, C. M., Shaw, D. S. & Hamm, P. B. (1986). The taxonomic structure of *Phytophthora megasperma*: Evidence for emerging biological species groups. *Transactions of the British Mycological Society* **87**, 557-573.

Hansen, E. M., Hamm, P. B., Shaw, C. G. & Hennon, P. E. (1988). *Phytophthora drechsleri* in remote areas of south east Alaska. *Transactions of the British Mycological Society* **91**, 379-384.

Hendrix, J. W. (1967). Light-cholesterol relationships in morphogenesis of *Phytophthora palmivora* and *P. capsici* sporangia. *Mycologia* **59**, 1107-1111.

Ho, H. H. (1982). Affinity groups among plant pathogenic species of *Phytophthora* in culture. *Mycopathologia* **79**, 141-146.

Ho, H. H. & S. C. Jong. (1986). A comparison between *Phytophthora cryptogea* and *P. drechsleri*. *Mycotaxon* **27**, 289-319.

Ho, H. H. & Jong, S. C. (1987). Critical problems in the taxonomy of *Phytophthora* species in culture. *Mycotaxon* **29**, 209-232.

Ho, H. H. & Jong, S. C. (1990). *Halophytophthora*, Gen. Nov., a new member of the family Pythiaceae. *Mycotaxon* **36**, 377-382.

Kaosiri, T. & Zentmyer, G. A. (1980). Protein, esterase and peroxidase patterns in the *Phytophthora palmivora* complex from cacao. *Mycologia* **72**, 988-1000.

Ko, W. H. (1978). Heterothallic *Phytophthora*: evidence for hormonal regulation of sexual reproduction. *Journal of General Microbiology* **107**, 15-18.

Ko, W. H., Chang, H. S., Su, H. J., Chen, C. C. & Leu, L. S. (1978). Peronophythoraceae, a new family of Peronosporales. *Mycologia* **70**, 380-384.

Leonian, L. H. (1925). Physiological studies on the genus *Phytophthora*. *American Journal of Botany* **12**, 444-498.

Leonian, L. H. (1934). Identification of *Phytophthora* species. *West Virginia Agricultural Experiment Station Bulletin* **262**, 1-36.

Merz, W. G., Burrell, R. G. & Gallegly, M. E. (1969). A serological comparison of six heterothallic species of *Phytophthora*. *Phytopathology* **59**, 367-370.

McNabb, S. A., Boyd, D. A., Belkhiri, A., Dick, M. W. & Klassen, G. R. (1987). An inverted repeat comprises more than three quarters of the mitochondrial genome in two species of *Pythium*. *Current Genetics* **12**, 205-208.

Mills, S. D., Förster, H. & Coffey, M. D. (1990). Taxonomic structure of *Phytophthora cryptogea* and *P. drechsleri* based on isozyme and mitochondrial DNA analysis. *Mycological Research* **95**, 31-48.

Newhook, F. J., Waterhouse, G. M. & Stamps, D. J. (1978). Tabular key to the species of *Phytophthora* de Bary. *Mycological Papers* **143**. Commonwealth Mycological Institute: Kew, U.K.

Nygaard, S. L., Elliott, C. K., Cannon, S. J. & Maxwell, D. P. (1989). Isozyme variability among isolates of *Phytophthora megasperma*. *Phytopathology* **79**, 773-780.

Old, K. M., Moran, G. R. & Bell, C. J. (1984). Isozyme variability among isolates of *Phytophthora cinnamomi* from Australia and Papua New Guinea. *Canadian Journal of Botany* **62**, 2016-2022.

Panabieres, F., Marais, A., Trentin, F., Bonnet, P., & Ricci, P. (1989). Repetitive DNA polymorphism analyses as a tool for identifying *Phytophthora* species. *Phytopathology* **79**, 1105-1109.

Ribeiro, O. K. (1978). *A Source Book of the Genus Phytophthora*. J. Cramer: Vaduz, Liechtenstein.

Rosenbaum, J. (1917). Studies of the genus *Phytophthora*. *Journal of Agricultural Research* **8**, 233-276.

Sansome, E. R. (1987). Fungal chromosomes with the light microscope. In *Evolutionary Biology of the Fungi*, (ed. A. D. M. Rayner, C. M. Brasier & D. Moore), pp. 97-113. Cambridge University Press.

Sansome, E. R. & Brasier, C. M. (1974). Polyploidy associated with varietal differentiation in the *megasperma* complex of *Phytophthora*. *Transactions of the British Mycological Society* **63**, 461-467.

Sansome, E. R., Brasier, C. M. & Griffin, M. J. (1975). Chromosome size differences in *Phytophthora palmivora*, a pathogen of cocoa. *Nature* **255**, 704-705.

Sansome, E. R., Brasier, C. M. & Hamm, P. B. (1991). *Phytophthora meadii* may be a species hybrid. *Mycological Research* **95**, 273-277.

Sansome, E. R., Brasier, C. M. & Sansome, F. W. (1979). Further cytological studies on the 'L' and 'S' types of *Phytophthora* from cocoa. *Transactions of the British Mycological Society* **73**, 293-302.

Shattock, R. C., Tooley, P. W. & Fry, W. E. (1986). Genetics of *Phytophthora infestans*: determination of recombination, segregation and selfing by isozyme analysis. *Phytopathology* **76**, 410-413.

Shaw, D. S. (1983). The cytogenetics and genetics of Phytophthora. In *Phytophthora: Its Biology, Taxonomy, Ecology and Pathology*, (ed. D. C. Erwin, S. Bartnicki-Garcia & P. H. Tsao), pp. 81-94. American Phytopathological Society: St. Paul, Minnesota.

Shaw, D. S. (1988). The *Phytophthora* species. *Advances in Plant Pathology* **6**, 27-51.

Shepherd, C. J. & Pratt, B. H. (1973). Separation of two ecotypes of *Phytophthora drechsleri* Tucker occurring in Australian native forests. *Australian Journal of Biological Science* **26**, 1095-1107.

Stamps, D. J., Waterhouse, G. M., Newhook, F. J. & Hall, G. S. (1990). Revised tabular key to the genus *Phytophthora*. *Mycological Papers* **162**. CAB International: Wallingford, U.K.

Tooley, P. W., Fry, W. E. & Villarreal Gonzalez, M. J. (1985). Isozyme characterization of sexual and asexual *Phytophthora infestans* populations. *Journal of Heredity* **76**, 431-435.

Tsao, P. H. & Alizadeh, A. (1988). Recent advances in the taxonomy and nomenclature of the so called '*Phytophthora palmivora*' MF4 occurring on cocoa and other tropical crops. In *Proceedings of the 10th International Cocoa Research Conference*, Santo Domingo, 17-23 May 1987, pp. 441-445.

Tsao, P. H. & Sisemore, D. J. (1978). Morphological variability in *Phytophthora parasitica* (*P. nicotianae*) isolates from citrus, tomato and tobacco. *Phytopathology News* **12**, 213 (abstract).

Tucker, C. M. (1931). Taxonomy of the genus *Phytophthora* de Bary. *Missouri Agricultural Experiment Station Research Bulletin* **153**, 1-208.

Waterhouse, G. M. (1963). Key to the species of *Phytophthora* de Bary. *Mycological Papers* **92**. Commonwealth Mycological Institute: Kew, U.K.

Waterhouse, G. M. (1970). Taxonomy in *Phytophthora*. *Phytopathology* **60**, 1141-1143.

Waterhouse, G. M., Newhook, F. J. & Stamps, D. J. (1983). Present criteria for classification of *Phytophthora*. In *Phytophthora: Its Biology, Taxonomy, Ecology and Pathology*, (ed. D. C. Erwin, S. Bartnicki-Garcia & P. H. Tsao), pp. 139-147. American Phytopathological Society: St. Paul, Minnesota.

Zentmyer, G. A. (1988). Origin and distribution of four species of *Phytophthora*. *Transactions of the British Mycological Society* **91**, 367-378.

Zentmyer, G. A., Kaosiri, T., Idosu, G. O. & Kellam, M. K. (1981). Morphological forms of *Phytophthora palmivora*. In *Proceedings of the 7th International Cocoa Research Conference, Douala, Cameroon*, ed. J. de Lafforest, pp. 291-295. Transla-Inter Ltd: London.

Chapter 9

Relationships between non-papillate, soilborne species of *Phytophthora*: root rot of raspberry

J. M. Duncan, D. M. Kennedy & P. H. Scott

Root rots of red raspberry (*Rubus idaeus* var. *idaeus*) caused by *Phytophthora* spp. have been known since the 1930's when there were widespread and severe outbreaks of a root rot in Scotland, supposedly caused by *P. citricola* (Waterston, 1937). A root rot caused by *P. erythroseptica* has also been recognised for many years in the Pacific Northwest of the United States and in British Columbia in Canada, but it is only within the last ten years that a root rot very similar to that seen in N.America has become the most serious disease problem encountered in raspberry production in many other parts of the world. Outbreaks have been confirmed in the United Kingdom, Eire (Duncan, Kennedy & Seemüller, 1987), Germany (Seemüller *et al.*, 1986), France (Nourrisseau & Baudry, 1987), Norway (Heiberg *et al.*, 1990) and Switzerland (Bolay & Lauber, 1989) with unconfirmed reports from several other countries; from New York, New Jersey and Ohio (Wilcox, 1987, 1989) outside of the Pacific Northwest in the United States; and in Australia, where *P. cryptogea* has been identified as the cause (Washington, 1988).

There is considerable evidence that many outbreaks have resulted from planting rooted canes from affected nurseries (Duncan *et al.*, 1987), with further spread occurring as a result of movement of soil carried by implements etc., and surface water run-off. Crop losses caused by the disease have been considerable and some plantations have been completely killed. Raspberry production on infested land has only been made possible through the use of fungicides and, in the Pacific Northwest, of cultivars with some resistance.

Isolation of *Phytophthora* species from raspberry

The availability of antimicrobials with little or no activity against *Phytophthora* spp. such as the thiabendazole fungicides, hymexazol and rifamycin, has improved the effectiveness of selective media for isolating slow-growing *Phytophthora* species from plant material (Montgomerie & Kennedy, 1983). Perhaps as a consequence, isolations from material displaying typical symptoms of root rot, often yield several species which require identification, and whose pathogenicities need to be determined. A num-

Table 9.1. *Phytophthora* species isolated worldwide from *Rubus*

Name	Group	Report	Host
P. cactorum	I[a]	Duncan *et al.*, 1987	raspberry
		Wilcox, 1987, 1989	raspberry
Phytophthora sp.	I?	Wilcox, 1989	raspberry
P. citricola	III	Waterston, 1937[b]	raspberry
		Wilcox, 1987, 1989	raspberry
		Duncan *et al.*, 1989	raspberry
P. syringae	III	Duncan *et al.*, 1987	raspberry
P. fragariae	V	McKeen, 1958	loganberry
		Nourrisseau & Baudry, 1987	raspberry
		Wilcox, 1989	raspberry
P. megasperma	V	Duncan *et al.*, 1987	raspberry
P. megasperma var. *megasperma*	V	Duncan *et al.*, 1987	raspberry
		Wilcox, 1989	raspberry
P. cambivora	VI	Duncan *et al.*, 1987	raspberry
P. cinnamomi	VI	Brien & Dingley, 1959	raspberry
P. cryptogea	VI	Washington, 1988	raspberry
		Wilcox, 1989	raspberry
P. drechsleri var. *drechsleri*	VI	Duncan *et al.*, 1987	raspberry
P. erythroseptica	VI	Converse & Schwartze, 1968	raspberry
		Seemüller *et al.*, 1986	raspberry
Phytophthora sp.	VI?	Wilcox, 1989	raspberry

[a]Groups according to the tabular key of Newhook, Waterhouse & Stamps (1978). Other unidentified isolates have been excluded because of the lack of good descriptions.

[b]Isolations do not appear to have been made.

ber of species have been reported as occurring on raspberry worldwide (see Table 9.1). Some are either non-pathogenic or only mildly pathogenic to raspberry, although in such cases pathogenicity is usually enhanced by waterlogging (Duncan & Kennedy, 1989). Isolates which are highly pathogenic without waterlogging usually have non-papillate sporangia, except *P. citricola* which can cause considerable damage in inoculation tests without the exacerbating effect of waterlogging (Duncan & Kennedy, 1988). The most pathogenic isolates have been assigned variously to *P. erythroseptica* (Converse & Schwartze, 1968; Seemüller *et al.*, 1986), *P. megasperma* (Duncan *et al.*, 1987), and *P. fragariae* (Nourrisseau & Baudry, 1987; Wilcox, 1989).

Table 9.2. Details of isolates referred to in this paper

Isolate	Details
A. Isolates highly pathogenic to raspberry	
(i) European isolates: representative of 71 isolates	
R6 (IMI 290879)	from raspberry, Oxfordshire, England
R17(IMI 296831)	from raspberry, Herefordshire, England
R25	from raspberry, Bodensee, Germany
R36	from raspberry, Munich, Germany
R40	from raspberry, Perthshire, Scotland
R61	from raspberry, Gloucestershire, England
R62	from raspberry, Gloucestershire, England
R95	from raspberry, Ross & Cromarty, Scotland
R98	from raspberry, Co. Waterford, Ireland
R100	from tayberry, S. Glamorgan, Wales.
R109	from raspberry, Perthshire, Scotland
R127	from raspberry, Njos, Norway
(ii) N. American isolates:	
M14 (ATCC 16184)[a]	from raspberry, Washington State
R130[b]	from raspberry, Vancouver, Washington State
B. P. cambivora	
R16 (IMI 296830)	from raspberry, Perthshire, Scotland
R48 (IMI 313725)	from raspberry, Perthshire, Scotland
R91	from raspberry, Perthshire, Scotland
C. P. drechsleri var. drechsleri	
R32 (IMI 303922)	from raspberry, Co Monaghan, Ireland
D. P. erythroseptica var. erythroseptica	
R144 (IMI 34684)	from potato, Ireland (type culture)
R146 (IMI 325820)	
R38 (IMI 303923)	from potato, Scotland
E. P. fragariae	
168 (IMI 278659)[c]	from strawberry, Kent, England
169 (ATCC 46092)	from loganberry (*Rubus loganobaccus*), British Columbia, Canada (McKeen, 1958)
293	from strawberry, Scotland
F. P. megasperma var. sojae	
R145 (IMI 131555)	from soybean
G. P. megasperma var. megasperma	
R24	from gooseberry, South Wales
R44	from raspberry, Perthshire, Scotland
R72	from raspberry, Perthshire, Scotland

[a]The original 'Canby' isolate of Converse & Schwartze (1968)

[b]One of sixteen isolates from red raspberry made during a visit in 1987 by the second author in co-operation with Dr P. R. Bristow of W. Washington Research and Extension Center, Puyallup, Washington.

[c]a single zoospore culture from a Hickman culture (H-4 of Montgomerie, 1967).

In studies where more than one species was isolated, the most pathogenic species was also the one most frequently isolated. Duncan *et al*. (1987) reported that 19 out of 35 isolates from Europe and the US belonged to 'pathogenic *P. megasperma*', a name which was given to distinguish highly pathogenic isolates from the different and less pathogenic *P. megasperma* var. *megasperma*. More recently, 46 out of 73 isolates belonged to this group (unpublished data). Wilcox (1989) identified 13 isolates as belonging to *P. fragariae* from a total of 33 isolates of seven species from raspberry. It is these common, highly pathogenic isolates with non-papillate sporangia which are the principal subject of this paper.

Comparison of isolates

A list of raspberry isolates used in comparative studies is given in Table 9.2. It includes the highly pathogenic isolates from raspberry (hereafter referred to as HPR-isolates), and isolates from raspberry which have been assigned to *P. cambivora*, *P. drechsleri*, and *P. megasperma* var. *megasperma*. Also included were isolates of *P. fragariae* from strawberry (Kennedy & Duncan, 1988) and loganberry (*Rubus loganobaccus*) (McKeen, 1958), *P. megasperma* var. *sojae* from soybean, and *P. erythroseptica* var. *erythroseptica* from potato.

Appearance on agar

The growth of a range of the isolates on V-8 agar, malt extract agar and a sucrose/asparagine agar is shown in Figs. 9.1 & 9.2. The HPR-isolates were quite distinct from all other species except *P. fragariae*. Their growth was generally uniform without petaloid or other patterns, and considerable quantities of aerial hyphae, particularly on malt extract and sucrose/asparagine agars produce colonies with a cottony appearance. They grew on malt extract agar although there was some variation between experiments, while isolates of *P. fragariae* from strawberry sometimes failed to grow on this medium (Ho & Jong, 1988). The HPR-isolates grew fairly slowly on all media, but usually slightly faster than isolates of *P. fragariae* (Fig. 9.1). The mean growth rates for 16 HPR-isolates on V-8 agar, french bean agar and sucrose asparagine agar at 20°C were 2, 2.5 and 1·7 mm per day respectively.

Cardinal temperatures

The cardinal temperatures of all HPR-isolates from raspberry were similar to those of *P. fragariae*; i.e. a minimum for growth of about 5°C, a maximum just below 30°C, and an optimum between 20°C and 25°C, similar, in fact, to those given by Wilcox (1989). The upper limit for growth of the isolate of *P. erythroseptica* from potato was above 30°C, and this species and *P. megasperma* var. *megasperma* grew much more rapidly at

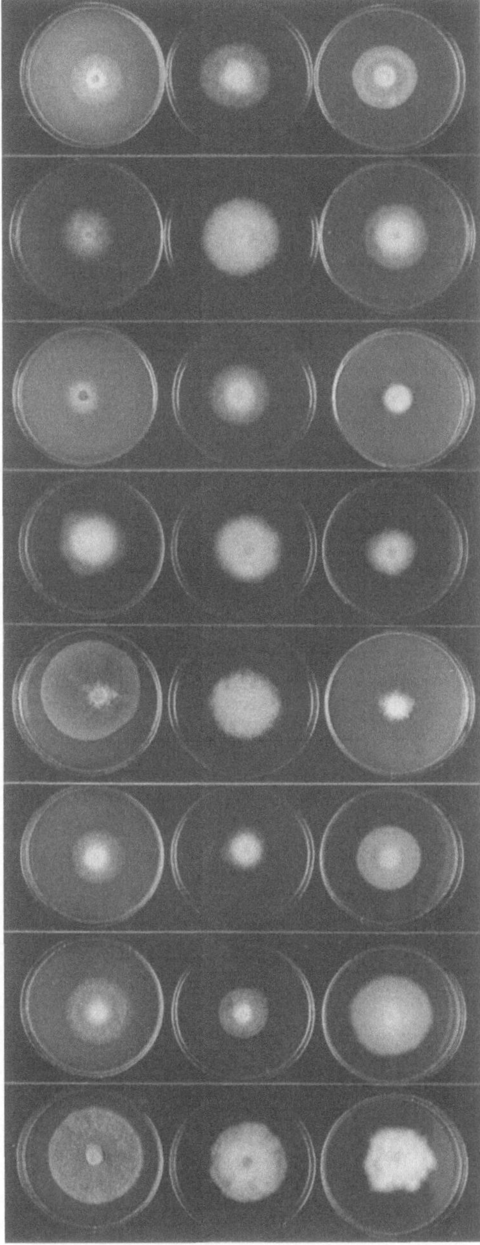

Fig. 9.1. Growth of isolates of *Phytophthora* at 20°C on V-8 agar, malt extract agar, and sucrose/asparagine agar. From the top: isolates highly pathogenic to raspberry R17, R25, R41, R100, M14; *P.fragariae*, 169 and 293; *P. megasperma* var. *sojae*, R145.

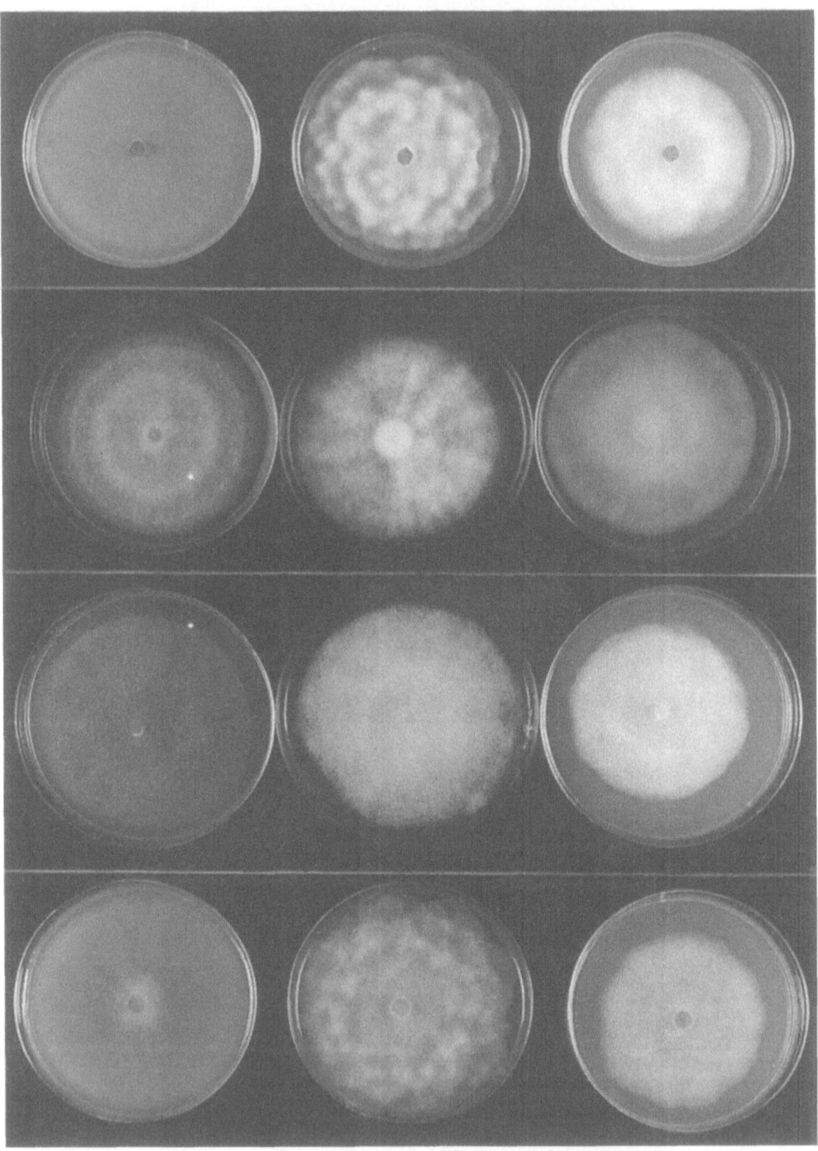

Fig. 9.2. Growth of isolates of *Phytophthora* at 20°C on V-8 agar, malt extract agar, and sucrose/asparagine agar. From the top: *P. erythroseptica* var. *erythroseptica*, R38; *P. cambivora*, R16; *P. megasperma* var. *megasperma* R24; *P. drechsleri* var. *drechsleri*, R32.

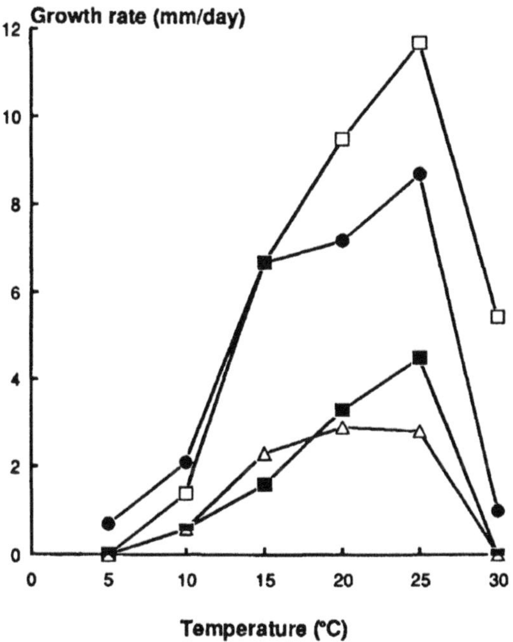

Fig. 9.3. Growth rates (mm/day) on V-8 agar at different temperatures: mean of seven European isolates highly pathogenic to raspberry (triangles); 'Canby' isolate, M14 (closed squares); mean of five isolates of *P. megasperma* var. *megasperma* (closed circles); *P. erythroseptica* var. *erythroseptica*, R38 (open squares).

most temperatures than any of the highly pathogenic isolates from raspberry (Fig. 9.3).

Sporulation

All HPR-isolates produced non-papillate sporangia and many of them displayed internal proliferation, but there were some differences. The original 'Canby' isolate of *P. erythroseptica* (M14) (Converse & Schwartze, 1968) tended to form new sporangia by sympodial branching rather than by internal proliferation and there was little 'nesting' of sporangia (Fig. 9.4a). All European HPR-isolates were very similar to *P. fragariae* in their patterns of sporulation with internal proliferation and 'nesting' (Fig. 9.4b), and limited observations on the more recent HPR-isolates from N. America indicated similar behaviour. Sporangia were large: for 16 European HPR-isolates the mean sizes were 59 x 40 μm, and for the 'Canby' isolate (M14) 58 x 42 μm, which compares with 60 x 38 μm for *P. fragariae* (Waterhouse, 1970), 47 x 32 μm for *P. erythroseptica*, and 71 x 38 μm for

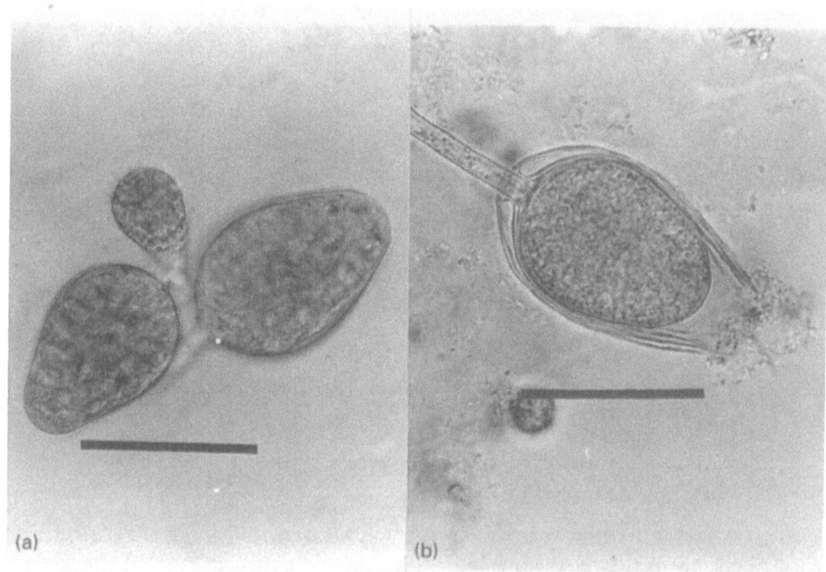

Fig 9.4. (a) Sporangia of M14 (the 'Canby' isolate) in water. Note sympodial branching. (b) Sporangia of R36, a raspberry isolate from Germany. Note 'nesting' of sporangia. Scale bars = 50 μm.

P. megasperma var. *megasperma*. The HPR-isolates cannot be separated, therefore, from *P. fragariae* on the basis of sporangial structure or size, although they usually sporulated about 1 day earlier than *P. fragariae* under the conditions used (Duncan *et al.*, 1987).

Oospore formation

Oospore formation in culture by the HPR-isolates was rarely abundant and in some cultures they were difficult to find e.g. R6. Typically, they were most abundant near the inoculum plug. Oogonia were yellow to golden-red in colour, and usually about 40 μm in diameter. The mean diameter for the oospores was 33 μm with walls 3 μm thick. The antheridia were approximately as wide as they were long (10 to 15 μm) and were predominantly amphigynous (Fig. 9.5), but a few paragynous forms occurred in cultures of most isolates. On complex media, such as french bean agar and oatmeal agar, they were sometimes found in close association

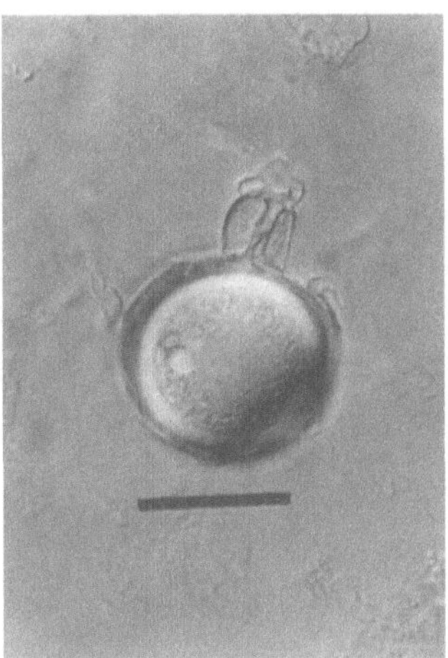

Fig. 9.5. Oospore of raspberry isolate R6 from England. Note typical amphigynous antheridium. Scale bar = 30 μm.

with the starch grains. In contrast, oospores were formed much earlier and were considerably more abundant in cultures of *P. erythroseptica* and *P. megasperma* var. *megasperma*, and the oogonial envelopes were either not coloured or only slightly yellow. The oospores of *P. erythroseptica* var. *erythroseptica* were similar in size to those of the raspberry isolates (32 μm diam.), while those of *P. megasperma* var. *megasperma* were much larger (45 μm diam.). *P. fragariae* produced few oospores, usually only in old cultures on complex media, and generally in close association with starch grains. They were often highly coloured (usually deep straw in colour), and deformed in shape. It was difficult to observe antheridial attachments in oospores formed in culture but, in recently infected strawberry roots when attached to still-developing oogonia, they were predominantly amphigynous and similar in size and shape to those of the raspberry isolates.

Electrophoretic patterns

Isolates were grown on a modified version of a glucose/yeast extract/peptone medium (Hall, Zentmyer & Erwin, 1969), and whole protein extracts in buffer were examined using SDS-polyacrylamide gel electrophoresis (SDS-PAGE). Gels were stained using a modification of a silver stain described by Blum, Beier & Gross (1987).

With the exception of *P. megasperma* var. *megasperma* (Fig. 9.6), all the species with non-papillate sporangia had similar protein patterns, yet there were differences which were generally consistent between species and could be used to separate them (Figs. 9.6 to 9.8). The three isolates of *P. fragariae*, which came from widely separated areas (Canada, UK, and Scotland, Table 9.2) and which had different virulences from one another,

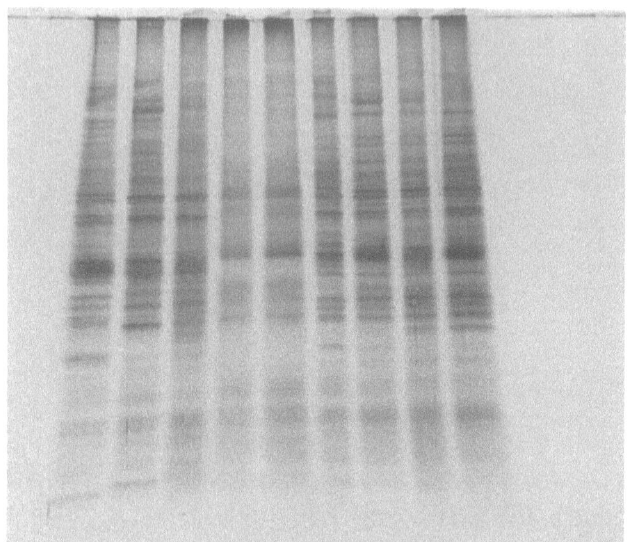

Fig. 9.6. Protein profiles of *Phytophthora* obtained with SDS-PAGE and silver staining. From left to right: 3 isolates of *P. erythroseptica* var. *erythroseptica*, R146, R144, R38; 2 isolates of *P. megasperma* var. *megasperma*, R44, R72; *P. megasperma* var. *sojae*, R145; 3 highly pathogenic isolates from raspberry, R100, M14, R36.

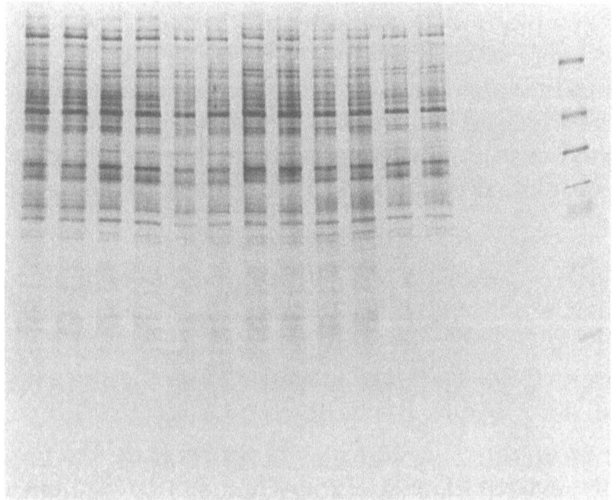

Fig. 9.7. Protein profiles of six isolates of *Phytophthora* highly pathogenic to raspberry, obtained with SDS-PAGE and silver staining. Each isolate is shown in duplicate. From left to right: R98, R95, R36, M14, R40, R61.

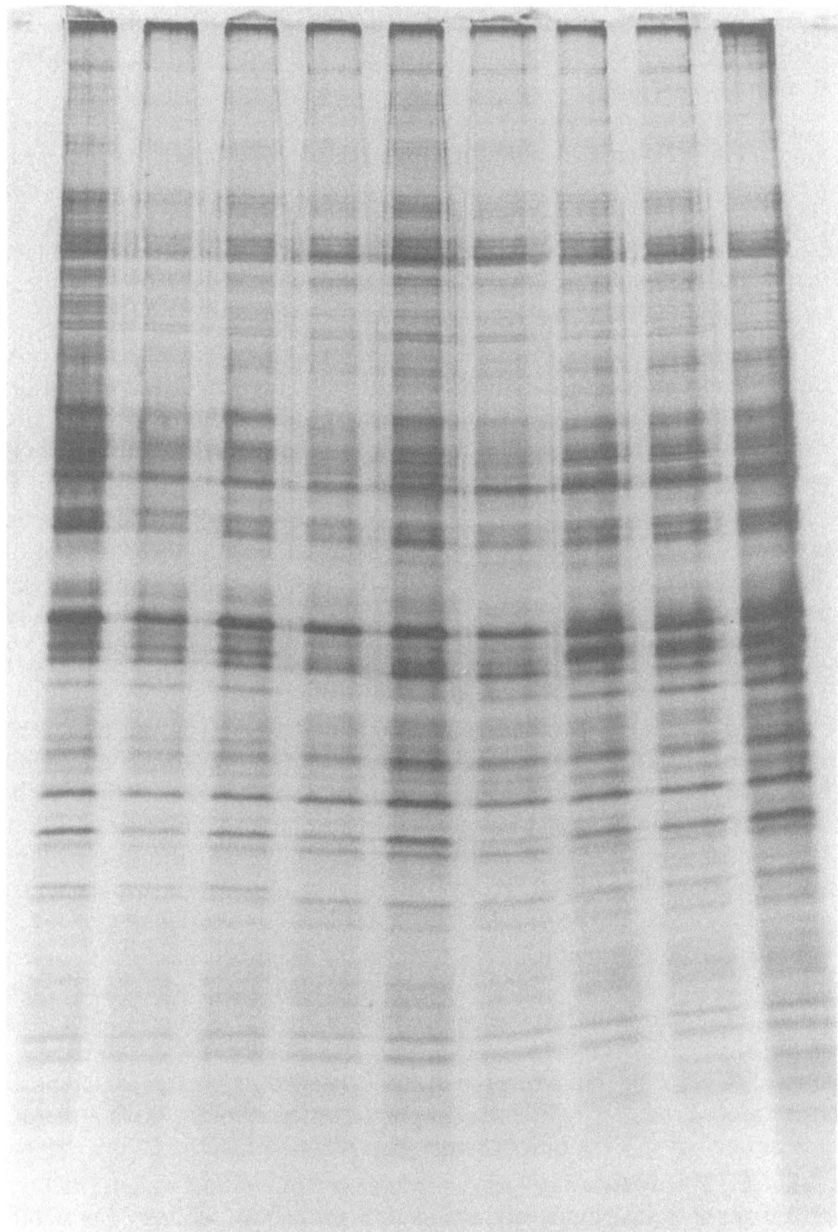

Fig. 9.8. Protein profiles of *Phytophthora* obtained with SDS-PAGE and silver staining. From left to right: 3 isolates of *P. fragariae* 169, 168, 293; 3 highly pathogenic isolates from raspberry R127, R109, R62; 3 isolates of *P. cambivora* R16, R48, R91.

had identical protein profiles. The Canadian isolate 169 (McKeen, 1958; Table 9.2), originally from loganberry, had the same pattern as the two strawberry isolates.

The HPR-isolates, regardless of origin, also formed a homogeneous group (Fig. 9.7), which could be distinguished readily from *P. megasperma* var. *sojae*, *P. erythroseptica* var. *erythroseptica* (Fig. 9.6), and *P. cambivora* (Fig. 9.8). One isolate, R100 from Tayberry (Table 9.2) differed from the other HPR-isolates in two or three bands (Fig. 9.6), but the 'Canby' isolate (M14) and the more recent HPR-isolates from N. America were identical to the European HPR-isolates.

The protein profiles of the HPR-isolates and *P. fragariae* were very similar and differed in only one or two bands, but the differences were consistent for all isolates examined. The results of the electrophoretic studies reinforced the view that the HPR-isolates formed a group on their own, albeit one very close to *P. fragariae*.

An exception to the general homogeneity of isolates within the same taxon was *P. erythroseptica* var. *erythroseptica*: each of the three isolates of this species had distinctive electrophoretic patterns (Fig. 9.6). One of the isolates was the type specimen of the species. A wider range of isolates of this species should be examined.

Pathogenicities

All HPR-isolates, whether from Europe or N. America, caused severe symptoms of root rot on red raspberry which had been inoculated with zoospore suspensions (Fig. 9.9). Rotting was equally severe with or without waterlogging. The Tayberry isolate (R100) was similar to all other HPR-isolates in its pathogenicity to raspberry. Neither it nor any other HPR-isolate caused root rot on Tayberry in inoculation experiments, unless the plants had been subjected to prolonged waterlogging.

No other species, not even the loganberry isolate (169) of *P. fragariae*, caused severe root rot on raspberry, although some symptoms were usually observed with the other species when the plants were subjected to waterlogging after inoculation (Duncan & Kennedy, 1989). All the isolates of *P. fragariae*, including the loganberry isolate (169), caused typical symptoms of red core on strawberry plants, but none of the HPR-isolates, nor isolates of species, other than *P. fragariae*, did so.

The HPR-isolates and *P. fragariae* have restricted host ranges: neither rotted apple fruits extensively, although slight rotting occurred occasionally after extended periods of incubation. In inoculation experiments *P. fragariae* infected members of the Rosaceae in the tribe Potentilleae (Converse & Moore, 1966; Pepin, 1967) to which both strawberry and raspberry belong. Pepin (1967), using the production of red steles and

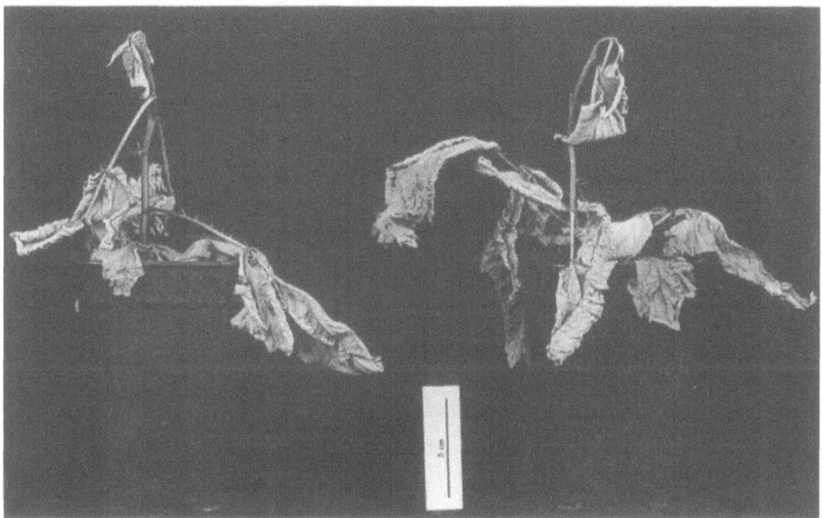

Fig. 9.9. Aerial symptoms of raspberry root rot on plants three weeks after inoculation with a zoospore suspension of a typical isolate of the highly pathogenic form, R6. The plant on the left was not waterlogged after inoculation; the plant on the left was waterlogged for four days.

---

oospores in roots as an indication of infection, managed to infect raspberry with *P. fragariae*, but in all instances infection rarely involved more than a few secondary rootlets (Table 9.3). There was no severe rotting of the type caused by the raspberry isolates. Our limited tests with one European HPR-isolate did not produce symptoms on apple seedlings, *Alchemilla* sp., *Filipendula hexapetala*, and *Poterium sanguisorba*, all of which are members of the Rosaceae, nor on the Potentilleae, *Potentilla argentea* and *Geum rivale*.

Both HPR-isolates and *P. fragariae* produced oospores in or very near to the steles of infected roots of their respective hosts. Other homothallic species attacking raspberry e.g. *P. cactorum*, produced oospores more evenly throughout the tissues of the root, most often in the cortex.

Nomenclature of the HPR-isolates

All the evidence suggests that the HPR-isolates i.e. non-papillate isolates of *Phytophthora* from Europe and N. America, which are highly pathogenic to raspberry, and which have been described variously as *P. erythroseptica* (Converse & Schwartze, 1968; Seemüller *et al.*, 1986),

Table 9.3. Severity of infection on members of the Rosaceae by six races of *P. fragariae*

Host species	Range of scores observed[a]
Geum coccineum	5-10
G. triflorum	4-10
G. macrophyllum	9-10
Dryas drummondii	5-9
Duchesnia indica	9-10
Potentilla anserina	5-10
P. argentea	3-10
P. glandulosa (2 var.)	6-10
P. gracilis	8-10
P. norvegica	8-10
P. erecta	5-10
Rubus parviflorus	8-10
red raspberry (14 cultivars)	6-8
strawberry cv Blakemore	3-5

[a]Scores go from 1 = most roots dead, to 10 = no infection; from Pepin (1967).

pathogenic *P. megasperma* (Duncan *et al.*, 1987), and *P. fragariae* (Wilcox, 1989), are the same. Their appearance on agar, growth rates, sporulation, oospore formation, electrophoretic patterns and pathogenicities support this view. All European isolates so far examined (Scottish, English, Welsh, Irish, German and Norwegian) fit into this group, and it is more than likely that the *Phytophthora* spp. causing root rot in France (Nourrisseau & Baudry, 1987) and Switzerland (Bolay & Lauber, 1989) also belong to it. Although we have not yet examined isolates from the last two countries, it is difficult to imagine that different species would be involved elsewhere in Europe, especially when the published etiology is so similar in both cases. The reported sizes and location of oospores in the steles of affected roots rather than the cortex, is further support for this view.

The HPR-isolates do not fit into *P. megasperma* var. *megasperma* nor *P. megasperma* var. *sojae*; in the former case there are differences in all characters examined, and in the latter, differences in sporangial size, antheridial type and protein profiles. Consequently, it appears that their assignment to *P. megasperma* by Duncan *et al.* (1987) was inappropriate. Neither can they be assigned to *P. erythroseptica* from which they also differ substantially in many respects. Converse & Schwartze (1968) did place their 'Canby' raspberry isolate in this species but only *sensu lato*, noting that the sporangia of their isolate were larger than those of *P. erythroseptica sensu strictu*. Their first preference had been to assign it to *P. fragariae* (Converse & Schwartze, 1965), but they did not do so because of its inability to cause red core symptoms on strawberry plants and because of a number of other differences, including lack of internally proliferating sporangia and strictly amphigynous antheridia. The tendency to sympodial branching of the sporangiophores in this isolate has been mentioned earlier, but it is not as pronounced in other N. American isolates. Likewise amphigyny, while not exclusive in most of the raspberry isolates, is predominant and may be so predominant in some isolates as to appear exclusive. The 'Canby' isolate is very similar to all other HPR-isolates in its protein profile and it is thus reasonable to place it in the same group.

The HPR-isolates and *P. fragariae* are similar in most respects, although there are some small but consistent differences between them. Ho & Jong (1988) listed various criteria which they thought distinguished *P. fragariae* from other species; the HPR-isolates satisfy many of them. However, unlike *P. fragariae* they do grow on malt extract agar, although this is not a definitive character and could not on its own preclude their inclusion in *P. fragariae*. There are also differences in numbers and form of the oospores in culture, but since oospore production is so scant and often aberrant in cultures of *P. fragariae*, it is difficult to make reliable comparisons. Thus, apart from small differences in protein profiles, the major difference is in the respective host ranges. Despite the isolation of a 'true' *P. fragariae* culture (169) by McKeen (1958) from loganberry, the HPR-isolates and *P. fragariae* cannot attack one another's hosts, except possibly where the soil is waterlogged. Wilcox (1989) and Nourrisseau & Baudry (1987) did not consider the small cultural differences, or the differences in host range sufficient to separate their HPR-isolates from *P. fragariae*. Yet although pathogenicity *per se* is not a valid criterion for separating taxa, it has in certain circumstances matched very closely other criteria, which are valid. Hansen *et al.* (1986) distinguished nine sub-groups of *P. megasperma* on the basis of various criteria, including protein profiles and karyotypes. Each of the sub-groups had a characteristic host range, and it was thought that they might represent emerging biological species. The same situation may apply to the HPR-isolates and *P. fragariae*, in which case the best course of action would be to consign the HPR-isolates to *P. fragariae*, while at the same time giving them some sub-specific epithet which would distinguish them from isolates which caused strawberry red core. In the meantime, comparisons between the two groups should be extended to include other criteria such as karyotypes, isozyme and restriction fragment length polymorphisms (Chapter 10).

Speculation on the origins of *P. fragariae* and the raspberry isolates

Accepting the close affinity of the HPR-isolates to *P. fragariae*, it is possible that one group evolved from the other, e.g. the raspberry isolates evolved in some red core infested site which had been planted subsequently to raspberries, and that selection in the surviving oospore population led to the emergence of strains highly pathogenic to raspberry. Raspberry and strawberry are more closely related than appearance and habit suggest; *P. fragariae* may infect raspberry to a very limited extent. However, this is probably too simplistic, since there is no apparent reason why, in gaining pathogenicity to raspberry, such isolates would have lost it to strawberry. Furthermore, experiments have shown a surprising lack of variation in the pathogenicity and, in particular, in the virulence of single oospore cultures of *P. fragariae* when compared with those of their parent cultures (Kennedy & Duncan, 1988).

It is now generally accepted that the cultivated strawberry (*Fragaria x ananassa*) arose from a cross of two New World species: *F. chiloensis* and *F. virginiana* (Wilhelm & Sagen, 1974). Clones of both species are resistant to Red Core (Scott & Lawrence, 1975), and in the case of *F. chiloensis* there are clones which react in a differential manner to the various races of the fungus (Pepin & Daubeny, 1966). Little is known about the resistance of old world strawberry species but all clones of *F. vesca* which have been tested are highly susceptible to the disease. Similarly, resistance to raspberry root rot is higher in raspberries derived from N. American red raspberry (*Rubus idaeus strigosus*) than in those derived from European red raspberry (*R. idaeus vulgaris*) (Bristow *et al*., 1988). All of which would suggest that *P. fragariae* and the HPR-isolates have their origin in the New World. Since both are undoubtedly 'cool temperate' species (Brasier, 1983), they could have arisen in the cooler, wetter parts of N. America, perhaps in the Pacific Northwest. *F. chiloensis* grows wild along cooler regions of the Pacific slope from Alaska to the tip of Chile, usually but not always very close to the shore (Wilhelm & Sagen, 1974).

The first reported outbreak of red core in N. America was in Illinois in 1930 (Anderson, 1940), but there is evidence that it could have been present earlier. When the strawberry breeding work of the University of California was moved in 1933 from the Santa Clara valley to Zyanthe, near Olympia in the Santa Cruz mountains, there were immediate outbreaks of a severe root rot which was later determined to be red core (Wilhelm & Sagen, 1974). At the time the disease was thought to be indigenous to the soil of the Olympia area (Wilhelm & Sagen, 1974), and perhaps other areas of California (Thomas, 1937), although the possibility that it was present but not manifest at the site in the warmer Santa Clara valley and was transferred on the plants to the new station, cannot be discounted. Further evidence for *P. fragariae* being indigenous to N. American soils was provided by Wooley (1955) who planted out certified disease-free strawberry plants in ostensibly undisturbed 'native' forest sites in Oregon. A number of the plants developed red core, suggesting that the fungus was present in the virgin soils. In this connection it is worth noting that Thomas (1935) reported that 'It was found in abundance on the roots of some planting stock shipped into California from Oregon'.

It is possible that both pathogens to strawberry and raspberry arose originally from a *Phytophthora* present in temperate areas of N. America or even S. America, the principal host range of which is in the Rosaceae, and members of the tribe Potentilleae in particular, which are well represented in the New World flora (Clark, 1976). *Potentilla* spp. have even been suggested as natural hosts for *P. fragariae* (Moore, Scott & Converse, 1964). Perhaps the two pathogens had their own preferred hosts within the tribe, namely *Fragaria* and *Rubus*, both of which were subsequently brought into cultivation by man. Both crops are propagated vegetatively, and there is ample evidence that this method of propagation has been responsible for the worldwide distribution of red core of strawberry and root rot of raspberry. It should be possible to test this hypothesis by attempting to isolate *Phytophthora* spp. from wild species of Rosaceae growing at virgin sites in N. America, and determine if they closely resemble the HPR-isolates and *P. fragariae*, but perhaps with different host ranges. Similar work in remote areas of Alaska, resulted in the isolation of *P. drechsleri* from undisturbed muskeg and forested areas (Hansen *et al.*, 1988).

Acknowledgements. We would like to acknowledge all those who have helped us with our work. In particular we would like to thank the staff of the CAB International Mycological Institute who have provided cultures on many occasions, Dr Pete Bristow who obtained material and financial assistance to allow the second author to visit the United States and to obtain fresh isolates from raspberry, and Dr Stephen Wilhelm of the University of California who provided copies of Dr Harold Thomas's early reports of Red Core on strawberries in California.

References

Anderson, H. W. (1940). Red stele root rot of strawberry. *Transactions of the Illinois State Horticultural Society* **74**, 383-393.

Blum, H., Beier, H. & Gross, H. J. (1987). Improved silver staining of plant proteins, RNA and DNA in polyacrylamide gels. *Electrophoresis* **8**, 93-99.

Bolay, A. & Lauber, H. -P. (1989). Un *Phytophthora*, cause d'un rapide dépérissement du Framboisier. *Revue Suisse de Viticulture, Arboriculture et Horticulture* **21**, 147-152.

Brasier, C. M. (1983). Problems and prospects in *Phytophthora* research. In *Phytophthora: Its Biology, Taxonomy, Ecology and Pathology*, (ed. D. C. Erwin, S. Bartnicki-Garcia & P. H. Tsao), pp. 351-363. The American Phytopathological Society: St. Paul, Minnesota.

Brien, R. M. & Dingley, J. M. (1959). A revised list of plant diseases recorded in New Zealand, fourth supplement, 1957-1958. *New Zealand Journal of Agricultural Research* **2**, 406-413.

Bristow, P. R., Daubeny, H. A., Sjulin, T. M., Pepin, H. S., Nestby, R. & Windom, G. E. (1988). Evaluation of *Rubus* germplasm for reaction to root rot caused by *Phytophthora erythroseptica*. *Journal of the American Society for Horticultural Science* **113**, 588-591.

Clark, L. J. (1976). *Wild flowers of the Pacific Northwest: from Alaska to Northern California*. (ed. J. G. Trelawny). Gray's Publishing Ltd: Sydney, BC, Canada.

Converse, R. H. & Moore, J. N. (1966). Susceptibility of certain *Potentilla* and *Geum* species to infection by various races of *Phytophthora fragariae*. *Phytopathology* **56**, 637-639.

Converse, R. H. & Schwartze, C. D. (1965). *Phytophthora* sp. from Washington pathogenic on roots of red raspberry. *Phytopathology* **55**, 503(abstract).

Converse, R. H. & Schwartze, C. D. (1968). A root rot of red raspberry caused by *Phytophthora erythroseptica*. *Phytopathology* **58**, 56-59.

Duncan, J. M. & Kennedy, D. M. (1988). *Phytophthora* root rot of raspberry. In *Annual Report of The Scottish Crop Research Institute*, 1987, pp. 94-95.

Duncan, J. M. & Kennedy, D. M. (1989). The effect of waterlogging on *Phytophthora* root rot of red raspberry. *Plant Pathology* **38**, 161-168.

Duncan, J. M., Kennedy, D. M. & Seemüller, E. (1987). Identities and pathogenicities of *Phytophthora* spp. causing root rot of red raspberry. *Plant Pathology* **36**, 276-289.

Hansen, E. M., Brasier, C. M., Shaw, D. S. & Hamm, P. B. (1986). The taxonomic structure of *Phytophthora megasperma*: evidence for emerging biological species groups. *Transactions of the British Mycological Society* **87**, 557-573.

Hansen, E. M., Hamm, P. B., Shaw III, C. G. & Hennon, P. E. (1988). *Phytophthora drechsleri* in remote regions of Southeast Alaska. *Transactions of the British Mycological Society* **91**, 379-384.

Hall, R., Zentmyer, G. A. & Erwin, D. C. (1969). Approach to taxonomy of *Phytophthora* through acrylamide gel electrophoresis of proteins. *Phytopathology* **59**, 770-774.

Heiberg, N., Duncan, J. M., Kennedy, D. M. & Semb, L. (1990). Raspberry root rots in Norway. *Proceedings of the Fifth International Rubus-Ribes Symposium*, In *Acta Horticulturae,* in press.

Ho, H. H. & Jong, S. C. (1988). *Phytophthora fragariae*. *Mycotaxon* **31**, 305-322.

Kennedy, D. M. & Duncan, J. M. (1988). Pathogenicity of *P. fragariae*. In *Annual Report of the Scottish Crop Research Institute*, 1987, pp. 115-116.

McKeen, W. E. (1958). Red stele root disease of the loganberry and strawberry caused by *Phytophthora fragariae*. *Phytopathology* **48**, 129-132.

Moore, J. N., Scott, D. H. & Converse, R. H. (1964). Pathogenicity of *Phytophthora fragariae* to certain *Potentilla* species. *Phytopathology* **54**, 173-176.

Montgomerie, I. G. (1967). Pathogenicity of British isolates of *Phytophthora fragariae* and their relationship with American and Canadian races. *Transactions of the British Mycological Society* **50**, 57-67.

Montgomerie, I. G. & Kennedy, D. M. (1983). An improved method of isolating *Phytophthora fragariae*. *Transactions of the British Mycological Society* **80**, 178-173.

Newhook, F. J., Waterhouse, G. M. & Stamps, D. J. (1978). Tabular key to the species of *Phytophthora* de Bary. *Mycological Papers* **143**. Commonwealth Mycological Institute: Kew, U.K.

Nourrisseau, J.-G. & Baudry, A. (1987). Un *Phytophthora* cause de dépérissement du framboisier en France. *Phytoma-Défense des cultures*, January 1987, pp. 39-41.

Pepin, H. S. (1967). Susceptibility of members of the Rosaceae to races of *Phytophthora fragariae*. *Phytopathology* **57**, 782-784.

Pepin, H. S. & Daubeny, H. A. (1966). Reactions of strawberry cultivars and clones of *Fragaria chiloensis* to six races of *Phytophthora fragariae*. *Phytopathology* **56**, 361-362.

Scott, D. H. & Lawrence, F. J. (1975). Strawberries. In *Advances In Fruit Breeding*, (ed. J. Janick & J. N. Moore), pp. 71-97. Purdue University Press: West Lafayette, Ind., U.S.A.

Seemüller, E., Duncan, J. M., Kennedy, D. M. & Riedl, M. (1986). *Phytophthora* sp. als Ursache einer Wurzelfale an Himbeere. *Nachrichtenblatt des Deutsches Pflanzenschutzdienstes* **38**, 17-21.

Thomas, H. E. (1935). Root rot studies. In *Report of the University of California College of Agriculture Agricultural Experiment Station*. Project No. 981.

Thomas, H. E. (1937). Root rot studies. In *Report of the University of California College of Agriculture Agricultural Experiment Station*. Project No. 981.

Washington, W. S. (1988). *Phytophthora cryptogea* as a cause of root rot of raspberry in Australia: resistance of raspberry cultivars and control by fungicides. *Plant Pathology* **37**, 225-230.

Waterhouse, G. M. (1970). The genus *Phytophthora* de Bary. *Mycological Papers* **122**. Commonwealth Mycological Institute: Kew, U.K.

Waterston, J. M. (1937). A note on the association of a species of *Phytophthora* with a 'Die-back' disease of the raspberry. *Transactions of the Botanical Society of Edinburgh* **32**, 251-259.

Wilcox, W. F. (1987). Minimize root rot losses. *Fruit Grower*, January 1987, 18-19.

Wilcox, W. F. (1989). Identity, virulence and isolation frequency of seven *Phytophthora* spp. causing root rot of raspberry in New York. *Phytopathology* **79**, 93-101.

Wilhelm, S. & Sagen, J. E. (1974). *A History of the Strawberry*. University of California,Division of Agricultural Sciences: California, USA.

Wooley, P. H. (1955). Physiologic specialization and etiology of red stele disease, *Phytophthora fragariae* Hickman of strawberries and attempts at control. Ph.D. Thesis, Oregon State College, USA.

Chapter 10

Variation in the species of the *Phytophthora megasperma* complex

Everett M. Hansen

Biological variation is at oncethe nemesis anda source of inspiration for mycologists. The problems of interpreting variation are well illustrated in the history of our understanding of *Phytophthora megasperma*. This group of organisms has been alternately split and lumped based on the isolates of the moment and the philosophy of the investigator. Is this one, highly variable species, or a conglomerate of discrete taxa? To answer we must have both a coherent species concept, and a grasp of the nature of variation in this context.

Variation in our mycological framework derives from at least three sources. Variation is the inevitable product of our observational and experimental techniques and the consequence of imperfect taxonomy. Variation also springs from the genetic systems of the organisms we study. Our response to this variation must be appropriate to its source.

The objectives of this Chapter are three-fold:

- to frame the debate about variation in *P. megasperma* in a historical and biological context;
- to summarize recent and unpublished evidence for segregated species within the complex; and
- to endorse a biological species concept for the genus that is both practical and consistent with phylogenetic nomenclature.

Variation and the nomenclature of *Phytophthora megasperma*

Drechsler (1931) recorded but was not troubled by the variation he observed in his new species, *P. megasperma*. He noted carefully the size distribution about the mean of oogonia and oospores, and commented that sporangia were '...too variable for profitable statistical metric treatment.' He described the range of conditions that affect size of *Phytophthora* structures including growth medium, temperature, and maturity, and quite sensibly chose to emphasize the sexual apparatus, '...distinguished by the unusually large dimensions of oogonium (47 μm) and oospore (41 μm).'

Tompkins, Tucker & Gardner (1936) are frequently credited (or blamed) for expanding the species concept of *P. megasperma* by including isolates from several hosts with smaller oogonia and oospores. It should be noted, however, that their measurements of Drechsler's own material were 5 μm smaller than his, and within the range of their isolates from other hosts. In fact, they made no note of their isolates being smaller, emphasizing instead the effects of cultural conditions on spore size. This sort of variation, really an artifact of non-standardized media and methods, continues to plague communication in our field, despite exhortations to improve techniques (Waterhouse, Newhook & Stamps, 1983).

Our taxonomic philosophies are another source of non-biological variation. Is one an adherent of the type concept or the biological species concept of taxonomy? Is morphology the only valid species character, or are differences in pathogenicity, growth or other behaviour also important? Depending on the answers to these questions, *P. megasperma* is either a single, highly variable species, or a cluster of much more uniform discrete species. This type of variation is a matter of definition rather than biology or technique.

The division of *P. megasperma* into two varieties, var. *sojae* and var. *megasperma*, based on spore size or pathogenicity has been widely accepted. Hildebrand (1959) emphasized specific pathogenicity to soybean in his new variety, while Waterhouse (Waterhouse & Blackwell, 1954; Waterhouse, 1963) included only morphology in her keys. Hildebrand's var. *sojae* was very uniform in morphology, behaviour, and the entire range of other characters reviewed below. Waterhouse's variety, on the other hand, encompassed isolates with a wider range of spore sizes and dramatic behavioural differences. This difference in philosophy has led to considerable taxonomic confusion.

The studies of Kuan & Erwin (1980) and Hansen & Hamm (1983) provide another example of the consequences of differing approach. They examined many of the same isolates and agreed that the system of two morphological varieties was unworkable, and that pathogenicity was an important factor. But they still came to different conclusions about the structure of the population. Kuan & Erwin (1980) looked at one morphological feature at a time and saw continuous variation whereas Hansen & Hamm (1983) looked at correlations between both morphological and physiological characters and saw discrete populations. In each case, starting assumptions dictated the appropriate statistical analysis, which in turn led to different conclusions.

Variation is also the real and natural product of growth and reproduction in all organisms. Genetic variation, the product of mutation and recombination, provides the fuel for the ongoing processes of natural

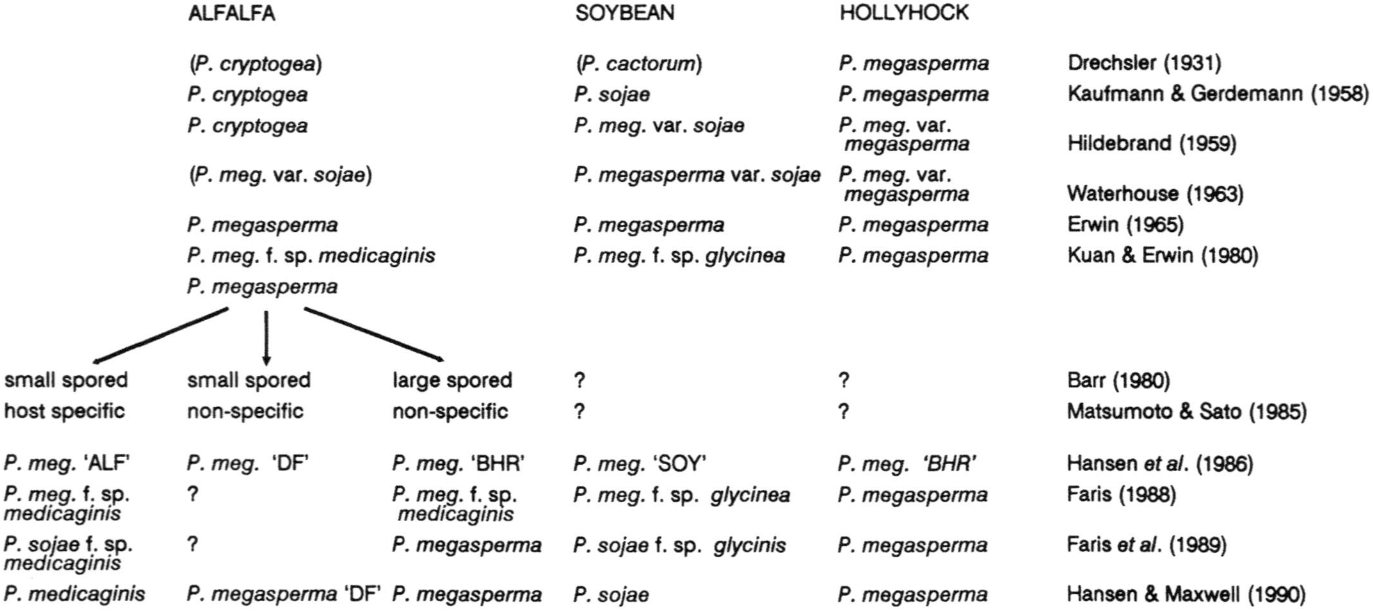

	ALFALFA		SOYBEAN	HOLLYHOCK	
	(*P. cryptogea*)		(*P. cactorum*)	*P. megasperma*	Drechsler (1931)
	P. cryptogea		*P. sojae*	*P. megasperma*	Kaufmann & Gerdemann (1958)
	P. cryptogea		*P. meg.* var. *sojae*	*P. meg.* var. *megasperma*	Hildebrand (1959)
	(*P. meg.* var. *sojae*)		*P. megasperma* var. *sojae*	*P. meg.* var. *megasperma*	Waterhouse (1963)
	P. megasperma		*P. megasperma*	*P. megasperma*	Erwin (1965)
	P. meg. f. sp. *medicaginis*		*P. meg.* f. sp. *glycinea*	*P. megasperma*	Kuan & Erwin (1980)
	P. megasperma				
small spored	small spored	large spored	?	?	Barr (1980)
host specific	non-specific	non-specific	?	?	Matsumoto & Sato (1985)
P. meg. 'ALF'	*P. meg.* 'DF'	*P. meg.* 'BHR'	*P. meg.* 'SOY'	*P. meg.* '*BHR*'	Hansen *et al.* (1986)
P. meg. f. sp. *medicaginis*	?	*P. meg.* f. sp. *medicaginis*	*P. meg.* f. sp. *glycinea*	*P. megasperma*	Faris (1988)
P. sojae f. sp. *medicaginis*	?	*P. megasperma*	*P. sojae* f. sp. *glycinis*	*P. megasperma*	Faris *et al.* (1989)
P. medicaginis	*P. megasperma* 'DF'	*P. megasperma*	*P. sojae*	*P. megasperma*	Hansen & Maxwell (1990)

Fig 10.1. Nomenclatural chronology of *Phytophthora megasperma* from three hosts.

selection and evolution. Pathogenicity was the primary focus of two reviews of genetic variation in *Phytophthora* species (Erwin *et al.*, 1963; Erwin, 1983); quite fitting for a genus of 'plant killers'. *P. megasperma* was not mentioned in the first article, but was featured as a problem of nomenclature in the second. Some isolates of *P. megasperma* are pathogenic only on soybean, others on alfalfa or on clover. Still others cause no significant disease on these leguminous hosts, but attack a broad range of mostly woody plants. These host preferences, of great practical importance to plant pathologists, have been the basis for several taxonomic proposals, including a segregated species (*P. sojae*; Kaufmann & Gerdemann, 1958), a variety (*P. megasperma* var. *sojae*; Hildebrand, 1959), and *formae speciales* (*P. megasperma* f. sp. *glycinea*, *medicaginis* and *trifolii*; Kuan & Erwin, 1980; Pratt, 1981; *P. sojae* f. sp. *medicaginis* and *glycinis*; Faris *et al.*, 1989). Within *P. megasperma* f. sp. *glycinea*, pathogenic to soybean, numerous cultivar-specific races of the pathogen have been defined. This fungus has become a model system for research on the physiological and genetic basis of host specificity. The fungus is unique in many features, and new isolates are unambiguously identified by simple tests. There is, however, no consensus on its name (Fig. 10.1).

Communication has been facilitated by designation of *formae speciales*, but problems remain. Firstly, host specificity is not absolute (Wilcox & Mircetich, 1987; Hansen & Hamm, 1983; Hamm & Hansen, 1981; Faris *et al.*, 1983, 1989). Three distinct taxa, all referable to *P. megasperma*, are naturally pathogenic on alfalfa, for example (Hansen, 1988). Secondly, *formae speciales* as generally used implies that isolates differ only in pathogenicity, and these host groups differ in many other characters as well. A combination of these problems led Faris and others (Faris *et al.*, 1983, 1986; Faris, 1985, 1988) to refer all *Phytophthora* isolates from alfalfa to *P. megasperma* f. sp. *medicaginis*, even as they demonstrated that they fell into two fundamentally different groups. Their new proposal (Faris *et al.*, 1989) remedies this problem, although it creates others (see below). In the end, taxonomy based on pathogenicity suffers the same failings as any based upon a single character.

Isolates of *P. megasperma* also vary in morphological andphysiological properties, even among the zoospore progeny of a single parent (Hamm & Hansen, 1982). Variation in oogonium (and oospore) size has been central to the controversy (Duncan, Kennedy & Scott, chapter 9). Waterhouse & Blackwell (1954) first recognized large-spored (42-52 μm) and small-spored (32-46 μm) forms, and Waterhouse (1963) later made this the basis of her var. *megasperma* ($>45\,\mu$m) and var. *sojae* ($<40\,\mu$m). As noted earlier, the inconsistency between Hildebrand's (1959) emphasis on pathogenicity and Waterhouse's reliance on morphology has created much of the confusion in the *megasperma* complex. This was recently accentuated by the proposal (Faris *et al.*, 1989) to reinstate *P. sojae* Kauf. & Gerd, not on the basis of pathogenicity as originally proposed or as suggested by Hansen & Maxwell (1990, see below), but on the basis of spore size. *P. sojae* (*sensu* Faris *et al.*, 1989) thus apparently includes small-spored isolates from alfalfa, soybean, Douglas fir, fruit trees and other hosts. The confusing nomenclature of *P. megasperma* is summarized in Fig. 10.1.

Not surprisingly, isolates vary in other fundamental characteristics besides pathogenicity. Sansome & Brasier (1974) showed that chromosome number differed among several isolates of *P. megasperma*, and Hansen *et al.* (1986) illustrated four karyotypes. They suggested that chromosome number was proportional to spore size within the *megasperma* complex and presented a compilation and measurement of variation in other morphological, physiological, and protein characters. There was a high degree of correlation between the characters (see also Hansen & Hamm, 1983). Their interpretation was not of a single highly variable species, but of a series of six discrete, relatively uniform populations, rather tentatively labelled as 'emerging biological species'.

Hansen & Maxwell (1990) have taken the next logical step, segregating the taxonomic species *P. sojae* Kaufmann & Gerdemann (= *P. megasperma* f. sp. *glycinea*), *P. medicaginis* sp. nov. (= *P. megasperma* f. sp. *medicaginis* in the sense of Kuan & Erwin (1980) but not Faris (1988)), *P. trifolii* sp. nov. (= *P. megasperma* f. sp. *trifolii*), and *P. megasperma* Drechs including the BHR, DF, and AC protein groups of Hansen *et al.* (1986). This restricted nomenclature will be used in the remainder of this paper, unless otherwise noted.

Recent evidence for biological species

Field populations

As stated by Grant (1977), 'The maintenance of separate character combinations under conditions of sympatry is a natural test, and our best criterion, of the species status of the population systems involved'. By this criterion, *P. sojae*, *P. medicaginis*, and *P. megasperma* are distinct species. In the most recent documentation, Nygaard *et al.* (1989) isolated both *P. sojae* and *P. medicaginis* from three different fields in Wisconsin that had been in a soybean-alfalfa-corn crop rotation for several years. Isolates not only maintained the distinctive growth characters of the respective species, but exhibited characteristic isozyme patterns for three enzyme systems, without evidence of variation or intermediate banding patterns. Other combinations of two species of the *megasperma* complex from the same crop are reported from New York and Oregon (Hansen, 1988), Japan (Matsumoto & Sato, 1985), and Ontario, Canada (Barr, 1980). In each case the unique morphological and physiological characters of the species were confirmed by total protein electrophoresis (Hansen, 1988; Faris *et al.*, 1986).

Hybridization barriers

It is difficult to demonstrate fertility barriers directly in homothallic fungi such as those of Section *Megasperma*. Layton & Kuhn (1988) bypassed the conventional difficulties, however, by using protoplasts of drug resistant mutants of *P. sojae* and *P. medicaginis* which were mixed under conditions that promoted plasmogamy. Heterokaryons, as identified by complementary growth on media containing two antimetabolite,s were readily formed in intra-specific mixtures but never in inter-specific ones.

Kuhn & Allen (1989) have provided other, less direct, evidence of the barriers to plasmogamy that separate these species. The DF form of *P. megasperma* carries a group of double stranded RNAs which are found in all DF isolates, but not in other groups of the complex (see also Chapter 17).

Isozyme patterns

Protein analysis is proving to be an increasingly valuable taxonomic tool. Proteins are gene products; they are a direct reflection of the functional portion of the fungus genome. The protein composition of an isolate can be visualized as a series of discrete bands on a gel after separation by electrophoresis and staining with suitable reagents. The procedures are inexpensive and no more time consuming or complicated than preparation of material for microscopy. Although the species of the *P. megasperma* complex are readily distinguished by their total protein electrophoretic patterns (Irwin & Dale, 1982; Hansen *et al.*, 1986; Faris *et al.*, 1986), the banding arrangements (40 to 70 distinct protein bands are regularly revealed) are too complex for detailed comparisons or phylogenetic analysis and individual proteins cannot be identified readily on total protein gels. Isozyme analysis overcomes these limitations by revealing the isomeric forms of one enzyme system at a time.

Two research groups have done detailed isozyme studies of *P. megasperma*, with complementary results (Nygaard *et al.*, 1989; Chapter 12). *P. medicaginis*, *P. sojae*, and *P. trifolii* are distinguished readily from one another, from *P. megasperma*, and from other *Phytophthora* species. There was virtually no intraspecific variation in the legume pathogens. All 224 isolates of *P. sojae* and 41 of 45 isolates of *P. medicaginis* as examined by Nygaard *et al.* (1989) were identical. Isolates of *P.megasperma*, in contrast, were split into several isozyme groups. Too few isolates have been examined for certainty, but both research groups have suggested that the variation within *P. megasperma sensu stricta* corresponds to the isolate groups (BHR, DF, AC) defined by Hansen *et al.* (1986) using total proteins and other characters.

DNA analysis

Direct analysis of both nuclear and mitochondrial DNA is now possible with several techniques. The challenge is to match the very sensitive techniques with appropriate statistical analysis to find patterns and phylogenetic relationships in the wealth of detail that is available. Förster has used RFLPs in mitochondrial DNA as a measure of similarity between isolates (Förster *et al.*, 1987, 1989; Chapter 11). Several types of analysis are possible but basically the patterns of fragment lengths for different isolates are compared as a measure of differences in DNA sequences. The results support those obtained with other methods. *P. medicaginis* and *P. sojae* are distinguished readily from each other and from other species, including *P. megasperma*. The latter seems to be comprised of several subgroups. The phylogenetic distances indicated between *P. sojae* and *P. medicaginis* are at least as great as those between either species and other *Phytophthora* species. Förster *et al.* (1989) concluded that soybean and alfalfa isolates 'are genetically distinct groups of fungi and represent biological species in which the occurrence of genetic exchange appears to be unlikely'.

Chromosome electrophoresis of the P. megasperma *complex*

Phytophthora megasperma, even with segregation of the legume pathogens, remains a variable species. This might be a consequence of the variation in chromosome number noted first by Sansome & Brasier (1974) and later elaborated by Hansen *et al.* (1986). Four karyotypes were distinguished, with oogonium size proportional to chromosome number. It was hypothesized that the karyotypes represented a polyploid series rather than unrelated sets of chromosomes. This explained the overriding similarities between isolates of the species.

Pulsed field gel electrophoresis (Howlett, 1990) provides a test of this hypothesis. In this technique DNA fragmentation is minimized and individual chromosomes, or at least chromosome size molecules of DNA, are separated by size in the gel. We expect that isolates with different numbers of copies of the same chromosome set should produce similar banding patterns in the gel.

Howlett (1989) has published the first images of chromosome gels of *P. megasperma*. She described nine chromosomal DNA bands from isolate 63 (Karyotype I, AC protein group), previously reported to have n = 13 to 14 chromosomes (Hansen *et al.*, 1986). Several large DNA bands remained unresolved under all conditions tested. Isolate 53, with n = 22

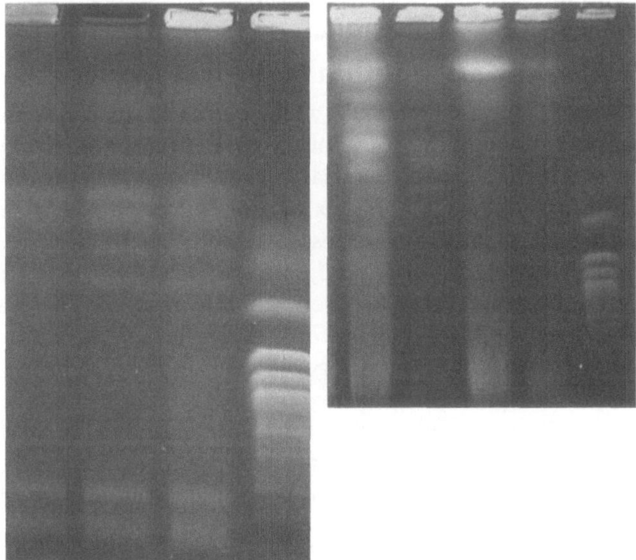

Figs. 10.2 & 10.3. Pulsed field gel electrophoresis of *Phytophthora* chromosomes. Fig. 10.2: Lanes 1, 2, & 3 represent three different protoplast preparations of isolate 63, *P. megasperma*. Lane 4 is *Saccharomyces cerevisiae*. Fig. 10.3. Lane 1 = Isolate number 78 of *P. megasperma*; Lane 2 = number 63; Lane 3 = number 62; Lane 4 = number 53; Lane 5 = *S. cerevisiae*.

to 24 (K III, BHR protein group) was electrophoretically very similar to isolate 63, with 9 DNA bands of about the same sizes (Howlett, 1989).

These preliminary results are being extended (E. Hansen, K. McCluskey & P. Hamm, unpublished). Banding patterns for different preparations of the same isolate are reproducible (Fig. 10.2). By comparison with DNA size standards, chromosomes range from about 2 Mb to 7 Mb, a similar size range to that observed in *Schizosaccharomyces pombe*. Sansome & Brasier (1974) were the first to note the remarkable range of chromosome sizes in this section. We have satisfactorily differentiated chromosomes of the isolates used by Howlett (1989, 1990), and also isolate 78 (K III, BHR protein group) with n = 23 to 27 chromosomes (Fig. 10.3). Our results are similar to Howlett's (1989); we find similar numbers of bands among the isolates. Other isolates examined to date show fewer bands, but resolution is still unsatisfactory, even with run times up to five days. There is a considerable amount of trial and error in establishing electrophoretic parameters (pulse duration and voltage) that give clear and unambiguous gel banding patterns.

Variation within biological species

Viewed as one complex species, *P. megasperma* in the old sense is undeniably variable. Even in the sense of the recent morphological definition of Faris *et al.* (1989), *P. sojae* is variable to the point of being unworkable. Viewed alternatively as a complex of isolated populations or discrete species, a different pattern emerges. *P. sojae sensu* Kaufmann & Gerdemann, *P. medicaginis*, *P. trifolii*, and perhaps the DF and AC groups of *P. megasperma*, are remarkable for their limited variation. These are sexual populations and not absolutely uniform in all characters, as shown by the common appearance of new pathogenic races of *P. sojae* or the variation among single zoospore progeny in several features (Hamm & Hansen, 1982). In character after character, however, including the very sensitive molecular measures, these species are very uniform. This is evidence in itself for the reproductive isolation of the populations, and for their status as species in the evolutionary sense.

The relative uniformity of these species is undoubtedly accentuated by their life styles. The species form homothallic, inbreeding populations. Asexual zoospores are responsible for most of the local spread and long distance transport is often via vegetative propagules carried in agricultural commerce. Host specialization is probably another force toward uniformity. Pathogenicity to a host species, in contrast to race/cultivar specificity, is evidently controlled by a number of interacting genes, and through epistatic effects may influence a significant portion of the genome. In a uniform host population the selection pressure for uniformity in the pathogen population must be strong. At the same time, the functional environment for the pathogen in a cultivated crop is relatively uniform, reducing the advantage of variability as a source of adaptation to changing environments.

Genetic uniformity may be a general feature of host-specialized species of *Phytophthora*. Species that fit this pattern include: *P. lateralis* (P. Hamm & E. Hansen, unpublished), *P. pseudotsugae* (Hamm & Hansen, 1983), *P. fragariae* on strawberries and its variant on raspberries (Chapter 9). The variability of the more omnivorous species stands in marked contrast, with *P. megasperma* as a case in point.

The variability of *P. megasperma* as defined here is in small part a consequence of its status as a residual species. Even with the DF isolates removed from consideration, however, the variability is striking. Mean oogonium diameters range from small-spored forms comparable in size to *P. sojae* to large-spored isolates similar to the type described by Drechsler (1931). The host range seems to be without limit. Chromosome number ranges from n = 12 to n = 30 or more. Isozyme and RFLP differences between isolates are especially striking. The possibility that this is still a grouping of unrelated forms must be considered seriously. There is other evidence, however, to support the contention that *P. megasperma*, too, is a biological species. Total protein profiles, while less uniform than *P. medicaginis* for example, still allow easy identification of a member of the species. Even isozyme and RFLP patterns suggest groupings of similar isolates. Spore size and chromosome number are probably both reflections of polyploidy. Perhaps most telling, isolates look similar in culture and behave similarly in the field.

P. megasperma is variable, but there are limits to that variation. Why is it different from the very uniform *P. sojae*, *P. medicaginis*, and *P.trifolii*? Perhaps this is a necessary adaption to successful parasitism on a wide range of unrelated hosts growing in disparate environments. Perhaps the tendency to heterothallism noted by Barr (1980) is involved. Perhaps this variation is the consequence of some cytological peculiarity that promotes chromosome duplications and deletions. Whatever the explanation, *P. megasperma* remains a most interesting species.

A species concept for *Phytophthora*

One reason that our taxonomy is unstable is the recent and rapid progress in *Phytophthora* genetics. The demonstration of diploidy in *Phytophthora* (Sansome, 1965) opened a new chapter that is just now being exploited. New knowledge and new tools are at last making a phylogenetic taxonomy of *Phytophthora* a realistic objective. Population genetics is not new, but its application to fungal taxonomy is. It has created a new definition of species: biological species are natural populations of organisms. They are groups of interbreeding populations, reproductively isolated from other such groups (Mayr, 1969). The definition was coined for mammals. A more general phrasing would be: species are populations of organisms that represent a common gene pool, reproductively isolated from other such populations by genetic barriers to gene flow. While the definition is short and the standards are reasonably clear, application of the concept to the real world is more complicated (Brasier, 1983). To be useful and accepted, a taxonomic system should satisfy at least two criteria: it should enhance our study of evolutionary biology, and aid communication between people interested in the organisms. The biological species concept clearly meets the first criterion and, if applied with a healthy measure of common sense, it will be an improvement over the current system for communication as well.

The objections to applying this standard to fungal taxonomy are usually phrased in terms of its impracticality, but in reality reflect the normal resistance to change expected from any traditional community. The challenges can be answered in turn:

Without the type specimen and the International Code of Nomenclature, chaos will overwhelm us! Answer: the two systems are not mutually exclusive. A designated specimen (and culture) as representative of the population is still a desirable reference. The Code prescribes rules for nomenclature, but does not provide a species definition for fungi; there is no conflict. With common sense the disruptions to established nomenclature will be minimal in most groups, because taxonomists have always tried to group fungi by natural relationships.

The new species will not be based on morphological criteria! Answer: true in some cases but, in most applications to date, morphological differences accompany genetic and physiological differences. Tradition, not the Code, requires morphological criteria for species. What is more important than morphology is practicality, especially in a genus noted for its lack of morphological features. While new molecular approaches may be unfamiliar to some taxonomists, many of these techniques require no more time or skill than preparing specimens for the microscope.

The biological species definition is applicable only to 'normal' biparental sexual systems! Answer: true again, in a narrow, classical sense, but Mayr (1969), who coined the definition of biological species, insisted that the key features were barriers to gene flow and the emphasis on populations. These are features common to all organisms. Once more, common sense rather than rigid adherence to narrow definitions is necessary. Botanists face many of the same challenges from unconventional reproductive systems, and they have, for the most part, accommodated the biological species concept.

We have insufficient data on population structure and genetic isolation to classify most fungi according to these criteria! Answer: we'll never have them if we don't start sometime. In many cases, including the homothallic species, we do not yet have the genetic tools to completely characterize species, but this should not prevent progress toward the goal. There is always one more test to run. In the end it still comes down to the personal judgement of the taxonomist, but hopefully that judgement will be framed by a widely accepted standard of species limits.

The biological species concept does put an extra burden on the taxonomist. Nomenclature and type can continue to follow the Code, but the species description must be based on a reasonable sample of the population and supported by reasonable genetic evidence of reproductive isolation. In practice, the latter requirement will usually be met in *Phytophthora* by evidence for a lack of intermediate forms in the two populations in question.

A consensus among taxonomists to accept the biological species norm for *Phytophthora* will not end the need for common sense and judgement in naming new species: the deciding factor in erecting a new species should be the practicality of the proposal. Does the population have a unique behaviour (pathogenicity or ecological niche)? Will communication be enhanced by the proposal? If we do not yet understand the behavioural significance of the observed population structure, then some form of subspecific recognition (Brasier & Rayner, 1987) without disruption of established species names is more appropriate.

Phytophthora megasperma provides a good example. Evidence that *P. sojae*, *P. medicaginis*, and *P. trifolii* represent isolated populations is convincing but certainly not complete. Specific recognition is justified by their unique pathogenic behaviour and as an aid to communication. All attempts to accommodate these populations within the established *P. megasperma* have led to confusion.

There is also evidence of morphotypes and correlated karyotypes within the population of *P. megasperma, sensu stricta*. Not only is the evidence that these represent isolated populations less compelling, but so far no behavioural differences have been associated consistently with the groups. Thus, continued non-hierarchical subspecific recognition (Brasier & Rayner, 1987) seems most appropriate.

There is indeed much to be learned about population structure and speciation in *Phytophthora* (Brasier, 1983; Hansen, 1987, 1988). Most species lack efficient aerial transport, but agricultural commerce has provided a world-wide distribution through vegetative propagules (Zentmyer, 1977). Genets and their ramets (Brasier & Rayner, 1987) commonly appear on opposite continents, establishing new populations. Mating types of heterothallic species are separated and assume an independent existence (Old, Moran & Bell, 1984). Selfing induced by a variety of stimuli may be as important a means of sexual reproduction as mating between A1 and A2 types (Brasier, 1972; Shaw, 1983). The importance, if any, of anastomosis and heterokaryosis or vegetative incompatibility is largely unevaluated. Ongoing investigations of sexual and presumed asexual populations of *P. infestans* (Chapter 13; Chapter 18) may provide a valuable model for genetic structure of populations under various conditions. Homothallic species are still more problematical. It is not clear that there is any mechanism for gene flow between individuals, yet in many cases, including *P. sojae*, *P. trifolii*, and *P. medicaginis*, species have maintained a remarkable uniformity. Among the possible explanations is a recent origin for the species mediated by inbreeding and homozygosity. Alternatively, some unknown mechanism may enforce a stasis on the genome (Slatkin, 1985), perhaps involving host specific pathogenicity and the consequent uniformity of environment and selection pressure.

Despite these problems, we can recognize *Phytophthora* species and the task is easier when we try to identify members of a population (biological species) in a particular ecological context, than when we are faced with matching an unknown isolate with a 'type' specimen or description. Species limits are usually discrete when we compare a full range of characters for populations of isolates. Where questions persist, we should suspect ongoing speciation, or ill-defined populations. The test of sympatry, where appropriate, is perhaps the most direct test of species status for two populations. Characteristics that represent a substantial portion of the genome are particularly helpful. These include protein electrophoresis, RFLPs, and behavioural features such as pathogenicity.

Achieving a natural classification has long been a goal of *Phytophthora* workers. Leonian (1934) stressed the 'variability of living things' and railed against the type concept. He anticipated the emphasis on populations with his definition of a species: 'a fluctuating group of organisms more or less loosely bound in a flexible orbit'. Waterhouse has taken a more conservative view. She has defended the formalities of the type method and the Rules of Botanical Nomenclature, but has always encouraged the testing of new species concepts and criteria (Waterhouse, 1970: Waterhouse *et al.*, 1983). Her *Key to the Species of Phytophthora* (1963) has become the standard reference for the genus, but it was never intended as a natural classification. That goal awaits the accumulation of knowledge and the acceptance of ideas.

The benefits of a biological species concept for *Phytophthora* taxonomy are substantial. It establishes an objective standard; one set by nature and evolution. The definition is flexible, subject to the various reproductive systems encountered in fungi. It recognizes speciation as an ongoing process, perhaps especially important in a genus of agricultural pathogens (Hansen, 1987, 1988; Brasier, 1987). It can readily accommodate a system of subspecific nomenclature as proposed by Brasier & Rayner (1987), which is essential in groups undergoing active speciation and when insufficient information is available to determine the species status of isolates. Finally, the biological species definition is practical because it embraces whole populations rather than individuals, and it emphasizes the behavioural differences between species that are so important to pathologists and ecologists.

References

Barr, D. J. S. (1980). Heterothallic-like reaction in the large-oospore form of *Phytophthora megasperma*. *Canadian Journal of Plant Pathology* **2**, 116-118.

Brasier, C. M. (1972). Observations on the sexual mechanisms in *Phytophthora palmivora* and related species. *Transactions of the British Mycological Society* **58**, 237-251.

Brasier, C. M. (1983). Problems and prospects in *Phytophthora* research. In *Phytophthora: Its Biology, Taxonomy, Ecology and Pathology* (ed. D. C. Erwin, S. Bartnicki-Garcia & P. H. Tsao), pp. 351-364. American Phytopathological Society: St. Paul, Minnesota.

Brasier, C. M. (1987). The dynamics of fungal speciation. In *Evolutionary Biology of the Fungi* (ed. A. D. M. Rayner, C. M. Brasier & D. Moore), pp. 231-260. Cambridge University Press.

Brasier, C. M. & Rayner, A. D. M. (1987). Whither terminology below the species level in fungi? In *Evolutionary Biology of the Fungi* (ed. A. D. M. Rayner, C. M. Brasier & D. Moore), pp. 379-388. Cambridge University Press.

Drechsler, C. A. (1931). A crown rot of hollyhock caused by *Phytophthora megasperma* n. sp. *Journal of the Washington Academy of Science* **21**, 513-526.

Erwin, D. C. (1965). Reclassification of the causal agent of root rot of alfalfa from *Phytophthora cryptogea* to *P. megasperma*. *Phytopathology* **55**, 1139-1143.

Erwin, D. C. (1983). Variability within and among species of *Phytophthora*. In *Phytophthora: Its Biology, Taxonomy, Ecology and Pathology* (ed. D. C. Erwin, S. Bartnicki-Garcia & P. H. Tsao), pp. 149-165. American Phytopathological Society: St. Paul, Minnesota.

Erwin, D. C., Zentmyer, G. A., Galindo, J. & Niederhauser, J. S. (1963). Variation in the genus *Phytophthora*. *Annual Review of Phytopathology* **1**, 375-396.

Faris, M. A. (1985). Variability and interaction between alfalfa cultivars and isolates of *Phytophthora megasperma*. *Phytopathology* **75**, 390-394.

Faris, M. A. (1988). Variability in growth of *Phytophthora megasperma* f. sp. *medicaginis* in relation to temperature. *Canadian Journal of Plant Pathology* **10**, 99-104.

Faris, M. A., Sabo, F. E. & Barr, D. J. S. (1983). Studies on *Phytophthora megasperma* isolates with different levels of pathogenicity on alfalfa cultivars. *Canadian Journal of Plant Pathology* **5**, 29-33.

Faris, M. A., Sabo, F. E., Barr, D. J. S. & Lin, C. S. (1989). The systematics of *Phytophthora sojae* and *P. megasperma*. *Canadian Journal of Botany* **67**, 1442-1447.

Faris, M. A., Sabo, F. E. & Cloutier, Y. (1986). Intraspecific variation in gel electrophoresis patterns of soluble mycelial proteins of *Phytophthora megasperma* isolated from alfalfa. *Canadian Journal of Botany* **64**, 262-265.

Förster, H., Kinscherf, T. G., Leong, S. A. & Maxwell, D. P. (1987). Molecular analysis of the mitochondrial genome of *Phytophthora*. *Current Genetics* **12**, 215-218.

Förster, H., Kinscherf, T. G., Leong, S. A. & Maxwell, D. P. (1989). Restriction fragment length polymorphisms of the mitochondrial DNA of *Phytophthora megasperma* isolated from soybean, alfalfa and fruit trees. *Canadian Journal of Botany* **67**, 529-537.

Grant, V. (1977). *Organismic Evolution*. Freeman & Co.: San Francisco.

Hamm, P. B. & Hansen, E. M. (1981). Host specificity of *Phytophthora megasperma* from Douglas-fir, soybean, and alfalfa. *Phytopathology* **71**, 65-68.

Hamm, P. B. & Hansen, E. M. (1982). Single-spore isolate variation: the effect on varietal designation in *Phytophthora megasperma*. *Canadian Journal of Botany* **60**, 2931-2938.

Hamm, P. B. & Hansen, E. M. (1983). *Phytophthora pseudotsugae*: a new species causing root rot of Douglas-fir. *Canadian Journal of Botany* **61**, 2626-2631.

Hansen, E. M. (1987). Speciation in *Phytophthora*: Evidence from the *Phytophthora megasperma* complex. In *Evolutionary Biology of the Fungi* (ed. A. D. M. Rayner, C. M. Brasier & D. Moore), pp. 325-338. Cambridge University Press.

Hansen, E. M. (1988). Speciation in plant pathogenic fungi. *Canadian Journal of Plant Pathology* **9**, 403-410.

Hansen, E. M. & Hamm, P. B. (1983). Morphological differentiation of host-specialized groups of *Phytophthora megasperma*. *Phytopathology* **73**, 129-134.

Hansen, E. M. & Maxwell, D. P. (1990). Species of the *Phytophthora megasperma* complex. *Mycologia*, in press.

Hansen, E. M., Brasier, C. M., Shaw, D. S. & Hamm, P. B. (1986). The taxonomic structure of *Phytophthora megasperma*: evidence for emerging biological species groups. *Transactions of the British Mycological Society* **87**, 557-573.

Hildebrand, A. A. (1959). A root and stalk rot of soybean caused by *Phytophthora megasperma* Drechsler var. *sojae* var. nov. *Canadian Journal of Botany* **37**, 927-957.

Howlett, B. J. (1989). An electrophoretic karyotype for *Phytophthora megasperma*. *Experimental Mycology* **13**, 199-202.

Howlett, B. J. (1990). Pulsed field gel electrophoresis as a method for examining phylogenetic relationships between organisms: its application to the genus *Phytophthora*. *Australian Systematic Botany* **3**, 75-80.

Irwin, J. A. G. & Dale, J. L. (1982). Relationships between *Phytophthora megasperma* isolates from chickpea, lucerne, and soybean. *Australian Journal of Botany* **30**, 199-210.

Kaufmann, J. J. & Gerdemann, J. W. (1958). Root rot of soybeans caused by *Phytophthora sojae* n. sp. *Phytopathology* **48**, 201-208.

Kuan, T.-L. & Erwin, D. C. (1980). Formae speciales differentiation of *Phytophthora megasperma* isolates from soybean and alfalfa. *Phytopathology* **70**, 333-338.

Kuhn, D. W. & Allen, B. (1989). Double-stranded RNA of *Phytophthora megasperma*. *Phytopathology* **79**, in press.

Layton, A. C. & Kuhn, D. N. (1988). Heterokaryon formation by protoplast fusion of drug-resistant mutants in *Phytophthora megasperma* f. sp. *glycinea*. *Experimental Mycology* **12**, 180-194.

Leonian, L. H. (1934). Identification of *Phytophthora* species. *West Virginia Agricultural Experiment Station Bulletin* 262.

Matsumoto, N. & Sato, T. (1985). Host non-specific *Phytophthora megasperma* with small oogonia from alfalfa-field soil. *Research Bulletin, Hokkaido National Agricultural Experiment Station* **143**, 75-84.

Mayr, E. (1969). *Principles of Systematic Zoology*. McGraw Hill: New York.

Nygaard, S. L., Elliott, C. K., Cannon, S. J. & Maxwell, D. P. (1989). Isozyme variability among isolates of *Phytophthora megasperma*. *Phytopathology* **79**, 773-780.

Old, K. M., Moran, G. F. & Bell, J. C. (1984). Isozyme variability among isolates of *Phytophthora cinnamomi* from Australia and Papua New Guinea. *Canadian Journal of Botany* **62**, 2016-2022.

Pratt, R. G. (1981). Morphology, pathogenicity and host range of *Phytophthora megasperma*, *P. erythroseptica*, and *P. parasitica* from arrowleaf clover. *Phytopathology* **71**, 276-282.

Sansome, E. R. (1965). Meiosis in diploid and polyploid sex organs of *Phytophthora* and *Achlya*. *Cytologia* **30**, 103-117.

Sansome, E. R & Brasier, C. M. (1974). Polyploidy associated with varietal differentiation in the *megasperma* complex of *Phytophthora*. *Transactions of the British Mycological Society* **63**, 461-467.

Shaw, D. S. (1983). The cytogenetics and genetics of *Phytophthora*. In *Phytophthora: Its Biology, Taxonomy, Ecology and Pathology* (ed. D. C. Erwin, S. Bartnicki-Garcia & P. H. Tsao), pp. 81-94. American Phytopathological Society: St. Paul, Minnesota.

Slatkin, M. (1985). Gene flow in natural populations. *Annual Review of Ecology and Systematics* **16**, 393-430.

Tompkins, C. M., Tucker, C. M. & Gardner, M. W. (1936). *Phytophthora* root rot of cauliflower. *Journal of Agricultural Research* **53**, 685-692.

Waterhouse, G. M. (1963). Key to the species of *Phytophthora* de Bary. *Mycological Papers* **92**.. Commonwealth Mycological Institute: Kew, U.K.

Waterhouse, G. M. (1970). The genus *Phytophthora* de Bary. *Mycological Papers*, **122**. Commonwealth Mycological Institute: Kew, U.K.

Waterhouse, G. M. & Blackwell, E. M. (1954). Key to the species of *Phytophthora* recorded in the British Isles. *Mycological Papers* **57**. Commonwealth Mycological Institute: Kew, U.K.

Waterhouse, G. M., Newhook, F. J. & Stamps, D. J. (1983). Present criteria for classification of *Phytophthora*. In *Phytophthora: Its Biology, Taxonomy, Ecology and Pathology* (ed. D. C. Erwin, S. Bartnicki-Garcia & P. H. Tsao), pp. 139-147. American Phytopathological Society: St. Paul, Minnesota.

Wilcox, W. F. & Mircetich, S. M. (1987). Lack of host specificity among isolates of *Phytophthora megasperma*. *Phytopathology* **77**, 1132-1137.

Zentmyer, E. A. (1977). Origin of *Phytophthora cinnamomi*: evidence that it is not an indigenous fungus in the Americas. *Phytopathology* **67**, 1373-1377.

Chapter 11

Approaches to the taxonomy of *Phytophthora* using polymorphisms in mitochondrial and nuclear DNA

Helga Förster & Michael D. Coffey

Current controversies in *Phytophthora* taxonomy (Chapters 8 and 10) are unlikely to be solved by traditional classification methods that are based on morphology. Few definitive characters are available and a high degree of morphological variability may be found within some species. Molecular methods can provide important new information to permit characterization of species. Analysis of total protein patterns (Kaosiri & Zentmyer, 1980; Erselius & Shaw, 1982; Erselius & de Vallavieille, 1984; de Vallavieille & Erselius, 1984) may reveal differences between isolates. However, the quantitative analysis of protein gels is very difficult since so many bands are present. Minor bands which could be of critical taxonomic value may be obscured by the presence of intense major bands. Also, biosynthesis of some proteins may be under environmental or developmental control and patterns may not remain constant.

DNA analysis, on the other hand, can provide a large number of characters that are not affected by changing environmental conditions. Through numerical data analysis, genetic differences can be quantified. Analysis of mitochondrial DNA (mtDNA) restriction fragment patterns provides a new and powerful tool in taxonomy and has been used for the classification of a variety of organisms including animals and 'higher' fungi (Birley & Croft, 1986; Taylor, 1986). It is a relatively small molecule, with sizes between 36·2 kb and 45·3 kb reported for *Phytophthora* (Förster *et al.*, 1987; Klimczak & Prell, 1984) and consequently the complete genome can be studied. In addition, there is no evidence for recombination between mitochondrial genomes. For example, protoplast fusion products between two strains of *P. capsici* (Lucas *et al.*, 1990) contained only the parental mtDNA types (Förster, unpublished). Furthermore single oospore progeny of crosses between different strains of *P. parasitica* (syn. *nicotianae*) only showed the parental mtDNA patterns whereas recombination of the nuclear genomes had occurred (Förster & Coffey, 1991). Consequently, relatedness is better estimated in comparisons of mtDNAs than in comparisons of nuclear DNAs.

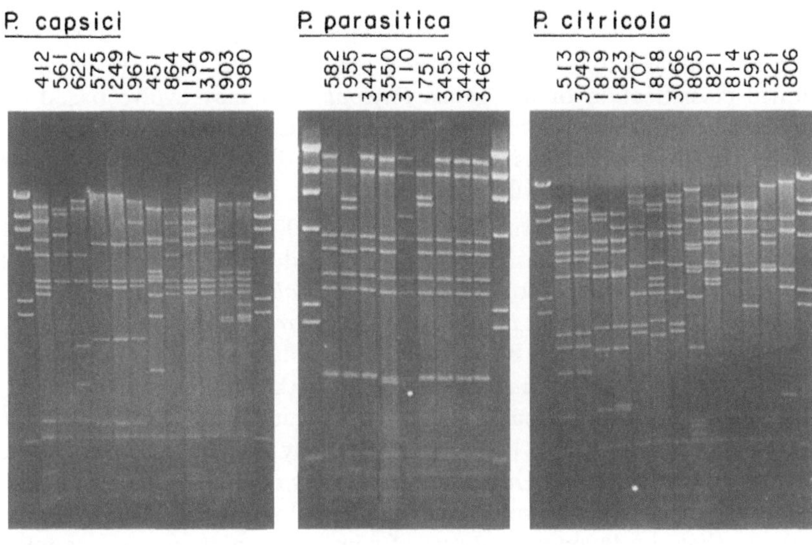

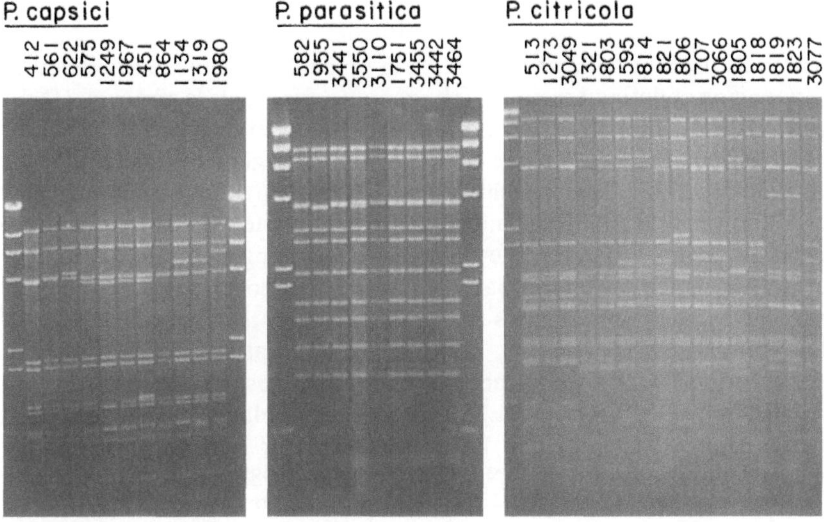

Fig. 11.1. Agarose gel electrophoresis of *P. capsici*, *P. parasitica*, and *P. citricola* mtDNAs digested with *Msp*I or *Hin*dIII. *Hin*dIII digested lambda DNA was used as a size marker. (From Förster, Oudemans & Coffey, 1990a).

The current classification system for *Phytophthora* (Newhook, Waterhouse & Stamps, 1978; Waterhouse, 1963) was designed for the identification of field isolates. Morphological groupings I-VI of species were not claimed to represent natural groupings. In the future a more natural classification of the genus *Phytophthora* may evolve based on molecular data which relate to the genetic variation within and between species. In this chapter, results which are based on these new molecular approaches to taxonomy will be presented.

To date, mtDNA diversity has been studied within eleven species of *Phytophthora*. Some of these species are considered to be relatively well characterized by morphological criteria (*P. palmivora*, *P. parasitica*, *P. cactorum*, *P. megakarya*, *P. infestans*, *P. citrophthora*, *P. capsici*, *P. citricola*) whilst others may be regarded as 'problem' species (*P. cryptogea*, *P. drechsleri*, *P. megasperma*). Isolates used in the studies were of worldwide origin and from a broad range of host plants. Most isolates used for mtDNA studies have also been investigated by isozyme analysis (Mills, Förster & Coffey, 1991; Oudemans & Coffey, 1991; Chapter 12). The results are discussed with reference to morphological data. Our evidence suggests that the taxonomic status of some *Phytophthora* species must change. A new species concept reflecting natural relationships should be accepted.

Methodology

Methods used to compare mtDNAs in taxonomic studies have involved sequencing of defined regions of the genome, for example ribosomal RNA genes (Gray, Sankoff & Cedergren, 1984; Köchel & Küntzel, 1982), or comparison of restriction site cleavage maps and gene orders (e.g. Bruns & Palmer, 1989; Taylor, Smolich & May, 1986). Such methods may eventually reveal additional information on possible mechanisms of molecular evolution. However, they are time-consuming and thus the number of isolates which can be compared is limited. Moreover, sequence analysis may involve genomic regions that evolve differently from other parts of the genome. In contrast, by comparison of total restriction fragment patterns generated by a number of endonucleases, restriction fragment length polymorphisms (RFLPs) from many isolates can be analyzed rather rapidly. Isolates can be prescreened for polymorphisms using mini-preparations of total DNA (Förster *et al.*, 1988) and can be put into groups of identical restriction fragment patterns. The groups can then be studied in more detail by purification of mtDNAs by caesium chloride gradient centrifugation (Förster *et al.*, 1987) and fragment patterns of the gels can be scored for common bands between isolates (Förster *et al.*, 1988; Fig. 11.1). If very different mtDNAs have to be compared and few bands appear to be in common between isolates, it may be very difficult,

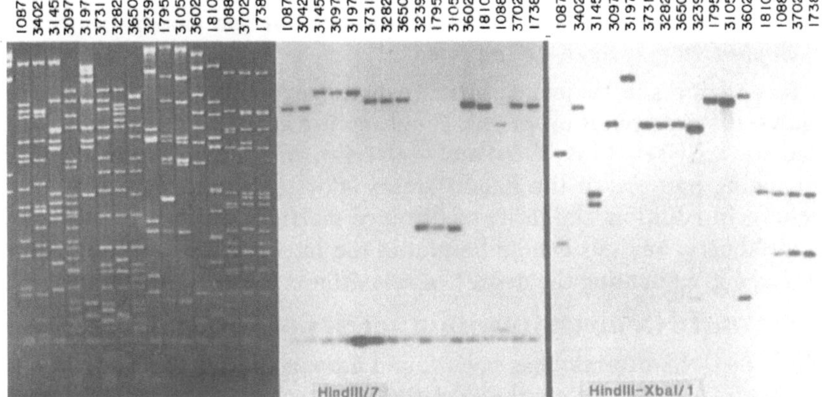

Fig. 11.2. Analysis of restriction fragment length polymorphisms in mtDNAs using hybridization techniques. Left: agarose gel electrophoresis of *Msp*I digested mtDNAs from various isolates of *P. cryptogea* and *P. drechsleri*. *Hin*dIII digested lambda DNA was used as a size marker (first lane). After Southern blotting, the membrane-bound DNA was hybridized with cloned mtDNA fragments *Hin*dIII/7 (middle) or *Hin*dIII-*Xba*I/I (right) from *P. megasperma* f. sp. *glycinea*. Polymorphisms in homologous DNA fragments are detected and scored more accurately than in the complex banding patterns of the total mtDNAs (From Mills *et al*., 1991).

or even impossible, to score the gels directly. In such cases a different approach can be used that is based on the hybridization with individual cloned mtDNA fragments (Mills *et al*., 1991). In fact, autoradiograms are much more easily scored than gels with complex banding patterns (Fig. 11.2).

Quantitative analysis of restriction fragment patterns is prone to several kinds of error: scoring of nonhomologous fragments of the same size as identical; lack of detection of small insertions or deletions, especially in larger DNA fragments; overestimation of divergence, if length mutations are scored several times in multiple digests; lack of differentiation of different mechanisms of mutation (site mutations, length mutations, rearrangements).

Furthermore, in comparisons of more distantly related isolates, fragment differences may be due to several mutational events. In general, mutation rates are known to be different in different regions of the genome. However, by using a number of restriction endonucleases and scoring many fragments, errors should be minimized and the relationships between isolates may then be considered to be good estimates. In addition, it is anticipated that the mutation mechanisms among isolates of *Phytophthora* may likely be the same.

Restriction site, sequence or restriction fragment length data can be analyzed by computer programs. Programs are available for phenetic and cladistic analyses (Fink, 1986) and both types may result in near identical branching patterns of the dendrograms (Förster, Oudemans & Coffey, 1990a). In addition, similarity or distance matrices which are calculated in a phenetic analysis can be helpful in the interpretation of the phenograms, e.g. estimating the degree of relatedness between isolates.

mtDNA diversity within species of *Phytophthora*

Much mtDNA diversity has been found among species of *Phytophthora*. However, it has to be emphasized that the degree of variation found is based on the researcher's selection of isolates. Conceivably, there may be other distinct isolates in additional geographical locations from where no cultures could be studied.

Our results are shown in phenograms based on mtDNA polymorphisms for isolates of *P. palmivora*, *P. parasitica*, *P. cactorum*, *P. megakarya*, *P. citrophthora*, *P. capsici*, *P. citricola*, *P. cryptogea* and *P. drechsleri* in Figs. 11.3–10. The phenograms were constructed with the NTSYS computer program using unweighted pair grouping with arithmetic averaging from Dice similarity coefficients. The analyses are based on 45–75 restriction fragments for each isolate. Details about the fungal isolates can be found in the tables below each figure (except for *P. megakarya* whose isolates are detailed in the figure legend).

Isolates of *P. palmivora* were shown to be quite uniform (Förster *et al.*, 1990a; Fig. 11.3, Table 11.1). Although different mtDNAs were found, they all shared a high degree of similarity. The mtDNA from an isolate of *P. arecae* was very similar to those of isolates of *P. palmivora*; the validity of the former as a separate species can be questioned.

Little mtDNA variation was also evident among isolates of *P. parasitica* (Förster *et al.*, 1990a; Fig. 11.4, Table 11.2) and the two proposed varieties, var. *parasitica* and var. *nicotianae* (Waterhouse, 1963) could not be differentiated. Comparative morphological studies by Tsao & Sisemore (1978) and Ho & Jong (1989) led to the same conclusion.

P. cactorum, too, was very uniform at the mtDNA level, although isolates came from diverse geographical locations and host plants (Fig. 11.5, Table 11.3). Most isolates were found to be identical; a few isolates

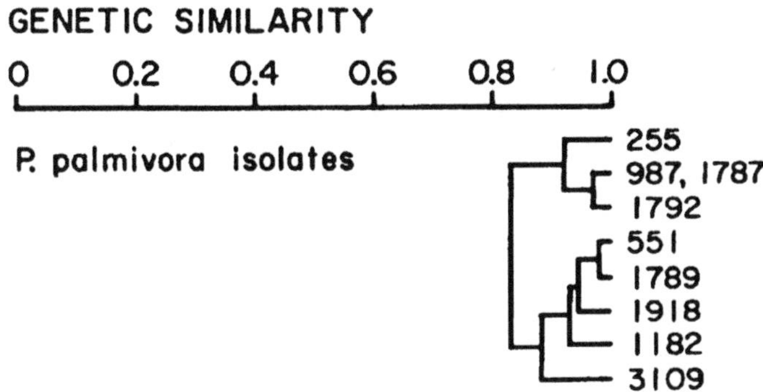

Fig. 11.3. Phenogram for *P. palmivora* isolates (From Förster, Oudemans & Coffey, 1990a).

Table 11.1. Isolates of *P. palmivora* and *P. arecae* used in the study.

Isolate	Host	Geographical origin
P255	*Theobroma cacao*	Costa Rica
P551	*Theobroma cacao*	Jamaica
P987	*Theobroma cacao*	Nigeria
P1182	*Morrenia odorata*	Florida
P1787	*Carica papaya*	Hawaii
P1789	*Dendrobium* sp.	Hawaii
P1792	*Cattleya* sp.	Hawaii
P1918	*Vanilla planifolia*	French Polynesia
P3109*	*Chamaedora* hybr.	Florida

**P. arecae*.

were polymorphic. However, all mtDNAs of *P. cactorum* shared a high degree of similarity. A number of the polymorphic isolates came from strawberry; some of these (isolates P6186, P6187) had characteristically pigmented oospores (Oudemans & Coffey, 1991). Otherwise, *P. cactorum* isolates were very similar in morphology (Oudemans & Coffey, 1991).

Little mtDNA variation was also evident among isolates of *P. infestans* (Chapters 17 and 18). Two quite similar mtDNA types were found which showed length mutations and rearrangements in a specific genomic region; other mutations involved single restriction sites.

In contrast to these species, isolates of *P. megakarya* from cacao could be separated into two distinct mtDNA subgroups (Förster *et al.*, 1990a;

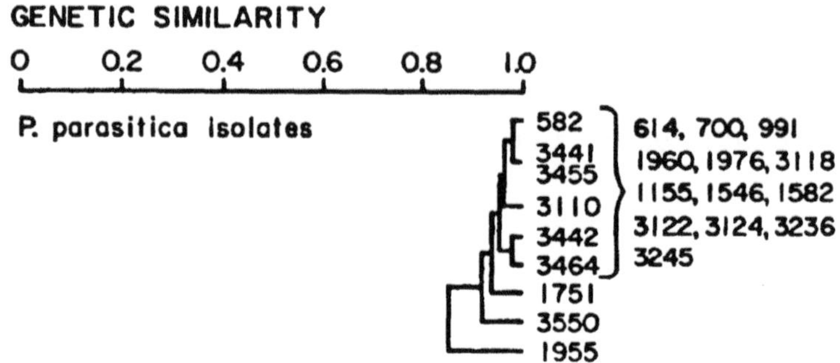

Fig. 11.4. Phenogram for *P. parasitica* isolates (From Förster, Oudemans & Coffey, 1990a).

Table 11.2. Isolates of *P. parasitica* (varieties '*nicotianae*' and '*parasitica*') used in the study.

Isolate	Host	Origin	Isolate	Host	Origin
var. *nicotianae*					
P582	*Nicotiana tabacum*	Kentucky	P1955	*Nicotiana tabacum*	S. Africa
P3110	*Nicotiana tabacum*	Kentucky	P3441	soil	Taiwan
P3550	*Dianthus* sp.	France			
var. *parasitica*					
P3442	*Fragaria vesca*	Japan	P3464	*Citrus* sp.	Thailand
variety unknown					
P614	*Theobroma cacao*	Nigeria	P700	*Lycopersicon esculentum*	Ponape
P991	*Citrus* sp.	California	P1155	*Citrus* sp.	Brazil
P1546	*Citrus* sp.	California	P1582	*Citrus* sp.	California
P1751	*Nicotiana tabacum*	Australia	P1960	*Nicotiana tabacum*	S. Africa
P1976	*L. esculentum*	California	P3118	*L. esculentum*	Brazil
P3122	*Citrus limon*	Brazil	P3124	*L. esculentum*	Brazil
P3236	*Citrus medica*	China	P3245	*Solanum melongena*	China
P3455	*Piper nigrum*	Malaysia			

Fig. 11.6). One group was formed by isolates from Nigeria, the other by those from Cameroon. The subgroups did not appear to be very closely related to each other. The type cultures of this species come from Nigeria and Cameroon, respectively, and no morphological differences have been reported (Brasier & Griffin, 1979). *P. megakarya* has only been found in West Africa; however, the host plant cacao is not native to this area. This

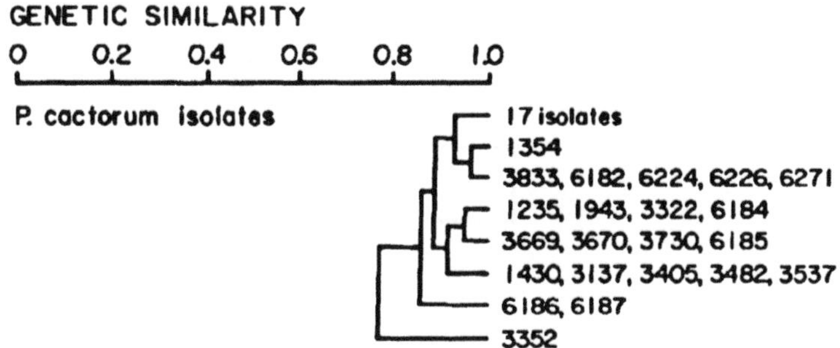

Fig. 11.5. Phenogram for *P. cactorum* isolates. The 17 identical isolates were: P472, P714, P1103, P1598, P1721, P1722, P1723, P1724, P3072, P3074, P3075, P3083, P3084, P3128, P3406, P3468, and P3634.

Table 11.3. Isolates of *P. cactorum* used in the study.

Isolate	Host	Origin	Isolate	Host	Origin
P472	*Pyrus communis*	California	P714	*Syringa* sp.	California
P1103	*Fagus sylvatica*	New York	P1235	*Raphiolepsis indica*	California
P1354	*Juglans* sp.	California	P1430	*Fragaria* sp.	W. Germany
P1598	*Malus* sp.	Netherlands	P1721	*Malus* sp.	New Zealand
P1722	*Malus* sp.	Maine	P1723	*Malus* sp.	S. Africa
P1724	*Malus* sp.	Poland	P1943	*Panax quincefolium*	Wisconsin
P3072	*Vitis vinifera*	S. Africa	P3074	*Cussonia paniculata*	S. Africa
P3075	*Malus* sp.	S. Africa	P3083	*Gypsophila* sp.	Mass. USA
P3084	*Trachymene caerulea*	Mass. USA	P3128	*Malus* sp.	
P3137	*Fragaria* sp.	W. Germany	P3322	cactus	
P3352	*Rubus idaeus*	Great Britain	P3405	*Fragaria* sp.	France
P3406	*Malus* sp.	W. Germany	P3468	*Juglans nigra*	France
P3482	*Aesculus hippocastanum*	Great Britain	P3537	*Fragaria vesca*	Great Britain
P3634	*Pyrus communis*	Canada	P3669	*Prunus* sp.	Michigan
P3670	*Malus* sp.	Michigan	P3730	*Malus* sp.	Michigan
P3833	*Malus* sp.	Arizona	P6182	*Rubus idaeus*	New York
P6184	*Prunus* sp.	New York	P6185	*Malus* sp.	New York
P6186	*Fragaria* sp.	New York	P6187	*Fragaria* sp.	New York
P6224	*Prunus armeniaca*		P6226	*Lilium candidum*	
P6271	*Prunus persica*	Oregon			

led Förster, Oudemans & Coffey (1990a) to speculate that the two *P. megakarya* subgroups evolved separately by founder events on unknown native plants.

Isolates of *P. citrophthora*, too, could be put into two distinct groups according to their mtDNA restriction fragment patterns which appear to be unrelated to each other (Förster *et al.*, 1990a; Fig. 11.7, Table 11.4).

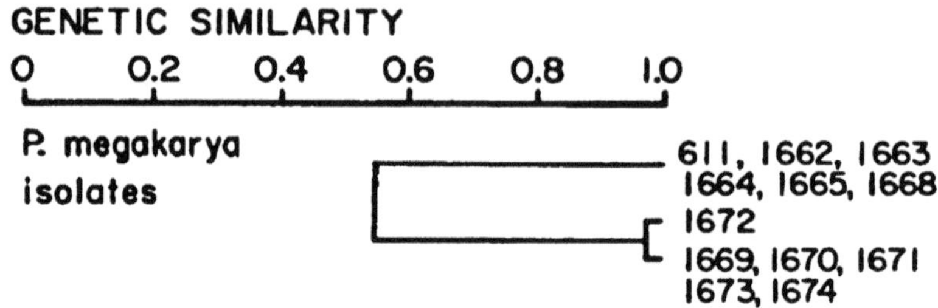

Fig. 11.6. Phenogram for *P. megakarya* isolates. The host for all isolates was *Theobroma cacao*. Isolates P611, P1662, P1663, P1664, P1665 and P1668 came from Nigeria, isolates P1669, P1670, P1671, P1672, P1673, and P1674 came from Cameroon (From Förster, Oudemans & Coffey, 1990a).

One group represents cacao isolates from Brazil, the only area to date where *P. citrophthora* has been found on that host. The other group is formed by citrus isolates from various locations worldwide. An additional kiwi group, clustering with the citrus isolates, could be differentiated by isozyme analysis (Chapter 12).

A higher degree of mtDNA diversity was found among isolates of *P. capsici* (Förster *et al.*, 1990a; Fig. 11.8, Table 11.5). A number of distinct and dissimilar types of mtDNA were present. No obvious correlation between pattern types and host or geographical origin was apparent. However, when many more isolates were included in an isozyme analysis, distinct subgroups could be differentiated (Chapter 12). MtDNA profiles of isolates formerly classified as *P. palmivora* MF4 were grouped with isolates of *P. capsici* from *Capsicum*, thus supporting their reclassification based on morphology (Brasier & Griffin, 1979; Tsao & Alizadeh, 1988).

P. citricola was found to be a very diverse species according to mtDNA restriction fragment analysis (Förster *et al.*, 1990a; Fig. 11.9, Table 11.6); many of the distinct mtDNA types do not appear to be closely related to one another. One distinct group of isolates from avocado (isolates P513, P1273, P3049, P3057 in Fig. 11.9) could be differentiated. Sixteen additional avocado isolates from four different groves in California were all identical and very similar to this group. This indicates that in *P. citricola*, too, distinct subgroups are present.

MtDNAs of *P. cryptogea* and *P. drechsleri* were also very diverse (Mills *et al.*, 1991; Fig. 11.10, Table 11.7). Nine distinct groups were found bearing little similarly to one another. In general, isolates classified as *P. cryptogea* and *P. drechsleri* were not found within the same group. Whereas

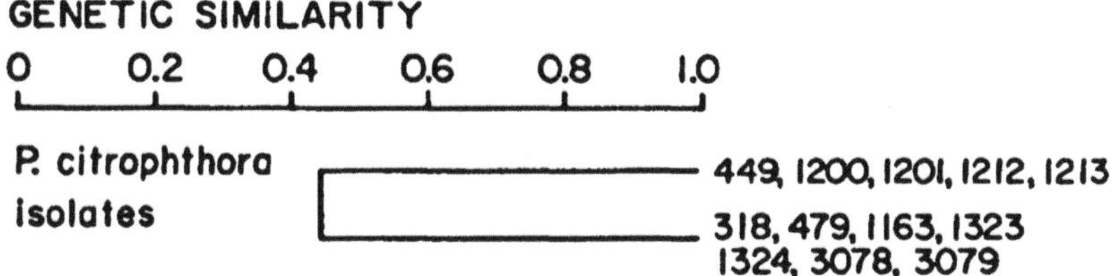

Fig. 11.7. Phenogram for *P. citrophthora* isolates (From Förster, Oudemans & Coffey, 1990a).

Table 11.4. Isolates of *P. citrophthora* used in the study.

Isolate	Host	Geographical origin
P449	*Theobroma cacao*	Brazil
P1200	*Theobroma cacao*	Brazil
P1201	*Theobroma cacao*	Brazil
P1212	*Theobroma cacao*	Brazil
P1213	*Theobroma cacao*	Brazil
P318	*Citrus* sp.	Australia
P479	*Citrus limon*	California
P1163	*Citrus* sp.	California
P1323	*Citrus* sp.	California
P1324	*Citrus* sp.	California
P3078	*Citrus limon*	S. Africa
P3079	*Citrus limon*	S. Africa

the host-specific *P. drechsleri* f. sp. *cajani* formed its own group, isolates of *P. cryptogea* f. sp. *begoniae* were found to be identical to isolates of *P. cryptogea* from several tree hosts. An isolate of *P. melonis* (P1746) had the same mtDNA restriction fragment pattern as isolates of *P. drechsleri* from cucumber and melon. The main criterion for separating *P. cryptogea* and *P. drechsleri* has been their temperature-growth response. Molecular data show that this separation is not justified, moreover, it suggests that the taxonomy of these species is much more complex than previously envisaged using morphological criteria (Bumbieris, 1974; Ho & Jong, 1986).

In *P. megasperma* the host-specific formae speciales *glycinea* and *medicaginis* formed distinct subgroups with very little mtDNA variability within

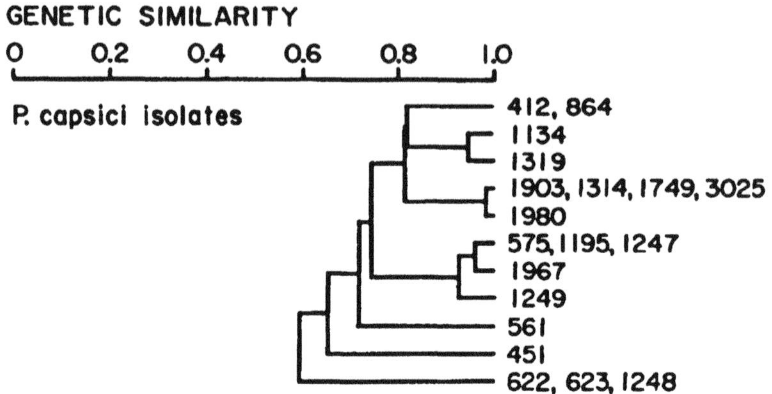

Fig. 11.8. Phenogram for *P. capsici* isolates (From Förster, Oudemans & Coffey, 1990a).

Table 11.5. Isolates of *P. capsici* used in the study.

Isolate	Host	Origin	Isolate	Host	Origin
P412 MF4	*Piper nigrum*	Guatemala	P1248	unknown	unknown
P451 MF4	*Theobroma cacao*	Brazil	P1249	*Spondias purpurea*	Costa Rica
P561 MF4	*Piper nigrum*	Brazil	P1314	*Capsicum annuum*	California
P575 MF4	*Theobroma cacao*	Mexico	P1319	*Capsicum annuum*	California
P622 MF4	*Theobroma cacao*	Brazil	P1749	*Cucumis sativus*	Taiwan
P623 MF4	*Theobroma cacao*	Brazil	P1903	*Gossypium* sp.	California
P864 MF4	*Theobroma cacao*	Cameroon	P1967	*Macadamia integrifolia*	Hawaii
P1134	*Capsicum annuum*	New Mexico	P1980	*L. esculentum*	California
P1195 MF4	*Theobroma cacao*	Mexico	P3025	*Capsicum annuum*	Chile
P1247	*Sechium edule*	Costa Rica			

each group (Förster *et al.*, 1989). Among isolates from other host plants many different mtDNA types were found (Förster *et al.*, 1988; Förster, unpublished). Some groups contained only isolates from a single host plant species (e.g. the Douglas fir group; the asparagus group). So far, no close relationships have been detected amongst most of these groups; more detailed analysis continues. Our results should shed light on the evolution of this complex species (Hansen *et al.*, 1986; Chapter 10) and provide a valuable new database for the renaming of proposed subgroups (Faris *et al.*, 1989; Chapter 10).

Little is known about mtDNA variation within species of 'higher' fungi. Usually only one or a few isolates from each species have been included in studies, or no quantitative analysis of restriction fragment data was

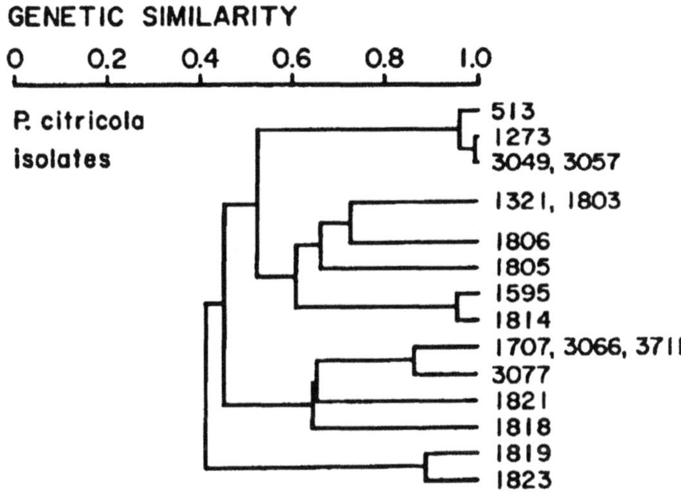

Fig. 11.9. Phenogram for *P. citricola* isolates (From Förster, Oudemans & Coffey, 1990a).

Table 11.6. Isolates of *P. citricola* used in the study.

Isolate	Host	Origin	Isolate	Host	Origin
P513	*Persea americana*	Mexico	P1818	*Medicago sativa*	S. Africa
P1273	*Persea americana*	California	P1819	*Curtisia dentata*	S. Africa
P1321	*Rubus idaeus*	California	P1821	*Ocotea bullata*	S. Africa
P1595	*Citrus sinensis*	Australia	P1823	*Olea capensis*	S. Africa
P1707	*Pinus* sp.	Australia	P3049	*Persea americana*	California
P1803	irrigation water	California	P3057	*Persea americana*	California
P1805	*Humulus lupulus*	California	P3066	*Pinus* sp.	California
P1806	*Prunus dulcis*	California	P3077	*Citrus* sp.	S. Africa
P1814	*Citrus limon*	S. Africa	P3711	*Persea americana*	California

carried out, e.g. in *Agaricus* (Hintz, Anderson & Horgen, 1989), *Aspergillus* (Kozlowski & Stepień, 1982) and *Armillaria mellea* (Jahnke, Bahnweg & Worrall, 1987). Also, dendrograms which are generated by different computer algorithms cannot be compared. Intraspecific comparisons within *Neurospora* were based on mtDNA restriction site cleavage map comparisons (Taylor *et al.*, 1986) and the amount of nucleotide substitutions was estimated. Consequently, these data cannot be compared with the analyses of *Phytophthora*. Thus, we are not yet able to compare the level of genetic diversity in *Phytophthora* with that in other fungal genera. On the other hand, ultrastructural and biochemical features (Burnett, 1987) as well as small-subunit ribosomal RNA sequence

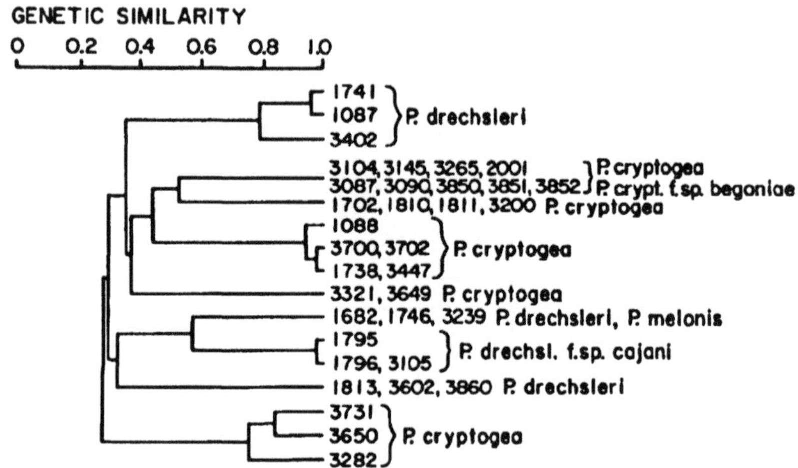

Fig. 11.10. Phenogram for *P. cryptogea*, *P. drechsleri*, and *P. melonis* isolates (From Mills *et al.*, 1991).

Table 11.7. Isolates of *P. cryptogea*, *P. drechsleri* and *P. melonis* used in the study.

Isolate	Host	Origin	Isolate	Host	Origin
Isolates of *P. cryptogea*					
P1088	*Aster* sp.		P1702	*Pseudotsuga menziesii*	Oregon
P1738	*L. esculentum*	Ireland	P1810	*Prunus avium*	California
P1811	*Juglans hindsii*	California	P2001	*Malus sylvestris*	Australia
P3087	*Pinus* sp.	Australia	P3090	*Melaleuca* sp.	Australia
P3104*	*Begonia elatior*	W. Germany	P3145*	*Begonia elatior*	W. Germany
P3200	*Pinus lambertina*	Oregon	P3265*	*Begonia elatior*	W. Germany
P3282	*Rubus* sp.	New York	P3321	*Prunus persica*	New York
P3447	*L. esculentum*	Channel Is.	P3649	*Malus pumila*	Arizona
P3650	*Malus pumila*	New York	P3700	*Asparagus officinalis*	California
P3702	*Asparagus officinalis*	California	P3731	*Prunus avium*	Michigan
P3850	*Actinidia deliciosa*	Australia	P3851	*Actinidia deliciosa*	California
P3852	*Actinidia deliciosa*	California			
Isolates of *P. drechsleri*					
P1087	*Beta vulgaris*	Idaho	P1682	*Cucumis melo*	Iran
P1741	*Solanum tuberosum*		P1795**	*Cajanus cajun*	India
P1796**	*Cajanus cajun*	India	P1813	*Prunus avium*	California
P3105**	*Cajanus cajun*	India	P3239	*Cucumis sativa*	China
P3402	*Beta vulgaris*	California	P3602	*Malus pumila*	Arizona
P3860	*Actinidia deliciosa*	California			
Isolate of *P. melonis*					
P1746	*Cucumis sativa*	Taiwan			

*f.sp. *begoniae*; **f.sp. *cajani*

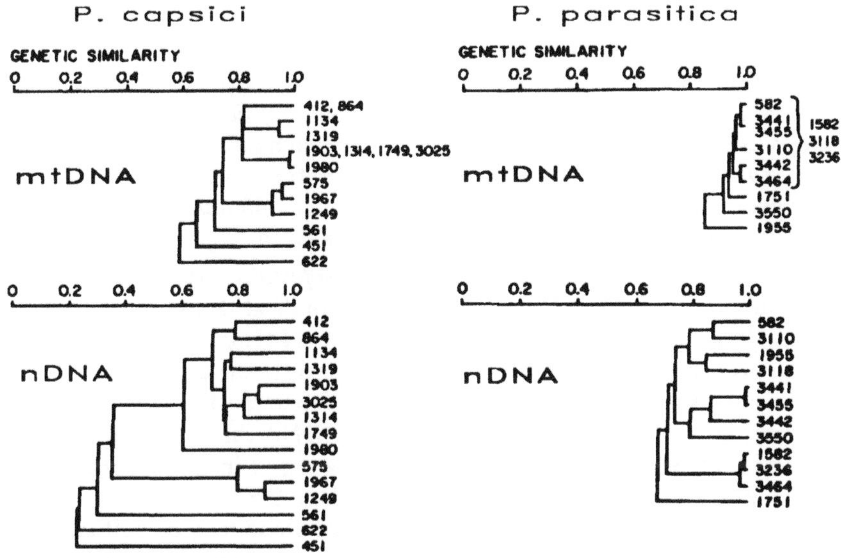

Fig. 11.11. Comparison of phenograms based on polymorphisms in mitochondrial (mtDNA) and nuclear (nDNA) from 15 isolates of *P. capsici* (left) and 12 isolates of *P. parasitica* (right). The phenograms based on mtDNA polymorphisms are essentially the same as in Figs. 11.4 and 11.8, except that some isolates have been excluded for which nuclear DNA RFLPs were not studied. Twenty-five and twenty-six single- or low copy DNA clones were used for the analysis for *P. capsici* and *P. parasitica*, respectively (From Förster, Oudemans & Coffey, 1990a).

analysis (Förster *et al.*, 1990b) indicate that the oomycetes form a different evolutionary line quite separate from the 'higher' fungi. To our knowledge no comparable data on mtDNA exist for the algal groups which are closely related to the oomycetes.

A comparison of mtDNA, nuclear DNA and isozymes

Comparisons between mitochondrial and nuclear DNA analyses have been carried out for *P. capsici* and *P. parasitica* (Förster *et al.*, 1990a). Polymorphisms in the nuclear DNAs were detected by hybridization with random cloned single or low copy DNA fragments from each species which were 500-4500 bp in size. Most of the cloned fragments did not hybridize to DNAs from other species, which may indicate that they were derived from non-coding or nonessential regions of the genome. It remains to be investigated if cDNA clones would be more useful for interspecies comparisons. Within both species it was evident that the degree of variation was much more pronounced among nuclear DNAs than it was among mtDNAs (Fig. 11.11). Again, this may be due to the cloning of nonessential DNA regions. It is known that non-coding DNA accumulates mutations much faster than coding DNA (Li, Luo & Wu, 1985). When the two dendrograms based on the two types of DNA analysis were compared, the branching patterns for *P. capsici* were found to be almost identical. This same comparison was much more difficult to make in *P. parasitica*, since there was so little variation among the mtDNAs. However, the two proposed varieties *nicotianae* and *parasitica*, were not distinguished, even using variation in nuclear DNA.

Most of the isolates, except those of *P. infestans*, were assessed for isozyme polymorphism as well as for mitochondrial DNA RFLPs (Mills *et al*., 1991; Oudemans & Coffey, 1991). It was found that both types of analyses correlated well. There were no major differences in the grouping of isolates or branching of the phenograms. Since most isozyme markers can be considered to be encoded by nuclear DNA this also confirms the finding that mtDNA and nuclear DNA data are in good agreement. Isozyme analysis is less time consuming than DNA analysis, thus for taxonomic purposes a detailed isozyme study alone may be sufficient to trace taxonomic relationships between isolates and species. For genetic studies, on the other hand, DNA markers are more conclusive; chromosomal and cytoplasmic inheritance can be differentiated, and a very large number of markers can be obtained by screening for polymorphisms.

A new species concept in the genus *Phytophthora*

From the above data it can be concluded that *Phytophthora* species have different genetic structures. Some species are quite uniform at the genetic level (*P. palmivora*, *P. parasitica*, *P. cactorum*, *P. infestans*), whilst other species are much more variable. However, within many of the variable species, rather uniform subgroups may be present which, in some cases, correlate with a specific host plant or geographic origin. In variable species which include many isolates with unique mtDNA patterns, and where only a few subgroups were found, additional more or less uniform subgroups may be found merely by screening more isolates. We conclude that most *Phytophthora* species probably have a clonal population structure. A very high degree of uniformity within a subgroup or within a species might be an indication of a recent evolutionary origin in which there was insufficient time for mutations to accumulate. Conversely, it is quite possible that there are selective constraints operating that maintain genetic uniformity. For example, Hansen (chapter 10) argues that narrow host specificity could be the main force determining uniformity within formae speciales of *P. megasperma*.

There was no correlation between intraspecific variability and homo- or heterothallism. However, it is unknown if homothallic species are inbreeding or outbreeding in their sexual reproduction. Moreover, even

Table 11.8. Dice similarity values within and between species of *Phytophthora*. The similarity values are based on mtDNA analyses (Förster, Oudemans & Coffey, 1990a; Förster, unpublished; Mills *et al.*, 1991).

Similarity values within species

P. palmivora	0·76 – 0·97	*P. capsici*	0·57 – 0·98
P. parasitica	0·82 – 0·98	*P. citricola*	0·32 – 0·96
P. cactorum	0·73 – 0·96	*P. cryptogea*	0·27 – 0·53
P. megakarya	0·56	*P. drechsleri*	0·23 – 0·58
P. citrophthora	0·66		

Similarity values between species

	palmivora	*parasitica*	*megakarya*	*citrophthora*	*capsici*
parasitica	0·34 – 0·35	–	–	–	–
megakarya	0·34 – 0·35	0·35 – 0·37	–	–	–
citrophthora	0·33 – 0·35	0·33 – 0·38	0·32 – 0·36	–	–
capsici	0·38 – 0·43	0·35 – 0·39	0·32 – 0·39	0·50 – 0·69	–
citricola	0·31 – 0·38	0·35 – 0·40	0·30 – 0·40	0·47 – 0·80	0·43 – 0·61

with heterothallic species the sexual cycle might occur infrequently, or not at all in the natural environment and asexual propagation may predominate. Brasier (1987) has proposed that clonal populations in heterothallic species might arise either by reduction of outcrossing ability or by loss of the sexual stage altogether, in response to severe ecological disturbances. Furthermore, in some heterothallic species, only a single mating type may be present in some populations so that only infrequent selfing is possible.

How should these apparently clonal subgroups be dealt with by the taxonomist? As mentioned above, subgroups are often very distinct and may be only distantly related. When similarity values from the phenetic analyses of subgroups within a species are compared, they are often in the same range as when different, well-characterized species are compared (Table 11.8). This may indicate that these subgroups represent different species. The question is where to draw the species borderline. How genetically different do subgroups have to be in order to give them species rank? Until we find out whether these subgroups represent reproductively isolated biological species, it is proposed to use the term 'molecular species' to define these subgroups of morphological species of *Phytophthora* which are shown not to be closely related by molecular techniques. Morphological species may consist of only one molecular species as in *P. palmivora*, *P. parasitica*, *P. cactorum* or *P. infestans*. In other cases several or many molecular species may exist within one morphological species.

Molecular species may represent what Wiley (1981) called evolutionary species. Evolutionary species were defined as single lineages of ancestor descendant populations which maintain their identities from other such lineages, and which have their own evolutionary tendencies and historical fates. However, until we determine if molecular species are biological species it would be premature to assign these subgroups new species names and further confuse the taxonomy of *Phytophthora*.

Conclusions

A high degree of morphological and physiological variation within some species of *Phytophthora* has been observed in the past. With the increasing use of new molecular methods in taxonomy, it is evident that some species are indeed very diverse at the molecular level. Knowledge of DNA variation provides information for a more accurate understanding of what comprises a species in *Phytophthora*. As a consequence, the taxonomic status of various *Phytophthora* species will have to be changed and a new species concept will have to be accepted in the future. Species are not necessarily always the genetically well-defined entities that taxonomists would like to see.

In the development of a more natural classification, some current species of *Phytophthora* will be retained (e.g. *P. palmivora*, *P. parasitica*), whilst others will have to be separated into new species and yet others will have to be discontinued due to redundancy. The genetic interrelationships between these species must be investigated in the future. In addition, sexual and protoplast compatibilities need to be investigated as well as molecular polymorphisms.

The existence of intraspecies diversity emphasizes the danger of assigning single isolates as type cultures. DNA studies show that there is no single isolate which is typical of *P. capsici*, *P. citricola*, *P. citrophthora*, *P. cryptogea* or *P. drechsleri*. The molecular based grouping of isolates may help to define subtle morphological, developmental or physiological properties that are characteristic of one group and that have been overlooked in the past.

Many cytoplasmic and nuclear markers are now available from molecular studies in a number of *Phytophthora* species and with a gene transformation system likely to be developed soon (chapters 19 & 20) there now lies ahead a golden opportunity for *Phytophthora* researchers. The genetics of *Phytophthora* was traditionally difficult to study due to the lack of appropriate markers. However, we believe it no longer remains a 'fungal geneticist's nightmare' (Shaw, 1983). These powerful new techniques of molecular biology should eventually solve the many genetic puzzles and provide a clearer understanding of mechanisms generating variation in *Phytophthora*.

References

Birley, A. J. & Croft, J. H. (1986). Mitochondrial DNAs and phylogenetic relationships. In *DNA Systematics*, ed. S.K. Dutta, pp. 107-137. CRC Press: Boca Raton, Florida, USA.

Brasier, C. M. (1987). The dynamics of fungal speciation. In *Evolutionary Biology of the Fungi*, eds. A. D. M. Rayner, C. M. Brasier & D. Moore, pp. 231-260. Cambridge University Press.

Brasier, C. M. & Griffin, M. J. (1979). The taxonomy of *Phytophthora palmivora* on cocoa. *Transactions of the British Mycological Society* **72**, 111-143.

Bruns, T. D. & Palmer, J. D. (1989). Evolution of mushroom mitochondrial DNA: *Suillus* and related genera. *Journal of Molecular Evolution* **28**, 349-362.

Bumbieris, M. (1974). Characteristics of two *Phytophthora* species. *Australian Journal of Botany* **22**, 655-660.

Burnett, J. H. (1987). Aspects of the macro- and micro-evolution of the fungi. In *Evolutionary Biology of the Fungi*, eds. A. D. M. Rayner, C. M. Brasier & D. Moore, pp. 1-15. Cambridge University Press.

de Vallavieille, C. & Erselius, L. J. (1984). Variation in protein profiles of *Phytophthora*: survey of a composite population of three species on *Citrus*. *Transactions of the British Mycological Society* **83**, 473-479.

Erselius, L. J. & de Vallavieille, C. (1984). Variation in protein profiles of *Phytophthora*: comparison of six species. *Transactions of the British Mycological Society* **83**, 463-472.

Erselius, L. J. & Shaw, D. S. (1982). Protein and enzyme differences between *Phytophthora palmivora* and *P. megakarya*: Evidence for self-fertilization in pairings of the two species. *Transactions of the British Mycological Society* **78**, 227-238.

Faris, M. A., Sabo, F. E., Barr, D. J. S. & Lin, C. S. (1989). The systematics of *Phytophthora sojae* and *P. megasperma*. *Canadian Journal of Botany* **67**, 1442-1447.

Fink, W. L. (1986). Microcomputers and phylogenetic analysis. *Science* **234**, 1135-1139.

Förster, H. & Coffey, M. D. (1990). Mating behavior of *Phytophthora parasitica*: evidence for sexual recombination in oospores using DNA restriction fragment length polymorphisms as genetic markers. *Experimental Mycology* **14**, 351-359.

Förster, H., Coffey, M. D., Elwood, H. & Sogin, M. L. (1990b). Sequence analysis of the small subunit ribosomal RNAs of three zoosporic fungi and implications for fungal evolution. *Mycologia* **8**, 306-312.

Förster, H., Kinscherf, T. G., Leong, S. A. & Maxwell, D. P. (1987). Molecular analysis of the mitochondrial genome of *Phytophthora*. *Current Genetics* **8**, 215-218.

Förster, H., Kinscherf, T. G., Leong, S. A. & Maxwell, D. P. (1988). Estimation of relatedness between *Phytophthora* species by analysis of mitochondrial DNA. *Mycologia* **80**, 466-478.

Förster, H., Kinscherf, T. G., Leong, S. A. & Maxwell, D. P. (1989). Restriction fragment length polymorphisms of the mitochondrial DNA of *Phytophthora megasperma* isolated from soybean, alfalfa, and fruit trees. *Canadian Journal of Botany* **67**, 529-537.

Förster, H., Oudemans, P. & Coffey, M. D. (1990). Mitochondrial and nuclear DNA diversity within six species of *Phytophthora*. *Experimental Mycology* **14**, 18-31.

Gray, M. W., Sankoff, D. & Cedergren, R. J. (1984). On the evolutionary descent of organisms and organelles: a global phylogeny based on a highly conserved structural core in small subunit ribosomal RNA. *Nucleic Acids Research* **12**, 5834-5852.

Hansen, E. M., Brasier, C. M., Shaw, D. S. & Hamm, P. B. (1986). The taxonomic structure of *Phytophthora megasperma*: evidence for emerging biological species groups. *Transactions of the British Mycological Society* **87**, 557-573.

Hintz, W. E. A, Anderson, J. B. & Horgen, P. A. (1989). Relatedness of three species of *Agaricus* inferred from restriction fragment length polymorphism analysis of the ribosomal DNA repeat and mitochondrial DNA. *Genome* **32**, 173-178.

Ho, H. H. & Jong, S. C. (1986). A comparison between *Phytophthora cryptogea* and *P. drechsleri*. *Mycotaxon* **27**, 289-319.

Ho, H. H. & Jong, S. C. (1989). *Phytophthora nicotianae* (*P. parasitica*). *Mycotaxon* **35**, 243-276.

Jahnke, K. -D., Bahnweg, G. & Worrall, J. J. (1987). Species delimitation in the *Armillaria mellea* complex by analysis of nuclear and mitochondrial DNAs. *Transactions of the British Mycological Society* **88**, 572-575.

Kaosiri, T. & Zentmyer, G. A. (1980). Protein, esterase, and peroxidase patterns in the *Phytophthora palmivora* complex from cacao. *Mycologia* **72**, 988-1000.

Klimczak, L. J. & Prell, H. H. (1984). Isolation and characterization of mitochondrial DNA of the oomycetous fungus *Phytophthora infestans*. *Current Genetics* **8**, 323-326.

Köchel, H. G. & Küntzel, H. (1982). Mitochondrial L-rRNA from *Aspergillus nidulans*: potential secondary structure and evolution. *Nucleic Acids Research* **10**, 4795-4801.

Kozlowski, M. & Stepień, P. P. (1982). Restriction enzyme analysis of mitochondrial DNA of members of the genus *Aspergillus* as an aid in taxonomy. *Journal of General Microbiology* **128**, 471-476.

Li, W. -H, Luo, C. -C. & Wu, C. -I. (1985). Evolution of DNA sequences. In *Molecular Evolutionary Genetics*, ed. R. J. MacIntyre, pp. 1-94. Plenum Press: New York & London.

Lucas, J. A., Greer, G., Oudemans, P. V. & Coffey, M. D. (1990). Fungicide sensitivity in somatic hybrids of *Phytophthora capsici* obtained by protoplast fusion. *Physiological and Molecular Plant Pathology* **36**, 175-187.

Mills, S. D., Förster, H. & Coffey, M. D. (1991). Taxonomic structure of *Phytophthora cryptogea* and *P. drechsleri* based on isozyme and mitochondrial DNA analyses. *Mycological Research* **95**, 31-48.

Newhook, F. J., Waterhouse, G. M. & Stamps, D. J. (1978). Tabular key to the species of *Phytophthora* de Bary. *Mycological Papers* **143**. Commonwealth Mycological Institute: Kew, U.K.

Oudemans, P. & Coffey, M. D. (1991). Isozyme comparisons within and among worldwide sources of three morphologically distinct species of *Phytophthora*. *Mycological Research* **95**, 19-30.

Shaw, D. S. (1983). The Peronosporales – a fungal geneticist's nightmare. In *Zoosporic Plant Pathogens. A Modern Perspective*, ed. S. T. Buczacki, pp. 86-121. Academic Press: London.

Taylor, J. W. (1986). Fungal evolutionary biology and mitochondrial DNA. *Experimental Mycology* **10**, 259-269.

Taylor, J. W., Smolich, B. D. & May, G. (1986). Evolution and mitochondrial DNA in *Neurospora crassa*. *Evolution* **40**, 716-739.

Tsao, P. H. & Alizadeh, A. (1988). Recent advances in the taxonomy and nomenclature of the so-called '*Phytophthora palmivora*' MF4 occurring on cocoa and other tropical crops. In *10th International Cocoa Research Conference Proceedings*, 17-23 May 1987, pp. 441-445. Santo Domingo.

Tsao, P. H. & Sisemore, D. J. (1978). Morphological variability in *Phytophthora parasitica* (*P. nicotianae*) isolates from citrus, tomato, and tobacco. *Phytopathology News* **12**, 213 (abstr.).

Waterhouse, G. M. (1963). Key to the species of *Phytophthora* de Bary. *Mycological Papers* **92**. Commonwealth Mycological Institute: Kew, U.K.

Wiley, E. O. (1981). *Phylogenetics*. Wiley: New York.

Chapter 12

Relationships between *Phytophthora* species: evidence from isozyme analysis

Peter Oudemans & Michael D. Coffey

The genus *Phytophthora* was erected by Anton de Bary (de Bary, 1876) and since that time over 40 species have been described (Newhook, Waterhouse & Stamps, 1978). Currently, the most widely accepted taxonomic scheme is that developed by Waterhouse (1963). Unfortunately, major unresolved problems still exist for identification of *Phytophthora* isolates to the species level. These include a lack of reliable diagnostic characters and a means to test their quality (Brasier, 1983). The framework developed for the taxonomy of the genus (Rosenbaum, 1917; Tucker, 1931; Leonian, 1934; Waterhouse, 1963; Newhook *et al.*, 1978) is based on morphological features of the sporangia, oospores, chlamydospores and hyphae as well as physiological characters such as cardinal temperature, nutritional requirements, and sensitivity to compounds such as malachite green. Several species exhibit good diagnostic characters, such as the coralloid mycelium of *P. cinnamomi* and the sympodially arranged papillate sporangia of *P. cactorum*, and these therefore represent good morphological species. However, an equal or greater number of species are defined by highly variable characters such as sporangial size, colony morphology, chlamydospore production, or temperature sensitivity. These may therefore either represent groups of cryptic species or, conversely, may be synonymous with other known species. Erwin (1983) cites several examples where morphological criteria have produced confusing and doubtful species designations, including the *P. megasperma* complex and the questionable varieties of *P. parasitica* (*P. nicotianae*).

One approach which has been used in *Phytophthora* systematics is the comparison of the electrophoretic patterns generated from total native or denatured soluble proteins. This technique has proven useful for demonstrating the identity of isolates belonging to a single *Phytophthora* species and confirming differences between species (Clare, 1963; Clare & Zentmyer, 1966; Gill & Powell, 1968; Gill & Zentmyer, 1978; Hall, Zentmyer & Erwin, 1969; Kaosiri & Zentmyer, 1980; Erselius & Shaw, 1982; Erselius & de Vallavieille, 1984; de Vallavieille & Erselius, 1984; Hansen *et al.*, 1986; Bielenin *et al.*, 1988). One major drawback with protein electrophoresis is that comparisons are based on a large number of unidentified protein bands and quantitative analysis can be difficult.

The use of isozyme electrophoresis has been advocated in taxonomic work by Avise (1975), Gottlieb (1981) and Ayala (1983) for animal, plant and insect systematics, respectively. Application of this approach to fungal systematics is as yet under exploited, although some promising studies have already been made (Old, Moran & Bell, 1984; Bonde *et al.*, 1984; Micales, Bonde & Peterson, 1986; Bosland & Williams, 1987; Stasz *et al.*, 1989; Mills, Förster & Coffey, 1991). Isozyme analysis, using starch gels, has one major advantage over total protein gel electrophoresis in that comparisons are based on a number of specific proteins which are each interpreted separately. In addition, the isozyme phenotype is consistent with a diploid organism and this has already been demonstrated in *Phytophthora* for certain enzymes (Shattock, Tooley & Fry, 1986; Spielman, McMaster & Fry, 1989; Chapter 16). Consequently, this approach can be used to quantify both genetic diversity within species and genetic distance between species (Nei, 1987).

Isozyme analysis

Isozyme analysis is based on cytochemical staining of specific enzymes separated by electrophoresis in a suitable medium, usually starch or acrylamide gels. Differences in the electrophoretic mobility of an enzyme can be attributed to mutations which result in an overall change in the net charge of the protein. Estimates of the probability of detecting such mutational events in starch gel electrophoresis range from 30 to 80% (Nei, 1987; Selander *et al.*, 1986). To date there are approximately 70 enzyme systems which have been used with various organisms (Harris & Hopkinson, 1976; Conkle *et al.*, 1982; Soltis *et al.*, 1983; Vallejos, 1983). For *Phytophthora* (Mills *et al.*, 1991; Oudemans & Coffey, 1991) approximately 30 enzyme systems have been screened; of 17 enzymes resolved, 7 stained for two putative loci, making a total of 24 identifiable loci (Table 12.1). To investigate the level of intraspecific variation present in a *Phytophthora* species, isolates are collected from as large a range of host plants and geographical locations as possible (Table 12.2). For each isolate, the data, expressed in the form of electrophoretic mobilities of several enzyme loci, are compared with those from all other isolates and this comparison is expressed in a distance matrix, such as Rogers (1972) genetic distance as modified by Wright (1978). The genetic distance ranges from zero to one, with zero indicating complete identity and one complete dissimilarity. Isolates can be grouped into electrophoretic types (ETs) with identical multilocus phenotypes. An unweighted pair grouping with mean averaging (UPGMA) cluster analysis (Sneath & Sokal, 1973) can be used to calculate the divergence among ETs of each species, and

Table 12.1. Enzyme stains for *Phytophthora*

Enzyme System	EC. number	Locus	No*
Oxidoreductases			
Diaphorase	1.6.4.3	DIA	2
Glucose-6-phosphate dehydrogenase	1.1.1.49	G6PDH	1
Isocitrate dehydrogenase	1.1.1.42	IDH	2
Lactate dehydrogenase	1.1.1.27	LDH	2
Malate dehydrogenase	1.1.1.37	MDH	2
Malic enzyme	1.1.1.40	ME	1
Phosphogluconate dehydrogenase	1.1.1.44	6PGD	2
Superoxide dismutase	1.15.1.1	SOD	1
Transferases			
Adenylate kinase	2.7.4.3	AK	1
Hexokinase	2.7.1.1	HEX	2
Phosphoglucomutase	2.7.5.1	PGM	1
Isomerases			
Glucosephosphate isomerase	5.3.1.9	GPI	1
Mannose-6-phosphate isomerase	5.3.1.8	M6PI	2
Triose phosphate isomerase	5.3.1.1	TPI	1
Others			
Aconitase	4.2.1.3	ACO	1
Peptidase	3.4.3.1	PEP	1
Fructose-1,6-diphosphatase	3.1.3.11	F16DP	1

*Number of putative loci observed.

is presented in the form of a dendrogram. In cases where more than one cluster is found per species, each one is treated as a subgroup. The dendrogram can be tested by using the cophenetic correlation to assess the accuracy with which the dendrogram reflects the original distance matrix. Interspecific comparisons are made using representative isolates from each species. Thus, interspecific comparisons are based upon the maximum diversity seen within a species and an evaluation can be made concerning the validity of each tested species on the basis of genetic distance.

Comparison of morphologically distinct species

In order to test the utility of isozyme analysis in *Phytophthora* taxonomy, it was necessary first to compare intraspecific variation with interspecific distances in a background of taxonomic certainty. For this purpose three easily identified species, *P. cactorum*, *P. cambivora* and *P. cinnamomi*

Table 12.2. A summary of the species and isolates subjected to isozyme analysis

Species & subgroup	Number of isolates	Number of loci	Number of ETs	Number of Hosts (most common)	Number of Countries (most common regions)
P. cactorum	47	18	2	16 (apple, strawberry)	10 (USA, Australia, Japan, Europe)
P. cambivora	25	18	8	6 (chestnut, apple)	5 (USA, Australia, Japan, Europe)
P. cinnamomi	81	18	8	22 (avocado, camellia)	15 (USA, Australia, S. Africa, China)
P. capsici/P. mexicana					
Group 1	50	18	14	9 (Capsicum, cacao)	14 (USA, Europe, Brazil)
Group 2	23	18	4	10 (cacao, black pepper)	7 (India, Hawaii, Brazil)
Group 3	11	18	1	1 (cacao)	1 (Brazil)
Total	84	18	19		
P. citrophthora					
Group 1	34	18	8	11 (citrus, kiwi)	13 (USA, Europe, Brazil)
Group 2	9	18	5	1 (cacao)	1 (Brazil)
Total	43	18	13		
P. palmivora/ P. arecae	106	17	7	20 (cacao, coconut)	19 (USA, Indonesia, West Africa)
P. megakarya					
Group 1	8	17	3	1 (cacao)	2 (Nigeria, Guinea)
Group 2	7	17	3	1 (cacao)	1 (Cameroon)
Total	15	17	6		
P. parasitica	60	19	18	9 (citrus, tobacco)	13 (USA, Australia, Fr. Polynesia)
P. botryosa	10	18	4	1 (rubber)	3 (Thailand, Malaysia, Vietnam)
P. meadii	33	18	13	1 (rubber)	2 (India, Sri Lanka)
P. katsurae	16	17	4	2 (coconut, Auracaria)	4 (Hawaii, Papua New Guinea)
P. heveae	14	17	5	6 (avocado, cacao)	4 (Guatemala, USA, Malaysia)

From Oudemans (1990) and Oudemans & Coffey (1991).

(Table 12.2) were compared (Oudemans & Coffey, 1991). *P. cactorum* is classified in group I of Waterhouse's (1963) taxonomic scheme since it produces papillate sporangia and paragynous antheridia. This species is identified by its subglobose sporangia which are produced in a sympodial manner and by the short pedicel. In contrast, *P. cinnamomi* and *P. cambivora* are placed in group VI of Waterhouse's (1963) scheme because they produce nonpapillate sporangia and amphigynous antheridia. *P. cambivora* is easily identified by the presence of bullate protuberances on the oogonial wall while *P. cinnamomi* can be distinguished by its unique coralloid mycelium. Furthermore, it has been hypothesized that *P. cambivora* and *P. cinnamomi* may be related since both produce a bicellular antheridium (Ashby, 1929; Royle & Hickman, 1964; Ho & Zentmyer, 1977; Ho, Zentmyer & Erwin, 1977).

Based on the banding patterns of 18 isozyme loci, *P. cactorum* proved to be the least diverse species with only two ETs and two polymorphic loci (Oudemans & Coffey, 1991). One of the ETs, CACT1, was the most common and widely distributed as opposed to CACT2 which consisted of isolates originating from strawberry roots in New York. Interestingly, the oospores of CACT2 were highly pigmented in contrast to those of CACT1 which were not. Thus, the divergence detected by isozymes is correlated with this character.

P. cambivora isolates were separated into 8 ETs, however one, CAMB1, was the most common worldwide. Five of the eight ETs were present in Australia, three in the USA, and one each was found in Europe and Japan. No distinctive morphological features were associated with any of the ETs. Eight ETs were also resolved in *P. cinnamomi*. Two of these, CINN4 and CINN5, were of A2 mating type whilst the other six ETs were A1. One enzyme, isocitrate dehydrogenase (IDH) differentiated the A1 and A2 isolates (Oudemans & Coffey, 1991). Earlier work by Arentz & Simpson (1986) indicated that isolates of the two mating types rarely co-existed in the native habitat of Papua New Guinea. Furthermore, no evidence was found in isozyme studies to indicate that hybridization had occurred between the two mating types (Old, Moran & Bell, 1984; Old, Dudzinski & Bell, 1988), with the exception of a single isolate from China which was A1 but possessed the A2 IDH isozyme phenotype (Oudemans & Coffey, 1991).

Interspecific comparisons among *P. cactorum*, *P. cambivora* and *P. cinnamomi* revealed extremely large genetic distances (Fig. 12.1). In addition, no evidence was obtained to suggest that *P. cambivora* and *P. cinnamomi* were related since the genetic distances seen among the three species were approximately equal.

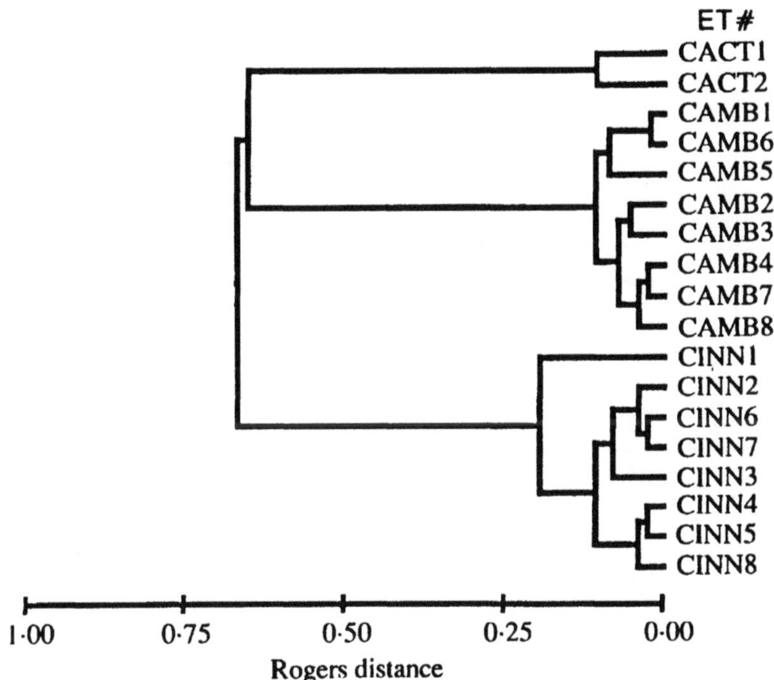

Fig. 12.1. A UPGMA cluster analysis of *Phytophthora cactorum* (CACT), *P. cambivora* (CAMB) and *P. cinnamomi* (CINN) isolates based on Modified Rogers Distance measure (Oudemans & Coffey, 1991).

An evaluation of the taxonomy of some species with papillate sporangia

P. palmivora

P. palmivora is a plurivorous pathogen widely distributed in tropical and subtropical areas of the world (Chee, 1974). Among the most notorious diseases incited by this pathogen is black pod of cacao which is distributed in most cacao growing areas (Zentmyer, 1988). In a survey of 100 isolates (Oudemans, 1990) only two of the 18 isozyme loci were variable and a total of 7 electrophoretic types were found (Table 12.2, Fig. 12.2). Generally, there was no correlation between ET and host or mating type. The highest level of isozyme diversity was found among isolates from southeast Asia which may indicate that the pathogen is native to that area.

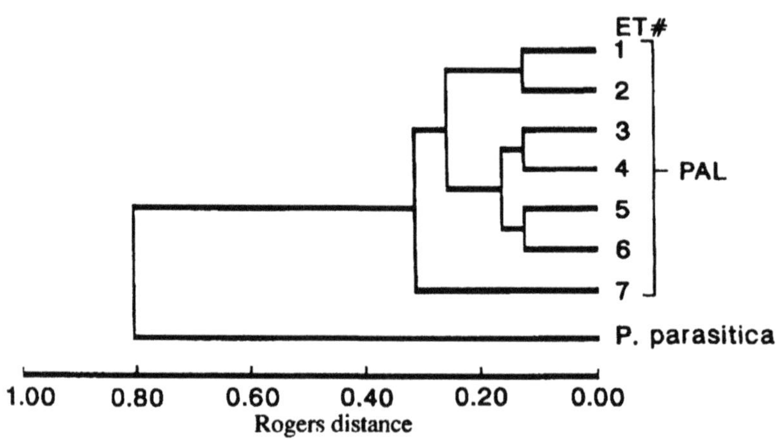

Fig. 12.2. Cluster analysis of *Phytophthora palmivora* (PAL).

This is consistent with the theory presented by Waterhouse (Zentmyer, 1988) based on the abundance of native hosts of *P. palmivora* in this region.

Interestingly, isolates described as *P. arecae* were identical to isolates of *P. palmivora*. The sample of isolates of *P. arecae*, though small, did include the only two available from the original host, *Areca catechu*. Unfortunately, no isolates from *Areca catechu* were available from the recent study of Sastry & Hegde (1987) in which *P. arecae* was reclassified as *P. meadii*. Tucker (1931) combined *P. palmivora*, *P. meadii* and *P. arecae* on morphological grounds, but Waterhouse (1963) maintained them as separate in her treatment of the genus. Based on isozyme data it appears that *P. arecae* and *P. palmivora* are most likely synonymous taxa and the minor morphological differences between them probably reflect a natural level of variation in the species.

P. parasitica

P. parasitica (syn. *P. nicotianae*) is a broad host range pathogen attacking conifers and other tree crops such as citrus, as well as a range of other herbaceous plants, many of which are economic hosts such as tobacco and tomato. Breda de Haan (1896) first described *P. nicotianae* from tobacco seedlings, however the causal organism was never isolated in pure culture

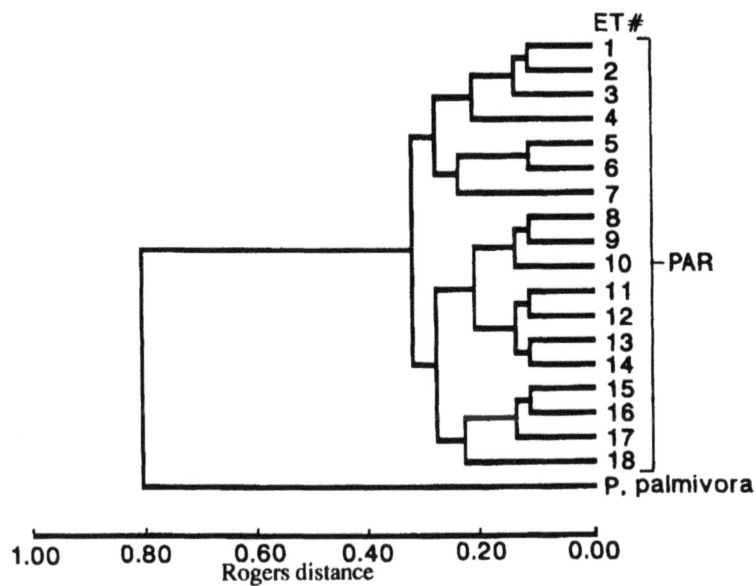

Fig. 12.3. Cluster analysis of *Phytophthora parasitica* (PAR)(Oudemans, 1990).

(Ashby, 1928). Dastur (1913) subsequently described *P. parasitica* as a pathogen of *Ricinus communis* in India. Ashby (1928) recognized the similarity between these two species and commented on their synonymity. In addition, he developed the first detailed classification system for this fungus and amended the description of *P. parasitica* to include isolates from tobacco (previously regarded as *P. nicotianae*) as well as other hosts such as tomato, rhubarb, *Polliar* sp., *Hevea* sp. and *Ricinus communis*. He created two groups of *P. parasitica* based on differences seen in diameters of oospores and oogonia. 'Microspora' referred to the isolates producing small (12 - 24 μm) oospores and 'Macrospora' to isolates producing larger (25 - 30 μm) oospores. Waterhouse (1963) preferred the older name *P. nicotianae* and recognized two morphological varieties: *P. nicotianae* var. *nicotianae* and *P. nicotianae* var. *parasitica*, which were in part based on Ashby's 'Macrospora' and 'Microspora' groups, respectively.

In our research we have attempted to test the various classification schemes through the use of both isozyme and morphometric analysis (Oudemans, 1990). Using a sample of 60 isolates, isozyme analysis resolved 18 ETs which in cluster analysis showed very limited genetic

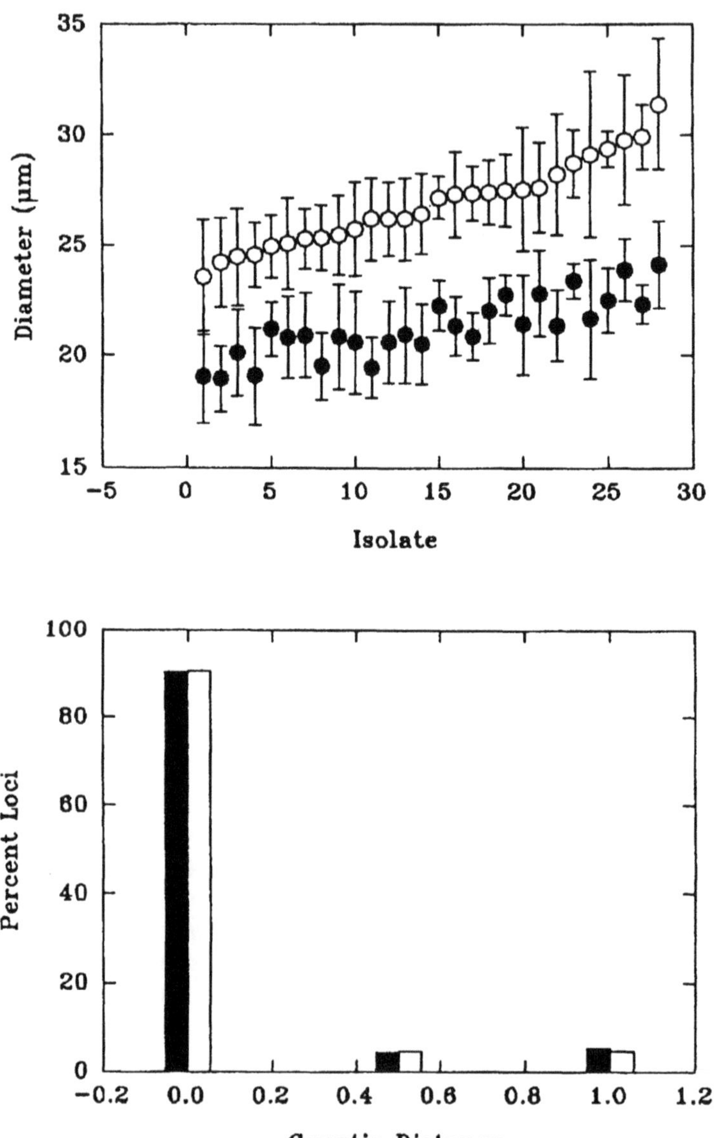

Fig. 12.4. Comparison of *P. parasitica* isolates to test for genetic divergence among isolates of the varieties *nicotianae* and *parasitica*. Top panel: a plot of oogonial (open circles) and oospore (closed circles) diameters (y-axis) across a range of *P. parasitica* isolates. Bottom panel: an analysis of the single locus effects within (filled bars) and among (open bars) two groups of ten isolates each. The two groups are taken from the figure above and represent minimum (isolates 1-10) and maximum (isolates 19-28) oogonial diameters (Oudemans, 1990).

divergence (Table 12.2, Fig. 12.3). To test the scheme developed by Waterhouse (1963), 29 isolates were examined for oospore and oogonial size. A plot of the dimensions of these structures revealed that no distinct size classes could be discerned (Fig. 12.4a). Therefore ten isolates from each of the two size extremes were used in further comparisons. A single locus comparison was made using an approach similar to that described by Ayala (1983). The percentage of loci yielding various genetic distance classes are plotted in a histogram. Since morphological divergence is often accompanied by some level of genetic divergence, it would be expected that the varieties of *P. nicotianae* designated by Waterhouse (1963) would have diverged significantly. The results of this analysis (Fig. 12.4b) clearly indicate that no genetic divergence exists between isolates with different oogonia sizes indicating that further classification into varietal subgroups is not justified.

P. meadii and *P. botryosa*

P. meadii and *P. botryosa* are both pathogens of rubber causing abnormal leaf fall and black stripe. In terms of distribution, *P. meadii* has been reported from India and Sri Lanka (Dantanarayana, Peries & Liyange, 1984; Liyanage & Wheeler, 1989) while *P. botryosa* has been reported from Thailand, Malaysia and Vietnam (Chee, 1969a). These two species are distinguished from one another on the basis of the size of sporangia and oogonia, *P. botryosa* producing smaller reproductive structures (Dantanarayana *et al.*, 1984). An important diagnostic feature of *P. botryosa* is the production of tightly clumped sporangia (Chee, 1969b; Dantanarayana *et al.*, 1984).

In an isozyme comparison of 43 isolates of these two species (Oudemans, 1990) a total of 17 ETs were found using 18 isozyme loci (Table 12.2). Cluster analysis revealed two clusters, one corresponding to the *P. meadii* isolates and another to the *P. botryosa* isolates (Fig. 12.5). However, the two clusters were separated by a much lower genetic distance than the isozymically unrelated species, *P. palmivora*. Of the 18 isozyme loci, *P. meadii* and *P. botryosa* shared 16 in common and 12 of these were not variable in the samples tested.

Despite the morphological differences observed between the two species (Dantanarayana *et al.*, 1984; Liyanage & Wheeler, 1989) they may be closely related. Based on their distribution, it is conceivable that speciation occurred as a result of geographic isolation. *P. meadii* and *P. botryosa* were probably present in southeast Asia prior to the relatively recent introduction of rubber (Baulkwill, 1989) which proved to be susceptible. One additional isolate (ET17) from a native host *Dioscorea* sp. in Malaysia (Tan, personal communication) clustered loosely with the rubber isolates of *P. meadii* and *P. botryosa*. Thus, further sampling of

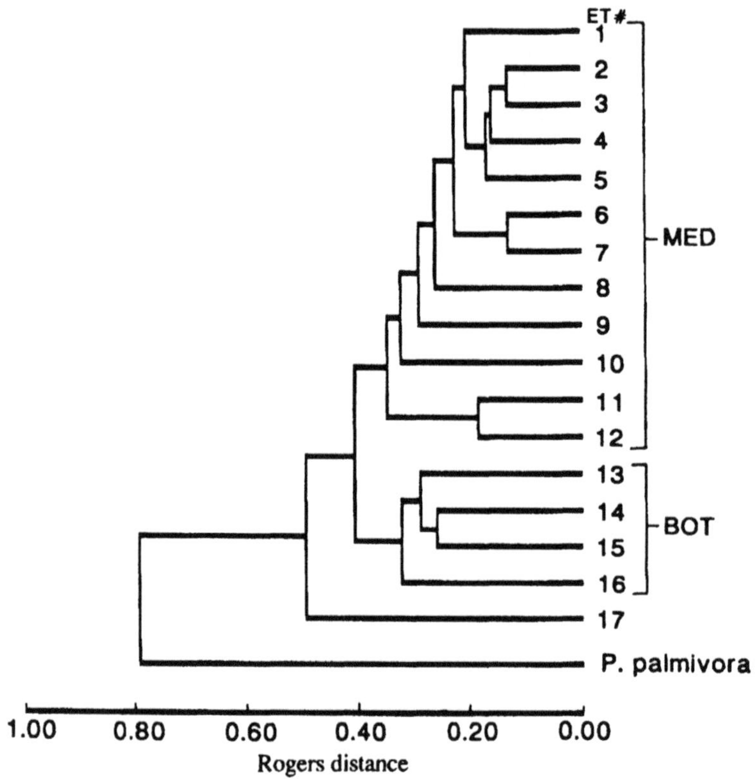

Fig. 12.5. Cluster analysis of *Phytophthora meadii* (MED) and *P. botryosa* (BOT)(Oudemans, 1990).

Phytophthora species in southeast Asia could help to establish whether a more genetically diverse native population of these species exists outside of the rubber crop.

P. megakarya

P. megakarya is a pathogen of cacao and is restricted to west Africa (Zentmyer, 1988). This species was described by Brasier & Griffin (1979) and corresponds to the older designation *P. palmivora* 'MF3' (Griffin, 1977). Isozyme analysis of 15 isolates revealed that two subgroups (Table 12.2, Fig. 12.6) were present and were separated by approximately the same genetic distance as *P. meadii* and *P. botryosa* (Oudemans, 1990). In addition, the two subgroups were geographically separated with MGK1 originating from Nigeria or Guinea and MGK2 from Cameroon. A similar pattern was seen in an analysis of restriction fragment length polymorph-

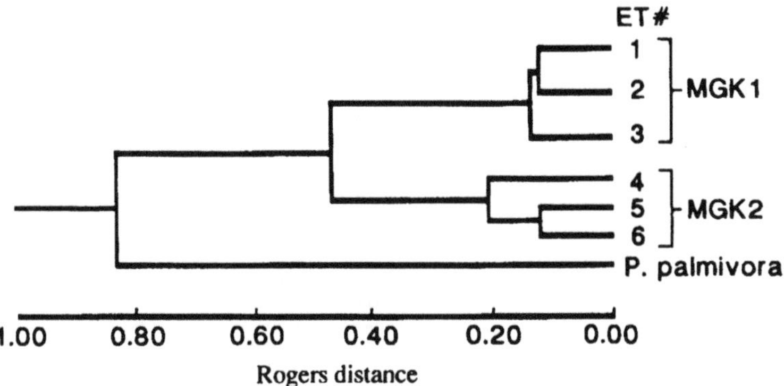

Fig. 12.6. Cluster analysis of *Phytophthora megakarya* (MGK) (Oudemans, 1990).

isms (RFLPs) of mtDNA using the same set of isolates, Chapter 11; Förster, Oudemans &Coffey, 1990). The pattern seen probably reflects a clinal variation of the isozyme phenotype which existed on a native host population prior to the introduction of cacao to West Africa (Förster *et al.*, 1990).

P. capsici

P. capsici was originally described as a pathogen of green pepper in New Mexico, USA (Leonian, 1922). Subsequently, it has been reported from a wide range of hosts and has a worldwide distribution (Stamps, 1985). Recently, the species has been redescribed to include isolates previously identified as *P. palmivora* 'MF4' (Tsao & Alizadeh, 1988).

In a survey of 84 isolates of *P. capsici*, 19 ETs were resolved using 18 isozyme loci (Oudemans, 1990). Cluster analysis revealed three distinct clusters which were designated CAP1, CAP2 and CAP3 (Table 12.2, Fig. 12.7). Unexpectedly, isolates previously identified as *P. palmivora* 'MF4' were distributed in all subgroups. The subgroup CAP1 was diverse for isozymes and contained isolates from hosts such as *Capsicum* sp., squash, tomato, cucumber and eggplant as well as isolates previously identified as *P. palmivora* 'MF4' from cacao and black pepper (*Piper nigrum*). In addition, one representative of *P. mexicana* also clustered with this subgroup. The second subgroup, CAP2, also contained a wide range of isolates from black pepper ('MF4' isolates), macadamia, papaya, and rubber. The isolates of this subgroup tended to be restricted in distribu-

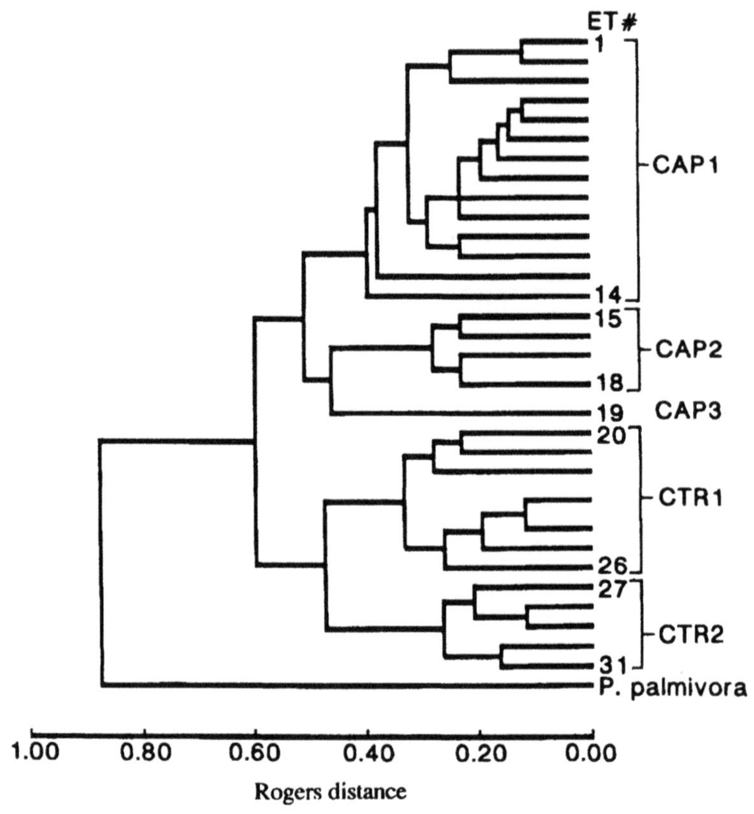

Fig. 12.7. Cluster analysis of *Phytophthora capsici* (CAP) and *P. citrophthora* (CTR) (Oudemans, 1990).

tion to more tropical latitudes. Uchida & Aragaki (1989) reported that isolates of *P. capsici* found in Hawaii were morphologically distinct from other isolates in terms of sporangia shape and size as well as in the production of chlamydospores. In an isozyme analysis, however, these same isolates were found to cluster with the CAP2 subgroup. It is possible that this subgroup could represent a distinct biological species; however, further morphological comparisons would be required to assess the stability of diagnostic characters, such as chlamydospore formation, using a group of isolates with a larger geographical range. The third subgroup, CAP3, was composed of a single ET and represented 'MF4' isolates derived strictly from cacao in Brazil. None of the other cacao isolates from Mexico, Central America or Cameroon occurred in this subgroup.

P. citrophthora

P. citrophthora was originally described as a pathogen of citrus in California, USA (Smith & Smith, 1906). This species has been described on several hosts including walnut, pistachio, almond and kiwi (Waterhouse & Waterston, 1964; Mircetich & Matheron, 1983; Conn, Gubler & Mircetich, 1988). Isolates obtained from cacao in Brazil were found to be morphologically indistinguishable from other *P. citrophthora* isolates (Kellam & Zentmyer, 1986). Isozyme analysis revealed that two genetically distinct subgroups exist among the isolates of this species (Table 12.2, Fig. 12.7). CTR1 contains all of the isolates derived from citrus, kiwi and various other hosts, whilst CTR2 contains only those from cacao in Brazil. Recently, Goodwin, Kirkpatrick & Duniway (1990) reported that the CTR1 isolates were reactive to two specific DNA probes developed for *P. citrophthora*, whereas CTR2 isolates were non-reactive. Using mtDNA RFLP analysis, Förster *et al.* (1990) also found that isolates from cacao (CTR2) were quite distinct from citrus isolates (CTR1). These results strengthen the argument that the groups CTR1 and CTR2 are genetically distinct.

Interspecific comparisons of papillate species

The interspecific comparisons (Fig. 12.8) of papillate species in Groups I and II of Waterhouse's (1963) scheme relied on information derived from 13 isozyme loci of which 11 were variable (Oudemans, 1990). The genetic distances among the species were generally very large and beyond the limit for use in a phylogenetic reconstruction of this group.

It is not entirely surprising that large genetic distances were seen between some papillate species in Group II since Waterhouse (1983) clearly indicated that her groupings in no way implied phylogeny. From the dendrogram (Fig. 12.8) there are at least 5 distinct lineages represented by *P. palmivora*, *P. megakarya*, *P. parasitica*, *P. meadii* and *P. capsici*. *P. cactorum*, which is classed in group I, forms a sixth lineage. However, within some of these lineages some pairs of species appeared to show higher levels of isozyme similarity. As mentioned earlier, *P. meadii* and *P. botryosa* demonstrated a high level of similarity. *P. heveae* and *P. katsurae* were previously considered very similar based on morphology (Stamps, 1985) and this was supported in the isozyme analysis. In addition, the subgroups of *P. capsici* and *P. citrophthora* also appeared to be related (Figs. 12.7, 12.8). Interestingly, Briard (unpublished) also demonstrated relatedness between *P. citrophthora* and *P. capsici* using sequence analysis of the 28s rRNA.

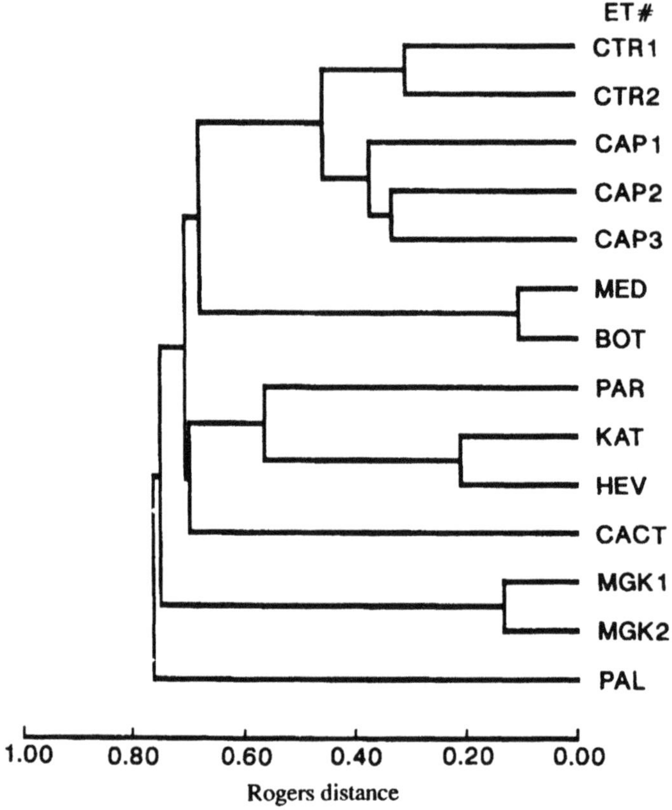

Fig. 12.8. Cluster analysis of ten species of *Phytophthora*. Included are *P. citrophthora* (CTR), *P. capsici* (CAP), *P. meadii* (MED), *P. botryosa* (BOT), *P. parasitica* (PAR), *P. katsurae* (KAT), *P. heveae* (HEV), *P. cactorum* (CACT), *P. megakarya* (MGK) and *P. palmivora* (PAL). Data from Oudemans (1990).

The large number of enzyme differences seen among species and subgroups underlines the possibility of using isozyme electrophoresis as a diagnostic aid. For instance, morphologically similar species such as *P. parasitica* and *P. palmivora* can be differentiated easily by this approach.

Conclusions

From isozyme analyses, some species commonly regarded as morphologically closely related, such as *P. parasitica* and *P. palmivora*, or *P. cambivora* and *P. cinnamomi*, were clearly differentiated as distinct genetic entities. Interspecific comparisons using this technique have also revealed instances where two species were conspecific, such as *P. palmivora* and *P. arecae* as well as *P. capsici* (CAP1) and *P. mexicana*. Many of the genetic relationships uncovered by isozyme analysis would not have been predicted on the basis of morphology alone. For example, *P. meadii* and *P. botryosa*, which are closely related by isozyme analysis, are considered quite distinct based on morphological considerations (Dantanarayana *et al*., 1984), but may be the result of fairly recent speciation events. Also, *P. capsici* and *P. citrophthora* which are also related genetically can be distinguished easily on morphological grounds (Zentmyer, 1988). Interspecific distances between unrelated species were very high and corresponded to generic level distances reported for other groups of organisms (Ayala, 1983). *Phytophthora* species in group II and group VI (Waterhouse, 1963) consist of several distinct and genetically divergent lines (Mills *et al*., 1991; Oudemans, 1990; Oudemans & Coffey, 1991). Unfortunately, a complete phylogenetic reconstruction was not possible since the genetic distances obtained using isozymes were too great.

Most importantly, isozyme analysis provides a rapid means of estimating the amount of genetic diversity within a species. Examination of genetic divergence below the species level revealed that isolates grouped into ETs with identical multilocus phenotypes. The ETs in turn formed clusters or subspecific groups which in some cases could represent cryptic species not readily identifiable using morphological comparisons. In particular, recent studies have revealed that *P. capsici*, *P. citrophthora* and *P. megakarya* are complex species (Förster *et al*., 1990; Chapter 11; Oudemans, 1990), clustering into subgroups which may represent distinct biological species that could reflect recent speciation events. Conversely, existing morphological taxa which demonstrate tight clustering patterns and little evidence of subspecific grouping such as *P. cactorum*, *P. cambivora*, *P. cinnamomi*, *P. palmivora* (*P. arecae*) and *P. parasitica* are also valid biological species.

References

Arentz, F. & Simpson, J. A. (1986). Distribution of *Phytophthora cinnamomi* in Papua New Guinea and notes on its origin. *Transactions of the British Mycological Society* **87**, 289-295.

Ashby, S. F. (1928). The oospores of *Phytophthora nicotianae* Br. De Haan, with notes on the taxonomy of *P. parasitica* Dastur. *Transactions of the British Mycological Society* **13**, 86-95.

Ashby, S. F. (1929). The production of sexual organs in pure cultures of *Phytophthora cinnamomi* Rands and *Blepharospora cambivora* Petri. *Transactions of the British Mycological Society* **14**, 260-263.

Avise, J. S. (1975). Systematic value of electrophoretic data. *Systematic Zoology* **23**, 465-481.

Ayala, F. J. (1983). Enzymes as taxonomic characters. In *Protein Polymorphism: Adaptive and Taxonomic Significance*, Systematics Association Special Volume No. 24, (ed. G. S. Oxford & D. Rollinson), pp. 3-26. Academic Press: New York.

Baulkwill, W. J. (1989). The history of natural rubber production. In *Rubber* (ed. C. C. Webster & W. J. Baulkwill), pp. 1-56. Longman Scientific and Technical: Essex, U.K.

Bielenin, A., Jeffers, S. N., Wilcox, W. F. & Jones, A. L. (1988). Separation by protein electrophoresis of six species of *Phytophthora* associated with deciduous fruit crops. *Phytopathology* **78**, 1402-1408.

Bonde, M., Peterson, G. L., Dowler, W. M. & May, B. (1984). Isozyme analysis to differentiate species of *Peronosclerospora* causing downy mildews of maize. *Phytopathology* **74**, 1278-1283.

Bosland, P. W. & Williams, P. H. (1987). An evaluation of *Fusarium oxysporum* from crucifers based on pathogenicity, isozyme polymorphism, vegetative compatibility, and geographic origin. *Canadian Journal of Botany* **65**, 2067-2073.

Brasier, C. M. (1983). Problems and prospects in *Phytophthora* research. In *Phytophthora: Its Biology, Taxonomy, Ecology, and Pathology* (ed. D. C. Erwin, S. Bartnicki-Garcia & P.H. Tsao), pp. 351-364. American Phytopathological Society: St. Paul, Minnesota.

Brasier, C. M. & Griffin, M. J. (1979). Taxonomy of '*Phytophthora palmivora*' on cocoa. *Transactions of the British Mycological Society* **72**, 111-143.

Breda de Haan, J. van (1896). De bibitziekte in de Deli-tabak veroorzaakt door *Phytophthora nicotianae*. *Mededeelingen uit'S Lands Plantentuin* **XV**, 1-107.

Chee, K. H. (1969a). Leaf fall due to *Phytophthora botryosa*. *Planters Bulletin of the Rubber Research Institute of Malaya* **104**, 190-198.

Chee, K. H. (1969b). Variability of *Phytophthora* species from *Hevea brasiliensis*. *Transactions of the British Mycological Society* **52**, 425-436.

Chee, K. H. (1973). Hosts of *Phytophthora palmivora*. In *Phytophthora Disease of Cocoa* (ed. P. H. Gregory), pp. 81-87. Longman: London.

Clare, B. G. (1963). Starch-gel electrophoresis of proteins as an aid to identifying fungi. *Nature* **200**, 803-804.

Clare, B. G. & Zentmyer, G. A. (1966). Starch-gel electrophoresis of proteins from species of *Phytophthora*. *Phytopathology* **56**, 1334-1335.

Conkle, M. T., Hodgkiss, P. D., Nunnally, L. B. & Hunter, S. C. (1982). *Starch gel electrophoresis of conifer seeds: a laboratory manual*. Technical Paper PSW-64. Pacific Southwest Forest and Range Experiment Station, Forest Service, U.S. Department of Agriculture: Berkeley, California.

Conn, K. E., Gubler, W. D. & Mircetich, S. M. (1988). *Phytophthora* spp. causing root and crown rot of kiwifruit in California. *Phytopathology* **78**, 1573 (abstract).

Dantanarayana, D. M., Peries, O. S. & Liyanage, A. de S. (1984). Taxonomy of *Phytophthora* species isolated from rubber in Sri Lanka. *Transactions of the British Mycological Society* **82**, 113-126.

Dastur, J. F. (1913). On *Phytophthora parasitica* nov. spec. A new disease of the castor oil plant. *Memoirs of the Department of Agriculture, India, Botanical Series* **5**, 177-231.

de Bary, A. (1876). Researches into the nature of the potato fungus - *Phytophthora infestans*. *Journal of the Royal Agricultural Society* **12**, 239-269.

de Vallavieille, C. & Erselius, L. J. (1984). Variation in protein profiles of *Phytophthora*: survey of a composite population of three species on citrus. *Transactions of the British Mycological Society* **83**, 473-479.

Erselius, L. J. & Shaw, D. S. (1982). Protein and enzyme differences between *Phytophthora palmivora* and *P. megakarya*: evidence for self fertilization in pairings of the two species. *Transactions of the British Mycological Society* **78**, 227-238.

Erselius, L. J. & de Vallavieille, C. (1984). Variation in protein profiles of *Phytophthora*: comparison of six species. *Transactions of the British Mycological Society* **83**, 463-472.

Erwin, D. C. (1983). Variability within and among species of *Phytophthora*. In *Phytophthora: Its Biology, Taxonomy, Ecology, and Pathology* (ed. D. C. Erwin, S. Bartnicki-Garcia & P.H. Tsao), pp. 149-165. The American Phytopathological Society: , St. Paul, Minnesota.

Förster, H., Oudemans, P. & Coffey, M. D. (1990). Mitochondrial and nuclear DNA diversity within six species of *Phytophthora*. *Experimental Mycology* **14**, 18-31.

Gill, H. S. & Powell, D. (1968). The use of polyacrylamide gel disc electrophoresis in delimiting three species of *Phytophthora*. *Phytopathologische Zeitschrift* **63**, 23-29.

Gill, H. S. & Zentmyer, G. A. (1978). Identification of *Phytophthora* species by disc electrophoresis. *Phytopathology* **68**, 163-167.

Goodwin, P. H., Kirkpatrick, B. C. & Duniway, J. M. (1990). Identification of *Phytophthora citrophthora* with cloned DNA probes. *Applied and Environmental Microbiology* **56**, 669-674.

Gottlieb, L. D. (1981). Electrophoretic evidence and plant populations. *Progress in Phytochemistry* **7**, 1-45.

Griffin, M. J. (1977). Cocoa *Phytophthora* workshop, Rothamsted Experimental Station, England, 24-26 May 1976. *Pest and News Summaries* (published by the Centre for Overseas Pest Research) **23**, 107-110.

Hall, R., Zentmyer, G. A. & Erwin, D. C. (1969). Approach to taxonomy of *Phytophthora* through acrylamide gel-electrophoresis of proteins. *Phytopathology* **59**, 770-774.

Hansen, E. M., Brasier, C. M., Shaw, D. S. & Hamm, P. B. (1986). The taxonomic structure of *Phytophthora megasperma*: evidence for emerging biological species groups. *Transactions of the British Mycological Society* **83**, 557-573.

Harris, H. & Hopkinson, D. A. (1976). *Handbook of Enzyme Electrophoresis in Human Genetics* (and Supplement, 1978). North Holland Publishing Company: Amsterdam.

Ho, H. H. & Zentmyer, G. A. (1977). Morphology of *Phytophthora cinnamomi*. *Mycologia* **69**, 701-713.

Ho, H. H., Zentmyer, G. A. & Erwin, D. C. (1977). Morphology of sex organs of *Phytophthora cambivora*. *Mycologia* **69**, 641-646.

Kaosiri, T. & Zentmyer, G. A. (1980). Protein, esterase, and peroxidase patterns in the *Phytophthora palmivora* complex from cacao. *Mycologia* **72**, 988-1000.

Kellam, M. K & Zentmyer, G. A. (1986). Morphological, physiological, ecological, and pathological comparisons of *Phytophthora* species isolated from *Theobroma cacao*. *Phytopathology* **76**, 159-164.

Leonian, L. H. (1922). Stem and fruit blight of peppers caused by *Phytophthora capsici* sp. nov. *Phytopathology* **12**, 401-408.

Leonian, L. H. (1934). Identification of *Phytophthora* species. *West Virginia Agriculture Experiment Station Bulletin* **262**, 2-36.

Liyanage, N. I. S & Wheeler, B. E. J. (1989). Comparative morphology of *Phytophthora* species on rubber. *Plant Pathology* **38**, 592-597.

Micales, J. A., Bonde, M. R. & Peterson, G. L. (1986). The use of isoenzyme analysis in fungal taxonomy and genetics. *Mycotaxon* **27**, 405-449.

Mills, S. D., Förster, H. & Coffey, M. D. (1991). Taxonomic structure of *Phytophthora cryptogea* and *P. drechsleri* based on isozyme and mitochondrial DNA analyses. *Mycological Research* **95**, 31-48.

Mircetich, S. M. & Matheron, M. E. (1983). *Phytophthora* root and crown rot of walnut trees. *Phytopathology* **73**, 1481-1488.

Nei, M. (1987). *Molecular Evolutionary Genetics*. Columbia University Press: New York, U.S.A.

Newhook, F. J., Waterhouse, G. M. & Stamps, D. J. (1978). Tabular key to the species of *Phytophthora* de Bary. *Mycological Papers* **143**. Commonwealth Mycological Institute : Kew, U.K.

Old, K. M., Moran, G. F. & Bell, J. C. (1984). Isozyme variability among isolates of *Phytophthora cinnamomi* from Australia and Papua New Guinea. *Canadian Journal of Botany* **62**, 2016-2022.

Old, K. M., Dudzinski, M. J. & Bell, J. C. (1988). Isozyme variability in field populations of *Phytophthora cinnamomi* in Australia. *Australian Journal of Botany* **36**, 355-360.

Oudemans, P. V. (1990). *Biochemical approaches to the taxonomy of Phytophthora*. Ph.D. Thesis, University of California, Riverside.

Oudemans, P. & Coffey, M. D. (1991). Isozyme comparison within and among worldwide sources of three morphologically distinct species of *Phytophthora*. *Mycological Research* **95**, 19-30.

Rogers, J. S. (1972). Measures of genetic similarity and distance. In *Studies in Genetics VII*, pp. 145-153. University of Texas: Austin, Texas, U.S.A.

Rosenbaum, J. (1917). Studies of the genus *Phytophthora*. *Journal of Agricultural Research* **8**, 233-276.

Royle, D. J. & Hickman, C. J. (1964). Observations on *Phytophthora cinnamomi*. *Canadian Journal of Botany* **42**, 311-318.

Sastry, M. N. L. & Hegde, R. K. (1987). *Phytophthora* associated with arecanut (*Areca catechu* Linn) in Uttara Kannada, Karnataka. *Current Science* **56**, 367-368.

Selander, R. K., Caugant, D. A., Ochman, H., Musser, J. M., Gilmour, M. N. & Whittam, T. S. (1986). Methods of multilocus enzyme electrophoresis for bacterial population genetics and systematics. *Applied and Environmental Microbiology* **51**, 873-884.

Shattock, R. C., Tooley, P. W. & Fry, W. E. (1986). Genetics of *Phytophthora infestans*: determination of recombination, segregation and selfing by isozyme analysis. *Phytopathology* **76**, 410-413.

Smith, R. E. & Smith, E. H. (1906). A new fungus of economic importance. *Botanical Gazette* **42**, 215-221.

Sneath, P. H. A. & Sokal, R. R. (1973). *Numerical Taxonomy: The Principles and Practice of Numerical Classification*. W.H. Freeman & Co.: San Francisco, California.

Soltis, D. E., Haufler, C. H., Darrow, D. C. & Gastony, G. J. (1983). Starch gel electrophoresis of ferns: a compilation of grinding buffers, gel and electrode buffers and staining schedules. *American Fern Journal* **73**, 9-27.

Spielman, L. J., McMaster, B. J. & Fry, W. E. (1989). Dominance and recessiveness at loci for virulence against potato and tomato in *Phytophthora infestans*. *Theoretical and Applied Genetics* **77**, 832-838.

Stamps, D. J. (1985). *Phytophthora capsici*. *Descriptions of Pathogenic Fungi and Bacteria*, No. 836. Commonwealth Mycological Institute: Kew, U.K.

Stasz, T. E., Nixon, K., Harman, G. E., Weeden, N. F. & Kuter, G. A. (1989). Evaluation of phenetic species and phylogenetic relationships in the genus *Trichoderma* by cladistic analysis of isozyme polymorphism. *Mycologia* **81**, 391-403.

Tsao, P. H. & Alizadeh, A. (1988). Recent advances in the taxonomy and nomenclature of the so-called '*Phytophthora palmivora*' MF4 occurring on cocoa and other tropical crops. *Proceedings, 10th International Cocoa Research Conference 17-23 May 1987, Santo Domingo*, 441-445.

Tucker, C. M. (1931). Taxonomy of the genus *Phytophthora* de Bary. *University of Missouri Agricultural Experiment Station Research Bulletin* **153**, 1-208.

Uchida, J. Y. & Aragaki, M. (1989). Comparison of pepper isolates of *Phytophthora capsici* from New Mexico to other solanaceous and non-solanaceous isolates. *Phytopathology* **79**, 1212 (abstract).

Vallejos, E. (1983). Enzyme activity staining. In *Isozymes in Plant Genetics and Breeding*, Part A (ed. S. D. Tanksley & T. J. Orton), pp. 469-516. Elsevier Science Publishers: Amsterdam.

Waterhouse, G. M. (1963). Key to the species of *Phytophthora* de Bary. *Mycological Papers* **92**. Commonwealth Mycological Institute: Kew, U.K.

Waterhouse, G. M. (1983). Present criteria for the classification of *Phytophthora*. In *Phytophthora, its Biology, Taxonomy, Ecology, and Pathology* (ed. D. C. Erwin, S. Bartnicki-Garcia & P.H. Tsao), pp. 139-147. American Phytopathological Society: St. Paul, Minnesota.

Waterhouse, G. M. & Waterston, J. M. (1964). *Phytophthora citrophthora*. *Commonwealth Mycological Institute Descriptions of Pathogenic Fungi and Bacteria*, No. 33.

Wright, S. (1978). *Evolution and the Genetics of Populations. Vol. 4: Variability within and among Natural Populations*. University of Chicago Press: Chicago, Illinois, U.S.A.

Zentmyer, G. A. (1988). Taxonomic relationships and distribution of species of *Phytophthora* causing black pod of cacao. *Proceedings, 10th International Cocoa Research Conference 17-23 May 1987, Santo Domingo*, 391-395.

Chapter 13

Variation in ploidy in *Phytophthora infestans*

P. W. Tooley & C. D. Therrien

While polyploidy is known to play an important role in the evolution of higher plants and animals (Stebbins, 1971; Lewis, 1980) its role in the evolution of fungi has been emphasized by some and dismissed by others (Maniotis, 1980). Although some have not considered polyploidy an important mechanism in fungal evolution (Burnett, 1975; Ullrich & Raper, 1977), others have provided evidence to the contrary (Rogers, 1973).

Difficulty in establishing the importance of polyploidy in fungal evolution stems largely from problems associated with chromosome studies (Maniotis, 1980). Specific problems in establishing karyotypes include the small size of nuclei and chromosomes, resistance of the cell wall to mechanical pressure necessary for spreading chromosomes, and uncertainty regarding the site of meiosis and the organization of chromosomes during mitosis (Maniotis, 1980). Due in part to these difficulties, very few comprehensive studies of chromosome numbers within species or among closely related genera have been performed to indicate the existence of polyploid series.

In spite of the above factors however, chromosome counts have provided clear evidence of polyploidy within certain genera, e.g. *Allomyces* (Emerson & Wilson, 1954; Rogers, 1973; Wilson, 1952) and multivalent formation at meiosis has provided evidence of polyploidy in *Cyathus stercoreus* (Lu, 1964) and *Xylaria curta* (Rogers, 1968).

Studies of chromosome number, DNA content and genome size in Oomycetes

In the Oomycetes numerous cytological studies have been performed and chromosome numbers estimated for many species (Maniotis, 1980). Likely polyploid series were found for several closely related species in the Saprolegniaceae and Pythiaceae (Win-Tin & Dick, 1975). Alternative methods have been employed to determine ploidy in Oomycetes. Bryant & Howard (1969) analyzed Feulgen-stained nuclei of *Saprolegnia terrestris* to identify meiosis in the life cycle.

Dodd, Horgen & Straus (1975) used DNA reassociation kinetics to estimate the genome size of *Achlya ambisexualis* at $4 \cdot 4 \times 10^7$ base pairs.

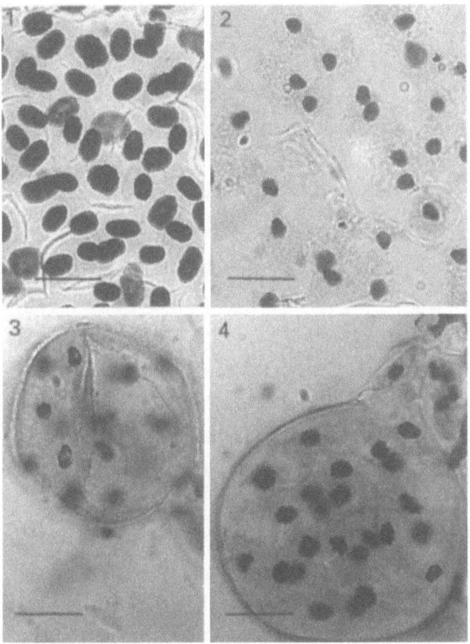

Fig. 13.1. Chick erythrocyte nuclei stained using the Feulgen reaction. The DNA content of the nuclei is ca. 2·41 pg per nucleus (Vaughn & Locy, 1969). Scale bar = 10 μm.

Fig. 13.2. Nuclei of zoospores of Japanese *Phytophthora infestans* isolate 1316 stained by the Feulgen reaction. DNA content estimated at 1·05 pg by comparison with chick erythrocyte nuclei. Scale bar = 10 μm.

Fig. 13.3. Feulgen stained nuclei inside oogonium of *P. infestans* produced in a cross between Mexican isolates 561 and 550 (both putative diploids). Scale bar = 10 μm.

Fig. 13.4. Feulgen stained nuclei in oogonium of *P. infestans* produced in a cross between Polish isolates 1125 (putative tetraploid) and 1157 (putative triploid). Scale bar = 10 μm.

In *Bremia lactucae*, Hulbert & Michelmore (1988) used DNA restriction fragment length polymorphisms to determine that many field isolates carried three or four alleles at a single locus and thus were hyperploids or heterokaryons. Reassociation kinetics, dot-blot genomic reconstructions and reverse genomic blots gave a consistent haploid genome size of 5×10^7 base pairs for *Bremia lactucae*; 65% of the nuclear DNA consisted of repetitive sequences (Francis, Hulbert & Michelmore, 1990). The above methods of genome size estimation could be applied in studies of polyploidy and genome evolution in the Oomycetes.

In the genus *Phytophthora*, a lively ploidy debate continued for years before it was generally accepted that members of this genus are vegetatively diploid and have only a short-lived haploid gametic phase (Shaw, 1983). Chromosome counts exist for many species in this genus (Brasier, 1983) and DNA-specific staining methods have been used to determine the place of meiosis in the life cycle of *P. drechsleri* (Mortimer & Shaw, 1975) and in measuring the DNA content in populations of *P. megasperma* (Rutherford & Ward, 1985) and *P. infestans* (Hansen *et al.*, 1986; Tooley & Therrien, 1987; Therrien *et al.*, 1989).

There is evidence from several taxonomic groups within *Phytophthora* that polyploidy may play a significant role in the evolution of this genus. Chromosome counts have shown that morphological species may be artificial and may contain subgroups of distinct karyotypes which are different biological species, e.g. *P. palmivora* and *P. megakarya* (Sansome, Brasier & Griffin, 1975; Sansome, Brasier & Sansome, 1979). Karyotypes within the aggregate species *P. megasperma* are highly variable and some polyploid groups, which also have high DNA contents, have been identified. It is not yet clear if these groups are closely related or belong to different biological species (but see Chapter 10). In a comprehensive study of *P. megasperma*, Hansen *et al.* (1986) found substantial variation in chromosome number correlated with DNA content within nine distinct subgroups of isolates. Two of the groups comprised at least four karyotypes, suggesting a polyploid series within closely related taxa.

Polyploidy in *Phytophthora infestans*

Sansome (1977) studied the karyotype of Welsh isolates of *P. infestans* by pairing A1 isolates with isolates of other heterothallic species. The chromosome numbers determined from meiotic metaphases of the Welsh isolates appeared to be approximately twice the number determined from a previous study of Mexican isolates (Sansome & Brasier, 1973). Sansome (1977) hypothesized that *P. infestans* may exist in nature outside Mexico in the tetraploid condition and that this may be the prevalent condition in temperate zones which may allow the fungus to adapt better to cooler conditions. A survey of isolates was urgently needed to determine how widespread polyploidy was in populations of *P. infestans*.

Feulgen cytophotometry of *P. infestans*

Such a survey was initiated by Tooley & Therrien (1987), who used Feulgen-DNA cytophotometry to determine the Feulgen-DNA content of zoospores of 23 Mexican and 18 non-Mexican isolates of *P. infestans*.

Absorption cytophotometry of the Feulgen-DNA dye complex has been used widely for over 30 years to determine relative amounts of DNA per nucleus in many organisms (Rasch, 1985). Feulgen cytophotometry had been applied previously to several fungi (Peabody, Motta & Therrien,

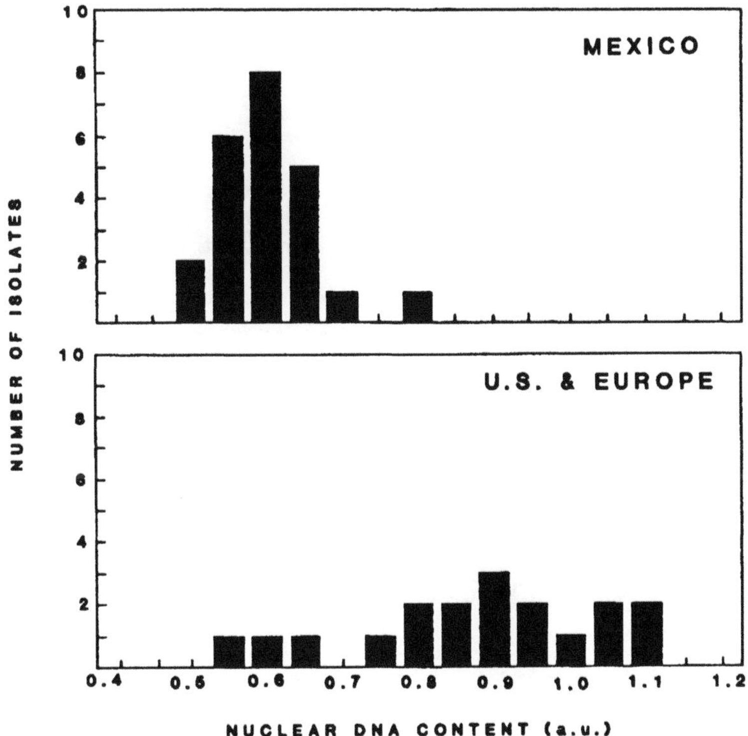

Fig. 13.5. DNA content (arbitrary units, a.u.) distribution in isolates of *P. infestans* from Mexico and from the United States and Europe (Tooley & Therrien, 1987).

1978; Therrien, 1966; Therrien & Yemma, 1974) including the Oomycete *Saprolegnia terrestris* (Bryant & Howard, 1969).

Zoospores of isolates of *P. infestans* collected from the field in Mexico, or obtained from the field or from culture collections in the US and Europe, were fixed at 5°C in 4% formalin in 0·25 M sucrose, post-fixed in 70% ethanol, and stained by the Feulgen reaction as modified by Therrien & Yemma (1974). Cytophotometric measurements were taken with a Zeiss Universal microscope equipped with a gradation interference filter monochromator and a PMI-1 photometer indicator. Patau's (1952) two wavelength method was used for all DNA calculations, and DNA values reported as arbitrary units (a.u.). Chick erythrocyte nuclei, with a nuclear DNA content of 2·41 pg (Vaughn & Locy, 1969), were used as an internal DNA standard (Dhillon, Berlyn & Miksche, 1977). Examples of Feulgen stained nuclei of chicken erythrocytes and *P. infestans* are shown in Figs. 13.1 to 13.4.

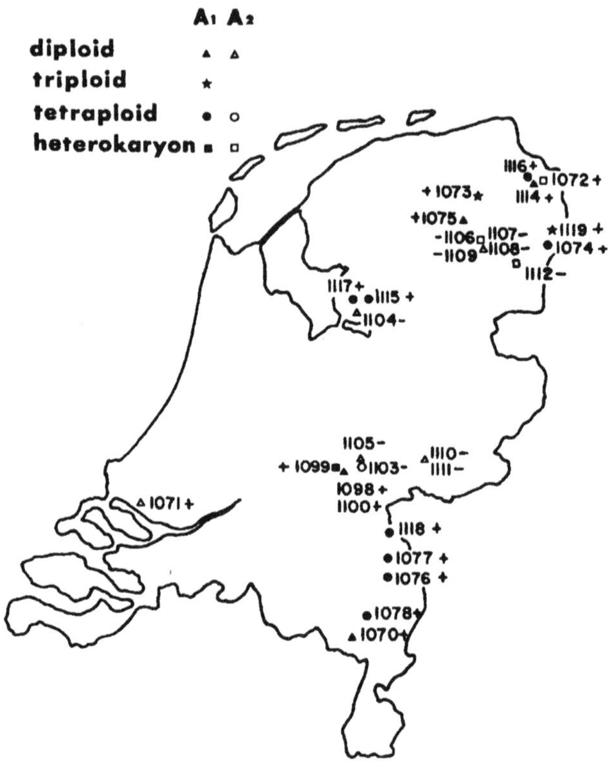

Fig. 13.6. Map of the Netherlands indicating locations where *P. infestans* isolates of both mating types and of differing ploidy levels were collected. + = isolate resistant to the fungicide metalaxyl; – = isolates sensitive to metalaxyl.

It was found that most Mexican *P. infestans* isolates contained about 50% of the DNA content of many non-Mexican isolates (Fig. 13.5) indicating that the former are probably diploid, in agreement with the earlier studies on chromosome number of Mexican material (Brasier & Sansome, 1975; Sansome & Brasier, 1973). In terms of absolute DNA content, the mean value for Mexican isolates was 0·52 pg DNA per nucleus. The non-Mexican group of isolates included those from the US and Europe and appeared to include isolates which were diploid, triploid, tetraploid, and aneuploid (Fig. 13.5; Tooley & Therrien, 1987). Two of the non-Mexican isolates with diploid DNA contents had been obtained from Southern California, and could have been of recent Mexican origin.

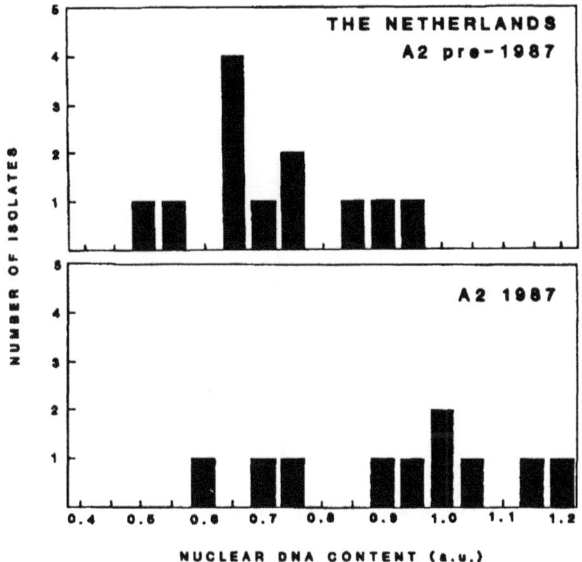

Fig. 13.7. DNA content (a.u.) distribution in isolates of A2 mating type of *P. infestans* from the Netherlands, collected prior to 1987 and in 1987 (Therrien *et al.*, 1989).

In a later study (Tooley, Therrien & Ritch, 1989), the DNA content of 34 isolates of *P. infestans* from Peru was examined using the same method. These isolates were all of A1 mating type and the spectrum of DNA content was very similar to that previously observed for isolates from the US and Europe. The DNA content of Peruvian isolates ranged from 0·53 to 1·22 a.u. (mean value 0·95 a.u.); the Peruvian population appeared to contain diploid, triploid, and tetraploid isolates.

DNA content in populations of *P. infestans* of both mating types

The recent discovery of the A2 mating type of *P. infestans* outside Mexico (e.g. Hohl & Iselin, 1984; Shaw *et al.*, 1985; Tantius *et al.*, 1986) poses new questions about the population biology of this fungus. For example, what effects would the A2 mating type have on variation in the field? If A1 isolates in an area were tetraploid and (possibly new) A2 isolates were diploid, sexual progeny might be absent or very rare. Such a barrier might effectively maintain isolation of the two mating types and prevent sexual reproduction. Thus, it is crucial to define ploidies of potentially sexual populations before breeding behaviour can be understood.

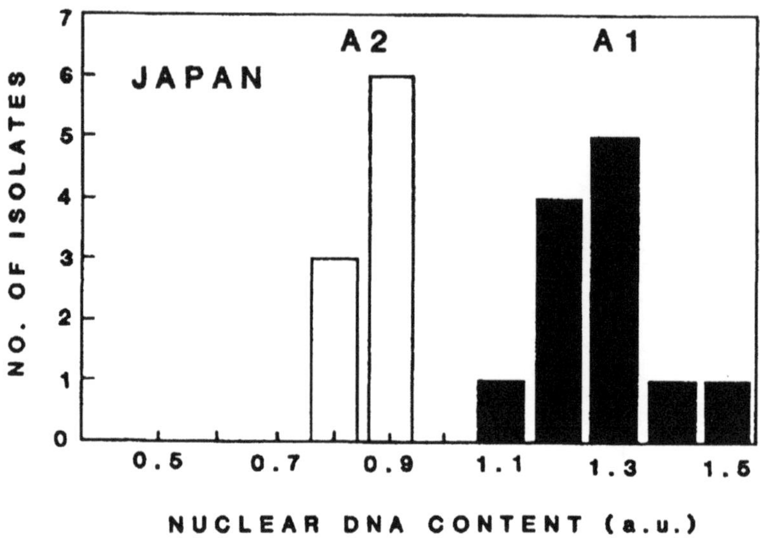

Fig. 13.8. DNA content (a.u.) distribution in isolates of A1 and A2 mating type of *P. infestans* from Japan, collected in 1988 (Therrien *et al.*, 1990).

Therrien *et al.* (1989) analyzed the DNA contents of 83 isolates from the Netherlands, where the A2 mating type was identified in 1981. It appeared that the Dutch population of *P. infestans* consisted largely of A1 mating type with a proportion of A2 mating type. To eliminate the possibility that DNA content differences could be caused by maintaining isolates for long periods in culture, fresh isolates of both mating types collected from the field in 1987, as well as isolates collected before 1987 and maintained in culture for various time periods, were examined. Fig. 13.6 shows the locations and presumed ploidy levels of the pre-1987 Dutch *P. infestans* isolates used in the study.

The measured DNA contents suggested that there was considerable variation in ploidy in the pre-1987 A1 and A2 isolates. Eleven (69%) of the A1 isolates were polyploid, while five (42%) of the A2 isolates were polyploid. Fifty-five isolates of both mating types collected during 1987 were also examined. Of the A1 isolates, 38 were tetraploid (84%), 5 were triploid, and 2 were diploid. Of the A2 isolates, 7 were tetraploid (70%), 1 was triploid and 2 were diploid.

Mean DNA contents must be interpreted with caution, since diploid-tetraploid heterokaryons may exist having intermediate mean

DNA contents similar to those of triploids. Of the pre-1987 A2s examined, 4 out of 12 isolates were interpreted, from their bimodal distributions of DNA contents, to be diploid-polyploid heterokaryons, while in the 1987 group of A2s there were no heterokaryons, but a higher frequency of tetraploids (Fig. 13.7). An increase in tetraploid A2 isolates in the Netherlands could be explained by selection if tetraploidy confers higher fitness compared with diploidy. Alternatively, tetraploid A2 strains could be entering the Netherlands from other regions in infected tubers or as windblown sporangia, causing an increased frequency of this phenotype.

DNA content of Japanese and Polish populations of *P. infestans*

Therrien *et al.* (1990) analyzed the DNA content of Japanese isolates of both mating types isolated from the field in 1988. Zoospores of 21 of the isolates were produced and subjected to Feulgen cytophotometry. The DNA contents of the isolates ranged from 0·84 to 1·46 a.u., with a mean of 1·1 a.u. However, isolates of different mating type fell into two distinct groups (Fig. 13.8). The A2 isolates had a mean DNA content of 0·87 a.u. (SD = 0·09), while the A1 isolates had a mean of 1·27 a.u. (SD = 0·03). Distinct differences in isozyme phenotype between the two mating types of this Japanese collection were also observed. The discrete, uniform groups formed by the two different mating types in Japan indicates absence of sexual reproduction in this population. The inferred ploidy difference between the two mating types could provide a barrier to mating; selfing and asexual reproduction only might be possible.

Fifty-three isolates from Poland were recently examined for nuclear DNA content; all except one were of A1 mating type. The Polish isolates showed a very wide range of DNA contents and appeared to encompass isolates of a wide range of ploidy (Fig. 13.9). Several Polish isolates appeared to be of higher ploidy, perhaps pentaploid or hexaploid. This 'population' could be experiencing rapid change and should be monitored closely in the future.

Conclusions

Polyploidy can provide an important mechanism for the maintenance of genetic variation, and thus has been considered an important evolutionary factor in various taxa (Lewis, 1980). Although polyploidy has been found to occur in true fungi (Rogers, 1973) and Oomycetes (Win-Tin & Dick, 1975), the biological relevance of polyploidy and its role in evolution of these groups remain obscure.

In *P. infestans*, polyploidy, as measured using Feulgen-DNA cytophotometry, appears to be quite common in populations outside Mexico (Table 13.1). Sansome (1977) speculated that polyploids could

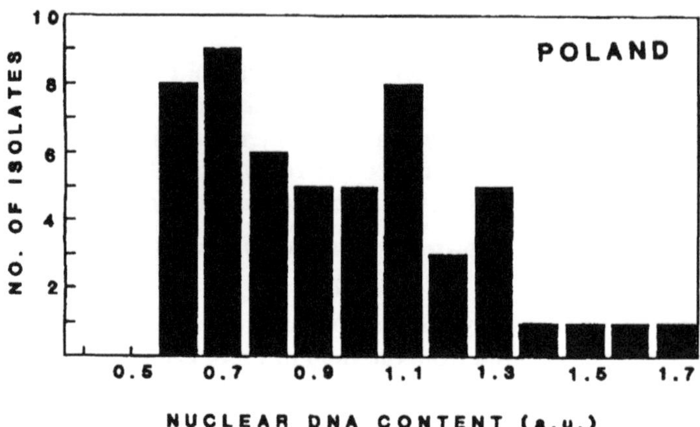

Fig. 13.9. DNA content (a.u.) distribution in Polish isolates of *P. infestans*.

have arisen by selection of autotetraploid nuclei that may occasionally arise during periods of rapid nuclear division. Conversely, if allopolyploid, they could have arisen by chromosome doubling in hybrids between *P. infestans* and other species (Sansome, 1977).

The presence of polyploidy may be advantageous to *P. infestans* in two ways. Sansome (1977) has speculated that tetraploid *P. infestans* may be better adapted to the cooler conditions found in temperate zones compared with their related diploids. Controlled studies have not been performed to evaluate this hypothesis. Secondly, since polyploids can harbour a wider array of virulence allele combinations compared with diploids, polyploidy could contribute to the development of the many virulence phenotypes of *P. infestans* observed (Sansome, 1977).

To evaluate the potential impact of polyploidy on *P. infestans* populations, more basic information is needed on the genetics of polyploidy in this organism. It has been shown that sexual reproduction may occur between *P. infestans* isolates of different ploidy. Hybrid progeny have been obtained from several presumed triploid x diploid crosses and segregation was observed for resistance to the fungicide metalaxyl (Spielman, Sweigard & Fry, unpublished). However, in other cases where viable progeny were not obtained in crosses, the parental

Table 13.1. Summary of ploidy level distribution in *P. infestans* populations determined using Feulgen cytophotometry.

P. infestans Origin	Mating type	Ploidy level 2N	3N	4N	H[a]	Total
Mexico	A1+A2	22	1	0	0	23
US & Europe	A1	3	6	8	1	18
Peru	A1	2	3	16	0	21
The Netherlands	A1+A2	16	8	54	5	83
	A1	7	7	46	1	61[d]
	A2	9	1	8	4	22[d]
Japan	A1+A2	0	9	10	2	21
	A1	0	0	10	2	12[d]
	A2	0	9	0	0	9[d]
Poland	A1 (+A2)[b]	9	16	21	7	53
Israel	A2 (+A1)[c]	3	1	2	0	6

[a]heterokaryon; [b]one or several A2 isolates only; [c]one or several A1 isolates only.
[d]derived from the total sample listed immediately above.

isolates were later found to differ in DNA content and were probably of different ploidy (Spielman, unpublished). Even if some viable progeny were obtained *in vitro* from such crosses, they may contain chromosomal abnormalities which prevent their survival in the field or their subsequent reproduction.

Isozyme banding patterns of *P. infestans* isolates, presumed from DNA content measurements to be polyploid, were identical to those of diploid Mexican isolates, indicating the presence of only two alleles per locus (Tooley, Fry & Villarreal Gonzalez, 1985). If autopolyploid, a single tetraploid individual could contain two copies of each of the two alleles, yielding a diploid isozyme banding pattern. Multivalents in the form of rings or chains were observed in polyploid *P. infestans* oogonia (Sansome, 1977), indicating possible autopolyploidy. Autopolyploidy was suggested by Win-Tin & Dick (1975) to occur in other Oomycetes as well, based on observations of multivalents at first metaphase (Sansome, 1963; Dick, 1969; Flanagan, 1970).

Further studies to determine the role of polyploidy in the population biology and genetics of *P. infestans* are needed. Key areas of inquiry may include investigations of the frequency of selfing versus outcrossing in crosses between parents of differing ploidy, and the viability, fitness and virulence of progeny from such crosses. Continual monitoring of ploidy shifts in A2 strains found in areas previously containing only A1 strains may shed light on the relationship between ploidy level and fitness of strains in temperate regions. Knowledge of the ploidy spectrum of mixed A1 and A2 populations will also help pathologists predict the likelihood that viable oospore progeny will be produced in those regions to serve as a major new source of over-wintering inoculum.

Improvements in cytogenetic methods to allow more rapid and accurate chromosome counts are needed for future studies of polyploidy in *P. infestans* and other Oomycetes. Additional work also needs to be performed in establishing the association between chromosome number and DNA content. Wittmann-Meixner, Weber & Bresinsky (1989) showed that DNA content (assessed using DAPI staining) was highly correlated with chromosome number for two *Pythium* species, but was poorly correlated for several *Achlya* species. It was shown for *Phytophthora megasperma* (Hansen *et al.*, 1986) that differences in DNA content of the various sub-groups were associated with the different chromosome numbers observed for the groups. Recent work with *P. infestans* (Whittaker, Shattock & Shaw, 1991) indicates that isolates with twice the DNA content had approximately twice the number of chromosomes at metaphase I.

Because of the small size of chromosomes in *Phytophthora*, counts can only give approximate karyotypes. However, accurate karyotypes have been produced for many eukaryotic microorganisms using pulsed field gel electrophoresis of chromosomal DNA molecules. Recently, such a karyotype was produced for an isolate of *P. megasperma* using a method which employed contour-clamped homogeneous electric fields (Howlett, 1989). However, the genome of *P. megasperma* is ca. 20% the size of that of *P. infestans* (Hansen *et al.*, 1986) so that separation of the presumably larger *P. infestans* chromosomes may not yet be possible using this method.

It is hoped that future studies will provide additional knowledge to assess the impact of polyploidy on the genetics, population biology and evolution of *P. infestans*. Perhaps this organism may then serve as a model for studies on the role of polyploidy in evolution of the Oomycetes and other taxa. Knowledge of the role of polyploidy in *P. infestans* biology will allow us better to understand at least one of the genetic mechanisms this organism possesses which makes it one of the world's most successful and destructive plant pathogens.

References

Brasier, C. M. (1983). Problems and prospects in *Phytophthora* research. In *Phytophthora: its Biology, Taxonomy, Ecology, and Pathology* (ed. D. C. Erwin, S.

Bartnicki-Garcia & P. H. Tsao) pp. 351-364. American Phytopathological Society: St. Paul, Minnesota, USA.

Brasier, C. M. & Sansome, E. (1975). Diploidy and gametangial meiosis in *Phytophthora cinnamomi*, *P. infestans*, and *P. drechsleri*. *Transactions of the British Mycological Society* **65**, 49-65.

Bryant, T. R. & Howard, K. L. (1969). Meiosis in the Oomycetes. I. A microspectrophotometric analysis of nuclear deoxyribonucleic acid in *Saprolegnia terrestris*. *American Journal of Botany* **56**, 1075-1083.

Burnett, J. H. (1975). *Mycogenetics*. John Wiley & Sons: New York and London.

Dhillon, S. S., Berlyn, G. P. & Miksche, J. P. (1977). Requirement of an internal standard for microspectrophotometric measurement of DNA. *American Journal of Botany* **64**, 660-670.

Dick, M. W. (1969). Morphology and taxonomy of the Oomycetes, with special reference to Saprolegniaceae, Leptomitaceae and Pythiaceae. I. Sexual reproduction. *New Phytologist* **68**, 751-775.

Dodd, J. G., Horgen, P. A., G Straus, N. A. (1975). Isolation and characterization of the deoxyribonucleic acid of the water mould, *Achlya ambisexualis*. *Cytobios* **13**, 31-36.

Emerson, R. & Wilson, C. M. (1954). Interspecific hybrids and cytogenetics and cytotaxonomy of *Euallomyces*. *Mycologia* **46**, 393-434.

Flanagan, P. W. (1970). Meiosis and mitosis in Saprolegniaceae. *Canadian Journal of Botany* **48**, 2069-2076.

Francis, D. M., Hulbert, S. H. & Michelmore, R. W. (1990). Genome size and complexity of the obligate fungal pathogen, *Bremia lactucae*. *Experimental Mycology* **14**, 299-309.

Hansen, E. M., Brasier, C. M., Shaw, D. S. & Hamm, P. B. (1986). The taxonomic structure of *Phytophthora megasperma*: evidence for emerging biological species groups. *Transactions of the British Mycological Society* **87**, 557-573.

Hohl, H. R. & Iselin, K. (1984). Strains of *Phytophthora infestans* with A2 mating type behaviour. *Transactions of the British Mycological Society* **83**, 529-531.

Howlett, B. J. (1989). An electrophoretic karyotype for *Phytophthora megasperma*. *Experimental Mycology* **13**, 199-202.

Hulbert, S. H. & Michelmore, R. W. (1988). DNA restriction fragment length polymorphism and somatic variation in the lettuce downy mildew fungus, *Bremia lactucae*. *Molecular Plant Microbe Interactions* **1**, 17-24.

Lewis, W. H. (ed.) (1980). *Polyploidy: Biological Relevance*. Basic Life Sciences **13**. Plenum Press: New York.

Lu, B. C. (1964). Polyploidy in the basidiomycete *Cyathus stercoreus*. *American Journal of Botany* **51**, 343-347.

Maniotis, J. (1980). Polyploidy in fungi. In *Polyploidy: Biological Relevance*, Basic Life Sciences **13** (ed. W. H. Lewis), pp. 163-192. Plenum Press: New York.

Mortimer, A. M. & Shaw, D. S. (1975). Cytofluorimetric evidence for meiosis in gametangial nuclei of *Phytophthora drechsleri*. *Genetical Research* **25**, 201-205.

Patau, K. (1952). Absorption microphotometry of irregular-shaped objects. *Chromosoma* **5**, 341-362.

Peabody, D. C., Motta, J. J. & Therrien, C. D. (1978). Cytophotometric evidence for heteroploidy in *Armillaria mellea*. *Mycologia* **70**, 487-498.

Rasch, E. M. (1985). DNA 'standards' and the range of accurate DNA estimates by Feulgen absorption microspectrophotometry. In *Advances in Microscopy* (ed. R. R. Cowden & F. W. Harrison), pp. 137-166. Alan R. Liss, Inc: New York.

Rogers, J. D. (1968). *Xylaria curta*: the cytology of the ascus. *Canadian Journal of Botany* **46**, 1337-1340.

Rogers, J. D. (1973). Polyploidy in fungi. *Evolution* **27**, 153-160.

Rutherford, F. S. & Ward, E. W. B. (1985). Estimation of relative DNA content in nuclei of races of *Phytophthora megasperma* f. sp. *glycinea* by quantitative fluorescence microscopy. *Canadian Journal of Genetics and Cytology* **27**, 614-616.

Sansome, E. (1963). Meiosis in *Pythium debaryanum* Hesse and its significance in the life history of the Biflagellatae. *Transactions of the British Mycological Society* **46**, 63-72.

Sansome, E. (1977). Polyploidy and induced gametangial formation in British isolates of *Phytophthora infestans*. *Journal of General Microbiology* **99**, 311-316.

Sansome, E. & Brasier, C. M. (1973). Diploidy and chromosomal structural hybridity in *Phytophthora infestans*. *Nature* **241**, 344-345.

Sansome, E., Brasier, C. M. & Griffin, M. J. (1975). Chromosome size differences in *Phytophthora palmivora*, a pathogen of cocoa. *Nature* **255**, 704-705.

Sansome, E., Brasier, C. M. & Sansome, F. W. (1979). Further cytological studies on the 'L' and 'S' types of *Phytophthora* from cocoa. *Transactions of the British Mycological Society* **73**, 293-302.

Shaw, D. S. (1983). The cytogenetics and genetics of *Phytophthora*. In *Phytophthora: Its Biology, Taxonomy, Ecology, and Pathology* (ed. D. C. Erwin, S. Bartnicki-Garcia & P. H. Tsao), pp 81-94. American Phytopathological Society: St. Paul, Minnesota, USA.

Shaw, D. S., Fyfe, A. M., Hibberd, P. G. & Abdel-Sattar, M. A. (1985). Occurrence of the rare A2 mating type of *Phytophthora infestans* on imported Egyptian potatoes and production of sexual progeny with A1 mating types from the U.K. *Plant Pathology* **34**, 551-556.

Stebbins, G. L. (1971). *Chromosomal Evolution in Higher Plants*. Edward Arnold, Ltd: London.

Tantius, P. M., Fyfe, A. M., Shaw, D. S. & Shattock, R. C. (1986). Occurrence of the A2 mating type and self fertile isolates of *Phytophthora infestans* in England and Wales. *Plant Pathology* **35**, 578-581.

Therrien, C. D. (1966). Microspectrophotometric measurement of nuclear deoxyribonucleic acid content in two myxomycetes. *Canadian Journal of Botany* **44**, 1667-1675.

Therrien, C. D., Daggett, S., Ritch, D. L., Sato, N. L., Spielman, L. J. & Tooley, P. W. (1990). Mating type, nuclear DNA content and isozyme composition of thirty-three isolates of *Phytophthora infestans* from Japan. *Phytopathology* **80**, 124 (abstr.).

Therrien, C. D., Ritch, D. L., Davidse, L. C,, Jespers, B. K. & Spielman, L. J. (1989). Nuclear DNA content, mating type and metalaxyl sensitivity of eighty-three isolates of *Phytophthora infestans* from The Netherlands. *Mycological Research* **92**, 140-146.

Therrien, C. D. & Yemma, J. J. (1974). Comparative measurements of nuclear DNA content in a heterothallic and a self-fertile isolate of the myxomycete *Didymium iridis*. *American Journal of Botany* **61**, 400-404.

Tooley, P. W., Fry, W. E. & Villarreal Gonzalez, M. J. (1985). Isozyme characterization of sexual and asexual *Phytophthora infestans* populations. *Journal of Heredity* **76**, 431-435.

Tooley, P. W. & Therrien, C. D. (1987). Cytophotometric determination of the nuclear DNA content of 23 Mexican and 18 non-Mexican isolates of *Phytophthora infestans*. *Experimental Mycology* **11**, 19-26.

Tooley, P. W., Therrien, C. D. & Ritch, D. L. (1989). Mating type, race composition, nuclear DNA content, and isozyme analysis of Peruvian isolates of *Phytophthora infestans*. *Phytopathology* **79**, 478-481.

Ullrich, R. C. & Raper, J. R. (1977). Evolution of genetic mechanisms in fungi. *Taxon* **26**, 169-179.

Vaughn, J. C. & Locy, R. D. (1969). Changing nuclear histone patterns during development. III. The deoxyribonucleic acid content of spermatogenic cells in the crab *Emerita analoga*. *Journal of Histochemistry and Cytochemistry* **17**, 591-600.

Whittaker, S. L., Shattock, R. C. & Shaw, D. S. (1991). Variation in DNA contents of *Phytophthora infestans* as measured by a microfluorimetric method using the fluorochrome DAPI. *Mycological Research* **95**, 602-610.

Wilson, C. M. (1952). Meiosis in *Allomyces*. *Bulletin of the Torrey Botanical Club* **79**, 139-159.

Win-Tin, & Dick, M, W. (1975). Cytology o£ Oomycetes. Evidence for meiosis and multiple chromosome associations in Saprolegniaceae and Pythiaceae, with an introduction to the cytotaxonomy of *Achlya* and *Pythium*. *Archives of Microbiology* **105**, 283-293.

Wittmann-Meixner, B., Weber, E. & Bresinsky, A. (1989). Different grades of correlation between relative nuclear DNA content, chromosome number and ploidy levels in fungi. *Opera Botanica* **100**, 267-274.

Chapter 14

Genetics of *Phytophthora infestans*: the Mendelian approach

D. S. Shaw & R. C. Shattock

Phytophthora infestans is rapidly becoming a model for the study of genetical and molecular interactions of a fungal pathogen with its host plant; the fungus is easily cultured *in vitro*, has a well defined sexual cycle, uninucleate asexual spores and what increasingly appear to be avirulence genes complementary to major resistance genes in both potato and tomato.

The renaissance of genetical work with *P. infestans* in the 1980s rapidly followed the coincidence of several advances. The A2 mating type was detected in many countries outside Mexico. The subsequent relaxation of quarantine restrictions allowed matings to be conducted in many laboratories for the first time and exchange of cultures carrying valuable markers became a reality. Improved methods for oospore germination were applied to *P. infestans*. Isozyme markers were identified, behaved as good Mendelian markers and were used to verify hybridisation and disomic inheritance (Shaw, 1991).

While rapid progress is being made using molecular methods (chapters 11 to 20), many problems in areas of basic genetics remain; some of these will be discussed in this chapter.

DNA contents of somatic nuclei and ploidy

Cytological work in the 1970s suggested that Mexican isolates were diploid but showed multiple association of certain chromosomes indicating translocation heterozygosity (Sansome & Brasier, 1973); some UK isolates were polyploid and possibly tetraploid (Sansome, 1977). Since then, DNA contents of nuclei from zoospores of isolates from many countries have been estimated (chapter 13). The uniformly low DNA contents (2C) of isolates from Central Mexico are believed to indicate diploidy; chromosome counts of such material at meiotic metaphase I show $2x=2n=24$ (upper estimate) (Sansome, 1977; Whittaker, Shattock & Shaw, 1991). However, a proportion of isolates from other countries has higher DNA contents. Isolates with around double the DNA content (4C) have $2x=4n=40$ (upper estimate) and are thus presumed to be tetraploid (Whittaker *et al.*, 1991), while isolates with even higher DNA contents are now known (chapter 13). Those with intermediate DNA contents may be aneuploids which carry extra copies of normal chromosomes or even super-numerary (B-type) chromosomes. How nuclei with different DNA contents came to exist in 'populations' of the fungus within the same geographic area and even within a single mycelium (heteroploidy) (Whittaker *et al.*, 1991) has yet to be determined. Perhaps pulsed-field gel electrophoresis of full length chromosomal DNA molecules (e.g. Howlett, 1989) will replace conventional cytological techniques which are limited by the small size of the chromosomes; variations in chromosome number in field and laboratory material could then be quickly and easily defined. So far, only limited separation of chromosomal sized DNA molecules has been achieved in *P. infestans*, probably due to the molecules being larger than the maximum size (*ca.* 12 Mb) which can at present be resolved (Carter, 1990; Fyfe & Whittaker, unpublished).

It is now clear that sexually compatible thalli of differing DNA content co-exist in the field and that *in vitro* matings between such isolates (2C x 4C) yield viable hybrid progeny with a continuous range of DNA contents (Whittaker, 1990). Fortunately, most of the progenies which have been analyzed genetically have had parents with DNA contents close to the minimum (i.e. presumed diploid). Nevertheless, even progeny of diploid parents might include trisomics resulting from variable disjunctions from multiple associations of meiotic chromosomes (cf. work with *P. drechsleri*, discussed below). Crosses of one or both parents having higher DNA content have been analyzed genetically but scarcity of appropriate markers has prevented patterns of inheritance being clearly defined. Until such information is available, parents with higher DNA contents should largely be avoided.

Sexual reproduction in *P. infestans*

The early work on the discovery of sexuality in the fungus suggested that the A2 mating type was confined to Mexico. However A2 was recovered in Switzerland in the early 1980s and has now been isolated from the field in most European countries, the Middle East, USSR, Japan and most recently in North America (Hohl & Iselin, 1984; Shaw 1987, 1991; Deahl *et al.*, 1990). This wide distribution suggests that it will be found wherever adequate surveys are made. Whether A2 has spread recently from Mexico to Europe and thence around the rest of the world (Chapters 3 and 15) or has been present in most populations for a long time is controversial. In England and Wales, where the most extensive sampling has been done, A2 was detected in small samples in 1981, 1982 and 1983 (Tantius *et al.*, 1986) and in larger samples during the four seasons 1985 – 88 (Shattock *et al.*, 1990). The frequency varied around 10%; there was no evidence of any increase in frequency of A2 over this period.

In many surveys, isolates bearing abundant oospores and few sporangia have been found. These, sometimes termed homothallic, sometimes self-fertile, have rarely been analyzed. Oosporic isolates from Mexico could transmit fertility to some of their zoospore progeny but also showed somatic segregation; other zoospore progeny were of single mating type (chapter 3). Several oosporic isolates from England and Wales have behaved differently. These are not homothallic, as single propagules (both single hyphal tips and single sporangia) segregated to yield either A1 or A2 mating types both of which were stable (Fyfe, unpublished). Thus these self-fertiles appear to be essentially a physical mixture of A1 and A2 hyphae which remain closely associated even after several years of subculture *in vitro*. Whether hyphal fusions and transient heterokaryons occur in such colonies has yet to be determined (Shaw, 1991).

Genetic analysis with isozyme and RFLP markers

Soon after clear isozyme polymorphisms became available (Chapter 15; Tooley *et al.*, 1985) their inheritance in oospore progeny was followed. Triple banded electromorphs of dimeric glucosephosphate isomerase and dimeric peptidase proved to be heterozygotes as they segregated as expected to produce the two single-banded phenotypes when a single parent was induced to self. Mendelian inheritance was also observed in various matings (A1 x A2) between triple-banded heterozygotes or between heterozygotes and homozygotes (Shattock *et al.*, 1987). Matings between different single-banded morphs produced a majority of triple-banded heterozygotes in F_1, as expected if hybridisation had taken place, but also a variable proportion of single-banded parental morphs. Since induced selfing of a single parent had been demonstrated, the single banded progeny were presumed to be the result of automixis of one or other parent (Shattock *et al.*, 1987). More recently, using mutants resistant to antibiotics, we have found that resistance may be transmitted to some progeny which, based on isozyme phenotype, appear to be selfs of the non-resistant parent (El-Refai, 1990). Either these progeny are not selfs or gametangia were formed on heterokaryotic/heteroplasmotic hyphae resulting from anastomosis of A1 and A2 hyphae (see somatic hybridisation below). Analysis using mitochondrial and additional nuclear markers is continuing.

The first analysis to involve single locus nuclear RFLPs of *P. infestans* (Carter, 1990) provides further insight. In a mating of diploid parents with glucosephosphate isomerase markers and many RFLP markers, most progeny had the triple-banded isozyme phenotype expected of hybrids. One single banded GPI phenotype, presumed to be a self, had all the

RFLP markers of the A1 parent and was probably formed by apomixis rather than by automixis (selfing). While most hybrids appeared to inherit one length marker allele at each locus from each parent, many of the Southern blots showed unequal density of the two bands of a presumed heterozygote. A few hybrids had inherited one allele from one parent but two alleles from the other at one locus. These anomalies could have resulted from trisomy similar to that involving segments bearing mating type genes in progenies of *P. drechsleri* (Mortimer, Shaw & Sansome, 1977; Sansome, 1980). Trisomy in this case is thought to result from the translocation heterozygosity of both parents. Translocation heterozygosity of this kind has been found in both mating types of *P. infestans* (Sansome, 1980).

These unexpected results may be typical of matings of *Phytophthora* spp., rather than merely the outcome of mating the particular (North American) parents chosen; analysis of the first matings of *P. nicotianae* (*parasitica*) involving RFLP markers has shown similar allele dosage variation but also absence of expected alleles in some progeny (Förster & Coffey, 1990). Clearly, many more matings of well marked parents must be analyzed with some urgency to explain these anomalies. the low viabilities and non-Mendelian ratios in F_2 and backcross progenies (e.g. Shattock, 1988; Chapter 15), often thought to involve inbreeding or lethal genes, may also become more readily understood.

Somatic hybridisation

It is now becoming increasingly clear that some sort of somatic hybridisation can take place in *Phytophthora* (Chapter 16). Co-cultivation of isolates of *P. infestans* differing in virulence to resistant potato differentials or in resistance to various antimetabolites has produced new phenotypes (reviewed by Shaw, 1983). Recently, similar somatic matings have been made using isozyme polymorphism, mating type and virulence as unselected markers in addition to the antimetabolite resistance markers used to select 'hybrids' resistant to two antimetabolites (El-Refai, 1990). 'Hybrid' growths were generated following fusion of parental protoplasts, but similar growths were also easily produced using inoculum from plates in which hyphae of the two parents were allowed to grow together. Hyphal tip propagations from 'hybrid' growths showed segregation of parental phenotypes suggesting that growths were heteroplasmon/heterokaryons. However, in some cases hybridity was transmitted through two generations of single zoospores and thus was unlikely to be due to heterokaryosis. Such recombinant single zoospore isolates had one parental phenotype for unselected markers but segregated the other parental phenotype in zoospore progeny (El-Refai, 1990). The transmission of mitochondrial markers and a range of nuclear

RFLPs is being followed in similar somatic matings in an attempt to clarify the roles of nuclear and mitochondrial genomes in what appears to be a complex interaction.

If heterokaryon/heteroplasmons and somatic hybrids are formed during co-cultivation of not only parents of a single mating type but also parents of both mating types then these could occur frequently following the hyphal interactions preceding sexual reproduction. This should be borne in mind when results from sexual matings are being interpreted.

The first evidence of somatic incompatibility has now been produced from co-cultivations of certain isolates of *P. infestans* from tomato and potato in the USSR. A zone of hyphal lysis between interacting colonies was observed (Gorborova *et al.*, 1989). The number and distribution of different vegetative incompatibility groups has yet to be determined.

The genetical determination of mating type

Heterothallism in *P. infestans* is a system of two mating or compatibility types, A1 and A2. Each of these mating types is essentially self sterile when grown in isolation but is bisexual, i.e. can form both antheridia and oogonia when suitably stimulated to mate or self-fertilise. Mating types of isolates of other heterothallic species of *Phytophthora* can be determined by pairing with opposite mating types of *P. infestans* in turn. Pheromones which stimulate sexual activity reciprocally seem to be released by each of a pair of compatible thalli (Ko, 1988). Similar molecules are probably produced by the same mating types of different heterothallic species resulting in the interspecies stimulation. Pheromone isolation and identification in *P. infestans* is in progress (Kemmitt, unpublished).

Progenies of *P. infestans* from matings segregate to give varying proportions of A1 and A2 mating types which often deviate from a 1:1 ratio. Backcross and F_2 progenies tend to deviate more than F_1 progenies (e.g. Shattock *et al.*, 1987). A few self-fertile progeny are often observed which usually revert to A1 or A2 during vegetative growth. Indeed germ sporangia from single germinating oospores can give rise to both A1 and A2 zoospore lines (Laviola & Gallegly, 1983). Analysis of self-fertile progeny in *P. drechsleri* suggested that they were aneuploid and possibly trisomic (Mortimer *et al.*, 1977; Sansome, 1980); cytological evidence for tertiary trisomy was found and led to the hypothesis that mating type genes are located on a chromosomal segment involved in a reciprocal translocation which inhibits mitotic crossing-over. Each parent in a mating, being heterozygous for the translocation, generates some disomic gametes and therefore trisomic progeny are not uncommon. It was further suggested that A2 might be heterozygous; this arrangement would ensure mating type segregation in progenies. It was thought that, at least in *P. infestans*, if A1 rather than A2 was heterozygous then rare somatic

Table 14.1. Numbers of progeny from single parent isolates induced to self using polycarbonate membranes

Isolate	Mating Type	Isozyme Genotype		Origin	Progeny							
					Mating Type		Isozyme Genotype					
							GPI				PEP	
		GPI	PEP		A1	A2	86/86	100/100	86/122	122/122	92/100	100/100
WF127[a]	A1	nt	92/100	New York State	1	0	-	-	-	-	1	0
40/34[a]	A1	nt	92/100	North Wales, UK	9	0	-	-	-	-	7	2
E14[b]	A2	100/100	nt	Egypt	0	28	-	28	-	-	-	-
550[b]	A2	86/86	nt	Mexico	0	8	8	-	-	-	-	-
10/5[b]	A1	86/122	nt	F_1 hybrid of 533 x 550	6	0	0	-	6	0	-	-
10/6[b]	A1	86/122	nt	F_1 hybrid of 533 x 550	1	0	0	-	0	1	-	-
16/4[b]	A2	86/122	nt	F_1 hybrid of 533 x 550	0	15	4	-	9	2	-	-
28/13[b]	A2	86/122	nt	F_1 hybrid of 533 x 550	0	1	0	-	1	0	-	-

GPI = glucosephosphate isomerase; PEP = peptidase. Data of [a]Shattock *et al*. (1986) and [b]Harrison (1990). nt = not tested.

segregation of A2 might occur but since all known isolates outside Mexico were A1, such a situation was unlikely. Both A1 and A2 isolates of *P. infestans* were shown to have multiple associations of chromosomes at meiotic metaphase (Sansome, 1980); further evidence, however, in support of this 'mating type complex' hypothesis is lacking.

The above hypothesis predicts that automixis of one mating type should yield 25% segregants of the other mating type and automixis of the other mating type should yield only the parental mating type. When selfing was induced by a compatible isolate separated by a porous polycarbonate membrane (Table 14.1), A1 isolates yielded only A1 progeny and A2 isolates yielded only A2 progeny. Expected segregations were observed when single parents were heterozygous for an isozyme marker (Shattock, Tooley & Fry, 1986; Harrison, 1990).

Thus, data available so far do not support the hypothesis that one mating type is a simple heterozygote for a mating type gene. However, selfs of *P. nicotianae* (syn. *parasitica*) produced in a similar manner show segregation of A2 from A1 parents and segregation of A1 from A2 parents (Ko, 1988). In this species also , growth of a single mating type in the presence of certain fungicides can result in self-fertile cultures from which lines of A1 and A2 mating type can be isolated. Chloroneb induces A1 capacity in A2 cultures whereas etridiazole (Truban) induces A2 capacity in A1 cultures (Ko, 1988). It is not known if these or other chemicals can induce mating type change in *P. infestans*.

It may be that mating type determination in *P. infestans* and *P.nicotianae* is different. Further data from hybrid and selfed progeny and from self-fertile progeny of parents with marked nuclear and mitochondrial genomes should allow further progress in our understanding of mating type determination.

The inheritance of resistance to phenylamide fungicides

Phenotypes resistant to the phenylamides were first detected soon after commercial release of metalaxyl in 1979, since when their frequency has increased in numerous countries (Chapters 22 & 23). Analysis of inheritance of resistance to metalaxyl (Shattock, 1988) suggested that a single incompletely dominant gene was involved, although backcross progeny showed unexpected frequencies of phenotypes for isozymes and metalaxyl sensitivity. In a recent study, inheritance of resistance in one mating with a sensitive strain was different when the cross was made on two different media (Fig. 14.1). Furthermore, a proportion of asexual progeny and some 'selfs' of the sensitive parent showed some resistance to metalaxyl (Harrison, 1990; Shaw, 1991). Further work with mitochondrial and nuclear RFLPs should determine if metalaxyl

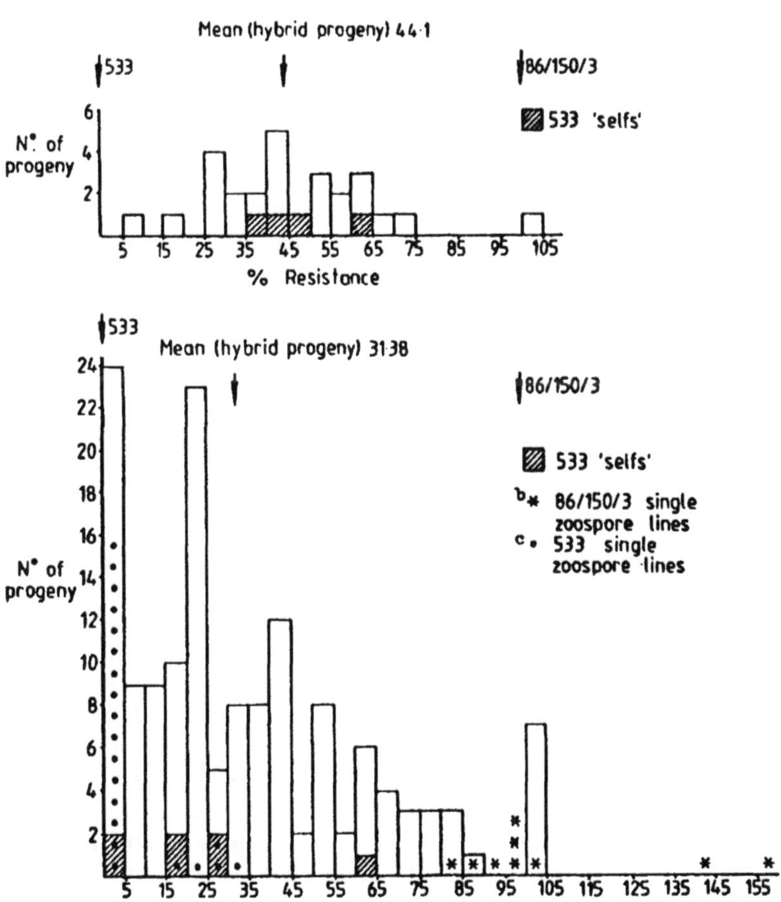

Fig. 14.1. Frequency of F_1 hybrid and 'selfed' single oospore progeny from a mating between a metalaxyl sensitive isolate (533) and a resistant isolate (86/150/3) on rye agar (top panel) and 10% V8 agar (bottom panel) assessed on rye agar with and without 10 μg ml^{-1} metalaxyl. % resistance = colony extension rate on rye agar with metalaxyl expressed as a percentage of that on rye agar alone. [b] single zoospore lines from isolate 86/150/3 assessed on rye agar; [c] single zoospore lines from isolate 533 assessed by dry weight assay using rye broth. Data from Harrison (1990).

resistance from the field can involve several genes or have a cytoplasmically inherited component.

The inheritance of specific virulence

Many physiologic races or virulence phenotypes of *P. infestans* have been identified each showing a different spectrum of compatibility and incompatibility on differential lines of tetraploid *Solanum tuberosum* carrying different single dominant genes for resistance to late blight. These fungal phenotypes were thought to carry single genes, complementary to host resistance genes, whose alleles determined virulence (compatibility) or avirulence (incompatibility). A gene-for-gene host pathogen system was thus suggested. The very first matings of *P. infestans* to be analyzed produced evidence of segregation and recombination of virulence determinants in F_1 progenies. Recent analyses (Al-Kherb, 1988; Spielman, McMaster & Fry, 1989; Spielman *et al.*, 1990; Fininsa, 1990, reviewed by Shaw 1991) have assessed large F_1 progenies from a series of matings of not only Mexican isolates but also isolates from UK, Egypt and California (e.g. Table 14.2). Some of the matings involved isozyme markers which identified hybrid progeny. F_2 and backcross progenies were raised from two matings. Results were largely in agreement with a basic gene-for-gene system having a single avirulence locus with dominant avirulence alleles or recessive virulence alleles determining the reaction of each differential line carrying a single resistance gene. Exceptional results deviating from the expected can be explained in a number of alternative ways. For example, host response may be influenced by allelic dose of the resistance gene, by expression of non-specific polygenic field resistance or by plant age-dependent expression of a resistance gene (Stewart, 1990). Progeny of low aggressiveness might be unable to express their virulence; some avirulence alleles to a specific resistance gene may be incompletely dominant or even recessive, or parents may possess different avirulence loci to the same resistance gene. Inhibitor genes may modify avirulence and segregate in the progeny to produce an excess of virulent phenotypes (Table 14.2; Norwood & Crute, 1984; Ilott, Hulbert & Michelmore, 1989). However, evidence for the definite involvement of recessive avirulent alleles or inhibitor loci remains unconvincing. Further work is needed to improve the assay for virulence and more data are needed from matings of multiply marked parents. Some evidence has been obtained for segregation of virulence during routine subculturing from a parent, Ca65, initially shown to have heterozygous avirulence to potato differential R10. Such segregation could have resulted following mitotic crossing over; evidence for similar instabilities of virulence phenotype has been detected in *Bremia lactucae* (Hulbert & Michelmore, 1988).

Table 14.2. Segregation of avirulent and virulent phenotypes among F_1, F_2 and backcross progeny of *Phytophthora infestans*.

Parents & progeny	Ratio of non-pathogenic (-) to pathogenic (+) on universal suscept r	Ratio of avirulent (-) to virulent (+) on R-gene differentials R1	R2	R3	R4	R5	R7	R8	R10	R11
Ca65[a]	+	+	-	-	+	-	-	-	-	-
E13	+	+	-	+	+	-	+	-	-	+
F_1b	7:136	0:119	99:37	57:79	5:114	133:3	49:70	70:60	104:32	51:68
F_1.34	+	+	-	+	+	-	+	+	-	+
F_1.48	+	+	-	+	+	-	+	+	+	+
F_2b	1:92	0:92	91:1	6:86	0:92	92:0	0:92	7:85	90:2	0:92
Ca65	+	+	-	-	+	-	-	-	-	-
F_1.48	+	+	-	+	+	-	+	+	+	+
Backcross[b]	0:50	0:50	38:12	0:50	0:50	50:0	0:50	15:35	23:27	0:50
565	+	-	+	+	+	-	+	-	-	-
550	+	+	+	+	+	+	+	-	-	+
F_1b	0:68	30:38	1:67	0:68	0:68	41:27	0:68	48:20	50:18	26:42
Ca65	+	+	-	-	+	-	-	-	-	+
550	+	+	+	+	+	+	+	-	+	+
F_1c	23:85	5:80	41:37	38:47	7:73	41:33	37:42	50:24	15:39	6:53

[a]Origins of isolates: Ca65 from California; E13 from Egypt; 550 & 565 from Mexico; F_1.34 & F_1.48 progeny of Ca65 x E13. Data from [b]Al-Kherb (1988) and [c]Fininsa (1990). An unique GPI marker in 550 allowed the hybrid nature of its F_1 progenies to be confirmed.

The *P. infestans*-tomato pathosystem offers an alternative for a model to study interactions at the genetic and molecular levels. Single resistance genes appear to be plentiful in some tomato lines, as differential responses to isolates of *P. infestans* are common (Al-Kherb, 1988). The development of pure lines of tomato with low field resistance and single genes for resistance is an attractive prospect. As the tomato genome is well characterized, it should be possible to analyze both sides of the pathosystem in detail and eventually understand the molecular basis of compatibility.

Conclusions

DNA polymorphisms in both the mitochondrial and the nuclear genome have now been shown to be common (Chapters 11, 17 & 18). Thus, plentiful genetic markers of chromosomes and cytoplasm are now available and will allow problems of basic genetics in both sexual and asexual cycles to be readdressed. The first, albeit fragmentary, linkage map of *P. infestans* exists (Carter, 1990); a genetical model system is in sight.

References

Al-Kherb, S. M. (1988). *The inheritance of host-specific pathogenicity in Phytophthora infestans*. Ph.D. Thesis, University of Wales, UK.

Carter, D. A. (1990). *The molecular genetics of Phytophthora infestans*. Ph.D. Thesis, University of London, UK.

Deahl, K. L., Goth, R. W., Young, R., Sinden, S. L. & Gallegly, M. E. (1990). Occurrence of the A2 mating type of *Phytophthora infestans* in the United States and Canada. *Phytopathology* **80**, 670 (abstract).

El-Refai, I.M. (1990). Somatic hybridization in *Phytophthora infestans*. Ph.D. Thesis, University of Tanta, Egypt.

Fininsa, C. (1990). Virulence in *Phytophthora infestans*: inheritance and variation. M.Phil. Thesis, University of Wales, UK.

Förster, H. & Coffey, M. E. (1990). Mating behavior of *Phytophthora parasitica*: evidence for sexual recombination in oospores using DNA restriction fragment length polymorphisms as genetic markers. *Experimental Mycology* **14**, 351-359.

Gorborova, Ye. V., Bagirova, S. F., Dolgova, A. V. & Dyakov, Yu. T. (1989). Vegetative incompatibility of *Phytophthora infestans* (Mont.) de Bary. *Proceedings of the Academy of Sciences of the USSR* **304**, 1245-1248.

Harrison, B. J. (1990). The genetics of *Phytophthora infestans*. Ph.D. Thesis, University of Wales, UK.

Hohl, H. R. & Iselin, K. (1984). Strains of *Phytophthora infestans* from Switzerland with A2 mating behaviour. *Transactions of the British Mycological Society* **83**, 529-531.

Howlett, B. J. (1989). An electrophoretic karyotype for the pathogenic fungus, *Phytophthora megasperma*. *Experimental Mycology* **13**, 199-202.

Hulbert, S. H. & Michelmore, R. W. (1988). DNA restriction fragment length polymorphism and somatic variation in the lettuce downy mildew fungus, *Bremia lactucae*. *Molecular Plant-Microbe Interactions* **1**, 17-24.

Ilott, T. W., Hulbert, S. H. & Michelmore, R. W. (1989). Genetic analysis of the gene-for-gene interaction between lettuce (*Lactuca sativa*) and *Bremia lactucae*. *Phytopathology* **79**, 888-897.

Ko, W. H. (1988). Hormonal heterothallism and homothallism in *Phytophthora*. *Annual Review of Phytopathology* **26**, 57-73.

Laviola, C. & Gallegly, M. E. (1983). Genetic recombination and mode of inheritance of pathogenic characters by *Phytophthora infestans* through sexual reproduction. In *Durable Resistance in Crops*, ed. F. Lamberti, J. M. Waller & N. A. Van Der Graff, pp. 339-345. Plenum Press: New York and London.

Mortimer, A. M., Shaw, D. S. & Sansome, E. R. (1977). Genetical studies of secondary homothallism in *Phytophthora drechsleri*. *Archives of Microbiology* **111**, 255-259.

Norwood, J. M. & Crute, I. R. (1984). The genetic control and expression of specificity in *Bremia lactucae* (lettuce downy mildew). *Plant Pathology* **33**, 385-400.

Sansome, E. (1977). Polyploidy and induced gametangial formation in British isolates of *Phytophthora infestans*. *Journal of General Microbiology* **99**, 311-316.

Sansome, E. (1980). Reciprocal translocation heterozygosity in heterothallic species of *Phytophthora* and its significance. *Transactions of the British Mycological Society* **74**, 175-185.

Sansome, E. & Brasier, C. M. (1973). Diploidy and chromosomal structural hybridity in *Phytophthora infestans*. *Nature* **241**, 344-345.

Shattock, R. C. (1988). Studies on the inheritance of resistance to metalaxyl in *Phytophthora infestans*. *Plant Pathology* **37**, 4-11.

Shattock, R. C., Tooley, P. W. & Fry, W. E. (1986). Genetics of *Phytophthora infestans*: characterisation of single oospore cultures from A1 isolates induced to self by intraspecific stimulation. *Phytopathology* **76**, 407-410.

Shattock, R. C., Tooley, P. W., Sweigard, J. A. & Fry, W. E. (1987). Genetic studies of *Phytophthora infestans*. In *Genetics and Plant Pathogenesis*, ed. P. R. Day & G. J. Jellis, pp. 175-185. Blackwell Scientific Publications: Oxford.

Shattock, R. C., Shaw, D. S., Fyfe, A. M., Dunn, J. R., Loney, K. H. & Shattock, J. A. (1990). Phenotypes of *Phytophthora infestans* collected in England and Wales from 1985 to 1988: mating type, response to metalaxyl and isoenzyme analysis. *Plant Pathology* **39**, 242-248.

Shaw, D. S. (1983). The Peronosporales: a fungal geneticist's nightmare. In *Zoosporic Plant Pathogens, A Modern Perspective*, ed. S. T. Buczacki, pp. 85-122. Academic Press: London.

Shaw, D. S. (1987). The breeding system of *Phytophthora infestans*: the role of the A2 mating type. In *Genetics and Plant Pathogenesis*, ed. P. R. Day & G. J. Jellis, pp. 161-174. Blackwell Scientific Publications: Oxford.

Shaw, D. S. (1991). Genetics. *Advances in Plant Pathology* **7**, 131-170.

Spielman, L. J., McMaster, B. J. & Fry, W. E. (1989). Dominance and recessiveness at loci for virulence against potato and tomato in *Phytophthora infestans*. *Theoretical and Applied Genetics* **77**, 832-838.

Spielman, L. J., Sweigard, J. A., Shattock, R. C. & Fry, W. E. (1990). The genetics of *Phytophthora infestans*: segregation of allozyme markers in F_2 and backcross progeny and the inheritance of virulence against potato resistance genes R2 and R4 in F_1 progeny. *Experimental Mycology* **14**, 57-69.

Stewart, H. E. (1990). The effect of plant age and inoculum concentration on expression of major gene resistance to *Phytophthora infestans* in detached potato leaflets. *Mycological Research* **94**, 823-826.

Tantius, P. M., Fyfe, A. M., Shaw, D. S. & Shattock, R. C. (1986). Occurrence of the A2 mating type and self-fertile isolates of *Phytophthora infestans* in England and Wales. *Plant Pathology* **36**, 578-581.

Tooley, P. W., Fry, W. E. & Villarreal-Gonzalez, M. J. (1985). Isozyme characterisation of sexual and asexual *Phytophthora infestans* populations. *Journal of Heredity* **76**, 431-435.

Whittaker, S. (1990). *Genetics of Phytophthora infestans: variation in DNA contents and ploidy*. Ph.D. Thesis, University of Wales, UK.

Whittaker, S., Shattock, R. C. & Shaw, D. S. (1991). Variation in DNA content of *Phytophthora infestans* as measured by a microfluorimetric method using the fluorochrome DAPI. *Mycological Research* **95**, in press.

Chapter 15

Isozymes and the population genetics of *Phytophthora infestans*

Linda J. Spielman

The classical and population genetics of *Phytophthora infestans*, the cause of late blight of potato and tomato, has undergone rapid development in the last several years. This is due to several methodological advances, one of which was the discovery of isozymes in *P. infestans* and their use as genetic markers. In this chapter I will focus on the application of isozyme methodology to the population genetics of *P. infestans*. This subject is especially interesting because the historical record of late blight outside of its probable centre of origin in central Mexico begins relatively recently, in the mid 1840's, with the first European and North American epidemics (Bourke, 1964; Chapter 2), and is more complete than the history of most other important diseases. After its appearance in Europe and North America, the disease was observed in many other regions, and by the end of World War II it had a worldwide distribution (Cox & Large, 1960; Miller & O'Brien, 1955). Since these early records of late blight often occurred shortly after the importation of European seed tubers (Cox & Large, 1960), it is likely that Europe served as the major distribution centre for *P. infestans* after its initial establishment there. Europe still serves as a major source of seed potatoes for many parts of the world, and thus may still be playing a pivotal role in the global dissemination of *P. infestans*. With isozymes as genetic tools, we are now able to start to determine how geographically separated *P. infestans* populations differ genetically, and we may learn more about the historical events which brought about the present distribution of late blight.

Although *P. infestans* has two mating types (A1 and A2), the A2 mating type was apparently initially limited to central Mexico, while Europe, the United States, Canada, the West Indies, South Africa, and presumably other regions where the fungus occurred, were apparently occupied exclusively by the A1 mating type (Galindo & Gallegly, 1960; Gallegly & Galindo, 1958; Shaw, 1983). However, in the early 1980's, A2 strains were found in Europe, Israel, and Egypt (Hohl & Iselin, 1984; Malcolmson, 1985; Shaw *et al.*, 1985; Therrien *et al.*, 1989; Cohen, McMaster, Sweigard & Fry, unpublished), and they have since been detected in China, Japan, and Brazil (Mosa *et al.*, 1989; Hohl, unpublished).

There are two commonly held hypotheses to account for the appearance of the A2 mating type outside of Mexico: 1) mutation from the A1 type, and 2) migration from an area inhabited by the A2 type, probably Mexico. These hypotheses can be investigated using the tools of population genetics. Mutation of A1 strains at the mating type locus (or loci) would produce A2 strains which would not differ from the indigenous population in other genetic traits, so no overall changes in the genetic make-up of the population would be expected. Likewise, if migrating strains were similar genetically to indigenous strains we would not expect to see differences between the pre- and post-migration populations. However, if migrating strains differed genetically from the indigenous population, a recent migration event might result in changes in the overall genetic profile of the population. Differences in the genetic composition of A1 and A2 isolates might or might not persist, depending on the amount of random mating which followed the migration. Of course, detection of population genetic shifts would require general knowledge of the genetic make-up of *P. infestans* populations, particularly pre- and post-migration populations and those which could have been the source of the migrants. This study has therefore been centred on the following questions:

- What is the genetic make-up and genetic structure of *P. infestans* populations in different parts of the world?
- What are the relationships among different *P. infestans* populations?
- What can we deduce about past migrations of *P. infestans*?

Isozyme methodology

Isozymes are ideal tools for this kind of investigation, because they are easily assayed, stable, codominant (which is important in a diploid organism like *P. infestans*), and generally under simple genetic control. Isozymes are defined as variant forms of an enzyme which share the same catalytic function, and allozymes are isozymes whose variant forms are coded at the same locus. Allozymes are therefore allelic to each other, and are ideal for population investigations, because the best analytical methods depend on knowledge of allelic frequencies and distributions.

Isozymes are assayed by means of starch gel electrophoresis, the methods for which have been standardized over the past 30 years (Shaw & Prasad, 1970; Siciliano & Shaw, 1976; Micales, Bonde & Peterson, 1986). Over 50 different enzyme systems have been tested in *P. infestans* using various extraction methods, buffers, running conditions, and staining recipes to optimize the resolution of each system (Spielman *et al.*, 1990). So far, 11 monomorphic enzymes and 18 polymorphic enzymes have been detected, but most of the latter were either so poorly resolved or genetically complex that it was not possible to deduce genotypes from the gel patterns observed. The two systems for which simple genetic control has been demonstrated are glucosephosphate isomerase (GPI, E.C. 5.3.1.9) and peptidase (PEP, E.C. 3.4.3.1) (Shattock, Tooley & Fry, 1986; Spielman *et al.*, 1990).

The enzymes commonly used in population studies are either monomeric or dimeric. For a monomeric enzyme, the gene product is the active enzyme unit, so homozygotes are one-banded and heterozygotes are two-banded. One of our polymorphic systems, mannosephosphate isomerase (MPI), is such an enzyme, and segregation of four electromorphs (93, 100, 105, and 111) is observed in crosses. However, the data suggest that there is more than one locus coding for MPI, and it is not possible to determine which alleles belong to which locus (Spielman *et al.*, 1990).

In dimeric enzymes, the initial gene products are inactive. Enzyme activity occurs only when two products, or subunits from a single gene, are joined. These subunits combine randomly, so if there are two different kinds present, as would be the case in a heterozygote, three kinds of active enzyme molecule are formed: two different homodimers and a heterodimer. Thus, homozygotes are single-banded, but heterozygotes are three-banded. Both GPI and PEP produce banding patterns on gels which are typical of dimeric enzymes and if two parents with different, single bands (homozygotes) are crossed, progeny which are all three-banded (heterozygotes) are obtained. If two identical heterozygotes are crossed, the three expected classes of progeny (1 homozygote : 2 heterozygote: 1 homozygote) are observed. Finally, if a heterozygote is crossed with a homozygote, the same two genotypes occur in the progeny in equal numbers. Unexpected classes do sometimes occur in low frequencies, probably as a result of selfing (Spielman *et al.*, 1990).

Unexpected ratios are sometimes observed in crosses between related parents (Spielman *et al.*, 1990). For instance, crosses between two heterozygous F_1 sibs have yielded excesses or deficiencies of expected homozygous classes, and in one backcross an expected class was absent. A likely explanation for these abnormalities is that inbreeding exposes deleterious recessive alleles linked to marker alleles. Nevertheless, the overall result of the crossing studies is that the electromorphs of GPI and PEP behave like Mendelian alleles.

It should be obvious from the foregoing discussion that the development of isozymes as genetic markers can be rather involved. Studies of *P. infestans* have so far yielded only two suitable polymorphic loci: *Gpi* with 7 alleles and *Pep* with 4 alleles. A few of the alleles we have detected are cryptic variants, isozymes which are distinguishable only

Table 15.1. Collections of *Phytophthora infestans* used in comparative studies.

Location	Dates of Collection	Number of Isolates	References[a]
United States/Canada	1979-1988	54	D, F
Central Mexico	1983-1988	71	D, F
Northwestern Mexico	1989	88	D
Peru	1984-1986	34	G
Netherlands	1984-1985	21	D
Poland	1988	44	E
United Kingdom	1988	39	C
Israel	1984-1986	9	A
Japan	1988	30	B

[a]References: (A) Cohen, McMaster, Sweigard & Fry, unpublished; (B) Sato, Spielman & Fry unpublished; (C) Shattock *et al.*, 1990; (D) Spielman, McMaster & Fry, unpublished; (E) Spielman, Sujkowski, McMaster & Fry, unpublished; (F) Tooley *et al.* 1985; (G) Tooley, Therrien & Ritch, 1989.

under certain running conditions. For example, on gels buffered with histidine at pH 7, three alleles of *Gpi* (*90*, *98*, and *100*) are indistinguishable, but the same buffer at pH 6 reveals the differing mobilities of these alleles (Spielman *et al.*, 1990).

Characterization of *P. infestans* populations

Isolates have been obtained from a number of populations of *P. infestans* in Europe, North America, South America, Israel and Japan (Table 15.1). These collections have been characterized for isozyme genotypes at *Gpi* and *Pep*, and the resulting data have been used to determine allelic (Table 15.2) and genotypic (Table 15.3) frequencies for each collection. Because the dates, sizes, and sampling methods of these collections were variable, rigorous statistical analyses are not warranted, but general comparisons of allelic and genotypic frequencies can give preliminary insight into the population genetics of *P. infestans*.

The central Mexican 'population' had greater allelic diversity (five *Gpi* alleles and three *Pep* alleles) than any other sample (Table 15.2). The lower allelic diversity in populations outside of central Mexico is consistent with the theory that central Mexico is the centre of origin of the *P. infestans*/*Solanum* pathosystem. Also supporting this theory is the

Table 15.2. Allelic frequencies in collections of *Phytophthora infestans*[a]

	Gpi allele							*Pep* allele		
	83	86	90	98	100	111	122	83	92	100
Central Mexico	0·01	0·15		0·03	0·71		0·11	0·01	0·18	0·82
NW Mexico		0·01			0·77	0·10	0·11		0·34	0·66
U.S.A/Canada		0·40			0·60				0·42	0·58
Peru		0·50			0·50				0·49	0·52
Netherlands			0·38		0·63				0·29	0·72
U.K.		0· 01			0·99				0·21	0·79
Poland		0·23	0·13		0·66				0·31	0·70
Israel					1·00					1·00
Japan		0·25			0·75				0·18	0·82

[a]Based on data from Shattock *et al.*, 1990; Spielman, Gu & Fry, unpublished; and Tooley *et al.*, 1985. Dates of collection and sample numbers are as given in Table 15.1.

finding that, for both loci, the most common allele in central Mexico was also the most common allele in all other populations. The differences in allele frequencies which exist between the central Mexican and the other populations could have arisen through founder effect, genetic drift, or selection. There are three alleles (*Gpi 83* and *98*, and *Pep 83*) which are so far known only from central Mexico, and three (*Gpi 90* and *111*, and *Pep 96*) which are found only outside of central Mexico (Tables 15.2 & 15.4). Whether these really are unique alleles can only be determined through further sampling, since our sample sizes were small enough to have missed rare alleles. If there are indeed unique alleles outside of central Mexico, they could have arisen by post-migration mutation.

The existence of the *Gpi 90* allele in the Netherlands may provide clues for tracking the recent movement of *P. infestans* from the Netherlands to other parts of the world. Thus, strains originating in the Netherlands could have been transferred to Poland, the only other location presently known to have the *Gpi 90* allele. Except for less diversity for mating type, the genotypes in the Polish sample are similar to those found in the Netherlands (Table 15.3). The small sample from Israel contained only the *Gpi 100* allele, but larger or more recent samples may very well reveal the *90* allele. Other parts of the world reported to contain the *Gpi 100* allele, such as Egypt (El-Korany, Sattar & Shaw, unpublished) and U.K. (Shattock *et al.*, 1990) may also harbour the *90* allele, since, as stated above, the two can only be distinguished from each other under certain gel conditions.

Table 15.3. Genotypic frequencies in some collections of *Phytophthora infestans*.

Genotypes *Gpi*	*Pep*	Mating Type	Central Mexico	NW Mexico	U.S.A. & Canada	Peru	Nether-lands	U.K.	Poland	Israel	Japan
83/100	92/100	A1	0·01								
86/86	100/100	A1	0·03								
		A2	0·01								
86/100	83/100	A1	0·01								
86/100	92/100	A1			0·68	0·97			0·45		0·33
		A2	0·01								
86/100	100/100	A1	0·08		0·11	0·03					0·17
		A2	0·04								
86/122	100/100	A1	0·07	0·02							
		A2	0·03								
90/100	92/100	A1					0·33		0·11		
		A2					0·14				
90/100	100/100	A1					0·14		0·14		
		A2					0·14				
98/98	92/100	A2	0·01								
98/98	100/100	A2	0·01								
100/100	92/92	A1	0·03		0·04						
		A2	0·01								
100/100	92/100	A1	0·07	0·68	0·17		0·10	0·21[b]	0·05		
		A2	0·13								0·03
100/100	100/100	A1	0·14				0·10	0·79[b]	0·20		
		A2	0·18				0·05		0·05	1·00	0·47
100/122	92/92	A1	0·01								
100/122	92/100	A1	0·01								
100/122	100/100	A1	0·03								
		A2	0·03								
100/111/122[c]	100/100	A2		0·30							
122/122	92/100	A1	0·01								

[a]Based on data from Shattock *et al.*, 1990; Spielman, Gu & Fry, unpublished; Tooley *et al.*, 1985. Dates of collection and sample numbers are as given in Table 15.1. [b]Although both mating types were found in the United Kingdom, their proportions within each dilocus isozyme genotype were not given (Shattock *et al.*, 1990). [c]100/111/122 designates a five-banded GPI phenotype which we believe to be the result of the presence of the three alleles *100*, *111*, and *122*.

Table 15.4. Genotypes found in early collections[a] of *Phytophthora infestans*.

Genotypes		Mating	Number of Isolates		
Gpi	*Pep*	Type	Netherlands (1980-1981)	Switzerland (1981)	Ireland/Wales (1980-1982)
86/100	92/100	A1	1	2	4
86/100	96/100	A1			1
86/100	100/100	A1	1		
90/100	92/100	A1	1		
		A2	1		
100/100	92/100	A1			1
		A2	1		
100/100	100/100	A1	4		
		A2	1		

[a]These collections were obtained from L. Davidse, H. Hohl, R. Shattock, and L. Sujkowski.

A look at some of the genotypic frequencies (Table 15.3) reveals that central Mexico contained more different allozyme genotypes at the two loci than any of the other populations, which, again, is consistent with a central Mexican centre of origin. Within each genotype, the mating types are distributed in roughly 1:1 proportions, as would be expected for a random mating population.

The northwestern Mexico collection (from the Los Mochis region on the west coast) yielded only three genotypes (Table 15.3), a much lower level of genotypic diversity than that found in the central Mexican sample. One of these genotypes (*Gpi 100/111/122*) is five-banded, unlike any of the patterns expected for a typical dimeric enzyme. If there are three alleles present, (*100*, *111*, and *122*) the superimposition of the three different heterozygote patterns could produce a five banded pattern. This is possible because of the existence of hyperploidy in *P. infestans* (Whittaker, Shattock & Shaw, 1991; Chapter 13). The *Gpi 111* allele in the normal heterozygote, *100/111* does occur in another population, and the three-allele hypothesis is being tested by genetic crosses and ploidy determinations.

Each of the three genotypes in the northwestern Mexican collection is limited to only one mating type (Table 15.3). Since sexual reproduction would have produced recombinants, this population is presumed to be asexual, although both mating types are represented. This apparent isolation of the two mating types may have been caused by differences in host preference among the genotypes since the isolates with the *Gpi 100/100, Pep 92/100* genotype all came from tomato and were A1, and both of the other genotypes were isolated only from potato.

The samples from Peru and the United States were also much less diverse than central Mexico for isozyme genotypes, and contained only A1 mating types. Although the United States sample had two rare genotypes not found in Peru, both populations appear to be dominated by the genotype: A1, *Gpi 86/100, Pep 92/100.*

The Japanese collection contained both mating types and four different dilocus allozyme genotypes (Table 15.3). As in northwestern Mexico, there was a complete association of genotype with mating type: all A1 strains were *Gpi 86/100*, and all A2 strains were *Gpi 100/100.* Thus, sexual reproduction appears to be absent in the Japanese population. Data for virulence phenotypes also suggests that the Japanese population is strongly clonal (Mosa *et al.*, 1989). Although all of the Japanese isolates came from potato, the two mating types do differ in DNA content (Chapter 13); this may explain the lack of recombination.

A different situation was found in the Netherlands. Although genotypic diversity was lower than that in central Mexico, it was higher than in the U.K., Americas or Japan. Three of the four di-locus isozyme genotypes included both A1 and A2 mating types, and the genotype frequencies are close to those expected from random mating, suggesting that sexual reproduction was occurring in the Netherlands.

All nine isolates from Israel were identical: A2, *Gpi 100/100, Pep 100/100.* Although the sample is small, this suggests that the population may have been relatively homogeneous when it was sampled, unlike most of the other populations so far examined.

Finally, I would like to present a hypothesis suggested by a comparison of the data discussed above (Table 15.3) with additional data for a few small European samples from the early 1980s (Table 15.4). Each of the samples from the early part of the decade contains isolates with the genotype: A1, *Gpi 86/100.* In the Netherlands sample of ten isolates, two carried this genotype, one with *Pep 92/100* and one with *Pep 100/100.* Two early Swiss isolates were A1, *Gpi 86/100, Pep 92/100.* And among six early isolates from Ireland and Wales, five were A1, *Gpi 86/100*, and of these, four were also *Pep 92/100.* In the more recent samples from the Netherlands and the U.K. described in Table 15.3, the A1, *Gpi 86/100* genotype is virtually absent; Shattock *et al.* (1990) reported that only one isolate in a total of 477 British isolates recently tested for GPI was *86/100*). The A1, *Gpi 86/100* genotype (with either *Pep 92/100* or *100/100*) also occurs commonly in the United States, Peru, and Japan, and, in fact, *Gpi 86/100* is never associated with the A2 mating type outside of central Mexico (Table 15.3). Poland also had A1 isolates with the *Gpi 86/100* genotype in 1988 (Table 15.3), although they appear to be on the decrease (data not shown).

These data suggest that a shift in genotypic frequencies may be taking place in Europe, and that this shift was initiated by a migration event(s) which introduced both new isozyme genotypes and A2 mating types. The A1s with genotype *Gpi 86/100*, *Pep 92/100* appear to have predominated in Europe before the migration, and may have been characteristic of the A1 populations worldwide before the introduction of new genotypes. In Europe, these indigenous genotypes are perhaps now being replaced by new ones (associated with both A1 and A2 mating types). In the Netherlands the replacement may have begun in the late 1970s, and appears now to be nearly complete. If the Netherlands was the initial site of introduction, as suggested by the early record of A2 mating types there, we would expect to see the most advanced situation there. The virtual absence of *Gpi 86/100* genotypes in recent samples from the U.K. suggests that there, also, replacement may be complete. The fact that the indigenous genotypes appear to have been completely replaced in both the Netherlands and the U.K. indicates that they may have been less fit than the introduced genotypes. If this is generally true, it will have implications for disease control in the affected regions. In samples from other locations, such as Japan and Poland, the original genotypes were found along with newer genotypes, although we are presently unable to tell how stable these situations are.

Conclusions

Investigations of isozyme diversity in populations of *P. infestans* have revealed a surprising amount of variation in both genetic make-up and genetic structure. The genetic structures of the populations were sometimes unexpected; some populations which contained both mating types had structures which indicated strictly clonal reproduction. The relative levels and patterns of diversity support the hypothesis that central Mexico was the centre of origin of the *P. infestans/Solanum* pathosystem. These data also support the hypothesis that the European A2 mating types resulted from migration. Comparison of collections from the early 1980s with more extensive later collections, has identified genotypes which appear to have been present in the original, pre-A2 European population, and other genotypes which appear to be replacing them. Some regions, such as Peru, remained uninvaded at the time of sampling, and in other regions, such as Poland and Japan, the indigenous and migrant populations were both present. There may be many other parts of the world where the original genotypes are still common, as well as areas where genetic changes associated with the introduction of new genotypes are now occurring. The hypothesis of population replacement should be tested by determining the mating types and isozyme genotypes of populations throughout the world. In some regions, it may be possible to observe the process of population replacement as it takes place, but this will require rapid action to sample and characterize diverse *P. infestans* populations. Restriction fragment markers promise to provide a more complete 'fingerprint' of these genotypes to assist the study of genetics of populations of *P. infestans* (Chapters 17, 18).

References

Bourke, P. M. A. (1964). Emergence of potato blight, 1843-46. *Nature* **203**, 805-808.

Cox, A. E. & Large, E. C. (1960). Potato blight epidemics throughout the world. *Handbook No. 174*, Agricultural Research Service, United States Department of Agriculture.

Gallegly, M. E. & Galindo, J. (1958). Mating types and oospores of *Phytophthora infestans* in nature in Mexico. *Phytopathology* **48**, 274-277.

Galindo, J. & Gallegly, M. E. (1960). The nature of sexuality in *Phytophthora infestans*. *Phytopathology* **50**, 123-128.

Hohl, H. R. & Iselin, K. (1984). Strains of *Phytophthora infestans* from Switzerland with A2 mating type behaviour. *Transactions of the British Mycological Society* **83**, 529-530.

Malcolmson, J. F. (1985). *Phytophthora infestans* A2 compatibility type recorded in Great Britain. *Transactions of the British Mycological Society* **85**, 531.

Micales, J. A., Bonde, M. R. & Peterson, G. L. (1986). The use of isozyme analysis in fungal taxonomy and genetics. *Mycotaxon* **27**, 405-449.

Miller, P. R. & O'Brien, M. (1955). Tomato late blight: its world distribution and present status. *Plant Disease Reporter*, **231**, 89.

Mosa, A. A., Kato, M., Sato, N., Kobayashi, K. & Ogoshi, A. (1989). Occurrence of the A2 mating type of *Phytophthora infestans* on potato in Japan. *Annals of the Phytopathological Society of Japan* **55**, 615-620.

Shattock, R. C., Tooley, P. W. & Fry, W. E. (1986). The genetics of *Phytophthora infestans*: identification of recombination, segregation and selfing by isozyme analysis. *Phytopathology* **76**, 410-413.

Shattock, R. C., Shaw, D. S., Fyfe, A. M., Dunn, J. R., Loney, K. H. & Shattock, J. A. (1990). Phenotypes of *Phytophthora infestans* collected in England and Wales from 1985 to 1988: mating type, response to metalaxyl and isoenzyme polymorphisms. *Plant Pathology* **39**, 242-248.

Shaw, C. R. & Prasad, R. (1970). Starch gel electrophoresis of enzymes - a compilation of recipes. *Biochemical Genetics* **4**, 297-320.

Shaw, D. S. (1983). The cytogenetics and genetics of *Phytophthora*. In *Phytophthora: Its Biology, Taxonomy, Ecology, and Pathology*, (ed. D. C. Erwin, S. Bartnicki-Garcia & P. H. Tsao), pp. 81-94. American Phytopathological Society: St. Paul, Minnesota.

Shaw, D. S., Fyfe, A. M., Hibberd, P. G. & Abdel-Sattar, M. A. (1985). Occurrence of the rare A2 mating type of *Phytophthora infestans* on imported Egyptian potatoes and production of sexual progeny with A1 mating types from the U.K. *Plant Pathology* **34**, 551-556.

Siciliano, M. J. & Shaw, C. R. (1976). Separation and visualization of enzymes on gels. In *Chromatographic and Electrophoretic Techniques*, ed. I. Smith, Vol. 2, pp. 185-209. Heinemann.

Spielman, L. J., Sweigard, J. A., Shattock, R. C. & Fry, W. E. (1990). The genetics of *Phytophthora infestans*: Segregation of allozyme markers in F2 and backcross progeny and the inheritance of virulence against potato resistance genes *R2* and *R4* in F1 progeny. *Experimental Mycology* **14**, 57-69.

Therrien, C. D., Ritch, D. L., Davidse, L. C., Jespers, B. K. & Spielman, L. J. (1989). Nuclear DNA content, mating type and metalaxyl sensitivity of eighty-three isolates of *Phytophthora infestans* from the Netherlands. *Mycological Research* **92**, 140-146.

Tooley, P. W., Fry, W. E. & Villareal Gonzalez, M. J. (1985). Isozyme characterization of sexual and asexual *Phytophthora infestans* populations. *Journal of Heredity* **76**, 431-435.

Tooley, P. W., Therrien, C. D. & Ritch, D. L. (1989). Mating type, race composition, nuclear DNA content, and isozyme analysis of Peruvian isolates of *Phytophthora infestans*. *Phytopathology* **79**, 478-481.

Whittaker, S., Shattock, R. C. & Shaw, D. S. (1991). Variation in DNA contents of *Phytophthora infestans* as measured by a microfluorometric method using the fluorochrome DAPI. *Mycological Research* **95**, 602-610.

Chapter 16

Parasexuality in *Phytophthora*?

David N. Kuhn

Parasexuality was broadly defined by Pontecorvo (1956) as the transfer of genetic material from one organism to another without meiosis or development of specialized sexual structures. Pontecorvo (1956) pointed out the importance of the parasexual cycle for plant pathogenic fungi, which often do not undergo a sexual cycle. The parasexual cycle was suggested as a possible means whereby new pathogenic phenotypes are produced by the recombination of pathogenic genotypes.

In the study of plant pathogenic fungi, the most common uses of the parasexual cycle are to follow double stranded RNA (dsRNA) from strain to strain and to determine vegetative compatibility groups as genetic markers (Adams *et al.*, 1987; Correll, Klittich & Leslie, 1987). In *Phytophthora*, heterokaryons have been used to study the inheritance of cultivar specific pathogenicity (Denward, 1970; Leach & Rich, 1969; Malcolmson, 1970; Layton & Kuhn, 1988b) and the opportunity for intraspecific genetic transfer *in planta* (Layton & Kuhn, 1990).

Hyphal anastomosis (Stephenson, Erwin & Leary, 1974), heterokaryon formation (Denward, 1970; Layton & Kuhn, 1988a; Leach & Rich, 1969; Long & Keen, 1977a; Malcolmson, 1970) and asexual karyogamy (Layton and Kuhn, 1988b) have been reported in *Phytophthora*. If a complete parasexual cycle exists in *Phytophthora*, its role in generating new pathogenic races should be investigated. Because all the stages of the parasexual cycle have not been documented in *Phytophthora*, the parasexual cycle will be outlined for a haploid fungus such as *Aspergillus*.

In haploid fungi (Pontecorvo, 1956; Fincham & Day, 1971; Burnett, 1975), the parasexual cycle involves the initial fusion of two hyphae to allow mixing of nuclei and cytoplasm (formation of heterokaryons and heteroplasmons). The heterokaryotic state is often maintained only under selective pressure, otherwise parental nuclei may segregate and give rise to sectors containing only one parental nuclear type. During sporogenesis cytoplasmic genetic elements may assort with their respective nuclei, assort with non-parental nuclei, become stable mixtures, or a single parental cytoplasmic type may predominate. At a low rate, nuclei of different genotype fuse and form a heterozygous diploid nucleus. The diploid nucleus and the conidium are larger to accommodate the two sets of chromosomes. In the diploid, somatic crossing over can occur between chromatids of homologous chromosomes (Wood & Kafer, 1969). Although most diploids are unstable, stable diploids have been isolated from the wild (e.g. Clarkson & Heale, 1985). Diploid nuclei, whether recombinant or not, are reduced at a low frequency to haploidy by non-disjunction. Reduction is not always rapid; a variety of aneuploids (di- and trisomics) is possible.

Although the parasexual cycle is widespread among true fungi, there are some barriers to parasexuality that appear to prevent indiscriminate interchange of genetic information between isolates. Isolates of a single species will often show 'vegetative incompatibility' which is the inability to form heterokaryons or undergo fusion with other isolates. Vegetative incompatibility is based on a number of nuclear genes and the most detailed studies have demonstrated that vegetative compatibility requires that isolates have identical alleles at a variety of loci (Fincham & Day, 1971; Burnett, 1975). A difference in alleles at even one locus causes incompatibility. Vegetative compatibility is useful in categorizing large collections of otherwise morphologically indistinguishable field isolates and, in some cases, is reliably correlated with pathogenicity, virulence or race (Correll, Klittich & Leslie, 1987).

Vegetative incompatibility is as widespread as the parasexual cycle among fungi. Caten (1971), in a classic study of heterokaryon formation in *Aspergillus* spp., found that vegetative incompatibility was more the rule than the exception in that very few isolates of his wild populations could form heterokaryons with other isolates. Caten speculated that vegetative incompatibility may be a fungal defence against exposure to harmful cytoplasmic elements such as mycoviruses, dsRNA or mitochondrial mutants.

Vegetative incompatibility barriers often prevent cell fusion and therefore, at least in some cases, can be overcome by first removing the cell wall. Preparation and fusion of protoplasts has been successful in a variety of fungi to overcome vegetative incompatibility and to prepare heterokaryons between different species (Borkhardt & Olson, 1983; Typas, 1983; Pingyan & Kaiying, 1987; Peberdy, 1989) or even between different genera (Groves & Oliver, 1984; Perez, Vallin, & Benitez, 1984).

Now that protoplasts can be used to overcome vegetative incompatibility, the parasexual cycle has been put to a variety of uses. It is particularly powerful in undefined genetic systems where multigenic characters are to be transferred from one isolate to another. Thus, the improvement of commercial fermentation yeasts (which are commonly genetically undefined) has been attempted by fusion of protoplasts of commercial isolates with protoplasts of *Saccharomyces cerevisiae* to introduce genes for growth on particular substrates (Spencer *et al.*, 1985).

Heterokaryon formation through protoplast fusion has also been used to study cytoplasmic genetic factors, their inheritance and their interaction with the nuclear genome. Of major interest has been the transfer of dsRNA from species to species. Since the natural passage of dsRNA from isolate to isolate requires hyphal anastomosis, transfer is normally limited to within a vegetative compatibility group (Day & Anagnostakis, 1973; Wood & Bozarth, 1973)). As a means of determining the phenotype associated with the dsRNA, transfer between vegetative compatibility groups, species (Pingyan & Kaiying, 1987) and genera (Perez, Vallin & Benitez, 1984) has been attempted by protoplast fusion. The replication and maintenance of mitochondria has been studied in similar protoplast fusion experiments (Borkhardt & Olson, 1983; Goodey & Bevan, 1983; Benitez *et al.*, 1984; Spencer *et al.*, 1985).

In the diploid yeast *Candida utilis*, the parasexual cycle was initially used to try to establish linkage groups for the species because no sexual cycle existed (Poulter *et al.*, 1981). Heterokaryon formation led to some intermixing of chromosomes in a single nucleus due to transfer of chromosomes from one nucleus to another rather than to karyogamy and loss of chromosomes (Sarachek, Rhoads & Schwarzhoff, 1981; Sarachek & Weber, 1984). Thus, in diploids like *Phytophthora*, the parasexual cycle may also be useful in transferring particular chromosomes from one isolate to another to form stable aneuploids.

Heterokaryon formation

Heterokaryon formation occurs naturally when two hyphae of different genotype fuse and the nuclei of the two hyphae are mixed. The mechanism by which hyphae fuse in true fungi is unknown. Heterokaryons can have a variety of forms (Burnett, 1975) depending on the type of fungi fusing. In fungi which are septate and uninucleate, heterokaryon formation may be only at the point of hyphal fusion if the nuclei cannot move from that point. In *Phytophthora*, which is coenocytic and multinucleate, hyphal fusion at any point could lead to a widespread mixing of nuclei and cytoplasm. After heterokaryon formation, homokaryotic sectors may form which regenerate the parental types. The cytoplasm of these parental types may vary from a mixture of cytoplasms to cytoplasm of one parent or the other. *Phytophthora* isolates with both nuclear and mitochondrial markers have not been used to study the process of hyphal fusion. Thus, the extent of nuclear and cytoplasmic mixing after fusion is unknown.

Production of protoplasts of Phytophthora *for heterokaryon formation*

To overcome vegetative incompatibility or to produce interspecific or intergeneric heterokaryons, protoplasts of the parental isolates are fused. The general requirements for protoplast production are an osmoticum to protect protoplasts from osmotic rupture after release from the cell wall, and a wall-removing enzyme. The variables in protoplast production are well described by Peberdy (1989). It is important to remember that there is no standard procedure for protoplast production. The best approach to obtain optimal yields of protoplasts is to experiment with different enzymes, osmotica, and developmental stages of the tissue. Most fungi have chitin in their cell walls and are thus digested by most commercially available cell wall-removing enzymes which contain chitinase. *Phytophthora* species do not have chitin in their cell walls and may be digested by cellulases normally used for plant protoplast production. The exact conditions and enzymes for optimal protoplast production vary from species to species and even from isolate to isolate. Layton & Kuhn (1988a) obtained good yields of protoplasts of *P. megasperma* from rapidly growing mycelial cultures or from zoospores which had been germinated for 24 hours; walls were digested by Driselase (Sigma) in mannitol. Campbell *et al.* (1989) and Kinghorn *et al.* (Chapter 19) have reported highly efficient production of protoplasts from sporangia and encysted zoospores of *P. infestans* digested with NovoZym 234 (Novo BioLabs) in $KCl/CaCl_2$. Because *Phytophthora* is coenocytic, protoplasts tend to form by protoplasm squeezing out through holes in the cell wall and then pinching off to form spheres. Nuclear content of protoplasts can vary from none to several (Layton & Kuhn, 1988a); thus, not all protoplasts will regenerate cell walls and form colonies.

Protoplast fusion can be driven by addition of polyethylene glycol and Ca^{2+} to a mixture of protoplasts from two parents. Polyethylene glycol reduces the volume of the medium and makes it energetically more favourable for membranes to fuse. Such fusion can occur between any cells and there appears to be no specificity due to membrane composition or cell source (Jones *et al.*, 1976; Kao *et al.*, 1974). Protoplast fusion can also be accomplished by alignment of cells in a mild electric field followed by brief pulses of current (Nea & Bates, 1987; Sonnenberg & Wessels, 1987). This electrofusion procedure is highly efficient but does require some special equipment. In both cases, fusion is random with the opportunity for fusion of parent A with parent A, parent B with parent B, A with B, or multiple events leading to large, multinucleate mixtures of A and B. In some cases, protoplasts can be stained differently, or protoplasts of different size or density can be fused to make identification of heterokaryons easier or to enrich a particular fraction for heterokaryons (Kamata & Nagata, 1987; Thomas & Rose, 1988). More usually, parents with two selectable mutations or phenotypes are used and heterokaryons that have regenerated their cell wall and grown to colonies can be selected on suitable media.

Mutant production in Phytophthora

To allow selection of true heterokaryons, isolates must be mutated and stable, nuclear mutations identified. Without such nuclear mutations, selection is not stringent enough to distinguish between true heterokaryons and physical mixtures of mycelia that may only rarely be heterokaryotic. In haploid fungi, selection of auxotrophs is not difficult and a variety of mutagens such as ultraviolet radiation (uv) or *N*-methyl-*N*'-nitro-*N*'-nitrosoguanidine (NTG) can be used, with approximately the same rate of mutation and range of mutant phenotypes observed for each mutagen. *Phytophthora* is believed to have a diploid mycelium and asexual spores (e.g. Long & Keen, 1977b; Shaw, 1983). Because it is difficult to obtain reasonable numbers of oospore progeny from most *Phytophthora* species, mutagenesis followed by generation of oospores to segregate potential mutations is not generally undertaken. Thus, isolation of mutants in *Phytophthora* has proved more difficult than in haploid fungi.

Mutagenesis using NTG has been most effective (Long & Keen, 1977b; Davidse, 1981; Joseph & Coffey, 1984; Layton & Kuhn, 1988a), although metalaxyl resistant mutants have also been produced by UV irradiation at essentially the same rate as with NTG (Davidse, 1981; Bruin & Edgington, 1982; Joseph & Coffey, 1984). Mutants must always be characterized for growth and sporulation before being used in heterokaryon formation experiments, due to the importance of zoosporogenesis in the analysis of heterokaryons. For example, Long & Keen (1977a) were able to form heterokaryons between auxotrophs but had difficulty in their analysis because one of the parents could not produce viable zoospores or oospores.

Some mutant phenotypes occur at a greater frequency than others (Long & Keen, 1977b; Davidse, 1981; Joseph & Coffey, 1984; Layton & Kuhn, 1988a). Mutation to metalaxyl resistance was easily selected with or without mutagenesis. Induced mutation to para-fluorophenylalanine or fluorotryptophan resistance was also easily selected (Layton & Kuhn, 1988a) at frequencies typical of a haploid rather than a diploid. Bruin & Edgington (1982) noted that mutation to metalaxyl resistance occurred at least 100 fold more frequently in UV irradiated *P. capsici* than in similarly treated *Peronospora parasitica.* Since the latter is an obligate parasite, mutant selection procedures are not strictly comparable. Nevertheless, there are several possible explanations.

Phytophthora species may only require mutation of a single allele to become metalaxyl or fluorophenylalanine resistant. Shattock's (1988) work suggested that resistance in *P. infestans* is due to a single semi-dominant allele. Alternatively, if resistance is recessive, wild types could already carry one resistance allele in the heterozygous condition and would need only one mutation in the other allele to become resistant. Another possibility is that monosomy of the chromosome carrying the resistance gene is common; again a single mutation only, would suffice. However, these alternative explanations are less likely as our own work on doubly-resistant heterokaryons argues strongly for dominant resistance. In the discussion which follows, the drug resistance markers used are assumed to be chromosomal rather than cytoplasmic. Thus, colonies exhibiting resistance markers from both parents are assumed to be heterokaryons and not heteroplasmons.

Heterokaryon production in Phytophthora *by protoplast fusion*

Layton & Kuhn (1988a) were able to isolate a number of metalaxyl and para-fluorophenylalanine mutants from two races of *P. megasperma* f.sp. *glycinea*; those with growth and sporulation most like the parent were used to produce heterokaryons. Protoplasts from metalaxyl and para-fluorophenylalanine resistant isolates were mixed and fused by polyethylene glycol and Ca^{2+} treatment. The fused protoplasts were allowed to regenerate for 24 hours and then putative heterokaryons were selected by overlaying with agar containing both metalaxyl and fluorotryptophan at levels inhibitory to either parent alone. Heterokaryons were produced at a rate of about 3×10^{-4}. Heterokaryons grew slowly in the presence of both drugs; control protoplasts of single parentage, fused or not, did not grow. Mixing of protoplasts without fusion produced no colonies able to grow in the presence of both drugs. Thus, in this experiment stringent selection for dual drug resistance made possible the isolation of true heterokaryons rather than cross-resistant physical mixtures of parental mycelia. The efficiency of heterokaryon production varied with mutation and isolate, as observed in other fungi (Typas, 1983). This may relate to nuclear and cytoplasmic incompatibility which prevents heterokaryon formation in some fungi even after successful protoplast fusion.

Layton & Kuhn (1988a) were unable to form heterokaryons between *P. megasperma* f.sp. *glycinea* mutants and a single isolate of *P. megasperma* f.sp. *medicaginis* resistant to metalaxyl. Failure may have been due to the availability of only one *P. megasperma* f.sp. *medicaginis* mutant for these experiments. In general, sexual and parasexual genetics in *Phytophthora* will be greatly aided by the availability of a large number of differing types of mutants.

Characterization of heterokaryons

Characterization of heterokaryons provides information on the mix of nuclei and cytoplasmic factors from both parents. In *Phytophthora*, heterokaryons can be characterized by sampling zoospore progeny. During zoosporogenesis, the existing nuclei and cytoplasm are transported to the sporangium where they are 'packaged' into uninucleate zoospores. Colonies grown from a single zoospore should contain identical nuclei. The nuclei, packaged singly into zoospores, should represent the nuclei present in the heterokaryon (if both parents were capable of producing viable zoospores, see Long & Keen, 1977a). Thus, by assessing the ratio of parental phenotypes found in the zoospore progeny of a heterokaryon, one can obtain information on its nuclear ratio.

The nuclear ratio in a heterokaryon can vary widely (Burnett, 1975). Layton & Kuhn (1988a) selected some heterokaryons with more than 20 times as many nuclei of one parent as the other which behaved exactly as those heterokaryons with roughly equal numbers of nuclei from each parent. Thus, neither heterokaryon phenotype nor segregation of zoospore progeny was a good indicator of nuclear ratio. However, it is important to note that because *Phytophthora* is coenocytic, even an extreme ratio might yield a doubly-resistant phenotype.

If the nuclei or mitochondria of the parents are shown to be significantly different by isozyme or restriction fragment length polymorphism (RFLP) analysis (Förster *et al.*, 1988; Chapters 11, 15, 17 and 18) the zoospore progeny of the heterokaryon can be analyzed for recombination and segregation of nuclear and cytoplasmic genotypes. If karyogamy has occurred in the heterokaryon, zoospore progeny expressing genes from both parents may appear. Long & Keen (1977a) observed only one putative recombinant among 1361 of the zoospore progeny of their heterokaryons. Layton & Kuhn (1988b) found 498 single zoospore colonies which expressed both parental drug resistance phenotypes among 2554 zoospore progeny from 11 different heterokaryons. Because both groups lacked isolates with multiple markers, neither group could determine whether stable tetraploids, aneuploids or a recombined diploid had been formed to bring both drug resistant traits into a single zoospore. If all zoospore progeny are resistant to both drugs, then the putative heterokaryon may simply be a mutation of either one or the other parent to double drug resistance. Layton & Kuhn (1988a) observed this in the fusions of *P. megasperma* f.sp. *glycinea* and *P. megasperma* f.sp.

medicaginis, where sufficient isozyme differences in the parents were available to make analysis possible.

Ideally, one would like to form heterokaryons between parents each containing multiple markers for both the nucleus and mitochondria, to allow detection of heteroplasmons, polyploids, recombinant diploids or aneuploids in zoospore progeny expressing both parental phenotypes. Due to the general lack of mutants of *Phytophthora*, this is rarely done (but see Lucas *et al.*, 1990). Isozymes could provide suitable markers to allow unambiguous identification of recombinants in zoospore progeny of heterokaryons. Unfortunately, the isolates of *P. megasperma* f.sp. *glycinea* used by Layton & Kuhn (1988a) were monomorphic for 19 enzymes.

Characterizing the phenotype of the heterokaryon

Once putative heterokaryons have been characterized by recovering both parental phenotypes in zoospore progeny, the virulence phenotype of the heterokaryons can be characterized. Layton & Kuhn (1988b) formed heterokaryons between mutants of *Phytophthora megasperma* f.sp. *glycinea* race 1 and race 3, resistant to metalaxyl or fluorotryptophan respectively. Heterokaryons were selected and maintained on plates with both metalaxyl and fluorotryptophan. The pathogenicity of the heterokaryons was tested by mycelial inoculation of hypocotyls of soybean differentials that could distinguish between race 1 and race 3 and also of soybean differentials which would be susceptible only to races other than 1 or 3 to detect novel pathogenicity phenotypes. Doubly drug resistant heterokaryons were recovered from the inoculated plants, which demonstrated that infection was due to the heterokaryon rather than to parental phenotypes segregating from the heterokaryon in the absence of selective pressure *in planta*. The race 1 phenotype was dominant to the race 3 phenotype as the majority of heterokaryons acted as race 1, independent of the race 1 to race 3 nuclear ratio. No novel races were observed. These experiments demonstrated that *Phytophthora* heterokaryons could be produced by protoplast fusion between races of *P. megasperma* f.sp. *glycinea* and that the race phenotype of the heterokaryon could be determined.

The real power of hyphal anastomosis or protoplast fusion is in the ability to mix the nuclear genomes and cytoplasmic components of two isolates and determine the effect on the phenotype of the heterokaryon/heteroplasmon (Layton & Kuhn, 1988b; Clark & Lott, 1989). Analysis of the heterokaryon/heteroplasmon phenotype requires no assumptions about the polygenic or monogenic nature of the genotype being studied or its location in the nucleus or cytoplasm. Thus, in the analysis of the inheritance of some traits, formation of heterokaryon/heteroplasmons may be preferable to sexual crossing because it does not disrupt particular combinations of alleles. Instead, the interaction of particular gene combinations from both parents may be observed.

Heterokaryon formation *in vivo*

It can be concluded that *Phytophthora* heterokaryons (or heteroplasmons) can be formed in the laboratory, but how often are *Phytophthora* heterokaryons found in nature? To address this question, Layton & Kuhn (1990) took advantage of the ease of detection of their metalaxyl and fluorotryptophan mutants of *Phytophthora megasperma* f.sp. *glycinea*. Plants susceptible to both mutants were inoculated with a mixture of metalaxyl and fluorotryptophan resistant zoospores and putative heterokaryotic mycelial colonies were reisolated from infected roots onto plates containing both drugs. Twenty-three double drug resistant colonies were observed from reisolations from 189 plants. Of these colonies, only one continued to grow when transferred from the initial isolation plate and root piece to a new double drug plate. Both parental drug resistance phenotypes were observed in the zoospore progeny, thus identifying this colony as a heterokaryon. No double drug resistant colonies were observed on controls where zoospores were mixed and plated directly onto selective medium. Although plants were inoculated with 20,000 zoospores of each mutant, it is not known what proportion actually penetrated the plant and so an estimate of the efficiency of heterokaryon formation *in planta* was not possible.

Heterokaryons formed *in planta* were detectable only because a positive selection for the doubly resistant phenotypes and against the parents was available. Without this selection, it would not be possible to determine if heterokaryons commonly form between races during a mixed infection, as the heterokaryon would have the race phenotype of one or the other of the parents as in the experiments of Layton & Kuhn (1988b).

Although it was not possible to calculate the frequency of heterokaryon formation in plants, the observation of one heterokaryon in the 189 plants tested raises the interesting question of whether there are environmental conditions that favour the production of heterokaryons. Two factors are particularly intriguing. Does growth in the plant lead to a more efficient alignment of hyphae of different isolates due to the likelihood that isolates that infect first may create a path for later infecting hyphae? Can increased hyphal alignment alone increase the rate of heterokaryon formation? Are there particular compounds or nutritional factors that increase the rate of heterokaryon formation? The environment inside the plant and the increased levels of extracellular wall degrading enzymes necessary to gain entrance into the plant cells may make hyphal fusion more likely. Limitation of certain necessary cofactors and the presence of a variety of sterols, which affect growth and sporulation of *Phytophthora*, may also play a role. We are interested in pursuing such questions to be able to predict how frequent heterokaryosis is in the field and judge its potential importance in generating new pathogenic races.

Many phenomena have been recorded in work with *Phytophthora* which could have their bases in parasexual events. For example, isolates of *P. megasperma* vary widely in chromosome number; increased chromosome number is correlated with increased oospore size (Hansen *et al.*, 1986; Chapter 10). Variations in DNA content in isolates of *P. infestans* suggest that polyploids and aneuploids are common (e.g. Chapter 13). Larger karyotypes in these species may have arisen through somatic fusion and parasexual karyogamy.

Variations among zoospore progeny from a single isolate (e.g. Hilty & Schmitthenner, 1962; Rutherford, Ward & Buzzell, 1985), high percentages of oospore abortion (Rutherford & Ward, 1985) and low viability of oospores (e.g. Shaw, 1983) are well known in *Phytophthora*. These phenomena might also be a result of parasexual events in the vegetative phase which come to light only after zoosporogenesis or after less frequent sexual reproduction.

Recent work with dsRNA as a marker argues for gene flow, not only within species but also between species; this transfer could have been asexual, i.e. parasexual. Molecules of dsRNA sharing sequence homology have been observed in *P. megasperma* isolates from widely separated geographical locations (Allen & Kuhn, unpublished). These dsRNAs also share homology with *P. infestans* dsRNA (Tooley, Hewings & Falkenstein, 1989). Transmission of dsRNA from one individual to another may be only by hyphal fusion, no infective virus particles or infective dsRNAs having yet been reported. Thus, some *P. infestans* and *P. megasperma* isolates may have, at some point, undergone hyphal fusion and the transfer of dsRNA. Such a transfer would fit well with the 'saltation' events seen in the co-evolution of plants and fungal pathogens (Dick, 1988; Hijwegen, 1988), where instead of a slow evolution to greater pathogenicity or wider host range, particular species seem to have gained the ability to grow on a new host plant in a single evolutionary leap. Horizontal transfer of genetic material between a successful pathogen and a potential pathogen has been suggested to explain such dramatic changes (Pirozynski, 1988; Pirozynski & Hawksworth, 1988).

Conclusions

In the laboratory, the parasexual cycle can be used to study the genetics of *Phytophthora*. Its potential in studying multigenic traits such as host range is great, as complex sets of alleles and cytoplasmic components can be combined.

The practical application of the parasexual cycle is limited by the lack of selectable mutants and mutants with multiple markers. Increased numbers of mutants will make possible the definition of vegetative compatibility groups and linkage groups in homothallic species and in species where oospore formation or germination is rare. Increased numbers of mutants from a variety of species will allow tests for interspecific heterokaryon formation and make the detailed analysis of karyogamy possible. Perhaps other laboratories would be willing to exchange mutants of value in this kind of study.

The level of natural heterozygosity or heterokaryosis and the range of ploidy in *Phytophthora* must be defined before the importance of the parasexual cycle in the generation of new pathogenic races or new pathogenic species can be determined. Heterokaryons can form *in planta* but the likelihood of survival of these heterokaryons or their contribution to the production of new pathogenic races is at present unknown.

References

Adams, G., Johnson, N., Leslie, J. F. & Hart, P. L. (1987). Heterokaryons of *Gibberella zeae* formed following hyphal anastomosis or protoplast fusion. *Experimental Mycology* **11**, 339-353.

Benitez, T., del Castillo, L., Aguilera, A. & Conde, J. (1984). Instability of *Saccharomyces cerevisiae* heterokaryons. *Current Genetics* **8**, 345-352.

Borkhardt, B. & Olson, L. W. (1983). Paternal inheritance of the mitochondrial DNA in interspecific crosses of the aquatic fungus *Allomyces*. *Current Genetics* **7**, 403-404.

Bruin, G. C. A. & Edgington, L. V. (1982). Induction of fungal resistance to metalaxyl by ultraviolet irradiation. *Phytopathology* **72**, 476-480.

Burnett, J. H. (1975). *Mycogenetics*. John Wiley & Sons: London.

Campbell, A. M., Moon, R. P., Duncan, J. M., Gurr, S. J. & Kinghorn, J. R. (1989). Protoplast formation and regeneration from sporangia and encysted zoospores of *Phytophthora infestans*. *Physiological and Molecular Plant Pathology* **34**, 299-307.

Caten, C. E. (1971). Heterokaryon incompatibility in imperfect species of *Aspergillus*. *Heredity* **26**, 299-312.

Clark, J. & Lott, T. (1989). Age heterokaryon studies in *Didymium iridis*. *Mycologia* **81**, 636-638.

Clarkson, J. M. & Heale, J. B. (1985). Heterokaryon compatibility and genetic recombination within a host plant between hop wilt isolates of *Verticillium albo-atrum*. *Plant Pathology* **34**, 129-138.

Correll, J. C., Klittich, C. J. R. & Leslie, J. F. (1987). Nitrate nonutilizing mutants of *Fusarium oxysporum* and their use in vegetative compatibility tests. *Phytopathology* **77**, 1640-1646.

Davidse, L. C. (1981). Resistance to acylalanine fungicides in *Phytophthora megasperma* f.sp. *medicaginis*. *Netherlands Journal of Plant Pathology* **87**, 11-24.

Day, P. R. & Anagnostakis, S. L. (1973). The killer system in *Ustilago maydis*: heterokaryon transfer and loss of determinants. *Phytopathology* **63**, 1017-1018.

Denward, T. (1970). Differentiation in *Phytophthora infestans* II. Somatic recombination in vegetative mycelium. *Hereditas* **66**, 35-48.

Dick, M. W. (1988). Coevolution in the heterokont fungi (with emphasis on the downy mildews and their angiosperm hosts). In *Coevolution of Fungi with Plants and Animals*, (ed. K. A. Pirozynski & D. L. Hawksworth), pp. 31-62. Harcourt Brace Jovanovich: London.

Fincham, J. R. S. & Day, P. R. (1971). *Fungal Genetics*. Blackwell Scientific Publications: Oxford.

Förster, H., Kinscherf, T. G., Leong, S. A. & Maxwell, D. P. (1988). Estimation of relatedness between *Phytophthora* species by analysis of mitochondrial DNA. *Mycologia* **80**, 466-478.

Goodey, A. R. & Bevan, E. A. (1983). Production and analysis of yeast cybrids. *Current Genetics* **7**, 69-72.

Groves, D. P. & Oliver, S. G. (1984). Formation of intergeneric hybrids of yeast by protoplast fusion of *Yarrowia* and *Kluyveromyces* species. *Current Genetics* **8**, 49-55.

Hansen, E. M., Brasier, C. M., Shaw, D. S., & Hamm, P. B. (1986). The taxonomic structure of *Phytophthora megasperma*: evidence for emerging biological species groups. *Transactions of the British Mycological Society* **87**, 557-573.

Hijwegen, T. (1988). Coevolution of Flowering Plants with Pathogenic Fungi. In *Coevolution of Fungi with Plants and Animals*, (ed. K. A. Pirozynski & D. L. Hawksworth), pp. 63-77. Harcourt Brace Jovanovich: London.

Hilty, J. W. & Schmitthenner, A. F. (1962). Pathogenic and cultural variability of single zoospore isolates of *Phytophthora megasperma* var. *sojae*. *Phytopathology* **52**, 869-872.

Jones, C. W., Mastrangelo, I. A., Smith, H. H., Liu, H. Z. & Meck, R. A. (1976). Interkingdom fusion between human (HeLa) cells and tobacco hybrid (GGLL) protoplasts. *Science* **193**, 401-403.

Joseph, M. C. & Coffey, M. D. (1984). Development of laboratory resistance to metalaxyl in *Phytophthora citricola*. *Phytopathology* **74**, 1411-1414.

Kamata, Y. & Nagata, T. (1987). Enrichment of heterokaryocytes between mesophyll and epidermis protoplasts by density gradient centrifugation after electric fusion. *Theoretical and Applied Genetics* **75**, 26-29.

Kao, K. N., Constabel, F., Michayluk, M. R. & Gamborg, O. L. (1974). Plant protoplast fusion and growth of intergeneric hybrid cells. *Planta* **120**, 215-227.

Layton, A. C. & Kuhn, D. N. (1988a). Heterokaryon formation by protoplast fusion of drug-resistant mutants in *Phytophthora megasperma* f.sp. *glycinea*. *Experimental Mycology* **12**, 180-194.

Layton, A. C. & Kuhn, D. N. (1988b). The virulence of interracial heterokaryons of *Phytophthora megasperma* f.sp. *glycinea*. *Phytopathology* **78**, 961-966.

Layton, A. C. & Kuhn, D. N. (1990) *In planta* formation of heterokaryons of *Phytophthora megasperma* f.sp. *glycinea*. *Phytopathology* **80**, 602-606.

Leach, S. S. & Rich, A. E. (1969). The possible role of parasexuality and cytoplasmic variation in race differentiation in *Phytophthora infestans*. *Phytopathology* **59**, 1360-1365.

Long, M. & Keen, N. T. (1977a). Evidence for heterokaryosis in *Phytophthora megasperma* var. *sojae*. *Phytopathology* **67**, 670-674.

Long, M. & Keen, N. T. (1977b). Genetic evidence for diploidy in *Phytophthora megasperma* var. *sojae*. *Phytopathology* **67**, 675-677.

Lucas, J. A., Greer, G., Oudemans, P. V. & Coffey, M. D. (1990). Fungicide sensitivity in somatic hybrids of *Phytophthora capsici* obtained by protoplast fusion. *Physiological and Molecular Plant Pathology* **36**, 175-187.

Malcolmson, J. F. (1970). Vegetative hybridity in *Phytophthora infestans*. *Nature* **225**, 971-972.

Nea, L. J. & Bates, G. W. (1987). Factors affecting protoplast electrofusion efficiency. *Plant Cell Reports* **6**, 337-340.

Peberdy, J. F. (1989). Fungi without coats - protoplasts as tools for mycological research. *Mycological Research* **93**, 1-20.

Perez, C., Vallin, C. & Benitez, J. (1984). Hybridization of *Saccharomyces cerevisiae* with *Candida utilis* through protoplast fusion. *Current Genetics* **8**, 575-580.

Pingyan, L. & Kaiying, C. (1987). Virus transmission through interspecies protoplast fusion in *Aspergillus*. *Transactions of the British Mycological Society* **89**, 73-81.

Pirozynski, K.A. (1988). Coevolution by horizontal gene transfer: a speculation on the role of fungi. In *Coevolution of Fungi with Plants and Animals*, (ed. K. A. Pirozynski & D. L. Hawksworth), pp. 247-268. Harcourt Brace Jovanovich: London.

Pirozynski, K. A. & Hawksworth, D. L. (1988). Coevolution of fungi with plants and animals: introduction and overview. In *Coevolution of Fungi with Plants and Animals*, (ed. K. A. Pirozynski & D. L. Hawksworth), pp. 1-29. Harcourt Brace Jovanovich: London.

Pontecorvo, G. (1956). The parasexual cycle in fungi. *Annual Review of Microbiology* **10**, 393-400.

Poulter, R., Jeffery, K., Hubbard, M. J., Shepherd, M. G. & Sullivan, P. A. (1981). Parasexual genetic analysis of *Candida albicans* by spheroplast fusion. *Journal of Bacteriology* **146**, 833-840.

Rutherford, F. S. & Ward, E. W. B. (1985) Evidence for genetic control of oospore abortion in *Phytophthora megasperma* f.sp. *glycinea*. *Canadian Journal of Botany* **63**, 1671-1673.

Rutherford, F. S., Ward, E. W. B. & Buzzell, R. I. (1985). Variation in virulence in successive single-zoospore propagations of *Phytophthora megasperma* f.sp. *glycinea*. *Phytopathology* **75**, 371-374.

Sarachek, A., Rhoads, D. D. & Schwarzhoff, R. H. (1981). Hybridization of *Candida albicans* through fusion of protoplasts. *Archives of Microbiology* **129**, 1-8.

Sarachek, A. & Weber, D. A. (1984). Temperature-dependent internuclear transfer of genetic material in heterokaryons of *Candida albicans*. *Current Genetics* **8**, 181-187.

Shattock, R. C. (1988). Studies on the inheritance of resistance to metalaxyl in *Phytophthora infestans*. *Plant Pathology* **37**, 4-11.

Shaw, D. S. (1983). The cytogenetics and genetics of *Phytophthora*. In *Phytophthora: Its Biology, Taxonomy, Ecology, and Pathology*, (ed. D. C. Erwin, S. Bartnicki-Garcia & P. H. Tsao), pp. 81-94. American Phytopathological Society: St. Paul, Minnesota.

Sonnenberg, A. S. M. & Wessels, J. G. H. (1987). Heterokaryon formation in the basidiomycete *Schizophyllum commune* by electrofusion of protoplasts. *Theoretical and Applied Genetics* **74**, 654-658.

Spencer, J. F. T., Spencer, D. M., Bizeau, C., Martini, A. V. & Martini, A. (1985). The use of mitochondrial mutants in hybridization of industrial yeast strains. Relative parental contributions to the genomes of interspecific and intergeneric yeast hybrids obtained by protoplast fusion, as determined by DNA reassociation. *Current Genetics* **9**, 623-625.

Stephenson, L. W., Erwin, D. C. & Leary, J. V. (1974) Hyphal anastomosis in *Phytophthora capsici*. *Phytopathology* **64**, 149-150.

Thomas, M. R. & Rose, R. J. (1988). Enrichment for *Nicotiana* heterokaryons after protoplast fusion and subsequent growth in agarose microdrops. *Planta* **175**, 396-402.

Tooley, P. W., Hewings, A. D. & Falkenstein, K. F. (1989). Detection of double-stranded RNA in *Phytophthora infestans*. *Phytopathology* **79**, 470-474.

Typas, M.A. (1983). Heterokaryon incompatibility and interspecific hybridization between *Verticillium albo-atrum* and *Verticillium dahliae* following protoplast fusion and microinjection. *Journal of General Microbiology*, **129**, 3043-3056.

Wood, H. A. & Bozarth, R. F. (1973). Heterokaryon transfer of viruslike particles and a cytoplasmically inherited determinant in *Ustilago maydis*. *Phytopathology* **63**, 1019-1021.

Wood, S. & Kafer, E. (1969). Effects of ultraviolet irradiation on heterozygous diploids of *Aspergillus nidulans* I. UV-induced mitotic crossing over. *Genetics* **62**, 507-518.

Chapter 17

DNA polymorphism in *Phytophthora infestans*: the Cornell experience

Stephen B. Goodwin

The population biology of *Phytophthora infestans* may be the most interesting of any *Phytophthora* species. This is partly ascribable to the prominent role played by this fungus in the Irish potato famine. Although other Phytophthoras certainly cause very destructive diseases, none has been so devastating to so many people over such a short period of time. Another factor adding to the interest in this fungus derives from its strongly heterothallic nature, and the fact that until recently only the A1 mating type was known anywhere outside of central Mexico. The potential for the introduction of the A2 mating type has lead to much speculation about the changes that may occur when it arrives in new locations. What really makes this fungus interesting for the population biologist is the historical record of its global spread. The fungus probably evolved in a relatively small area in central Mexico, from which it was disseminated through the activities of human beings in the last 150 years. The extremely destructive nature of this disease ensured that it was not long ignored after its arrival in a new location, providing an excellent written record of the initial migration of the fungus outside Mexico (see chapter 3). This written record is a boon for population geneticists and evolutionary biologists because it may be possible to document the effects of limited, potentially recurring, migration events on the genetic structure of populations.

However for this to occur it is necessary that the requisite genetic markers be available. Although *P. infestans* has been the subject of the most intensive genetic studies of any *Phytophthora* species, genetic analyses in this organism are still in their infancy. Isozymes have proved to be useful markers in many fungi, but for *P. infestans* only three systems have been studied in detail; in only two of these have the genetics been worked out unambiguously (Chapter 15). Genetic studies have also been carried out on the mating type locus, and although mating type seems to segregate in a 1:1 ratio in most crosses, the exact genetic basis for this trait remains elusive (Chapter 14). Drug resistances and a few other markers also exist (Shaw, 1988), yet these may not occur in natural populations and clearly the number of markers available is not sufficient for thoroughly understanding the population genetics and evolution of this organism.

For these reasons the development and mapping of molecular markers is imperative for *P. infestans*. Because these markers are needed to address basic questions in both genetics and population biology, both mitochondrial and nuclear markers must be used. Mitochondrial DNA variation may be more useful than nuclear DNA variation when studying the migration events that have undoubtedly played an important role in the recent evolution of this species, and this molecule may also be useful for inferring phylogenetic relationships among *Phytophthora* species (Chapter 11). As *P. infestans* can reproduce both asexually and sexually in the same population, it would be advantageous to develop a method for distinguishing between asexual and sexual progeny. The recently developed technique of DNA fingerprinting with probes for moderately repetitive DNA may be useful for this purpose. Genetically mapped, single copy nuclear DNA probes are also needed for studying such phenomena as linkage disequilibrium in *P. infestans* populations. Mapping is one of the first steps in the development of molecular markers for population genetics and is necessary for interpreting the observed patters of variation in natural populations. Due to the broad scope of potential applications for these markers, three topics relating to the development and use of molecular markers in *P. infestans* will be discussed in this chapter: mitochondrial DNA restriction site mapping; single-copy nuclear DNA RFLP probes; and DNA fingerprinting.

Mitochondrial DNA restriction site mapping

There are several reasons for undertaking molecular mapping studies on the mitochondrial genome of *P. infestans*. The size of the mitochondrial genome in Oomycetes is relatively small compared to those in the true fungi and plants, ranging from 36 kb in *P. infestans* to 70 kb in *Pythium diclinum* (reviewed in McNabb & Klassen, 1988). The largest *Phytophthora* mitochondrial DNA (mtDNA) is the 45 kb reported for *P. megasperma* f. sp. *glycinea* (Förster *et al.*, 1987). The mtDNA of Oomycetes and other fungi has a very different buoyant density in a CsCl-bisbenzimide gradient compared to nuclear DNA, and this property allows the mtDNA to be purified relatively easily (Garber & Yoder, 1983). Because of this ease of purification and small size, physical mapping of restriction enzyme sites on the mitochondrial genomes of the Oomycetes is a straightforward process. The mitochondrial genomes of many organisms are believed to be evolving rapidly (Brown, George & Wilson, 1979) and this could be important when studying a species such as *P. infestans*, which has expanded its range from a small area, probably in central Mexico, to a panglobal distribution in a very short span of evolutionary time. Another advantage of mtDNA is that it is not normally recombined during sexual reproduction and therefore will be passed intact from parents to progeny regardless of any rearrangements that may occur to the nuclear genome. This property of the mtDNA makes it a very useful marker for migration events. If a previously unknown mtDNA type is introduced to a new location it will serve as a marker for the clonal progeny of that mtDNA, whether or not sexual reproduction is occurring. These properties of the mtDNA make it ideal both for studying population genetics and for elucidating the evolutionary relationships among *Phytophthora* species.

Previous work on the mtDNA of *P. infestans* has yielded a restriction map consisting of the sites for 11 different restriction enzymes (Klimczak & Prell, 1984), and mitochondrial DNA mapping in other *Phytophthora* species has yielded some interesting evolutionary insights (Chapter 11). However, some of the restriction enzyme sites, such as *Eco*RI, which were mapped in *P. infestans*, were not mapped in the other species, and some of the enzymes which gave very diagnostic restriction patterns in the other species were not mapped in *P. infestans*. This is unfortunate because it seriously limits the comparisons that can be made among species. One facet of my work has been to extend the mtDNA restriction map of *P. infestans* to include those enzymes shown to be useful in other *Phytophthora* species, as well as to screen for additional potentially informative enzymes. Enzyme sites which have been partially or totally mapped so far include *Kpn*I, *Sac*I, *Eco*RV, *Hae*III, *Hin*dIII, and *Hha*I. This work has already yielded some interesting results. Based on the Klimczak & Prell (1984) restriction map, Förster *et al.* (1988) postulated the occurrence of three regions of homology among the mitochondrial genomes of *P. infestans*, *P. megasperma* and *P. nicotianae*. One of these regions contained the large and small ribosomal RNA genes, which have not been localized on the *P. infestans* map. If these regions really are conserved among all three species then it should be possible to predict, based on the consensus map for *P. megasperma* and *P. nicotianae*, the locations of sites on the *P. infestans* map for enzymes that were not studied previously. Of particular interest are the enzymes *Hae*III, *Hin*dIII, and *Hha*I, each of which cuts the *P. infestans* mtDNA many times. When added to the map, the sites for these additional enzymes were not informative for two of the putative conserved regions but were very diagnostic for the third region. This region is shown in detail in Fig. 17.1. The top part of Fig. 17.1 shows the sites on the consensus map for *P. megasperma* and *P. nicotianae* (Förster *et al.*, 1988), while the bottom shows the corresponding map of *P. infestans* with six previously unmapped sites for the enzymes *Hae*III, *Hin*dIII, and *Hha*I. These additional sites were located on the *P. infestans* map almost exactly where expected. This confirms the highly conserved nature of this region and proves it probably contains the large and small ribosomal RNA genes. Differences within this region may therefore be useful for making phylogenetic comparisons

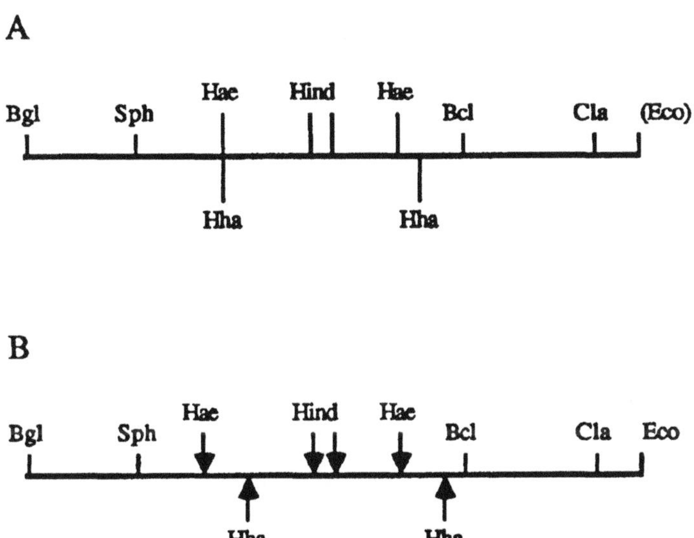

Fig. 17.1. Detailed restriction map of the mitochondrial DNA region containing the small ribosomal RNA gene in *Phytophthora*. A = the consensus map for *P. megasperma* and *P. nicotianae* (from Förster *et al.*, 1988). B = the *P. infestans* map. Previously unmapped restriction sites are shown by arrows.

among species, as suggested by Förster *et al.* (1988). An advantage of mapping the mitochondrial genome is that any changes which occur to it can be located easily. So far, four variant forms of the mitochondrial genome have been discovered in a search for mtDNA variation in isolates from many parts of the world using the enzyme *Eco*RI. Additional variation has been found using other enzymes (chapter 18) but only the *Eco*RI variants will be discussed here. The most common mtDNA form, which is the one mapped by Klimczak & Prell (1984), has been designated as form A. This form is characterized by ten bands when cut with the restriction enzyme *Eco*RI. A variant of this, form B, was first reported among isolates from Israel (Goodwin & Fry, 1988), but has since been found to have a worldwide distribution. Carter *et al.* (1990 and Chapter 18) have designated A and B forms as Types I and II respectively. This form is about 1·7 kb larger than the A form and has apparently undergone some rearrangements (Fig. 17.2). The three largest *Eco*RI fragments in Klimczak & Prell's original map have been reduced to two, and a small (about 300 base pair) fragment not reported earlier has also been lost in

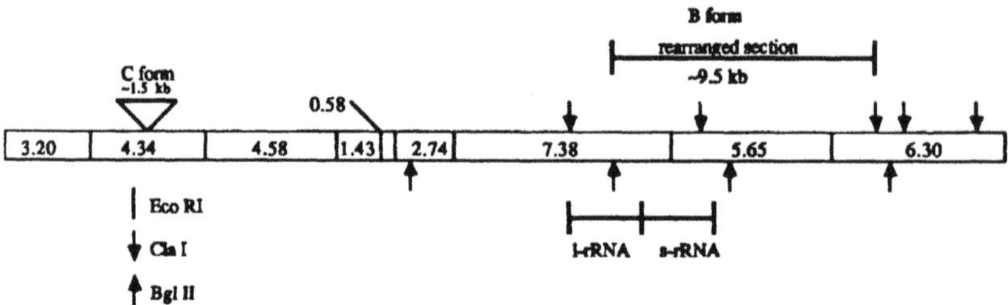

Fig. 17.2. Simplified restriction map of the A form of the *P. infestans* mitochondrial genome showing the probable locations of the changes to the B and C forms. Fragment sizes are as given by Klimczak & Prell (1984). Probable locations of the large and small ribosomal RNA genes (based on Förster *et al.*, 1988) are indicated.

the B form. The endpoints of the putative rearranged section have been localized between a *Bgl*II site on one side and a *Cla*I site on the other. This form is characterized by the loss of two *Eco*RI sites and one *Cla*I site, so it cannot be due to a simple insertion. A minimum of three mutations are required to get from the A to the B form. The C form of the mtDNA has recently been found in one isolate from Peru. This form differs from the A form in that it appears to have an insertion of about 1·5 kb to the fifth-largest *Eco*RI fragment (Fig. 17.2). The C form appears to be the result of a single insertion with no additional rearrangements.

The 'D' form of the mtDNA has recently been found in isolates from central Mexico and Peru. The nature of this change is complex; it appears to be a combination of the A and C forms, with two additional bands of unknown origin that are not found in other isolates. One of these bands may have been found previously in a mitochondrial DNA sample of an isolate from central Mexico yet did not appear in subsequent preparations of the same isolate. At this time it is not entirely clear what the bands represent; one possibility is that they are not of mitochondrial origin but, instead, are attributable to plasmids that have some similarity to mitochondrial sequences.

The occurrence of the different *P. infestans* mtDNA forms has revealed an interesting geographical pattern in over 170 isolates tested (Table 17.1), particularly because the A and B forms are so different and no intermediate forms have been found. So far, the B form has not been found among 33 isolates tested from the Toluca Valley area of central Mexico, yet is more common than the A form in samples from both northeastern

Table 17.1. Numbers of isolates of mitochondrial DNA genotypes by location.

	mtDNA genotype			
Location	A	B	C	D
Mexico				
Northeast	13	20	–	–
Northwest	7	24	–	–
Central	22	–	–	11
Ireland	1	–	–	–
Poland	13	8	–	–
The Netherlands	12	2	–	–
Israel	–	8	–	–
Japan	8	8	–	–
Peru	1	–	1	13
United States	1	–	–	–

and northwestern Mexico. The B form is most often associated with isolates that have genotypes *100/100* and *100/100* for the isozymes glucosephosphate isomerase and peptidase respectively and is found with both mating types (Table 17.2). Interestingly, the B form has not yet been found among isolates with the *86/100* and *92/100* isozyme genotype for glucosephosphate isomerase and peptidase; this is the genotype that may have predominated in areas outside Mexico prior to the 1980's (Chapter 15). Two conclusions may be drawn from these data. The strong association between mitochondrial DNA form and other genetic markers in some locations suggests that sexual reproduction is probably not occurring within these populations even though both mating types are common. This fully agrees with the conclusions reached earlier for these same populations using isozyme markers (Chapter 15). The second conclusion is that mtDNA may be a useful marker for reconstructing recent global migration events. It has been suggested that the recent appearance of the A2 mating type outside Mexico may have been due to the importation of potatoes into Europe in the late 1970's from northern Mexico (Turkensteen, unpublished). Because the B form is very common in northern Mexico, yet not found in central Mexico (Goodwin & Fry, 1988), it is entirely plausible that the B form mtDNA and the A2 mating type were introduced into Europe concurrently. It must be emphasized that these are preliminary results based on somewhat limited sampling, but so far the pattern is suggestive.

Table 17.2. Nuclear (isozyme) genotypes associated with the B form of the mitochondrial DNA by location.

Location	Mating type	GPI	PEP
Mexico			
Northeast	A1	100/100	100/100
	A2	100/100	100/100
	A2	100/111/122	100/100
Northwest	A1	100/100	92/100
Poland	A1	90/100	100/100
	A1	100/100	100/100
	A2	100/100	100/100
The Netherlands	A1	90/100	92/100
Israel	A2	100/100	100/100
Japan	A2	100/100	92/100
	A2	100/100	100/100

GPI = glucosephosphate isomerase; PEP = peptidase

Polymorphisms of nuclear DNA

Molecular mapping of the mitochondrial genome involves the physical mapping of individual restriction sites. This can be done whether or not there is any variation at these sites and the whole process, while rather time consuming, is relatively straightforward. Finding and mapping nuclear DNA markers is a much more complicated, long-term objective. Instead of a physical map of restriction sites, a genetic map of restriction fragment length polymorphisms (RFLPs) is produced. Mapping cannot proceed until suitable polymorphisms are located. In spite of the large commitment in time and resources required to complete a genomic RFLP map, there are several compelling reasons for proceeding with such a project. One immediate benefit is that once the genetic data are available, mapped probes can be used for other purposes. There is a need for independent, well-characterized loci in *P. infestans* for population genetics studies. The existence of gametic phase disequilibrium among loci can be an indicator of the evolutionary forces acting on populations (Hedrick, 1983), yet disequilibrium values cannot be interpreted accurately unless the linkage relationships among loci are known. Variations in ploidy may have important ramifications for the population biology of *P. infestans* (Chapter 13). Molecular markers will facilitate much more exact studies on the nature of ploidy variation, but will only be useful once the inheritance of each allele at a locus has been determined. A long-term goal is to facilitate map-based cloning of important genes such as virulence and fungicide resistance. Basic techniques involved in RFLP mapping have been reviewed elsewhere (Michelmore & Hulbert, 1987). What follows is a general outline of a strategy for developing nuclear DNA probes in *P. infestans*. The immediate goals of this project are three: to generate single-copy probes for RFLP mapping and population genetics; to identify multicopy probes that may be useful for other purposes such as DNA fingerprinting; and to elucidate the organization of the genome of *P. infestans*. RFLP mapping is also being pursued by Carter *et al.* (chapter 18).

Sources of probes

Three sources of probes can be used for developing RFLP markers: random genomic clones; copy DNA (cDNA) clones; and heterologous probes of DNA cloned from other organisms. So far only random genomic clones have been studied in *P. infestans*. A library was made by digesting total genomic DNA to completion with *Eco*RI, using DNA that had been purified previously in a caesium chloride/bisbenzimide gradient to remove most mitochondrial sequences. After size fractionation in an agarose gel, DNA fragments between 0·5 and 2·5 kb were purified and ligated into the vector BlueScript. Only small inserts were used in an attempt to avoid repetitive DNA. The resulting recombinant molecules were transformed into *E. coli* strain DH5α; approximately 150 recombinant clones were selected at random from this library for further analysis. Because of the high repetitive DNA content of *P. infestans* (see below), a cDNA library is currently being made to ensure that a large number of single-copy clones will be available for RFLP analysis.

Identification of RFLPs

A small number of *P. infestans* isolates was chosen for screening the clones from the library. The objective was to choose isolates representing a wide range of genetic diversity. For this reason isolates from central Mexico were chosen because this area is believed to represent the centre of origin and diversity of *P. infestans* (Chapter 3). Another consideration was to include isolates that would be compatible in crosses. Because many crosses among potential parents in this fungus are unsuccessful (Spielman, unpublished), and it is impossible to predict which parents will produce a successful cross, only proven parents were chosen. Based on these criteria, seven different *P. infestans* isolates, among which four successful crosses had already been produced, were chosen for analysis. Large progeny sizes were available for each of these four crosses, and they had already been scored for segregation of mating type, isozyme, and some virulence phenotypes (Table 17.3). For the screening procedure,

Table 17.3. Mexican isolates screened for DNA polymorphisms, and the crosses in which they have been used

Cross	No. of progeny	Markers segregating
510 × 543	124	Mating type, virulence to R2[a], R3, R4 and Ph1[b]
510 × 562	142	Mating type, GPI[c], PEP[d], virulence to R1, R2, R4 and Ph1
568 × 575	146	Mating type, GPI[c], PEP[d], virulence to R2, R3, R4 and Ph1
580 × 618	112	Mating type, GPI[c], PEP[d], virulence to R2, R3, R4 and Ph1

[a]virulence to foliage of potato differential carrying major gene R2 (etc) for resistance.

[b]virulence to foliage of tomato differential carrying major gene Ph1 for resistance.

[c]isozymes of glucosephosphate isomerase.

[d]isozymes of peptidase

total genomic DNA from all seven parents was digested to completion with the restriction enzyme *Eco*RI, each was blotted to 20 replicate filters and the blots screened with individual clones from the library. Randomly selected clones, which would include both unique and multicopy sequences, were chosen for screening to get an idea of genome composition as well as to identify suitable polymorphic probes. Radioactive ^{32}P-dCTP was incorporated into each clone using the random hexamer method (Feinberg & Vogelstein, 1984) and the clones were hybridized to the genomic blots for 20 hours at 65°. Each filter was then washed three times for ten minutes each in 2 × SSC, 0·1% SDS, in 1 × SSC, 0·05% SDS, and in 0.5 × SSC, 0·025% SDS at 65° and placed on X-ray film. Following autoradiography, each clone was characterized as single copy, moderately repetitive, or highly repetitive. Highly-repetitive clones were those that hybridized to more than 20 intense bands in each isolate, gave a dark smear of hybridization with no discrete bands or a combination of the two. Moderately-repetitive clones hybridized to from four to 20 bands, while those that were categorized as single copy hybridized to from one to three bands. Clones that gave a single intense band of hybridization with other faintly hybridizing bands also were characterized as single copy. Among the first 60 clones tested, 11% were highly repetitive, 57% were moderately repetitive and about one-third were single copy. Most of the repetitive clones (about 85%), but only about 10% of the single-copy clones, were polymorphic on the seven parental isolates tested.

One of the reasons for screening parents that had been involved in four different crosses initially was to determine whether one of the crosses would be more informative than the others for generating segregation data. However, based on the band-counting method of Nei & Li (1979) there was not much difference among the four crosses studied; the proportion of shared bands ranged from 80 to 83 percent. Thus from the standpoint of DNA variation it probably does not make too much difference which parents are chosen for a cross. The choice of a cross can then be made based on segregations for markers of interest, such as virulence. However, unexpected complications can arise in a genetically uncharacterized organism such as *P. infestans*. For example, nothing is known about major chromosomal rearrangements such as inversions and translocations in *P. infestans*. For these reasons preliminary RFLP segregation data are being obtained for two different crosses, to be sure that segregations are normal before the final choice of a primary cross is made. So far, segregations for a limited number of RFLP markers in the two different crosses have behaved according to Mendelian expectations. To achieve one of the long-term goals of the RFLP mapping project – mapping all of the virulence genes – undoubtedly several crosses will be required.

Methods for generating RFLP segregation data and constructing RFLP maps have been described elsewhere (Botstein *et al*., 1980). Many of the single-copy clones identified so far have not been polymorphic based on the results obtained by digesting the DNA with a single restriction enzyme; the next phase of this project is to search for polymorphisms within the regions hybridizing to these monomorphic clones by digesting the DNA with a number of different restriction enzymes. Eight restriction enzymes have been chosen for this purpose and it is anticipated that a high proportion of the clones can be utilised. The high proportion of repetitive DNA is in agreement with C_0t curve analyses and genome reconstruction experiments in other *Phytophthora* species where the repetitive DNA fraction was 50% or higher (Tyler & Mao, unpublished). This high repetitive DNA content will complicate chromosome walking strategies and may make map-based cloning of virulence genes extremely difficult.

DNA fingerprinting

One of the early benefits of the somewhat conservative strategy for generating RFLP probes outlined above is that several probes have been identified that are candidates for DNA fingerprinting. Each of these probes hybridizes to a number of highly variable bands. Furthermore, all seven isolates gave a unique banding pattern when tested with these clones and the number of bands identified with each probe was 20 or more. DNA fingerprinting is a relatively new technique that was first developed for humans (Jeffreys, Wilson & Thein, 1985) but has been shown to be useful in other mammals and birds (reviewed in Burke, 1989). So far it has been used only infrequently in plants (Dallas, 1988) and fungi (Scherer & Stevens, 1988; Hamer *et al.*, 1989), and never in an Oomycete. The technique involves identifying specific DNA sequences that occur many times at different locations throughout the genome. When total DNA is cut with a restriction enzyme, separated on an agarose gel, blotted and probed with one of these sequences, many bands will be visible on an autoradiogram indicating that the sequence is present within many fragments of different size. Such a pattern can be used as a 'DNA fingerprint' (Jeffreys *et al.*, 1985) to identify individual isolates. Asexual progeny from a given isolate will all have an identical DNA banding pattern. Because each asexual clonal line of the fungus can be identified unequivocally, this technique has the potential to lead to greater advances in our understanding of the population biology of *P. infestans* than any other finding since the discovery of the A2 mating type.

Three things must be known about the banding patterns produced by these probes before they can be used as DNA fingerprints:

(i) individual bands must show Mendelian behaviour;

(ii) the individual loci must be unlinked; and

(iii) the banding patterns of an individual must be stable through mitotic divisions.

Due to the complex nature of these patterns a complete genetic analysis may be difficult or impossible, at least initially. However, Mendelian segregation conditions will be satisfied if for a cross between two parents, one of which has a given band and one of which does not, about half of the progeny have the band (Jeffreys *et al.*, 1988). Linkage can be tested by looking for cosegregation among bands. For *P. infestans*, somatic stability can be demonstrated by showing that individual patterns are stable through single zoospores. These tests have been carried out for two moderately repetitive DNA probes isolated from *P. infestans* (Goodwin & Fry, 1990; Goodwin *et al.*, unpublished). Over 60 single-zoospore isolates showed banding patterns that were identical to the original isolates, so these patterns do appear to be mitotically stable. Segregation data were obtained for 13 out of 23 variable bands for probe RG57 and, in each case except one, the patterns were consistent with normal Mendelian expectations; the one exception was probably due to a sampling phenomenon. Two bands cosegregated, but the remainder did

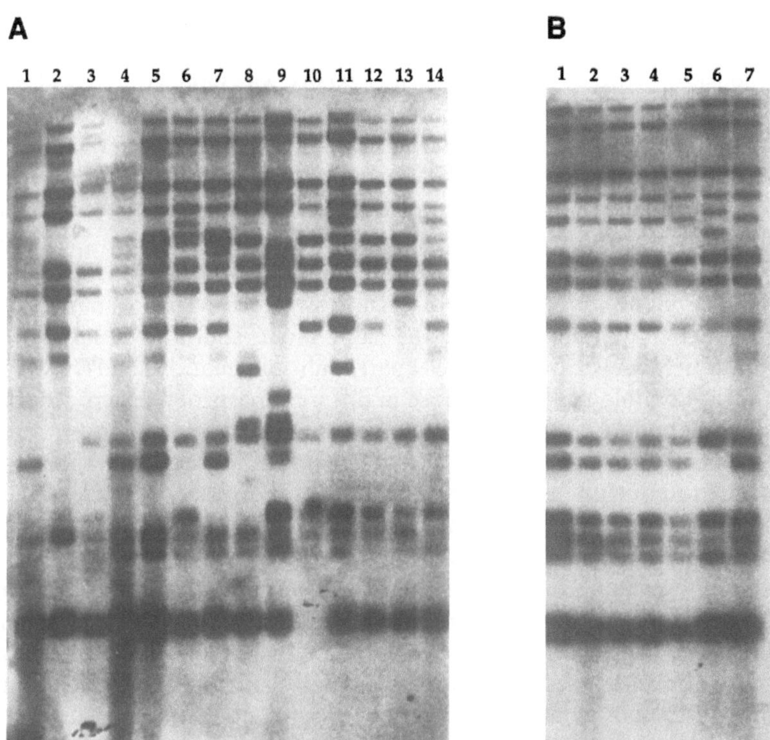

Fig. 17.3. Typical blots of DNA from Mexican isolates probed with RG57. Each lane contains approximately 2 μg of total genomic DNA digested to completion with *Eco*RI. A = 14 central Mexican isolates showing the large amount of variation detected with this probe. B = 7 isolates from northwestern Mexico showing two of the three banding patterns found at this location.

not appear to be tightly linked. Probe RG7 gave similar results, except that there was a larger number of cosegregating bands.

Probe RG57 has been used to study the genetic diversity of over 200 isolates from a worldwide culture collection. This probe revealed at least 25 different bands, 23 of which have been variable among the isolates tested so far (Goodwin & Fry, 1990; Goodwin *et al.*, unpublished) and the separation distances among the bands are such that each band can be scored unambiguously. Isolates obtained from different locations in Mexico showed different levels of genetic diversity. In central Mexico almost every isolate was unique, as expected for a sexually reproducing organism in its centre of origin. In contrast, in northwestern Mexico there were only three different banding patterns among over 40 isolates tested, indicating that reproduction in this location was almost entirely clonal. In northeastern Mexico the pattern of diversity was intermediate between these two extremes, with 15 clones found among 26 isolates tested. The same pattern was also evident in the frequencies of individual bands. All 25 bands occurred in central Mexico, but only 18 in northeastern Mexico and 13 in northwestern Mexico. Typical blots from northwestern and central Mexico are shown in Fig. 17.3. Note the large amount of variation in central Mexico and the lesser variation in northwestern Mexico, where only three clones predominated. In northwestern Mexico, many fields were genetically homogeneous; the same clone could be isolated from different fields several kilometres apart. This information sheds new light on the epidemiology of this disease in different parts of Mexico. Another interesting point is that there were only three isozyme genotypes present in both northeastern and northwestern Mexico (Chapter 15), which might lead one to conclude that the levels of genetic diversity in each of these areas were similar. However, the DNA data revealed that this was not the case and illustrates the resolving power provided by these markers. A more limited sample of isolates from the United States and Europe has also been studied. So far there has been a very high level of genetic diversity in the Netherlands, comparable to that observed in central Mexico. The amount of diversity was less in Poland and the United States where particular clonal lines were predominant. The high degree of resolution provided by these markers will add greatly to our understanding of the population biology and epidemiology of *P. infestans*.

Whether or not these probes provide true genetic fingerprints is debatable, because the number of bands and their degree of polymorphism is not as high as those reported in some other organisms. However, because there are fewer bands, each one can be scored unambiguously, providing extremely high resolution. With 20 or more independent loci the probability of two unrelated individuals having the same banding pattern is extremely low. The fact that only a limited number of clones was identified in certain areas and that the identities of these clones also correlated well with differences recognized for mating type, mtDNA, host range, and isozymes (unpublished data) provides additional evidence that these markers are somatically stable.

Conclusions

Recent advances in the development of molecular genetic markers for *P. infestans* are ushering in a new era of discovery. Mitochondrial DNA variation provides markers that may prove effective in following the migration of *P. infestans* around the world and for monitoring the effectiveness of quarantine restrictions. Determining the complete map and full extent of changes to the mitochondrial genome in the A, B, C, and D forms is a high priority for the future. More intensive sampling of *P. infestans* populations worldwide must be completed soon to provide a baseline against which future changes can be assessed. A thorough check of all extant isolates obtained before the late 1970's could give a clearer picture of what existed prior to the most recent introductions. It would also be desirable to determine the mode of inheritance of mtDNA in *P. infestans*. One way to do this may be to look for cosegregation of mtDNA with known cytoplasmic markers such as double-stranded RNA.

A preliminary RFLP map of the *P. infestans* genome based on randomly cloned genomic sequences is now foreseeable. It is to be hoped that as particular genes are isolated from *Phytophthora* species, their map locations will also be determined. A suitable cDNA library is currently under construction and including these clones in the RFLP mapping process is a high priority. Studying the inheritance of virulence to all of the known potato and tomato resistance genes is also a high priority. Due to the difficulties involved in this kind of work, this area should receive more attention now so that it does not become a bottleneck in the future. It is hoped that mapping in both nuclear and mitochondrial genomes will allow the relative rates of evolution of the two types of markers to be compared. Although mtDNA sequence evolution is rather rapid in animals (Brown *et al*., 1979) it is very much slower in plants (Palmer & Herbon, 1988) and its rate in Oomycetes is unknown.

DNA fingerprinting has the potential to be a very useful technique both for RFLP mapping and for understanding population biology. For the future, it will be important to determine the organization of these sequences in the *P. infestans* genome. The ability to identify individual clones of the fungus will permit a greater understanding of the population biology of this organism, and for asexually reproducing populations may render other markers, such as isozymes, obsolete. Furthermore, it may be possible to determine the origin of new phenotypes in asexually reproducing populations. A migrant carrying a previously unknown banding pattern should be easily distinguishable from a newly-arisen mutant. The large number of loci that can be studied with these probes may permit recently developed techniques for measuring gene flow among populations (Slatkin, 1981, 1985) to be applied to *P. infestans*. These markers will also permit theories of parasexual recombination to be tested rigorously.

International cooperation can greatly accelerate the RFLP mapping process. Some laboratories may be primarily concerned with cloning particular genes. Libraries or cloned sequences made during this type of endeavour may provide suitable probes for RFLP mapping. A mechanism must be devised by which these and other markers can be exchanged among investigators to facilitate genetic mapping studies in the future.

References

Botstein, D., White, R. L., Skolnick, M. & Davis, R. W. (1980). Construction of a genetic linkage map in man using restriction fragment length polymorphisms. *American Journal of Human Genetics* **32**, 314-331.

Brown, W. M., George, M. & Wilson, A. C. (1979). Rapid evolution of animal mitochondrial DNA. *Proceedings of the National Academy of Sciences, USA* **76**, 1967-1971.

Burke, T. (1989). DNA fingerprinting and other methods for the study of mating success. *Trends in Ecology and Evolution* **4**, 139-144.

Carter, D. A., Archer, S. A., Buck, K. W., Shaw, D. S. & Shattock, R. C. (1990). Restriction-fragment length polymorphisms of mitochondrial DNA of *Phytophthora infestans*. *Mycological Research* **94**, 1123-1128.

Dallas, J. F. (1988). Detection of DNA 'fingerprints' of cultivated rice by hybridization with a human minisatellite DNA probe. *Proceedings of the National Academy of Sciences, USA* **85**, 6831-6835.

Feinberg, A. P. & Vogelstein, B. (1984). A technique for radiolabelling DNA restriction endonuclease fragments to high specific activity. *Analytical Biochemistry* **137**, 266-267.

Förster, H., Kinscherf, T. G., Leong, S. A. & Maxwell, D. P. (1987). Molecular analysis of the mitochondrial genome of *Phytophthora*. *Current Genetics* **12**, 215-218.

Förster, H., Kinscherf, T. G., Leong, S. A. & Maxwell, D. P. (1988). Estimation of relatedness between *Phytophthora* species by analysis of mitochondrial DNA. *Mycologia* **80**, 466-478.

Garber, R. C. & Yoder, O. C. (1983). Isolation of DNA from filamentous fungi and separation into nuclear, mitochondrial, ribosomal, and plasmid components. *Analytical Biochemistry* **135**, 416-422.

Goodwin, S. B. & Fry, W. E. (1988). Mitochondrial DNA polymorphism in *Phytophthora infestans*. *Phytopathology* **78**, 1542 (abstract).

Goodwin, S. B. & Fry. W. E. (1990). DNA fingerprinting in *Phytophthora infestans*. *Phytopathology* **80**, 1006 (abstract).

Hamer, J. E., Farrall, L., Orbach, M. J., Valent, B. & Chumley, F. G. (1989). Host species-specific conservation of a family of repeated DNA sequences in the genome of a fungal plant pathogen. *Proceedings of the National Academy of Sciences, USA* **86**, 9981-9985.

Hedrick, P. W. (1983). *Genetics of Populations*. Jones & Bartlett: Boston.

Jeffreys, A. J., Royle, N. J., Wilson, V. & Wong, Z. (1988). Spontaneous mutation rates to new length alleles at tandem-repetitive hypervariable loci in human DNA. *Nature* **332**, 278-281.

Jeffreys, A. J., Wilson, V. & Thein, S. L. (1985). Hypervariable 'minisatellite' regions in human DNA. *Nature* **314**, 67-73.

Klimczak, L. J. & Prell, H. H. (1984). Isolation and characterisation of mitochondrial DNA of the oomycetous fungus *Phytophthora infestans*. *Current Genetics* **8**, 323-326.

McNabb, S. A. & Klassen, G. R. (1988). Uniformity of mitochondrial DNA complexity in Oomycetes and the evolution of the inverted repeat. *Experimental Mycology*, **12**, 233-242.

Michelmore, R. W. & Hulbert, S. H. (1987). Molecular markers for genetic analysis of phytopathogenic fungi. *Annual Review of Phytopathology*, **25**, 383-404.

Nei, M., & Li, W. (1979). Mathematical model for studying genetic variation in terms of restriction endonucleases. *Proceedings of the National Academy of Sciences, USA* **76**, 5269-5273.

Palmer, J. D. & Herbon, L. A. (1988). Plant mitochondrial DNA evolves rapidly in structure, but slowly in sequence. *Journal of Molecular Evolution* **28**, 87-97.

Scherer, S. & Stevens, D. A. (1988). A *Candida albicans* dispersed, repeated gene family and its epidemiologic applications. *Proceedings of the National Academy of Sciences, USA* **85**, 1452-1456.

Shaw, D. S. (1988). The *Phytophthora* species. *Advances in Plant Pathology* **6**, 27-51.

Slatkin, M. (1981). Estimating levels of gene flow in natural populations. *Genetics* **99**, 323-335.

Slatkin, M. (1985). Rare alleles as indicators of gene flow. *Evolution* **39**, 53-65.

Chapter 18

DNA polymorphisms in *Phytophthora infestans*: the UK experience

D. A. Carter, S. A. Archer, K. W. Buck, D.S. Shaw & R.C. Shattock

Genetic polymorphisms are defined as variations in inheritable traits. Morphological and nutritional traits have been supplemented by those identified as enzyme differences, i.e. isozymes or enzymes which share catalytic activity but vary in electrophoretic mobility or antigenic reactivity. Recombinant DNA techniques now allow such variations to be traced back to the DNA level, where they occur as altered nucleotide sequences. As a consequence of the redundancy of the genetic code and the presence of non-coding sequences, 'silent' DNA modifications can occur whereby base substitution does not result in an alteration of amino acid sequence; the same techniques can be used to detect these polymorphisms which would go unnoticed in protein analysis. Some of the non-expressed sequences may be more free from the functional constraints imposed on coding DNA and may therefore yield even higher levels of polymorphisms than is seen in genes that specify functional products.

Polymorphisms in ribosomal, mitochondrial and nuclear DNA have been employed in population studies in a large range of organisms. Ribosomal DNA (rDNA) genes are heavily constrained by a strict conservation of the products they encode, and variations are assayed by direct sequence analysis. Extensive population phylogenies have been constructed based on sequence polymorphisms in 16S rDNA (e.g. Pace, Olsen & Woese, 1986). Ribosomal DNA of *P. infestans* has yet to be examined.

The most straightforward approach to analyzing polymorphisms in mitochondrial and nuclear DNA is by comparison of digestion profiles produced by type II restriction endonucleases. These enzymes recognise specific sequences of nucleotides and will cleave double stranded DNA wherever such sequences occur, producing a range of lengths of linear DNA known as restriction fragments. Changes to a recognition site by base substitution, by loss or gain of nucleotides or by its deletion, mean that the restriction enzyme is no longer able to cleave at that site, causing a single restriction fragment to be produced from the two original fragments flanking the altered site. Likewise, new sites may be generated by

Table 18.1. *Phytophthora infestans* isolates.

Isolate	Origin	Host[1]	mtDNA type	Mating type	Ploidy[2]	GPI[3]	Peptidase[3]	Virulence[4]
Ca65	USA (California)	T	IIb	A1	2C	100-100	92-100	1,4,7,11
E13a	Egypt	P	IIa	A2	2C	100-100	100-100	1,3,4,7,8,10,11
550	Mexico	S	Ia	A2	2C	86-86	100-100	1,2,3,4,5,7,8,10,11
533	Mexico	P	Ia	A1	2C	122-122	92-100	1,2,3,4,7,10,11
7-85	Sweden	P	Ia	A1	nt	100-100	100-100	1,2,11
WF127	USA (New York)	P	Ib	A1	4C	86-100	92-100	2,3,4,7
83-5	UK (Gwynedd)	P	Ia	A1	nt	nt	nt	1,2,3,4,7,10,11
86-126-1A	UK (Cambridge)	P	Ib	A1	3C	86-100	92-100	nt
K1067	Ireland	P	Ib	A1	nt	86-100	92-100	nt
86-127-2	UK (Gwynedd)	P	Ia	SF	nt	100-100	92-100	nt
86-150-3	UK (South Wales)	P	Ia	A2	2C	100-100	100-100	nt
87-182-9	UK (Norfolk)	P	IIa	A2	2C	100-100	100-100	nt
E14C2	Egypt	P	IIa	A2	2C	100-100	100-100	1,4,7,10,11
565	Mexico	P	Ia	A1	2C	100-122	92-100	2,3,4,7
Ismailia B	Egypt	P	Ia	A1	2C	100-100	100-100	1,4,7,10,11
84-19	Spain	P	Ia	A1	4C	100-100	92-100	1,3,4,7,8,10,11
K1098	The Netherlands	P	Ia	A1	2C	100-100	100-100	3,4,7
NR-132	Switzerland	P	Ia	A2	nt	86-100	100-100	nt
CG3	New Zealand	P	Ia	A1	nt	100-100t	100-100	1,2,3,4,7,11
12-87	Brazil	P	Ib	A1	2C	86-100	92-100	3,4,7
644-86	Brazil	P	IIa	A2	nt	100-100	100-100	1,3,4,5,7,11
87-34-4	UK (Dyfed)	P	Ia	A1	nt	100-100	100-100	nt
87-244-1	UK (Avon)	T	Ia	A1	nt	100-100	100-100	nt
87-205-2	UK (Cornwall)	P	Ia	A2	nt	100-100	100-100	nt

[1]Host plant from which isolate was obtained, P = potato, T = tomato, S = *Solanum* sp. [2]Ploidy data from S. L. Whittaker, R. C. Shattock & D. S. Shaw (1991). [3]Alleles at glucosephosphate isomerase (Gpi-1) and peptidase loci from Tooley *et al.* (1985) and unpublished results of S. L. Whittaker, R. C. Shattock & D. S. Shaw; nt = not tested. [4]Virulence, as assayed by ability to grow on potatoes carrying the appropriate resistance genes (S. Al-Kherb, D. S. Shaw & R. C. Shattock, unpublished results). SF = self fertile.

point mutations, which will cause an increase in the number of restriction fragments. Restriction fragment sizes can also be changed by insertions or deletions of DNA between two recognition sites. All such changes are collectively denoted restriction fragment length polymorphisms, or RFLPs.

RFLPs were first used in genetic analysis in 1974. Linkage of temperature-sensitive mutations of adenovirus to specific restriction fragment length differences was used to locate the mutations on a physical map of the restriction fragments (Grodzicker *et al.*, 1974). RFLPs were used to show the maternal inheritance of mammalian mitochondrial DNA (Hutchinson *et al.*, 1974) and the Mendelian inheritance of ribosomal DNA in yeast (Petes & Botstein, 1977). The possibility of using RFLPs for construction of a genetic linkage map was first proposed by Botstein *et al.* (1980) in the context of human genetics. Since then, RFLPs have found increasing application in studies of a wide range of organisms and have proved to be particularly useful in studies where other heritable traits are infrequent.

Population and genetic studies of *P. infestans* have traditionally been hampered by the lack of easily scored heritable traits. Variation in fungal morphology may be unstable and can be influenced by external environmental factors. Other markers which have been used, such as mating type and virulence, may likewise show instability and do not always demonstrate classical Mendelian inheritance as expected of nuclear chromosomal genes. Isozyme markers are not generally affected by environmental factors, show Mendelian inheritance and are codominant, allowing homozygotes to be distinguished from heterozygotes. Some isozymes have been utilised in genetic studies of *P. infestans* (e.g. Tooley, Fry & Villarreal Gonzalez, 1985; Shattock, Tooley & Fry, 1986; Chapter 15); however, the range available and the level of polymorphisms revealed by each isozyme is limited and has only allowed a rudimentary analysis of variation among isolates of *P. infestans* (Chapter 14). RFLPs, which are theoretically much more numerous, should allow a far more extensive analysis.

RFLPs of mitochondrial DNA

Examination of DNA polymorphisms in mitochondrial genomes has been applied to a diverse range of organisms (Avise *et al.*, 1979; Sederoff *et al.*, 1981; Latorre, Moya & Ayala, 1986; Cann, Stoneking & Wilson, 1987; Ball *et al.*, 1988). The size range of mitochondrial DNA (mtDNA) varies greatly in different organisms, but generally it is small enough to allow polymorphisms to be visualised by enzyme cleavage and gel electrophoresis alone, with the whole mitochondrial genome viewed in its entirety. As they are generally inherited uniparentally and only a single

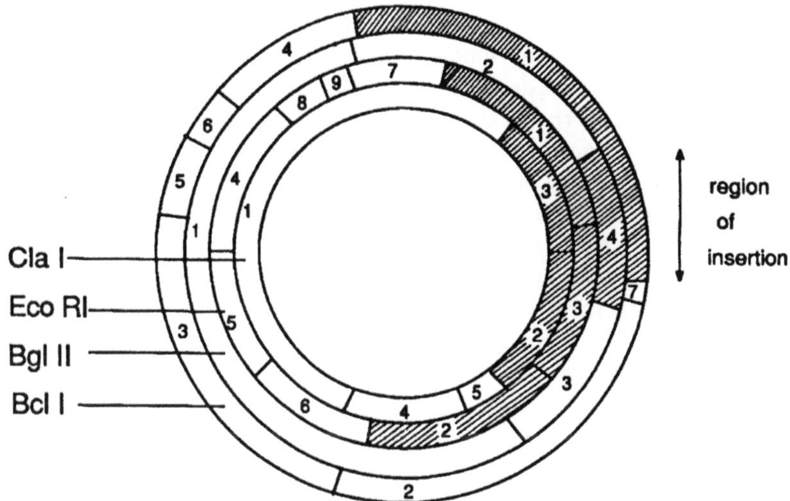

Fig. 18.1. Restriction map of *Phytophthora infestans* type I mtDNA, adapted from Klimczac & Prell (1984) to indicate the region of the 2 kb insertion in Type II mtDNA (Carter *et al.*, 1990). The shaded blocks identify fragments which are altered in the inserted strains.

mtDNA species usually occurs in a given organism, mitochondrial genomes are essentially haploid, making analysis simpler than that of nuclear DNA in diploid organisms. While variability in mtDNA may not be high enough to distinguish every individual it is frequently sufficient for studies of populations and higher taxa. In fungi, mtDNA polymorphisms have been utilised to assess relationships between genera (Hoeben & Clark-Walker, 1986), between species (Weber *et al.*, 1986; Anderson, Petsche & Smith, 1987; Croft, 1987; Förster *et al.*, 1987; Chapter 11) and between isolates of a single species (Garber & Yoder, 1984; Hulbert & Michelmore, 1988; Förster *et al.*, 1989; Bates, Buck & Brasier, 1990). Mitochondrial DNA was previously isolated from *P. infestans* by Klimczac & Prell (1984) and a physical map was obtained using a number of restriction enzymes. It was shown to be a circular molecule of 36·2 kb with a GC content of 22·4%, but analysis of five isolates differing in virulence did not reveal any RFLPs among them. Recently we have extended both the number and global range of isolates investigated (see Table 18.1) and used additional restriction enzymes (Carter *et al.*, 1990). RFLP analysis initially divided the isolates into two groups. The first group, which contained 19 of the 24 isolates examined, gave identical mtDNA restriction fragment patterns to those obtained by Klimczac & Prell using the same restriction enzymes. The remaining five isolates from Brazil, California, Egypt and the United Kingdom had mtDNA which contained an insertion of approximately 2 kb. This was localised on the Klimczac & Prell map by examining which of the restriction fragments had been altered in the mitochondrial genomes containing the insertion (Fig. 18.1). The insertion occurred in the same location in the mtDNA of all five of these isolates, suggesting a common origin, despite their global distribution (Table 18.1). Alteration of an *Eco*R1 site distal to the insertion site suggested that some sequence changes in addition to the insertion had occurred. The altered *Eco*R1 site was not present in any of the mtDNAs lacking the insertion and therefore probably was created either during or following the insertion event. The insertion was present in four out of eight isolates of the A2 mating type, but in only one out of 15 A1 isolates; thus it may have arisen in the A2 population with subsequent transfer to the A1 population.

Mitochondrial DNA lacking the insertion was classified as Type I, and mtDNA containing the insertion as Type II. This polymorphism now appears to be identical to that reported by Goodwin & Fry (1988; Chapter 17) whose forms A and B correspond to our Types I and II respectively. These two mitochondrial genome types were further subdivided using *Msp*I and *Hha*I respectively. *Msp*I digestion revealed an additional site in four of the 19 Type I mtDNAs. The four isolates which contain this mitochondrial genome are heterozygous at both the *Gpi*-1 and *Pep*-1 loci, a combination not seen in any of the other 20 isolates. They thus appear to form a closely related group, although they have a wide geographical distribution (Table 18.1). This conservation of genotype suggests that members of this group have not readily exchanged genetic material with *P. infestans* belonging to other groups. Mitochondrial DNA with the extra *Msp*I site was designated TypeIb; that lacking the site was designated Type Ia. An additional *Hha*I site was found in one of the five Type II mtDNAs; this was designated Type IIb and the remaining four were designated Type IIa (see Table 18.1).

RFLPs of nuclear DNA

Endonuclease digestion of nuclear DNA (nDNA) produces a large range of restriction fragment sizes which are seen as a smear after agarose gel electrophoresis and ethidium bromide staining. Therefore hybridisation with cloned radiolabelled sequences is necessary to distinguish differences in sizes of specific restriction fragments from nuclear DNA. Fig. 18.2 illustrates how a cloned sequence is used to detect polymorphism in nDNA from different individuals.

Early work on nuclear RFLPs used clones containing genes of known functions (Jeffreys, 1979; Rotwein *et al.*, 1981; Ullrich *et al.*, 1980). Clones from cDNA libraries with inserts of unknown gene sequences have also been exploited. These can be particularly useful for detecting single copy sequences when the level of repetitive DNA is high but have the disadvantage that their insert sequences may be more highly conserved than other single copy sequences and signals produced with digests of nuclear DNA may be weak because of the presence of introns (Helentjaris, 1987). Most commonly, 'anonymous' clones are constructed which contain random inserts of digested genomic DNA. These inserts may detect single copy or repetitive DNA sequences. Single copy sequences are desirable in genetic mapping work, where the RFLPs they detect can be assigned to a single locus and linkage groups can be established in a manner analogous to that using any other polymorphic genetic marker. RFLPs can also be detected if the probe hybridises to a sequence repeated in tandem a variable number of times and an enzyme cutting outside the repeated sequence is used. If such a probe is specific for repeats occurring at a single locus (a locus-specific probe) it is useful for mapping but if it hybridises to repeats at many loci (a multi-locus probe) it reveals a many-banded profile allowing 'typing' or 'fingerprinting'. Both kinds of probe have been much used in the population biology of man (e.g. Jeffreys, Wilson & Thein, 1985), birds (e.g. Burke, 1989) and many other organisms. Dispersed repetitive DNA sequences which often produce highly variable hybridisation profiles have also been used for typing. O'Dell *et al.* (1989) identified polymorphisms in such sequences in the fungal pathogen, *Erysiphe graminis*. We now report the use of anonymous nDNA clones to characterise further the 24 isolates of *P. infestans* from many countries, previously analyzed for mtDNA polymorphisms (Carter *et al.*, 1990). Polymorphism at the nDNA level was found to be sufficiently high to allow preliminary grouping of the isolates based on shared hybridisation patterns using probes for single copy sequences. The probes have also allowed us to detect increases in ploidy levels and to examine the frequency of heterozygosity.

Preparation of nDNA

Table 18.1 shows the isolates used (from the University College of North Wales collection) and details of their phenotype. Methods used for growing the fungus, extracting DNA and endonuclease digestion were as previously published (Carter *et al.*, 1990). CsCl gradients with *bis*-benzimide were used to separate the more dense nDNA from the less dense mtDNA (Garber & Yoder, 1983). A second cycle of centrifugation in CsCl then dialysis at 4°C against TE buffer was used to purify the nDNA further.

Restriction of nDNA

Preparations of nDNA were readily digested with a number of restriction endonucleases. When digests were electrophoresed in agarose gels and the gels stained with ethidium bromide, smears were detected

1. Possible Alleles at Hybridising Locus

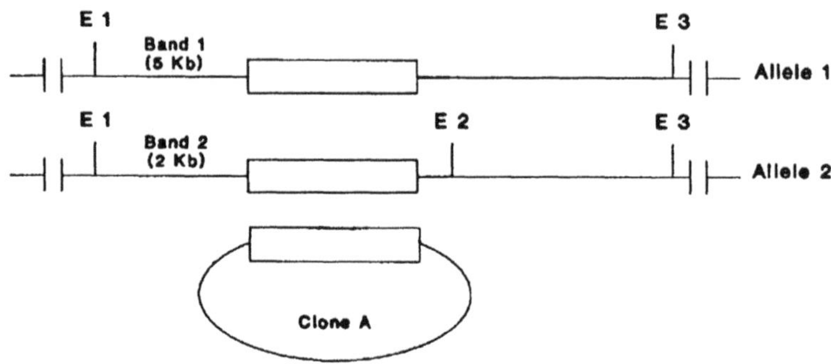

E1, E2 & E3 - sites for restriction enzyme 'E'

E2 is polymorphic

▭ - hybridising regions

2. Use of Radiolabelled Clone A to Detect RFLPs

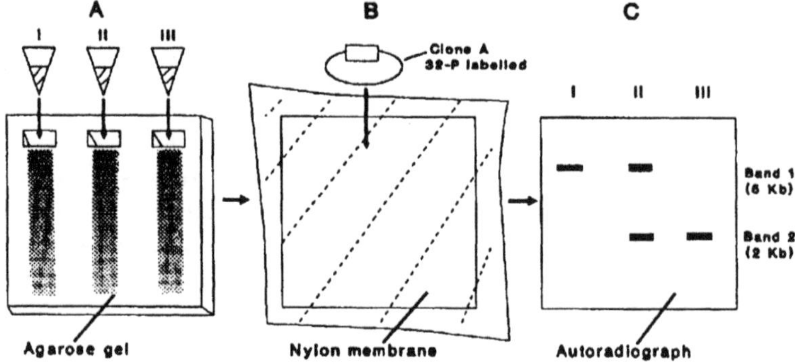

A: DNA from three diploid individuals digested with enzyme 'E' is size-fractionated by gel electrophoresis

B: DNA is transferred by capillary action to a nylon membrane Blot is hybridised with a radiolabelled clone A

C: Autoradiography reveals restriction fragments hybridising with clone A

Fig. 18.2. (Above and on facing page) Detection and interpretation of RFLPs in *Phytophthora infestans* nuclear DNA.

3. Interpretation

INDIVIDUAL	PHENOTYPE	GENOTYPE						
		E1	E2	E3	/	E1	E2	E3
I	Band 1	+	-	+	/	+	-	+
II	Band 1 & Band 2	+	-	+	/	+	+	±
III	Band 2	+	+	±	/	+	+	±

+ : Presence of site

- : Absence of site

± : Site is outside hybridising region therefore presence or absence cannot be determined.

Fig. 18.2. (Continued from facing page) Detection and interpretation of RFLPs in *Phytophthora infestans* nuclear DNA.

together with several discrete bands (Fig. 18.3). In *Hin*dIII digests, the average length of fragments constituting the smear was longer than that in the *Msp*I digests, reflecting a lower frequency of sites for *Hin*dIII (recognises a six base sequence) than for *Msp*I (recognises a four base sequence). The discrete banding patterns detected in the *Hin*dIII gels, which were different from those previously detected in *Hin*dIII digests of mtDNA (Carter *et al.*, 1990), were conserved among the 24 isolates and probably arose from repetitive DNA sequences in the nDNA. However, the high molecular weight bands detected in the *Msp*I digests corresponded in size to the bands previously detected in *Msp*I digests of mtDNA of these isolates (Carter *et al.*, 1990) and indicate that preparations of nDNA can be contaminated with mtDNA despite purification by two cycles of centrifugation in caesium chloride density gradients containing *bis*-benzimide. The mtDNA bands were visible in *Msp*I digests of nDNA preparations owing to a depletion of C + G residues in mtDNA relative to nDNA, so that *Msp*I, which has a recognition sequence of CCGG, cuts the mtDNA less frequently than the nDNA.

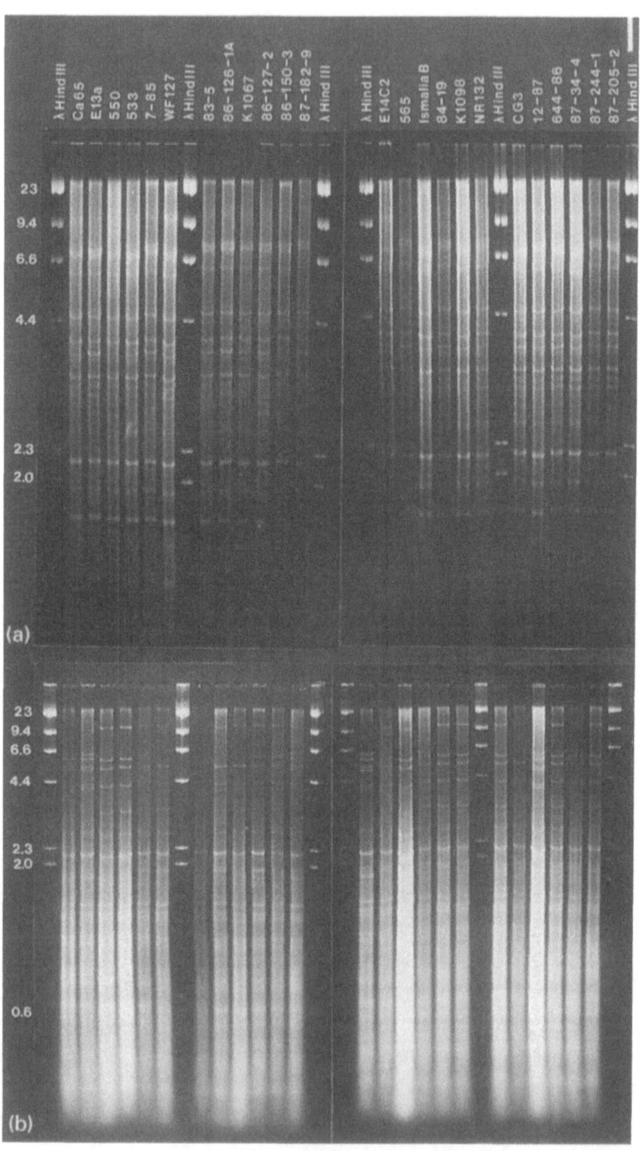

Fig. 18.3. Agarose gel electrophoresis of nDNA of 24 isolates digested with *Hin*dIII (a) or *Msp*I (b). The sizes (in kb) of marker DNAs (phage λ DNA digested with *Hin*dIII) are shown in this and subsequent figures on the side of the gel.

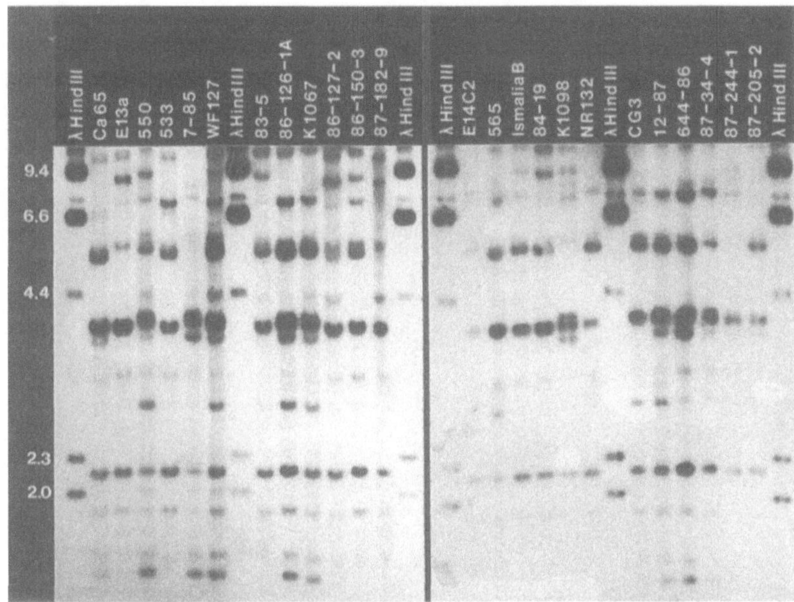

Fig. 18.4. Southern blot of *Bgl*II-digested nDNA from 24 isolates, hybridised with pIN-1.

Production and characterization of probes to detect polymorphisms in nDNA

Ligation of *Hin*dIII-digested nDNA from isolate 550 into the *Hin*dIII site of pUC19 and transformation of *Escherichia coli* DH5 (Cobianchi & Wilson, 1987; Hanrahan, 1985) resulted in a large number of recombinant plasmids with inserts ranging in size from approximately 500 bp to 7 kb. 160 of these were dot-blotted and probed with radiolabelled nDNA of *P. infestans*. Clones producing an intense signal following autoradiography should contain inserts of repetitive DNA, whereas those producing a very faint signal should have inserts of single or low copy number sequences. These latter clones were further analyzed by using each as a probe on a blot from a gel of nDNA of isolate 550 digested with *Hin*dIII. If the insert sequence, which should not contain a *Hin*dIII site, is present as a single copy in the 550 genome, it should detect a single band, provided that 550 is homozygous at this locus, or two bands if heterozygous. Additional bands indicate the clone contains a sequence repeated within the genome.

Polymorphisms in repetitive sequences

The recombinant plasmid pIN-1, which had an insert size of 3·5 kb gave a faint signal on the dot-blot, and a considerable number of hybridising bands on the blot of 550, indicating that it contained DNA of a moderately

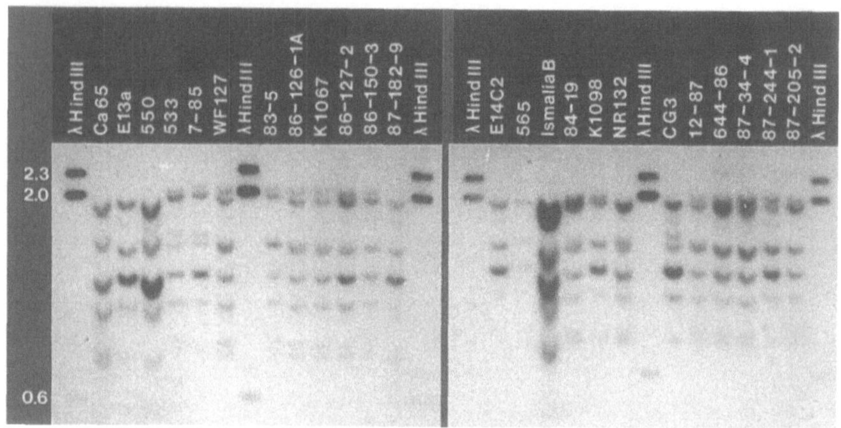

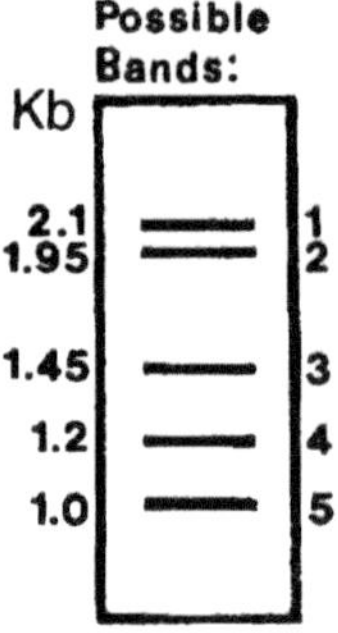

Fig. 18.5. (above) Southern blot of *Msp*I-digested nDNA from 24 isolates, hybridised with pIN-41; the diagram alongside shows the size and numbering scheme of the bands detected in the blot.

---

repetitive nature. This clone was initially selected to screen for polymorphisms among *Bgl*II-digested nDNA of the 24 isolates of *P. infestans*. Fig. 18.4 shows the resulting autoradiograph. A high degree of nuclear DNA polymorphism among the 24 isolates was revealed. This could be useful for 'typing' closely related isolates or for assessing variation within a limited geographical area.

pIN-41 which contained a 0·5 kb insert, gave a less complex hybridisation pattern with *Msp*I-digested nDNA indicating a low copy number sequence. This clone allowed a preliminary grouping of the isolates based on shared fragments (Fig. 18.5, Table 18.2).

Polymorphisms in unique sequences

More informative than the low copy number probes, however, were those which detected only a single sequence within the *P. infestans*

Table 18.2. Grouping of isolates according to *Msp*I restriction fragments which hybridise with clone pIN-41.

Hybridising bands*	Isolates
2 + 4 + 5	Ca65, E13a, 550, 87-182-9, E14C2, 565, Ismailia B
1 + 2 + 4 + 5	86-126-1A, K1067, 86-127-2, 86-150-3, 84-19, K1098, 12-87, 644-86, 87-34-4, 87-244-1, 87-205-2, WF127
1 + 2 + 3 + 4	83-5
2 + 3 + 4 + 5	NR132, CG3
1 + 2 + 3 + 4 + 5	533, 7-85

*The bands are numbered as in Fig. 18.5.

genome. These enabled a further grouping of the isolates and allowed putative genotypes at the hybridising locus to be constructed. They also provided information on the level of heterozygosity and on the existence of polyploidy or heterokaryosis among the isolates.

pIN-2 contains a 1·8 kb insert of single copy nDNA, and revealed a number of polymorphisms when hybridised with *Msp*I or *Bgl*II digests of the 24 isolates. Hybridisation with *Msp*I-digested nDNA (Fig. 18.6) gave three main bands, each of which could exist either independently or in combination. An explanation for the generation of these bands in terms of different combinations of *Msp*I sites is given in Fig. 18.6b. Possible bands with their sizes in kb are given on the left, and are numbered from the largest (slowest running on the gel) to the smallest (fastest running). The region of the DNA with which the clone hybridises is shown schematically in the centre of the diagram. The open box represents the region on the chromosome homologous to the pIN-2 insert sequence. The polymorphic *Msp*I sites within and flanking this sequence are designated M1 to M4. Band 3 is 1·7 kb, slightly smaller than the insert which must therefore contain at least one *Msp*I site. The remaining 0·1 kb probably hybridises with fragments too small to blot efficiently, as only faint bands were seen at the bottom of the autoradiograph. In band 2, the M2 site has been lost and the radiolabelled insert now detects a fragment of 2·1 kb. Likewise the band 1 phenotype of 2·7 kb is seen when both the M2 and M3 sites are lost.

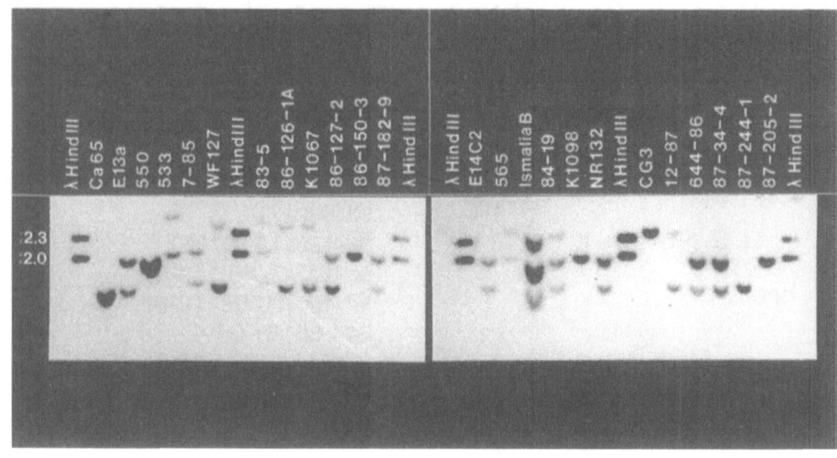

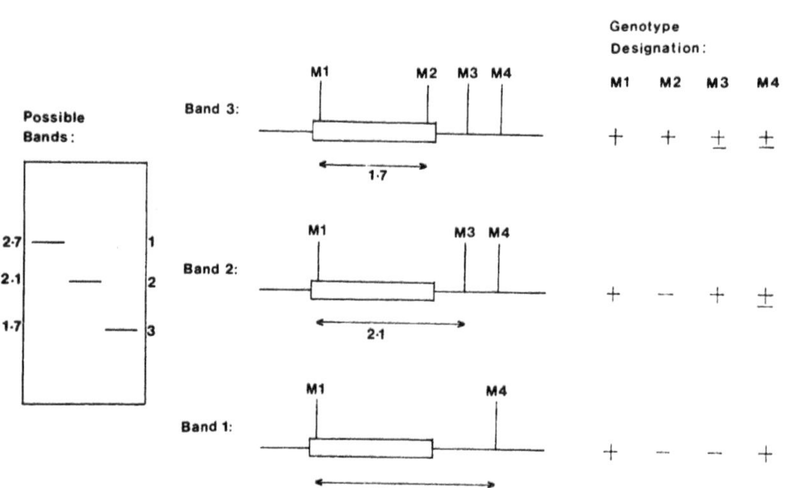

Fig. 18.6. Top: Southern blot of *Msp*I-digested nDNA from 24 isolates, hybridised with pIN-2 (clone 2); the bottom panel shows a diagrammatic model to explain the formation of the bands detected in the blot above. Only the first five higher molecular weight bands have been considered. Genotype designations for each of the band phenotypes are given on the right of the diagram, expressed as the presence (+) or absence (–) of the *Msp*I sites. Sites outside the region hybridising with the radiolabelled probe which cannot be determined are designated +.

Table 18.3. Grouping of isolates according to *Msp*I restriction fragments which hybridise with pIN-2.

Hybridising bands	Isolates	Proposed genotype
homozygous		
1	CG3	+−−+/+−−+
2	550, 86-150-3, K1098, 87-205-2	+−+ +/+−+ +
3	Ca65, 87-244-1	+ + + +/+ + + +
heterozygous		
1 + 2	565	+−−+/+−+ +
1 + 3	WF127, 86-126-1A, K1076, 12-87	+−−+/+ + + +
2 + 3	E13a, E14C2, 7-85, 86-127-2, 87-182-9, NR132, 644-86, 87-34-4	+−+ +/+ + + +
2·8 kb band + 2	533	+−−−+/+−+ +
'hyperploid'		
1 + 2 + 3	83-5, Ismailia B, 84-19	+−−+/+−++/++++

As *P. infestans* is a diploid organism, isolates can give rise to a single band (homozygous), or two bands (heterozygous) in any combination. All possible combinations of bands are seen among the 24 isolates, with the majority being heterozygous (see Table 18.3). Isolate 533 gives rise to band 2 plus a unique band slightly larger than band 1, which could be the result of loss of the M4 site in one chromosome of this isolate. Three of the isolates produce all three bands on the autoradiograph. These may have a ploidy level greater than two, a duplication or else contain intact nuclei from more than one source and be functioning as stable heterokaryons. DAPI-fluorescence of single nuclei of isolate 84/19 is now known to be twice that of most other isolates (Whittaker, Shattock & Shaw, 1991) indicating that this isolate is polyploid and probably tetraploid. Ismailia B has a DNA content typical of a diploid; polyploidy but not aneuploidy can thus be ruled out for this isolate. The DNA content of 83–5 has yet to be determined. Hulbert & Michelmore (1988) noted the occurrence of more than two alleles at a given locus in a number of isolates of *Bremia lactucae* and designated such isolates 'hyperploids'; this term will also be used here to denote isolates of *P. infestans* with more than two alleles.

When pIN-2 was used as a probe against *Bgl*II-digested nDNA from the 24 isolates of *P. infestans*, a more complex pattern resulted (Fig. 18.7).

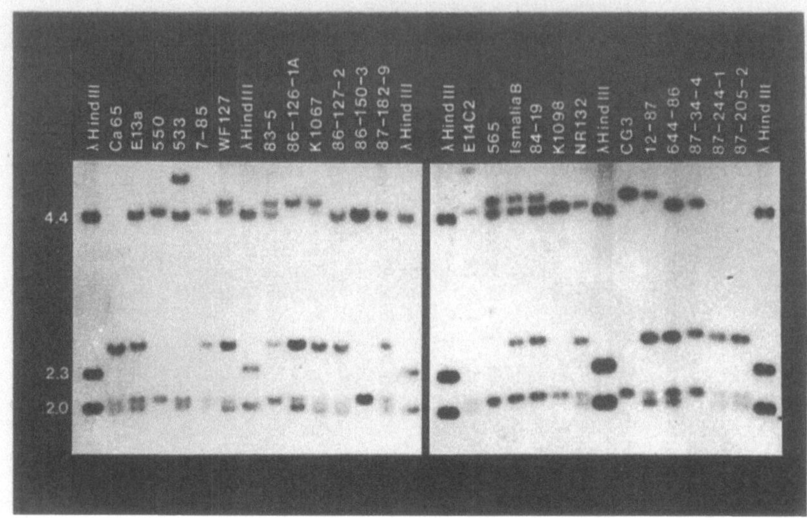

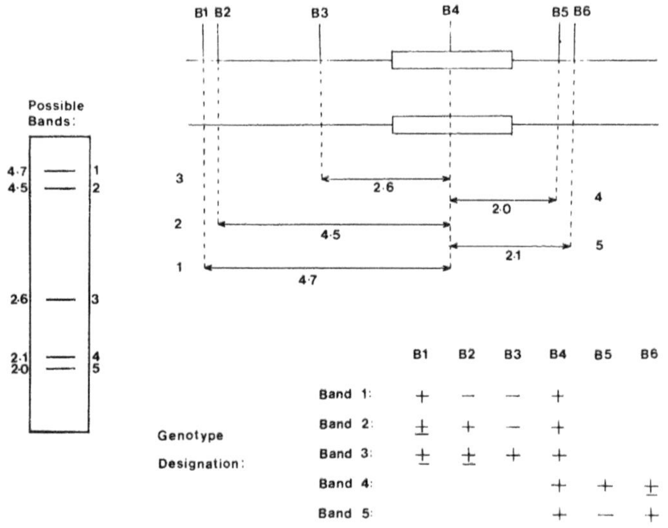

Fig. 18.7. Top: Southern blot of *Bgl*II-digested nDNA from 24 isolates, hybridised with pIN-2 (clone 2); the bottom panel shows a diagrammatic model to explain the formation of the bands detected in the blot above.

Table 18.4. Grouping of isolates according to *Bgl*II restriction fragments which hybridise with pIN-2.

Hybridising bands	Isolates	Proposed genotype[1,2]
homozygous		
1 + 4	CG3	+−−+++/+−−+++
2 + 4	550, 86-150-3, K1098	++−+++/++−+++
heterozygous		
3 + 4 + 5	Ca65, 87-244-1, 87-205-2	++++−+/++++++
1 + 2 + 4	565	+−−+++/++−+++
2 + 3 + 4	87-34-4	++−+++/++++++
1 + 3 + 4 + 5	12-87, K1067, 86-126-1A	+−−+−+/++++++ OR +−−+++/++++−+
2 + 3 + 4 + 5	E13a, 7-85, 86-127-2, 87-182-9, NR132, 644-86	++−+++/++++−+ OR ++−+−+/++++++
5·4kb band + 2 + 4 + 5	533	(+)−−+++/++−+−+ OR
5·6kb band + 2 + 4 + 5	E14C2	(+)−−+−+/++−+++
hyperploid		
1 + 2 + 3 + 4	Ismailia B, 84-19, 83-5	+−−+++/++−+++/ ++++++
1 + 2 + 3 + 4 + 5	WF127	+−−+−+/++−+−+/ ++++−+/++++++

[1]For isolates 533 and E14C2, (+) indicates that site B1 has been displaced upstream of the region shown in Fig. 18.7b.

[2]There are several possibilities for the genotype of isolate WF127. The one proposed is based on the relative intensities of the bands (Fig. 18.7).

Table 18.5. Grouping of isolates according to *Msp*I restriction fragments which hybridise with pIN-29.

Hybridising bands*	Isolates
1 + 3 + 4 + 5	Ca65, E13a, 533, 7-85, 83-5, 86-127-2, 86-150-3, 565, 87-182-9, E14C2, Ismailia B, 84-19, K1098, CG3, 87-34-4, 87-205-2
1 + 4 + 5	550, NR132, 87-244-1
1 + 3 + 5	644-86
1 + 2 + 3 + 4 + 5	WF127, 86-126-1A, K1067, 12-87

*Bands are numbered in order of decreasing size (see Fig. 18.8)

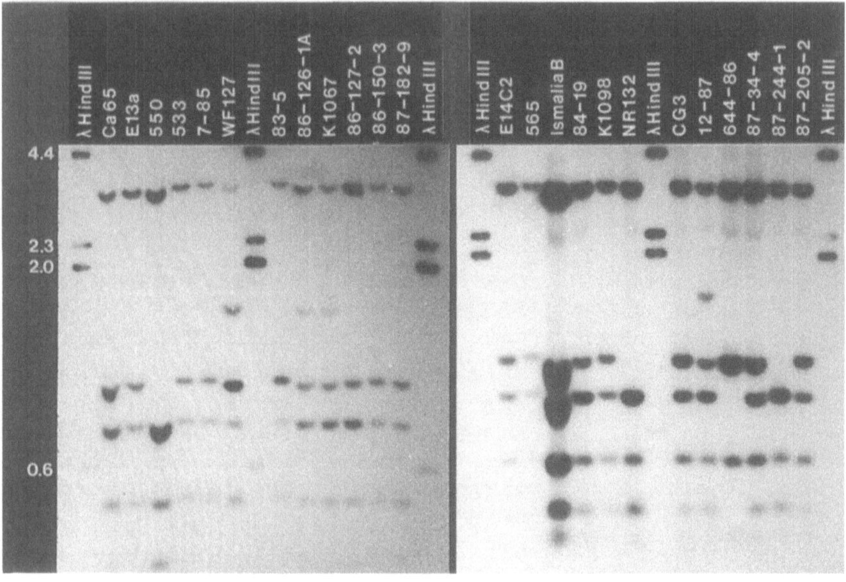

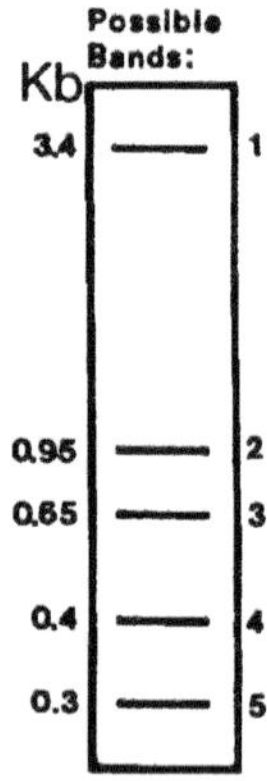

Fig. 18.8. Top: Southern blot of *Msp*I-digested nDNA from 24 isolates, hybridised with pIN-29; the diagram alongside shows the size and numbering scheme of the bands detected in the blot above.

Isolates are again grouped according to hybridisation patterns (Table 18.4). In this model, there is a *Bgl*II site (B4) which is conserved among all the 24 strains. The observed bands can be generated by polymorphisms at sites B2, B3 and B5. Isolates have been designated homozygous when only two bands occur and heterozygous when three or four bands occur. Again heterozygosity predominates for this clone/enzyme combination. 533 and E14C2 each have a novel band larger than band 1, which probably results from polymorphisms at and beyond B1 respectively. Isolates Ismailia B, 84-19 and 83-5 produced evidence for three alleles and WF127

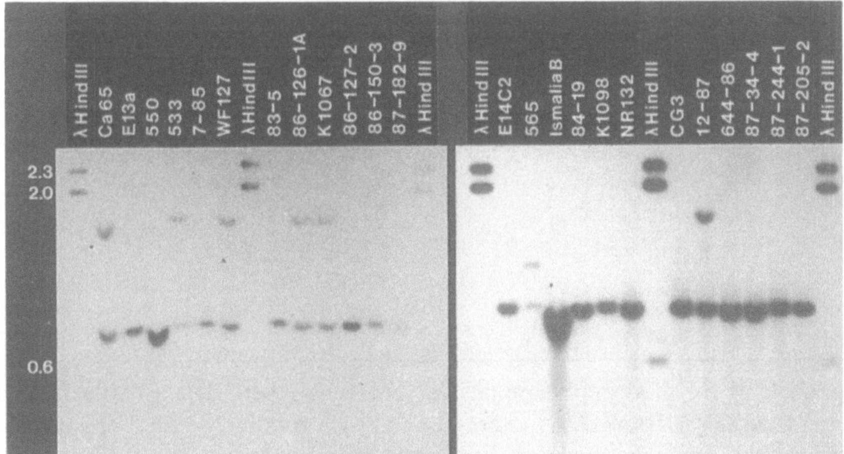

Fig. 18.9. Southern blot of *Msp*I-digested nDNA from 24 isolates, hybridised with pIN-11.

---

for four alleles; WF127 is known to have double the normal DNA content per nucleus (Tooley & Therrien, 1987).

The results of hybridising *Msp*I-digested nDNA with two other single copy clones, pIN-29, and pIN-11, are shown in Figs. 18.8 and 18.9 and the corresponding groupings of isolates based on common banding patterns are listed in Tables 18.5 and 18.6 respectively. Several bands were detected by pIN29, suggesting that the insert in this clone contains a number of *Msp*I sites. Seventeen of the 24 isolates gave a single band with pIN-11, indicating that they were homozygous at this locus. The remaining seven isolates (86-126-1A, K1067, 12-87, WF127, 533, 565 and Ca65) appeared to be heterozygous at this locus, giving rise to an additional band of 1·6 kb (1·1 kb in the case of 565).

Relatedness of the 24 isolates

Analysis of the banding patterns and groupings (Figs. 18.4 to 18.9, Tables 18.2 to 18.6) shows that isolates 86-126-1A, K1067 and 12-87 are indistinguishable, using four different anonymous probes which represent at least four different nuclear loci. Moreover, they often shared RFLPs not found in any of the other 20 isolates, e.g. a 1·5 kb band in *Msp*I-digests detected by pIN-29 (Fig. 18.8). Hence the results support the suggestion (Carter et al., 1991;p Chapter 15) based on their unique heterozygous

Table 18.6 Grouping of isolates according to *Msp*I restriction fragments which hybridise with pIN-11.

Hybridising bands	Isolates
3	E13a, 550, 7-85, 83-5, 86-127-2, 86-150-3, 87-182-9, E14C2, Ismailia B, 84-19, K1098, NR132, CG3, 644-86, 87-34-4, 87-244-1, 87-205-2
1 + 3	Ca65, 533, WF127, K1067, 86-126-1A, 12-87
1 + 2	565

genotypes for glucosephosphate isomerase and peptidase (Table 18.1) and their Type Ib mtDNA characterised by a relatively rare *Msp*I polymorphism, that these isolates share a common origin and have undergone little or no recombination with other genotypes, despite their global distribution. WF127, the only other isolate studied which has type Ib mtDNA and the same double-heterozygous isozyme genotype, differs from the other three isolates only in a nDNA polymorphism detected with pIN-2 and virulence on an additional differential (Table 18.1). Hence it is likely to be derived from other isolates in this unique group by mutation, rather than by recombination with a different strain.

The only other isolates which were indistinguishable on the basis of nDNA polymorphisms were E13a (from Egypt) and 87-182-9 (from the UK). These isolates were both A2 mating type with the relatively rare type IIa mtDNA. They are clearly very closely related and probably share a common origin.

The finding of two groups of closely related isolates (the 86-126-1A, K1067, 12-87, WF127 group and the E13a, 87-182-9 group), each of which contains isolates obtained from widely separated geographical regions (Table 18.1), emphasises the likelihood of spread as a result of the export/import of infected tubers. It is probably relevant that Egypt has obtained much of its potato seed from the UK and ships large quantities of ware to the UK.

The degree of nDNA polymorphism detected with the five clones was sufficiently high to enable each of the remaining 18 isolates to be distinguished from each other i.e. each of these isolates is unique. By extending this study, it should be possible, based on proportions of nDNA polymorphisms in common, to determine the degree of relatedness between isolates. Combined with studies of mtDNA RFLPs and the frequency of occurrence of the A1 and A2 mating types, it should be feasible to assess the importance of the sexual cycle in the epidemiology of this important pathogen.

Conclusions

This preliminary survey on isolates from diverse geographical locations indicates the usefulness of RFLP markers in both mitochondrial and nuclear DNA for studies of populations of *P. infestans*. Variation in the mtDNA appears to be quite limited but where it occurs it provides a simple and precise method of following a polymorphism through a population. Mitochondrial DNA polymorphisms also provide a marker for cytoplasmic genes, and can be used to investigate phenotypes that do not show classical Mendelian inheritance. Isolating purified mtDNA is laborious; however, fragments detecting site polymorphisms can be cloned, radiolabelled and used to detect mtDNA polymorphisms in restriction digests of total DNA. Alternatively, polymorphisms detected by *Msp*1 and *Hha*1, and possibly other enzymes with GC-rich restriction sites, can be found in restriction digests of total DNA by ethidium bromide staining.

A high degree of polymorphism at the nuclear level is seen among the isolates of *P.infestans* included in this survey, such that single copy sequence probes alone can be used to distinguish between most of the isolates. More closely related isolates could be examined using probes that contain repetitive DNA sequences which are generally non-coding and therefore more polymorphic than the non-repetitive sequences. Such probes could be useful in a field situation where the spread of disease is thought to have arisen from a limited number of original pathotypes. They could also be used to study the spread of novel phenotypes, for example increased virulence or fungicide resistance, through a population.

Clones containing single copy sequences give useful preliminary evidence for the existence of hyperploidy. Two out of four hyperploid isolates are now known to have nuclei with larger DNA contents; further cytological work is necessary to identify how the increased DNA is organised in the other two. The single copy probes have also demonstrated a considerable level of heterozygosity in the isolates included in this survey. This has been exploited for genetic mapping, where heterozygosity in the parents allows segregation of alleles in the F_1 population (Hulbert *et al.*, 1988), avoiding the need for F_2 or backcrosses where inbreeding may cause problems (Chapter 15). A rudimentary genetic map of *P. infestans* using RFLP markers has now been prepared (Carter, 1990). Pulse field gel electrophoresis has been initiated to achieve separation of whole chromosomes of *P. infestans*. Radiolabelled clones can then be hybridised to blots of the chromosomes and thus can be placed directly into linkage groups. Parental isolates have been chosen with as diverse virulence phenotypes as possible, so that once linkage of the virulence genes with the arbitrary DNA markers is established the genes themselves can be isolated and their functions determined.

References

Anderson, J. B., Petsche, D. M. & Smith, M. L. (1987). Restriction fragment length polymorphisms in biological species of *Armillaria mellea*. *Mycologia* **79**, 69-76.

Avise, J. C., Gibun-Davidson, C., Laerm, J., Patton, J. M. C. & Lansman, R. A. (1979). Mitochondrial DNA clones and matriarchial phylogeny within and among geographic populations of the pocket gopher *Geomys pientis*. *Proceedings of the National Academy of Sciences of the USA* **76**, 6694-6698.

Ball, R. M., Freeman, S., James, F. C., Bermingham, E. & Avise, J. C. (1988). Phylogeographic population structure of Red-winged Blackbirds assessed by mitochondrial DNA. *Proceedings of the National Academy of Sciences of the USA* **85**, 1558-1562.

Bates, M. R., Buck, K. W. & Brasier, C. M. (1990). Molecular variation in the Dutch elm disease fungus. In *Molecular Evolution* (UCLA Symposium, March 1989), (ed. M. T. Clegg & S. J. O'Brien), pp. 171-178. Alan R. Liss, Inc.: New York.

Botstein, D., White, R., Skolnick, M. & Davis, R. W. (1980). Construction of a genetic linkage map in man using restriction fragment length polymorphisms. *American Journal of Human Genetics* **32**, 314-331.

Burke, T. (1989). DNA fingerprinting and other methods for the study of mating success. *Trends in Ecology and Evolution* **4**, 139-144.

Cann, R. L., Stoneking, M. & Wilson, A. C. (1987). Mitochondrial DNA and human evolution. *Nature* **325**, 31-36.

Carter, D. A. (1990). DNA polymorphisms as genetic markers in *Phytophthora infestans*. PhD Thesis, University of London, UK.

Carter, D. A., Archer, S. A., Buck, K. W., Shaw, D. S. & Shattock, R. C. (1990). Restriction fragment length polymorphisms of mitochondrial DNA of *Phytophthora infestans*. *Mycological Research* **94**, 1123-1128.

Croft, J. H. (1987). Genetic variation and evolution in *Aspergillus*. In *Evolutionary Biology of the Fungi*, (ed. A. D. M. Rayner, C. M. Brasier & D. Moore), pp. 311-323. Cambridge University Press.

Cobianchi, G. F. & Wilson, S. H. (1987). Enzymes for modifying and labelling DNA and RNA. In *Guide to Molecular Cloning Techniques*, (ed. S. L. Berger & A. R. Kimmel), pp. 94-110. Academic Press: New York.

Förster, H., Kinscherf, T. G., Leong, S. A. & Maxwell, D. P. (1987). Molecular analysis of the mitochondrial genome of *Phytophthora*. *Current Genetics* **12**, 215-218.

Förster, H., Kinscherf, T. G., Leong, S. A. & Maxwell, D. P. (1989). Restriction fragment length polymorphisms of the mitochondrial DNA of *Phytophthora megasperma* isolated from soybean, alfalfa and fruit trees. *Canadian Journal of Botany* **67**, 529-537.

Garber, R. C. & Yoder, O. C. (1983). Isolation of DNA from filamentous fungi and separation into nuclear, mitochondrial, ribosomal and plasmid components. *Analytical Biochemistry* **135**, 416-422.

Garber, R. C. & Yoder, O. C. (1984). Mitochondrial DNA of the filamentous ascomycete *Cochliobolus heterostrophus*. *Current Genetics* **8**, 621-628.

Goodwin, S. B. & Fry, W. E. (1988). Mitochondrial DNA polymorphism in *Phytophthora infestans*. *Phytopathology* **78**, 1542 (abstract).

Grodzicker, T., Williams, J., Sharp, P. & Sambrook, J. (1974). Physical mapping of temperature sensitive mutations of adenoviruses. *Cold Spring Harbor Symposium on Quantitative Biology* **39**, 439-446.

Hanahan, D. (1985). Techniques for transformation of *E. coli*. In *DNA Cloning, a Practical Approach*, Vol. 1., (ed. D. M. Glover), pp. 109-135. IRL Press Ltd.: xxxxxxx.

Helentjaris, T. (1987). A genetic linkage map for maize based on RFLPs. *Trends in Genetics* **3**, 217-221.

Hoeben, P. & Clark-Walker, G. D. (1986). An approach to yeast classification by mapping mitochondrial DNA from *Dekkera/Brettanomyces* and *Eeniella* genera. *Current Genetics* **10**, 371-379.

Hulbert, S. H., Ilott, T. W., Legg, E. J., Lincoln, S. E., Lander, E. S. & Michelmore, R. W. (1988). Genetic analysis of the fungus *Bremia lactucae*, using restriction fragment length polymorphisms. *Genetics* **120**, 947-958.

Hulbert, S. H. & Michelmore, R. W. (1988). DNA restriction fragment length polymorphism and somatic variation in the lettuce downy mildew fungus, *Bremia lactucae*. *Molecular Plant-Microbe Interactions* **1**, 17-24.

Hutchinson, C. IIIrd, Newbold, J., Potter, S. & Edgell, M. (1974). Maternal inheritance of mammalian mitochondrial DNA. *Nature* **251**, 536.

Jeffreys, A. J. (1979). DNA sequence variants in the ^{a}Y-, ^{A}Y-, δ and β-globin genes of man. *Cell* **18**, 1-10.

Jeffreys, A. J., Wilson, V. & Thein, S. L. (1985). Individual-specific 'fingerprints' of human DNA. *Nature* **316**, 76-79.

Klimczac, L. J. & Prell, H. H. (1984). Isolation and characterisation of the mitochondrial DNA of the oomycetous fungus *Phytophthora infestans*. *Current Genetics* **8**, 323-326.

Latorre, A., Moya, A. & Ayala, F. J. (1986). Evolution of mitochondrial DNA in *Drosophila subobscura*. *Proceedings of the National Academy of Sciences of the USA* **83**, 8649-8653.

O'Dell, M., Wolfe, M. S., Flavell, R. B., Simpson, C. G. & Summers, R. W. (1989). Molecular variation in populations of *Erysiphe graminis* on barley, oats and rye. *Plant Pathology* **38**, 340-351.

Pace, N. R., Olsen, G. J. & Woese, C. R. (1986). Ribosomal RNA phylogeny and the primary lines of evolutionary descent. *Cell* **45**, 325-336.

Petes, T. K. & Botstein, D. (1977). Simple Mendelian inheritance of the reiterated ribosomal DNA of yeast. *Proceedings of the National Academy of Sciences of the USA* **74**, 5091-5095.

Rotwein, P., Chyn, R., Chirgwin, J., Cordell, B., Goodman, H. M. & Permutt, M. A. (1981). Polymorphism in the 5'-flanking region of the human insulin gene and its possible relation to type 2 diabetes. *Science* **213**, 1117-1120.

Sederoff, R. R., Levings, C. S. IIIrd, Timothy, D. H. & Hu, W. W. L. (1981). Evolution of DNA sequence organisation in mitochondrial genomes of *Zea*. *Proceedings of the National Academy of Sciences of the USA* **78**, 5953-5957.

Shattock, R. C., Tooley, P. W. & Fry, W. E. (1986). Genetics of *Phytophthora infestans*: determination of recombination, segregation, and selfing by isozyme analysis. *Phytopathology* **76**, 410-413.

Tooley, P. W., Fry, W. E. & Villarreal Gonzalez, M. J. (1985). Isozyme characterisation of sexual and asexual *Phytophthora infestans* populations. *Journal of Heredity* **76**, 431-435.

Tooley, P. W. & Therrien, D. (1987). Cytophotometric determination of the nuclear DNA content of 23 Mexican and 18 non-Mexican isolates of *Phytophthora infestans*. *Experimental Mycology* **11**, 19-26.

Ullrich, A., Dull, T. J., Gray, A., Brosius, J. & Sures, I. (1980). Genetic variation in the human insulin gene. *Science* **209**, 612-615.

Weber, C. A., Hudspeth, M. E. S., Moore, G. P. & Grossman, L. I. (1986). Analysis of the mitochondrial and nuclear genomes of two Basidiomycetes, *Coprinus cinereus* and *Coprinus stercorarius*. *Current Genetics* **10**, 515-525.

Whittaker, S. L., Shattock, R. C. & Shaw, D. S. (1991). Variation in DNA content of *Phytophthora infestans* as measured by a microfluorimetric method using the fluorochrome DAPI. *Mycological Research* **95**, 602-610.

Chapter 19

Gene structure and expression in *Phytophthora infestans* and the development of gene-mediated transformation

James R. Kinghorn, Richard P. Moon, Shiela E. Unkles & James M. Duncan

Recombinant DNA technology should help to elucidate the mechanisms underlying pathogenicity and virulence in *Phytophthora infestans*. An important tool in this technology is gene-mediated transformation. Efficient gene transfer systems have been developed for many ascomycete fungi and for representatives of several other groups of filamentous fungi (for reviews see Ballance, 1986; Gurr, Unkles & Kinghorn, 1987; Hynes, 1986; Ramsbosek & Leach, 1987; Mishra, 1985; Fincham, 1989; Unkles, 1989). So far, transformation has only been reported for one member of the Oomycetes, *Achlya ambisexualis*, (Manavathu *et al*., 1988). These results provide encouragement for successful gene manipulation of other oosporic fungi including the potato late blight pathogen. Gene transfer systems require three essential components: a protoplasting and regeneration system; a selectable marker; and a functional promoter and terminator.

Protoplasting and regeneration systems

The production of large yields of protoplasts and their regeneration at high frequencies is central to the development of a gene mediated transformation system. Although protoplasts of *P. cinnamomi* and *P. nicotianae* (syn. *parasitica*) were first produced over two decades ago (Bartnicki-Garcia & Lippman, 1966, 1967), it is only comparatively recently that protoplast formation and subsequent regeneration into macroscopic colonies have been reported for *P. infestans* (Pesti & Ferenczy, 1979; Campbell *et al*., 1989) and *P. parasitica* (Jahnke, Leipoldt & Prell, 1987). This section describes the principal factors found to be important for protoplast formation and regeneration from mycelium, sporangia and encysted zoospores of *P. infestans*. Results presented from our laboratories refer mainly to *P. infestans* isolate ATCC 48720 (virulence phenotype 0, mating type A1) and ATCC 36609 (virulence phenotype 3.4, mating type A2). The term 'protoplast' describes digested cells which (a) lyse in water, (b) fuse in polyethylene glycol solutions and

(c) lack fluorescence in Calcofluor White solutions (Hess & Leipoldt, 1979)(cyst and sporangial protoplasts) or have diminished fluorescence in such solutions, localised to small patches of the cell surface (mycelial protoplasts). The formation and regeneration of protoplasts from cysts and sporangia is illustrated in Fig. 19.5 (opposite p. 306).

Age of culture

The age of mycelium is an important factor influencing protoplast yield and viability for many fungi (Peberdy, 1979). Also, it is generally accepted that higher transformation efficiencies are achieved using protoplasts derived from younger mycelium of 'model' filamentous fungi such as *Aspergillus nidulans*, rather than using protoplasts derived from older mycelium. It therefore seemed logical to prepare protoplasts from young mycelium of *P. infestans* for use in transformation experiments. We found that 36–48 h mycelium grown from a relatively heavy inoculum of sporangia in 40 ml pea broth (Campbell *et al.*, 1989) in 250 ml conical flasks gave good yields of protoplasts (approximately 3×10^6 protoplasts per 40 ml; Table 19.1). Weight for weight, older mycelium (4 to 8 day) tended to yield fewer protoplasts in a given period of digestion, and regeneration frequencies were lower (Moon, unpublished).

Cell wall degrading enzymes

It has been found that 80-90% of the dry weight of cell walls of *Phytophthora* spp. consists of β-linked glucose polymers (glucans) with small amounts of other carbohydrate. These glucans are of two types, a non-cellulosic fraction (β-1,3 linked) which is highly branched (β-1,6 linked) and a smaller, cellulosic, fraction (β-1,4 linked). The ratio of non-cellulosic to cellulosic glucans varies between 2:1 and 10:1 by dry weight, according to species and cell type (Bartnicki-Garcia & Wang, 1983). Obviously, efficient digestion of such cell wall material requires selection of an appropriate enzyme preparation. *Phytophthora* protoplasts have been produced using enzymes prepared from *Streptomyces* (*P. cinnamomi* and *P. nicotianae*; Bartnicki-Garcia & Lippman, 1966), the snail *Helix pomatia* (*P. infestans*; Pesti & Ferenczy, 1979), and the commercial preparation NovoZym 234 (Novo Biolabs) from *Trichoderma* sp. (*P. nicotianae*; Jahnke *et al.*, 1987, and *P. infestans*; Campbell *et al.*, 1989). We assessed three commercial enzyme preparations which possessed enzymic activities capable of hydrolysing some of the types of bond involved. These were Lyticase (Sigma Chemical Company; β-1,3 glucanases from *Arthrobacter*), Cellulase (BDH; endo-1,4-β glucanase from *Trichoderma*), and NovoZym 234 (β-1,3 glucanase, laminarinase, and chitinase among other enzyme activities from *Trichoderma*). Sporangia and encysted zoospores were digested with these enzymes in 1 M KCl solutions at room temperature (20–25°C). Only

Table 19.1. Summary of optimal protoplasting and regeneration conditions.

	Cell type		
	Encysted zoospores	Sporangia	Mycelium
Protoplasting osmoticum	0·63 M KCl + 0·2 M $CaCl_2$	0·63 M KCl + 0·2 M $CaCl_2$	0·35 M $CaCl_2$
NovoZym 234 concentration (batch 1961) mg ml^{-1}	50	20	10
Digestion period (h)	2	24	1·5
Yield	50%	70%	3·6 x 10^6 per culture
Regeneration osmoticum	0·8 M sorbitol	1·0 M sorbitol	0·8 M mannitol
Regeneration %age	90	80	27

NovoZym 234 (batch no. 1961) completely protoplasted encysted zoospores and sporangia of *P. infestans* isolate ATCC 48720 whilst Lyticase and Cellulase were not effective. Combining NovoZym 234 and Lyticase did not yield protoplasts from these cell forms more efficiently than NovoZym 234 alone; combining NovoZym 234 with Cellulase actually reduced the rate of digestion of walls of sporangia and cysts, possibly as a result of inhibitors in the Cellulase preparation. Therefore, no synergistic effect was observed and consequently we used only NovoZym 234 in further studies. No experiments of this kind were carried out on mycelia. As has been found with other fungi, the efficiency of protoplasting varied according to the batch of NovoZym 234 used. A recent batch (no. 2416) did not protoplast sporangia of isolate ATCC 48720 even after 48 h incubation although protoplast formation from mycelia and cysts was not significantly different from that with batch no. 1961.

Osmotic stabilizers

We tested various mineral salts [KCl, NaCl, KNO_3, $NaNO_3$ and $Ca(NO_3)_2$] and sugar alcohols (sorbitol and mannitol) as osmotic stabilizers for protoplasts from cysts and sporangia. Our results showed that optimum concentrations of mineral salts generally gave higher yields of protoplasts whilst sugar alcohols resulted in slower and incomplete digestion of the cell walls. Although protoplast yields were similar for KCl and NaCl, subsequent regeneration frequencies were much higher for protoplasts produced in KCl; 1 M KCl was optimal for the number of

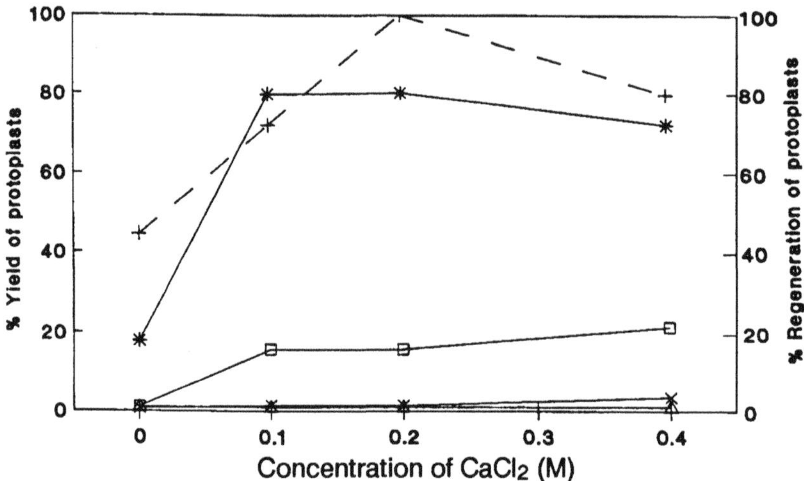

Fig. 19.1. Effect of $CaCl_2$ on production and regeneration of sporangial protoplasts of *P. infestans*. Sporangia were treated for 24 h at 25°C with 20 mg ml^{-1} NovoZym 234 in osmotica containing various concentrations of $CaCl_2$ and KCl, keeping the osmolality isotonic to 1 M KCl. The percentage of sporangia yielding protoplasts after 24 h is shown (+). Regeneration of these protoplasts after 2 days at 17°C in variously amended pea broths is also shown: regeneration in pea broth + 1 M sorbitol alone (asterisk), sorbitol + 0·1 M $CaCl_2$ (open square), sorbitol + 0·2 M $CaCl_2$ (×), 0·4 M $CaCl_2$ (open triangle). Similar results were obtained with cyst protoplasts (from Campbell *et al.*, 1989).

---

regenerated protoplasts obtained from the original suspension of sporangia or cysts (data not shown). Protoplasts were regenerated in liquid media containing sorbitol at optimal concentrations of 0·8 M and 1 M for cyst and sporangial protoplasts respectively (Fig. 19.5d & h). Very low regeneration frequencies were observed in media with 1 M KCl as osmoticum. Pesti & Ferenczy (1979) and Jahnke *et al.* (1987) found 0·35 M $CaCl_2$ optimal for the formation of protoplasts from mycelium of *P. infestans* and *P. nicotianae* respectively. Optimal regeneration frequencies of 8% for *P. infestans* and about 19% for *P. nicotianae* were obtained using 0·4 M mannitol + 0·1 M $CaCl_2$ as the regeneration osmoticum. In our experiments, digestion of two day old mycelium of isolate ATCC 48720 in 0·35 M $CaCl_2$ yielded protoplasts with optimal regeneration frequencies of 25-30% when 0·8 M mannitol was used as the regeneration osmoticum (Table 19.1).

Table 19.2. Minimum inhibitory concentrations (μg ml^{-1}) of various drugs suppressing colony establishment from encysted zoospores (Z) or mycelial plug (M) inocula of three *P. infestans* isolates.

	Isolate					
	ATCC 48720		I259		I230	
Drug	M	Z	M	Z	M	Z
G418	4	4	4	2	4	1
Hygromycin B	100	50	5	10	100	25
Oligomycin C	20	20	2	<1	20	20
Phleomycin	300	150	<2	<2	200	40

Hygromycin B (Calbiochem, USA), G418 (Gibco, UK), phleomycin (Cayla, France) as filter-sterilised aqueous solutions and oligomycin C (Sigma, UK) (dissolved in absolute ethanol) were added to molten pea broth agar at 55°C. The minimum inhibitory concentration of a drug was defined as the minimum concentration which prevented measurable fungal radial growth from a hyphal plug or which prevented macroscopic growth from encysted zoospores, after 12-15 days incubation at 18°C (Moon & Gould, unpublished).

Pesti & Ferenczy (1979) and Jahnke *et al.* (1987) found that $CaCl_2$ in regeneration media increased protoplast regeneration frequencies. This observation led us to investigate the effect of $CaCl_2$ on regeneration of protoplasts from sporangia, cysts and mycelium of isolate ATCC 48720. Contrary to their results, we found that 0·1 to 0·4 M $CaCl_2$ in the regeneration medium reduced regeneration frequencies of cyst and sporangial protoplasts by at least 75% (Fig. 19.1). Viability of mycelial protoplasts was reduced by 66% in liquid and solid media amended with 0·4 M mannitol + 0·1 M $CaCl_2$. This may be an effect of salts in general as we find regeneration frequencies of mycelial and cyst protoplasts are substantially reduced unless they are washed free of KCl or $CaCl_2$ with a solution such as 0·9 M mannitol before being added to regeneration media. However, digesting cysts and sporangia in $CaCl_2$/KCl solutions isotonic to 1 M KCl dramatically increased subsequent viability in sorbitol-emended regeneration media (Fig. 19.1). Maximum viabilities of over 90% were obtained using 0·2 M $CaCl_2$ + 0·63 M KCl as the protoplasting osmoticum (Table 19.1). Digesting young mycelium in this osmoticum rather than in 0·35 M $CaCl_2$ was slower.

Regeneration of protoplasts

The method of plating protoplasts of *P. infestans* on solid media can affect viability markedly. We spread aliquots of protoplasts gently onto wet, freshly made and poured pea broth agar (Campbell *et al.*, 1989). The inoculated plates were not allowed to dry so that protoplasts regenerated in a film of liquid on the agar surface. Drying of plated protoplasts or use of dry or old medium reduced regeneration frequencies significantly. Only fresh medium was used for regenerating protoplasts in liquid culture. Jahnke *et al.* (1987) reported similar findings with *P. nicotianae*. Pesti & Ferenczy (1979) failed to obtain colonies when protoplasts of *P. infestans* mycelia were plated out in thin overlays at 43°C. We have employed this technique with sporangial protoplasts using 0·5% low melting point agarose (BRL), inoculated near the gelling temperature (about 30°C). In this way we obtained colonies from protoplasts of sporangia, although regeneration frequencies were not determined. Establishment of macroscopic colonies was not significantly affected if mycelial protoplasts were allowed to regenerate for 24 h on thin agar plates before being overlaid with non-osmotically stabilized 0·7% conventional agar medium poured when near the gelling temperature of about 40°C.

Selection systems

Development of a transformation system for *P. infestans* using dominant selectable marker genes sandwiched between promoters and terminators from other filamentous fungi has been attempted in a number of research laboratories including our own. Since *P. infestans* isolates may exhibit diploidy, polyploidy or aneuploidy and many grow poorly on minimal media we considered that systems which depend on the repair of auxotrophy or a growth defect would be difficult to implement. Consequently, a system based on dominant, selectable genes which confer resistance to antibiotics such as hygromycin B, phleomycin, G418 (geneticin), and oligomycin C (Table 19.2) was thought more likely to be successful. Before employing such markers it was necessary to verify that the selective agent was toxic to *P. infestans*. The minimum inhibitory concentrations (MIC) of G418, hygromycin B, oligomycin C and phleomycin were determined for both encysted zoospores and mycelial cells using three isolates of *P. infestans*, including ATCC 48720. The results presented in Table 19.2 show that the MIC was influenced by the cell type and isolate. It was also influenced by protoplast formation and treatment. Nevertheless, the low MICs for G418, hygromycin B, oligomycin C and phleomycin suggested that they could be useful selective agents for transformation. Chloramphenicol and methotrexate, on the other hand, seemed less suitable candidates since relatively high concentrations were required to retard growth significantly (data not shown). Oligomycin C, although inhibitory, has not yet been used as a selective agent since a mutant allele is normally required for the transformation vector (Ward, Wilkinson & Turner, 1986); attempts to

Table 19.3. Heterologous expression systems tested.

Plasmid	Promoter	Reference
(a) *hph* gene and hygromycin B		
pH1S	*C. heterostrophus* promoter 1	2
pH1B	*C. heterostrophus* promoter 1	2
pH25	*C. heterostrophus* promoter 2	2
pDH25	*A. nidulans trpC*	2, 3
pAN7-1	*A. nidulans gpdA*	1, 2, 3
pHL1	*U. maydis hsp70*	2, 3
pPS29	*C. acremonium* IPNS	2
pROH1	*C. acremonium* IPNS	1
pLG90	*S. cerevisiae* CYC1	1
pRD1	*P. megasperma actA*	1
pTPH	*D. melanogaster Hsp*70 (minimal)	3
(b) *kan*^{903} gene and G418		
pBC10	*E. coli*	1
pEB10	*E. coli*	1
pJL2	*E. coli*	1
pETM2	CAMV 35S	2
pSV2neo	SV40 (early)	1
(c) *ble* gene and phleomycin		
pAN8-1	*A. nidulans gpdA*	1
pUT332	*S. cerevisiae* CYC1	1
pUT701	*Streptomyces*	1

References: 1 = St. Andrews University and Scottish Crop Research Institute (UK); 2 = O. C. Yoder, Cornell University (USA); 3 = T. Peterson, Max-Planck Institute, Koln (Germany).

generate an oligomycin-resistant strain were unsuccessful (Gould & Moon, unpublished). Finally, we failed to generate an orotidine 5'-phosphate decarboxylase (*pyr*) auxotrophic mutant by selection for resistance to 5-fluoroorotic acid (Gould, unpublished), which we take to support our view that transformation of *P. infestans* based on auxotrophic selection is unlikely to be successful.

Functional promoters and terminators

Heterologous gene expression cassettes

Heterologous gene expression cassettes derived from a number of organisms, including filamentous fungi, have been used in these studies (Table

19.3). Expression systems have taken advantage of the promoters and terminators of several genes. Such constructs have been used by several research groups in attempts to transform protoplasts of *P. infestans*. Surprisingly perhaps, to date there has been no unequivocal evidence for the successful generation of transformants. Even the 'universal' pAN7-1 vector which has been used to transform a wide range of filamentous fungi has been unsuccessful with *P. infestans*. Transformation procedures employing lithium acetate or $CaCl_2$/PEG treatments or electroporation have been equally unsuccessful. It is worth noting that all three cell types of the fungus have been used in transformation experiments in our laboratory. Putative transformants have been observed on selective media by many research groups but, disappointingly, probes have failed to detect vector sequences in Southern blots of such transformants.

Some recent work by Peterson and colleagues has revealed the ability of zoospores to take up exogenous plasmid DNA; Southern analysis suggests that some of this DNA was concatamerised and was chromosomally integrated. Unfortunately, neither stable nor transient expression of the heterologous genes carried on such vectors was observed (Peterson, unpublished).

The reason(s) for failure to detect transformation is unclear. Potential problems include: (1) the possibility that *P. infestans* requires its own expression signals; (2) the failure of plasmid DNA to enter protoplasts; (3) incorrect methylation of introduced DNA with consequent digestion

Fig. 19.2. (Facing page) Heterologous hybridisations to *P. infestans* genomic DNA. *P. infestans* genomic DNA from strains ATCC 48720 and E13a (a) and (b) or strain ATCC 48720 alone and *A. nidulans* Glasgow wild type strain (c) and (d) were digested, fragments separated by electrophoresis in 0·7% agarose and Southern blotted onto nylon filters. Digests (a) and (b) B, *Bam* HI; H, *Hin*dIII; (c) and (d) 1, *A. nidulans*, *Eco*RI; 2, *P. infestans*, *Eco*RI; 3, *P.infestans*, *Bam* HI; 4, *P. infestans*, *Hin*dIII. Hybridisation conditions for oligonucleotide probes were: pre-hybridisation overnight at 30°C (a) or 43°C (b) in 5 x SSPE, 0·5% skimmed milk, 1% SDS, 0·1% $Na_4P_2O_7.10\ H_2O$, 250 μg ml^{-1} sonicated herring sperm DNA, followed by overnight hybridisation in the same solution with ^{32}P end-labelled probe. Washing was in 5 x SSC, 0·1% SDS, 0·1% $Na_4P_2O_7.10\ H_2O$, (a) or 1 x SSC, 0·1% SDS, 0·1% $Na_4P_2O_7.10\ H_2O$, (b). Hybridisation conditions for DNA fragment probes were pre-hybridisation for 3 h at 52°C in the above solution followed by overnight hybridisation in the same solution to probe ^{32}P-labelled using random hexanucleotide primers (Amersham). Washing was in 5 x SSC, 0·1% SDS, 0·1% $Na_4P_2O_7.10\ H_2O$, followed, if necessary, by 3 x SSC, 0·1% SDS, 0·1% $Na_4P_2O_7.10\ H_2O$. Probes (a) alkaline protease mixed 20-mer oligonucleotides; (b) β-tubulin mixed 18-mer oligonucleotides; (c) actin *A. nidulans actA* 0·83 kb *Nco*I-*Kpn*I fragment (Fidel *et al.*, 1988); (d) glyceraldehyde 3-phosphate dehydrogenase, a 1 kb *Eco*RI-*Bam*HI fragment of pCNDA51-a (Punt *et al.*, 1988). Arrows indicate the positions of faint bands.

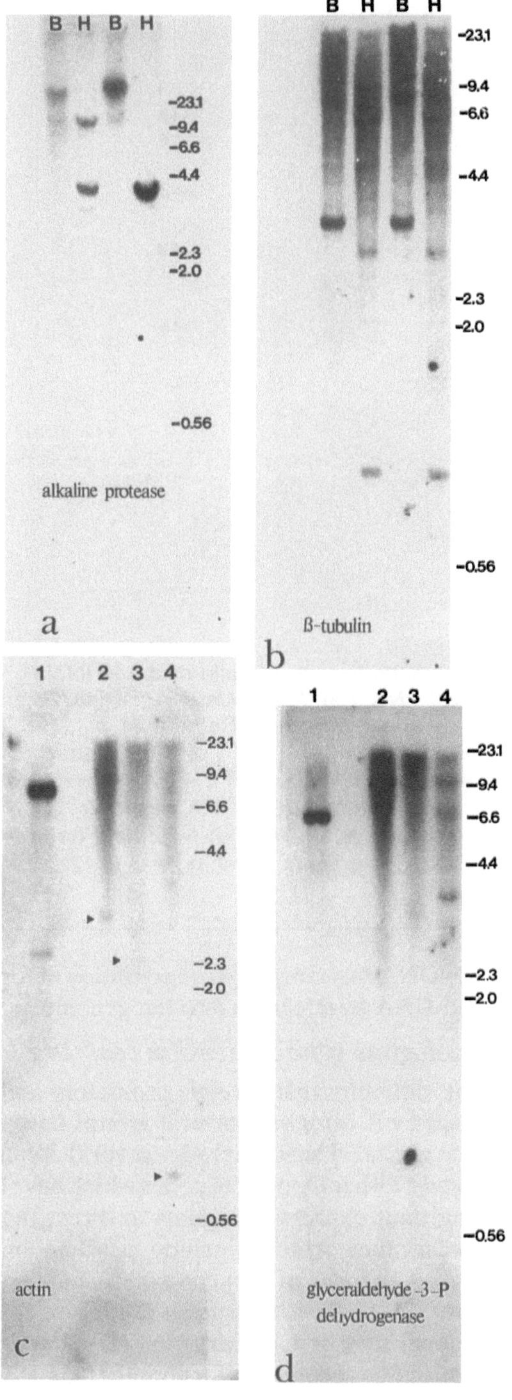
B H B H
-23.1
-9.4
-6.6
-4.4
-2.3
-2.0
-0.56
alkaline protease
a
B H B H
-23.1
-9.4
-6.6
-4.4
-2.3
-2.0
-0.56
ß-tubulin
b
1 2 3 4
-23.1
-9.4
-6.6
-4.4
-2.3
-2.0
-0.56
actin
c
1 2 3 4
-23.1
-9.4
-6.6
-4.4
-2.3
-2.0
-0.56
glyceraldehyde-3-P
dehydrogenase
d

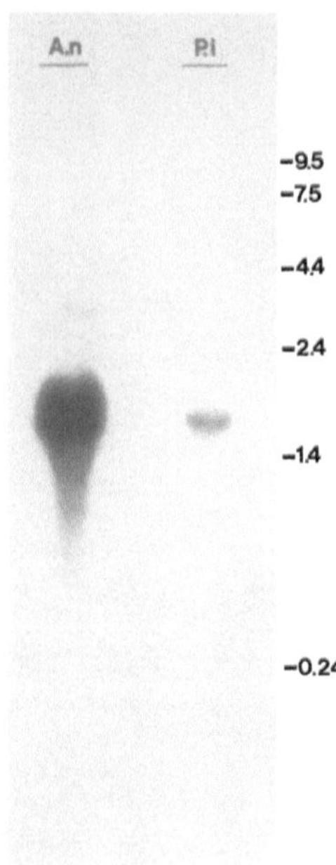

Fig. 19.3. Heterologous hybridization of actin probe to total *P. infestans* RNA (lane labelled P.i). Total RNA from *P. infestans* ATCC 48720 and *A. nidulans biA1* was prepared by the method of Cathala *et al.* (1983), run on 1% formaldehyde agarose (Davis *et al.*, 1986) and transferred to a nylon filter. Pre-hybridisation at 42°C for 3 h in 20 x SSPE, 20% de-ionised formamide, 1% SDS, 250 μg ml^{-1} sonicated herring sperm DNA was followed by hybridisation overnight to the *A. nidulans* actin 0·83 kb *NcoI-KpnI* fragment ^{32}P-labelled using random hexanucleotide primers. Washing was at 42°C in 1 x SSC, 0·1% SDS, 0·1% $Na_4P_2O_7.10\ H_2O$.

---

or cleavage of foreign DNA sequences; (4) degradation of foreign protein; (5) failure of plasmid DNA to integrate into the genome.

Homologous gene expression cassettes

To circumvent the difficulty that foreign promoters and terminators may not be recognised by *P. infestans*, several groups have attempted to isolate homologous genes. These include several heatshock genes (Peterson, unpublished). Other Oomycete genes which have been isolated with the aim of using their expression signals to direct the synthesis of bacterial antibiotic resistance proteins include histidine and tryptophan biosynthetic genes, *his3* and *trp1* of *P. nicotianae* (Chapter 20), the heat shock protein gene, *hsp70* of *Bremia lactucae* (Judelson & Michelmore, 1989) and the actin gene, *actA* of *P. megasperma* (Dudler,1990). In terms of the use of homologous promoter-terminator sequences to drive

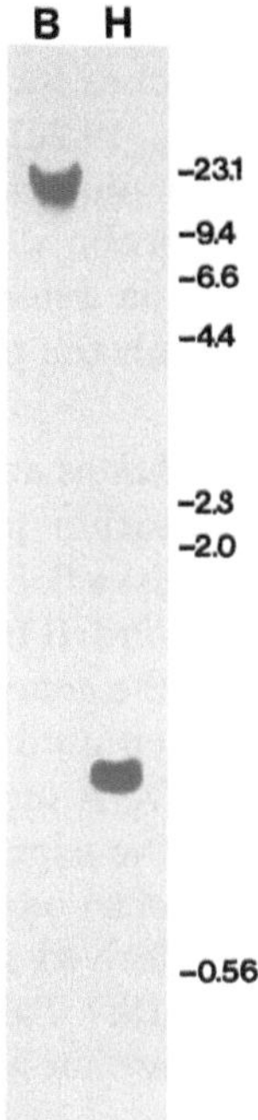

Fig. 19.4. Back-hybridization of *actA* to *P. infestans* genomic DNA. *P. infestans* ATCC 48720 genomic DNA was digested with *Bam* HI (B) or *Hin*dIII (H), fragments separated by electrophoresis in 0·7% agarose and transferred to a nylon filter. Pre-hybridisation for 3 h and hybridisation overnight, in the solution detailed in the legend to Fig. 19.3, was at 65°C. The probe was the 0·8 kb *Hin*dIII fragment of *P. infestans actA*, ^{32}P-labelled using random hexanucleotide primers. Washing was at 65°C in 0·2 x SSC, 0·1% SDS, 0·1% $Na_4P_2O_7.10\ H_2O$.

---

heterologous expression, it may be important to choose highly expressed genes with a strong promoter such as those from the glycolytic pathway, e.g. glyceraldehyde 3-phosphate dehydrogenase or phosphoglycerate kinase. In *Saccharomyces cerevisiae* the latter promoter directs expression of heterologous genes with a 500-fold greater efficiency than a tryptophan synthetase promoter (Mellor *et al.*, 1985).

We have used gene fragments from other fungi and synthetic oligonucleotides as probes to detect and isolate specific genes from a genomic library of *P. infestans*. Partial (*Sau* 3A) digests of genomic DNA were size-selected and cloned in a λ DASH II vector (Stratagene) to give 3×10^5 clones with inserts averaging 17 kb. Colony hybridization on nylon transfers under relaxed conditions identified clones with sequences similar to the probes. The oligonucleotide probes were derived from suitable highly conserved regions observed by comparison of known amino acid sequence data of a protein from several different species.

Hybridisation was observed to probes for alkaline protease (Fig.19.2a), β-tubulin (Fig. 19.2b), actin (Fig. 19.2c), glyceraldehyde 3-phosphate dehydrogenase (Fig. 19.2d) and phosphoglycerate kinase (data not shown). In addition, the actin probe derived from *A. nidulans* also hybridised to a mRNA of 1·8 kb on a northern blot of total RNA of *P. infestans* (Fig. 19.3). At present, only the putative actin clones have been studied further (see below).

Cross-hybridisation of the *A. nidulans* actin probe to DNA of six clones identified two different hybridisation patterns. Five clones showed specific hybridisation to two probes: a 0·8 kb *Hin*dIII fragment and a 20 kb *Bam* HI fragment. The 0·8 kb *Hin*dIII fragment probably corresponds to a *Hin*dIII band of similar size in the genomic Southern blot (Fig. 19.2c). This 0·8 kb HindIII fragment, when used as the probe, back hybridised to a band of this size in genomic DNA digested with *Hin*dIII (Fig. 19.4) as well as to a mRNA species of the expected size (1·8 kb) (Fig. 19.6). This cloned gene was designated *actA* and inserted in pUC13 for DNA sequencing. A comparison of its derived amino acid sequence with that of *A. nidulans actA* is shown in Fig. 19.7. Taken together with the Southern and northern blot data, the evidence shows that we have isolated clones of a *P. infestans* actin gene.

The sixth putative actin-containing clone encompassed a 20 kb *Hin*dIII and a 3 kb *Bam* HI fragment which hybridised strongly to the *A. nidulans actA* probe. The 3 kb *Bam* HI was subcloned into pUC13 and found to back-hybridise to fragments of the expected size in Southern blots of *Bam* HI digested genomic DNA. Preliminary DNA sequence data confirmed similarity to *A. nidulans actA*. Hybridization of the 3 kb *Bam* HI fragment

Fig. 19.5. (Facing page) Formation and subsequent regeneration of cyst (a-d) and sporangial (e-h) protoplasts. (a) A zoospore about to undergo encystment. (b) The same cell 3 min. later. The cyst begins to fluoresce as the cell wall is synthesised. (c) Encysted zoospores after 1·5 h digestion with NovoZym 234 (Batch number 1961). Small patches of fluorescence remain on the cell surface but are apparent only when viewed under UV light alone. (Note the relatively intact sporangial cell walls). (d) Protoplasts of encysted zoospores 1 h after removal of NovoZym 234. The cell wall is quickly regenerated but this does not, in itself, indicate whether the cell has the ability to develop further into a macroscopic colony. (e) Sporangia before digestion with Novozym 234. (f) Sporangia after 24 h digestion. Note that the cell walls are completely digested away. (g) Sporangial protoplast after removal of the NovoZym 234. Regeneration and germination are relatively rapid. (h) Substantial growth is established after 48 h. Note: protoplast yield from cysts and sporangia was always found to be less than 100% due to some cell disintegration. Cells in (a) to (d), (g) and (h) were stained with Calcofluor White PMS (American Cyanamid Co., U.S.A.) and viewed under combined visible and UV light [(a) - (d) & (g)] or UV light alone (h). Cells in (e) and (f) were unstained and viewed in visible light only.

Transformation in *Phytophthora infestans*

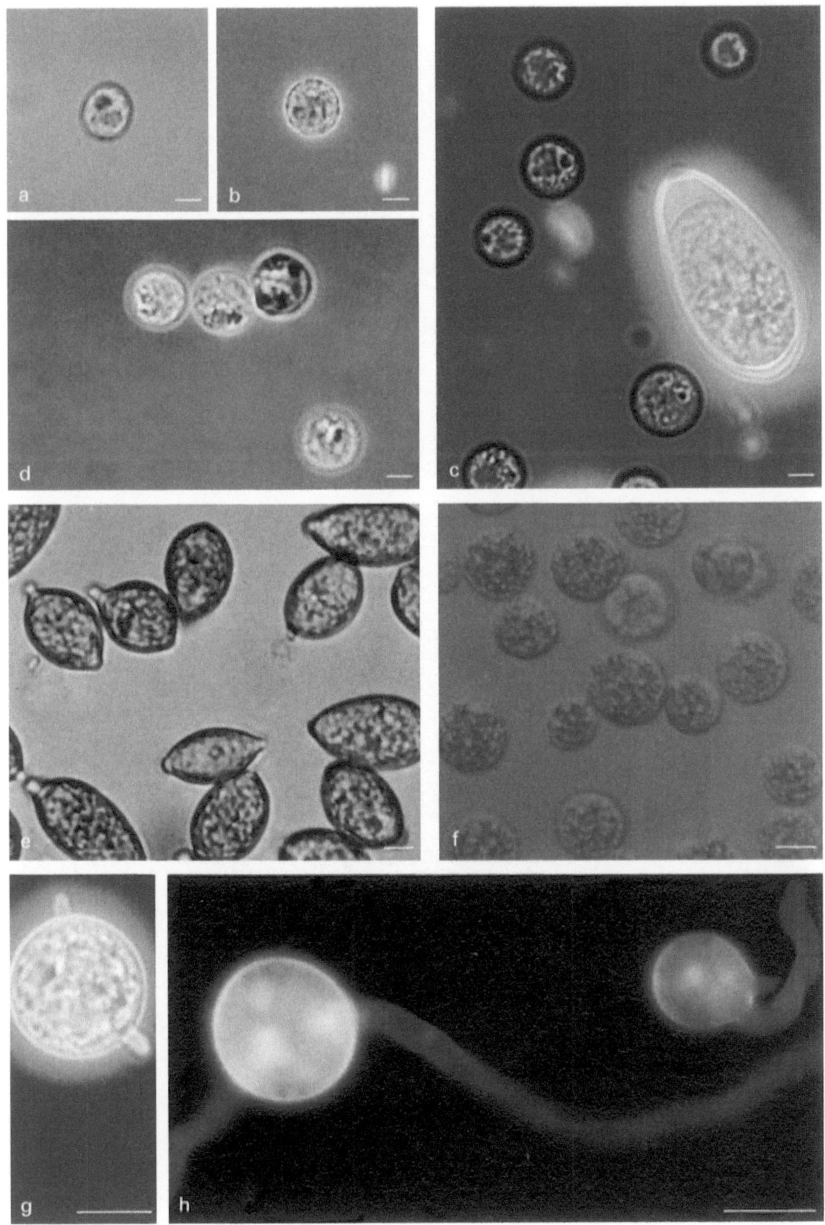

(*Facing p.* 306)

This plate is available for download in colour from www.cambridge.org/9780521189767

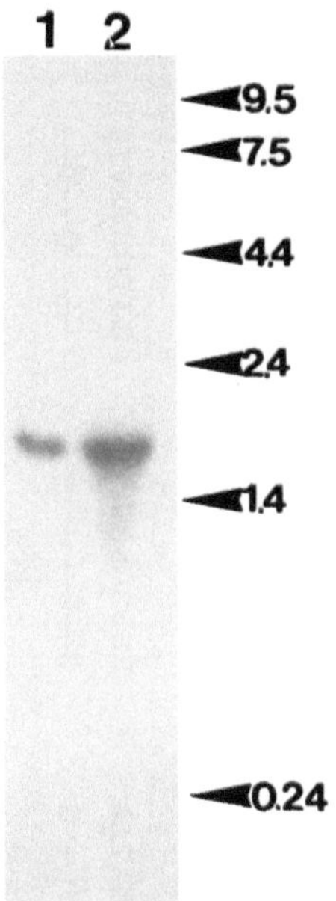

Fig. 19.6. Homologous hybridization of *actA* to total *P. infestans* RNA. Total RNA from *P. infestans* ATCC 48720 was prepared by the method of Cathala *et al.* (1983). 10 μg (lane 1) and 40 μg (lane 2) run on 1% formaldehyde agarose (Davis *et al.*, 1986) and transferred to a nylon filter. Pre-hybridisation at 42°C for 3 h in 20 x SSPE, 50% de-ionised formamide, 1% SDS, 250 μg ml^{-1} sonicated herring sperm DNA was followed by hybridisation overnight to the 0·8 kb *Hin*dIII fragment of *P. infestans actA* ^{32}P-labelled using random hexanucleotide primers. Washing was at 42°C in 0·1 x SSC, 0·1% SDS, 0·1% $Na_4P_2O_7.10H_2O$.

to total RNA (data not shown) revealed a transcript of the expected size (1·8 kb) but only after a long exposure time (20 times longer than with the 0·8 kb *Hin*dIII fragment probe). This clone was designated *actB*. Assuming that the probes were roughly of equivalent specific activity, our results suggest that the two transcripts were expressed at different levels. However, the very weak signal produced by the *actB* probe may alternatively represent cross-hybridisation to the *actA* transcript. That there appear to be (at least) two genes *actA* and *actB* does, of course, have an important bearing on the use of transcriptional signals for expression of a heterologous bacterial gene.

Finally, the identification of two genes would suggest that there may be multiple actin genes. We are repeating Southern blot experiments with *actA* and *actB* at low stringency to determine if multiple bands are

```
                                                                      50
P. infestans        ALVVD  NGSGMCKAGF  AGDDAPRAVF  PSIVGRPKHL  GIMVGMDQKD
A. nidulans         ---I-  ----------  ----------  -------R-H  ---I--G---
                                                                     100
P. infestans   AYVGDEAQSK  RGVLTLKYPI  EHGIVTNWDD  MEKIWHHTFY  NELRVAPEEH
A. nidulans    S---------  --I---R---  ---V------  ----------  ----------
                                                                     150
P. infestans   PVLLTEAPLN  PKANRERMTQ  IMFETFNVPA  MYVNIQAVLS  LYASGRTTGC
A. nidulans    --------I-  --S---K---  -V--------  F--S------  ---------I
                                                                     200
P. infestans   VLDSGDGVSH  TVPIYEGYAL  PHAIVRLDLA  GRDLTDYMMK  ILTERGYSFT
A. nidulans    --------T-  V------F--  ----S-V-M-  -------L--  ----------
                                                                     250
P. infestans   TTAEREIVRD  IKEKLTYIAL  DFDQEMKTAA  ESSGLEKSYE  LPDGNVIVIG
A. nidulans    ----------  -----C-V--  --E--IQ--S  Q--S------  ----Q--T--
                                  270
P. infestans   NERFRTPEVL  FQPSLIGKEA
A. nidulans    -----A-KA-  ----VL-L-A
```

Fig. 19.7. Comparisons of predicted amino acid sequences (in conventional one-letter code; see Sambrook, Fritsch & Maniatis, 1989) for *P. infestans* and *A. nidulans* actin proteins (Fidel, Doonan & Morris, 1988). Amino acids are indicated for the *A. nidulans* only where they differ from those in the *P. infestans* sequence. Dashes represent amino acids identical to those of *P. infestans* actin. The *P. infestans* sequence is inferred from the 0·8 kb *Hin*dIII fragment of the *actA* gene. It should be noted that the *P. infestans* amino acid sequence start corresponds to residue 5 of the *A. nidulans* sequence and ends at residue 270.

observed. Actin gene families are common in other eukaryotes (Hightower & Meagher, 1986). A recent paper (Dudler, 1990) reports the cloning and nucleotide sequence of the *P. megasperma* actin gene. Unlike *P. infestans* (this communication) there is only one actin gene in *P. megasperma* but, as in *P. infestans*, the *P. megasperma* gene lacks introns which is rather unusual for actin genes (Dudler, 1990). The protein from *P. megasperma* is significantly diverged from other actins, including those of filamentous fungi.

Conclusions

Phytophthora infestans appears to be more difficult to transform than most other filamentous fungi. This has been the frustrating experience of several research groups over the last few years. One obvious possibility is that *P. infestans* may be unable to recognise heterologous expression signals. Several *Phytophthora* genes have been isolated and preliminary DNA sequences obtained. Such studies should provide information (and surprises) regarding the relationships between *Phytophthora* and other lower eukaryotes. Now that several *P. infestans* genes have been isolated, their promoters and terminators will hopefully ensure the expression of introduced selectable markers.

Acknowledgements We thank the Agriculture and Food Research Council for a Link Award between St. Andrews University and the Scottish Crop Research Institute, Invergowrie. RPM acknowledges the support of the Science and Engineering Research Council through an SERC-CASE studentship. We also wish to thank former colleagues Alison Campbell, Debra Gould and Sarah Jane Gurr for their research contributions and Edward Campbell for sequence comparison work. We acknowledge receipt of unpublished data and helpful comments from a number of *Phytophthora* colleagues and of genes for use as probes from R. Morris (Rutgers University, U.S.A.), P. Punt (TNO, Netherlands) and S. M. Kingsman (Oxford University, U.K.).

References

Ballance, D. J. (1986). Sequences important for gene expression in filamentous fungi. *Yeast* **2**, 229-236.

Bartnicki-Garcia, S. & Lippman, E. (1966). Liberation of protoplasts from the mycelium of *Phytophthora*. *Journal of General Microbiology* **42**, 411-416.

Bartnicki-Garcia, S. & Lippman, E. (1967). Enzymic digestion and glucan structure of hyphal walls of *Phytophthora cinnamomi*. *Biochimica et Biophysica Acta* **136**, 533-543.

Bartnicki-Garcia, S. & Wang, M. C. (1983). Biochemical aspects of morphogenesis in *Phytophthora*. In *Phytophthora: its Biology, Taxonomy, Ecology and Pathology* (ed. D. C. Erwin, S. Bartnicki-Garcia & P. H. Tsao), pp. 121-137. The American Phytopathological Society: St. Paul, Minnesota, U.S.A.

Campbell, A. M., Moon, R. P., Duncan, J. M., Gurr, S. J. & Kinghorn, J. R. (1989). Protoplast formation and regeneration from sporangia and encysted zoospores of *Phytophthora infestans*. *Physiological and Molecular Plant Pathology* **34**, 299-307.

Cathala, G., Savouret, J. -P., Mendez, B., West, B. L., Karin, M., Martial, J. S. & Baxter, J. D. (1983). A method for isolation of intact translationally active ribonucleic acid. *DNA* **2**, 329-335.

Davis, L. G., Dibner, M. D. & Batley, J. F. (1986). *Basic Methods in Molecular Biology.* Elsevier: New York.

Dudler, R. (1990). The single copy actin gene of *Phytophthora megasperma* encodes a protein considerably diverged from any other known actin. *Plant Molecular Biology* **14**, 415-422.

Fidel, S., Doonan, J. H. & Morris, N. R. (1988). *Aspergillus nidulans* contains a single actin gene which has unique intron locations and encodes an α-actin. *Gene* **70**, 283-293.

Fincham, J. R. S. (1989). Transformation in fungi. *Microbiological Reviews* **53**, 148-170.

Gurr, S. J., Unkles, S. E. & Kinghorn, J. R. (1987). The structure and organization of nuclear genes of filamentous fungi. In *Gene Structure in Eukaryotic Microbes* (ed. J. R. Kinghorn), pp. 99-139. IRL Press: Oxford.

Hess, D. & Leipoldt, G. (1979). Regeneration of shoots and roots from isolated mesophyll protoplasts of *Nemesia strumosa*. *Biochemie und Physiologie der Pflanzen* **174**, 411-417.

Hightower, R. C. & Meagher, R. B. (1986). The molecular evolution of actin. *Genetics* **114**, 315-332.

Hynes, J. M. (1986). Transformation of filamentous fungi. *Experimental Mycology* **10**, 1-8.

Jahnke, K. -D., Leipoldt, G. & Prell, H. H. (1987). Studies on the preparation and viability of *Phytophthora parasitica* spheroplasts. *Transactions of the British Mycological Society* **89**, 213-220.

Judelson, H. S. & Michelmore, R. W. (1989). Structure and expression of a gene encoding heat-shock protein Hsp70 from the Oomycete fungus *Bremia lactucae*. *Gene* **79**, 201-217.

Manavathu, E. K., Suryanarayana, K., Hasnain, S. E. & Leung, W. (1988). DNA-mediated transformation in the aquatic filamentous fungus *Achlya ambisexualis*. *Journal of General Microbiology* **134**, 2019-2028.

Mellor, J., Dobson, M. J., Roberts, N. A., Kingsman, A. J. & Kingsman, S. M. (1985). Factors affecting heterologous gene expression in *Saccharomyces cerevisiae*. *Gene* **33**, 215-226.

Mishra, N. C. (1985). Gene transfer in fungi. *Advances in Genetics* **21**, 73-180.

Peberdy, J. F. (1979). Fungal protoplasts: isolation, reversion and fusion. *Annual Review of Microbiology* **33**, 21-39.

Pesti, M. & Ferenczy, L. (1979). Formation and regeneration of protoplasts from *Phytophthora infestans*. *Acta Phytopathologica Academia et Scientiarum Hungaricae* **14**, 1-4.

Punt, P. J., Dingemanse, M. A., Jacobs-Meijsing, B. J. M., Pouwels, P. H. & van den Hondel, C. A. M. J. J. (1988). Isolation and characterisation of the glyceraldehyde-3-phosphate dehydrogenase gene of *Aspergillus nidulans*. *Gene* **69**, 49-57.

Ramsbosek, J. & Leach, D. (1987). Recombinant DNA in filamentous fungi: progress and prospects. *CRC Critical Reviews in Biotechnology* **6**, 357-393.

Sambrook, J., Fritsch, E. F. & Maniatis, T. (1989). *Molecular Cloning: A Laboratory Manual*. Cold Spring Harbor Press: Cold Spring Harbor.

Ward, M., Wilkinson, B. & Turner, G. (1986). Transformation of *Aspergillus nidulans* with a cloned, oligomycin-resistant ATP synthase subunit 9 gene. *Molecular and General Genetics* **202**, 265-270.

Unkles, S. E. (1989). Fungal biotechnology and the nitrate assimilation pathway. In *Molecular and Genetic Aspects of Nitrate Assimilation*, (ed. J. L. Wray & J. R. Kinghorn), pp. 341-363. Oxford University Press: Oxford.

Chapter 20

Towards transformation in *Phytophthora nicotianae*

Hermann H. Prell, Petr Karlovský & Günther Bahnweg

Infection and invasion of a plant by a pathogenic fungus proceed in multistep processes with contributions from both plant and fungus. One promising approach in analyzing the fungal contribution to this poorly understood process is the identification of the different fungal genes involved. Molecular genetics offers theoretical approaches and experimental techniques for identification, isolation and analysis of such genes *in vivo* as well as *in vitro*. The most important requirement is a transformation system to allow transfer of genetic material back into the fungus. Only then can genes important in pathogenesis be identified from the phenotype which is expressed in the transformed fungus.

Several years ago when we decided to establish a transformation system for pathogens of the genus *Phytophthora*, molecular genetics of this genus had not begun. Thus we had to start from the very beginning. The initial step towards our goal was the investigation of the DNA of *Phytophthora* which would provide a source of DNA sequences for the construction of transformation vectors.

Characterization of *Phytophthora* DNA

We started our experiments using *P. infestans*, but later switched to *P. nicotianae* (syn. *parasitica*), which grows much faster in a simple, chemically defined medium (Henninger, 1963).

DNA was isolated from young mycelium grown in liquid medium after inoculation either with a zoospore suspension, with minced hyphae, or with a suspension of enzymatically generated protoplasts (Leipoldt, unpublished). Mycelium was broken up in a mortar under liquid nitrogen, or after lyophilization. The DNA was selectively purified by CTAB (cetyltrimethylammonium bromide) and centrifuged in a bisbenzimide-CsCl density gradient, essentially as described by Klimczak & Prell (1984). After centrifugation, four DNA bands of different density were observed (Fig. 20.1) and each band was withdrawn separately from the gradient. The following four DNA fractions were obtained from *P. infestans* (GC content in mol percent shown in brackets): mitochondrial (mt)DNA (24%); ribosomal (r)DNA (about 50%); nuclear or chromosomal (n)DNA (54%); and as a very faint band, satellite (sat)DNA (about 68%). Many other *Phytophthora* species gave essentially the same pattern of

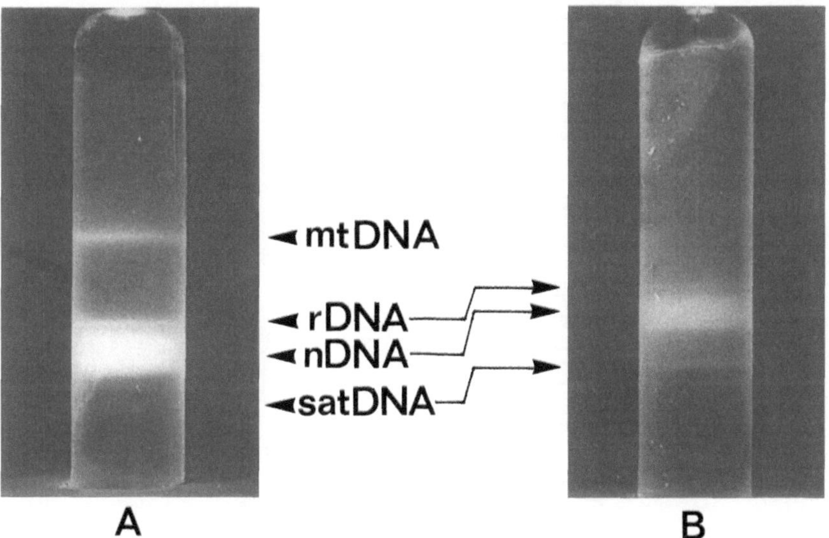

Fig. 20.1 Bisbenzimide-CsCl density gradient centrifugation of DNA isolated from *Phytophthora infestans*. In panel A total DNA is separated (42000 rpm, 40 hours) in four density fractions. From top to bottom: mt-, r-, n- and satDNA (the lowest DNA band is very faint). In panel B, recentrifugaion of DNA collected from the lower part of four centrifuge tubes in order to enrich for satDNA (42000 rpm, 40 hours). From top to bottom: r-, n- and satDNA.

---

banding with only minor variations in sharpness. However, some species lacked the satDNA band.

Using gel electrophoresis we recently discovered plasmid (pl)DNA in three *Phytophthora* species. However, only in one species (see below) was the copy number sufficient for an easy separation from contaminating nDNA fragments.

The mtDNA represents an independent DNA species found only in mitochondria (Klimczak & Prell, 1984). Restriction enzyme analysis revealed a circular molecule of about 36 kb for both *P. infestans* and *P. nicotianae*. The mtDNA restriction maps of both species are rather similar (Fig. 20.2). Comparisons among maps of different *Phytophthora* species reveal varying degrees of similarity which can be used for taxonomic characterization (de Cock & Bahnweg, unpublished; Chapter 11).

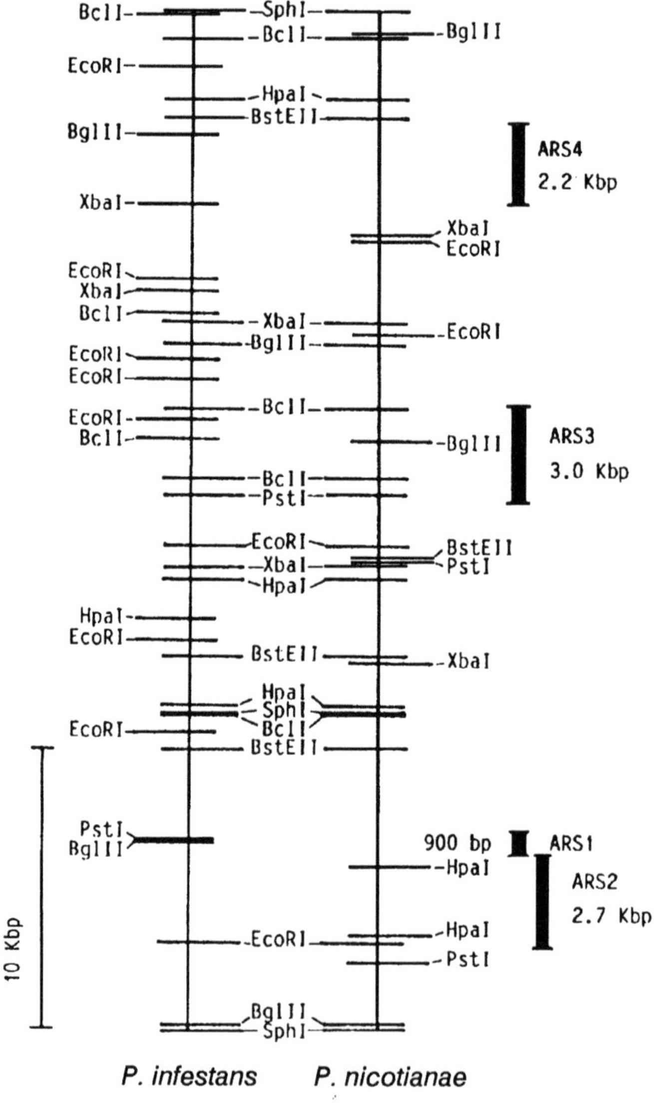

Fig. 20.2 Restriction maps of mt DNAs of *P. infestans* and *P. nicotianae*. Circular maps were opened at one of the *Sph*I sites. Note the 19 sites common to both molecules. Cloned *Hin*dIII-fragments displaying ars-activity in yeast were localized by Southern hybridization (Mühlbauer, 1988).

The r- and satDNA of *Phytophthora* represent portions of repetitive DNA within the high molecular weight DNA. This was demonstrated by electrophoretic separation of undigested DNA according to molecular weight and by Southern hybridization with cloned probes of yeast rDNA and *P. infestans* satDNA. As with mtDNA, the rDNA restriction maps from different species are very similar (Fig. 20.3). Differences, which were found to occur only in the spacer regions, can be used to quantify relationships among *Phytophthora* species (de Cock & Bahnweg, unpublished).

The discovery of satDNA was quite unexpected. This DNA fraction appeared in bisbenzimide-CsCl density gradients of *P. infestans* as a very faint band of high buoyant density and high GC content. Its restriction by *Bam*HI yielded three fragments of 6·7, 3·6, and 3·5 kb in equal molarity and one fragment of 6·5 kb in at least two-fold molarity (Möller & Bahnweg, unpublished). These four '*Bam*HI-families', which belong to a class of middle repetitive sequences, were cloned and further analyzed (Möller, 1989; Möller & Bahnweg, unpublished). The different families cross-hybridize because they harbour a set of particular, heterologous sequences, at least some of which are present in every family.

Digestion of genomic DNA of *P. infestans* with *Eco*RI, which does not cut satDNA, and subsequent electrophoresis and hybridization with the four cloned *Bam*HI fragments yielded hybridizing DNA only in the high molecular weight region above 25 kb. This suggests that satDNA sequences are clustered in only a few places in the genome.

Restriction analysis of satDNA and comparison with the restriction maps of cloned 6·5 kb and 6·7 kb fragments proved that copies of these *Bam*HI fragments are arranged as head-to-tail repeats. Quantitative dot blot hybridization of different amounts of the 6·5 kb fragment with defined amounts of total genomic DNA of *P. infestans* revealed a copy number of approximately 15 copies per haploid genome. Several cloned copies of the 6·5 kb repeat display some variations in their restriction patterns.

Electrophoresis and Southern hybridization of *Bam*HI, *Hin*dIII and *Eco*RI digested DNA showed that sequences homologous to the cloned 6·5 kb family of *P. infestans* are also present in about 26 of 53 *Phytophthora* species investigated. These were present in restriction fragments of varying sizes in the different species. Interestingly, all so called marine and aquatic *Phytophthora* species belong to the group showing no homology to this particular probe. Recently, these 'marine' species have been excluded from the genus *Phytophthora* by Ho & Jong (1990).

We called the most prominent DNA fraction of the bisbenzimide-CsCl gradient nDNA (Fig. 20.1). Restriction and electrophoresis suggests that

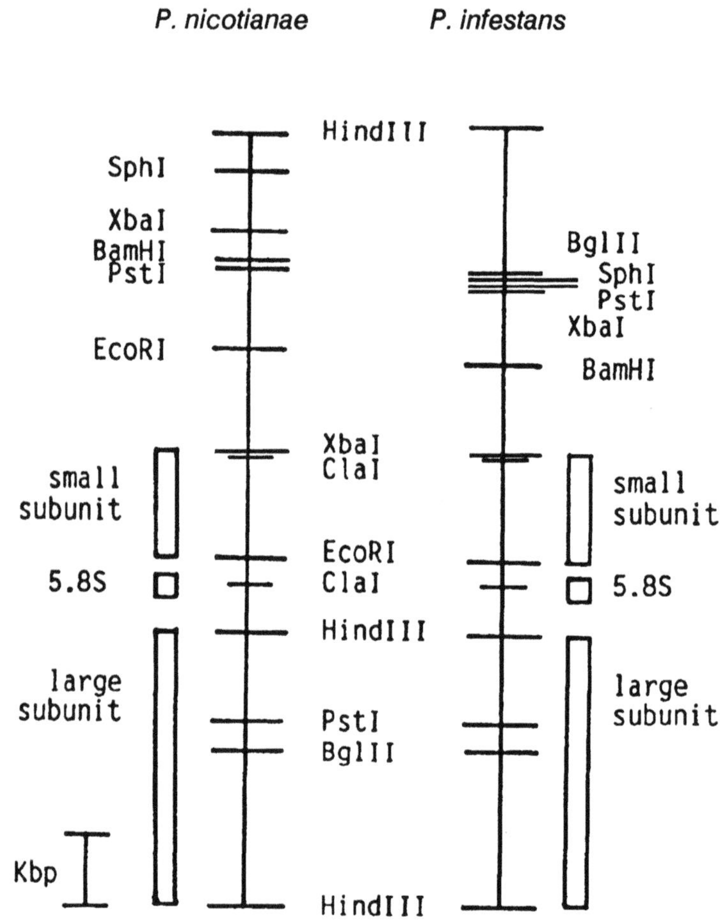

Fig. 20.3 rDNA repeat units of *P. nicotianae* and *P. infestans*. Genes for the large, the small, and the 5·8S rRNA subunits share common restriction sites, whereas differences occur only in the spacer between the large and small subunits. Restriction fragments carrying genes are identical in size; spacers of the two species differ by approximately 100 bp in length. *Cla*I restriction sites were not completely mapped and only those occurring in the rRNA genes are shown.

this fraction contains mainly unique chromosomal DNA. However, some bands within the smear of restriction DNA fragments indicate that it also contains repetitive DNA in a proportion not yet determined, as well as contaminating r- and satDNA. A complexity of the nDNA fraction of $2{\cdot}3 \times 10^7$ base pairs for the haploid genome as measured by DNA renaturation kinetics (Mühlbauer, 1988) corresponds to the complexity estimated for *Neurospora* DNA (Krumlauf & Marzluf, 1980).

Uptake of DNA into *Phytophthora*

After conversion of mycelium of *P. nicotianae* into protoplasts by enzymatic digestion with NovoZym 234 (Jahnke, Leipoldt & Prell, 1987) DNA is taken up in the presence of polyethyleneglycol (10% PEG 4000 for 1 min) or after electroporation (3 μF/250 V). However, PEG is highly toxic to *Phytophthora*, and only 10^{-4} to 10^{-3} of the protoplasts are able to regenerate in soft agar to form a mycelium. However, using electroporation, up to 10% of protoplasts regenerate. Plasmid DNA (pBR322 or its derivatives) taken up after both kinds of treatment is fairly stable. It can be traced by Southern hybridization for five days in regenerating protoplasts. During this period several nuclear divisions take place.

The construction of vectors

Transformation of fungal protoplasts is usually accomplished by using bacterial plasmids carrying a selectable marker gene (Fincham, 1989). Additional DNA fragments or genes to be transferred into recipient protoplasts may be inserted into the vector. Recovery of a stable transformed clone requires not only DNA uptake but also expression of the marker gene of the vector and the hereditary transmission of the vector, or its marker gene to progeny mycelium. Sequences which are homologous (i.e. derived from *Phytophthora* DNA) should be recognised by the recipient cells and thus 'fully adapted' plasmids carrying such signals should yield stable transformants. If, however, a vector is 'partly adapted' and carries heterologous signals from an unrelated organism, transformation is less likely.

For the construction of vectors we used derivatives of pBR322 from *Escherichia coli*, furnished with the neomycin resistance coding region derived from transposon Tn5. If this genetic marker, *neo*, is supplied with recognisable expression sequences it should be dominantly expressed in *P. nicotianae*. Selection of transformants would be performed with the neomycin analogue G418, which is highly toxic to *P. nicotianae*: $0{\cdot}5\ \mu$g ml^{-1} G418 in liquid medium and 10 μg ml^{-1} in agar medium completely inhibits growth (see also Table 19.2, p. 299).

In order to identify a promoter recognized by *P. nicotianae*, we first tested 'partly adapted' vectors, with a heterologous promoter fused to the

5' end of the *neo* coding region. The following promoters were tested: Herpes simplex virus thymidine kinase promoter; SV40 early promoter; nopaline synthetase promoter from *Agrobacterium tumefaciens*. The *neo* gene of Tn903 with its own bacterial promoter was also used. Some of the vectors were tested both with and without an additional insert of *P. nicotianae* rDNA to aid transmission of the vector (see below). We also constructed derivatives of the *E. coli* plasmid pUC19 carrying random DNA fragments from *P. nicotianae* as well as fragments able to promote neomycin resistance expression in *E. coli*. All these vectors yielded resistant colonies of *P. nicotianae* in numbers 2-3 times above the spontaneous background level. However, many of these neomycin resistant isolates exhibited a tolerance level only three times above that of sensitive wild type, grew slowly and finally stopped growth after a few transfers. These discouraging results did not indicate whether the failure of these 'partly adapted' vectors was caused by insufficient expression of the *neo* gene, or by faulty hereditary transmission of vector DNA, or by both. To overcome this problem we decided to construct 'fully adapted' vectors furnished with DNA inserts isolated from *P. nicotianae* itself. Accordingly, an expression signal derived from a cloned gene of *P. nicotianae* was fused to the coding region of the *neo* gene. This construct, after insertion into a vector was employed in transformation experiments to identify *P. nicotianae* DNA sequences facilitating stable hereditary transmission of G418 resistance.

Isolation of signal sequences for gene expression

For expression of the *neo* gene in *P. nicotianae*, it is necessary that its coding sequence is flanked by a promoter at its 5' end and a polyadenylation signal at its 3' end. We constructed a gene library of *P. nicotianae* in the expression vector pUC19 to identify both of these signal sequences. Plasmids were selected from the library which could complement the amino acid requiring *E. coli* mutants *trp*C1117, defective in N-(5'-phosphoribosyl) anthranilate isomerase (PRAI) and the mutant *his*B463, defective in imidazole glycerophosphate dehydratase (IGPS). The inserts (about 7 kb) were active in complementation regardless of their orientation with respect to the *lac* promoter of the vector or of the presence of high levels of *lac* repressor. This indicated that transcription starts from within the *Phytophthora* DNA insert and not from the *lac* promoter in the plasmid. The *trp*-complementing fragment (TRP1) was then shortened to 2·3 kb. The complementing region within the fragment was localized using insertional mutagenesis by *in vitro* insertion of the Omega interposon (Prentki & Krisch, 1984) at 62 randomly distributed sites within the fragment. Subsequent deletion of most of the Omega DNA left behind a 10 bp insertion harbouring a *Bam*HI restriction site. If such an insertion

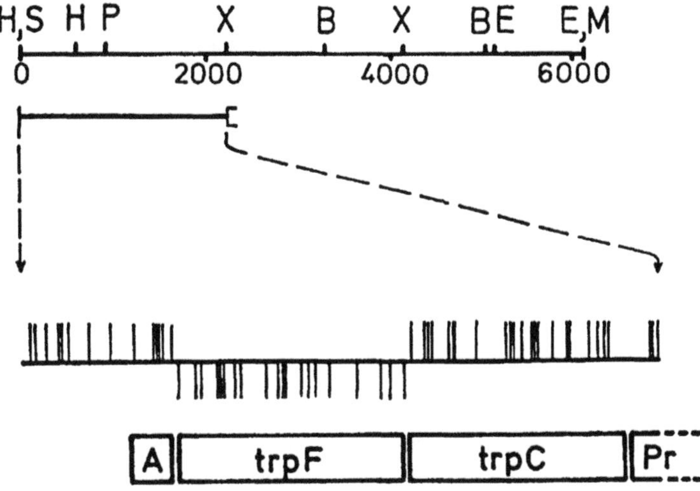

Fig. 20.4 Map of the cloned DNA fragment of *P. nicotianae*. H = *Hin*dIII, S = *Sal*I, P = *Pst*I, X = *Xho*I, B = *Bam*HI, E = *Eco*RI, M = *Sma*I). The scale is in bp. This fragment was shortened to 2·3 kb and mutated by *in vitro*-generated insertions: those insertions which did not abolish the complementing activity are shown above the line, inactivating insertions below the line. Pr = promoter region, trpC = C-domain, trpF = F-domain, A = polyadenylation site. The trpF region comprises 820 bp.

mutation occurred within the coding region it could cause a reading frame shift, preventing formation of a functional gene product. Mutated plasmids, each carrying one insertion mutation, were mapped for the location of the insertion and tested for ability to complement *E. coli trp*C1117. In this way, 21 insertions within a region of about 820 bp were shown to prevent complementation (Fig. 20.4). Thus, this region is the complementing sequence coding for PRAI.

Applying sucrose gradient profiles for the identification of tryptophan synthesizing enzymatic activities, Hütter & DeMoss (1967) were able to show that the domains for IGPS (domain C) and PRAI (domain F) in *Saprolegnia* and *Pythium* are located on one bifunctional polypeptide. By analogy with other fungi, the domain F should be localized at the carboxyl terminus of the polypeptide. Accordingly, in *P. nicotianae* one could

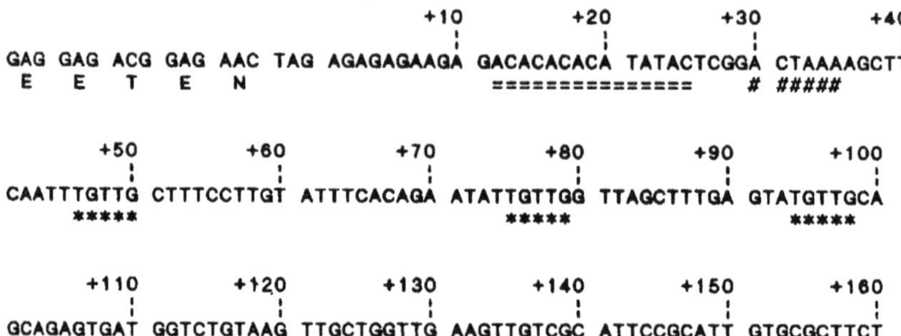

Fig. 20.5 Part of the TRP1 nucleotide sequence. 3'-end of the F domain coding region of *P. nicotianae* with polyadenylation signal (####), three TG-rich domains (****) and alternating purine-pyrimidine sequence (= = = =).

expect the polyadenylation signal to map just at the 3' end of the PRAI coding region. To verify this assumption and to obtain some general information on gene structure in *Phytophthora*, the TRP1 gene and its flanking regions were sequenced. A long reading frame was identified starting from outside the complementing region (PRAI; domain F) and continuing through it from right to left (Fig. 20.4). Just at the 3' end of PRAI we found a typical polyadenylation signal, an ACTAAA sequence 30 bp downstream from the TGA stop codon and three copies of the GT-rich sequence TGTTG further downstream (Fig 20.5). Surprisingly, the ACTAAA signal is preceded by the sequence ACACACACATA-TAC. Such alternating purine-pyrimidine tracts have been found in proximity to other eukaryotic genes and in promoters of archebacteria. They adopt a distorted, noncanonical helix structure *in vitro* and *in vivo* (McClelan, Palecek & Lilley, 1986; Palecek, Rašovská & Boublíková, 1988). One such sequence was found in the common termination region of counter-directionally transcribed TRP1 and inorganic pyrophosphatase (IPP) genes in *Kluyveromyces lactis* (Stark & Milner, 1989). Further sequencing downstream of TRP1 and analysis of other genes of *P. nicotianae* ought to reveal whether some analogy with the situation in *K. lactis* does exist (i.e. another gene converging with TRP1 on a short intergenic space) or whether polyadenylation in *P. nicotianae* requires very specific structural properties within its signal site.

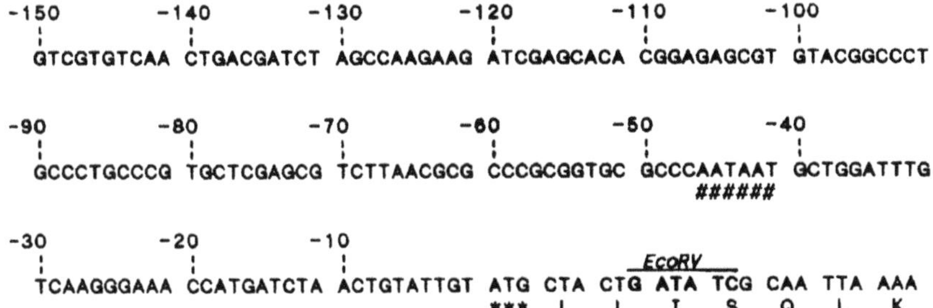

Fig. 20.6 Part of the TRP1 nucleotide sequence of *P. nicotianae*. 5'-end of the C domain coding region, translation start codon (***) and upstream sequence. The putative TATA box is shown (####).

The amino acid sequence of the putative protein coded by the long reading frame encompassing PRAI was compared with known sequences of tryptophan biosynthetic enzymes. By this means the two domains were identified: N-(5'-phosphoribosyl) anthranilate isomerase (PRAI, F domain) on the carboxy-terminus and indole glycerophosphate synthase (IGPS, C domain) on the amino-terminus.

The C-domain of the TRP1 product is much shorter than that of other known IGPS enzymes. Sequence alignment revealed two deletions near the amino-terminus of the domain. An in-frame TAA stop codon 9 bp upstream of the presumed translation start, ATG, strongly suggests that this ATG is indeed the start codon for the reading frame of TRP1 (Fig. 20.6). However, sequence motifs unambiguously corresponding to canonical promoter sequences (TATA box, CAAT box) have not yet been identified within the 184 bp upstream of this translation start codon. Further sequencing and S1 nuclease mapping of mRNA should reveal the promoter site. Interestingly, this promoter in *P. nicotianae*, or the accompanying translation start, is not recognized by *E. coli*, since the 7 kb long PRAI complementing fragment from the gene library was not able to complement *E. coli trp*C670, which is defective in IGPS (domain C). Therefore, the expression in *E. coli* of the TRP1 gene of *P. nicotianae*, leading to complementation of the *trp*C1117 mutation, must start from a site far downstream of the transcription start in *P. nicotianae*. This site is probably within the coding sequence for domain C.

As our sequencing data revealed, glutamine amidotransferase is not coded for by TRP1 of *P. nicotianae*. This is in contrast to all other

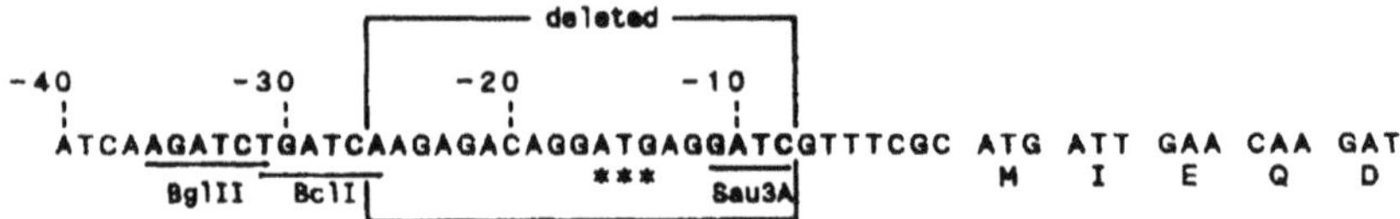

Fig. 20.7 The 5'-end of the Tn5 *neo* coding region and its upstream nucleotide sequence. The ATG (***) in the region between the true translation start codon of *neo* and the upstream *Bgl*II site was removed by deleting the fragment bordered by the compatible restriction sites *Bcl*I and *Sau*3A.

---

eukaryotic organisms for which TRP1 sequences are known (Hütter & Niederberger, 1986). The structure of the *Phytophthora* TRP1 gene is the first demonstration of a new gene organization within the tryptophan pathway among fungi; this organization has only been identified in the Enterobacteriaceae. Our finding is in accordance with activity profiles obtained by sucrose gradient analysis in *Saprolegnia* and *Pythium* (Hütter & DeMoss, 1967). Since bifunctional proteins have proven to be very stable characters (Ahmad & Jensen, 1988a, b), this finding supports results which indicate a great evolutionary distance between Oomycetes and other fungi (Gunderson *et al*., 1987).

In order to create a marker for selection of *Phytophthora* transformants, we joined the promoter region of TRP1 to the 5' end of the *neo* coding region from Tn5 and the polyadenylation signal of TRP1 to its 3' end. The product of *neo*, aminoglycoside 3'-phosphotransferase type II, confers resistance to the antibiotic G418. The isolation of regulatory sequences from TRP1 was accomplished by employing the *Bam*HI sites artificially introduced in the course of insertional mutagenesis (see above).

We constructed two fusions. A translation fusion was made using the *Eco*RV sites located in the fourth codon of TRP1 (Fig. 20.6) and a modified *neo* gene. The presumed product should have seven extra aminoacyls at its amino-terminus, which will probably not interfere with its activity (Reiss, Sprengel & Schaller, 1984). For transcriptional fusions, the first codons of TRP1 were removed by *Bal*31 digestion and the fragment containing the promoter was ligated to the filled-in *Bgl*II site in front of the *neo* gene (Reiss *et al*., 1984). In order to prevent an undesirable translation initiation at the ATG between a *Bgl*II site and the true start of the gene, a 19 bp fragment (*BclI*/*Sau*3A) in the promoter region of *neo* was removed; thus the *Bgl*II site was positioned only 11 bp upstream of the start codon (Fig. 20.7). Both fusions are currently being tested for transformation in *Phytophthora* and will be used in transient expression experiments.

We have also constructed a cosmid gene library of *P. nicotianae* containing DNA inserts of about 40 kb. From this library we isolated TRP1 by DNA-hybridization. We found identical restriction patterns for both the TRP1 gene isolated by complementation and that from the cosmid bank. This excludes gross sequence rearrangements during selection in *Escherichia coli*. Currently, we are isolating other TRP genes from the cosmid library in order to analyze the genetic structure of the tryptophan biosynthetic pathway in *P. nicotianae*.

Isolation of sequences for hereditary transmission

Stable transformation depends either on integration of introduced DNA into the chromosomal DNA of the recipient cell or on autonomous replication of the vector within the transformed cell and its transmission to all progeny. In each case, success depends upon recognition by the transformed cell of DNA sequences present in the vector.

Integrative vectors are circular plasmids, which partially or completely integrate into the chromosomal DNA by recombination (Hinnen, Hicks & Fink, 1978; Fincham, 1989). The inserted vector DNA is then passively replicated together with the chromosomal DNA. Vector integration can occur at homologous and, more frequently, at heterologous sites. In the latter case, however, many small sites of homology dispersed within the chromosomal DNA may provide for recombinational integration (Fincham, 1989; Orr-Weaver & Szostak, 1985). Homologous recombination depends on the presence in the vector of DNA sequences from the recipient organism. Thus integration could be promoted by inserts of repetitive DNA (rDNA or satDNA) or of unique DNA, such as the TRP1 gene. In the latter case a detailed analysis of the mode of integration would be feasible.

Autonomously replicating vectors require an origin of replication recognized by the recipient cell. Autonomously replicating sequences (ars) from the yeast *Saccharomyces cerevisiae* have been inserted into bacterial plasmid vectors. These vectors replicate in yeast cells where the ars is recognised as an origin of replication (Beggs, 1978; Struhl *et al.*, 1979; Stinchcomb *et al.*, 1980). So far, ars from mtDNA have provided the most successful replicating vectors (Stahl, Leitner & Esser, 1987) but it has also been shown that an ars isolated from nDNA of *Aspergillus nidulans* promotes high frequency transformation via vector integration

(Ballance & Turner, 1985). A search was made for suitable ars in *P. nicotianae*.

We identified four different ars located in mtDNA of *P. nicotianae* (Fig. 20.2). After reduction to its minimal size, one of these ars was sequenced (ars3; 246 bp). In its AT-rich sequence (85 mol percent AT) a consensus sequence for ars (5'-[A/T]TTTAT[A/G]TTT[A/T]-3') was identified as well as segments with a strongly asymmetrical purine-pyrimidine distribution which are also believed to be necessary to initiate replication. Constructs of each of these putative ars with a neomycin gene and expression signals are being tested for replication in *Phytophthora*.

Native plasmids of *Phytophthora* species should be a much more promising source for the isolation of an origin of replication. Alternatively they could even serve as backbone for the construction of a self-replicating vector. *P. iranica* contains a linear plasmid of 3·8 kb present in about 50 copies per diploid genome which is easily isolated (de Cock & Bahnweg, unpublished). Restriction enzyme mapping revealed two *Xho*I-sites, approximately 60 to 80 bp from either end. These were used to ligate the plasmid without its ends into pUC18 of *E. coli*. The cloned *P. iranica* plasmid served as a hybridization probe. No homologies to plasmids of other species of *Phytophthora* or *Pythium*, or to mitochondrial restriction fragments of *P. iranica*, were detected. Further investigation is needed to find out how to employ the plasmid or parts of it to construct a self replicating transformation vector for *P. nicotianae*.

References

Ahmad, S. & Jensen, R. A. (1988a). The phylogenetic origin of the bifunctional tyrosine-pathway protein in the enteric lineage of bacteria. *Molecular Biological Evolution* **5**, 282-297.

Ahmad, S. & Jensen, R. A. (1988b). New prospects for deducing the evolutionary history of metabolic pathways in prokaryotes: aromatic biosynthesis as a case-in-point. *Origins of Life* **18**, 41-57.

Ballance, D. J. & Turner, G. (1985). Development of a high-frequency transforming vector for *Aspergillus nidulans*. *Gene* **36**, 321-331.

Beggs, J. D. (1978). Transformation of yeast by a replicating hybrid plasmid. *Nature* **275**, 104-109.

Fincham, J. R. S. (1989). Transformation in fungi. *Microbiological Reviews* **53**, 148-170.

Gunderson, J. H., Elwood, H., Ingold, A., Kindle, K. & Sogin, M. L. (1987). Phylogenetic relationships between chlorophytes, chrysophytes, and oomycetes. *Proceedings of the National Academy of Sciences, USA* **84**, 5823-5827.

Henninger, H. (1963). Zur Kultur von *Phytophthora infestans* auf vollsynthetischen Nährsubstraten. *Zeitschrift für allgemeine Mikrobiologie* **3**, 126-135.

Hinnen, A., Hicks, J. B. & Fink, G. R. (1978). Transformation of yeast chimeric ColE1 plasmid carrying LEU2. *Proceedings of the National Academy of Sciences, USA* **75**, 1929-1933.

Ho, H. H. & Jong, S. C. (1990). *Halophytophthora*, gen. nov., a new member of the Pythiaceae. *Mycotaxon* **36**, 377-382.

Hütter, R. & DeMoss, J. A. (1967). Organization of the tryptophan pathway: a phylogenetic study of the fungi. *Journal of Bacteriology* **94**, 1896-1907.

Hütter, R. & Niederberger, P. (1986). Tryptophan biosynthetic genes in eukaryotic microorganisms. *Annual Review of Microbiology* **40**, 55-77.

Jahnke, K.-D., Leipoldt, G. & Prell, H. H. (1987). Studies on preparation and viability of *Phytophthora parasitica* spheroplasts. *Transactions of the British Mycological Society* **89**, 213-220.

Klimczak, L. J. & Prell, H. H. (1984). Isolation and characterization of mitochondrial DNA of the oomycetous fungus *Phytophthora infestans*. *Current Genetics* **8**, 323-326.

Krumlauf, R. & Marzluf G. A. (1980). Genome organization and characterization of the repetitive and inverted repeat DNA sequences in *Neurospora crassa*. *Journal of Biological Chemistry* **255**, 1138-1145.

McClelan, I. A., Palecek, E. & Lilley, D. M. I. (1986). (A-T)n tracts embedded in random sequence DNA – formation of a structure which is chemically reactive and torsionally deformable. *Nucleic Acid Research* **14**, 9291-9309.

Möller, E. M. (1989). Repetitive DNA-Sequenzen von *Phytophthora infestans* (Montagne) de Bary: Ihre Charakterisierung und Verwendung zur Konstruktion von DNA-Sonden für diagnostische und taxonomische Zwecke. Ph.D. thesis, University of Göttingen, Germany.

Mühlbauer, E. (1988). Molekulargenetische Untersuchungen an *Phytophthora* spp. Experimentelle Arbeiten zum Aufbau eines Transformationssystems. Dissertation, University of Göttingen, Germany.

Orr-Weaver, T. L. & Szostak, J. W. (1985). Fungal recombination. *Microbiological Reviews* **49**, 33-58.

Palecek, E., Rašovská, E. & Boublíková, P. (1988). Probing of DNA polymorphic structure in the cell with osmium tetroxide. *Biochemical and Biophysical Research Communications* **150**, 731-738.

Prentki, P. & Krisch H. M. (1984). *In vitro* insertional mutagenesis with a selectable DNA fragment. *Gene* **29**, 303-313.

Reiss, B., Sprengel, R. & Schaller, H. (1984). Protein fusions with the kanamycin resistance gene from transposon Tn5. *EMBO Journal* **3**, 3317-3322.

Stahl, U., Leitner, E. & Esser, K. (1987). Transformation of *Penicillium chrysogenum* by a vector containing a mitochondrial origin of replication. *Applied Microbiology and Biotechnology* **26**, 237-241.

Stark, M. J. R. & Milner, J. S. (1989). Cloning and analysis of the *Kluyveromyces lactis* TRP1 gene: a chromosomal locus flanked by genes encoding inorganic pyrophosphatase and histone H3. *Yeast* **5**, 35-50.

Stinchcomb, D. T., Thomas, M., Kelly, J., Selker, E. & Davis, R. W. (1980). Eucaryotic DNA segments capable of autonomous replication in yeast. *Proceedings of the National Academy of Sciences, USA* **77**, 4559-4563.

Struhl, K., Stinchcomb, D. T., Scherer, S. & Davis, R. W. (1979). High-frequency transformation of yeast: autonomous replication of hybrid DNA molecules. *Proceedings of the National Academy of Sciences, USA* **76**, 1035-1039.

Chapter 21

Potato late blight: forecasts and disease suppression

W. E. Fry & M. A. Doster

Fungicides have figured prominently in the suppression of potato late blight ever since the discovery of Bordeaux mixture during the late 19th century. Use of fungicides has been very effective, particularly in comparison with the frustratingly unstable disease suppression associated with specific host plant resistance. The need for strategies giving effective, stable disease suppression is particularly important in locations where outbreaks of this explosive and destructive disease are a persistent threat. Certainly, the experience in Europe and the United Kingdom during the 1980s is an example. In these and other locations, successful potato production sometimes depends on very frequent applications of fungicide.

Although the frequent application of fungicide has been a successful disease suppression strategy, the economic costs can be high and there may be negative side effects, such as environmental and food contamination and fungicide resistance. Thus, scientists and growers have long sought more effective and efficient techniques for using fungicides. In this paper, we discuss techniques for timing fungicide applications (including forecasts) and use of host resistance as part of an integrated strategy which minimizes fungicide use. We then consider approaches for delivering this information to growers.

Forecasts

Initial applications

Forecasts for potato late blight have been in development for well over sixty years; and their use has aided disease management. The most successful approach has been to identify that time during the season when fungicide applications should be initiated. The most highly publicized techniques are those reported from Europe and the United States (Beaumont, 1947; Hyre, 1954; Krause, Massie & Hyre, 1975; Schrödter & Ullrich, 1966; van Everdingen, 1926; Wallin, 1962). Each of these techniques identifies the time of first occurrence of late blight as a function of weather; they have been developed empirically, and are surprisingly accurate in an extremely complex system.

The appearance of late blight in foliage when the source of inoculum is an infected seed tuber is a rare event (Doster, Sweigard & Fry, 1989), but in the northeastern USA, if foliar late blight is going to appear from infected seed tubers, it typically appears very soon after the time predicted by BLITECAST (Krause *et al.*, 1975; Doster *et al.*, 1989). The BLITECAST prediction is usually restricted to a single field. Protectant fungicide applications made more than several days prior to appearance of inoculum are essentially wasted, and in many situations, very early applications are not needed and can be eliminated. Timing the initial spray is a component of late blight forecasts that is becoming increasingly adopted by growers in the United States, particularly those involved with integrated pest management programmes.

Non-systemic fungicide application frequency

Although many late blight forecasts predict only the initiation of regular applications, some also address the frequency of applications throughout the season. A system developed by Wallin (1962) and adapted into BLITECAST by Krause *et al.* (1975), recommends the frequency of applications subsequent to the initial one. Additionally, our group at Cornell devised a forecast which incorporated the effects of both host resistance and the dynamics of fungicide weathering (Fry, Apple & Bruhn, 1983). Unfortunately, when we analyzed the benefits of adjusting the frequency of fungicide sprays according to the weather (relative to a standard weekly schedule), we were unable to demonstrate improved disease suppression or a reduction in the amount of fungicide used (Fohner, Fry & White, 1984). These are the two criteria by which we evaluated forecasts. We subsequently reasoned that such forecasts should not be expected to provide benefits because they are based on previous weather and recommend the application frequency of non-systemic fungicides which have no influence on *Phytophthora infestans* after a germ tube has entered a leaf. However, it appeared to us that adjustment of fungicide frequency might be possible if a systemic fungicide with post-infection efficacy were used, or if forecasts of future weather (e.g. Vincelli & Lorbeer, 1989) could be used.

Application frequency and metalaxyl

We then began development of a disease forecast for timing the frequency of fungicide sprays, but using the systemic fungicide metalaxyl which has some post-infection efficacy. Our approach was to use a complex simulation model which accurately reflected late blight development as a function of host resistance and weather, the deposition and redistribution of a protectant fungicide (chlorothalonil), and the deposition and dynamics of the systemic fungicide metalaxyl (Bruhn & Fry, 1981, 1982a, b; Milgroom & Fry, 1988a). In this model, chlorothalonil only influences the establishment of new lesions, whereas metalaxyl also inhibits sporulation from and expansion of established lesions. The model had already been validated and had provided insight into disease management. We expected to identify a candidate forecast system through analysis of simulation experiments, and then to evaluate this system in a series of field experiments. At this time we have just completed the simulation experiments. Because of the importance of metalaxyl resistance in *P. infestans*, results of the simulation experiments must be interpreted using at least two criteria: disease suppression, and suppression of metalaxyl resistance.

A trial and error approach was used to evaluate candidate forecasts using metalaxyl (Doster & Fry, unpublished). All simulation experiments assumed that the season was 12 weeks in duration (typical for northeastern USA), that the first infections appeared in the fourth week of the season, and that metalaxyl (0·20 kg a.i. ha^{-1}) was always applied in a mixture with chlorothalonil (1·61 kg a.i. ha^{-1}). Obviously, the levels of disease reflect the magnitude of the 'initializing' parameters (the number of initial lesions and the time of their appearance). Thus the absolute levels of disease are not critical to our evaluation, but the relative performance of the various strategies is important. All simulations were done with fifty years of diverse weather. The results of using any particular strategy are presented as the mean of the area under the disease progress curve (AUDPC), and the mean final disease rating for the 50 years. We identified the effects of various forecast strategies on disease caused by metalaxyl-sensitive isolates (Doster & Fry, unpublished). The reference treatment in all cases was four applications of the commercial product for northeastern USA (metalaxyl: chlorothalonil mixture on a 14 day schedule) with the initial application on day 28.

Our first general approach was to evaluate various weather-dependent strategies. One strategy was to apply metalaxyl after a specified number of 'favourable days' (a favourable day being one which favours late blight development). We defined a favourable day as having a severity value [*sensu* Krause *et al*., 1975] greater than zero. Severity values are a function of temperature and periods of high relative humidity (≥ 90%). A severity value of one is associated with high relative humidity periods of 16, 13 or 10 hours at temperatures of 7·2, 11·7 or 15·1°C respectively (Krause *et al*., 1975). A severity value of two is associated with high relative humidity periods of 19, 16 or 13 hours at these temperatures. The maximum severity value for a single day is four, and no severities accumulate if the temperature is below 7·2 or above 26·6° (Krause *et al*., 1975). From our analysis, we learned that application of the metalaxyl mixture after eight favourable days was slightly more effective than the reference treatment (Table 21.1) and used slightly fewer applications. Other mixture/application strategies

Table 21.1. Weather dependent application strategies using a metalaxyl:chlorothalonil mixture.

Rule	No. of applications	Area under disease progress curve mean	max	Final disea se(%)	Proportio n of refer- ence do se
14-day interval (reference)	4·0	1·3	2·7	7·6	1·00
after 8 favourable days	3·9	1·0	2·8	5·5	0·97
after 9 favourable days	3·6	1·7	6·1	10·9	0·89
after 12 favourable days	2·8	4·6	12·2	31·1	0·71
after 2 consecutive favourable days					
minimum interval = 8	4·0	1·0	6·3	6·2	1·00
minimum interval = 9	3·7	1·3	6·3	8·3	0·94
minimum interval = 14	2·8	3·3	9·0	22·6	0·71
after 14 severity values	4·0	0·8	2·2	4·6	1·00
after 17 severity values	3·5	1·8	6·1	11·3	0·87
after 20 severity values	3·1	3·2	9·2	21·7	0·79

Results shown are the means of simulation of 50 years of diverse weather (Doster & Fry, unpublished). Severity values are defined in Krause *et al.* (1975).

involving favourable days used less fungicide, but suppressed disease less well than the reference treatment (Table 21.1).

A second strategy was to apply a metalaxyl mixture after two consecutive favourable days, but with different minimum intervals before recording favourable days. When the minimum interval was eight days, the same number of metalaxyl applications was made as in the reference treatment, but there was slightly better disease suppression on average. However, the variance was larger (permitting a larger maximum AUDPC), indicating a greater risk associated with this treatment than with the reference treatment. When the minimum interval was nine days, 94% as much fungicide was used as in the reference treatment, but disease suppression was very nearly identical to that when the minimum interval was eight days (Table 21.1).

A third strategy was to apply metalaxyl after a given number of severity values (as defined by Krause *et al.*, 1975). Applications made after the accumulation of 14 severity values used the same amount of fungicide as used in the reference treatment, but suppressed disease more effectively. Applications made after the accumulation of 17 or 20 severity values used

Table 21.2. Decision formula strategies.

Decision formula Make an application if	No. of applications	Area under disease progress curve mean	max	Final disease (%)	Proportion of reference dose
0·2× SV-0·03× day≥ 1·0	4·0	0·6	2·0	3·8	1·01
0·14× SV-0·03× day≥ 1·0	3·2	2·6	9·4	19·1	0·79
0·10× SV-0·01× day≥ 1·0	3·6	1·1	3·2	7·6	0·90
0·05×SV-0 ·04×day+0·20SI ≥1	4·0	0·5	1·0	3·0	1·01
0·15×SV-0 ·03×day+0·03SI ≥1	3·7	0·9	2·8	6·3	0·93
0·15×SV-0 ·03×day+0·02SI ≥1	3·6	1·0	3·2	7·2	0·89
0·05×SV-0 ·05×day+0·18SI ≥1	3·1	1·8	3·1	13·5	0·78
14-day interval (reference)	4·0	1·3	2·7	7·6	1·00

Results shown are the means of simulations of 50 years of diverse weather (Doster & Fry, unpublished). SI = spray interval, SV = severity value.

less fungicide, but permitted more disease than occurred with the standard reference treatment (Table 21.1).

Several of these weather-dependent strategies used fungicide more effectively than the standard reference treatment. It appears that a weather-dependent application strategy might permit a decrease of nearly 10% in fungicide use while suppressing disease as well as the standard reference treatment. However, we were somewhat disappointed that this simulation analysis of various weather- dependent application strategies did not identify a strategy which allowed larger reductions in fungicide amounts relative to the standard reference treatment.

Our next general approach was to evaluate various decision formulae which depended on severity values, the number of days since emergence, and the interval since the last application. In this analysis, several tactics suppressed disease better with the same amount of fungicide as in the standard reference treatment, whilst others suppressed disease as well as the standard, but with less fungicide. With the better of these tactics, less than 90% of the fungicide used in the standard reference treatment was required and disease was suppressed to similar levels (Table 21.2); so the decision formula approach appeared more effective than the simpler forecasts.

Our third general approach was to investigate various fungicide timing tactics that were independent of weather. One of the most effective of these was to apply the metalaxyl mixture at one-half the recommended dosage, but to make six applications during the season rather than four.

Table 21.3. Weather independent strategies.

Treatment Dosage	days of application	No. of applications	Area under disease progress curve mean	max	Final disease (%)	Proportion of reference dose
1·00	28, 42, 56, 70	4·0	1·3	2·7	7·6	1·00
0·50	28, 38, 48, 58, 68, 78	6·0	0·4	0·9	1·9	0·75
1·00	28, 38, 48, 58	4·0	0·2	0·4	1·1	1·00
0·75	28, 38, 48, 58	4·0	0·3	0·7	2·3	0·75
1·00	28, 38, 52	3·0	1·1	2·7	18·4	0·75
1·00	28, 46, 56	3·0	1·4	3·0	11·1	0·75

Results shown are the means of simulations of 50 years of diverse weather (Doster & Fry, unpublished).

With this tactic, only 75% as much fungicide was used as in the standard reference treatment, but disease suppression was better than in the standard reference treatment (Table 21.3). Other effective weather-independent tactics concentrated applications during the middle portion of the season. The most effective way to enhance fungicide efficacy and to suppress disease was to apply metalaxyl every ten days beginning on the 28th day at three-quarters of the recommended dosage for a maximum of four applications (treatment 4, Table 21.3). This treatment used only three-quarters as much fungicide as the standard reference treatment but also suppressed disease much more effectively. This treatment is particularly attractive because lower dosages of metalaxyl will decrease the selection for resistance (Milgroom & Fry, 1988b; Milgroom, Levin & Fry, 1989).

Selection for metalaxyl resistance

We used the simulation model to predict the metalaxyl resistance selection associated with various metalaxyl application strategies (Doster, Milgroom & Fry, unpublished). We investigated the influence of various numbers of mixture applications on resistant and sensitive strains separately. This is particularly appropriate because the model describes focal development. The rationale was that late blight is often a focal disease and the population of *P. infestans* within a focus may be clonal. As one might expect, greater numbers of metalaxyl sprays select more strongly for resistant individuals than do fewer sprays. Conversely, increasing numbers of metalaxyl sprays strongly suppress sensitive individuals. However, it turns out that the timing of applications within the season is quite important (Doster, Milgroom & Fry, unpublished). For sensitive strains, applications made during the middle of the season are very much more effective in suppressing disease than those made at the beginning or at the end. For resistant types, applications made at the end of the season are associated with the lowest severities of disease caused by resistant individuals. This occurs because the level of protectant fungicide in the mixture is lower than that recommended for applications of the protectant alone. Thus, because metalaxyl has no influence on the resistant individuals, mixture applications made at the beginning or during the middle of the season partially release the resistant types from the suppressive influence of the protectant fungicide.

The timing of metalaxyl mixtures (whether used alone or integrated with a programme of protectant sprays) influences resistant and sensitive individuals differently. The choice of timing technique will depend on the proportion of resistant individuals in a population. When resistance is very rare, mixture applications should be applied during the middle of the season. However, if resistance is common, the metalaxyl mixture we tested should be used at the end of the season or not at all.

Plant resistance

Integration of plant resistance into a potato late blight management programme has been a goal of growers and plant pathologists for more than 100 years (Chapter 25). This goal has been partially achieved in some situations, but it has also been frustrated by the ability of the fungus to overcome specific resistances. As far as we know, specific resistances to *P. infestans* rarely (if ever) contribute to long term, stable disease suppression in agroecosystems. Conversely, levels of general, nonspecific resistance are usually insufficient alone for adequate late blight suppression. However, even the modest levels of general resistance available in commercial potato cultivars can contribute to overall disease suppression in an integrated management strategy. Relative to some of the most susceptible cultivars grown commercially in the northeastern United States, some of the most resistant cultivars have disease suppressive effects approximately equivalent to the effects of half of the recommended dosage of protectant fungicide applied on a weekly schedule to the susceptible cultivars (Fry, 1978). These resistances are incorporated into management recommendations made to commercial growers (Halseth *et al.*, 1987). For example, if a grower applies fungicide on a weekly basis to the most susceptible cultivars, then application every other week to the most resistant cultivars is recommended (Halseth *et al.*, 1987). Unfortunately, most of the commercial cultivars in the northeastern United States and certainly the most popular ones, are fairly susceptible to potato late blight.

In addition to suppressing overall disease, resistant cultivars also help to limit the selection for metalaxyl resistance. When metalaxyl is used on a population containing some resistant individuals, it selects very strongly for resistant biotypes. Selection is most rapid when the growth rate of the population is rapid (Milgroom & Fry, 1988b). Because general resistance slows pathogen growth rates, it also reduces selection for metalaxyl resistance. Unfortunately, the same mechanism also slows reversion to sensitivity in the absence of selection for metalaxyl resistance.

Some authors have questioned whether general resistance is likely to be stable and/or durable. Some have suggested that certain pathogen biotypes are specifically adapted to certain host genotypes (Chapter 25). Several reports have supported the concept of adaptation to specific host genotypes in *P. infestans* (Caten, 1974; Jeffrey, Jinks & Grindle, 1962; Jinks & Grindle, 1963; Latin, Mackenzie & Cole, 1981). On the other hand, other reports have not found evidence for adaptation (James & Fry, 1983; Paxman, 1963; van der Plank, 1971). In an effort to address the role of environment in this interaction, Parker (1989) investigated the general resistance of four commercial potato cultivars in two widely distant locations but against the same fungal genotypes. Parker (1989) found that the relative resistance ranking of several commercial cultivars was the same in Central New York State USA as it was in Toluca, Central Mexico. However, one cultivar, Alpha, was much more resistant in New York than in Toluca. This different response was apparently a plant reaction because assessments were done with the same pathogen isolates in both locations, and because other cultivars had similar resistances in both locations.

At this point, we believe that if the pathogen population can adapt to specific host genotypes, the magnitude of adaptation is small and is unlikely to cause a significant problem over the short term in practical disease management. (However, there appears to be no *a priori* reason why such adaptation could not occur.) Thus it appears that fungicides can be adjusted to complement host resistance.

Delivery systems

Many factors influence a grower's decision to adopt new technology. Changing economic, social and environmental pressures are among those causing potato growers to evaluate more carefully their use of fungicides. New technology will be adopted if the benefits are clearly visible and if the new technology can be implemented readily.

Computer-assisted information delivery systems appear to capture the attention of growers and plant pathologists. The use of a main frame computer to process BLITECAST recommendations was one of the exciting selling points of this program in the mid-1970's. More recently, BLITECAST and early blight suppression information have been incorporated into a computerized set of recommendations at the University of Wisconsin. The computer program has been used by growers in the upper Midwest, and the program has been used on more than 10,000 ha (Stevenson, unpublished).

The next step in computer-assisted delivery programs is probably the development of expert systems for potato production. A prototype system is CUPIDS (Cornell University Potato Integrated Decision support System). The prototype currently contains recommendations only for the suppression of potato early blight and potato late blight, but the system is designed to accommodate recommendations for insect and weed suppression as well. The recommendations are still quite simple and conservative. For example, if early blight is likely to be a more important problem than late blight, then only the early blight suppression recommendations are presented (Pelletier, 1987). It appears to us that computer-assisted disease management systems will be easier to use than printed recommendations. Such programs should enable plant pathologists to identify the most appropriate management strategies in complex systems. It is precisely for these complex systems that computer-assisted information delivery systems are most appropriate.

Conclusions

Although progress in potato late blight management and forecasting has been slow, significant insight and advances have been achieved. Several useful decision rules concerning the initiation of fungicide applications have been developed in different locations around the world. Dosages of protectant fungicides can be adjusted to complement even modest levels of general resistance in commercial potato cultivars. Further improvements in the effectiveness of systemic fungicides appear likely, and departures from fixed-interval/fixed-dosage calendar spray schedules will be facilitated with a new generation of computer-based decision support systems. Our experiences with fungicides and host resistances during the past 100 years emphasize the need for an integrated approach for potato late blight suppression.

References

Beaumont, A. (1947). The dependence on the weather of the dates of outbreak of potato late blight epidemics. *Transactions of the British Mycological Society* **31**, 45-53.

Bruhn, J. A. & Fry, W. E. (1981). Analysis of potato late blight epidemiology by simulation modeling. *Phytopathology* **71**, 612-616.

Bruhn, J. A. & Fry, W. E. (1982a). A statistical model of fungicide deposition on potato foliage. *Phytopathology* **72**, 1301-1305.

Bruhn, J. A. & Fry, W. E. (1982b). A mathematical model of the spatial and temporal dynamics of chlorothalonil residues on potato foliage. *Phytopathology* **72**, 1306-1312.

Caten, C. E. (1974). Intra-racial variation in *Phytophthora infestans* and adaptation to field resistance for potato blight. *Annals of Applied Biology* **77**, 259-270.

Doster, M. A., Sweigard, J. A. & Fry, W. E. (1989). The influence of host resistance and climate on the initial appearance of foliar late blight of potato from infected seed tubers. *The American Potato Journal* **66**, 227-233.

Fohner, G. R., Fry, W. E. & White, G. B. (1984). Computer simulation raises questions about timing protectant fungicide application frequency according to a potato late blight forecast. *Phytopathology* **74**, 1145-1147.

Fry, W. E. (1978). Quantification of general resistance of potato cultivars and fungicide effects for integrated control of potato late blight. *Phytopathology* **68**, 1650-1655.

Fry, W. E., Apple, A. E. & Bruhn, J. A. (1983). Evaluation of potato late blight forecasts modified to incorporate host resistance and fungicide weathering. *Phytopathology* **73**, 1054-1059.

Halseth, D. E., Sieczka, J. B., Tingey, W. M. & Zitter, T. A. (1987). *Cornell recommendations for potato production*. Cornell University, Ithaca, New York, U.S.A.

Hyre, R. A. (1954). Progress in forecasting late blight of potato and tomato. *Plant Disease Reporter* **38**, 245-253.

James, R. V. & Fry, W. E. (1983). Potential for *Phytophthora infestans* populations to adapt to potato cultivars with rate-reducing resistance. *Phytopathology* **73**, 984-988.

Jeffrey, S. T. B., Jinks, J. L. & Grindle, M. (1962). Intra-racial variation in *Phytophthora infestans* and field resistance to potato blight. *Genetica* **32**, 323-338.

Jinks, J. L. & Grindle, M. (1963). Changes induced by training *Phytophthora infestans*. *Heredity* **18**, 245-264.

Krause, R. A., Massie, L. B. & Hyre, R. A. (1975). BLITECAST; a computerized forecast of potato late blight. *Plant Disease Reporter* **59**, 95-98.

Latin, R. X., Mackenzie, D. R. & Cole, Jr., H. (1981). The influence of host and pathogen genotypes on the apparent infection rates of potato late blight epidemics. *Phytopathology* **71**, 82-85.

Milgroom, M. G. & Fry, W. E. (1988a). A model for the effects of metalaxyl on potato late blight epidemics. *Phytopathology* **78**, 559-565.

Milgroom, M. G. & Fry, W. E. (1988b). A simulation analysis of the epidemiological principles for fungicide resistance management in pathogen populations. *Phytopathology* **78**, 565-570.

Milgroom, M. G., Levin, S. A. & Fry, W. E. (1989). Population genetics theory and fungicide resistance. In *Plant Disease Epidemiology*, Vol. II, (ed. K. J. Leonard & W. E. Fry), pp. 340-367. McGraw Hill: New York, U.S.A.

Parker, J. M. (1989). *Stability of disease expression in the potato late blight system*. Ph.D. Thesis, Cornell University, Ithaca, New York, U.S.A.

Paxman, G. J. (1963). Variation in *Phytophthora infestans*. *European Potato Journal* **6**, 14-23.

Pelletier, J. R. (1987). *Computer simulation of cultivar resistance in fungicide effects on epidemics of potato early blight*. Ph.D. Thesis, Cornell University, Ithaca, New York, U.S.A.

Schrödter, H. & Ullrich, J. (1966). Further research on the biometeorology and epidemiology of *Phytophthora infestans* (Mont.) de Bary. A new proposal for the solution of problems of epidemiological prognosis. *Phytopathologische Zeitschrift* **56**, 265-278.

van Everdingen, E. (1926). Het verband tusschen de weergesteldheid en de aardappelziekte *Phytophthora infestans*. *Tijdschrift over Plantenziekten* **32**, 129.

van der Plank, J. E. (1971). Stability of resistance to *Phytophthora infestans* in cultivars without R genes. *Potato Research* **14**, 263-270.

Vincelli, P. C. & Lorbeer, J. W. (1989). Blight alert: a weather-based predictive system for timing fungicide applications on onion before infection periods of *Botrytis squamosa*. *Phytopathology* **79**, 493-498.

Wallin, J. R. (1962). Summary of recent progress in predicting late blight epidemics in the United States and Canada. *American Potato Journal* **39**, 306-312.

Chapter 22

Current problems in the chemical control of late blight: the Northern Ireland experience

Louise R. Cooke

Potato late blight, caused by *Phytophthora infestans* remains one of the most intractable problems for potato producers in Ireland. Within Northern Ireland, the potato is the most important agricultural crop in monetary terms, with production valued at £20 million in 1987 (Anon., 1988), with £5 million and £15 million due to seed and ware respectively. Unfortunately, the moist, mild climate which is well-suited to potato production is equally favourable for development of potato blight. Within Northern Ireland, foliage infection often builds up too late in the season to cause yield reductions. However, sporulation of *P. infestans* on leaves and stems shortly before haulm destruction, accompanied by high rainfall, can result in significant infection of tubers. Depending on the rate of symptom development and the extent of secondary bacterial infection, this may produce rotting of tubers in the ground before lifting, losses in store or infection in the progeny crop if the stock is used for seed.

From the introduction of Bordeaux and Burgundy mixtures until the mid 1970's, the potato crop was protected from blight by frequent application of non-systemic fungicides. These compounds function by killing the pathogen on the plant surface before infection occurs, but have little curative activity. Their effectiveness is thus dependent on maintaining a complete cover of the foliage; under conditions of rapid plant growth or high rainfall this may be impossible to achieve.

The introduction of the first systemic fungicides with good activity against Oomycete pathogens appeared to open up new possibilities for control of potato blight (Schwinn, 1983). The translocation, persistence and curative activity of the phenylamides permitted the use of longer spray intervals and offered greater flexibility, vital in a country where spray operations are frequently rendered impossible by rain. Within Northern Ireland, metalaxyl was the first phenylamide introduced and remains the most widely used, although benalaxyl, cyprofuram, ofurace and oxadixyl have also been marketed for control of potato blight. In this paper, experience of potato blight control with the phenylamides in Northern Ireland over the last ten years will be related to the occurrence of resistance and its influence on disease control.

The initial occurrence of phenylamide resistance

Metalaxyl was first used to control potato blight on a commercial scale in the UK in 1978. Whereas in other European countries, a wettable powder (Ridomil 25 WP) containing metalaxyl as sole active ingredient was marketed, in the UK the product approved for use on potatoes (Fubol 58 WP) was a mixture of metalaxyl with the dithiocarbamate protectant, mancozeb. The benefits of improved control of blight, particularly in the tubers, and longer spray intervals (Smith, 1979), resulted in metalaxyl formulations rapidly gaining popularity with potato growers. However, in the summer of 1980 there was a sudden failure of disease control on metalaxyl-treated crops in the Republic of Ireland and the Netherlands, which was shown to be due to selection of phenylamide-resistant strains of *P. infestans* (Dowley & O'Sullivan, 1981; Davidse *et al.*, 1981).

In Northern Ireland, control of potato blight by the commercial metalaxyl + mancozeb formulation remained excellent throughout the 1980 season. Nonetheless, infected tubers from the 1980 crop were shown to contain phenylamide-resistant strains of *P. infestans* (Cooke, 1981). Concern at this first detection of resistance in *P. infestans* within the UK led the manufacturers of metalaxyl (Ciba-Geigy) to withdraw the recommendation for its use in Northern Ireland in 1981. At the same time, metalaxyl formulations were also withdrawn from the Republic of Ireland. Subsequently, in 1985 metalaxyl formulations were re-introduced to both Northern Ireland and the Republic of Ireland. Levels of usage of phenylamide formulations are discussed in a later section.

The incidence of phenylamide resistance, 1981-89

In Northern Ireland, the incidence of phenylamide-resistant strains of *P. infestans* has been monitored with the assistance of the Department of Agriculture's Potato Inspectors who provided samples of infected foliage from seed crops in all potato-growing areas of the Province. Isolates obtained from individual crops were tested for their ability to sporulate on metalaxyl-treated detached leaflets (1981-83) or on discs floating on metalaxyl solutions (1984-89) and classified as resistant if they sporulated on a 100 mg metalaxyl l^{-1} treatment (Cooke, 1986a). This test is non-quantitative and isolates which yield a mixture of resistant and sensitive sporangia will be detected as resistant (Sozzi & Staub, 1987). A limited number of isolates has also been tested using a quantitative technique, and of ten resistant isolates tested, eight had >80% resistant sporangia, whilst the remaining two had 72% and 41% respectively (Walker, 1990).

In the period 1981 to 1985, the proportion of isolates including phenylamide-resistant strains fluctuated from season to season (Fig. 22.1), always being less than 50%; no consistent trend was observed

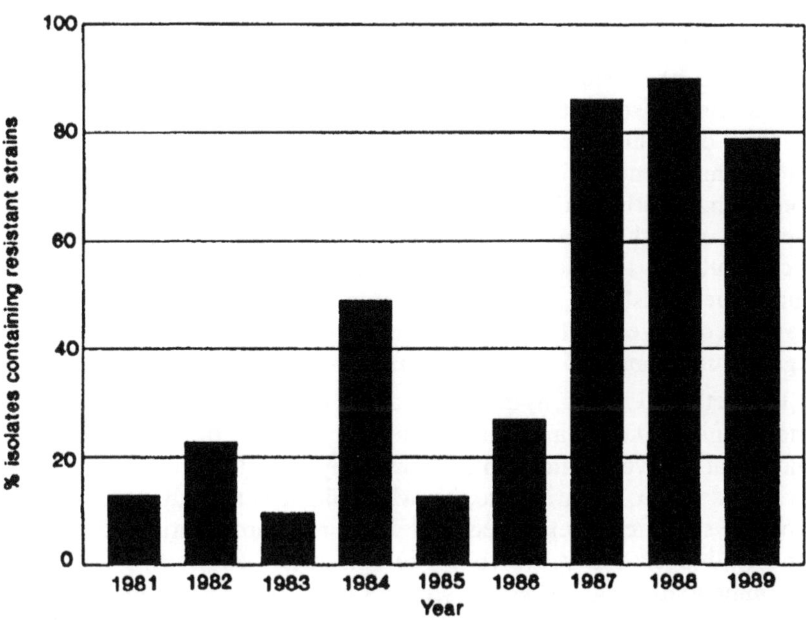

Fig. 22.1. Proportion of viable isolates of *Phytophthora infestans* containing phenylamide-resistant strains in Northern Ireland, 1981-1989.

(Cooke, 1986a). In contrast, in 1981 in the Republic of Ireland, following the failure of metalaxyl to control *P. infestans* the previous year, phenylamide-resistant strains were detected in 75% of isolates, but this figure declined to 21% in 1982 and 6% in 1983 (Dowley & O'Sullivan, 1985).

From 1983 onwards, there is a close similarity in the level of detection of phenylamide-resistant strains in Northern Ireland and the Republic. In both cases, the occurrence of phenylamide-resistant strains increased dramatically in the seasons following 1985, reaching 86% in Northern Ireland (Cooke, 1988) and 76% in the Republic of Ireland in 1987 (Dowley & O'Sullivan, unpublished). In Northern Ireland, this level of resistance was maintained in 1988 and only slightly reduced in 1989 (Fig. 22.1).

Thus, after five years, when the incidence of phenylamide resistance appeared under control, the situation changed abruptly. Before considering the factors which may have contributed to this sudden escalation in the occurrence of resistant strains, the strategy which was adopted in an attempt to minimise the build up of resistance will be outlined.

Strategy for use of phenylamides on potatoes

Since 1980, phenylamides have only been marketed for use on potatoes in formulations also containing dithiocarbamate fungicides. In the UK, the amount of dithiocarbamate present in most formulations containing phenylamides now gives a rate equivalent to that used in products containing a dithiocarbamate alone. However, the strategy adopted for the use of such formulations in the UK is different from that used successfully in the Netherlands. This strategy, developed by the UK Fungicide Resistance Action Committee (FRAC) Phenylamide Group, was encouraged within Northern Ireland by the Department of Agriculture from 1985 when phenylamides were again widely marketed.

In Northern Ireland, as in the rest of the UK, the emphasis has been on encouraging phenylamide use from the start of the season up to a maximum of five applications, followed by use of a protectant, usually based on fentin. Curative use, i.e. when blight is already present in the crop, is specifically excluded and label recommendations have been modified over the years to make this more explicit. There is no recommendation against use on seed potatoes. In the Republic of Ireland, the strategy has been broadly similar to that in the UK, except that a maximum of three phenylamide applications is currently recommended, which is in line with the international FRAC group limit of two to four applications per season (Urech & Staub, 1985). In contrast, in the Netherlands, the anti-resistance strategy involves use of phenylamides on ware crops only, to reduce the risk of survival of resistant strains of *P. infestans* in infected seed, up to two consecutive applications being permitted only when conditions are favourable to the spread of blight (Davidse, 1985; Chapter 23).

Usage of phenylamides in Northern Ireland

In the years from 1981 to 1983, the Department of Agriculture for Northern Ireland and the agrochemical manufacturers recommended that phenylamides should not be used on potato crops in Northern Ireland. Despite this, some growers continued to obtain and use metalaxyl + mancozeb from Great Britain. Accurate data on usage for these years are not available, but it is probable that 20 to 30% of the seed crop received some phenylamide applications. In 1985, metalaxyl + mancozeb was re-introduced to Northern Ireland and by this time, formulations containing cyprofuram, ofurace and oxadixyl were also available. In this year, over 50% of the seed crop area was treated with phenylamides and this level was maintained in 1986 and 1987 (Fig. 22.2). In 1988, the crop

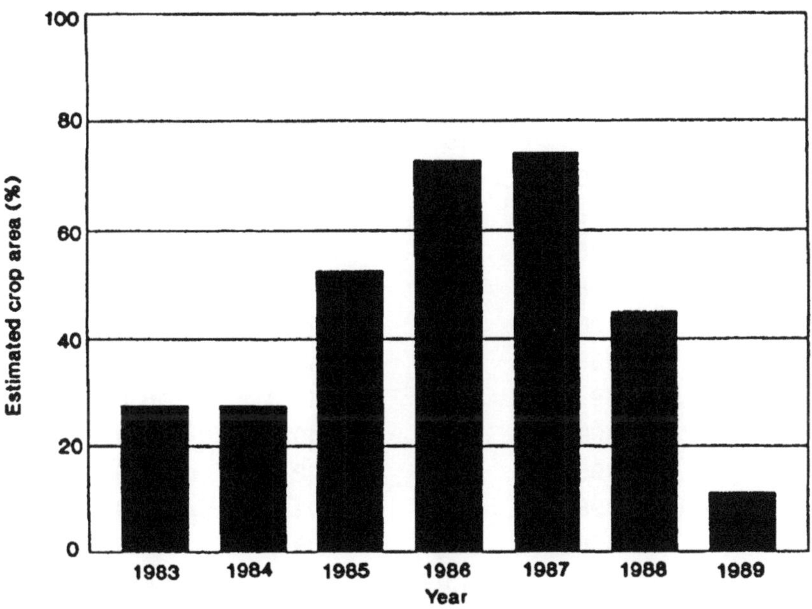

Fig. 22.2. The proportion of seed potato crops treated with phenylamides in Northern Ireland, 1983-1989.

area receiving phenylamide treatments was reduced mainly as a result of grower reaction to problems with blight control experienced in 1987.

Phenylamides were also re-introduced to the Republic of Ireland in 1985. In 1986 and 1987 respectively, 79% and 83% of farms surveyed used phenylamide fungicides, but in 1988 as in Northern Ireland a smaller proportion (63%) received phenylamides (Dowley & O'Sullivan, unpublished).

Undoubtedly, the increased crop area treated with phenylamides from 1985 onwards was one factor which encouraged the build-up of resistance. However, even before 1985 in some parts of Northern Ireland, most notably Counties Antrim and Londonderry, a substantial proportion of the seed crop was regularly treated with phenylamides. It thus appears unlikely that increased treated crop area was solely responsible for the increase in phenylamide resistance.

Meteorological conditions 1981-88

The weather pattern during the summer months has a profound influence on the incidence of potato blight and thus on the size of the *P. infestans*

Table 22.1. Meteorological data for July and August, 1981-1988, County Londonderry, Northern Ireland

			Temperature (°C)*	
Month	Year	Rainfall (mm)	minimum	maximum
July	1981	75·3	10·1	16·2
	1982	15·7	11·0	18·2
	1983	26·5	12·3	20·2
	1984	54·6	11·2	17·7
	1985	115·7	10·8	16·7
	1986	84·3	10·9	16·3
	1987	89·2	10·7	17·5
	1988	121·5	10·0	16·0
August	1981	48·4	11·3	18·0
	1982	107·9	10·7	16·9
	1983	42·4	12·2	19·6
	1984	52·6	11·9	19·0
	1985	207·3	9·9	15·5
	1986	113·8	8·3	14·7
	1987	96·9	10·7	17·1
	1988	112·9	10·5	16·5

*mean daily figures.

population and the number of occasions on which selection for resistance can occur. It also indirectly influences the number of fungicide applications made to individual crops, since the frequency of application is increased when conditions favour the rapid spread of blight (Chapter 21).

The amount of rainfall in the late June to August period is of crucial importance in determining the incidence of foliage infection in Northern Ireland and its impact is modified by the prevailing temperatures which influence the rate of development of the pathogen. The overnight temperature is generally too low for spread of infection before the second half of June; the first field outbreaks are seldom observed before June 20. However, if rainfall is high in July, infection can spread extremely rapidly through the foliage and if wet weather continues into August, there is a risk of tuber infection. In Northern Ireland, seed crops are subject to statutory dates for destruction of the haulm, which are always before the end of August for maincrop cultivars. Rainfall in September thus only influences tuber infection in ware crops.

In 1982, 1983 and 1984, the weather in July was warm and dry (Table 22.1), and in each of these seasons the incidence of blight infection was low. In contrast, in 1985 very high rainfall occurred throughout June, July and August, resulting in widespread foliage infection and severe tuber infection in some crops (Anon., 1985). In 1986, 1987 and 1988, rainfall was again above average during July and August. Above average temperatures in July 1987 produced conditions particularly favourable for the rapid spread of *P. infestans*. Thus, over the four years from 1985 to 1988, the numbers of *P. infestans* sporangia and zoospores exposed to phenylamides in Northern Ireland were undoubtedly far greater than in earlier years.

It seems probable that this succession of summers with weather extremely conducive to blight was the over-riding factor in encouraging selection for phenylamide resistance in the *P. infestans* population.

The similarity in resistance increases in Northern Ireland and the Republic of Ireland tends to lend support to this view. Whether a strategy of phenylamide use similar to that adopted in the Netherlands (Chapter 23) might have prevented the build up of resistance in Ireland cannot now be definitely ascertained, but given the infection pressures experienced in the 1985-88 seasons this appears unlikely. A policy of recommending two phenylamide applications when conditions favoured the spread of infection would have been extremely difficult to implement since such conditions prevailed for long periods during each of the four seasons and would almost certainly have resulted in curative use of these products. It is unclear if this would have led to selection pressures in favour of resistance greater than those actually experienced.

The impact of phenylamide resistance on control of potato blight

From the phenylamide resistance survey results reported above(Fig. 22.1), it is clear that the level of resistant strains in the Northern Ireland population has increased markedly over the past few years. It is therefore important to determine how disease control with phenylamide-containing formulations is influenced by the level of resistance. All phenylamides are now marketed as formulations also containing dithiocarbamate protectants, so even were the population totally resistant, disease control would not fail. In the worst scenario, it would decline to the level achieved by the dithiocarbamate alone, but this discounts the possibility of a synergistic interaction between the components of the mixture (Chapter 24).

The impact of resistance on disease control was investigated in a series of field trials. Where the initial inoculum (applied to rows of unsprayed infector plants adjacent to the sprayed plots) was either phenylamide-sensitive or a mixture of phenylamide-sensitive and

-resistant isolates, products containing a phenylamide plus a 'full rate' of mancozeb consistently out-performed mancozeb alone in terms of foliage blight control (Cooke, 1986a; Cooke & Norrie, 1987). However, when the initial inoculum was phenylamide-resistant, there was no difference in foliage blight control between metalaxyl + mancozeb and mancozeb alone applied at the same intervals (Cooke & Little, 1989). There was no evidence of any interaction between the two components of the mixture in this situation.

The influence of resistance on prevention of tuber infection by phenylamides has also been investigated. When metalaxyl was first introduced, excellent tuber blight control was one of its most outstanding features (Smith, 1979), even when *P. infestans* infected the foliage of tuber blight susceptible cultivars. This was in part due to its ability to protect tubers directly from infection by phenylamide-sensitive strains of *P. infestans* (Bruin, Edgington & Ripley, 1982). A trial in which healthy tubers were inoculated immediately after lifting, showed that whereas very few tubers from metalaxyl-treated plants became infected by a phenylamide-sensitive isolate of *P. infestans*, a phenylamide-resistant isolate infected them as successfully as it did tubers from untreated plants (Cooke, 1986b).

Thus, when phenylamide-resistant strains dominate the *P. infestans* population, the benefits from the use of phenylamides in terms of foliage and tuber blight control are substantially reduced. However, because they are now used only in mixtures with dithiocarbamates, a gradual erosion of performance is observed, not a sudden loss of disease control. The effect on disease control is comparable to that of the quantitative changes in sensitivity of pathogen populations to demethylation-inhibiting fungicides or dodine, as opposed to the dramatic impact of the qualitative development of resistance to a phenylamide or benzimidazole used alone (Skylakakis, 1985), although its cause is quite different. Consequently, it is extremely difficult to determine at what point the substantially greater cost of using a product containing a phenylamide is no longer economically justified.

Survival of phenylamide-resistant *Phytophthora infestans*

If further restrictions on the use of phenylamides were implemented or if the crop area treated declined substantially, is it likely that the incidence of phenylamide-resistant strains of *P. infestans* would decrease? The answer to this question is dependent on the comparative fitness of phenylamide-resistant and -sensitive strains in the absence of phenylamide usage. The marked decline in the occurrence of resistant strains in the Republic of Ireland between 1981 and 1983 (Dowley & O'Sullivan, 1985) may have resulted from reduced sporulation compared with that of the sensitive strains (Dowley, 1987). In isolates obtained from Northern Ireland, no differences in sporulation capacity were detected (Walker, 1990), but phenylamide-resistant isolates generally grew faster than sensitive ones at low temperature (Walker & Cooke, 1988). This may lead to a proportionately greater loss of resistant strains over the winter, due to increased rotting of infected tubers.

However, trials to date have failed to demonstrate if this influences the incidence of resistant strains in the following season's crop. In contrast, in Israel, Kadish & Cohen (1988) found that phenylamide-resistant field isolates of *P. infestans* produced larger lesions than phenylamide-sensitive isolates and competed successfully with the latter on intact plants even in the absence of metalaxyl application. These authors concluded that this high fitness of the phenylamide-resistant isolates provided a possible explanation for the prevalence of resistant strains within the *P. infestans* population in Israel and for the severity of infection in field crops.

It seems possible that when phenylamide-resistant strains were first selected from the population, particularly where selection was due to the use of products containing metalaxyl alone, resistance was not necessarily combined with other traits conferring a high level of fitness. Resistance to phenylamides conferred such a strong competitive advantage that it over-rode any minor defects as long as phenylamide usage continued, but once usage ceased such inadequate strains were soon lost. Now, after nine years of selection within the resistant sub-population, resistance and fitness have probably combined to produce strains as fit as the wild-type. There seems, therefore, no reason to anticipate an abrupt decline in resistant strains, even if all use of phenylamides ceased at once.

Conclusions

In the years before 1986, the results of annual surveys for phenylamide-resistance within the *P. infestans* populations in Northern Ireland and the Republic of Ireland suggested that the strategy of using phenylamides only as pre-pack mixtures with protectant fungicides was proving successful in limiting further selection for resistance. Since then the situation has altered radically and phenylamide-resistant strains are now detected in $>70\%$ of isolates tested. The major factor responsible for this change was probably the succession of four years (1985-1988) with summer weather very conducive to the rapid spread of *P. infestans* which produced many more opportunities for selection in favour of resistance. Data from field trials and surveys suggest that with phenylamide-resistant strains of *P. infestans* in Ireland at their current levels, the benefits of using products containing phenylamides are much reduced. Even if all phenylamide usage now ceased, an abrupt decline in the occurrence of phenylamide-resistant strains is unlikely since no marked defects in their fitness have been detected.

What lessons can be learned from this experience which may be of use in the future for the control of potato blight or the management of resistance? Firstly, strategies which appear to combat the build-up of resistance when infection levels are moderate may prove ineffective when subjected to high pressures. Such high infection pressure is particularly liable to occur with air-borne pathogens which are strongly influenced by prevailing conditions. Second, the role of tuber-borne infections in initiating blight epidemics may have been underestimated. If other sources of overwintering of *P. infestans* can be controlled, then breaking the cycle of selection for resistance by restricting the use of systemic fungicides to ware crops (as occurs in the Netherlands) will have a major impact, particularly if seed and ware crops are sufficiently distant for spread of aerial inoculum between them to be minimal. Unfortunately, for potato production in Ireland, perhaps the most important attribute for any new fungicide intended to control *P. infestans* is the ability to prevent tuber infection particularly in seed crops. A fungicide with a ware only recommendation would be of limited value.

Finally, that mixtures of systemic and non-systemic fungicides are not well-suited to combating resistance, even if in this case no other option is available. Although applied as a mixture, the two fungicides become separated once they contact the plant, both spatially (by uptake and translocation) and in time (by preferential loss of one fungicide). Once this occurs, selection of strains resistant to one of the fungicides becomes feasible. The problem of the use of two fungicides with different persistences as a mixture is theoretically easily resolved, by using an application interval dictated by the less persistent partner, as is now advocated for phenylamide/mancozeb mixtures (Urech & Staub, 1985). However, this may still be difficult to achieve in practice, if weather conditions do not permit sufficiently frequent applications or if growers are recommended to use a shorter interval than that previously advocated. The use of a mixture of a systemic and a non-systemic fungicide presents a further problem, since parts of the plant, such as new foliage, may be protected by the systemic alone. In a recent study, when a phenylamide + mancozeb formulation was applied to potato foliage by conventional high volume spraying, the undersides of the leaves were shown to be protected by the phenylamide alone and were readily infected by phenylamide-resistant strains of *P. infestans* (Marquinez Ramirez, 1989). To obviate the situation where parts of the plant are protected by the systemic fungicide only, application techniques which optimise cover are needed. In the case of mixtures containing metalaxyl, however, the potato tuber provides a further site at which selection in favour of resistance in the presence of the phenylamide alone can occur.

For the future, rather than attempting to reduce the selection for resistance to compounds with an inherently high resistance risk, it would be better to identify fungicides with a lower risk. This might be achieved by developing fungicides which bind to the active sites of fungal enzymes or co-factors, acting as pseudo-substrates, as occurs in the case of the demethylation-inhibiting fungicides binding to cytochrome P-450 (Nyfeler & Huxley, 1986). With such compounds, the development of resistance due to binding site modifications is much less probable and other mechanisms of resistance generally seem less likely to lead to a qualitative loss of activity. However, if a fungicide produces inhibition by binding to an enzyme at a site distinct from the active site, there appears a much greater potential risk of a binding site change without loss of enzymic activity. Whether this applies to metalaxyl will not be clear until the binding site is elucidated (Chapter 23).

I thank Messrs G. Little, S. Norrie and T. Currie and Miss D. Moore for their assistance in this work, and Miss R. Marquinez Ramirez and Miss A. S. L. Walker (postgraduate students) for the use of some of their data. Mr L. J. Dowley (Oak Park Research Centre, Teagasc) kindly allowed me to quote from his results obtained in the Republic of Ireland. Members of the UK FRAC Phenylamide Group funded vacation students who assisted in resistance monitoring, 1982-1988.

References

Anonymous, (1985). Plant Pathology Research Division. Technical Work: Potatoes-Blight. *Annual Report on Research and Technical Work of the Department of Agriculture for Northern Ireland, 1985.*

Anonymous, (1988). Statistical Review of Northern Ireland Agriculture. *Department of Agriculture for Northern Ireland. Economics and Statistics Division.*

Bruin, G. C. A., Edgington, L. V. & Ripley, B. D. (1982). Bioactivity of the fungicide metalaxyl in potato tubers after foliar sprays. *Canadian Journal of Plant Pathology* **4**, 353-356.

Cooke, L. R. (1981). Resistance to metalaxyl in *Phytophthora infestans* in Northern Ireland. *Proceedings, 1981 British Crop Protection Conference, Pests and Diseases* **2**, 641-649.

Cooke, L. R. (1986a). Acylalanine resistance in *Phytophthora infestans* in Northern Ireland. *Proceedings, 1986 British Crop Protection Conference, Pests and Diseases* **2**, 507-514.

Cooke, L. R. (1986b). Plant Pathology Research Division. Potato Diseases–Potato late blight (*Phytophthora infestans*). *Annual Report on Research and Technical Work of the Department of Agriculture for Northern Ireland, 1986*, p. 136.

Cooke, L. R. (1988). Beating Potato Blight. *Agriculture in Northern Ireland* **2**, 8-9.

Cooke, L. R. & Little, G. (1989). The control of a phenylamide-resistant potato blight infection by fungicide formulations. *Tests of Agrochemicals and Cultivars No 10, Annals of Applied Biology* **114** (supplement), 174-175.

Cooke, L. R. & Norrie, S. (1987). A comparison of the control of potato blight by acylalanine-containing formulations in the presence of acylalanine-resistant inoculum. *Tests of Agrochemicals and Cultivars No 8, Annals of Applied Biology* **110** (supplement), 76-77.

Davidse, L. C. (1985). Resistance to acylalanines in *Phytophthora infestans* in The Netherlands. *EPPO Bulletin* **15**, 403-409.

Davidse, L. C. Looijen, D., Turkensteen, L. J. & van der Wal, D. (1981). Occurrence of metalaxyl resistant strains of *Phytophthora infestans* in Dutch potato fields. *Netherlands Journal of Plant Pathology* **87**, 65-68.

Dowley, L. J. (1987). Factors affecting the survival of metalaxyl-resistant strains of *Phytophthora infestans* (Mont.) de Bary in Ireland. *Potato Research* **30**, 473-475.

Dowley, L. J. & O'Sullivan, E. (1981). Metalaxyl-resistant strains of *Phytophthora infestans* (Mont.) de Bary in Ireland. *Potato Research* **24**, 417-421.

Dowley, L. J. & O'Sullivan, E. (1985). Monitoring metalaxyl resistance in populations of *Phytophthora infestans*. *Potato Research* **28**, 531-534.

Kadish, D. & Cohen, Y. (1988). Competition between metalaxyl-sensitive and metalaxyl-resistant isolates of *Phytophthora infestans* in the absence of metalaxyl. *Plant Pathology* **37**, 558-564.

Marquinez Ramirez, R. (1989). Effects of cultivar, pathogen strain and fungicide on the development of potato late blight. M.Sc. thesis, The Queen's University of Belfast, U.K.

Nyfeler, R. & Huxley, P. (1986). The application of modern methods in chemical fungicide research. *1986 British Crop Protection Conference, Pests and Diseases* **1**, 207-215.

Schwinn, F. J. (1983). New developments in chemical control of *Phytophthora*. In *Phytophthora: Its Biology, Taxonomy, Ecology and Pathology*, ed. D. C. Erwin, S. Bartnicki-Garcia & P. H. Tsao, pp. 327-334. The American Phytopathological Society: St. Paul, Minnesota.

Skylakakis, G. (1985). Two different processes for the selection of fungicide-resistant sub-populations. *EPPO Bulletin* **15**, 519-525.

Smith, J. M. (1979). The use of metalaxyl for the control of downy mildew diseases. *Proceedings, 1979 British Crop Protection Conference, Pests and Diseases* **1**, 331-339.

Sozzi, D. & Staub, T. (1987). Accuracy of methods to monitor the sensitivity of *Phytophthora infestans* to phenylamide fungicides. *Plant Disease* **71**, 422-425.

Urech, P. A. & Staub, T. (1985). The resistance strategy for acylalanine fungicides. *EPPO Bulletin* **15**, 539-543.

Walker, A. S. L. (1990). Comparison of the fitness of phenylamide-sensitive and phenylamide-resistant strains of *Phytophthora infestans*. Ph.D. thesis, The Queen's University of Belfast, U.K.

Walker, A. S. L. & Cooke, L. R. (1988). The survival of phenylamide-resistant strains of *Phytophthora infestans* in potato tubers. *Proceedings, 1988 Brighton Crop Protection Conference, Pests and Diseases* **1**, 353-358.

Chapter 23

Phenylamides and *Phytophthora*

L. C. Davidse, G. C. M. van den Berg-Velthuis, B. C. Mantel & A. B. K. Jespers

The discovery and development of the phenylamide fungicides for the control of plant diseases caused by the Peronosporales has been one of the major contributions of the agrochemical industry to agriculture in the 1980s. The high inherent activity of the phenylamides, their protective and curative properties against all Peronosporales, their rapid uptake by plant tissue and high acropetal systemicity made these compounds excellent products to combat a group of devastating pathogens (Schwinn & Staub, 1987). A serious threat, however, to the performance of the phenylamides is the development of resistance in populations of target organisms. Resistance to phenylamides developed shortly after their introduction in *Phytophthora infestans* and *Plasmopara viticola*. Subsequently, resistance has been reported for many other pathogens. To cope with resistance, use strategies have been developed in which the ability of the compounds to retard an epidemic is optimized at the lowest selection pressure possible (Urech, 1988).

This chapter describes two aspects of the phenylamides. In the first part the biochemical basis of their antifungal activity and the mechanism of resistance in target fungi is discussed. The second part focuses on the contribution of metalaxyl to controlling late blight in the Netherlands and the strategy to manage resistance that has been followed in this country.

Mechanism of action of phenylamides

Identification of the phenylamide target site

The mechanism of action of the phenylamides has been most intensively studied using metalaxyl as a representative member of this group. Metalaxyl specifically inhibits incorporation of [5,6-^{3}H]uridine into RNA by living mycelia of *Phytophthora infestans* and *P. megasperma* f. sp. *medicaginis* (Davidse *et al.*, 1988). Incorporation of [methyl-^{3}H]thymidine into DNA and [^{14}C]phenylalanine into protein is affected to a lesser extent. Since metalaxyl at low concentrations does not affect the uptake of precursors or the conversion of nucleosides into nucleotides, inhibition of RNA synthesis must be responsible for the observed effects on uridine incorporation. At metalaxyl concentrations that are fully in-

Table 23.1. Effect of metalaxyl on endogenous RNA polymerase activity in nuclei isolated from phenylamide-sensitive (S) and phenylamide-resistant (R) strains[a] of *P. megasperma* f.sp. *medicaginis* (*Pmm*) and *P. infestans* (*Pinf*).

Concentration of metalaxyl in assay mixture (μg ml^{-1})	Polymerase activity as % of control in nuclei from:			
	Pmm S	*Pmm* R	*Pinf* S	*Pinf* R
0·1	68	100	68	96
1	56	99	60	95
10	51	100	57	88

[a]For methods and origin of strains, see Davidse *et al.* (1988)

Table 23.2. Effects of metalaxyl and α-amanitin on endogenous RNA polymerase activity of nuclei isolated from a phenylamide-sensitive strain of *P. megasperma* f. sp. *medicaginis*.

Inhibitor(s) added	Concentration (μg ml^{-1})	Activity as % of control
metalaxyl	10·0	64
α-amanitin	0·1	76
α-amanitin	5·0	72
α-amanitin	100·0	71
α-amanitin	250·0	78
metalaxyl + α-amanitin	10·0 + 5·0	38
metalaxyl + α-amanitin	10·0 + 100·0	32
metalaxyl + α-amanitin	10·0 + 250·0	42

hibitory to growth, a complete inhibition of uridine incorporation does not occur. In *P. megasperma* f. sp. *medicaginis*, uridine incorporation is reduced to *ca* 20 % of the control value; in *P. infestans* it is reduced to *ca* 60 %. Apparently only part of the cellular RNA synthesis is sensitive to phenylamides, indicating a specific inhibition of one of the RNA polymerases. Uridine incorporation by phenylamide-resistant strains of both *Phytophthora* spp. is not affected by metalaxyl.

In vitro studies with crude, cell free extracts or partially purified RNA polymerases failed to demonstrate any inhibitory effect on RNA synthesis (Davidse, Hofman & Velthuis, 1983). Endogenous RNA polymerase activity of isolated nuclei, however, appeared to be sensitive (Table 23.1).

Table 23.3. Binding of [^{3}H]metalaxyl[a] in cell-free mycelial extracts[b].

Source of extract	K_d-value[c] (μM)	Binding capacity[c] (pmol mg^{-1} protein)
P. megasperma f. sp. *medicaginis* S[d]	0·17	5·4
P. megasperma f. sp. *medicaginis* R	∞	none
P. infestans S	0·07	3·0
P. infestans R	∞	none

[a] Specific activity 551 mCi mmol^{-1}.

[b] S100 fractions prepared according to Swanson & Holland (1983) containing 24·5, 23·0, 15·8 and 16·2 mg protein ml^{-1} for *P. megasperma* f. sp. *medicaginis* S, *P. megasperma* f. sp. *medicaginis* R, *P. infestans* S and *P. infestans* R, respectively.

[c] As determined by plotting binding data in a Scatchard plot.

[d] S, phenylamide sensitive strain; R, phenylamide resistant strain.

α-Amanitin, a specific inhibitor of RNA polymerase II, was also inhibitory in this assay (Table 23.2). The effects of metalaxyl and α-amanitin are additive, indicating interference of metalaxyl with an RNA polymerase activity different from the α-amanitin sensitive polymerase II. Analysis of radiolabelled RNA formed *in vivo* during incubation with metalaxyl showed that synthesis of polyadenylated RNA (polyA RNA) which represents most of the messenger RNA (mRNA) was less sensitive than that of total RNA, the majority of which is ribosomal RNA (rRNA). Fractionation of *in vivo* labelled total RNA by agarose gel electrophoresis showed that, in the presence of metalaxyl, neither high molecular weight rRNA species is labelled. These results indicate that metalaxyl selectively inhibits rRNA synthesis by interference with the activity of the RNA polymerase I-template complex. In resistant strains the target site has been changed resulting in insensitivity.

Characterization of the phenylamide target site

To substantiate the idea that a target site change is responsible for phenylamide resistance, binding studies were performed with [^{3}H]metalaxyl and cell-free mycelial extracts of sensitive and resistant strains. Binding was determined with a Sephadex G25 gel filtration assay. Specific binding of metalaxyl occurred only in extracts of sensitive strains (Table 23.3; Davidse, 1988, 1990). Other phenylamides competitively inhibited [^{3}H]metalaxyl binding (Table 23.4), indicating that the mechanism of action of the different phenylamides is similar.

Although binding activity could easily be demonstrated in crude cell-free mycelial extracts of phenylamide-sensitive *Phytophthora* strains,

Table 23.4. Inhibition of [^{3}H]metalaxyl binding[a] in a cell-free mycelial extract[b] by phenylamides.

Compound	Binding as % of control[c] in the presence of the test compound at the indicated concentration (μg ml^{-1})			
	0·01	0·1	1	10
metalaxyl	94	53	13	3
cyprofuram	110	109	78	24
benalaxyl	94	49	18	4
oxadixyl	93	86	51	14

[a] determined at a [^{3}H]metalaxyl (specific activity 551 mCi mmol^{-1}) concentration of 0·50 μM.

[b] S-35 fraction of *P. megasperma* f. sp. *medicaginis* S prepared according to Swanson & Holland (1983) containing 17·5 mg protein ml^{-1}.

[c] binding activity 2·1 pmole [^{3}H]metalaxyl mg protein^{-1}.

---

binding activity decreased considerably when the extracts were subjected to Heparin-Sepharose affinity chromatography, a first step in the purification of RNA polymerases (Hammond & Holland, 1983). The binding assay, therefore, cannot be used to follow purification of the phenylamide receptor. An alternative approach would be to label the receptor covalently with a photoaffinity probe. Purification of the labelled complex can then be followed either by a radio-assay or an immunoassay, which respectively require a radiolabelled photoaffinity probe or antiserum that selectively recognizes covalently bound phenylamide residues.

A suitable photoaffinity probe would be 3-azidometalaxyl (Figure 23.1), since substituents like amino, nitro and chloro at the 3-and 5-position do not cause a drastic change in biological activity of the compound (Hubele *et al.*, 1983). This compound, therefore, was synthesized and its biological properties were determined. Azidometalaxyl inhibited mycelial growth of the phenylamide-sensitive strain of *P. megasperma* f. sp. *medicaginis* (EC_{50} = 4 μg ml^{-1}), whereas the compound was almost inactive against growth of the phenylamide-resistant strain. Like metalaxyl, azidometalaxyl inhibited endogenous RNA polymerase activity in isolated nuclei, although at least a ten-fold higher concentration was required to achieve a similar level of inhibition (Table 23.5). Azidometalaxyl bound to the phenylamide receptor, as indicated by its ability to compete with [^{3}H]metalaxyl (Table 23.6). Hence azidometalaxyl is a suitable photoaffinity probe for the receptor. Photoaffinity labelling experiments up till

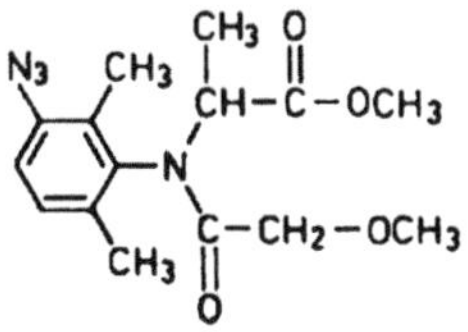

Fig. 23.1. Structure of 3-azidometalaxyl.

Table 23.5. Effect of metalaxyl and azidometalaxyl on endogenous RNA polymerase[a] activity of nuclei isolated from a phenylamide-sensitive strain of *P. megasperma* f. sp. *medicaginis*.

Inhibitor	Polymerase activity as % of control in the presence of the inhibitor at a concentration (μg ml^{-1}) of:		
	0·1	1·0	10·0
metalaxyl	63	53	52
azidometalaxyl	92	67	57

[a] determined according to Davidse *et al.*, 1983.

Table 23.6. Inhibition of [^{3}H]metalaxyl binding[a] in a cell-free mycelial extract[b] by azidometalaxyl.

Binding as % of control[c] in the presence of the test compound at the indicated concentration (μg ml^{-1})			
0·01	0·1	1	10
88	86	55	24

[a] determined at a [^{3}H]metalaxyl (specific activity 551 mCi mmol^{-1}) concentration of 0·59 μM.

[b] S-100 fraction of *P. megasperma* f. sp. *medicaginis* S prepared according to Swanson & Holland (1983) containing 26·3 mg protein ml^{-1}.

[c] binding activity 2·8 pmole [^{3}H]metalaxyl mg protein^{-1}.

now have failed to demonstrate specifically labelled protein bands in a Western blot using rabbit antibodies raised against a formaldehyde-conjugated 3-aminometalaxyl-bovine serum albumin complex and goat antirabbit-IgG conjugated alkaline phosphatase as second antibody to detect phenylamide-bound residues. Identical labelling patterns were observed using mycelium of a sensitive and a resistant strain of *P. megasperma* f. sp. *medicaginis* as sources of the crude protein extract which indicatedn onspecific labelling. Preincubation of the extracts with or without metalaxyl prior to photolabelling with azidometalaxyl and including dithiothreitol as a suppressor of nonspecific labelling in the extracts,

Table 23.7. Availability of phenylamides and phenylamide mixtures during 1979-1989 in the Netherlands and instructions for use.

Year	Products available	Instructions for use
1979	Ridomil 25 WP[a]	biweekly
1980	Ridomil 25 WP	biweekly
	Ridomil Speciaal 58 WP[b]	biweekly
1981	none	
1982	none	
1983	none	
1984	Ridomil Speciaal 58 WP[c]	for use in tankmix with full rate of preventive compounds, maximum of two applications
1985	Ridomil Delta 47 WP[d]	maximum of two applications in critical situations only, no use on seed potatoes
1986	same	same
1987	same	same
1988	same	same
1989	same	same, except for N.E. Netherlands

[a] recommended rate: 1 kg ha^{-1} = 250 g metalaxyl ha^{-1}.

[b] recommended rate: 2 kg ha^{-1} = 200 g metalaxyl + 960 g mancozeb ha^{-1}.

[c] from August 18 onwards.

[d] recommended rate: 2.5 kg ha^{-1} = 250 g metalaxyl + 690 g maneb + 230 g fentin acetate ha^{-1}.

also did not result in differentially labelled protein bands. Fractionation of labelled extracts that were preincubated with or without metalaxyl on Heparin-Sepharose to enrich the nucleic acid bound RNA polymerases also failed to result in a differential labelling pattern of the enriched fraction.

Managing phenylamide resistance in late blight in the Netherlands

The population dynamics of phenylamide-resistant strains of *P. infestans* in the Netherlands have been intensively studied since 1980 when resistance first appeared after almost exclusive use of metalaxyl (Davidse, 1985; Davidse, Danial & Van Westen, 1983; Davidse, Hellinga & Straathof, 1987; Davidse *et al.*, 1981, 1989). After metalaxyl was unavailable in the period 1981–1983, it was reintroduced in August 1984 in a mixture with

Table 23.8. Phenylamide resistance in samples[a] of *P. infestans* collected during the potato growing season in the Netherlands during 1986-1989.

Year	Number of sites sampled	% sites treated with Ridomil Delta 47 WP	% sites with resistant strains
1986	61	2	0
1987	148	49	16
1988	208	48	43
1989	123	7	46

[a] Samples were obtained from blighted leaflets or tubers from fields, allotment gardens, volunteer plants and dump sites. Usually one bulk isolate obtained from up to 6 blight lesions was tested in the leaf disc assay (Davidse *et al.*, 1981).

protectant fungicides for limited use on ware and starch potato crops. A maximum of two applications per season and only in critical situations was recommended (Table 23.7). Monitoring results of the population from 1986 to 1989 are presented in Table 23.8, together with data on the use of metalaxyl at the sites sampled.

Although in 1986 the late blight fungus was found on May 22 at a dump site, the disease did not become widespread until the end of the growing season because of the relatively dry summer. In total, 61 samples were analyzed, none of which contained resistant strains. The use of metalaxyl mixtures in 1986 was very limited. Potato growers relied on preventive fungicides for late blight control.

In 1987, the weather throughout the summer was very conducive to the development of late blight. The disease was already widespread by mid-June and remained present during the whole growing season. The severity of the epidemic is best reflected by the fact that warnings for infection periods were broadcast 22 times by the Advisory Service after consulting the KNMI (Royal Netherlands Meterological Institute). Spraying, however, was not always possible due to heavy rainfall. Ridomil Delta 47 WP was intensively used, mostly up to the recommended two times, in alternation with preventive products. Resistant strains were found throughout the summer at a low frequency, and increased in frequency towards the end of the growing season.

In 1988, the extremely favourable weather conditions for blight development in July initiated a severe epidemic. Resistant strains were found at 43% of the sites sampled (Table 23.8). Application of Ridomil Delta resulted in an increase in resistance. Sixty-eight percent of the treated

Table 23.9. Effect of Ridomil Delta 47 WP application on the frequency of resistant strains.

Year	Number and spray history of sites[a] and % sites with resistant strains							
	+ Ridomil Delta		− Ridomil Delta		Untreated		Unknown	
	no.	%	no.	%	no.	%	no.	%
1987	75	15	28	11	14	7	31	19
1988	98	68	53	15	22	23	32	28
1989	8	88	51	45	47	45	17	29

[a] including fields, allotment gardens, volunteer plants and dump sites.

fields yielded resistant isolates, whereas only 15% of fields treated with preventive compounds yielded resistant isolates (Table 23.9).

In 1989, the disease was not severe in most regions, but due to an intensive monitoring campaign a sufficient number of isolates was obtained to evaluate the situation. Use of Ridomil Delta was very limited (Table 23.8). The frequency of resistance did not decline and even in fields not treated with Ridomil Delta resistance could be detected easily (Table 23.9).

In the years before 1988, most of the fields where resistance was detected were located in a region in the north-eastern part of the country (Table 23.10; Davidse *et al.* 1989). In this region potatoes are predominantly grown for the starch industry. Metalaxyl mixtures have been used in this region almost continuously since 1979 and it is a public secret that these mixtures were sprayed more than twice per season at a rate half of that recommended ($2 \cdot 5$ kg ha^{-1}) in a tankmix with preventive compounds on both the ware and seed potato crop, which are usually grown together on the same farm. This practice is different from that in other regions where growers of table potatoes and potatoes for processing buy most of their seed potatoes from the specialized seed potato growers via the trading houses. Since Ridomil Delta is not registered for use on seed potatoes in the Netherlands, the blight population in the table and processing crop is less exposed to the selection pressure of metalaxyl than the population in the starch potato crop. This explains the regional difference in the occurrence of resistance up toand including 1987. In 1988, however, resistant strains were also frequently found in other regions (Table 23.10). For example, in three out of eight other regions more than 50% of the fields sampled yielded resistant strains. Resistance declined in these regions in 1989, but in N.E. Netherlands resistance increased further, 78% of the samples containing resistant strains. Intensive sampling of 11 fields

Table 23.10. Location of sites that yielded resistant isolates

Year	Region	Total number of sites[a] sampled	Sites with resistant strains number	% of total
1987	N.E. Netherlands	46	14	30
	Other regions	91	8	9
1988	N.E. Netherlands	67	33	49
	Other regions	137	56	41
1989	N.E. Netherlands	46	36	78
	Other regions	79	21	27

[a] including fields, allotment gardens, volunteer plants and dump sites.

in this region yielded 125 single lesion isolates, 111 of which (89%) were resistant.

A comparison of the frequency of metalaxyl-resistant strains in blighted seed tubers harvested in 1988 in various regions (Table 23.11) with the resistance level for N.E. Netherlands and other regions (Table 23.10) lends support to the idea that the higher level of resistance in N.E. Netherlands is caused by the higher frequency of resistance in seed-borne inoculum. Seventy six percent of the samples from seed lots harvested in this area in 1988 contained resistant strains and 78% of the field samples collected in 1989 contained resistant strains. The corresponding figures for other regions were respectively 31 and 27% (Table 23.10 & 23.11).

Thus, the advice given early in 1989 to the growers of starch potatoes not to use Ridomil Delta (see Table 23.7) proved correct. The advice had been based on the high frequency of resistance in samples of seed lots to be used to plant the 1989 crop in this region and on the assumption that in this area the initial inoculum originates from seed tubers. For the table potato crop and crops for processing in other regions, it had been assumed that if the 1989 epidemic started from seed borne inoculum the frequency of resistance would be of the order of 10-20% (Table 23.11). In this situation, metalaxyl would be expected to make a fair contribution to the performance of Ridomil Delta. Hence the recommendation for the use of Ridomil Delta on these crops was maintained at a maximum of two applications. However, since applications of Ridomil Delta proved to be unnecessary in 1989 it may be expected that the level of resistance did not increase further. Hence the use of Ridomil Delta in the production area of table potatoes and potatoes for processing can still be advised in the Netherlands in 1990.

Table 23.11. Phenylamide resistance in samples of *P. infestans* collected from seed potato lots harvested in 1988 in the starch potato production area (N.E. Netherlands) and other regions.

Region	Lots sampled	% lots with resistance	Isolates	% resistant isolates
N.E. Netherlands	25[a]	76	133	41[c]
Other regions	35[b]	31	137	12

[a] seed lots for starch potato crops (n = 16) and seed lots for table and processing potato crops (n = 9). [b] seed lots for table and processing potato crops. [c]value differs significantly from that for other regions ($P < 0{\cdot}05$; χ^2-test).

Concluding remarks

Identification of the phenylamide receptor in the RNA polymerase I-template complex proved difficult. Photoaffinity labelling, a traditional biochemical approach to characterize receptors, particularly those that are labile during purification, has not yet been successful. A molecular biological approach by cloning the phenylamide resistance gene cannot be followed yet, because a transformation system for *Phytophthora* species is still lacking (Chapters 19, 20) and the gene cannot be selected in established transformation systems because the higher fungi used in these systems are naturally insensitive to phenylamides

The extensive monitoring programme on phenylamide resistance in *P. infestans* carried out in the Netherlands during the period 1980–89 produced a wealth of information on the complex dynamics of resistance and the response of potato growers to the recommended resistance management strategy.

Growers generally followed the strategy in which Ridomil Delta is applied only twice in consecutive treatments and in critical situations only. In addition, both the price of the product and the quantity of the product made available by the supplier contributed to its limited use. An inherent risk of this strategy is the application of the product at too late a stage. In that case a phenylamide-sensitive population may not be reduced sufficiently in size and may cause damage because it is not adequately controlled by the preventive compounds in the mixture, and/or *de novo* genesis of resistant strains may occur because the size of the population has exceeded a critical level.

The strategy, on the other hand, prevents the use of the metalaxyl mixture in cases where preventive products are expected to give adequate disease control, thereby reducing the selection pressure. In view of the history of phenylamide resistance in the Netherlands, the strategy of limited use in critical situations only was chosen. It is obvious that intensive monitoring should be an integrated part of this strategy to avoid application at too late a stage. A further requirement is that growers should be well educated.

Convincing seed potato growers to refrain from using Ridomil Delta as another part of the strategy to manage resistance, is more difficult. Although growers understand the rationale, they nevertheless use the product in case of emergency, knowing that it is unlikely that the offence will be discovered. In order to be fully effective, separate production of seed in regions where no table or processing crops are grown, is also necessary. In the Netherlands this situation more or less occurs in the northern part of the province of Friesland where seed potatoes account for 73% of the total potato acreage. Nevertheless, the ban on the use of Ridomil Delta in seed potatoes has contributed to keeping the selection pressure low.

In retrospect it can be stated that the combined efforts of the supplier of metalaxyl, the research team at the university, and the Advisory Service to develop a strategy to combat phenylamide resistance in *P. infestans* has led to an optimal usage of metalaxyl in late blight control in the Netherlands. Ten years after resistance developed, metalaxyl mixtures can still be advised for use against late blight up to a maximum of two applications in table potatoes and potatoes for processing. Time will tell whether this will continue to be possible. If the frequency of phenylamide resistance increases further, the ability of the fungus to generate highly pathogenic resistant mutants apparently cannot be counteracted by any strategy, and only total withdrawal of the product can be recommended.

Acknowledgements. This research was financially supported by Ligtermoet Chemie BV, Roosendaal and Ciba Geigy Ltd, Basel, Switzerland. Thanks are due to advisors and regional specialists of the Advisory Service for Crop Protection, inspectors of the NAK, technical and sales staff of the agrochemical industries and Cebeco-Handelsraad, contractors and potato growers, who collected late blight samples and to W. E. Fry for comments on the manuscript.

References

Davidse, L. C. (1985). Resistance to acylalanines in *Phytophthora infestans* in the Netherlands. *EPPO Bulletin* **15**, 403-409.

Davidse, L. C. (1988). Phenylamide fungicides: Mechanism of action and resistance. In *Fungicide Resistance in North America*, (ed. C. J. Delp), pp. 63-65. American Phytopathological Society: St. Paul, Minnesota.

Davidse, L. C. (1990). Biochemical basis of resistance to phenylamide fungicides. In *Managing Resistance to Agrochemicals*, (ed. M. B. Green, H. M. LeBaron & W. K. Moberg), pp. 215-223. American Chemical Society: Washington, USA.

Davidse, L. C., Danial, D. L. & Van Westen, C. J. (1983). Resistance to metalaxyl in *Phytophthora infestans* in the Netherlands. *Netherlands Journal of Plant Pathology* **89**, 1-20.

Davidse L. C., Gerritsma, O. C. M., Ideler, J., Pie, K. & Velthuis, G. C. M. (1988). Antifungal modes of action of metalaxyl, cyprofuram, benalaxyl and oxadixyl in phenylamide-sensitive and phenylamide-resistant strains of *Phytophthora megasperma* f. sp. *medicaginis* and *Phytophthora infestans*. *Crop Protection* **7**, 347-355.

Davidse, L. C., Hellinga, J. H. & Straathof, H. J. M. (1987). Resistance to phenylamides in *Phytophthora infestans* in the Netherlands; can the problem be solved? In *Systemic Fungicides and Antifungal Compounds*, (ed. H. Lyr & C. Polter), pp. 185-190. Tagungs Bericht 253. Akademie der Landwirtschaftswissenschaften der DDR: Berlin, GDR.

Davidse, L. C., Henken, J., Van Dalen, A., Jespers, A. B. K. & Mantel, B. C. (1989). Nine years of practical experience with phenylamide resistance in *Phytophthora infestans* in the Netherlands. *Netherlands Journal of Plant Pathology* **75** (Supplement 1), 197-213.

Davidse L. C., Hofman A. E. & Velthuis G. C. M. (1983). Specific interference of metalaxyl with endogenous RNA polymerase activity in isolated nuclei from *Phytophthora megasperma* f. sp. *medicaginis*. *Experimental Mycology* **7**, 344-361.

Davidse, L. C., Looijen, D., Turkensteen, L. J. & Van der Wal, D. (1981) Occurrence of metalaxyl-resistant strains of *Phytophthora infestans* in the Netherlands. *Netherlands Journal of Plant Pathology* **87**, 65-68.

Hammond, C. I. & Holland M. J. (1983) Purification of yeast RNA polymerases using heparin agarose affinity chromatography. *Journal of Biological Chemistry* **258**, 3230-3241.

Hubele, A., Kunz, W., Eckhardt, W. & Sturm, E. (1983). The fungicidal activity of acylalanines. In *Pesticide Chemistry, Human Welfare and the Environment*, Vol 1: *Synthesis and Structure-Activity Relationships*, (ed. P. Doyle & T. Fujita), pp. 233-242. Pergamon Press: Oxford, UK.

Schwinn, F. J. & Staub, T. (1987). Phenylamides and other fungicides against Oomycetes. In *Modern Selective Fungicides*, (ed. H. Lyr), pp. 259-273. VEB Gustav Fischer Verlag and Longman Group UK, Ltd: Jena, GDR and London.

Swanson, M. E. & Holland, M. J. (1983) RNA polymerase I-dependent selective transcription of yeast ribosomal DNA. *Journal of Biological Chemistry* **258**, 3242-3250.

Urech, P. A. (1988). Phenylamide resistance management strategies. In *Fungicide Resistance in North America*, (ed. C. J. Delp), pp. 74-75. American Phytopathological Society: St. Paul, Minnesota.

Chapter 24

Synergism between fungicides for control of *Phytophthora*

U. Gisi

Fungicides are combined in mixtures for three main reasons: (1) to widen the spectrum of antifungal activity; (2) to exploit synergistic interactions between the fungicides by which the overall activity can be increased or the amounts used reduced without loss of activity; (3) to delay the selection of resistant strains. In the past, more emphasis was given to spectrum enlargement; only recently, synergistic interactions and their impact on resistance risk have been studied in more detail (Staub & Sozzi, 1983; Gisi, Binder & Rimbach, 1985; Grabski & Gisi, 1987; Samoucha & Gisi, 1987a). At the same time, mathematical models became available describing the dynamics of pathogen populations under the selection pressure of fungicide mixtures (Levy, Levi & Cohen, 1983; Levy & Strizyk, 1985; Josepovits & Dobrovolszky, 1985). Numerous examples of synergism among fungicides for *Phytophthora* control under laboratory and field conditions have been reported for mixtures between contact fungicides and cymoxanil or phenylamides (Gisi *et al.*, 1985; Samoucha & Cohen, 1986, 1988; Samoucha, Ben-Hador & Cohen, 1989). Nevertheless, very little information is available about the mechanism(s) of synergism, although several hypotheses have been postulated partly based on data from other types of fungicides (De Waard, 1987). In this paper, a review is presented of test methods, greenhouse and field results, possible mechanisms and significance of synergistic interactions of fungicide mixtures for *Phytophthora* control and their impact on the build-up of resistance.

Definitions and test methods

Fungicide mixtures which produce activity that corresponds to the sum of the effects of the components alone are said to show additive interactions. Synergism is defined as simultaneous action in which the total response of an organism to the mixture is greater than the sum of the responses to the individual components. There are several methods for estimation of the extent of interactions of fungicides in mixtures, for example those of Abbott and of Wadley. The expected efficacy of a mixture, expressed as percent control ($\% \ C_{exp}$) can be predicted by the Abbott formula (Levy *et al.*, 1986):

$$\% C_{exp} = A + B - (AB/100)$$

in which A and B are the control levels given by the single fungicides. For calculating synergism, the ratio (SF, synergy factor) between the observed experimental efficacy of the mixture (C_{obs}) and the expected efficacy of the mixture (C_{exp}) is computed:

$$SF = C_{obs}/C_{exp}$$

If SF is 1, interactions are additive, if SF is higher than 1, they are synergistic. The Abbott approach is very simple and interactions can be studied with as few as three test elements, i.e. A alone, B alone and A + B. However, the accuracy of this test method is doubtful at levels of control by the single components higher than about 70%, because SF values rapidly decrease towards 1·0 (Levy *et al.*, 1986). If A and B alone provide 70% control each, the expected control level of the mixture A + B is 91% and synergistic interactions are only present if the observed control level is higher than 91%. A more reliable but also more labour-intensive method is the approach of Wadley (Gisi *et al.*, 1985, Levy *et al.*, 1986) in which dose-response curves of A alone, B alone and A + B are constructed. With a logit-log transformation, the dose-response curves are converted to linear regressions which can be used for calculating EC-values ('equally effective concentrations') for different levels of control, e.g. EC_{50}, EC_{90}. When a and b are the absolute amounts of the components in the mixture, the expected effective concentration (EC_{exp}) providing a given control level can be calculated. For example, for 90% control:

$$EC_{90exp} = (a + b)/(a/EC_{90A} + b/EC_{90B})$$

By dividing the expected and the observed experimental EC_{90}, the interactions between the two fungicides A and B in the mixture can be calculated; if this value is 1, interactions are additive, if it is higher than 1, interactions are synergistic. Unlike the Abbott method, the Wadley approach allows estimation of the interactions at any fungicide concentration. In addition, the Wadley approach can be used to define the optimum ratio of the components in the mixture to achieve the highest control levels with the minimum amounts of the components (Fig. 24.1); this ratio represents maximum synergy levels (Levy *et al.*, 1986).

Greenhouse results

Dose-response curves were established for the phenylamide oxadixyl (A), the dithiocarbamate mancozeb (B) and for the cyanoacetamideoxime cymoxanil (C) using *Phytophthora infestans* on potatoes or tomatoes. Disease control achieved by the single components and by the different mixtures was expressed as EC_{90}, which was used to construct an isobologram for A + B mixtures with ratios between 1 + 1 and 1 + 32 (Fig. 24.1). Maximum levels of synergy were found at ratios of A + B = 1 + 7.

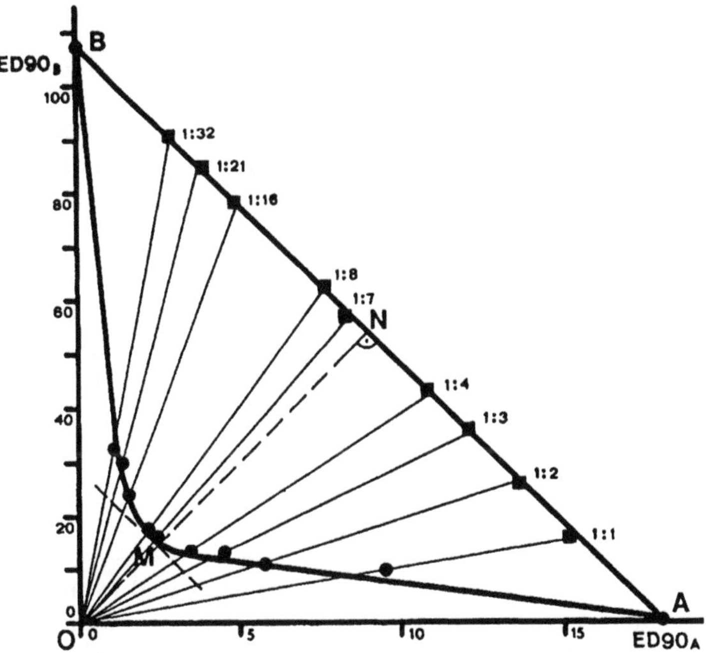

Fig. 24.1. Isobologram for the two fungicides A (oxadixyl) and B (mancozeb) active against *Phytophthora infestans* on potato plants, giving a theoretical (ANB) and an experimental (AMB) isobole for A+B mixtures with ratios between 1+1 and 1+32. The level of synergy can be measured by the ratio R = ON/OM which is maximal for the mixture A+B = 1+7 (Levy *et al.*, 1986).

Similarly, mixtures of B + C, A + C, and A + B + C were tested and maximum synergy levels evaluated. Fungicide concentrations could be reduced significantly for all components in the mixtures and for both phenylamide-sensitive and -resistant strains (Table 24.1). The efficacy of the three-way mixture A + B + C = 1 + 7 + 0·4 was little influenced by phenylamide sensitivity of pathogen strains, whereas A + B was much less effective against resistant than against sensitive strains. All mixtures showed synergistic interactions of varying degrees depending on the components and the strains (Table 24.1). When sensitive and resistant strains were compared, synergy levels increased for all cymoxanil-containing mixtures (B + C, A + C, A + B + C), but decreased for A + B. Mathematical models proposed by Levy & Strizyk (1985) suggest that selection for resistance is very slow if synergistic interactions in a mixture

Table 24.1. Fungicidal activities (EC_{90} mg l^{-1}) of oxadixyl (A), mancozeb (B), cymoxanil (C) alone and in mixture and synergy ratio of mixtures against sensitive and phenylamide-resistant strains of *Phytophthora infestans* on tomato (Grabski & Gisi, 1987).

	EC_{90} (mg l^{-1})		Synergy ratio	
Fungicides	sensitive	resistant	sens.	res.
A	34	>5000	–	–
B	776	495	–	–
C	48	23	–	–
A:B = 1:7	69 (9+60)*	406 (51+355)	3·0	1·4
B:C = 7:0·4	134 (127+7)	51 (48+3)	3·2	4·6
A:C = 1:0·4	22 (16+6)	41 (29+12)	1·7	2·0
A:B:C = 1:7:0·4	74 (9+62+4)	63 (8+53+3)	2·4	4·2

*Figures in parentheses are concentrations of the individual components in the mixture.

are equally high or higher for the resistant compared to the sensitive subpopulation (Fig. 24.2). Therefore, two-way mixtures of phenylamides plus contact fungicides may not delay the build-up of resistant strains sufficiently. This assumption was confirmed by greenhouse and field experiments (Staub & Sozzi, 1983; Cohen & Samoucha, 1990). Phenylamides combined with other contact fungicides of the dithiocarbamate or phthalimide groups, with copper- or fentin-based fungicides, with chlorothalonil, but also with the systemics fosetyl-Al, propamocarb or dimethomorph showed synergistic interactions (Gisi *et al*., 1985). The phenomenon of synergism is not restricted to phenylamide mixtures; there are significant synergistic interactions between mancozeb and cymoxanil as well as mancozeb and fosetyl-Al (Gisi *et al*., 1985).

Field results

Under field conditions, the construction of dose-response curves is very labour-intensive and requires specially designed field trials. In addition, several fungicide applications are normally necessary to control an ongoing epidemic in the field. If interactions between fungicides are to be investigated under field conditions, it is essential to apply the components alone and in the mixture at identical rates and intervals. As an example of many field trials performed in different countries over several years (Sandoz trials), Table 24.2 illustrates interactions between oxadixyl, mancozeb and cymoxanil which gave synergy ratios between 1·7 and 5·3; cymoxanil-containing mixtures had higher synergism for resistant than for sensitive strains. Dosages yielding 90% disease control were reduced

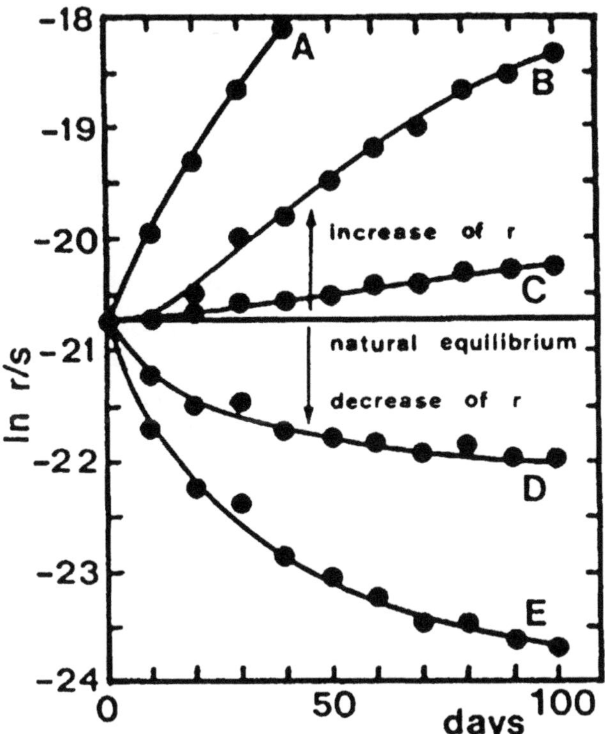

Fig. 24.2. Change in the proportion of resistant (r) and sensitive (s) strains when treated with a mixture of fungicides showing different levels of synergy. Quotient of synergy ratio for r and s: <0·3 (A), 1 (B), 3 (C), 6 (D), 9 (E). Redrawn from Levy & Strizyk (1985).

---

significantly in the mixtures. The three-way mixtures A + B + C were not much affected by different sensitivities of the strains, whereas the two-way mixture A + B was clearly less effective against resistant than sensitive strains. At recommended rates and spray intervals, the fungicides allowed the following levels of attack (% foliage blighted) in an Israeli field trial containing a fully resistant population: control plot, 95%; A + B, 58% (14 day interval); C 46 % and B 36 % (both at 7 day interval); B + C, 34 % (14 day interval) and 7% (7 day interval); A + B + C, 10 % (14 day interval)(Samoucha *et al.*, 1989).

Table 24.2. Fungicidal activities (EC_{90} g ha^{-1}) of oxadixyl (A), mancozeb (B), cymoxanil (C) alone and in mixture and synergy ratio of mixtures against sensitive and phenylamide-resistant strains of *Phytophthora infestans* on potato under field conditions (Samoucha & Cohen, 1989).

	EC_{90} (g ha^{-1})		Synergy ratio	
Fungicides	sensitive	resistant	sens.	res.
A	60	>2000	–	–
B	760	1130	–	–
C	350	310	–	–
A:B = 1:7	100 (12+88)*	550 (69+481)	3·2	2·3
B:C = 7:1·75	220 (177+43)	260 (209+51)	2·8	2·9
A:B:C = 1:7:0·4	180 (21+150+9)	260 (31+217+12)	1·7	4·2
A:B:C = 1:7:2	110 (11+77+22)	150 (15+105+30)	2·8	5·3

*Figures in parentheses are dosages of the individual components in the mixture. Recommended rates for *P. infestans* control in many countries are A:B:C = 200 + 1400 + 80 g ha^{-1}.

Table 24.3. Duration of activity (% control) of mixtures of oxadixyl (A), mancozeb (B), cymoxanil (C) against phenylamide-resistant *Phytophthora infestans* on tomato under greenhouse conditions (Samoucha & Gisi, 1987b).

	Disease control (%) days after application					
Fungicides*	1	4	6	8	9	14
A:B = 1:7	100	66	41	23	9	nt
B:C = 7:0·4	100	100	100	92	17	nt
A:C = 1:0·4	100	100	99	94	25	nt
A:B:C = 1:7:0·4	100	100	100	100	100	97

*A: 160 mg l^{-1}; B: 1120 mg l^{-1}; C: 64 mg l^{-1}.; nt = not tested.

Table 24.4. Duration of activity (days) for 50% and 90% control of phenylamide-resistant *Phytophthora infestans* on field grown potato treated with oxadixyl (A), mancozeb (B), cymoxanil (C) alone and in mixtures (Samoucha, Levy & Cohen, 1988).

Fungicides	Duration (days) of	
	>50% control	>90% control
B	9	5
C	7	4
A:B = 1:7	10	5
B:C = 7:1·75	14	8
A:B:C = 1:7:1·75	16	11

A (oxadixyl) was not tested alone. Fungicides were used at the recommended rates.

Significance of synergism

Mixtures of fungicides with synergistic interactions have two main advantages: (i) increase of the overall fungicidal activity; and (ii) activity against sensitive and resistant strains.

Overall activity

The recommended field rate of a product is normally not adapted to each disease situation. Thus, the rate has to be high enough to control disease when fungal attack is severe. A mixture with synergistic interactions between the components gives more reliable disease control in most cases. Reduced rates of the components can be used in the mixture and thus less chemicals are consumed over the whole season. Synergistic mixtures offer a longer duration of activity as compared to single components. Under greenhouse conditions and against phenylamide-resistant strains of *P. infestans*, A + B lost its efficacy rapidly, whereas the mixtures A + C and B + C gave about 90% control for 8 days and A + B + C for more than 14 days (Table 24.3). Under field conditions, B + C and A + B + C provided 90% control for 8 and 11 days, respectively, whereas that of the single components lasted only for about 4 to 5 days (Table 24.4).

Resistance

There are two aspects of synergistic interactions of fungicides in mixtures in regard to resistance:

- control of populations containing resistant strains
- decrease of selection pressure.

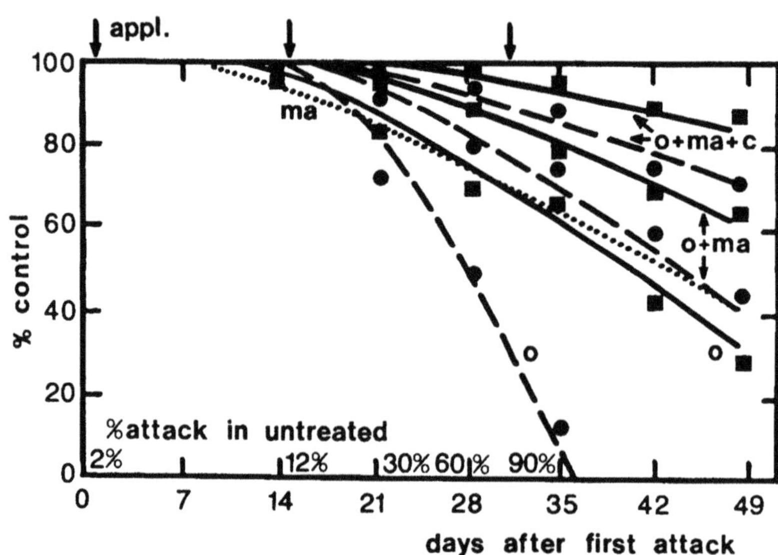

Fig. 24.3. Efficacy of oxadixyl and various mixtures against *Phytophthora infestans* populations containing initially 10% (solid lines) and 100% (dashed lines) resistant sporangia, respectively, on potatoes treated three times (arrows) at 14 day intervals (mancozeb 7 days, dotted line) in the field (Sandoz trials). o = oxadixyl, ma = mancozeb, c = cymoxanil.

Under field conditions, phenylamides alone give limited control of resistant populations; the control is significantly enhanced in synergistic mixtures (Fig. 24.3). The higher the proportion of the phenylamide-resistant subpopulation, the higher the contribution of the non-phenylamide components in the mixture. More importantly, the use of mixtures of components with synergistic interactions decreases the selection pressure for resistance. For full control of sensitive *Phytophthora* populations, about 250 to 300 g ha^{-1} oxadixyl are needed, whereas in the mixture with mancozeb, 200 g oxadixyl give even better levels of disease control (Sandoz field trials). This reduction of the 'at risk' fungicide also reduces the rate of build-up of resistant strains. Staub & Sozzi (1983) using growth chamber experiments, showed that a mixture of metalaxyl + mancozeb delayed the build-up of resistance compared to metalaxyl alone. Similarly, the addition of cymoxanil to oxadixyl and mancozeb reduced resistance build-up much more than the two-way mixture oxadixyl + mancozeb and

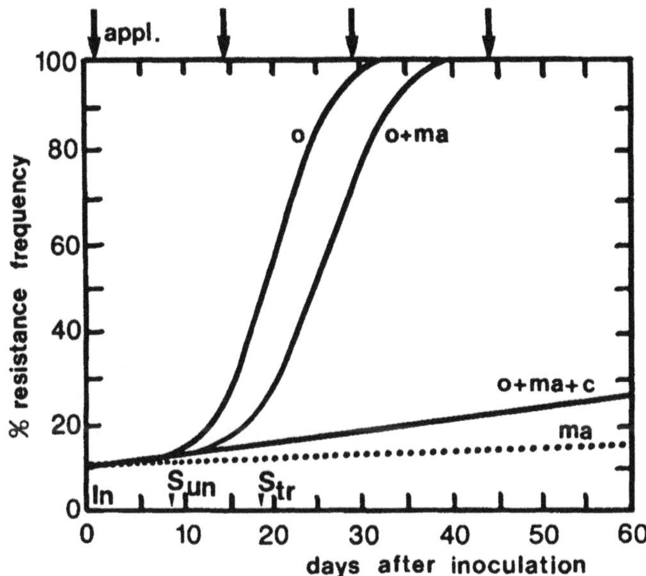

Fig. 24.4. Percent resistance frequency of a *Phytophthora infestans* population initially containing 10% resistant sporangia developing on field grown potatoes treated four times (arrows) with oxadixyl (o), mancozeb (ma), oxadixyl + mancozeb (o + ma), or oxadixyl + mancozeb + cymoxanil (o + ma + c). In = inoculation; S_{un} = first symptoms on untreated plants; S_{tr} = first symptoms on treated plants (redrawn from Cohen & Samoucha, 1990).

the increase of the resistant subpopulation was almost nil (Fig. 24.4). These results confirm the predictions of Levy & Strizyk (1985, shown in Fig. 24.2) that fungicide mixtures with synergistic interactions that are higher against resistant than sensitive strains are very effective in delaying the build-up of resistance. The synergistic interactions in the three-way mixture oxadixyl (A) + mancozeb (B) + cymoxanil (C) may be attributed to all four possible combinations, A + B, B + C, A + C, and A + B + C; all contribute to the overall synergistic effect. Comparative experiments indicated that oxadixyl produced higher synergy levels than other phenylamides when mixed with other fungicides (Gisi *et al.*, 1985).

Mechanisms of synergism

Very little is known about the physiological and biochemical mechanisms of synergism between phenylamides and other fungicides. Nevertheless, there are several possible hypotheses:

- increased uptake and binding of fungicides to the site of action may lead to higher concentrations in the target cell;
- decreased biodegradation of fungicides in fungal and plant cells may prolong activity;
- different sites of action of fungicides both in the fungal cell (biochemical mechanism) and in the life cycle of the fungus (pathogenic mechanism) may decrease the aggressiveness of the fungus.

Increased uptake and binding

In binding experiments, Davidse (1988; Chapter 23) showed that phenylamide resistance in *Phytophthora megasperma* f. sp. *medicaginis* and *P. infestans* was due to a change at the target site (RNA polymerase). Recently, Bashan, Levy & Cohen (1989) demonstrated decreased uptake of phenylamides into a resistant isolate of *P. infestans*. As in other classes of fungicides, a mixture of active ingredients with different sites of action may:

- stimulate the passive uptake of single components;
- inhibit the energy-dependent efflux of single components that consequently accumulate within the cells;
- affect membrane functioning or uridine uptake (De Waard & Van Nistelrooy, 1982; Davidse, 1984).

Decreased biodegradation

When grape leaves were treated with a mixture of the phenylamide fungicide benalaxyl and mancozeb, the amount of benalaxyl inside the leaves was higher than when benalaxyl was applied alone (Gozzo, Pizzingrilli & Valcamonica, 1988). In addition, benalaxyl was degraded in the leaves less rapidly in the presence of mancozeb. Although the nature of the metabolites was unaffected,their amounts in the plant were markedly reduced. Since degradation pathways in plants and fungi may be similar, increased levels and a longer duration of fungicidal activity not only of phenylamides, but also of cymoxanil may occur if they are applied together with dithiocarbamates or other fungicides inhibiting oxidases. Many field results indicate longer duration of activity of mixtures between phenylamides or cymoxanil and contact fungicides.

Decreased aggressiveness

Different types of fungicides may affect different sites in the metabolic pathway and also different sites in the life cycle of the target fungi. Most contact fungicides (e.g. mancozeb) remain, after application, on the plant surface or in the cuticle; systemic fungicides penetrate into the tissue and are translocated according to their physico-chemical properties. During the infection process, the fungus may be faced with several consecutive sites of inhibition if the fungicides are applied together, e.g. spore germination and initial penetration by mancozeb, formation of haustoria by phenylamides, hyphal growth and sporulation by cymoxanil. The synergistic interactions may be simply a summation of these sites. Alternatively, the fungus may be weakened by the contact fungicide at the germination stage, decreasing its aggressiveness and reducing the likelihood of successful infection; the systemic fungicide can then inhibit the fungus at concentrations which are normally sublethal (Samoucha & Cohen, 1986). The combination of the two effects may result in synergistic interactions. Preliminary experiments showed that respiration inhibitors produced inhibitory effects similar to those of mancozeb on the infection of potato tuber discs by sporangia of *P. infestans*, supporting this hypothesis (Bashan, Levy & Cohen, 1988).

Conclusion

Synergistic interactions between fungicides can be measured *in vitro* as well as under greenhouse and field conditions using either the Abbott or the Wadley approach. Synergistic mixtures allow a decrease of the amount of active ingredients without reducing the overall activity. The duration of activity is longer in most cases when such mixtures are used. Decreased dosages of fungicides in synergistic mixtures lower the risk of selecting for resistant strains. There is still much uncertainty about the mechanisms of synergism between phenylamides, cymoxanil and contact fungicides.

References

Bashan, B., Levy, Y. & Cohen, Y. (1988). Synergistic interactions between respiration inhibitors and phenylamide fungicides in controlling *Phytophthora infestans*. *Phytoparasitica* **16**, 83 (abstract).

Bashan, B., Levy, Y. & Cohen, Y. (1989). Toxicity and incorporation of metalaxyl and oxadixyl to phenylamide-sensitive and resistant isolates of *Phytophthora infestans*. *Mededelingen Faculteit Landbouwwetenschappen Rijksunivesiteit Gent* **54**, 696-705.

Cohen, Y. & Samoucha, Y. (1990). Competition between oxadixyl- sensitive and -resistant field isolates of *Phytophthora infestans* on fungicide-treated potato crops. *Crop Protection* **9**, 15-20.

Davidse, L. C. (1984). Antifungal activity of acylalanine fungicides and related chloracetanilide herbicides. In *Mode of Action of Antifungal Agents*, ed. A. P. J. Trinci & J. F. Ryley, pp. 239-255. Cambridge University Press.

Davidse, L. C. (1988). Phenylamide fungicides: mechanism of action and resistance. In *Fungicide Resistance in North America*, ed. C. Delp, pp. 63-65. American Phytopethological Society: St. Paul, Minnesota.

De Waard, M. A. (1987). Synergism and antagonism in fungicides. In *Modern Selective Fungicides*, ed. H. Lyr, pp. 355-365. Gustav Fischer Verlag: Jena, & Longman Group (UK) Ltd.: London.

De Waard, M. A. & Van Nistelrooy, J. G. M. (1982). Antagonistic and synergistic activities of various chemicals on the toxicity of fenarimol to *Aspergillus nidulans*. *Pesticide Science* **13**, 279-286.

Gisi, U., Binder, H. & Rimbach, E. (1985). Synergistic interactions of fungicides with different modes of action. *Transactions of the British Mycological Society* **85**, 299-306.

Gozzo, F., Pizzingrilli, G. & Valcamonica, C. (1988). Chemical evidence of the effects of mancozeb on benalaxyl in grape plants as possible rationale for their synergistic interaction. *Pesticide Biochemistry and Physiology* **30**, 136-141.

Grabski, C. & Gisi, U. (1987). Quantification of synergistic interactions of fungicides against *Plasmopara* and *Phytophthora*. *Crop Protection* **6**, 64-71.

Josepovits, G. & Dobrovolszky, A. (1985). A novel mathematical approach to the prevention of fungicide resistance. *Pesticide Science* **16**, 17-22.

Levy, Y. & Strizyk, S. (1985). Une modélisation du développement de souches résistantes aux fongicides systémiques comptetenu du mode d'application, de l'adaptabilité génétique du champignon et de la synergie entre fongicides. *EPPO Bulletin* **15**, 519-525.

Levy, Y., Levi, R. & Cohen, Y. (1983). Buildup of a pathogen subpopulation resistant to a systemic fungicide under various control strategies: A flexible simulation model. *Phytopathology* **73**, 1475-1480.

Levy, Y., Benderly, M., Cohen, Y., Gisi, U. & Bassand, D. (1986). The joint action of fungicides in mixtures: Comparison of two methods for synergy calculations. *EPPO Bulletin* **16**, 651-657.

Samoucha, Y. & Cohen Y. (1986). Efficacy of systemic and contact fungicide mixtures in controlling late blight in potatoes. *Phytopathology* **76**, 855-859.

Samoucha, Y. & Cohen Y. (1988). Synergistic interactions of cymoxanil mixtures in the control of metalaxyl-resistant *Phytophthora infestans* of potato. *Phytopathology* **78**, 636-640.

Samoucha, Y. & Cohen, Y. (1989). Field control of potato late blight by synergistic fungicidal mixtures. *Plant Disease* **73**, 751-753.

Samoucha, Y. & Gisi, U. (1987a). Use of two- and three-way mixtures to prevent buildup of resistance to phenylamide fungicides in *Phytophthora* and *Plasmopara*. *Phytopathology* **77**, 1405-1409.

Samoucha, Y. & Gisi, U. (1987b). Systemicity and persistence of cymoxanil in mixture with oxadixyl against *Phytophthora* and *Plasmopara*. *Crop Protection* **6**, 393-398.

Samoucha, Y., Ben-Hador, G. & Cohen, Y. (1989). Chemical control of late blight epidemics incited by metalaxyl-resistant *Phytophthora infestans* in potato crops in Israel. *Phytoparasitica* **17**, 185-195.

Samoucha, Y., Levy, R. S. & Cohen, Y. (1988). Efficacy over time of cymoxanil mixtures in controlling late blight in potatoes incited by a phenylamide-resistant isolate of *Phytophthora infestans*. *Crop Protection* **7**, 210-215.

Staub, T. & Sozzi, D. (1983). Recent practical experience with fungicide resistance. *10th International Congress of Plant Protection* **2**, 591-598.

Chapter 25

Development and evaluation of blight resistant potato cultivars

L. J. Dowley, E. O'Sullivan & H. W. Kehoe

Late blight, caused by *Phytophthora infestans* has been a worldwide problem for almost 150 years. The main potato grown in Ireland from 1810 was the cultivar Lumper and the failure of this blight susceptible cultivar in 1846 was largely responsible for the Irish famine. Since then, resistance to late blight has been considered a priority in many breeding programmes.

Major gene resistance

Salaman (1911), one of the first to study blight resistance in the potato, discovered that crosses between *Solanum tuberosum* and *S. demissum* were highly resistant to blight. It was later confirmed that this type of hypersensitive resistance was controlled by a series of major (R) genes (Black, 1952). An international scheme for the nomenclature of the different physiological races of *P. infestans* and of genes controlling immunity in *S. demissum* derivatives was proposed by Black *et al.* (1953).

In nature most physiological races arise by mutation (Gallegly, 1968; Toxopeus, 1956; Peterson & Mills, 1953). Sexual reproduction can also be an important source of physiological race development (Pristou & Gallegly, 1956). Some variation may be generated by a parasexual cycle (Leach & Rich, 1969), while some could also be under cytoplasmic control (Caten & Jinks, 1968). The relative importance of these different sources of variation is not clear (see Chapter 14). The recent confirmation of the occurrence of the A2 mating type in the United Kingdom (Malcolmson, 1985) may suggest a greater role for sexual reproduction in producing genetic diversity, however its precise role under natural conditions in Northern Europe is not yet clear. Since hypersensitive resistance based on one or a few genes is now considered futile (Simmonds, 1970), more emphasis is currently placed on field resistance (Umaerus, 1983).

Field resistance

The importance of field resistance has been demonstrated by Niederhauser (1962; Chapter 3) and Niederhauser, Cervantes & Servin (1954); this type of polygenic resistance can be found in many cultivars and certain wild species including *S. demissum* and *S. stoloniferum*

(Toxopeus, 1964). Field resistance is made up of a number of different types of resistance (Thurston, 1971), which may include inhibition of zoospore germination, appressorium formation, penetration, tissue invasion and sporulation. Hodgson (1962) has found that potato cultivars can express all or none of these types of resistance and has suggested that each of these resistance mechanisms is under separate genetic control. Because field resistance is thought to be based on a number of different components which can be present in varying degrees, it is difficult to develop a single laboratory test which accurately determines field resistance.

Factors affecting field resistance

Lapwood (1961a & b) has shown that resistance is positively correlated with late maturity and that the lower leaves are more susceptible than the upper. Susceptibility has also been found to increase with increasing plant age (Peterson & Mills, 1953), while leaf susceptibility also increases with increased physiological age (Dowley, 1975). Seasonal variation in resistance has been confirmed by Grainger (1956) who associated this with variation in total carbohydrate content. Field resistance can be reduced by decreasing day length and light intensity (Umaerus, 1959). The importance of temperature and wetting duration has been stressed by Rotem, Cohen & Putter (1971). Spore concentration is also important and the ED_{50} has been used as a measure of resistance (Lapwood & McKee, 1966).

Many naturally occurring chemicals have been associated with resistance to *P. infestans* in the potato. These include carbohydrates (Grainger, 1956), sugars (Dowley, 1975), protein (Grummer, 1956) and phenols (Tomiyama, 1963). Enzymes such as peroxidase (Umaerus, 1959) and polyphenoloxidase (Tomiyama, 1963), as well as phytoalexins and auxins (Fehrmann & Diamond, 1967) have also been associated with resistance. In response to infection, many of these chemicals appear in minute quantities or their rates of accumulation may differ slightly between resistant and susceptible cultivars. As a result, measurement of such chemicals may not, on their own, be very reliable as a test for field resistance.

The main problem associated with field resistance is that it is strongly correlated with late maturity (Lapwood, 1961a; Toxopeus, 1964) and this is not acceptable in a new potato variety. High levels of foliage resistance are usually associated with high levels of tuber resistance. This association, however, is not consistent and therefore separate tests must be carried out to determine the resistance of foliage and tubers.

Methods of resistance testing

It is generally agreed that hypersensitive resistance is most easily assessed using detached leaves (Toxopeus, 1954). However, Umaerus (1963) has questioned the use of detached leaves in testing for field resistance and although good results have been obtained by Hodgson (1962), Lapwood (1961a) and Malcolmson (1969), most programmes include a field test.

A field test to determine tuber resistance was developed by Howard, Langton & Jellis (1976) using guard rows of a susceptible cultivar inoculated with a complex race. Lapwood (1967) developed a laboratory test based on tuber inoculation while Langton (1972) developed a method based on the rate at which the fungus can grow through a tuber slice. However, Lacey (1966) has shown that the distribution of tubers within the ridge varies with the cultivar. As this also affects the degree of infection with tuber blight, laboratory tests may not accurately reflect field resistance.

The potato breeding programme at Oak Park Research Centre was initiated in 1961 with blight resistance as a major objective. Over a 20 year period there has been a change in virulence phenotype within *P. infestans* in Ireland from simple to very complex races (O'Sullivan & Dowley, 1983). As a result, potato breeding in Ireland has concentrated on field resistance without any deliberate attempt to eliminate R-genes. Sources of resistance have included resistant cultivars and semi-commercial seedlings together with *Solanum demissum* and *S. stoloniferum* derivatives.

Seedlings are grown under normal field conditions from the time the initial tubers are produced. During the next five years of agronomic evaluation, seedlings are also exposed to natural infection with late blight. As there is a blight epidemic in Ireland in most years, at least some natural selection takes place during these five years. Formal testing for resistance to both foliage and tuber blight starts when seedlings reach the eighth year of evaluation. Testing is continued for a minimum of four years. Foliage blight resistance testing is based on a field test which relies on natural infection. Each plot consists of 25 tubers in each of four rows and each cultivar is replicated six times in a randomised block design. A row of Arran Victory is planted between each plot in an attempt to equalise inoculum pressure. Blight assessments, on the centre two rows, are made every seven days based on the B.M.S. Key (Cox & Large, 1960). Resistance assessments are based on the rate of defoliation from the time of initial infection to 75% defoliation under unsprayed conditions. Defoliation for each seedling is then compared with known control cultivars in each of the three main maturity groups. The final score for each seedling is given on a 1-9 scale where 9 represents the greatest level of resistance. Physiological race indicators are also planted within the trial area so that the sequence of appearance of the different physiological races can be monitored. Races carrying virulence to the R2 gene are the last to appear in the field each year, indicating that the R2 gene may confer some benefit in a breeding programme.

Tuber resistance is assessed for each seedling at harvest. This score is merely a guide to field resistance as field results may be conditioned by cryptic error, maturity class and timing of weather patterns. This field test is complemented by a laboratory test where 50 tubers of each seedling are inoculated with a spore suspension of 5×10^3 sporangia ml^{-1}. Prior to inoculation (dipping for 1 min), the sporangial suspension is incubated for 3 h at 10°C to promote zoospore formation. Following inoculation, the tubers are incubated for 10 days at 18°C and >95% R.H. The tuber resistance score, on a 1-9 scale as for foliage resistance, is derived from the percentage of tubers infected.

All seedlings with a rating less than 5 are discarded unless they have other compensating commercial traits. The emphasis on blight resistance will always be conditioned by the availability of alternative control measures.

Stability of field resistance

The question of stability of field resistance has been debated for many years with no definite outcome, although the consensus of opinion seems to favour stability. However, isolates of *P. infestans* have been reported to increase in aggressiveness by serial passage through resistant potato cultivars (Reddick & Mills, 1938). Potato isolates can also increase in aggressiveness towards the tomato in the same way (Mills, 1940).

This process of adaptation might make resistant cultivars more susceptible over time (Jeffery, Jinks & Grindle, 1962; Toxopeus, 1956). However, Niederhauser (1962) states that this process would be very slow while Toxopeus (1956) suggested that it may be very limited due to the high number of mutations required and emphasised the importance of tuber resistance in preventing the carry-over of resistance.

The blight resistance screening programme at Oak Park has produced blight resistance ratings for some of the more important Irish cultivars (Table 25.1). It can be seen that cv. Cara has a high level of both foliage and tuber resistance. This cultivar accounts for nearly 90% of the total seed potato export trade from Ireland.

During 1988, high levels of tuber blight were recorded in a number of Donegal-grown crops of Cara following handling and storage. In all cases, foliage blight levels had been low and no obvious explanation could be

Table 25.1. Blight resistance of some maincrop potato cultivars

Cultivar	Foliage Resistance	Tuber Resistance
Cara	7	7
Kerr's Pink	4·5	3
King Edward	3	2
Pentland Crown	5	4
Pentland Dell	4	1·5
Pentland Ivory	4	7
Record	5·5	7

Entries show resistance on a scale of 1-9, where 9 represents the greatest resistance.

found from examination of cultural practices or weather patterns. In general, the dry matter content in the Donegal tubers was very low suggesting a possible link between susceptibility and low carbohydrate content. Different seed stocks of Cara and isolates of *P. infestans* were examined in an effort to determine if the apparent change in tuber blight resistance was due to changes in the host, pathogen or the environment.

Infection of tubers

P. infestans was isolated from four different stocks of Donegal-grown Cara (isolates B, C, D & H) and from one stock of Carlow-grown Kerr's Pink (isolate I). The isolates were maintained on tuber slices of Carlow-grown Cara on moist filter paper in petri-dishes at 18-20°C, a day-length of 12 h and a light intensity of 150 lux. Inoculum was prepared from 7-day old cultures and standardised at $2 \cdot 5 \times 10^3$ or 5×10^3 sporangia ml^{-1}. Zoospore formation was induced by incubation for 3 h at 10°C prior to inoculation. Tubers were inoculated by dipping for 1 min in a spore suspension and incubated in plastic boxes lined with moist peat at 20°C in the dark. Each treatment was replicated 2 to 4 times in a randomized block design and each replicate consisted of 50 tubers. Disease assessments were carried out after a minimum incubation period of 10 days.

In a preliminary experiment, the four Donegal isolates were inoculated onto Carlow-grown tubers of Kerr's Pink and Cara. There were no differences in the aggressiveness of these isolates and, as expected, Kerr's Pink was more susceptible than Cara.

To determine if infection could be transmitted to healthy tubers during a dry handling process, two sporulating tubers infected with isolate C or isolate H were placed in a sealed plastic container with twenty healthy

Table 25.2. Infection of healthy tubers of cv. Cara following dry handling with blight-infected sporulating tubers

Isolate	Infection (%) after incubation for: 10 days	21 days	35 days
C	100	100	100
H	40	60	100

Table 25.3. Tuber infection of cvs. Kerr's Pink and Cara grown at different locations and inoculated with isolates of *P. infestans* from Donegal

Isolate	Tuber infection (%) Carlow Kerr's Pink	Carlow Cara	Donegal Cara
B	84	41	100
C	100	45	100
D	67	26	94
H	87	37	99
Mean	84·5	37·3	98·3
L.S.D. (5%) = 2·6			

Donegal Cara tubers. The container was then rotated ten times to distribute the inoculum. The tubers were then placed in plastic boxes lined with moist peat and incubated at 15°C and 95% R.H. in the dark. Each treatment was replicated four times. Blight assessments were carried out after incubation for 10, 21 and 35 days.

Within 10 days of inoculation, profuse sporulation was observed on the surface of infected Cara tubers. Examination of stored tubers in Donegal also revealed that the fungus was sporulating on infected tubers of a number of different varieties under normal storage conditions. Healthy tubers developed high levels of infection very soon after handling (Table 25.2). This showed for the first time that potato tubers could become infected during an in-store dry handling process. This change in the epidemiology of the disease could be due to changes in the host, pathogen or environment (Chapter 21).

Evidence for increased aggressiveness in P. infestans

In a further experiment, the four isolates from Donegal were inoculated onto Carlow-grown Kerr's Pink and Cara and onto Donegal-grown Cara. The Donegal Cara developed significantly more infection than the Carlow Cara while there was little difference between the Carlow Kerr's Pink and

Table 25.4. Tuber infection in cvs. Kerr's Pink and Cara grown at the same locations and inoculated with different isolates of *P. infestans* from Donegal and Carlow

	Tuber infection (%)	
Isolate	Kerr's Pink	Cara
C (Donegal)	86	22
H (Donegal)	88	22
I (Carlow)	98	0
L.S.D. (5%) = 17·8		

the Donegal Cara (Table 25.3). The percentage infection in the Carlow Cara was considered to be much higher than normal. This would suggest a change in both the host and the pathogen, although the apparent change in the host could be due to variation in dry matter between sites.

When isolates C & H from Donegal and isolate I from Carlow were inoculated onto Carlow-grown Kerr's Pink and Cara, the Donegal isolates proved much more aggressive on Cara than was the Carlow isolate (Table 25.4). Little difference was observed between the isolates on Kerr's Pink. These results suggest that in nature, strains of *P. infestans* can develop which are cultivar-specific and which can overcome some but not all of the field resistance in the tubers. Experiments which will determine if changes have taken place in the host have not yet been completed.

Conclusions

Field resistance to *P. infestans* is currently the most common form of resistance being incorporated into new potato cultivars. This type of resistance has been considered to be stable under field conditions. Results from experiments conducted with the resistant cultivar Cara confirm that there is variation in the pathogen which can erode some of the tuber resistance of this cultivar. Despite this fact, field resistance will remain the most important form of resistance available to the breeder. The complex nature of field resistance together with the potential for variation within the pathogen makes long term field testing for blight resistance an essential component of cultivar assessment.

It has also been established that *P. infestans* can produce viable sporangia on infected tubers of different varieties during storage and that this inoculum can infect healthy tubers during a dry handling process. This new development has important implications for post-harvest tuber blight control.

For the future it would be desirable to develop standard laboratory and field tests to measure field resistance in new cultivars together with the introduction of an international set of blight resistance indicator cultivars for each maturity class. More research is also required to develop useful rapid biochemical techniques for identifying blight resistance in breeding lines. To this end, further work on carbohydrates, especially sugars may be productive. It is also important to monitor more isolates for possible changes in virulence towards specific cultivars.

References

Black, W. (1952). A genetical basis for the classification of strains of *Phytophthora infestans*. *Proceedings of the Royal Society of Edinburgh, series B* **65**, 36-51.

Black, W., Mastenbroek, C., Mills, W. R. & Peterson, L. C. (1953). A proposal for an international nomenclature of races of *Phytophthora infestans* and of genes controlling immunity in *Solanum demissum* derivatives. *Euphytica* **2**, 173-179.

Caten, C. E. & Jinks, J. L. (1968). Spontaneous variability of single isolates of *Phytophthora infestans*. I. Cultural variation. *Canadian Journal of Botany* **46**, 329-348.

Cox, A. E. & Large, E. C. (1960). *Potato blight epidemics throughout the world*. Agricultural Handbook No. 174. US Department of Agriculture: Washington, D.C.

Dowley, L. J. (1975). *Ontogenetic and environmental predisposition of tomato foliage to infection with races of Phytophthora infestans (Mont.) de Bary*. M.S. Thesis, University of New Hampshire.

Ferhmann, H. & Diamond, A. E. (1967). Studies on auxins in the *Phytophthora* disease of the potato tuber. I. Role of indole-acetic-acid in pathogenesis. *Phytopathologische Zeitschrift* **59**, 83-100.

Gallegly, M. E. (1968). Genetics of pathogenicity of *Phytophthora infestans*. *Annual Review of Phytopathology* **6**, 375-396.

Grainger, J. (1956). Host nutrition and attack by fungal parasites. *Phytopathology* **46**, 445-456.

Grummer, G. (1956). The relation between the protein metabolism of cultivated plants and their susceptibility to parasitic fungi. *Reviews of Applied Mycology* **35**, 115-116 (abstr.).

Hodgson, W. A. (1962). Studies on the nature of partial resistance in the potato to *Phytophthora infestans*. *American Potato Journal* **39**, 8-13.

Howard, H. W., Langton, F. A. & Jellis, G. J. (1976). Testing for field susceptibility of potato tubers to blight (*Phytophthora infestans*). *Plant Pathology* **25**, 13-14.

Jeffery, S. I. B., Jinks, J. L. & Grindle, M. (1962). Intraracial variation in *Phytophthora infestans* and field resistance to potato blight. *Genetica* **32**, 323-364.

Lacey, J. (1966). The distribution of healthy and blighted tubers in potato ridges. *European Potato Journal* **9**, 86-98.

Langton, F. A. (1972). The development of a laboratory method of assessing resistance of potato tubers to late blight (*Phytophthora infestans*). *Potato Research* **15**, 290-301.

Lapwood, D. H. (1961a). Laboratory assessments of the susceptibility of potato haulm to blight (*Phytophthora infestans*). *European Potato Journal* **4**, 117-128.

Lapwood, D. H. (1961b). Potato haulm resistance to *Phytophthora infestans*. III. Lesion distribution and leaf destruction. *Annals of Applied Biology* **49**, 704-716.

Lapwood, D. H. (1967). Laboratory assessments of the susceptibility of potato tubers to infection by blight (*Phytophthora infestans*). *European Potato Journal* **10**, 127-135.

Lapwood, D. H. & McKee, R. K. (1966). Dose response relationships for infection of potato leaves with zoospores of *Phytophthora infestans*. *Transactions of the British Mycological Society* **49**, 679-686.

Leach, S. S. & Rich, A. E. (1969). The possible role of parasexuality and cytoplasmic variation in *Phytophthora infestans*. *Phytopathology* **59**, 1360-1365.

Malcolmson, J. F. (1969). Factors involved in resistance to blight (*Phytophthora infestans* (Mont.) de Bary) in potatoes and assessment of resistance using detached leaves. *Annals of Applied Biology* **64**, 461-468.

Malcolmson, J. F. (1985). *Phytophthora infestans*: A2 compatibility type recorded in Great Britain. *Transactions of the British Mycological Society* **85**, 531.

Mills, W. R. (1940). *Phytophthora infestans*. Adaptive parasitism of *Phytophthora infestans*. *Phytopathology* **30**, 17.

Niederhauser, J. S. (1962). Evaluation of multigenic 'field resistance' of the potato to *Phytophthora infestans* in 10 years of trials at Toluca, Mexico. *Phytopathology* **52**, 746.

Niederhauser, J. S., Cervantes, J. & Servin, L. (1954). Late blight in Mexico and its implications. *Phytopathology* **44**, 406-408.

O'Sullivan, E. & Dowley, L. J. (1983). Physiological specialisation in strains of *Phytophthora infestans* sensitive and resistant to metalaxyl. *Irish Journal of Agricultural Research* **22**, 105-107.

Peterson, L. C. & Mills, W. R. (1953). Resistance of some American varieties to late blight of potatoes. *American Potato Journal* **30**, 65-70.

Pristou, R. & Gallegly, M. E. (1956). Differential reaction of potato hosts to foreign and domestic potato physiological races of *Phytophthora infestans*. *American Potato Journal* **33**, 287-295.

Reddick, D. & Mills, W. (1938). Building up virulence in *Phytophthora infestans*. *American Potato Journal* **15**, 29-34.

Rotem, J., Cohen, Y. & Putter, J. (1971). Relativity of limiting and optimum inoculum loads, wetting durations and temperatures for infection by *Phytophthora infestans*. *Phytopathology* **61**, 275-278.

Salaman, R. N. (1911). Studies in potato breeding. In *Proceedings, IVth International Congress of Genetics*, p. 397.

Simmonds, N. W. (1970). In *Annual Report of the Scottish Plant Breeding Station for 1969*, p. 18.

Thurston, H. D. (1971). Relationship of general resistance: Late blight of potato. *Phytopathology* **61**, 620-626.

Tomiyama, K. (1963). Physiology and biochemistry of disease resistance in plants. *Annual Review of Phytopathology* **1**, 295-324.

Toxopeus, H. J. (1954). Leaf testing as a method of genetical analysis of immunity from *Phytophthora infestans*. *Euphytica* **3**, 233-240.

Toxopeus, H. J. (1956). Reflections on the origin of new physiological races of *Phytophthora infestans* and the breeding for resistance in potatoes. *Euphytica* **5**, 221-237.

Toxopeus, H. J. (1964). Treasure digging for blight resistance in potatoes. *Euphytica* **13**, 206-222.

Umaerus, V. (1959). Relationship between peroxidase activity in potato leaves and resistance to *Phytophthora infestans*. *American Potato Journal* **36**, 124-131.

Umaerus, V. (1963). Field resistance to late blight in potatoes. In *Recent Plant Breeding Research, Svalof 1946-61*, ed. E. Akerberg, A. Hagberg, G. Olsson & O. Tedin, pp. 223-245. John Wiley & Sons: New York.

Umaerus, V., Umaerus, M., Erjefalt, L. & Nilsson, B. A. (1983). Control of *Phytophthora* by host resistance: problems and progress. In *Phytophthora: Its Biology, Taxonomy, Ecology and Pathology*, (ed. D. C. Erwin, S. Bartnicki-Garcia & P. H. Tsao), pp. 315-326. American Phytopathological Society: St. Paul, Minnesota.

Chapter 26

Microbial suppression of *Phytophthora cinnamomi*

A. Ruth Finlay & A. R. McCracken

Phytophthora cinnamomi (Rands) is an extremely virulent pathogen with a very wide host range. It was first described by Rands (1922) on cinnamon trees (*Cinnamomum burmannii* Blume) in Sumatra. Since then it has been reported as causing disease on nearly 1000 host plants in at least 67 countries. *P. cinnamomi* is soil-borne and attacks the small absorbing roots of primarily woody plants. Major outbreaks of the disease have included jarrah dieback of eucalyptus in Western Australia and pineapple root rot in California (Newhook, 1970). In the United Kingdom the Hardy Nursery Stock Industry suffers major economic losses through *P. cinnamomi* infection of *Chamaecyparis lawsoniana* and various species of *Cupressus*, *Rhododendron*, *Erica* and *Calluna* (Evans & Melville, 1984). Symptoms of the disease include root and foot rot, wilting, reduction in leaf size, leaf discolouration and death of infected plants. *P. cinnamomi* can be disseminated in flowing water, on infected nursery stock, on equipment or by movement of infested soil or gravel.

There are many factors which influence the disease impact by this pathogen. These include high soil moisture (Reeves, 1975), high temperatures which favour sporulation and growth of the fungus (Grant & Byrt, 1984; Shearer, Shea & Deegan, 1987) and plant water status which affects host susceptibility.

There are a number of approaches to the control of *P. cinnamomi*. These include production of clean nursery stock, use of resistant varieties and chemical, physical and biological control (Chapter 28). Only limited success has been achieved in the development or selection of host plants resistant to *P. cinnamomi*. There are also problems associated with attempts to control this pathogen chemically. Systemic fungicides, such as the acylalanines, can inhibit chlamydospore production and zoospore release (Margot, 1980), but resistance can develop rapidly. Furthermore, there are no commercially available fungicides which destroy oospores of *P. cinnamomi*.

For these and other considerations there has been a growing interest and recognition of the opportunities for biological control of *P. cinnamomi*, the mechanisms of which can be grouped into three categories:

- (i) reduction of pathogen inoculum by antagonistic microorganisms, either resident or introduced (Baker & Cook, 1974);
- (ii) protection of plant surfaces against infection;
- (iii) management of physiological incompatibility between host and pathogen (Cook, 1981).

Three biological control mechanisms may operate in the rhizosphere or infection court against the pathogen:

- (i) amensalism – where the biological control agent produces either volatile or non-volatile antibiotics or toxic metabolic by-products which inhibit growth of the pathogen; this antibiosis probably occurs in microhabitats where nutrients are abundant;
- (ii) parasitism and predation – involving active contact between microorganisms resulting in either the degradation of hyphal walls or mycophagy of whole propagules (Barnett & Binder, 1973; Boosalis, 1964);
- (iii) competition – involving demand by two or more microorganisms for the same limited resource which may be nutrients, oxygen or space.

None of these strategies is likely to be mutually exclusive, indeed a successful antagonist will probably employ several different strategies (Whipps & Lynch, 1986).

A positive correlation has been demonstrated between lysis of hyphae of various species of *Phytophthora* and populations of *Pseudomonas*, *Bacillus* and *Streptomyces* spp. (Malajczuk, 1983). Actinomycetes are dominant in *Phytophthora* suppressive soils (Broadbent, Baker & Waterworth, 1971; Malajczuk & McComb, 1979) and antifungal activity has been mainly attributed to antibiotics although autolysis of fungal hyphae in the presence of *Streptomyces* spp. may also have been due to nutrient deprivation (Hsu & Lockwood, 1969).

Among the fungi, *Epicoccum purpurascens* has been reported as causing marked inhibition of *P. cactorum in vitro* (Gilmore, 1986) and as an active antagonist of *P. cinnamomi* (Brown, Finlay & Ward, 1987). Control of *P. cinnamomi* by basidiomycete species, particularly mycorrhizal fungi, has shown some promise. Ectomycorrhizas reduced infection by *P. cinnamomi* of a wide range of tree hosts although the exact mechanism of root protection is unknown (Malajczuk, 1979; Marais & Klotze, 1976; Marx, 1973; Ross & Marx, 1972).

Only limited success has been achieved in inducing resistance in plants to *Phytophthora* diseases using cross-protection by inoculation with incompatible races or with pathogens of other hosts. When two Australian forest species, *Banksia grandis* and *Casuarina fraserana* were pre-inoculated with *Pythium irregulare* to protect against infection by *Phytophthora cinnamomi*, a degree of cross-protection was achieved, possibly due to a hypersensitive reaction occurring on inoculated roots reducing their susceptibility to *P. cinnamomi* (Borstel, 1979). Similarly, protection against *P. cinnamomi* root rot in *Persea* spp. has been obtained by prior inoculation with other *Phytophthora* spp. Strong local protection against *P. citricola* and *P. cinnamomi* infection of *Persea indica* was achieved by prior inoculation with a non-pathogenic citrus isolate of *P. nicotianae* (*parasitica*) (Cohen & Coffey, 1984; Dolan & Coffey, 1984; Dolan, Cohen & Coffey, 1986).

We examined a wide range of microorganisms shown to have antagonistic properties against *P. cinnamomi*. Some investigations were also carried out on the potential of using other *Phytophthora* species as cross-protectants against *P. cinnamomi*.

Interaction between *Phytophthora cinnamomi* and other microorganisms

The isolates of *P. cinnamomi* used throughout were as follows: A, B and C obtained from the CAB International Mycological Institute (IMI 129910, IMI 157800, IMI 70473); D and E were isolated from infected roots of *Berberis* spp.; isolates F and G from *Calluna* spp.; isolates H and I from *Cupressocyparis leylandii*; and J from *Chamaecyparis lawsoniana* cv. *Ellwoodii*.

Actinomycetes

Three *Streptomyces* spp., *S. lavendulae* (Waks. & Curt.) Waksman & Henrici, *S. viridifaciens* Gourevitch & Lein, and an unidentified species (isolate IsS) were tested for antagonism against a range of *P. cinnamomi* isolates. Each antagonist was placed separately on a corn meal agar (CMA) plate as spots at 3 out of 4 points symmetrically around and 2 cm away from a central inoculum plug taken from a 2 day old culture of *P. cinnamomi*. After incubation at 24°C for 5 days, inhibition was visually assessed and the mycelia of *P. cinnamomi* were examined microscopically.

All three *Streptomyces* spp. were inhibitory to all of the *P. cinnamomi* isolates examined. There were large zones of inhibition between the actinomycete colonies and *P. cinnamomi*. Growth of *P. cinnamomi* on the side free from an actinomycete colony was also retarded, compared to that on the control plates of *P. cinnamomi* alone.

The production of toxin by *Streptomyces* spp. was assayed using a cup-plate method in which culture filtrates of each of the *Streptomyces* spp. grown in yeast-maltose (YM) broth for 5 days at 23°C were pipetted into wells cut in CMA 2 cm from a central inoculum of *P. cinnamomi*. All inhibited the growth of the fungus. *S. viridifaciens* was the most inhibitory to all of the *P. cinnamomi* isolates tested although there was a large difference in the sensitivity of isolates. *Streptomyces* sp. IsS had the least inhibitory effect on *P. cinnamomi* (Fig. 26.1).

Basidiomycetes

The interaction between ten basidiomycete species and nine isolates of *P. cinnamomi* was examined on agar plates. The micro-organisms used were *Agaricus arvensis* Schaeff. ex Secr. S. Lange non Cooke, *Langermannia gigantea* (Batsch ex Pers.) Rostk., *Marasmius oreades* (Bolt. ex Fr.), *Panaeolus foenisecii* (Pers. ex Fr.) Schroeter, two isolates of *Psilocybe semilanceata* (Fr. ex Secr.) Kummer, *Stropharia semiglobata* (Batsch. ex Fr.) Quel., all of which were associated with grassland. Other organisms used were *Thelephora terrestris* (Ehrh.) Fr., isolated from mycorrhizal roots of *Tsuga canadensis* 'Jeddeloh', *Ceraceomyces sublaevis* (Bres. Julich) isolated from flax and a species of basidiomycete (isolate

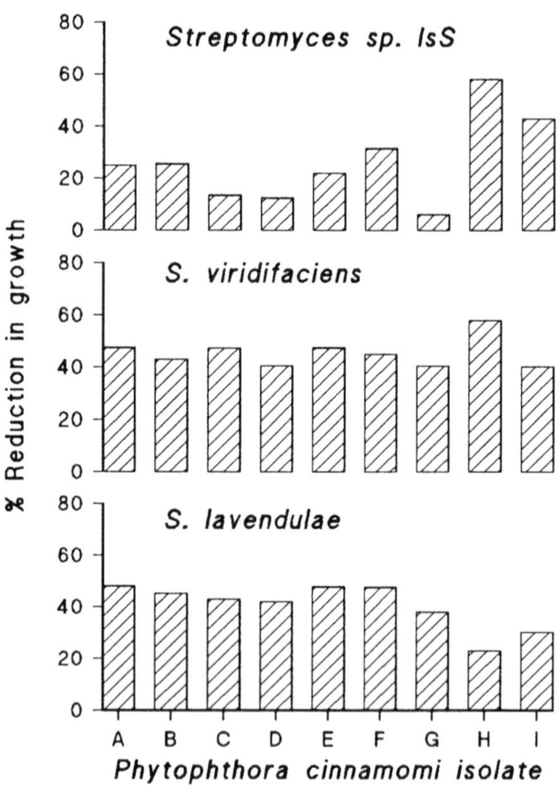

Fig. 26.1 Inhibition of growth of nine isolates of *Phytophthora cinnamomi* by three *Streptomyces* species.

Table 26.1. Inhibition of radial growth of *Phytophthora cinnamomi* by a range of basidiomycete species in opposition culture on modified Melin-Norkran's Agar

	Reduction (%) in radial growth of *P. cinnamomi* [a] Basidiomycete species B1 to B5 [b]				
P. cinnamomi isolate	B1	B2	B3	B4	B5
A	18·3	55·2	26·7	31·4	46·3
B	9·6	60·3	0	0	25·9
C	0	68·8	4·7	0	42·6
D	22·9	57·2	0	38·2	54·8
E	5·9	46·1	9·1	1·4	31·2
F	0	61·7	0	0	43·1
G	16·5	59·5	11·1	7·1	36·6
H	28·7	61·3	59·6	48·7	60·8
I	5·8	56·4	55·5	44·6	46·0
J	15·9	53·9	19·0	24·5	28·6
LSD (*P*=0·05)	11·4	8·0	8·2	6·6	6·7

[a]Mean (arcsin transformation) of 3 replicates for each *P. cinnamomi*/basidiomycete combination.

[b]Key to basidiomycete species: B1 = *Agaricus arvensis*, B2 = *Langermannia gigantea*, B3 & B4 = *Psilocybe semilanceata*, B5 = *Stropharia semiglobata*.

---

no. RS1) previously identified as mycoparasitic on *P. cinnamomi*. All of these fungi were grown on modified Melin-Norkran's medium (Marx, 1969) for 22 days at 23°C. A plug of *P. cinnamomi* was then placed on the agar 1 cm away from the leading edge of the basidiomycete colony. After a further 5 days incubation at 23°C the effect on radial growth of *P. cinnamomi* was measured. Five of the basidiomycetes tested (*A. arvensis*, *L. gigantea*, *S. semiglobata* and both isolates of *P. semilanceata*) were inhibitory to at least some of the *P. cinnamomi* isolates. The most effective against the widest range of isolates was *L. gigantea* (Table 26.1). One of the isolates of *Psilocybe semilanceata* was particularly inhibitory to some *Phytophthora cinnamomi* isolates (almost 60% reduction in growth) while having no effect on others. Culture filtrates of two of the most inhibitory basidiomycete species, *L. gigantea* and *S. semiglobata*, growing in malt broth were assayed for anti-fungal activity using a cup-plate method (Brown, Finlay & Ward, 1987) but failed to inhibit *P. cinnamomi* colony growth.

Other organisms.

Four species of *Mortierella*, *M. bainieri* Costantin, *M. echinosphaera* Plaats-Niterink, *M. gamsii* and *M. nana* were, along with *Myrothecium verrucaria*, tested for their effect on the radial growth of *P. cinnamomi*. Also tested was an isolate of *Trichoderma harzianum* Rifae and four isolates of *Trichoderma* species obtained from composted Sitka Spruce bark. All co-cultures were examined for evidence of hyphal interactions.

None of the *Mortierella* or *Trichoderma* species formed zones of inhibition but rather grew through the *P. cinnamomi* colonies. In co-culture with *Mortierella bainieri*, *P. cinnamomi* hyphae produced multiple short branches. Hyphae of *M. nana* and *M. echinosphaera* formed close associations with *P. cinnamomi* although no similar association was noted with *M. gamsii*.

Myrothecium verrucaria caused some slight inhibition and as it approached *P. cinnamomi* it induced numerous swellings on the hyphae of several isolates. When the two colonies intermingled the antagonist coiled and caused distortion of the *P. cinnamomi* hyphae. There was also some evidence of possible invasion of chlamydospores. Disease suppression *in vivo* by *Myrothecium* has been reported recently by Coffey (Chapter 28).

Hyphae of *Trichoderma* spp. showed characteristic coiling around the hyphae of *P. cinnamomi* while *T. harzianum* penetrated the hyphae. When plugs of agar were removed from co-cultures and stimulated to produce sporangia, few sporangia were formed. However, one of the cultures of *T. harzianum* isolated from composted bark had a significant stimulatory effect on the number of sporangia formed (Table 26.2). This could increase rather than decrease inoculum potential.

Interaction between *Epicoccum purpurascens* and *Phytophthora cinnamomi*

Four isolates of *E. purpurascens* inhibited *P. cinnamomi* growth the latter mycelium becoming swollen and stunted with regions of protoplasmic degeneration (Brown, Finlay & Ward, 1987). Where colonies intermingled the hyphae of *E. purpurascens* coiled around and penetrated the *P. cinnamomi* hyphae. Cellulase and β-1,3-glucanase activities were detected in cultures of *E. purpurascens* grown on hyphal wall material of *Phytophthora* spp. as the sole carbon source. Volatile antibiotics produced by *E. purpurascens* (Martin, 1963) on *P. cinnamomi* elicited varied responses. During the first three days of inoculation, two out of the ten *P. cinnamomi* isolates were significantly inhibited while three were significantly stimulated. On further incubation the inhibitory effect declined, but two isolates continued to be stimulated (Table 26.3).

Table 26.2. Effect of *Trichoderma harzianum* on subsequent sporangium production by *P. cinnamomi*.

Treatment	Mean number of sporangia in zone P	I	O[a]	LSD (P=0·05)
P. cinnamomi is. A	60·8	–	–	
				26·7
P. cinnamomi is. A & *T. harzianum*	104·2	17·6	0	
P. cinnamomi is. B	22·7	–		
				11·5
P. cinnamomi is. B & *T. harzianum*	28·2	7·4	0	

[a]plugs taken from opposing cultures: P = *Phytophthora* region; I = *interaction zone*; 0 = opponent colony.

When *E. purpurascens* was grown in two selective media, six antifungal compounds were detected (Brown, Finlay & Ward, 1987). In a sucrose/casamino acid medium, four compounds, epicorazines A and B and two (X and Y) of unknown identity (designated the epicorazine fraction), were produced at an early stage of the growth of the organism. These products were not detected in a glucose/ammonium phosphate medium (GAP). The compound flavipin was formed in both media but was produced preferentially in GAP. A third unidentified antifungal compound was detected in both media at a late growth stage.

Flavipin extracted from cultures of *E. purpurascens* grown in GAP inhibited radial growth of all *P. cinnamomi* isolates tested but to different extents (Table 26.4), although in dual cultures, isolates had varied in their response to *E. purpurascens*. Zoospores of *P. cinnamomi* were 15 times more sensitive at pH 5 to flavipin (ED_{50} 0·53 μg ml^{-1}) than *Botrytis allii* conidia (ED_{50} 7·8 μg ml^{-1}) (Brown, Finlay & Ward, 1987). Colony growth of *P. cinnamomi* was also inhibited by the compounds present in the 'epicorazine fraction' (Table 26.4).

Experiments were carried out to test the effect of *E. purpurascens* on *P. cinnamomi* infection of *Lupinus albus* and *Erica vagans*. Initially, *L. albus* cv. Vladimir seedlings were grown in a hydroponic system in the presence and absence of *E. purpurascens*. The fungus did not exert any

Table 26.3. Inhibition (%) of colony growth of *Phytophthora cinnamomi* isolates caused by volatile antibiotics produced by *Epicoccum purpurascens* (I3) after 3 and 7 days incubation.

	Reduction in colony growth (%)	
P. cinnamomi isolate	3 Day	7 Day
A	9·2	3·0
B	14·9*	-24·7
C	1·0	-8·6
D	24·1*	-4·8
E	-32·6	-9·2
F	-43·8*	-30·0*
G	-6·4	0·02
H	-45·2*	-14·4
I	-15·2	-9·3
J	-40·7*	-22·4*

*Effect significant at $P = 0 \cdot 05$. Negative values indicate stimulation of *P. cinnamomi* by *E. purpurascens*.

significant effect on either shoot length or dry weight. However, tap roots were significantly longer and mean root dry weights correspondingly greater in the presence of *E. purpurascens* (Table 26.5). Although mycelium was not evident on microscopic examination of the roots, *E. purpurascens* was recovered from non-sterilised roots. There were no corresponding differences in shoot and root lengths and weights of plants grown in compost amended and not amended with *E. purpurascens*. Two-day-old germinated seedlings of *L. albus* were planted in sterilised compost which was then drenched with a spore suspension of *E. purpurascens*. Three days later, plants were challenged with a zoospore suspension of *P. cinnamomi*. After a further three weeks the presence of *P. cinnamomi* on roots and in soil was determined by plating root segments on P_{10} VPH selective medium (Tsao & Ocana, 1969) and by using a baiting technique (McCracken, 1985). Although no effect on the sizes of plants was observed, no *P. cinnamomi* was recovered from either lupin roots or soil samples inoculated with *E. purpurascens*. Rooted cuttings of *Erica vagans* transplanted into a 1:1 peat/sand mixture were drenched with a spore suspension of *E. purpurascens*. After seven days, plants were challenged with *P. cinnamomi* zoospores and above-ground symptoms were monitored at monthly intervals (Finlay, 1987). No *P. cinnamomi* infections were recorded in either of two experiments.

Table 26.4. Inhibition of mycelial growth of *Phytophthora* spp. by partially purified flavipin and by the 'epicorazine fraction'

Phytophthora sp.	Flavipin conc. causing 80% reduction in colony growth (μg ml^{-1})	Reduction (%) in colony growth by 'Epicorazine fraction'
P. cinnamomi (is. A)	50·5 (5·02)[a]	57·8 (4·61)
P. cinnamomi (is. B)	40·2 (4·51)	52·9 (4·14)
P. cinnamomi (is. C)	75·5 (5·52)	58·5 (4·89)
P. cactorum	30·1 (2·53)	32·1 (2·72)
P. citricola	186·3 (15·62)	42·8 (3·17)
P. cryptogea	125·1 (11·47)	27·2 (1·91)
P. syringae	32·3 (2·09)	35·6 (2·43)

[a]Standard errors ($P = 0 \cdot 05$) are shown in parentheses. Data from Brown, Finlay & Ward (1987).

Table 26.5. Effect of *Epicoccum purpurascens* on growth of *Lupinus albus* in a hydroponic system.

	E. purpurascens present	*E. purpurascens* absent
Shoot length (cm)	10·4^{a}	10·9^{a}
Tap root length (cm)	33·2^{a}	24·0^{c}
Shoot dry weight (g)	0·301^{a}	0·305^{a}
Root dry weight (g)	0·087^{a}	0·075^{b}

Values followed by the same letter are not significantly different at $P = 0 \cdot 05$, a-b denotes difference at $P = 0 \cdot 05$ and a-c difference at $P = 0 \cdot 001$.

Cross protection

All of the *Phytophthora* species tested in paired confrontations on potato dextrose agar were able to affect the growth of *P. cinnamomi* (Table 26.6). The degree of protection of glasshouse grown *Erica vagans* against infection by *Phytophthora cinnamomi* was examined using isolates of *P. cactorum*, *P. citricola*, *P. cryptogea*, *P. syringae* and *Phytophthora* species PA, from a wood rot, and PB and PC from apple rots. Vermiculite infested with these organisms was mixed with sterilised 1:1 peat/sand mixture at a rate of 10 cm^3 of inoculum per 1000 cm^3 mixture. Six-month old cuttings

Table 26.6. Interaction *in vitro* between *Phytophthora cinnamomi* and opposing *Phytophthora* species.

	Interaction category*	Mean colony radius (mm) *P. cinnamomi*	Opposing species
P. cactorum	M	25·2	10·4
P. citricola	MC	21·3	17·5
P. cryptogea 1	MC	25·1	12·4
P. cryptogea 2	O	19·7	24·9
P. syringae	MC	26·2	10·1
Phytophthora is. PA	O	11·7	49·2
Phytophthora is. PB	MC	24·3	11·7
Phytophthora is. PC	MC	25·1	10·9
P. cinnamomi	O	32·5	32·5
LSD (0·05)		1·0	0·9

*interaction categories: M = mutual inhibition; MC = mutual inhibition with clear zone; O = overgrowth.

---

of *E. vagans* were transplanted into the inoculated potting mixture. One month later a *P. cinnamomi* suspension (8×10^4 zoospores ml^{-1}, 10 ml per pot) was applied to half the treated plants and half the control plants. A further 2 months later, above-ground symptom expression was assessed. A small random sample of plants from each treatment was selected for destructive analysis. Disease severity was again assessed after 3, 5 and 7 months. A final sample was taken for destructive analysis 8 months after inoculation. Roots of plants were washed under running water. Sections of roots were surface sterilised and plated on P_{10}VPH selective medium. Stem base segments were similarly surface sterilised and plated on the same selective medium. The presence or absence of *P. cinnamomi* was noted 4 days later. The baiting technique was used to assess the presence of the pathogen in the soil.

None of the *E. vagans* plants treated with other *Phytophthora* species were significantly protected from infection by the pathogen, *P. cinnamomi*. General trends after 2 months indicated that plants grown in compost infested with *P. cactorum* or *P. cryptogea* prior to challenge with *P. cinnamomi* exhibited greater disease severity than control plants exposed to *P. cinnamomi* alone. Most *Phytophthora* species tested were only mildly pathogenic or non-pathogenic to *E. vagans* in the absence of *P. cinnamomi*; *P. cryptogea*, however, was pathogenic.

Table 26.7. Effect of prior inoculation with several *Phytophthora* species on infection of *Erica vagans* by *Phytophthora cinnamomi*.

	Disease Index at indicated month after *P. cinnamomi* challenge		
	2	3	7
P. cactorum + Pc*	2·3	2·7	3·4
P. cactorum	0	0	1·2
P. citricola + Pc	1·7	2·7	3·1
P. citricola	0·1	0	0·3
P. cryptogea + Pc	2·5	2·7	4·1
P. cryptogea	1·7	1·3	1·4
P. syringae + Pc	0·5	1·2	1·9
P. syringae	0·1	0·1	1·1
Phytophthora is. PA + Pc	1·0	1·8	2·5
Phytophthora is. PA	0·1	0	0·9
Phytophthora is. PB + Pc	0·9	1·0	1·6
Phytophthora is. PB	0·3	0·2	0·4
Phytophthora is. PC + Pc	0·8	1·4	3·2
Phytophthora is. PC	0·1	0·1	0·7
P. cinnamomi	1·8	2·2	2·4
Control	0	0	0·2
LSD (0·05)	0·9	1·2	1·4

*Pc = *Phytophthora cinnamomi*. Above-ground symptoms were monitored using a 0 to 5 Disease Index (Finlay, 1987).

Pre-inoculation with *P. syringae* and the *Phytophthora* sp. isolates PA, PB or PC reduced the disease rating on plants 2 months after challenges with *P. cinnamomi* compared to challenged plants grown in uninfested compost. This trend was still evident after 3 months although isolates A and C afforded less protection than earlier in the experiment. After 7 months most 'protected' plants showed severe wilting or death (Table 26.7). The recovery of *P. cinnamomi* from roots 2 months after treatment was variable; the pathogen was, however, recovered from all diseased roots from all plants exposed to other *Phytophthora* species prior to challenge. Highest percentage root infection was recorded from plants challenged with *P. cinnamomi* alone and the lowest from plants which had been previously exposed to *P. cryptogea*. Recovery of *P. cinnamomi* from soil, using pine needle baits, was lower than direct recovery from roots. The highest recovery (48% of baits infected) was from compost containing

Table 26.8. Reisolation of *Phytophthora cinnamomi* from *Erica vagans* 2 months and 8 months after 'protection' with *Phytophthora* species.

	% roots infected with *P. cinnamomi*		% pine needle baits infected with *P. cinnamomi*	
	2 months	8 months	2 months	8 months
P. cactorum + Pc	50·3	0	23·6	0
P. citricola + Pc	49·9	9·2	23·7	0
P. cryptogea + Pc	25·6	0	27·5	0
P. syringae + Pc	64·3	23·9	30·0	43·3
Phytophthora is. PA + Pc	52·1	22·5	21·0	0
Phytophthora is. PB + Pc	61·3	27·1	37·1	20·9
Phytophthora is. PC + Pc	54·8	11·3	33·4	22·5
P. cinnamomi only	75·0	0	48·1	0
Control	0	0	0	0
LSD (0·05)	51·9	34·4	35·9	36·0

Pc = *Phytophthora cinnamomi.*

only *P. cinnamomi* and the least from soil to which PA had been added (Table 26.8).

Conclusions

Only in a few instances have *Phytophthora* diseases been controlled successfully using biological and/or cultural methods (Shea & Broadbent, 1983). In particular, diseases induced by *P. cinnamomi* are especially difficult to control by these methods because of the wide host range and adaptive life cycle of the pathogen. Limitations and problems associated with the use of chemicals, however, combined with considerable economic losses to the horticultural industry caused by *P. cinnamomi* encourage further investigation of potential biocontrol methods.

In the present study, cross-protection of the woody host, *E. vagans* against *P. cinnamomi* using other *Phytophthora* spp. showed no potential as a control method. Much more promising results were obtained from the *in vitro* experiments with fungi and actinomycetes antagonistic to *P. cinnamomi*. Subsequent introduction of some of those antagonists into soil gave partial control of infection of host plants. Although the *Streptomyces* species were very inhibitory to the growth of *P. cinnamomi*, their potential as biocontrol agents in soil was considered limited since actinomycetes are more prevalent in dry soils (Chen & Griffin, 1966), producing greater quantities of antibiotics at low water potential (Wong & Griffin, 1974), conditions which are unfavourable for *P. cinnamomi* pathogenesis. Several of the basidiomycetes examined were antagonistic to *P. cinnamomi*; a range of different mechanisms of antagonism including the secretion of antibiotics was demonstrated. Although no β-1,3-glucanase production was detected from *Stropharia semiglobata* or *Langermannia gigantea*, it is known that many of the Agaricales produce cellulases (Wood & Fermor, 1985) and these may have some role in breaking down *P. cinnamomi* cell walls. The apparent mycoparasitic attack by an unidentified basidiomycete species (RB) and by *Ceraceomyces sublaevis* indicated that some basidiomycetes may be able to reduce *P. cinnamomi* inoculum. There was however, no evidence of penetration and destruction of chlamydospores. All of the basidiomycetes examined had slow growth rates which may have restricted their competitive ability.

Similarly, although *Myrothecium verrucaria* was capable of antagonising *P. cinnamomi* in opposition culture by antibiosis and mycoparasitism, it grew slowly and it seems unlikely that soil infected with this organism could maintain sufficiently high population levels to obtain long term disease control. Munnecke (1984) successfully introduced *M. verrucaria* into fumigated soils where populations remained high for approximately 6 months. However, the populations then rapidly declined and although the rate of infestation by *P. cinnamomi* was slower in treated plants, eventually disease incidence was as great as in the non-treated plots.

The fungal antagonist showing most potential was *Epicoccum purpurascens*. Fungal interaction between *P. cinnamomi* and *E. purpurascens* in culture resulted in extreme antagonism via antibiosis, parasitism involving destruction of *Phytophthora* hyphae after contact, and the production of a wide range of antifungal compounds. *E. purpurascens* enhanced root growth of lupins in hydroponic culture, but not in soil. The fungus grew saprophytically in soil very well, but did not seem to preferentially colonise the rhizosphere. It is probable, therefore, that *E. purpurascens* suppresses the saprophytic stage of *P. cinnamomi* but cannot prevent infection once zoospores reach the root surface.

The potential use of *E. purpurascens* or other fungi antagonistic to *P. cinnamomi* in producing container-grown woody ornamentals would appear to be rather limited. *P. cinnamomi* is a particularly virulent and opportunistic pathogen with a large host range. The successful introduction and maintenance of high populations of a suitable antagonist would be very difficult. A recent development in container media is the use of composted tree bark, to replace peat. This material is reported to be suppressive to many pathogens including *P. cinnamomi* (Hoitink, Schmitthenner & Herr, 1975; Hoitink, 1980). At least some of the suppressive activity of composted bark is due to its microbial populations

(Nelson & Hoitink, 1983). Effective colonisation of a bark substrate by introduced antagonists may endow it with suppressive properties or enhance those which already exist. Unfortunately, *E. purpurascens* does not colonise this substrate very quickly or effectively (Finlay, 1987). Perhaps the best prospects for biological control and microbial suppression of *P. cinnamomi* and other root pathogens of container grown woody plants may be the use of introduced antagonists along with appropriate composted tree barks.

Acknowledgement The work described here was made possible by funding provided by the Department of Agriculture for Northern Ireland.

References

Baker, K. F. & Cook, R. J. (1974). *Biological control of plant pathogens*. W. H. Freeman & Co.: New York.

Barnett, H. L. & Binder, F. L. (1973). The fungal host-parasite relationship. *Annual Review of Phytopathology* **11**, 273-292.

Boosalis, M. G. (1964). Hyperparasitism. *Annual Review of Phytopathology* **2**, 363-376.

Borstel, P. J. (1979). Cross protecting West Australian native plants against *Phytophthora cinnamomi* Rands using *Pythium irregulare* Buism. Honours dissertation, University of Western Australia, Perth.

Broadbent, P., Baker, K. F. & Waterworth, Y. (1971). Bacteria and actinomycetes antagonistic to fungal root pathogens in Australian soils. *Australian Journal of Biological Science* **24**, 925-944.

Brown, A. E., Finlay, A. R. & Ward, J. S. (1987). Antifungal compounds produced by *Epicoccum purpurascens* against soil-borne plant pathogenic fungi. *Soil Biology and Biochemistry* **19**, 657-664.

Chen, A. W. & Griffin, D. M. (1966). Soil physical factors and the ecology of fungi. V. Further studies in relatively dry soils. *Transactions of the British Mycological Society* **49**, 419-426.

Cohen, Y. & Coffey, M. D. (1984). Protecting *Persea indica* seedlings from *Phytophthora citricola* by a prior inoculation with three other *Phytophthora* species. *Phytopathology* **74**, 807 (abstract).

Cook, R. J. (1981). Biological control of plant pathogens: Overview. In *Biological Control in Crop Production* (ed. G. C. Papavizas), Beltsville Agricultural Research Centre, Maryland, Symposium No. 5, pp. 23-44. Granada: London.

Dolan, T. E. & Coffey, M. D. (1984). Biocontrol of *Phytophthora cinnamomi* on *Persea indica* and *P. americana* by prior inoculation with *Phytophthora parasitica*. *Phytopathology* **74**, 807 (abstract).

Dolan, T. E., Cohen, Y. & Coffey, M. D. (1986). Protection of *Persea* species against *Phytophthora cinnamomi* and *P. citricola* by prior inoculation with a citrus isolate of *P. parasitica*. *Phytopathology* **76**, 194-198.

Evans, E. J. & Melville, S. C. (1984). Damage and loss caused by disease in hardy ornamentals. In *Plant Disease: Infection, Damage and Loss* (ed. R. K. S. Wood & G. J. Jellis), pp. 267-272. Blackwell Scientific Publications: Oxford.

Finlay, A. R. (1987). Microbial suppression of *Phytophthora cinnamomi*. Ph.D. Thesis, The Queen's University of Belfast, U.K.

Gilmore, C. (1986). *Epicoccum purpurascens*, a potential antagonist of soil borne plant pathogenic fungi. HNC Project, University of Ulster at Jordanstown.

Grant, B. R. & Byrt, P. N. (1984). Root temperature effects on the growth of *Phytophthora cinnamomi* in the roots of *Eucalyptus marginata* and *E. calophylla*. *Phytopathology* **74**, 179-184.

Hoitink, H. A. J. (1980). Composted bark, a lightweight growth medium with fungicidal properties. *Plant Disease* **64**, 142-147.

Hoitink, H. A. J., Schmitthenner, A. F. & Herr, L. J. (1975). Composted bark for control of root-rot in ornamentals. *Ohio Reporter* **60**, 25-26.

Hsu, S. C. & Lockwood, J. L. (1969). Mechanisms of inhibition of fungi in agar by streptomycetes. *Journal of General Microbiology* **57**, 149-158.

McCracken, A. R. (1985). The isolation of *Phytophthora* species from apple orchard soils. *Record of Agricultural Research (Department of Agriculture, N. Ireland)* **33**, 31-36.

Malajczuk, N. (1979). Biocontrol of *Phytophthora cinnamomi* in eucalypts and avocados in Australia. In *Soilborne Plant Pathogens* (ed. B. Schippers & W. Gams), pp. 635-652. Academic Press: London.

Malajczuk, N. (1983). Microbial antagonism to *Phytophthora*. In *Phytophthora, its Biology, Taxonomy, Ecology and Pathology* (ed. D. C. Erwin, S. Bartnicki-Garcia & P. H. Tsao), pp. 197-218. The American Phytopathological Society: St Paul, Minnesota.

Malajczuk, N. & McComb, A. J. (1979). The microflora of unsuberized roots of *Eucalyptus calophylla* R. Br. and *Eucalyptus marginata* Donn. ex Sm. seedlings grown in soil suppressive and conducive to *Phytophthora cinnamomi* Rands. I. Rhizosphere bacteria, actinomycetes and fungi. *Australian Journal of Botany* **27**, 235-254.

Marais, L. J. & Klotze, J. M. (1976). Ectomycorrhizae of *Pinus patula* as biological deterrents to *Phytophthora cinnamomi*. *South African Forestry Journal* **99**, 35-39.

Margot, P. (1980). Influence of soil applications of metalaxyl (Ridomil®) on *Phytophthora cinnamomi* and *P. nicotianae* var. *parasitica* causing root and collar rot of avocado and citrus. In *The Second Southeast Asian Symposium and Plant Disease in the Tropics*. Program and Abstracts, 102.

Martin, C. M. (1963). Studies on the nature and effects of volatile fungal metabolites. PhD Thesis, Glasgow University.

Marx, D. H. (1969). The influence of ectotrophic mycorrhizal fungi on the resistance of pine roots to pathogenic infections. I. Antagonism of mycorrhizal fungi to root pathogenic fungi and soil bacteria. *Phytopathology* **59**, 153-163.

Marx, D. H. (1973). Growth of ectomycorrhizal shortleaf pine seedlings in soil with *Phytophthora cinnamomi*. *Phytopathology* **63**, 18-23.

Munnecke, D. E. (1984). Establishment of micro-organisms in fumigated avocado soil to attempt to prevent reinvasion of the soils by *Phytophthora cinnamomi*. *Transactions of the British Mycological Society* **83**, 287-294.

Nelson, E. B. & Hoitink, H. A. J. (1983). The role of microorganisms in the suppression of *Rhizoctonia solani* in container media amended with composted hardwood bark. *Phytopathology* **73**, 274-278.

Newhook, F. J. (1970). *Phytophthora cinnamomi* in New Zealand. In *Root Diseases and Soil Borne Pathogens* (ed. T. A. Tousson, R. V. Bega & P. E. Nelson), pp. 173-176. University of California Press: Berkeley.

Rands, R. D. (1922). Streepkanker van kaneel verooraakt door *Phytophthora cinnamomi* n. sp. *Medelingen van het Institut voor Planterzickten* **54**, 1-53.

Reeves, R. J. (1975). Behaviour of *Phytophthora cinnamomi* Rands in different soil and water regimes. *Soil Biology and Biochemistry* **7**, 19-24.

Ross, E. W. & Marx, D. H. (1972). Susceptibility of sand pine to *Phytophthora cinnamomi*. *Phytopathology* **62**, 1197-1200.

Shea, S. R. & Broadbent, P. (1983). Developments in cultural and biological control of *Phytophthora* diseases. In *Phytophthora: Its Biology, Taxonomy, Ecology and Pathology* (ed. D. C. Erwin, S. Bartnicki-Garcia & P. H. Tsao), pp. 335-350. The American Phytopathological Society: St Paul, Minnesota.

Shearer, B. L., Shea, S. R. & Deegan, P. M. (1987). Temperature-growth relationships of *P. cinnamomi* in the secondary phloem of roots of *Banksia grandis* and *Eucalyptus marginata*. *Phytopathology* **77**, 661-665.

Tsao, P. H. & Ocana, G. (1969). Selective isolation of species of *Phytophthora* from natural soils on an improved antibiotic medium. *Nature* **223**, 636-638.

Whipps, J. M. & Lynch, J. M. (1986). The influence of the rhizosphere on crop productivity. In *Advances in Microbial Ecology*, vol 9 (ed. K. C. Marshall), pp. 187-244. Plenum Press: New York.

Wong, P. T. W. & Griffin, D. M. (1974). Effect of osmotic potential on Streptomycete growth, antibiotic production and antagonism to fungi. *Soil Biology and Biochemistry* **6**, 319-325.

Wood, D. A. & Fermor, T. R. (1985). Nutrition of *Agaricus bisporus*. In *The Biology and Technology of the Cultivated Mushroom* (ed. P. B. Flegg, D. M. Spencer & D. A. Wood), pp. 43-61. Wiley-Interscience: Chichester.

Chapter 27

Chemical and biological control of *Phytophthora* species in woody plants

Olaf K. Ribeiro & Robert G. Linderman

The impact of *Phytophthora* spp. on many economically important crops worldwide has been well documented (Gregory, 1983). Recent literature reviews indicate that diseases caused by *Phytophthora* species have become an important concern in production of many agricultural, horticultural and forest plant species.

Phytophthora spp. infecting many woody plant species kill young and old plants alike. Most often, infections occur during propagation or juvenile stages and symptoms of disease appear later, when plants are in late stages of production or are in the final planting site. This fact dictates that control measures should, if at all possible, be preventative in nature, whether they are changes in cultural practices, application of chemicals, or involve biological control strategies. Where this has not been possible, however, there is a need to develop control strategies to reduce disease regarding severity on infected plants as well as spread to healthy plants.

Cultural controls of *Phytophthora* diseases have traditionally involved use of resistant or tolerant plants, such as tree fruit rootstocks, but in most cases such plants are not available or are not used because they are horticulturally unacceptable. Maintenance of optimal soil drainage is always advocated to reduce incidence and spread of these diseases, but that is not always practiced by nurserymen or landscapers. Minor levels of infection become severe under poorly drained conditions. Usually, this leads to the use of chemicals to prevent or slow down the disease. Biological control strategies are in their infancy except for the use of some potting mixes with claimed suppressiveness to diseases caused by *Phytophthora* spp. and other pathogens (Chapters 26, 28).

The structure and mode of action of systemic chemicals with efficacy against *Phytophthora* spp. has been reviewed in detail by Schwinn (1983) and Cohen & Coffey (1986). Biological control strategies were reviewed by Linderman *et al*. (1983), as well as in the comprehensive treatment of biological control by Cook & Baker (1983). Our discussion will therefore be limited to theoretical and practical control strategies of *Phytophthora* spp. on woody plants other than citrus, avocado, and tropical and sub-tropical plants since these are discussed by Coffey (chapter 28).

Chemical control

Nursery stock

A wide variety of woody plant species are grown in nurseries, either in containers or in ground beds. Propagation of these plants is either by rooting of vegetative cuttings or by seed. Field-grown seedlings or rooted cuttings are often grown in fumigated soil or in soil-less potting mixes. Rootstocks are produced in stoolbeds. Any of these nursery production systems has multiple opportunities for the introduction of inoculum of *Phytophthora* species, which leads to the production of *Phytophthora*-contaminated nursery grown plants. The realization that this happens has become a major concern to nurserymen who produce plants for planting into orchards or in landscape settings, or for grafting and/or growing-on, often in other nurseries. We have documented large losses in field plantings of fruit trees, in many species of woody ornamentals, and in conifer plantations due to planting of *Phytophthora*-infected nursery stock. This situation is well illustrated by the work of Julis, Clayton & Sutton (1978) and Jeffers & Aldwinckle (1988) who showed that there is widespread contamination of apple tree nursery stock by *P. cactorum* and *P. cambivora*. Stocks from all nurseries sampled in the U.S., Canada and Europe were contaminated with one or both of these species. Unbudded clonal rootstocks were infected more than nursery trees ready for field planting. That rootstocks can be infected in stoolbeds was recently confirmed by Tidball & Linderman (1990).

Control of *Phytophthora* spp. in stoolbed and liner nurseries is difficult because (a) plants are densely grown, making early detection difficult, (b) intensive cropping occurs at the same site for several years, allowing for the build-up of inoculum, (c) high levels of water and fertilizers are used to promote rapid growth, resulting in succulent but often stressed plants, (d) incipient *Phytophthora* infections can rarely be detected visually, (e) *Phytophthora*-contaminated irrigation and run-off water may repeatedly reinfest areas or move inoculum from one area of a nursery to others, (f) chemicals able to control *Phytophthora* are few and difficult to deliver in a consistent and effective manner, (g) containers for propagation, growing, and handling are readily contaminated and frequently re-used without effective means of disinfestation, and (h) clonally-propagated stocks may be infected and transfer inoculum or infections to new propagules. There are advantages, however, to nursery production with respect to potential control of *Phytophthora* diseases because of the unique growing and handling practices and the high value of the crops allowing for innovative, even expensive control procedures. The key is applying control measures in a timely, preventative way before infections are established.

Prophylactic treatments

An increasing number of woody plant nurseries apply prophylactic treatments of metalaxyl or fosetyl-Al to control *Phytophthora* diseases, either alone or alternately. The problem of pathogen resistance (Utkhede, 1988) to these chemicals in this type of management strategy is not a serious concern with container-grown nursery plant material that is shipped within 1-2 years. However, resistance could build-up in field-grown plant systems or in stool beds. Several instances have occurred in the Pacific Northwestern U. S. where fungicide-treated trees in the nursery succumbed in the orchard during the first growing season. Over 50% loss was recorded in some field plants of young apple and ornamental trees (Ribeiro, unpublished data). The chemicals had not completely eradicated the pathogen, possibly due to pathogen resistance, or simply incomplete eradication. Some chemicals are known to be only fungistatic (e.g. metalaxyl), not fungicidal. Furthermore, it is nearly impossible to deliver chemicals to all infection sites.

Chemical eradication

Eradication of *Phytophthora* infections on woody plants is difficult if not impossible if the plants are in soil or potting mixes in containers. Bareroot stock, on the other hand, can be treated by chemical root dips. Dipping rootstocks to eradicate *Phytophthora* infections or contamination was first reported by Brown & Hendrix (1980) who treated young apple rootstocks, contaminated with *P. cactorum*, with chemicals such as captafol, copper sulphate, PCNB, ethazole (terrazole 35WP), maneb and benomyl. Captafol reduced *P. cactorum* infection levels the most. Copper sulphate was unacceptable because of phytotoxicity at effective levels. More recently, Jeffers & Wilcox (1986) attempted to eradicate *Phytophthora* spp. from infected apple rootstocks by dipping dormant plants for 10 or 60 min in 1 mg ml^{-1} solutions of metalaxyl, captafol, mancozeb, $Cu(OH)_2$, or 1·05% sodium hypochlorite. The last was phytotoxic, and none of the other chemicals entirely eradicated *Phytophthora* spp. Nonetheless, treated plants did not show symptoms, whereas water-treated control plants did. Unfortunately, treated trees were not evaluated in later years to see if they remained disease free.

Recently, Tidball & Linderman (1990) demonstrated the efficacy of dipping apple rootstocks for 10 min in high concentrations of metalaxyl or fosetyl-Al to reduce *Phytophthora* infections or contamination. The least disease occurred if treatments with either chemical were applied after removal from stoolbeds and before cold storage, although they also reported carry-over effects of those chemicals applied in the field before harvest. None of the dip treatment chemical dosages were phytotoxic, even though they exceeded label recommendations.

Nursery containers

Growing containers are frequently re-used for several years, and are thus a significant source of *Phytophthora* inoculum as contamination from previously infected plants. Thorough washing followed by a soak in some disinfectant such as sodium hypochlorite is the routine followed by most nurserymen. While this treatment may be largely effective, it sometimes fails, even with smooth plastic containers. It is frequently ineffective with styroblock containers which are moulded from expanded polystyrene beads which have 100-150 cells/block and are widely used in forest nurseries. With continued use and re-use, these blocks develop minute cracks, crevices and holes in which infected root pieces and fungal spores can accumulate. Major losses have occurred in some conifer nurseries when contaminated blocks were used. Efforts to sanitize them have been successful only with toxic chemicals that pose a hazard to employees. Less toxic chemicals have been less successful, as have been trials to use sophisticated block washing machines, dip tanks, and pressure washing.

Field-grown plants

Phytophthora disease control on plants grown in field nurseries or orchards has been variable. Plants that exhibit advanced symptoms of *Phytophthora* infections are rarely saved by any type of treatment.

Several reports detailing efforts to control *Phytophthora* infections in apple trees have been reviewed by Utkhede (1987a, b). In general, copper-based products, as well as fungicides such as maneb, mancozeb and ferbam, have not controlled disease adequately on infected trees. On the other hand, the systemic fungicides metalaxyl and fosetyl-Al, applied singly or in combination, have been effective in many woody plant nursery and field situations.

Utkhede (unpublished) found that metalaxyl and fosetyl-Al reduced collar rot disease of apple when applied twice a year at the onset of symptoms. Severely affected trees were not significantly influenced by any of the treatments. Similar results were obtained by Ribeiro (unpublished) and by Valdebenito-Sanhueza (unpublished). Valdebenito-Sanhueza found no significant differences using fungicide combinations of metalaxyl, mancozeb and captafol applied to apple trees severely infected with collar rot and root rot.

A long term effort to control *P. cactorum* in apple orchards was recently concluded by Anliker (unpublished) in an orchard that was naturally infested with the pathogen but showed no initial symptoms of root or collar rot. The apple cultivars (Red Delicious and Golden Delicious) grafted on *Phytophthora*-susceptible EMLA.106 and EMLA.104 rootstocks were planted in 1977; tests began in 1980 and were concluded in 1988. Metalaxyl was applied at different times of the year, either as a collar drench, dripline spray, or row spray. Measuring several parameters, Anliker found (unpublished) that metalaxyl collar-drenched in the fall was most effective. Ribeiro, however, found no correlation of population levels of the pathogen in soil with symptom expression or degree of infection observed.

The efficacy of chemical treatments for the control of *Phytophthora* spp. in apple trees was also reported by Ellis, Ferree & Madden (1986) and Orlikowski, Leoni-Ebeling & Schmidle (1986). The latter workers found that preventative drenches of metalaxyl either reduced or prevented *Phytophthora* disease development for 6 weeks; fosetyl-Al and metalaxyl + fosetyl-Al at various concentrations gave control for 8 to 18 weeks. Fungicide treatments after infection had occurred or was apparent, reduced but did not eradicate the disease.

Christmas trees

In recent years, Christmas tree plantings of Noble fir (*Abies procera*), Douglas fir (*Pseudotsuga menziesii*), Shasta fir (*A. magnifica* var. *shastensis*), and Grand fir (*A. grandis*) have exhibited increasing levels of infection by *Phytophthora* spp. Ribeiro (unpublished) has identified *P. cactorum*, *P. cambivora*, *P. cinnamomi*, *P. cryptogea*, *P. drechsleri*, *P. megasperma*, and *P. pseudotsugae* on trees in the Pacific Northwestern United States. Fungicide trials by Chastagner (unpublished) indicated that treatments in the plantation fields were ineffective, possibly because several *Phytophthora* spp. were present. In young Douglas fir plantations where only one *Phytophthora* sp. was present, metalaxyl treatments were effective (Ribeiro, unpublished data).

Ornamentals

Most woody ornamentals, both in the nursery and in landscape plantings, have responded to treatments with metalaxyl or fosetyl-Al followed by application of high-phosphate fertilizer in the form of monoammonium phosphate (Ribeiro, unpublished data). However, due to the present environmental concerns, alternative control measures are needed, including utilization of beneficial microbes (such as mycorrhizal fungi, fungal and bacterial biocontrol agents) and soil amendments.

Fumigation

Fumigation with broad spectrum chemicals is often used to eliminate weeds and soilborne pathogens from soils to be planted. Fumigation with methyl bromide with or without chloropicrin is an extreme measure, since it affects virtually all microflora. If plants introduced into these treated soils are infected by or contaminated with *Phytophthora* spp., the lack of competition from soil antagonists can lead to a very rapid increase and spread of *Phytophthora* populations. Also, trees planted in fumigated soils may be stunted for the first year or two, presumably due to a lack of beneficial mycorrhizas (Linderman, 1987) and/or microbial activity. Fumigation of heavy clay soil is often ineffective because the fumigant does not penetrate deeply enough to reach *Phytophthora* propagules that were formed in infected roots at greater depths in the soil.

Vapam (metham sodium) may be a better alternative to methyl bromide fumigation because it can be used in restricted areas where live plant material is present, and it effectively kills root pathogens but does not affect as much of the microflora (Ridge & Theodorou, 1972). For example, Linderman (1987) showed that VA mycorrhizal fungi were not affected by Vapam in nursery soil (at levels that eliminated *Pythium*), compared to their complete elimination by methyl bromide. Tree species grown in soil treated with methyl bromide were stunted and variable in the absence of mycorrhizas; growth was uniform in Vapam-treated soil.

Treatment of irrigation water

Many water systems (rivers and ponds) in agricultural areas are contaminated with propagules of *Phytophthora* spp. This is true in ornamental production areas where recycled water is used and must be treated either chemically or by filtration to prevent continual re-contamination of growing areas.

In the Pacific Northwestern United States, chlorine injectors have been used successfully to reduce *Phytophthora* and *Pythium* levels in water. These systems are calibrated to deliver the desired dosage of chlorine needed in relation to the treatment time, either in the holding tank or in the irrigation line. Such a system is presently being used to treat irrigation water for 12 ha of a 20 ha arboretum near Seattle, WA. The water is pumped from a holding pond that is heavily contaminated with *Phytophthora* spp., and 1 μg ml^{-1} chlorine is injected into the line at a rate monitored by computer. Since the installation of the system 3 years ago, the incidence of *Phytophthora* disease has dropped to negligible levels, and thus has considerably reduced the reliance on fungicides. Plants with *Phytophthora* diseases before the system was installed were rhododendrons, yews, heaths, heathers, firs, cedar and hemlocks; now only an occasional rhododendron exhibits disease symptoms.

Ultraviolet (uv) radiation can also be used to treat large volumes of water. Equipment is available that can treat over 4500 l min^{-1}, but it is expensive and water must be clarified by filtration before uv sterilization. A small unit was successfully tested in an apple orchard, but the long term effectiveness and cost/benefit of such units in agricultural and horticultural enterprises have not been determined. The same could be said for other methods of water treatment, such as with ozone or by filtration.

Cultural and biological control

Fungicide applications alone rarely cure infected woody plants (Ellis *et al.*, 1986; Orlikowski *et al.*, 1986; Utkhede, 1987a, b). Fungicide treatments supplemented by various organic amendments have been suggested. Organic amendments may significantly affect soilborne diseases, but the effects may depend on the kind and state of decomposition (Linderman, 1989), ranging from stimulation in early stages of decomposition to disease suppression as decomposition approaches humification. The effect of various organic substances, as well as other cultural and biological controls of *Phytophthora* diseases have been discussed previously by Shea & Broadbent (1983). They did not, however, address the effects of humic substances, a subject of interest to us.

Humic substances

Amendment of soil with humus can influence the physical, chemical and biological systems therein (Linderman, 1989). Humic and fulvic acids from organic substances can significantly influence plants in a variety of ways, including enhanced seed germination, improved root growth and enhanced nutrient uptake. In these and other ways, humic acids enhance plant growth. Schisler & Linderman (1989) reported on the effects of high humic-content organic materials on soil microorganisms, including mycorrhizal fungi and soilborne pathogens, in several soil types. Responses were highly dependent on the soil type and rates of application, making prediction of effects extremely difficult. They did demonstrate, however, that in one soil heavily infested with root rot/damping off pathogens (species of *Fusarium* and *Pythium*) all humic materials reduced damping off. Extracts of amended soils stimulated, had no effect, or reduced sporangium production by *Phytophthora cinnamomi*.

In young apple orchards, the addition of a highly decomposed hypnum peat (also used by Schisler & Linderman, 1989) into the planting hole enhanced root growth and reduced the incidence of *Phytophthora* root rot (Ribeiro unpublished data). Large scale trials are presently being conducted in a newly planted orchard in virgin soil using that hypnum peat, either placed in the planting hole (1:1 mixture), applied as a surface mulch, or incorporated into the top 45 to 60 cm of soil. After 1 year, no trees in any of the peat-treated plots show symptoms of *Phytophthora* infections, whereas several trees in the untreated areas now exhibit symptoms. Ellis *et al.*, (1986) found that a 1:1 mixture of composted hardwood bark and field soil placed in the planting hole resulted in significantly fewer infected apple trees than those planted in non-amended soil. However, amended soil had significantly more infected trees than soils treated with metalaxyl or captafol.

Bacterial and fungal biocontrol agents

Many soil bacteria may exert inhibitory effects on species of *Phytophthora*, but few reports exist on specific bacteria that inhibit these pathogens in woody plant species. Utkhede (1987) reported good control of *P. cactorum* infections of apple trees by the application of a bacterial antagonist, *Enterobacter aerogenes* (strain B8). After 2 years of a 5-year study, B8 reduced disease incidence compared to non-treated trees, but was no better than applying fungicides alone. The mode of action was suggested to be production of antibiotics based on high activity *in vitro*. Other reports discuss activity of bacterial biocontrol agents against *Phytophthora* spp. (Yona, 1988), but little is known about the mode of action *in vivo*. It is even possible that some chemical control measures may indirectly involve such antagonists. Malajczuk *et al.* (1983) reported a marked interaction between lytic soil microorganisms and metalaxyl added to soil. They suggested caution in applying such materials that could reduce populations of microbes that could lyse propagules of *Phytophthora*.

Trichoderma spp. are becoming well-known for their capacities to control soilborne pathogens. The first successful application of *Trichoderma* to control *P. cactorum* on apple was by Valdebenito-Sanhueza (1987). She used naturally infested soil first treated with 3% formaldehyde and then, after 2 weeks, placed in large pots. Half the pots received *T. viride* inoculum produced on autoclaved sorghum grain (200 g inoculum/hole). One week later, apple seedlings were planted in all pots, and one week after that, pots with and without *Trichoderma* treatment were inoculated with *P. cactorum*. *Trichoderma* reduced mortality from natural or artificial inoculation to zero, and no infections have occurred on those trees planted into the field. Most untreated trees became diseased after planting out. This work emphasizes the benefit of establishing biocontrol agents in advance of exposure to the pathogen in soil.

Gees & Coffey (1989; Chapter 28) have reported promising activity of *Myrothecium roridum* for the control of *P. cinnamomi* on *Persea indica* in greenhouse trials, using natural or artificial inoculum. This fungal biocontrol agent exhibited good rhizosphere competence on *Persea*, but its efficacy on other hosts is known.

In current studies by Linderman, apple rootstocks are being pre-treated with fungal and bacterial biocontrol agents to eradicate incipient infections and protect roots from subsequent infections while stocks are in cold storage and upon transplant into the field.

Mycorrhizal fungi have been studied for their potential to reduce plant responses to various environmental stresses, including root infection caused by *Phytophthora* species. There is little doubt that both ectomycorrhizas and VA mycorrhizas on various woody plant species can affect plant growth under nutrient stress conditions, but some confusion and disagreement exists on whether or not mycorrhizas reduce *Phytophthora* diseases. Clearly, ectomycorrhizas can deter infection by these pathogens by various means, including production of a fungal mantle that protects root tips, production of antibiotic materials that inhibit the pathogen, and by selectively changing the rhizosphere populations to increase levels of antagonistic microflora (Marx, 1972; Malajczuk, 1988; Duchesne, Ellis & Peterson, 1989). With VA mycorrhizas, however, the picture is not so clear. Several studies (reviewed by Dehne, 1982 and Schenck, 1989) suggest that VA mycorrhizal fungi do not reduce infections by *Phytophthora* spp., but other studies on different hosts suggest the contrary (Bärtschi, Gianinazzi-Pearson & Vegh, 1981). Linderman and his colleagues are currently studying the potential for biocontrol with VA mycorrhizal fungi on apple rootstocks, inoculated singly or in combination with fungal or bacterial biocontrol agents. The basis for that approach is that mycorrhizal fungi and antagonistic microbial associates may function in biocontrol in combination within the rhizosphere (Linderman, 1988). That idea was supported by the work of Meyer & Linderman (1986) showing reduced sporangium production by *P. cinnamomi* in soil around roots with VA mycorrhizas compared to that around non-mycorrhizal roots.

Conclusion

The data presented on chemical control make it obvious that fungicide treatments alone will not cure *Phytophthora*-infected plants. There is thus a need to re-evaluate traditional methods presently advocated for control of these diseases. The several forms of survival structures of *Phytophthora* spp. make their control a challenging proposition. A successful offensive against *Phytophthora* diseases of woody plants must take into account a number of interacting factors including the following:

- Number of infecting species and host ranges. Do the species involved produce long term survival structures (e.g. oospores, chlamydospores)? Is the host range of the species involved large enough to permit survival in surrounding plants and weeds?
- Debris/organic matter. Will the organic matter present prevent the applied chemical or biological control agent from reaching its target site? Do the pathogen propagules overwinter in debris present in the field?

- Soil type. Heavy clay soils increase the time for applied control agents to penetrate the soil profile and reach target sites; sandy soils on the other hand, result in rapid percolation of applied chemicals and biologicals through the soil profile and often results in leaching if heavy irrigation or rainfall follows applications.
- Microbial activity in the soil profile. Are there microbial antagonists to *Phytophthora* present in the soil? Are the microorganisms present capable of rapid breakdown of the applied chemical or are they antagonistic to introduced biological control agents?
- Application of biocontrol agents. Are the methods developed for the application of biological control agents suitable to permit the introduced agents to reach their target sites without loss of activity? (See Chapters 26 and 28.)
- Inoculum sources. Recurring infections can be prevented by monitoring possible sources of contamination such as irrigation water, re-used containers, and nursery stock.
- Resistant plants. The use of resistant plant material can enhance chemical and biological control.
- Environmental stresses. Elimination of plant stress factors, such as excessive salts, poor nutrition, flooding effects, etc., that enhance *Phytophthora* infections can reduce disease incidence and severity.

Novel approaches using biological agents currently known to have efficacy against *Phytophthora* may offer better possibilities for long term control than using only chemicals. Approaches using natural soilborne microorganisms are both intellectually and environmentally appealing. The present environmental concerns on pesticide use make it imperative that we examine systems that offer a more scientific basis for integrated pest management systems. Perhaps lower levels of chemicals can be used in combination with biological control agents. It is imperative that with perennial woody plant species, inoculation with beneficial microbes be done early in the production cycle so that they may become established and maintained on the roots prior to transplant to the permanent growing site (Linderman, 1986). Otherwise, the problem, as with chemicals, of not being able to deliver the control agent to the site of activity may be repeated. Prevention of *Phytophthora* diseases on woody host plants appears to be the only approach that makes long term control feasible.

References

Bärtschi, H., Gianinazzi-Pearson, V. & Vegh, I. (1981). Vesicular-arbuscular mycorrhiza and root rot disease (*Phytophthora cinnamomi* Rands) development in *Chamaecyparis lawsoniana* (Murr.) Parl. *Phytopathologische Zeitschrift* **103**, 213-218.

Brown, E. A. & Hendrix, F. F. (1980). Efficacy of *in vitro* activity of selected fungicides for control of *Phytophthora cactorum* collar rot of apple. *Plant Disease* **64**, 310-312.

Cohen, Y. & Coffey, M. D. (1986). Systemic fungicides and the control of oomycetes. *Annual Review of Phytopathology* **24**, 311-338.

Cook, R. J. & Baker, K. F. (1983). *The Nature and Practices of Biological Control of Plant Pathogens*. APS Press: St. Paul, Minnesota.

Dehne, H. W. (1982). Interaction between vesicular-arbuscular mycorrhizal fungi and plant pathogens. *Phytopathology* **72**, 1115-1119.

Duchesne, L. C., Ellis, B. E. & Peterson, R. L. (1989). Disease suppression by the ectomycorrhizal fungus *Paxillus involutus*: contribution of oxalic acid. *Canadian Journal of Botany* **67**, 2726-2730.

Ellis, M. A., Ferree, D. C. & Madden L. V. (1986). Evaluation of metalaxyl and captafol soil drenches, composted hardwood bark soil amendments, and graft union placement on the control of apple collar rot. *Plant Disease* **70**, 24-26.

Gees, R. & Coffey, M. D. (1989). Evaluation of a strain of *Myrothecium roridum* as a potential biocontrol agent against *Phytophthora cinnamomi*. *Phytopathology* **79**, 1079-1084.

Gregory, P. H. (1983). Some major epidemics caused by *Phytophthora*. In *Phytophthora: Its Biology, Taxonomy, Ecology and Pathology*, (ed. D. C. Erwin, S. Bartnicki-Garcia & P. H. Tsao), pp. 271-278. American Phytopathological Society: St. Paul, Minnesota.

Jeffers, S. N. & Aldwinckle, H. S. (1988). *Phytophthora* crown rot of apple trees: sources of *Phytophthora cactorum* and *P. cambivora* as primary inoculum. *Phytopathology* **78**, 328-335.

Jeffers, S. N. & Wilcox, W. F. (1986). Preplant root dips for apple rootstocks naturally infected with *Phytophthora* species. *Phytopathology* **76**, 1105 (abstract).

Julis, A. J., Clayton, C. N. & Sutton, T. B. (1978). Detection and distribution of *Phytophthora cactorum* and *P. cambivora* on apple rootstocks. *Plant Disease Reporter* **62**, 516-520.

Linderman, R. G. (1986). Managing rhizosphere microorganisms in the production of horticultural crops. *HortScience* **21**, 1299-1302.

Linderman, R. G. (1987). Response of shade tree seedlings, VA mycorrhizal fungi, and Pythium to soil fumigation with Vapam and methyl bromide. In *Proceedings of the 7th North American Conference on Mycorrhizae*, ed. D. M. Sylvia, L. L. Hung & J. H. Graham, p. 30. Institute of Food and Agricultural Sciences, University of Florida, Gainesville, Florida, USA.

Linderman, R. G. (1988). Mycorrhizal interactions with the rhizosphere microflora; The mycorrhizosphere effect. *Phytopathology* **78**, 366-371.

Linderman, R. G. (1989). Organic amendments and soilborne diseases. *Canadian Journal of Plant Pathology* **11**, 180-183.

Linderman, R. G., Moore, L. W., Baker, K. F. & Cookesy, D. A. (1983). Strategies for detecting and characterizing systems for biological control of soilborne plant pathogens. *Plant Disease* **67**, 1058-1064.

Malajczuk, N. (1988). Interaction between *Phytophthora cinnamomi* zoospores and micro-organisms on non-mycorrhizal and ectomycorrhizal roots of *Eucalyptus marginata*. *Transactions of the British Mycological Society* **90**, 375-382.

Malajczuk, N., Broughton, T. J., Campbell, N. A. & Litchfield, R. T. (1983). Interaction between the fungicide metalaxyl and soil microorganisms on survival of *Phytophthora cinnamomi*. *Annals of Applied Biology* **103**, 57-61.

Marx, D. H. (1973). Mycorrhizae and feeder root diseases. In *Ectomycorrhizae. Their Ecology and Physiology*, ed. G. C. Marks & T. T. Kozlowski, pp. 351-382. Academic Press: New York.

Meyer, J. R. & Linderman, R. G. (1986). Selective influence on populations of rhizosphere or rhizoplane bacteria and actinomycetes by mycorrhizas formed by *Glomus fasciculatus*. *Soil Biology and Biochemistry* **18**, 191-196.

Orlikowski, L. B., Leoni-Ebeling, M. & Schmidle, A. (1986). Efficacy of metalaxyl and fosetyl-Al in the control of *Phytophthora cactorum* on apple trees. *Zeitschrift für Pflanzenkrankheiten und Pflanzenschutz* **93**, 202-209.

Ridge, E. H. & Theodorou, C. (1972). The effect of soil fumigation on microbial recolonization and mycorrhizal infection. *Soil Biology and Biochemistry* **4**, 295-305.

Schenck, N. C. (1989). Vesicular-arbuscular mycorrhizal fungi and the control of fungal root diseases. In *Innovative Approaches to Plant Disease Control*, ed. I. Chet, pp. 179-191. John Wiley & Sons: New York.

Schisler, D. A. & Linderman, R. G. (1989). Influence of humic-rich organic amendments to coniferous nursery soils on Douglas-fir growth, damping-off and associated soil microorganisms. *Soil Biology and Biochemistry* **21**, 403-408.

Schwinn, F. G. (1983). New developments in chemical control of *Phytophthora*. In: *Phytophthora: Its Biology, Taxonomy, Ecology and Pathology*, (ed. D. C. Erwin, S. Bartnicki-Garcia & P. H. Tsao), pp. 327-334. American Phytopathological Society: St. Paul, Minnesota.

Shea, S. R. & Broadbent, P. (1983). Developments in cultural and biological control of *Phytophthora* diseases. In *Phytophthora: Its Biology, Taxonomy, Ecology and Pathology*, (ed. D. C. Erwin, S. Bartnicki-Garcia & P. H. Tsao), pp. 335-350. American Phytopathological Society: St. Paul, Minnesota.

Tidball, C. J. & Linderman, R. G. (1990). *Phytophthora* root and stem rot of apple rootstocks from stool beds. *Plant Disease* **74**, 141-146.

Utkhede, R. S. (1987a). Chemical and biological control of crown and root rot of apple caused by *Phytophthora cactorum*. *Canadian Journal of Plant Pathology* **9**, 295-300.

Utkhede, R. S. (1987b). Control of crown rot (*Phytophthora cactorum*) of apple trees with the systemic fungicides metalaxyl and fosetyl-Al. *Pesticide Science* **19**, 289-295.

Utkhede, R. S. (1988). *In vitro* selection of strains of *Phytophthora cactorum* resistant to metalaxyl. *Journal of Phytopathology* **122**, 35-44.

Valdebenito-Sanhueza, R. M. (1987). Uso de formaldeído e *Trichoderma* para previnir a recolonização de solo por *Phytophthora* em pomares do macieras. In *Anais da II Reduniao Sobre Controle Biologico de Doencas de Plantas*, p. 55. Campinas Funação Cargill.

Yona, I. Pinkas, Y., Elad, Y. & Chet, I. (1988). Biological control of *Phytophthora* root rot in *Persea indica* seedlings. *Phytoparasitica* **16**, 67.

Chapter 28

Strategies for the integrated control of soilborne *Phytophthora* species

Michael D. Coffey

The majority of *Phytophthora* species are soilborne pathogens causing damping off, root rots, collar or crown rots, stem rot, foliar blights, pod or fruit rots, or even bud rot (Table 28.1). The exceptions are those rare *Phytophthora* diseases, such as late blight of potatoes and tomatoes, that have a predominant above-ground etiology. The soil acts as an important reservoir of inoculum for many *Phytophthora* species, including those that cause their principal damage to aerial parts of the host plant. A classical example of the latter is *P. megakarya* which causes black pod of cocoa in Nigeria (Gregory, 1983). Disease is frequently initiated by rain droplets containing zoospores of the pathogen; these motile spores, released from sporangia produced on infected rootlets, are present on the soil surface. Rain splash disperses the zoospore inoculum at least 0·5 m into the canopy of the cocoa tree to initiate primary infections (Gregory *et al*., 1984). *Phytophthora* species may exist as dynamic colonizers of host root tissues producing sporangia which can produce multitudes of zoospores (Zentmyer & Erwin, 1970). Additionally, they can survive as oospores (Schmitthenner, 1985), chlamydospores (Weste & Vithanage, 1979), or sporangia (Old, Oros & Malafant, 1984), and spore viability is greatly extended in decaying root fragments (Old *et al*., 1984), or other plant debris in the soil. *Phytophthora* species are therefore highly adapted to their soil environment, and once established are extremely difficult to eradicate. In fact, Cook & Baker (1983) regard the spread of *P. cinnamomi* through sale of contaminated nursery stock as 'a particularly dangerous type of environmental pollution'.

Traditional approaches to the control of many soilborne pathogens, including *Phytophthora* species, have involved planting in well-drained sites and avoidance of infested soils where possible. Unfortunately, such cultural (protective husbandry) methods have limited usefulness since pathogens such as *Phytophthora* are frequently introduced either with planting material or *via* contaminated soil or water; most *Phytophthora* species have the remarkable ability to subsist as dormant propagules in relatively low densities in soil (e.g. 1 to 3 colony forming units g^{-1} dry wt soil for *P. cinnamomi*), yet in the presence of a host can build up rapidly

Table 28.1. Soilborne *Phytophthora* species and some diseases they cause.

Pathogen	host (disease)
Root rots	
P. capsici	green pepper, cucumber
P. cinnamomi	avocado, walnut
P. citrophthora	citrus
P. cryptogea	tomato
P. drechsleri	cucumber, melon
P. fragariae	strawberry (red stele)
P. megasperma	soybean
P. megasperma	lucerne
P. megasperma	cherry, apple
Foot, collar (crown) and stem (trunk) rots	
P. cactorum	apple
P. cambivora	apple, cherry
P. capsici	black pepper
P. capsici	green pepper
P. citricola	avocado, walnut
P. citrophthora	citrus
P. cryptogea	tomato
P. megasperma	cherry
P. megasperma (*P. sojae*)	soybean
P. palmivora	cocoa, durian
P. nicotianae (syn.. *parasitica)*	citrus, tobacco (gummosis, black shank)
P. syringae	almond
Foliage or blossom blights	
P. cactorum	rhododendron (dieback)
P. capsici	black pepper (*Piper nigrum*)
P. capsici	green pepper (*Capsicum annuum*)
P. capsici	macadamia
P. nicotianae (syn. *parasitica)*	tomato
Fruit, tuber and pod rots	
P. capsici	green pepper
P. capsici	cocoa (black pod)
P. citrophthora	cocoa (black pod)
P. citrophthora	citrus (brown rot)
P. erythroseptica	potato (pink rot)
P. megakarya	cocoa (black pod)
P. palmivora	cocoa (black pod)
P. palmivora	coconut (premature nutfall)
P. nicotianae (syn. *parasitica)*	tomato (buckeye rot)
P. syringae	apple
Bud rot	
P. katsurae	coconut
P. palmivora	coconut, arecanut

by producing repeated cycles of sporangia from which multitudes of infective zoospores can be released under saturated soil conditions.

In more recent times, soil fumigation using general biocides such as methyl bromide or aerated steaming of soil, has proved useful in horticultural operations raising potted plants. The same technologies have sometimes been applied to the production of nursery fruit trees destined for field planting, such as avocado (Coffey, 1987). However, fumigation has found little application with most field soils due to its high cost. In the rare instances where it has been attempted in the field, such as with *P. cinnamomi* on avocados in California, results have generally been disappointing. This is in part related to the difficult soil profiles often encountered in a field situation and the need to fumigate to sufficient depth, typically 5 to 10 cm, in order to inactivate the pathogen population. But perhaps more importantly, fumigation can rarely eradicate *Phytophthora* in such field situations, and the pathogen can rapidly build up in a soil now largely devoid of microbial competitors. However, soil fumigation with methyl bromide-chloropicrin mixtures did revolutionise production of one Californian crop – the strawberry. Introduction of such fumigation technology in the 1960s for the first time permitted dependable high yields of strawberries, largely due to the temporary control of *Verticillium* wilt, as well as other soilborne pathogens including *Ceratobasidium* and *Phytophthora fragariae* (Wilhelm & Paulus, 1980). The method has been successful largely because strawberry is treated as an annual crop in California and, most importantly, because it is a high value crop which has been successfully subjected to intensive breeding for high yield potential and good marketing qualities.

For the majority of *Phytophthora* problems, effective control practices have been much slower to evolve. Selection and breeding for resistance or tolerance has been advocated. In most instances, however, suitable germ plasm has been difficult to find, particularly with tree crops where the subsequent progress has been slow due to the time frames of 15 to 20 years needed to evaluate resistance in such perennial crops. In rare cases, resistance has been much easier to detect and diverse types of germ plasm have been found in available accessions. A classic example is that of resistance to *P. megasperma* f. sp. *glycinea* (*P. sojae*) in soybean (Schmitthenner, 1985). *Phytophthora* root rot (PRR) of soybean first became a problem in the midwest USA in the early 1950s; at almost the same time, useful sources of resistance were detected in some cultivars which were deployed in midwest soybean fields, but races of the pathogen capable of overcoming such resistance soon emerged.

A major advance occurred in the 1970s with the development of several new classes of systemic fungicide capable of controlling Oomycete pathogens, especially *Pythium*, *Phytophthora* and downy mildews (Bruin & Edgington, 1983; Cohen & Coffey, 1986; Schwinn, 1983). Amongst these fungicides, two classes have emerged which possess extremely useful properties against various soilborne *Phytophthora* species; the phenylamides, exemplified by metalaxyl (Schwinn, 1983) and the phosphonates, particularly fosetyl-aluminium and potassium phosphonate (Coffey & Ouimette, 1989). Metalaxyl possesses excellent mobility in soils (Cohen & Coffey, 1986) and good activity against different *Phytophthora* species. The majority of the soilborne species against which metalaxyl is used, including *P. megasperma* f. sp. *glycinea*, *P. cinnamomi*, *P. nicotianae* (*P. parasitica*) and *P. palmivora* are very sensitive (Coffey & Bower, 1984a). Fosetyl-aluminium, and particularly its breakdown product, phosphonic acid (Cohen & Coffey, 1986; Ouimette & Coffey, 1988, 1989), are also quite active against some *Phytophthora* species, notably *P. cinnamomi*, *P. citricola*, *P. citrophthora* and the majority of isolates of *P. palmivora* (Coffey & Bower, 1984b; Coffey & Joseph, 1985; Coffey & Ouimette, 1989). Interestingly, buffered phosphonic acid possesses very little activity against certain other species, notably *P. infestans* and *P. megasperma* f. sp. *glycinea* (Coffey & Bower, 1984b; Coffey & Ouimette, 1989).

Fosetyl-aluminium, or more specifically phosphonic acid (the latter usually formulated as potassium phosphonate), possesses remarkable systemic properties, being translocated downwards to the roots of a plant after foliar application (Coffey & Ouimette, 1989; Fenn & Coffey, 1984; Ouimette & Coffey, 1989). The unique ambimobility of phosphonate (Ouimette & Coffey, 1990) has proven a decisive factor in the use of phosphonate fungicides for effective control of several soilborne Phytophthoras. As an example, previous chemical control methods, including use of metalaxyl soil applications, had proved uneconomic and relatively ineffective for suppression of *P. cinnamomi* on susceptible rootstocks of large bearing avocado trees (Coffey, 1987; Darvas, Toerien & Milne, 1984; Pegg & Whiley, 1987). Trunk injections of a phosphonate fungicide, fosetyl- aluminium or potassium phosphonate, proved more economical and effective than previous methods for control of PRR on avocados (Coffey, 1987; Darvas *et al.*, 1984; Pegg *et al.*, 1985).

Precisely because valuable new chemicals and application technologies are finally becoming available, which are relatively economical and effective for control of several important soilborne Phytophthoras, there has been an increasing awareness in the last decade of the need to develop integrated control approaches for disease management (Coffey, 1987; Pegg & Whiley, 1987; Schmitthenner, 1985). Firstly, introduction of additional measures, including various cultural practices and use of moderately resistant or tolerant germ plasm, may reduce the levels of fungicides required without diminishing the control achieved (Schmitthenner, 1985). Secondly, reduced usage of such fungicides should help to prevent their premature failure, either due to enhanced biodegradation in soils (Bailey & Coffey, 1985; Pegg & Whiley, 1987) or due to emergence of resistant strains (Bower & Coffey, 1985; Cohen & Coffey, 1986; Davidse, 1982). It has been suggested that the emergence of fungicide-resistant strains of soilborne *Phytophthora* pathogens is unlikely because of their inherently different epidemiology, as compared to foliar pathogens such as *P. infestans* (Schwinn, 1983). However, the rapidity with which virulent strains of *P. megasperma* f. sp. *glycinea* appeared to overcome resistance in soybean cultivars should be a reminder that such concepts are not necessarily supported by actual experiences in the field. Constant use of a site-specific fungicide such as metalaxyl (Davidse, Hofman & Velthuis, 1983) can be expected to lead eventually to development of genetically stable resistance. Bruin & Edgington (1981) clearly demonstrated this resistance phenomenon in the laboratory with *P. capsici*. More recently, Bower & Coffey (1985) demonstrated that the fungicidal activity of fosetyl-aluminium and potassium phosphonate, as well as metalaxyl, could all be overcome using chemical mutagenesis to produce fungicide-resistant strains of *P. capsici*. Moreover, such fungicide-resistant strains have proved to be environmentally fit and capable of surviving as long as wild type strains in soil (Lucas, Bower & Coffey, 1990).

The remainder of this chapter will concentrate on approaches to control of two exclusively soilborne *Phytophthora* problems: *P. cinnamomi* on avocados and *P. megasperma* f. sp. *glycinea* on soybeans. In addition, two problems which are partially soilborne but also have a very important aerial phase in their pathology, *Phytophthora* species on cocoa and *P. capsici* on black pepper, will be considered. Such disease problems are especially troublesome since control strategies that effectively suppress the soil phase of the disease, may have little or no impact on the aerial phase, and *vice versa*.

Specific strategies for integrated control

Phytophthora *root rot of avocados*

The avocado (*Persea americana*) exists as three horticultural races primarily based on fruit characters: Mexican (var. *drymifolia*), Guatemalan and West Indian. The industry began in the early part of this century but did not expand dramatically until the 1960s. In California, a problem known as avocado decline or dying back was recorded in the 1920s but did not receive any serious attention until the 1940s. Wager, a South American plant pathologist, was the first to diagnose the cause of the problem in California and South Africa as *P. cinnamomi* (Wager,

1942). However, Tucker (1928) had previously isolated *P. cinnamomi* from diseased avocados in Puerto Rico.

The root system, especially the fine feeder roots, of avocado is extremely susceptible to *P. cinnamomi*. No selection of *Persea americana* has been found that has high resistance to the disease. Due to the availability of relatively cheap land and water in California, the area devoted to this high value crop, colloquially known as 'green gold', increased from 8000 ha in 1969 to approximately 32000 ha in 1985. Traditionally, the main control measures were to make new plantings in sites free of disease using clean nursery stock. However, even this approach had its failures, due to accidental introduction of the pathogen, either on the trees, or through flooding or movement of contaminated soil from other ranches with the disease problem.

The practical possibilities of integrated control of PRR of avocados evolved in the 1980s with the commercial availability of moderately resistant or tolerant rootstocks such as 'Duke 7' (Coffey, 1987) and anti-Oomycete fungicides such as metalaxyl and fosetyl-aluminium (Cohen & Coffey, 1986). Worldwide, the four components of integrated control of PRR of avocados are:

- hygiene and sanitation
- cultural and biological control
- use of tolerant rootstocks, and
- fungicides.

Hygiene and sanitation are key factors since such practices aim to delay development and spread of PRR for as long as possible. In the nursery, avocado trees are raised in well-drained potting mixes which have been fumigated with methyl bromide or treated with aerated steam to free them of pathogens. Care is taken to reduce the possibilities for accidental introduction of the pathogen by providing disinfectant footbaths and tyre dips for vehicles.

A key cultural component consists of planting trees on mounds (Coffey, 1987) or ridges (Pegg & Whiley, 1987), especially on heavier soils and where the site is prone to flooding. This feature not only improves drainage but also serves to reduce soil compaction. In Queensland, Australia, mulching and cover cropping is practiced to maintain the high organic matter typical of the rain forest soils in which avocados are planted (Pegg & Whiley, 1987). This not only serves to maintain a large population of antagonistic microorganisms but improves soil structure, facilitating drainage. Companion cropping with citrus in California, bananas in New South Wales and coffee in Costa Rica has been practiced to remove excessive soil moisture particularly in heavier soils, and perhaps to slow down spread of the pathogen. These intercropping practices have not been critically evaluated, though in the case of bananas, their removal leads to rapid loss of the avocados due to PRR (Pegg & Whiley, 1987).

The search for rootstocks with resistance to *P. cinnamomi* has gone on continuously since the 1950s (Coffey, 1987). Several distant relatives of the avocado in another subgenus *Eriodaphne* (Kopp, 1964), including *Persea borboni*, *P. pachypoda* and *P. cinerascens*, were found to be resistant. Unfortunately, these species were not graft-compatible with avocado. Moderate resistance or tolerance has been found in a few selections of avocado (*Persea americana*) including Duke 7 and G6 (Zentmyer, 1980), Barr Duke and Thomas (Coffey, 1987; Gabor, Guillemet & Coffey, 1990) and Velvick (Pegg & Whiley, 1987). Ideally, such rootstocks should be established in disease-free sites since they are produced as cuttings and lack the initial vigour of seedling rootstocks. Recently, additional resistance has been found in two graft-compatible near relatives to *Persea americana*: *Persea steyermarkii* and *P. schiedeana* (Coffey, 1987; Gabor & Coffey, 1990b). Additionally, three natural hybrids of *Persea americana* × *Persea schiedeana* have been discovered which perform as moderately resistant rootstocks under greenhouse (Gabor & Coffey, 1990a) and field conditions (Gabor *et al.*, 1990). Interestingly, these rootstocks (G755a, b, c) possess some physiological resistance to *Phytophthora cinnamomi*, restricting lesion development as compared to both susceptible and tolerant *Persea americana* rootstocks (Dolan & Coffey, 1986; Gabor & Coffey, 1990a). *Persea schiedeana* and *P. steyermarkii* selections and also possess this type of physiological resistance (Gabor & Coffey, 1990b). In contrast, no evidence of physiological resistance has been found in the majority of avocado (*Persea americana*) rootstocks with field resistance (Gabor *et al.*, 1990). Selections such as Duke 7 (Kellam & Coffey, 1985), Thomas (Gabor & Coffey, 1990a) and Velvick (Pegg & Whiley, 1987) are tolerant primarily because they can produce new feeder roots more efficiently and in larger numbers than can susceptible rootstocks. Current levels of tolerance in these rootstocks allow establishment of avocados in well-drained sites where *Phytophthora cinnamomi* is either absent during the establishment phase, or where fungicides such as metalaxyl or fosetyl-aluminium are used (Coffey, 1987).

Fungicides such as metalaxyl and fosetyl-aluminium became available in the 1980s. With avocados on clonally propagated, tolerant rootstocks both fungicides have performed well, allowing establishment of trees even on heavier soils, providing cultural care was good (Coffey, 1987). In addition, they have proved very economical, with annual costs per tree less than one-tenth the purchase cost of a single avocado tree on a tolerant rootstock. However, metalaxyl is not economical for use on larger trees,

5 years or older, because it has to be applied as a soil treatment and progressively larger amounts are required (Coffey, 1987). Biodegradation is also a factor since many soils in California, South Africa and Australia exhibit enhanced abilities to metabolize the fungicide after one or two years' usage (Bailey & Coffey, 1985, 1986; Pegg & Whiley, 1987). In contrast, the phosphonate fungicides fosetyl-aluminium and potassium phosphonate have proved quite economical and retained their relative efficacy on large trees over several years. The most effective and efficient method of application of these two fungicides has been through use of trunk injections using plastic syringes (Coffey, 1987; Darvas *et al.*, 1984; Pegg *et al.*, 1985).

The prognosis for effective control of avocado root rot now looks exceedingly bright. Choice of a good planting site, use of mounds or ridges especially on heavier soils or poorly drained sites, tolerant rootstocks and judicious use of fungicides appear to be key components in integrated control.

Phytophthora *root rot of soybean*

P. megasperma f. sp. *glycinea* (*P. sojae*, Faris *et al.*, 1989) is the only severe *Phytophthora* disease of a major animal feed crop in the U.S.A. (Schmitthenner, 1985). The disease was first observed in 1948 in Indiana and by 1954 the causal agent had been identified as a *Phytophthora* species. First erroneously named *P. cactorum*, it was later described as a new species *P. sojae* (Kaufmann & Gerdemann, 1958). Unfortunately, this early species description, though probably correct in the light of recent research (chapter 10), was not generally accepted on broad morphological grounds and the pathogen was re-classified as *P. megasperma* var. *sojae* (Hildebrand, 1959). Subsequently, Kuan & Erwin (1980) recognized the host-specific nature of this soybean pathogen and named it *P. megasperma* f. sp. *glycinea*. Recent biochemical and molecular studies have supported the unique genetic nature of the pathogen (Faris *et al.*, 1989; Förster *et al.*, 1989; Hansen *et al.*, 1986; chapter 10) and Faris *et al.* (1989) went so far as to suggest renaming the pathogen *P. sojae*.

Unlike with avocado and *P. cinnamomi*, resistance to *P. megasperma* f. sp. *glycinea* was readily found in existing soybean accessions. In fact, many different alleles and seven loci expressing hypersensitive resistance to the pathogen have been detected and incorporated into new cultivars. Despite the availability of much useful resistant germ plasm, the pathogen has advanced progressively and now damages about 6·5 million ha in the USA and Canada (Schmitthenner, 1985). Since races capable of overcoming all major genes for resistance are known to exist already, the continuing use of major genes as a primary method of control appears vulnerable. Consequently, alternative strategies based on integrated control principles are now being evaluated (Schmitthenner, 1985; Schmitthenner & Van Doren, 1985).

Moderate resistance or tolerance to *P. megasperma* f. sp. *glycinea*, largely unrelated to major gene resistance, has been identified in some breeding lines. Tolerance is defined as 'the relative ability to survive root infection, either natural or artificial, without showing severe symptom development such as death, stunting, or yield loss' (Walker & Schmitthenner, 1984b). Since heritability is quite high (Walker & Schmitthenner, 1984b) it would appear feasible to incorporate such tolerance into commercial cultivars. Tolerance also appears much more stable than resistance and some lines have been grown for over 15 years without emergence of pathogen races with increased virulence to such hosts. The main stumbling block is that under severe disease situations the yield of tolerant lines is only about 75% of that of a resistant cultivar.

Fungicides have been evaluated for their efficacy against PRR of soybean. Metalaxyl is moderately effective when used as a seed dressing, though cultivars with low tolerance are not protected after seedling emergence (Schmitthenner, 1985). In addition, there is some evidence that metalaxyl may reduce yields, though it has yet to be determined whether this is cultivar- or environment-specific. Metalaxyl soil treatment is highly effective against PRR using highly tolerant, but not susceptible cultivars (Anderson & Buzzell, 1982), though the higher amounts employed make this method much more expensive.

Various cultural factors have been evaluated for their role in limiting PRR. Of these, improving impeded drainage through tiling appears the most significant (Schmitthenner, 1985). Spring tillage also resulted in somewhat higher yields with high-tolerant cultivars as compared to zero-tillage treatments. In contrast, planting on ridges or beds and rotation with another crop had little impact on PRR severity under Ohio conditions (Schmitthenner, 1985; Schmitthenner & Van Doren, 1985).

Two integrated control packages have been proposed. In one option, metalaxyl soil treatment is combined with the use of high-tolerant cultivars. The principal disadvantage is the high cost of metalaxyl. The most comprehensive option involves complete tillage, tile drainage, high-tolerant cultivars and a metalaxyl seed treatment (Schmitthenner & Van Doren, 1985). The main disadvantage is the cost of tiling, while some growers prefer to plant soybeans without tillage.

Use of major gene resistance remains a primary control strategy. However, the availability of tolerant breeding lines which appear more durable in the face of pathogenic diversity, provides soybean breeders with new opportunities. Improvements in the levels of tolerance appear possible through recurrent selection (Walker & Schmitthenner, 1984c).

Additional security might be achieved by incorporating major resistance genes into a tolerant genetic background. The fungicide metalaxyl appears to be an invaluable component of integrated control but should be used sparingly so as to minimize potential problems with enhanced soil biodegradation and development of fungicide-resistant strains.

Black pod and stem canker of cocoa

Phytophthora species attacking cocoa include *P. palmivora*, worldwide, *P. megakarya* in West Africa and *P. capsici* and *P. citrophthora* in Brazil (Zentmyer, 1988). In 1985, worldwide losses due to black pod were estimated at £1540 million (Evans & Prior, 1987).

Cultural control is widely practiced in an attempt to reduce black pod incidence. Approaches include regular harvesting of pods, which are produced throughout the year, and canopy thinning to improve air circulation and reduce relative humidity. Removal of infected pods has also been advocated, though this does increase the risk of spreading the disease (Evans & Prior, 1987). Of these cultural measures, regular harvesting of pods plays the most significant role, since it reduces the damage due to spread of the pathogen to other pods and bark, where flower cushion infections and stem cankers may develop.

Phytophthora species, particularly *P. megakarya* (Gregory *et al*., 1984) and *P. palmivora* (Gregory, 1983), are present as a primary reservoir in the soil. Rain splash disperses the zoospore inoculum into the lower part of the tree. Additional rain splash and also invertebrate vectors such as ants serve to move the inoculum within the canopy of the cocoa tree. Once this secondary aerial phase of the disease is well established, control by cultural practices becomes ineffectual.

Where economic considerations permit their use, fungicides represent the principal weapon against *Phytophthora* on cocoa. Copper fungicides, particularly cuprous oxide and cupric oxychloride formulations, are widely used as foliar sprays (Evans & Prior, 1987). Their effectiveness is dependent on the timing and frequency of spraying and the amount of rainfall. Trends in fungicide usage are towards increasing the copper content whilst decreasing the number of applications (Pereira, 1985) and the possible use of flowable copper formulations with low-volume spray equipment (Fulton, 1989).

Control of stem canker has been achieved using metalaxyl combined with an insecticidal paint to control bark-damaging insects (Prior & Smith, 1982). However, the method is extremely labour-intensive since young cankers are difficult to detect. Metalaxyl combined with cuprous oxide has also been used as a foliar spray for control of black pod (McGregor, 1982, 1984). Recently, trunk injections of potassium phosphonate have proved potentially effective in controlling black pod and stem canker (Anderson, Middleton & Guest, 1989). The economics and feasibility of using fungicides and especially the newer systemic compounds, depends on the type of cocoa operation (plantation company or smallholder), the yield potential and value of the crop, the disease severity, and the amount of rainfall (McGregor, 1983; Prior, 1984; Evans & Prior, 1987). One conclusion is clear, however; high yields are only likely to be obtained using regular chemical treatment (Prior, 1984).

The majority of the world's cocoa is highly susceptible to black pod disease. Breeding efforts with cocoa, in common with many other perennial crops, are by necessity slow. In the case of black pod, efforts are further hampered by the lack of reliable and rapid, small-scale tests for resistance. Furthermore, resistance to stem canker may not confer resistance to black pod. Resistance to bark infection seems to be more common (Evans & Prior, 1987). However, useful resistance to black pod has been detected in Trinitario and Upper Amazon clones (Evans & Prior, 1987). Breeding programmes are further complicated by the fact that there are three important species, *P. capsici*, *P. megakarya* and *P. palmivora*, attacking cocoa in different parts of the world (Zentmyer, 1988).

Prospects for more effective control of black pod and stem canker look quite bright although progress will be slow. Incorporation of useful resistance to both canker and black pod into high yielding cultivars will be a critical step. New systemic materials (metalaxyl, potassium phosphonate) should only be used to reduce initial disease incidence, and be followed by well-timed applications of copper fungicides.

Foot rot or quick decline of black pepper

Phytophthora capsici is the cause of a serious foot rot in the principal countries where black pepper (*Piper nigrum*) is produced: India, Indonesia, Malaysia and Brazil (de Waard, 1986). The causal agent, originally described as *P. palmivora* MF4 and more recently as *P. capsici* (Kasim & Prayitno, 1979), has now been confirmed as *P. capsici* following a taxonomic revision of the latter species (Tsao & Alizadeh, 1988). Isozyme analysis of isolates of *P. capsici* from worldwide sources has revealed that there are three distinct populations or subgroups designated CAP1, CAP2 and CAP3 (chapter 12). Isolates from black pepper belong to CAP1 and CAP2. No evidence of hybridization between CAP1 and CAP2 has been found, despite the fact that both subgroups can be present on black pepper in the same country. It has not yet been determined whether the two types of *P. capsici* differ in their pathology on black pepper; isolates of CAP2 often form chlamydospores whereas CAP1 types never do (Mchau & Coffey, unpublished).

Black pepper originates from the wet tropical jungles of the Western Ghats in Kerala, India. It is primarily a smallholder crop and foot rot is the major limiting factor in its production (de Waard, 1986). The origins of *P. capsici* are unknown but perhaps the widespread distribution of *Capsicum* spp. in India may have played a role since these are also hosts of this pathogen.

P. capsici is a soilborne fungus; it attacks the roots, collars, aerial stems, leaves, branches and fruit spikes of the black pepper vine. The fungus is spread by water and by soil cultivation operations in the pepper garden. The roots and stem collar are most adversely affected leading to rapid wilting and death of the entire vine. Runner shoots, being in close contact with the ground, can also become infected at the onset of a monsoon or rainy period. These lesions, which occur on both shoots and young leaves, act as primary foci for secondary spread of the pathogen via rain splash (Nambiar & Sarma, 1982; Ramachandran *et al*., 1988). The disease spreads out from vine to vine in a non-random, centrifugal fashion (Anandaraj, Ramachandran & Sarma, 1990).

Cultural practices are widely practiced in an attempt to control foot rot (de Waard, 1986). Disease-free cuttings should be raised in nurseries using fumigated soil mixtures (Sarma *et al*., 1988). Sanitation measures include removal of infected vines and pruning of runner and hanging shoots to remove sources of inoculum. A legume and grass cover can also be planted to reduce chances of rain splash of inoculum (Holliday & Mowat, 1963; Ramachandran, Sarma & Anandaraj, 1990). Provision of good drainage may also reduce disease incidence and vines may be grown on raised beds or mounds. Minimum tillage and restriction of movement of personnel in the garden will aid in reducing inadvertent spread of the soilborne pathogen on mud or infected plant debris. Finally, where pepper production utilises shade trees as in India, shade pruning can be practiced to reduce humidity around the vines (Ramachandran *et al*., 1990).

Chemical control is expensive for smallholders and experiences with traditional fungicides such as Bordeaux mixture have often been inconsistent (de Waard, 1986; Ramachandran *et al*., 1990). In India, Bordeaux mixture is used as a 10% paste to treat the stem collar and as a 1% spray for the vine. Poor results are partly due to heavy rainfalls and also because the main growth of the vine occurs during the monsoon season when new unprotected foliage is continually being produced. In addition, many farmers use such fungicides as curative treatments (Ramachandran *et al*., 1990).

Research with the newer systemic fungicides, particularly metalaxyl and fosetyl-aluminium, used either as a soil drench or foliar spray, has demonstrated their efficacy against *P. capsici* (Ramachandran & Sarma, 1985; Ramachandran *et al.*, 1990; Sarma, Ramachandran & Anandaraj, 1990). An interesting observation in India was that the insecticides endosulfan and quinalphos, used to control the flea beetle (*Longitarsus nigripennis*), were also active against *P. capsici in vivo* (Ramachandran & Sarma, 1988). The possibility exists that reduced levels of a fungicide such as metalaxyl may be required when used with such insecticides (Ramachandran *et al.*, 1990).

A search for rootstock resistance or tolerance has been actively pursued, especially in India since black pepper originates from Kerala state. Several wild relatives, including *Piper columbrinum*, *P. hispidum* and *P. arifolium* are resistant. Attempts at using *Piper columbrinum*, though initially quite successful, resulted in graft union problems by the fifth year, and a 40 to 80% loss by the seventh year (de Waard, 1986). Perhaps more promising is the discovery that several cultivars of black pepper possess some tolerance to the disease (Sarma, Nambiar & Nair, 1982; Sarma *et al.*, 1990). One of these cultivars, Karimunda, is a high yielder and may provide a key component in an integrated control programme.

At present, the prognosis for control of *P. capsici* on black pepper appears more bleak than that for the previous three examples of integrated control that have been described. In part this is because black pepper is almost exclusively a smallholder crop and consequently financial resources to allow technology transfer, plus low prices for the commodity, discourage changes in traditional farm practices (de Waard, 1986). In addition, *P. capsici* is highly successful as a soilborne pathogen, thrives in the tropical monsoon climates where black pepper is grown and is capable of attacking its host at its most vulnerable point, the stem collar.

Future developments in integrated control

Socio-economic and political factors

Undoubtedly, the most useful developments that could occur in integrated control of debilitating soilborne diseases of the type described would be availability of effective means to encourage growers and farmers to adopt such progressive measures. In many cases, financial backing is insufficient to educate farmers effectively, especially smallholders. Furthermore, unless growers are provided with concrete proof of the success of the new methods, they are unlikely to take financial risks especially in a situation where profits are virtually non-existent. Kendrick (1988) points out that integrated control 'is not a panacea for pest and disease control' but rather 'it is an ecological approach to maintaining plant health'. He goes on to suggest that effective transmission of such philosophies will require radical changes in research management, costly and complex implementation procedures and even critical changes in the

Table 28.2. General strategies for integrated control.

Disease-free seed or stock	
Planting site ideally free of major pathogens	
Hygiene and sanitation	
Modification of soil environment:	
	mounding or ridging
	improved drainage
	organic amendments
	fumigation
Preplant applications of:	
	systemic fungicides
	biological antagonists
Tolerant or resistant germ plasm	
Postplant applications of systemic fungicides	
Cultural practices (protective husbandry):	
	mulching
	pruning
	intercropping
	irrigation regimes
	fertilizer practices
	removal of diseased plants

profession of plant pathology itself. Further, he suggests that widespread adoption of integrated control practices will only occur once the 'profession gains a foothold in our private sector economy'. The era of the professional 'plant doctor' may be on the horizon in those countries where agriculture is largely business and profit oriented. Perhaps the critical factor for change will be the extent of the impact on agriculture of new legislation restricting usage of pesticides.

It is also perhaps ironic that current integrated control practices largely mimic many facets of traditional farming practice. For example, use of landraces, which are nothing more or less than tolerant cultivars, is frequently an important component of traditional methods. The current vogue of 'sustainable agriculture' involves use of different cultural practices to maintain and stabilize crop production. Mulching and the use of raised beds, ridges or mounds also figure prominently in many types of traditional agriculture (Thurston, 1990). Traditional farmers have successfully managed disease problems, often completely unaware of the specific nature of the pathogens, for millennia relying mainly on both sustainable and labour-intensive cultural practices (Thurston, 1990). Until recently, efforts at improving smallholder agriculture rested largely on the introduction of intensive temperate-climate technologies into the tropics. Frequently, results have been disastrous, including poor adaptation or susceptibility to disease of introduced cultivars and lack of pesticide efficacy due to a variety of factors. There is now an increasing awareness that traditional smallholder agriculture must be at the basis of future developments and advances in integrated control practices in the tropics.

Prospects for improved control

Integrated control consists of a large number of individual components (Table 28.2). Many of these are already widely exploited in agricultural and horticultural practices worldwide. However, there are two specific areas where further improvements are possible: development of more tolerant germ plasm and improved use of fungicides.

Experience with soybean (Schmitthenner, 1985) highlights the limitations of utilizing race-specific resistance against soilborne *Phytophthora* species. Perhaps fortuitously, this type of resistance has not been available for incorporation into many other crop plants (e.g. avocado, black pepper). Increasingly, emphasis in both annual and perennial crops has been on the development of tolerant rootstocks (Coffey, 1987) or cultivars (Sarma *et al.*, 1988a; Walker & Schmitthenner, 1984a). Such screening for tolerance is time-consuming and laborious. Recent progress in the development of rapid, small-scale tests for detecting and measuring tolerance (Gabor & Coffey, 1991; Sarma *et al.*, 1990; Walker & Schmitthenner, 1984a) in avocado, black pepper and soybean promise acceleration in availability of new tolerant germ plasm. Even with such advances, the ultimate tests have to be maintained in the field over many years. Knowledge of productivity, commodity quality, susceptibility or resistance to other pests and diseases, and other agronomic characters have to be evaluated. Ultimately, however, improved tolerance to important diseases offers the best hope of making integrated control attractive to the farmer. This is particularly so with the smallholder, since 'disease resistant varieties offer control with no input from the farmer' (Evans & Prior, 1987).

Fungicides are also a key component of many approaches to integrated control of *Phytophthora* (Table 28.2). Ideally, they are used to establish a crop in the absence of disease, though their more frequent usage is in an existing, often severe disease situation. The new 'curative' fungicides metalaxyl and particularly the two phosphonates, fosetyl-aluminium and potassium phosphonate, may open up new opportunities for more effective control. Potentially, they allow farmers to reduce disease levels, sometimes in a dramatic way, in a serious disease situation (Darvas *et al.*, 1984; Anderson *et al.*, 1989). New methods, including trunk injection or foliar application of phosphonates, as well as low-volume or ultra low-volume application of aerial sprays, may provide further successes in control of other *Phytophthora* diseases. Because of their ambimobile systemic properties, phosphonates are especially useful for control of sensitive *Phytophthora* species such as *P. palmivora*, both above and below ground. Perhaps the only danger is that success with these new chemicals may dissuade governments and scientists from aggressively pursuing new strategies based on integrated control! I believe this would be a mistake since such site-specific compounds are vulnerable as fungicide resistant strains of the pathogen may develop (Cohen & Coffey, 1986).

Biological control

Biological control, which may already be an important component (Pegg & Whiley, 1987) of some practices (e.g. mulching, organic amendments), remains the 'Cinderella' of integrated control. This is largely because insufficient effort has been made to identify the microbial components present in suppressive soils. Further, private enterprise has to date shown little interest in investment and development of such practices. This is partly because the technologies to deliver microbial agents to the target sites in the soil are largely in their infancy. Moreover, such biological control agents may be quite specific with respect to a pathogen, host or soil environment (Linderman *et al.*, 1983) and thus their market potential could be limited. Other factors which have sometimes inhibited commercial development of biological control methods by the plant protection industry are fears that they may compete directly with synthetic pesticides and that patent or product protection may be difficult.

Despite these various obstacles, some progress is being made towards identifying potentially active anti-*Phytophthora* microbial agents. Very recently, *Trichoderma* and *Gliocladium* species were identified in small scale pot tests for their ability to suppress symptoms of *P. cactorum* on apple seedlings (Smith, Wilcox & Harman, 1990). Although these tests were conducted only in pasteurized soil mixes, they provide some encouragement that biocontrol agents may eventually play a role in practical control of *Phytophthora* species.

In our own research, we have identified a strain of *Myrothecium roridum* that is active in suppressing *P. cinnamomi* on highly susceptible seedlings of *Persea indica*, a relative of avocado (Gees & Coffey, 1989). In addition to efficacy in various potting mixes, it was also very active in three *Phytophthora*-conducive field soils supplemented with additional inoculum of *P. cinnamomi*. The fungal antagonist could be recovered from actively growing root tips four weeks after inoculation.

In the above studies, the biological control agents were isolated either from field soils suppressive to *Phytophthora* (Gees & Coffey, 1989), suppressive to another Oomycete *Aphanomyces euteiches*, or from a conducive soil where *Phytophthora* had not developed (Smith *et al*., 1990). Another factor that could facilitate their use in the integrated control of *Phytophthora* is that *Gliocladium*, *Myrothecium* and *Trichoderma* are unlikely to be inhibited by either metalaxyl or the phosphonate fungicides.

Future progress in biological control will require the development of appropriate and economic formulations of microbial agents. Shelflife of the product will be a key factor, as will the ability to produce it on a large scale in an economic manner. Finally, the product will have to be consistent in its performance in different soils, environments and cultural conditions. Perhaps the best opportunities for effective biocontrol are where plants can be raised in fumigated mixes, such as with avocado, and where the important phase of the disease is in the soil.

Conclusions

Integrated control can be compared to politics or economics, in that there are not only diametrically opposing viewpoints, there are also many shades of opinion. In this review I have attempted to present strategies for four important *Phytophthora* problems. Obviously, emphasis on specific components of a strategy will depend on the advisor, but will vary according to conditions, not only due to differences in agriculture and pathology, but also specific economic and political factors. Nevertheless, it is quite apparent that even disease case-histories as different as avocado root rot in agribusiness-oriented California and foot rot of black pepper in smallholder India have many parallels. The need for high-yielding lines with superior tolerance to *Phytophthora* is common to avocado, cocoa, black pepper and soybean. The need for more effective and less expensive methods of delivering 'curative' fungicides is also universal. There is an increasing emphasis on cultural practices including the use of mounds or ridges, cover crops and pruning practice and removal of infected plant material, since these favour better crop performance and tend to reduce disease incidence. A possible hidden advantage of integrated control is the synergistic effect of combining practices. With soybean root rot, this is well illustrated by the use of tolerant cultivars together with metalaxyl (Anderson & Buzzell, 1982; Schmitthenner, 1985). Biological control is not an obvious component of *Phytophthora* suppression, although it is very much a factor in the use of cover crops and mulches which form an important part of avocado root rot control in Queensland (Pegg & Whiley, 1987). At a time when there is an increasing emphasis on reducing pesticide usage, integrated control may come of age. It certainly offers potential advantages for both smallholder agriculture and agribusiness. However, much will depend on the will of government, administrators and research leaders to provide the complex framework allowing successful technological transfer to the farmer. In the final analysis we should not underestimate the logistic and economic difficulties of the task. We should realize that whilst integrated control is an ecologically more acceptable form of agriculture, it is initially expensive to develop and will always remain relatively complex to put into successful practice.

References

Anandaraj, N., Ramachandran, N. & Sarma, Y. R. (1990). Epidemiology of foot rot disease of black pepper (*Piper nigrum* L.) in India. In *Proceedings of the International Pepper Community Workshop on Joint Research for the Control of Black Pepper Diseases*, 27-29 October 1988, Panaji, Goa, (ed. Y. R. Sarma). National Research Centre for Spices: Calicut, India.

Anderson, R. D., Middleton, R. M. & Guest, D. I. (1989). Development of a bioassay to test the effect of phosphorous acid on black pod of cocoa. *Mycological Research* **93**, 110-112.

Anderson, T. R. & Buzzell, R. I. (1982). Efficacy of metalaxyl in controlling *Phytophthora* root and stalk rot of soybean cultivars differing in field tolerance. *Plant Disease* **66**, 1144-1145.

Bailey, A. M. & Coffey, M. D. (1985). Biodegradation of metalaxyl in avocado soils. *Phytopathology* **75**, 135-137.

Bailey, A. M. & Coffey, M. D. (1986). Characterization of microorganisms involved in accelerated biodegradation of metalaxyl and metolachlor in soils. *Canadian Journal of Microbiology* **32**, 562-569.

Bower, L. A. & Coffey, M. D. (1985). Development of laboratory tolerance to phosphorous acid, fosetyl-Al, and metalaxyl in *Phytophthora capsici*. *Canadian Journal of Plant Pathology* **7**, 1-6.

Bruin, G. C. A. & Edgington, L. V. (1981). Adaptive resistance in Peronosporales to metalaxyl. *Canadian Journal of Plant Pathology* **3**, 201-206.

Bruin, G. C. A. & Edgington, L. V. (1983). The chemical control of diseases caused by zoosporic fungi – a many-sided problem. In *Zoosporic Plant Pathogens, A Modern Perspective*, ed. S. T. Buczacki, pp. 193-232. Academic Press: London.

Coffey, M. D. (1987). *Phytophthora* root rot of avocado: an integrated approach to control in California. *Plant Disease* **71**, 1046-1052.

Coffey, M. D. & Bower, L. A. (1984a). *In vitro* variability among isolates of six *Phytophthora* species in response to metalaxyl. *Phytopathology* **74**, 502-506.

Coffey, M. D. & Bower, L. A. (1984b). *In vitro* variability among isolates of eight *Phytophthora* species in response to phosphorous acid. *Phytopathology* **74**, 738-742.

Coffey, M. D. & Joseph, M. C. (1985). Effects of phosphorous acid and fosetyl-Al on the life cycle of *Phytophthora cinnamomi* and *P. citricola*. *Phytopathology* **75**, 1042-1046.

Coffey, M. D. & Ouimette, D. G. (1989). Phosphonates: antifungal compounds against Oomycetes. In *Nitrogen, Phosphorus and Sulphur Utilization by Fungi*, ed. L. Boddy, R. Marchant & D. J. Read, pp. 107-129. Cambridge University Press: Cambridge, UK.

Cohen, Y. & Coffey, M. D. (1986). Systemic fungicides and the control of Oomycetes. *Annual Review of Phytopathology* **24**, 331-338.

Cook, R. J. & Baker, K. (1983). *The Nature and Practice of Biological Control of Plant Pathogens*. American Phytopathological Society: St. Paul, Minnesota, USA.

Darvas, J. M., Toerien, J. C. & Milne, D. L. (1984). Control of avocado root rot by trunk injection with phosethyl-Al. *Plant Disease* **68**, 691-693.

Davidse, L. C. (1982). Acylalanines: resistance in downy mildews, *Pythium* and *Phytophthora* spp. In *Fungicide Resistance in Crop Protection*, ed. J. Dekker & S. G. Georgopoulos, pp. 118-127. Pudoc: Wageningen, The Netherlands.

Davidse, L. C., Hofman, A. E. & Velthuis, G. C. M. (1983). Specific interference of metalaxyl with endogenous RNA polymerase activity in isolated nuclei from *P. megasperma* f. sp. *medicaginis*. *Experimental Mycology* **7**, 344-361.

de Waard, P. W. F. (1986). Current state and prospective trends of black pepper (*Piper nigrum* L.) production. *Outlook on Agriculture* **15**, 186-195.

Dolan, T. E. & Coffey, M. D. (1986). Laboratory screening technique for assessing resistance of four avocado rootstocks to *Phytophthora cinnamomi*.*Plant Disease* **70**, 115-118.

Evans, H. C. & Prior, C. (1987). Cocoa pod diseases: causal agents and control. *Outlook on Agriculture* **16**, 35-41.

Faris, M. A., Sabo, F. E., Barr, D. J. S. & Lin, C. S. (1989). The systematics of *Phytophthora sojae* and *P. megasperma*. *Canadian Journal of Botany* **67**, 1442-1447.

Fenn, M. E. & Coffey, M. D. (1984). Studies on the *in vitro* and *in vivo* antifungal activity of fosetyl-Al and phosphorous acid. *Phytopathology* **74**, 606-611.

Förster, H., Kinscherf, T. G., Leong, S. A. & Maxwell, D. P. (1989). Restriction fragment length polymorphisms of the mitochondrial DNA of *P. megasperma* isolated from soybean, alfalfa, and fruit trees. *Canadian Journal of Botany* **67**, 529-537.

Fulton, R. H. (1989). The cacao disease trilogy: black pod, Monilia pod rot, and witches'-broom. *Plant Disease* **73**, 601-603.

Gabor, B. K. & Coffey, M. D. (1990). Quantitative analysis of the resistance to *Phytophthora cinnamomi* in five avocado rootstocks under greenhouse conditions. *Plant Disease*, **74**, 882-885.

Gabor, B. K. & Coffey, M. D. (1991). Rapid methods for evaluating resistance to *Phytophthora cinnamomi* in avocado rootstocks. *Plant Disease* **75**, in press.

Gabor, B. K., Guillemet, F. B. & Coffey, M. D. (1990). Comparison of field resistance to *Phytophthora cinnamomi* in twelve avocado rootstocks. *HortScience* **25**, 1655-1656.

Gees, R. & Coffey, M. D. (1989). Evaluation of a strain of *Myrothecium roridum* as a potential biocontrol agent against *Phytophthora cinnamomi*. *Phytopathology* **79**, 1079-1084.

Gregory, P. H. (1983). Some major epidemics caused by *Phytophthora*. In *Phytophthora: Its Biology, Taxonomy, Ecology, and Pathology*, ed. D. C. Erwin, S. Bartnicki-Garcia & P. H. Tsao, pp. 271-278. American Phytopathological Society: St. Paul, Minnesota, USA.

Gregory, P. H., Griffin, M. J., Maddison, A. C. & Ward, M. R. (1984). Cocoa black pod: a reinterpretation. *Cocoa Growers Bulletin* **35**, 5-22.

Hansen, E. M., Brasier, C. M., Shaw, D. S. & Hamm, P. B. (1986). The taxonomic structure of *Phytophthora megasperma*: evidence for emerging biological species groups. *Transactions of the British Mycological Society* **87**, 557-573.

Hildebrand, A. E. (1959). A root and stalk rot of soybeans caused by *Phytophthora megasperma* var. *sojae* var. nov. *Canadian Journal of Botany* **37**, 927-957.

Holliday, P. & Mowat, W. P. (1963). Foot rot of *Piper nigrum* L. (*Phytophthora palmivora*). *Phytopathological Paper no. 5*. Commonwealth Mycological Institute: Kew, UK.

Kasim, R. & Prayitno, S. (1979). Factors influencing growth and sporulation of *Phytophthora capsici* from *Piper nigrum* L. *Pemberitaan Lembaga Penelitian Tanaman Industri* (Indonesia) **34**, 41.

Kaufmann, M. J. & Gerdemann, J. W. (1958). Root and stem rot of soybeans caused by *Phytophthora sojae* n. sp. *Phytopathology* **48**, 201-208.

Kellam, M. K. & Coffey, M. D. (1985). Quantitative comparison of the resistance to *Phytophthora* root rot in three avocado rootstocks. *Phytopathology* **75**, 230-234.

Kendrick, Jr., J. B. (1988). A viewpoint on integrated pest management. *Plant Disease* **72**, 647.

Kopp, L. E. (1966). A taxonomic revision of the genus *Persea* in the western hemisphere (*Persea* - Lauraceae). *Memoirs of the New York Botanical Garden* **14**, 1-120.

Kuan, T. -L. & Erwin, D. C. (1980). Formae speciales differentiation of *Phytophthora megasperma* isolates from soybean and alfalfa. *Phytopathology* **70**, 333-338.

Linderman, R. G., Moore, L. W., Baker, K. F. & Cooksey, D. A. (1983). Strategies for detecting and characterizing systems for biological control of soilborne plant pathogens. *Plant Disease* **67**, 1058-1064.

Lucas, J. A., Bower, L. A. & Coffey, M. D. (1990). Fungicide resistance in soil-borne *Phytophthora* species. *EPPO Bulletin* **20**, 199-206.

McGregor, A. J. (1982). A small-scale screening technique for evaluating fungicides against *Phytophthora palmivora* pod rot of cocoa. *Annals of Applied Biology* **101**, 25-31.

McGregor, A. J. (1983). Experiments on the profitability of chemical black pod control in Papua New Guinea. *Tropical Pest Management* **29**, 129-136.

McGregor, A. J. (1984). Comparison of cuprous oxide and metalaxyl with mixtures of these fungicides for control of *Phytophthora* pod rot of cocoa. *Plant Pathology* **33**, 81-87.

Nambiar, K. K. N. & Sarma, Y. R. (1982). Some aspects of epidemiology of foot rot of black pepper. In *Proceedings of the Workshop on Phytophthora Diseases of Tropical Cultivated Plants*, ed. K. K. N. Nambiar, pp. 225-232. Central Plantation Crops Research Institute: Kasaragod, India.

Old, K. M., Oros, J. M. & Malafant, K. W. (1984). Survival of *Phytophthora cinnamomi* in root fragments in Australian forest soils. *Transactions of the British Mycological Society* **82**, 605-613.

Ouimette, D. G. & Coffey, M. D. (1988). Quantitative analysis of organic phosphonates, phosphonate, and other inorganic anions in plants and soil by using high-performance ion chromatography. *Phytopathology* **78**, 1150-1155.

Ouimette, D. G. & Coffey, M. D. (1989). Phosphonate levels in avocado (*Persea americana*) seedlings and soil following treatment with fosetyl-Al or potassium phosphonate. *Plant Disease* **73**, 212-215.

Ouimette, D. G. & Coffey, M. D. (1990). Symplastic entry and phloem translocation of phosphonate. *Pesticide Biochemistry and Physiology* **38**, 18-25.

Pegg, K. G. & Whiley, A. W. (1987). *Phytophthora* control in Australia. *South African Avocado Growers' Association Yearbook* **10**, 94-96.

Pegg, K. G., Whiley, A. W., Saranah, J. B. & Glass, R. J. (1985). Control of *Phytophthora* root rot of avocado with phosphorus acid. *Australasian Plant Pathology* **14**, 25-29.

Pereira, J. L. (1985). Chemical control of *Phytophthora* pod rot of cocoa in Brazil. *Cocoa Growers' Bulletin* **36**, 23-38.

Prior, C. (1984). Approaches to the control of diseases of cocoa in Papua New Guinea. *Journal of Plant Protection in the Tropics* **1**, 39-46.

Prior, C. & Smith, E. S. C. (1982). Control of bark canker in cocoa with metalaxyl. *Tropical Pest Management* **28**, 45-48.

Ramachandran, N. & Sarma, Y. R. (1985). Efficacy of three systemic fungicides in controlling *Phytophthora* infection of black pepper. *Indian Phytopathology* **38**, 160-162.

Ramachandran, N. & Sarma, Y. R. (1988). Fungitoxic effect of endosulfan and quinalphos on *Phytophthora palmivora*, the foot rot pathogen of black pepper. *Annals of Applied Biology* **112** (Supplement: Tests of Agrochemicals and Cultivars No. 9), 26-27.

Ramachandran, N., Sarma, Y. R. & Anandaraj, N. (1990). Management of *Phytophthora* infections in black pepper. In *Proceedings of the International Pepper Community Workshop on Joint Research for the Control of Black Pepper Diseases*, 27-29 October 1988, Panaji, Goa, (ed. Y. R. Sarma). National Research Centre for Spices: Calicut, India.

Ramachandran, N., Sarma, Y. R., Anandaraj, N. & Abraham, J. (1988b). Effect of climatic factors on *Phytophthora* leaf infection in black pepper grown in arecanut-black pepper mixed cropping system. *Journal of Plantation Crops* **16**, 110-118.

Sarma, Y. R., Nambiar, K. K. N. & Nair, M. K. (1982). Screening of black pepper (*Piper nigrum* L.) and *Piper* species against *Phytophthora palmivora*. In *Proceedings of the Workshop on Phytophthora Diseases of Tropical Cultivated Plants*, ed. K. K. N. Nambiar, pp. 242-247. Central Plantation Crops Research Institute: Kasaragod, India.

Sarma, Y. R., Ramachandran, N. & Anandaraj, M. (1990). Status of black pepper diseases in India. In *Proceedings of the International Pepper Community Workshop on Joint Research for the Control of Black Pepper Diseases*, 27-29 October 1988, Panaji, Goa, (ed. Y. R. Sarma). National Research Centre for Spices: Calicut, India.

Sarma, Y. R., Ramachandran, N., Anandaraj, M. & Ramana, K. V. (1988). Disease management in black pepper. *Indian Cocoa, Arecanut and Spices Journal* **11**, 123-127.

Schmitthenner, A. F. (1985). Problems and progress in control of *Phytophthora* root rot of soybean. *Plant Disease* **69**, 362-368.

Schmitthenner, A. F. & Van Doren, Jr., D. M. (1985). Integrated control of root rot of soybean caused by *P. megasperma* f. sp. *glycinea*. In *Ecology and Management of Soilborne Plant Pathogens*, eds. C. A. Parker, A. D. Rovira, K. J. Moore, P. T. Wong & J. F. Kollmorgen, pp. 263-266. American Phytopathological Society: St. Paul, Minnesota, USA.

Smith, V. L., Wilcox, W. F. & Harman, G. E. (1990). Potential for biological control of *Phytophthora* root and crown rots of apple by *Trichoderma* and *Gliocladium* spp. *Phytopathology* **80**, 880-885.

Schwinn, F. J. (1983). Recent developments in chemical control of *Phytophthora*. In *Phytophthora: Its Biology, Taxonomy, Ecology, and Pathology*, ed. D. C. Erwin, S. Bartnicki-Garcia & P. H. Tsao, pp. 327-334. American Phytopathological Society: St. Paul, Minnesota, USA.

Thurston, H. D. (1990). Plant disease management practices of traditional farmers. *Plant Disease* **74**, 96-102.

Tsao, P. H. & Alizadeh, A. (1988). Recent advances in the taxonomy and nomenclature of the so-called '*Phytophthora palmivora*' MF4 occurring on cocoa and other tropical crops. In *10th International Cocoa Research Conference Proceedings*, 17-23 May 1987, pp. 441-445. Santo Domingo.

Tucker, C. M. (1928). Report of the plant pathologist. *Porto Rico Agricultural Experiment Station Report* **1927**, 25-27.

Wager, V. A. (1942). *Phytophthora cinnamomi* and wet soil in relation to the dying-back of avocado trees. *Hilgardia* **14**, 519-532.

Walker, A. K, & Schmitthenner, A. F. (1984a). Comparison of field and greenhouse evaluations for tolerance to *Phytophthora* rot in soybean. *Crop Science* **24**, 487-489.

Walker, A. K. & Schmitthenner, A. F. (1984b). Heritability of tolerance to *Phytophthora* rot in soybean. *Crop Science* **24**, 490-491.

Walker, A. K. & Schmitthenner, A. F. (1984c). Recurrent selection for tolerance to *Phytophthora* rot in soybean. *Crop Science* **24**, 495-497.

Weste, G. & Vithanage, K. (1979). Survival of chlamydospores of *Phytophthora cinnamomi* in several non-sterile, host-free forest soils and gravels at different soil water potentials. *Australian Journal of Botany* **27**, 1-9.

Wilhelm, S. & Paulus, A. O. (1980). How soil fumigation benefits the California strawberry industry. *Plant Disease* **64**, 264-270.

Zentmyer, G. A. (1980). *Phytophthora cinnamomi* and the diseases it causes. *Phytopathological Monograph* No. **10**. American Phytopathological Society: St. Paul, Minnesota, USA.

Zentmyer, G. A. (1988). Taxonomic relationships and distribution of species of *Phytophthora* causing black pod of cacao. In *10th International Cocoa Research Conference Proceedings*, 17-23 May 1987, pp. 391-395. Santo Domingo.

Zentmyer, G. A. & Erwin, D. C. (1970). Development and reproduction of *Phytophthora*. *Phytopathology* **60**, 1120-1127.

Index

A

B

C

D

E

H

M

Q

R

S

T

U

V

W

X

Y

Z

www.ingramcontent.com/pod-product-compliance
Ingram Content Group UK Ltd.
Pitfield, Milton Keynes, MK11 3LW, UK
UKHW042207080726
473066UK00007B/291

* 9 7 8 0 5 2 1 1 8 9 7 6 7 *

ACCESO GRATIS *a la Lectura en la Nube*

Para visualizar el libro electrónico en la nube de lectura envíe junto a su nombre y apellidos una fotografía del código de barras situado en la contraportada del libro y otra del ticket de compra a la dirección:

En un máximo de 72 horas laborables le enviaremos el código de acceso con sus instrucciones.

La visualización del libro en **NUBE DE LECTURA** excluye los usos bibliotecarios y públicos que puedan poner el archivo electrónico a disposición de una comunidad de lectores. Se permite tan solo un uso individual y privado.

ASPECTOS CIVILES Y PROCESALES DE LA LEY DE VIVIENDA

ASPECTOS CIVILES Y PROCESALES DE LA LEY DE VIVIENDA

ANTONIO GÁLVEZ CRIADO
DIRECTOR

tirant lo blanch
Valencia, 2025

En caso de erratas y actualizaciones, la Editorial Tirant lo Blanch publicará la pertinente corrección en la página web www.tirant.com.

© TIRANT LO BLANCH
EDITA: TIRANT LO BLANCH
C/ Artes Gráficas, 14 - 46010 - Valencia
TELFS.: 96/361 00 48 - 50
FAX: 96/369 41 51
Email: tlb@tirant.com
www.tirant.com
Librería virtual: www.tirant.es
Depósito legal: V-4093-2024
ISBN: 978-84-1071-815-9

Si tiene alguna queja o sugerencia, envíenos un mail a: *atencioncliente@tirant.com*. En caso de no ser atendida su sugerencia, por favor, lea en *www.tirant.net/index.php/empresa/politicas-de-empresa* nuestro procedimiento de quejas.

Responsabilidad Social Corporativa: http://www.tirant.net/Docs/RSCTirant.pdf

Autores

MARÍA TERESA ALONSO PÉREZ
Catedrática de Derecho civil. Universidad de Zaragoza

VÍCTOR BASTANTE GRANELL
Profesor Titular de Derecho civil. Universidad de Almería

PAULA CASTAÑOS CASTRO
Profesora Titular de Derecho civil. Universidad de Málaga

MARÍA DOLORES CERVILLA GARZÓN
Catedrática de Derecho civil. Universidad de Cádiz

JOSÉ MANUEL DE TORRES PEREA
Profesor Titular de Derecho civil. Universidad de Málaga

ANTONIO GÁLVEZ CRIADO
Profesor Titular de Derecho civil. Universidad de Málaga

ISABEL L. MARTENS JIMÉNEZ
Doctora en Derecho. Universidad de Málaga

JESÚS MARTÍN FUSTER
Profesor Ayudante Doctor de Derecho civil. Universidad de Málaga

JOSÉ MANUEL MARTÍN FUSTER
Profesor Ayudante Doctor de Derecho civil. Universidad de Málaga

MARÍA LUISA MORENO-TORRES HERRERA
Catedrática de Derecho civil. Universidad de Málaga

ANA MARÍA RODRÍGUEZ TIRADO
Profesora Titular de Derecho procesal. Universidad de Cádiz

ISABEL ZURITA MARTÍN
Catedrática de Derecho civil. Universidad de Cádiz

Índice

Abreviaturas

AAPP:	Administraciones Públicas
BOE:	Boletín Oficial del Estado
BOJA:	Boletín Oficial de la Junta de Andalucía
BOPV:	Boletín Oficial del País Vasco
CC:	Código civil
CCAA:	Comunidades Autónomas
CCCat:	Código civil de Cataluña
CE:	Constitución Española
CEDH:	Convenio Europeo de Derechos Humanos
CGPJ:	Consejo General del Poder Judicial
CNMC:	Comisión Nacional de los Mercados y la Competencia
DA:	Disposición adicional
DF:	Disposición final
DT:	Disposición transitoria
INE:	Instituto Nacional de Estadística
IPREM:	Indicador público de renta de efectos múltiples
LAU:	Ley de arrendamientos urbanos
LCCI:	Ley de contratos de crédito inmobiliario
LDV:	Ley por el derecho a la vivienda
LEC:	Ley de enjuiciamiento civil
LO:	Ley orgánica
LOE:	Ley de ordenación de la edificación

LPH:	Ley de propiedad horizontal
ODS:	Objetivos de Desarrollo Sostenible
RD:	Real decreto
STC:	Sentencia del Tribunal Constitucional
STS:	Sentencia del Tribunal Supremo
TC:	Tribunal Constitucional
TEDH:	Tribunal Europeo de Derechos Humano
TRLGDCU:	Texto refundido de la Ley general para la defensa de los consumidores y usuarios
TRLSRU:	Texto refundido de la Ley de suelo y rehabilitación urbana
TS:	Tribunal Supremo
UE:	Unión Europea

Presentación

MARÍA LUISA MORENO-TORRES HERRERA

En el tema de la vivienda, que constituye hoy una de las mayores preocupaciones de los ciudadanos españoles, se encuentran implicados diversos sectores del ordenamiento jurídico, lo que explica el variado contenido de la Ley 12/2023, de 24 de mayo, por el derecho a la vivienda. Dentro de este texto legal, que quienes lo han promovido insisten en presentar como la primera ley de vivienda postconstitucional de ámbito estatal, como si ese simple dato debiera determinar su aceptación entusiasta, se encuentran normas de muy diversa naturaleza, dictadas al amparo de diferentes títulos competenciales.

La presente obra no contiene un estudio completo de la Ley por el derecho a la vivienda, sino que se centra exclusivamente, como bien refleja su título, en los aspectos civiles y en los aspectos procesales. El objetivo fundamental que persigue es el análisis de las normas de Derecho privado presentes en la ley, lo que ha exigido a su vez el tratamiento de importantes cuestiones procesales (aunque estas sin ánimo de exhaustividad), íntimamente relacionadas con aquellas, como las modificaciones de la Ley de enjuiciamiento civil referidas a los desahucios de los arrendatarios en situación de vulnerabilidad o al desalojo de quienes ocupan indebidamente una vivienda.

No precisa apenas justificación la publicación de una obra que contiene un análisis jurídico referido a una ley nueva, pero quizás sí que requiera alguna explicación el hecho de que se trate de un análisis parcial. Varias razones pueden invocarse en este sentido. Por una parte, desde una primera y rápida lectura del texto legal llama la atención el tratamiento poco riguroso que reciben los aspectos jurídico-privados, claramente descuidados incluso desde el punto de vista puramente terminológico. Por otra parte, instituciones centrales del Derecho civil, como la propiedad, la posesión o el contrato, parecen haber quedado afectadas (aunque quizás no del todo conscientemente) por algunos de los preceptos de la Ley.

Estas primeras impresiones, y el convencimiento de la necesidad de contar con expertos ius privatistas y procesalistas para comprobar en su caso las hipótesis de partida, me llevaron a tomar la iniciativa de reunir a un grupo de especialistas para acometer la tarea cuyos resultados se recogen en estas páginas. La implicación y buen hacer de cada uno de ellos, su espíritu de colaboración y su capacidad para trabajar en equipo, unidos a su cualificación, han resultado decisivos para que este proyecto llegue a buen puerto. No menos importante ha sido el papel del director de la obra, el profesor Antonio Gálvez Criado, quien ha liderado este proyecto con la sabiduría y humildad que lo caracterizan. Quiero dejar expresa constancia de mi agradecimiento a todos y cada uno de los participantes.

Las actividades del equipo, incluyendo la difusión de los resultados a través de esta obra colectiva, han sido financiadas por la Universidad de Málaga, que cuenta con un plan propio de investigación. En concreto, en el año 2022 resolvió, en una convocatoria competitiva, financiar la constitución de la "Red temática para el estudio de los aspectos civiles y procesales de la Ley de vivienda" (Referencia: D5-2022_03). Las actividades de esta red se iniciaron antes de la aprobación de la Ley, que es bien sabido que sufrió un importante retraso respecto de las previsiones iniciales, hasta que, finalmente, fue publicada a final de la precedente legislatura.

Si bien, y como se ha apuntado, ha habido una estrecha colaboración entre los autores, lo que se estima imprescindible para la coherencia y unidad de una obra que no es una mera suma de trabajos individuales, cada uno de ellos es autor intelectual exclusivo de su propio capítulo, redactado desde la máxima libertad. Ello explica que pueda encontrarse, sobre ciertas cuestiones, disparidad de criterio, lo que no hace sino enriquecer el resultado.

Pero en lo que no hay disparidad alguna es en el rigor con el que se tratan los temas ni tampoco en los objetivos generales que se persiguen en cada uno de los capítulos. Se ha cuidado muy especialmente que, en cada uno de ellos, los lectores puedan identificar con claridad tres grandes cuestiones: cuáles son las respuestas que ha dado el legislador, qué aspectos de esas respuestas son dudosos, discutibles o criticables, y cuál habría debido ser la postura o solución del legislador estatal a cada problema.

Es conveniente que el lector sea consciente de que lo que encontrará en estas páginas es un análisis jurídico y no valoraciones ideológicas o políticas, que fácilmente están a su alcance en otros lugares. Se han querido evitar juicios superfluos o escasamente fundados, optándose por un análisis sosegado, que dio comienzo durante la fase de tramitación parlamentaria y tiene en cuenta, por supuesto, todo el marco normativo implicado y también el proyecto de ley y no únicamente el texto definitivo. Es muy oportuno que esta publicación se cierre cuando la ley de la que trata tiene más de un año de vigencia, lo que ha permitido, además, disponer ya de algunas publicaciones científicas que abordan sus contenidos, incluyendo la muy reciente obra colectiva "El derecho a la vivienda en tiempos de incertidumbre", dirigida por Mate, Hernández y Alonso. Sin embargo, la que ahora se presenta es la única centrada, estrictamente, en los aspectos de Derecho privado tal y como se contienen en la Ley.

El libro se presenta plenamente actualizado, no solo porque en su elaboración se ha utilizado la bibliografía especializada existente hasta la fecha, sino también, como no, porque incorpora las últimas reformas legislativas, como las llevadas a cabo por los Reales Decreto-ley 6/2023 y 8/2023. Igualmente, se ha tenido en cuenta la Sentencia del Pleno del Tribunal Constitucional n. 79/2024, de 21 de mayo, mediante la cual se resuelve el recurso de inconstitucionalidad 5491-2023, interpuesto por el Consejo de Gobierno de la Junta de Andalucía en relación con diversos preceptos de la Ley 12/2023. Mediante ella se resuelve el primero de los muchos recursos de inconstitucionalidad interpuestos contra la Ley por el derecho a la vivienda, pero están por ver las soluciones de los restantes. Como ha escrito recientemente Nasarre Aznar, es difícil valorar hoy qué sucederá con la Ley. Pero ello no significa, en absoluto, que las reflexiones que se recogen en las páginas que siguen sean útiles solo en el momento presente o que vayan a perder sentido una vez que se vayan despejando algunas incertidumbres. Muy por el contrario, están llamadas a permanecer en el tiempo, habida cuenta del planteamiento al que responden.

En efecto, sea cual sea la postura del Tribunal Constitucional respecto de las complejas competencias legislativas en materia de vivienda, y sean cuales sean las decisiones de las Comunidades autónomas en relación a los muchos e importantes temas que la ley estatal deja en sus manos, mantendrán su utilidad las reflexiones que esta obra contiene sobre aspectos tan importantes como los siguientes: el concepto legal de vivienda, la no discriminación en materia de vivienda, la naturaleza del derecho a una vivienda digna y adecuada, el contenido del derecho de propiedad sobre la vivienda, los deberes de información de los vendedores o arrendadores de vivienda, la configuración jurídica de los administradores de fincas, la intervención legislativa en materia de arrendamientos, la tutela del poseedor de la vivienda, el papel de la conciliación e intermediación en conflictos con los ocupantes de la vivienda o la protección del arrendatario vulnerable.

Todas estas cuestiones se estudian con profundidad en la obra, a propósito, ciertamente, del articulado de la Ley 12/2023, pero con amplitud de miras y sin ceñirse exclusivamente a lo en ella preceptuado. Los temas aparecen tratados en el mismo orden en el que aparecen en el texto legal y se completan con un importante capítulo que contiene las reflexiones de conjunto sobre el Derecho privado en la Ley. Frente a lo que pudiera pensarse, este último capítulo no es una síntesis o resumen de lo anterior, sino el lugar en el que se ofrece, de la mano de un autor distinto, una visión de conjunto sobre los aspectos jurídico-privados de la Ley. Es decir, que el análisis monográfico de cada uno de los temas, convenientemente anunciados en el título que se ha dado a cada uno de los capítulos, se enriquece y complementa con la reflexión global. Pero esta no sustituye a aquellos ni viceversa.

Creo, sinceramente, que el análisis jurídico que aquí se presenta resulta necesario. Obviamente, no es definitivo ni concluyente, ni lo pretende. Nada lo es en Derecho, y mucho menos cuando lo que está en juego es, nada más ni nada menos, que el derecho de todos los españoles a una vivienda digna y adecuada que, según el art. 47 de la Constitución, han de promover todos los poderes públicos sin excepción. El estudio realizado por los doce autores que participan en este libro constituye una aportación valiosa para diseñar en el futuro soluciones legislativas en materia de vivienda, referidas a aspectos concretos. Pero, además, demuestra que el legislador no debe prescindir de los técnicos y que, con independencia de la orientación ideológica a la que responda la política de vivienda, la decisión sobre las medidas legislativas más adecuadas en cada caso y la redacción de las leyes que las establezcan no puede hacerse prescindiendo de quienes poseen los conocimientos necesarios y la visión de conjunto del ordenamiento jurídico; y que, en el caso concreto de la vivienda, resulta imprescindible la aportación de los expertos en Derecho privado.

En Málaga, a 16 de julio de 2024

Capítulo I

Las definiciones de vivienda en la ley por el derecho a la vivienda

ISABEL L. MARTENS JIMÉNEZ
Doctora en Derecho
Universidad de Málaga

I. INTRODUCCIÓN

"Ley por el derecho a la vivienda" es el título elegido para la Ley 12/2023 (LDV) que persigue, según su propia Exposición de Motivos, la "protección del derecho a acceder a una vivienda digna y adecuada y a su disfrute", siguiendo el dictado del artículo 47 de nuestra Constitución. Este precepto, sin embargo, no define qué debe entenderse por "vivienda digna y adecuada" lo que quizás, y en opinión de gran parte de nuestra doctrina, ha supuesto un problema de materialización efectiva de este derecho en la práctica, en perjuicio de la sociedad. 45 años después de la publicación de la Constitución ve la luz la primera ley estatal de desarrollo de este derecho y de este concepto jurídico indeterminado.

Precisamente, por ser la "vivienda" el objeto sobre el que recae la ley, la primera pregunta que le surge a cualquier jurista –y probablemente al lego en Derecho– al enfrentarse a la Ley por el derecho a la vivienda es: ¿qué es una "vivienda" y cuándo responde a los criterios de la dignidad y adecuación a los que se refiere la Constitución española?

Quizás se deba a ello la decisión del legislador de incluir en la ley un artículo, el tercero, destinado a definir ciertos términos que tendrán relevancia en el texto. De todos los términos elegidos por el legislador, centraremos el presente análisis únicamente en aquellos que se refieran a la "vivienda" o a los tipos o características de esta "vivienda".

Dicho análisis perseguirá el objetivo de concluir sobre el alcance de las definiciones, su utilidad y su impacto respecto del resto del articulado, y las medidas que contiene la ley, y otras normas de nuestro ordenamiento jurídico que regulen o se refieran a la vivienda como objeto de Derecho.

En cuanto a la metodología seguida para completar este análisis, se plantearán tres preguntas concretas que titulan los siguientes apartados de este capítulo, y que parten de una cuestión inicial: la propia necesidad de incluir un listado de definiciones en la ley[1] y la finalidad perseguida por el legislador. Dado que en la tramitación parlamentaria y en la propia ley no se incluye una explicación de esta decisión del legislador, proponemos a través de las tres preguntas, tres finalidades que podrían haber motivado al legislador a incluir este listado de definiciones.

Estas tres finalidades son: delimitar el ámbito de aplicación de la ley; unificar conceptos incluyendo los avances doctrinales y legislativos de ámbito nacional e internacional; y/o lograr coherencia interna en el texto y facilitar la aplicación de las medidas.

---

1 López y López (2022, p. 29, nota n.º 20) critica esta técnica que tilda de "ajena a nuestra tradición", explicando que es propia –y apropiada– de la tradición jurídica anglosajona, pero que en nuestro ordenamiento, en el que no existen las restricciones del ordenamiento anglosajón, y la interpretación extensiva y la aplicación analógica son posibles, "las definiciones legislativas pueden acabar convirtiéndose en material jurídico deleznable". Finaliza su crítica afirmando que la escribe "con la absoluta convicción de que, ante un legislador aluvional y sin demasiados conocimientos técnicos, estos riesgos son ciertos, tanto como todos (y no somos pocos) los que los denunciamos somos *voces quae clamant in deserto*".

Como decimos, ninguna de estas finalidades se menciona por el legislador, y partimos de la hipótesis de que el propio legislador ignoró la importancia y utilidad que podría haber tenido este precepto, haciendo un uso inútil y contraproducente de la herramienta definitoria que tenía a su disposición.

El análisis necesario para responder a las tres preguntas anunciadas será el camino para llegar a una conclusión, confirmando la hipótesis inicial planteada, y para proponer mejoras.

II. ¿SIRVEN LAS DEFINICIONES DE VIVIENDA PARA DELIMITAR EL ÁMBITO DE APLICACIÓN DE LA LEY?

La importancia de fijar el ámbito de aplicación de una ley es evidente, pero con mucha mayor intensidad en este supuesto. La razón es obvia: el problema competencial. No es este capítulo el lugar adecuado para analizar el conflicto competencial que ha generado la LDV, pero sí las consecuencias que supone respecto de la decisión del legislador estatal de haber incluido definiciones en el texto de la ley.

En concreto, haciendo buen uso de la competencia que le brinda el artículo 148.1.3.ª CE, y a la vista del vacío legal, muchas CCAA han acometido la fundamental tarea de definir y regular no pocos aspectos relacionados con la vivienda. Pues bien, el artículo 3.º LDV también establece una serie de definiciones que ya están presentes en dichas leyes autonómicas, pero con un contenido distinto al reflejado en aquellas.

Surge, por ende, un problema de coordinación entre las definiciones autonómicas y estatales, al que el artículo 3.º LDV cree dar solución mediante una breve indicación al inicio del precepto de prevalencia de la definición autonómica. La dificultad está en el efecto práctico que tiene esa indicación.

Este "problema de coordinación" al que hemos hecho referencia, puede explicarse con un ejemplo: la definición de "vivienda"[2].

Según el art. 3.a) LDV, la "vivienda" es: "edificio o parte de un edificio de carácter privativo y con destino a residencia y habitación de las personas, que reúne las condiciones mínimas de habitabilidad exigidas legalmente, pudiendo disponer de acceso a espacios y servicios comunes del edificio en el que se ubica, todo ello de conformidad con la legislación aplicable y con la ordenación urbanística y territorial"; sin embargo, conforme al art. 3.a) de la Ley 18/2007, de 28 de diciembre, del derecho a la vivienda de Cataluña, es: "toda edificación fija destinada a que residan en ella personas físicas o utilizada con este fin, incluidos los espacios y servicios comunes del inmueble en el que está situada y los anexos que están vinculados al mismo, si acredita el cumplimiento de las condiciones de habitabilidad que fija la presente ley y la normativa que la desarrolle y cumple la función social de aportar a las personas que residen en ella el espacio, las instalaciones y los medios materiales necesarios para satisfacer sus necesidades personales ordinarias de habitación".

Las definiciones no son iguales, y por indicación del apartado primero del art. 3 LDV, debe prevalecer la definición de

---

2 De hecho, el CGPJ en su Informe Jurídico al anteproyecto LDV (n.º 71, p. 28) es especialmente crítico con la inclusión en la ley de una definición para el término "vivienda", por cuanto que "conceptuar con alcance estatal el *factum* 'vivienda' cuando éste es el término que sirve a la delimitación exclusiva de la competencia autonómica resulta de nuevo cuestionable en la medida en que redefine desde la norma estatal el ámbito de las competencias autonómicas". Asimismo, el problema de coordinación no se limita a la relación norma autonómica/norma estatal, sino que incluso está presente entre normas estatales. Sirva como ejemplo la divergencia entre las definiciones de la LDV y la LAU, en sus arts. 2 y 7.

la norma autonómica, pero entonces, ¿quiere decir la ley que sus medidas se aplican a las viviendas se definan estas como se definan en cada comunidad autónoma? ¿o solo se aplican en aquellas CCAA en la que no haya definiciones o en las que estás sean coincidentes?[3] ¿qué alcance tiene realmente la expresión "a los efectos de lo dispuesto en esta ley"?

A la vista, por tanto, de este gran problema que plantea la LDV, y que fue advertido al legislador desde prácticamente todos los ámbitos –académico[4], político[5], judicial[6]...–, resul-

---

[3] El numeral 50 del Informe jurídico del CGPJ al Anteproyecto de la LDV destaca precisamente este problema: "... esa 'expropiación' de la regulación autonómica por el legislador estatal generará, sin duda, una situación de gran inseguridad jurídica al resultar ciertamente difícil determinar qué norma habrá de aplicarse, la estatal o la autonómica, cuando no concuerden completamente entre sí".

[4] Muy clarificador a estos efectos son las intervenciones de Sergio Nasarre Aznar y Trayter Jiménez que se pueden leer en el Diario de Sesiones del Congreso de los Diputados, n.º 651, sesión n.º 27, del 25 de abril de 2022, pp. 1-70. El primero de los catedráticos, el Dr. Nasarre Aznar, considera que uno de los grandes problemas del artículo 3 LDV (a la fecha de su intervención era el PLDV) es el "conflicto de los conceptos" (p. 4). En similares términos se pronuncia Trayter Jiménez para quien las definiciones del PLDV chocaban con las definiciones ya recogidas en normas autonómicas.

[5] Es este un aspecto destacado por la mayoría de los partidos políticos durante la tramitación parlamentaria, junto con el conflicto competencial; al haber sido definidos los términos contenidos en el art. 3 LDV en las CCAA, los problemas de coordinación entre unas y otras definiciones generan una indefensión indeseada. En el Diario de sesiones del Congreso de los Diputados publicado el 27 de mayo de 2022, n.º 89-3, se recogen 33 enmiendas al art. 3 LDV, una señal del franco desacuerdo de diversos partidos políticos al contenido de este artículo.

[6] A estos efectos, dispone el CGPJ en el numeral 70 de su Informe jurídico al Anteproyecto de la LDV que "... resulta difícil asumir

ta incomprensible que el legislador no haya creído oportuno abordar el problema de la coordinación con las normas ya existentes[7], a los efectos de evitar duplicidades, divergencias; en definitiva, más dudas de derecho que soluciones de hecho.

Lo lógico, cuando surgen dudas sobre la aplicabilidad de las medidas recogidas en una ley, es consultar el artículo que delimita su ámbito de aplicación. No es posible, sin embargo, en este supuesto, por cuanto que, como se puede observar en el índice de la ley, el Título Preliminar de la LDV, compuesto por seis artículos, no incluye un artículo que determine a qué situaciones de hecho deben aplicarse las medidas incluidas en la ley.

Ya hemos visto, que el artículo 3.º LDV, lejos de aclarar a qué situaciones se aplican las medidas de la ley, genera mayores dudas de aplicación. En cuanto al artículo 1.º LDV determina el "objeto de la ley" que, como bien es sabido, no sirve para determinar el ámbito de aplicación de la misma, sino para señalar cuál es su fin, por qué y para qué fue creada, un resumen, si se quiere, de la explicación ofertada en la Exposición de Motivos. El artículo 2.º de la LDV se

---

que las definiciones que contiene lo sean a los limitados efectos de la presente Ley, pues pasan a ser 'conceptos jurídicos', integrados ahora en el ordenamiento positivo, cuya coordinación con otras leyes civiles no está clara y que en algunos casos propician la exclusión de tales definiciones por absolutamente innecesarias".

7 En este sentido, Prats Albentosa (2022, p. 168), refiriéndose al régimen de la propiedad abordado por las medidas de la LDV, entiende que aquel "no puede abordarse desde un criterio estanco de división competencial Estado/CCAA, sino desde la cooperación entre las Cortes Generales y las Asambleas Legislativas de las CCAA", y ello, por cuanto que ese régimen de propiedad tiene por objeto la "vivienda", y la competencia sobre esta materia es autonómica, conforme al art. 148.1.3.ª CE.

titula "fines de las políticas públicas de vivienda", o lo que es lo mismo: una forma más extensa y detallada de repetir lo ya indicado en el artículo 1.º. Los artículos 4.º "servicios de interés general", 5.º "acción pública" y 6.º "principio de igualdad y no discriminación en la vivienda" tampoco solventan el problema de la indeterminación del ámbito de aplicación de la ley.

Resulta absolutamente incomprensible, decíamos, no solo que no se incluyera en la ley una disposición general destinada a aclarar el ámbito de aplicación de la ley, sino que ni siquiera se facilitara, ante tal omisión, la delimitación por otra vía, como podría haber sido el artículo que alberga las definiciones.

Incomprensible, en efecto, por cuanto que el legislador, naturalmente, no es ajeno a la estructura propia de las normas, siendo así que en la Resolución de 28 de julio de 2005, de la Subsecretaría, por la que se da publicidad al Acuerdo del Consejo de Ministros, de 22 de julio de 2005, por el que se aprueban las Directrices de técnica normativa, se establece que: "Las disposiciones generales son aquellas que fijan el objeto y ámbito de aplicación de la norma, así como las definiciones necesarias para una mejor comprensión de algunos de los términos en ella empleados. Deberán figurar en los primeros artículos de la disposición y son directamente aplicables, en cuanto forman parte de la parte dispositiva de la norma".

La importante función de delimitación competencial, reservada normalmente a esa disposición denominada "ámbito de aplicación", no queda salvada porque el legislador diga vagamente al inicio de cada precepto: "...en el ámbito de competencias del Estado..." (art. 1 LDV), "...en el ámbito de sus respectivas competencias..." (art. 2 LDV), "...en tanto no entren en contradicción con las reguladas por las administraciones competentes en materia de vivienda..." (art. 3 LDV), puesto que, justamente, la duda generalizada es qué competencia corresponde al Estado en materia de vivienda[8], sobre todo, a la vista de una ley muy extensa que parece abarcarla toda[9].

Pero, si no se utiliza la propia ley para aclarar justamente qué dimensiones son las que le corresponden al Estado y a las que deberán aplicarse estas medidas y cómo debe funcionar ese "marco normativo coherente, estable y seguro", al que hace

---

8 De hecho, el legislador, en su propia Exposición de Motivos ya apunta al problema competencial cuando dice que: "La vivienda resulta ser, así, un bien esencial de rango constitucional que presenta múltiples dimensiones. Desde esta perspectiva, el propio Tribunal Constitucional ha reiterado que no constituye un título competencial autónomo, sino que puede recaer bajo distintos títulos competenciales estatales o autonómicos dependiendo de cuál sea el enfoque y cuáles los instrumentos regulatorios utilizados en cada caso por el legislador. Dicha complejidad competencial es clara consecuencia de las distintas dimensiones constitucionales que presenta la vivienda" y más aún: "Conforme al artículo 148.3 de la Constitución, todas las Comunidades Autónomas tienen asumida en sus Estatutos de Autonomía, sin excepción, la competencia plena en materia de vivienda. A diferencia del Estado, que sólo puede incidir, con distinto alcance y sobre la base de títulos competenciales diversos, en la política de vivienda, los legisladores autonómicos pueden formular completos programas normativos de la acción pública en la materia. Uno y otras están abocados, por tanto, a articular sus respectivas actuaciones de modo que puedan desplegarse en un marco normativo coherente, estable y seguro que haga posible la realidad del derecho reconocido en el artículo 47 de la Constitución en ejecución de las medidas, acciones, planes y programas correspondientes y la igualdad básica de todos los españoles en relación con dicho derecho".

9 Prats Albentosa (2022, p. 169) llama la atención sobre la admisión en el propio Proyecto de la LDV de la falta de competencia del legislador estatal para regular determinadas materias, que le lleva a frecuentes remisiones a la legislación autonómica. Por ello, el autor concluye que ello lleva a dudas sobre la idoneidad de las premisas en las que se basa la propia elaboración del Proyecto y sobre la competencia del Estado en cuanto al contenido que se propone en aquel.

referencia la Exposición de Motivos de la LDV, difícilmente se estará facilitando la aplicación de la misma[10].

Nuestra respuesta a la primera pregunta, esto es, si sirven las definiciones de vivienda para delimitar el ámbito de aplicación de la ley, es negativa. La LDV no delimita su ámbito de aplicación y las definiciones no sitúan al aplicador de la ley, quién se enfrentará al problema de coordinación entre la LDV y las normas estatales previas a la LDV que se refieran a la vivienda como objeto de derecho, y las normas autonómicas en materia de vivienda.

III. ¿SIRVEN LAS DEFINICIONES DE VIVIENDA PARA UNIFICAR CONCEPTOS INCORPORANDO LOS AVANCES DOCTRINALES Y LEGISLATIVOS EN LA MATERIA?

Habiendo contestado negativamente a la primera pregunta planteada y descartando ya como finalidad del artículo de las definiciones la de delimitar el ámbito de aplicación de la ley, procedemos a analizar si las definiciones sirven para unificar los conceptos en materia de vivienda, y si el legislador ha tomado en consideración todos los novedosos estudios y avances que se han logrado ya no solo en el ámbito internacional y europeo, sino incluso en el nacional[11].

---

10 En este sentido, Sainz-Cantero Caparrós (2023, p. 638), sobre las definiciones manifiesta que: "… se han observado contradicciones y deficiencias técnicas importantes sobre la normativa proyectada, a pesar de que la vocación armonizadora de la normativa estatal respecto de las normativas autonómicas justificaría y daría sentido a estas definiciones…".

11 Así lo trató de explicar Sergio Nasarre Aznar durante la tramitación parlamentaria (Diario de Sesiones del Congreso de los Diputados,

En concreto, nuestro análisis se dividirá en cuatro apartados: la vivienda adecuada e inadecuada, la vivienda habitual y secundaria, la infravivienda y el sinhogarismo, y el hogar. Además de hacer una comparativa entre la propia terminología empleada en la LDV frente a la utilizada en el ámbito internacional y nacional, veremos si el legislador ha tomado en consideración el trabajo desarrollado por los expertos nacionales e internacionales al establecer definiciones en materia de vivienda.

1. Vivienda adecuada e inadecuada (o más bien: derecho a una vivienda adecuada)

Una crítica común en la doctrina a la LDV es que no se entiende la segregación en las definiciones entre la "vivienda", la "vivienda digna y adecuada" y la "infravivienda". Lo cierto es que una vivienda o es digna y adecuada o es infravivienda[12], según veremos a continuación.

---

n.º. 651, sesión n.º 27, del 25 de abril de 2022, p. 4) "Y, por último, el conflicto de conceptos. Como ya les he dicho, lo más importante de la norma es que haya unos buenos conceptos y que nos entendamos todos en lo que es vivienda, en lo que es infravivienda, etcétera, pero el problema es que cada comunidad autónoma tiene su propio concepto de cada uno de estos. Como mínimo, yo creo que el proyecto de ley tiene que aspirar a ser inspirador, y para eso tendría que haber recogido las últimas novedades de esos conceptos, novedades doctrinales publicadas en 2021 y 2022, donde después de cientos de páginas en tesis doctorales y en libros, a nivel internacional y a nivel nacional, se llega a un concepto de vivienda, de infravivienda o de sinhogarismo mucho más completo de lo que han previsto allí, que en muchas ocasiones es una copia de la Ley del Suelo, y quiero decir que una ley de vivienda no tiene nada que ver con el suelo".

12 Nasarre Aznar (2022, p. 13) muestra su sorpresa ante esa distinción que hace la LDV entre la vivienda y la vivienda digna y adecuada, enfatizando que técnicamente, si la vivienda no es digna y adecuada, no puede

No obstante, parece que el problema va más allá de una mera segregación entre "vivienda" y "vivienda digna y adecuada". En efecto, si el legislador en ambos casos se hubiera limitado a definir la vivienda como objeto de derecho, no tendría sentido, como bien dice la doctrina, esa separación entre "vivienda" y "vivienda digna y adecuada".

Pero no es esto lo que hace el legislador, por el contrario, parece definir la vivienda como objeto de derecho en el art. 3.a) LDV, y la vivienda digna y adecuando en cuanto a derecho de los ciudadanos en el art. 3.c) LDV. Como se verá a lo largo de este apartado, en la mayoría de las definiciones que se refieren a la "vivienda adecuada", lo que realmente se está definiendo no es la vivienda, sino el derecho.

Entrando ya en el análisis de la adecuación de la definición a los avances en la materia, en el ámbito internacional son varios los instrumentos jurídicos que se refieren o incluso regulan el derecho a la vivienda de los ciudadanos[13]. No obstante, destaca

---

ser vivienda, sino que será "infravivienda". Más aún, aclara que, en tal caso, directamente no "debería ampararse legalmente de ningún modo (art. 47 CE)". En iguales términos, Argelich Comelles (2023, p. 88).

13 Obviamente, no es el PIDESC el único texto internacional que hace referencia a la vivienda, pero sí es el más claro y por ello hemos decidido desarrollarlo aquí. Aun así, podemos mencionar los siguientes textos que, a semejanza del anterior, no es que sean muy precisos en su literalidad, pero que, a través de la interpretación de bien las observaciones generales en el caso del primero, o bien de los tribunales competentes, sirven para acotar los conceptos que deben ser estudiados en materia de vivienda. Así, entre otros, el art. 25 de la Declaración Universal de los Derechos Humanos proclamada por la Asamblea General de las Naciones Unidas en París, el 10 de diciembre de 1948 en su Resolución 217 A (III), dispone que "Toda persona tiene derecho a un nivel de vida adecuado que le asegure, así como a su familia, la salud y el bienestar, y en especial la alimentación, el vestido, la vivienda, la asistencia

el art. 11 del PIDESC[14], no tanto en cuanto al contenido de este artículo, sino a la interpretación y explicación contenida

---

médica y los servicios sociales necesarios; tiene asimismo derecho a los seguros en caso de desempleo, enfermedad, invalidez, viudez, vejez u otros casos de pérdida de sus medios de subsistencia por circunstancias independientes de su voluntad. 2. La maternidad y la infancia tienen derecho a cuidados y asistencia especiales. Todos los niños, nacidos de matrimonio o fuera de matrimonio, tienen derecho a igual protección social" y a su vez el art. 34 de la Carta de Derechos Fundamentales de la Unión Europea, en su apartado 3.º: "Con el fin de combatir la exclusión social y la pobreza, la Unión reconoce y respeta el derecho a una ayuda social y a una ayuda de vivienda para garantizar una existencia digna a todos aquellos que no dispongan de recursos suficientes, según las modalidades establecidas por el Derecho comunitario y las legislaciones y prácticas nacionales".

14 Según el Instrumento de Ratificación de España del Pacto Internacional de Derechos Económicos, Sociales y Culturales, hecho en Nueva York el 19 de diciembre de 1966, el art. 11 del PIDESC dispone que: "1. Los Estados Partes en el presente Pacto reconocen el derecho de toda persona a un nivel de vida adecuado para sí y su familia, incluso alimentación, vestido y vivienda adecuados, y a una mejora continua de las condiciones de existencia. Los Estados Partes tomarán medidas apropiadas para asegurar la efectividad de este derecho, reconociendo a este efecto la importancia esencial de la cooperación internacional fundada en el libre consentimiento.
2. Los Estados Partes en el presente Pacto, reconociendo el derecho fundamental de toda persona a estar protegida contra el hambre, adoptarán, individualmente y mediante la cooperación internacional, las medidas, incluidos programas concretos, que se necesiten para:
a) Mejorar los métodos de producción, conservación y distribución de alimentos mediante la plena utilización de los conocimientos técnicos y científicos, la divulgación de principios sobre nutrición y el perfeccionamiento o la reforma de los regímenes agrarios, de modo que se logre la explotación y la utilización más eficaces de las riquezas naturales;
b) Asegurar una distribución equitativa de los alimentos mundiales en relación con las necesidades, teniendo en cuenta los problemas que se plantean tanto a los países que importan productos alimenticios como a los que los exportan".

en la Observación General n.º 4. Gracias a esta OG, existe una enumeración de los requisitos que debe cumplir una vivienda para ser considerada “adecuada”. No obstante, y quizás este sea el aspecto destacable y que debería haber trasladado el legislador español a su LDV, el propio Comité admite que “la adecuación viene determinada en parte por factores sociales, económicos, culturales, climatológicos, ecológicos y de otra índole”, pero que, aun así, es “posible identificar algunos aspectos de ese derecho que deben ser tenidos en cuenta a estos efectos en cualquier contexto determinado”.

Seguidamente realiza una enumeración de estos aspectos, desarrollándolos:

a) seguridad jurídica de la tenencia[15]: destacando el Comité que independientemente del tipo de tenencia, todas las personas deben gozar de cierto grado de seguridad, sobre todo, frente al desahucio, hostigamiento u otras amenazas[16];

---

15 “Seguridad jurídica de la tenencia. La tenencia adopta una variedad de formas, como el alquiler (público y privado), la vivienda en cooperativa, el arriendo, la ocupación por el propietario, la vivienda de emergencia y los asentamientos informales, incluida la ocupación de tierra o propiedad. Sea cual fuere el tipo de tenencia, todas las personas deben gozar de cierto grado de seguridad de tenencia que les garantice una protección legal contra el desahucio, el hostigamiento u otras amenazas. Por consiguiente, los Estados Partes deben adoptar inmediatamente medidas destinadas a conferir seguridad legal de tenencia a las personas y los hogares que en la actualidad carezcan de esa protección consultando verdaderamente a las personas y grupos afectados”.

16 Pérez Daudi (2022, p. 339) haciendo un recorrido por los Dictámenes del Comité DESC, destaca que el derecho a la vivienda digna se vulnera cuando no se ofrece al ciudadano una alternativa habitacional en caso de desahucio y lanzamiento. Es una manifestación, por tanto, del criterio de seguridad jurídica en la tenencia,

b) disponibilidad de servicios, materiales, facilidades e infraestructura[17], esto es, acceso a servicios indispensables para la salud y la nutrición, entre otros;

c) gastos soportables[18]: de manera que los ciudadanos no tengan que renunciar a la satisfacción de otras necesidades básicas;

d) habitabilidad[19]: como punto central se refiere nuevamente el Comité a la salud, en el sentido de que la habi-

---

que debe ser tenido en cuenta a la hora de establecer una definición o los parámetros para delimitar su contenido.

17 "Disponibilidad de servicios, materiales, facilidades e infraestructura. Una vivienda adecuada debe contener ciertos servicios indispensables para la salud, la seguridad, la comodidad y la nutrición. Todos los beneficiarios del derecho a una vivienda adecuada deberían tener acceso permanente a recursos naturales y comunes, a agua potable, a energía para la cocina, la calefacción y el alumbrado, a instalaciones sanitarias y de aseo, de almacenamiento de alimentos, de eliminación de desechos, de drenaje y a servicios de emergencia".

18 "Gastos soportables. Los gastos personales o del hogar que entraña la vivienda deberían ser de un nivel que no impidiera ni comprometiera el logro y la satisfacción de otras necesidades básicas. Los Estados Partes deberían adoptar medidas para garantizar que el porcentaje de los gastos de vivienda sean, en general, conmensurados con los niveles de ingreso. Los Estados Partes deberían crear subsidios de vivienda para los que no pueden costearse una vivienda, así como formas y niveles de financiación que correspondan adecuadamente a las necesidades de vivienda. De conformidad con el principio de la posibilidad de costear la vivienda, se debería proteger por medios adecuados a los inquilinos contra niveles o aumentos desproporcionados de los alquileres. En las sociedades en que los materiales naturales constituyen las principales fuentes de material de construcción de vivienda, los Estados Partes deberían adoptar medidas para garantizar la disponibilidad de esos materiales".

19 "Habitabilidad. Una vivienda adecuada debe ser habitable, en sentido de poder ofrecer espacio adecuado a sus ocupantes y de protegerlos

tabilidad supone que el espacio que ocupen las personas les proteja de factores externos;

e) asequibilidad[20]: a diferencia de los gastos soportables, que se circunscriben al momento ya de uso de la vivienda, este criterio de la asequibilidad: se refiere al momento inicial de acceso a la vivienda;

---

del frío, la humedad, el calor, la lluvia, el viento u otras amenazas para la salud, de riesgos estructurales y de vectores de enfermedad. Debe garantizar también la seguridad física de los ocupantes. El Comité exhorta a los Estados Partes a que apliquen ampliamente los Principios de Higiene de la Vivienda preparados por la OMS, que consideran la vivienda como el factor ambiental que con más frecuencia está relacionado con las condiciones que favorecen las enfermedades en los análisis epidemiológicos; dicho de otro modo, que una vivienda y unas condiciones de vida inadecuadas y deficientes se asocian invariablemente a tasas de mortalidad y morbilidad más elevadas".

20 "Asequibilidad. La vivienda adecuada debe ser asequible a los que tengan derecho. Debe concederse a los grupos en situación de desventaja un acceso pleno y sostenible a los recursos adecuados para conseguir una vivienda. Debería garantizarse cierto grado de consideración prioritaria en la esfera de la vivienda a los grupos desfavorecidos como las personas de edad, los niños, los incapacitados físicos, los enfermos terminales, los individuos VIH positivos, las personas con problemas médicos persistentes, los enfermos mentales, las víctimas de desastres naturales, las personas que viven en zonas en que suelen producirse desastres, y otros grupos de personas. Tanto las disposiciones como la política en materia de vivienda deben tener plenamente en cuenta las necesidades especiales de esos grupos. En muchos Estados Partes, el mayor acceso a la tierra por sectores desprovistos de tierra o empobrecidos de la sociedad, debería ser el centro del objetivo de la política. Los Estados deben asumir obligaciones apreciables destinadas a apoyar el derecho de todos a un lugar seguro para vivir en paz y dignidad, incluido el acceso a la tierra como derecho".

f) lugar[21] : destacando la importancia para el adecuado desarrollo de las personas que la ubicación de las viviendas permitan a sus habitantes el acceso a servicios esenciales, sin incurrir por ello en excesivos gastos por desplazamientos;

g) adecuación cultural[22].

Además de estos criterios[23], se refiere el Comité a los "demás derechos" que recogen los Pactos y otros instrumentos internacionales aplicables[24].

---

21 "Lugar. La vivienda adecuada debe encontrarse en un lugar que permita el acceso a las opciones de empleo, los servicios de atención de la salud, centros de atención para niños, escuelas y otros servicios sociales. Esto es particularmente cierto en ciudades grandes y zonas rurales donde los costos temporales y financieros para llegar a los lugares de trabajo y volver de ellos puede imponer exigencias excesivas en los presupuestos de las familias pobres. De manera semejante, la vivienda no debe construirse en lugares contaminados ni en la proximidad inmediata de fuentes de contaminación que amenazan el derecho a la salud de los habitantes".

22 "Adecuación cultural. La manera en que se construye la vivienda, los materiales de construcción utilizados y las políticas en que se apoyan deben permitir adecuadamente la expresión de la identidad cultural y la diversidad de la vivienda. Las actividades vinculadas al desarrollo o la modernización en la esfera de la vivienda deben velar por que no se sacrifiquen las dimensiones culturales de la vivienda y porque se aseguren, entre otros, los servicios tecnológicos modernos".

23 Tal y como destaca Caballé Fabra (2020, p. 216), lo que no hace el Comité en la OG n.º 4º, es aclarar si se deben dar todos estos criterios para que la vivienda sea adecuada, o si puede seguir siéndolo aun en caso de ausencia de alguno de ellos. Es este un aspecto que el legislador estatal sí debería afrontar, sobre todo, a la vista del reparto competencial, en unos casos asumiendo la responsabilidad de su determinación y en otros, dejándolo en manos de las CCAA.

24 En esta línea, al estudiar la legislación autonómica, propone Pisarello (2009, p. 5) que "Esta conexión entre el derecho a la vivienda y otros derechos sociales y civiles clásicos –especialmente visible en algunos

En iguales términos y más recientemente, la ONU en la Nueva Agenda Urbana publicada por el Programa de las Naciones Unidas para los Asentamientos Humanos (ONU-Habitat) explica que: "la vivienda adecuada, asequible y bien ubicada es una vía para el disfrute de varios derechos humanos, incluidos los derechos al trabajo, la salud, la seguridad social, el voto, la privacidad y la educación", y caracteriza a continuación dicha vivienda adecuada conforme a los criterios mencionados (seguridad de la tenencia, disponibilidad de servicios, materiales, instalaciones e infraestructura, asequibilidad, habitabilidad, accesibilidad, localización y adecuación cultural)[25].

Asimismo, el Relator Especial de las Naciones Unidas ofrece una breve definición de "derecho a una vivienda adecuada" ("el derecho humano a una vivienda adecuada es el derecho de todo hombre, mujer, joven y niño a tener un hogar y una comunidad seguros en que puedan vivir en paz y dignidad")

---

estatutos de autonomía– hace que el carácter 'digno y adecuado' de la misma deba definirse a partir de su relación con el resto de derechos tutelados por el ordenamiento y con las diferentes declaraciones y convenios sobre derechos humanos ratificados por el Estado español (art. 10.2), desde la CSE o el CEDH hasta el propio PIDESC". Asimismo, en la definición que recoge el Relator Especial de las Naciones Unidas en su Informe sobre una vivienda adecuada como elemento integrante del derecho a un nivel de vida adecuado, de 13 de febrero de 2008 (A/HRC/7/16, p. 5) tiene en cuenta precisamente esas "interrelaciones que existen entre el derecho a una vivienda adecuada y otros derechos humanos afines, como el derecho a la alimentación, el agua, la salud, el trabajo, la tierra, los medios de vida, la propiedad y la seguridad de la persona, así como a la protección contra el trato inhumano y degradante, la no discriminación y la igualdad de géneros, son la base de la realización del derecho a una vivienda adecuada".

25 El documento se puede consultar en https://unhabitat.org/sites/default/files/2021/10/nueva-agenda-urbana-ilustrada.pdf [fecha de última consulta: 27 de mayo de 2024].

para, seguidamente, enumerar todos los criterios que deben servir para conceptualizarla.

En concreto son catorce los elementos que enumera en el Informe sobre una vivienda adecuada como elemento integrante del derecho a un nivel de vida adecuado, de 13 de febrero de 2008 (A/HRC/7/16, p. 6): seguridad de la tenencia; bienes y servicios públicos; bienes y servicios ambientales (incluidos la tierra y el agua); asequibilidad (incluido el acceso a la financiación); habitabilidad; accesibilidad (física); ubicación; adecuación cultural; garantía frente a la expoliación; información, capacidad y creación de capacidad; participación y posibilidad de expresión; reasentamiento; medio ambiente seguro; seguridad (física) y privacidad.

Pues bien, a diferencia de la práctica propia de los textos internacionales que hemos mencionado anteriormente, no opta el legislador en la LDV por establecer unos criterios que ayuden a definir y delimitar el derecho a una vivienda adecuada, sino que por el contrario, dispone en su art. 3.c) LDV: "la vivienda que, por razón de su tamaño, ubicación, condiciones de habitabilidad, accesibilidad universal, eficiencia energética y utilización de energías renovables y demás características de la misma, y con acceso a las redes de suministros básicos, responde a las necesidades de residencia de la persona o unidad de convivencia en condiciones asequibles conforme al esfuerzo financiero, constituyendo su domicilio, morada u hogar en el que poder vivir dignamente, con salvaguarda de su intimidad, y disfrutar de las relaciones familiares o sociales, favoreciendo el pleno desarrollo y la inclusión social de las personas".

Lejos de establecer unos criterios que ayuden a determinar en qué supuestos y condiciones una vivienda será adecuada para que el ciudadano ejercite su derecho, se establece una definición tipo y limitada con referencia a aspectos físicos, materiales y a requisitos administrativos, remisión a otras normas del ordenamiento y una vaga mención a otros principios como el libre desarrollo de la personalidad.

Es una definición insuficiente y no incorpora las novedades internacionales, pero a mayor abundamiento, no se inspira en las normas del ordenamiento interno que sí se adecuan más a estos avances y al uso de criterios para definir.

En concreto, sirve de ejemplo la definición recogida la Ley 1/2010, de 8 de marzo, reguladora del derecho a la vivienda en Andalucía.

En el art. 2 de esta ley destaca un aspecto que la LDV no ha incluido en la definición y es advertir de que los requisitos que enumera suponen un mínimo, es decir, que esos aspectos *tienen* que concurrir para que la vivienda se califique de "digna y adecuada", pero *no* son los únicos. En particular, exige que sea una edificación fija y habitable, accesible y de calidad.

En cuanto a la calidad, se desarrolla en el art. 3, siendo especialmente destacable que los niveles de calidad y parámetros mínimos exigibles a la vivienda, que deben ser desarrollados reglamentariamente, se adecuarán a la diversidad de unidades familiares, contribuyendo a la cohesión social y velando por la mejora a través de las innovaciones tecnológicas. Asimismo, se menciona expresamente la "información" en el acceso a la vivienda.

Una lectura pausada de estos dos artículos –2 y 3– de la ley andaluza, muestra una estrategia bien distinta a la estatal y es la de establecer unos parámetros mínimos que permiten calificar la vivienda como "adecuada" y "digna", asumiendo que no es factible fijar una definición cerrada, como no es factible prever todas las situaciones, necesidades y posibilidades que se pueden dar en la práctica en el ámbito de la vivienda.

Nuestra doctrina se pronuncia en línea con la utilización de unos criterios mínimos para definir este derecho. Así, por ejemplo, Molina Roig (2018, pp. 133 y 134) diferencia entre el elemento interno[26] y el externo[27]; y Caballé Fabra (2020, pp. 187-191) diferencia tres elementos: el económico y financiero, el estructural y el simbólico, destacando dentro de este último especialmente el denominado "factor X"[28] o "factor moral que vincula dicho inmueble a quien lo habita"[29].

2. *Vivienda habitual y vivienda secundaria*

El legislador utiliza los términos "vivienda" "residencia" y "domicilio" como sinónimos a lo largo de toda la LDV, a pesar de que en el art. 3.i) LDV solamente defina la "residencia habitual" como: "la vivienda que constituye el domicilio permanente de la persona que la ocupa y que puede acreditarse a través de los datos obrantes en el padrón municipal u otros medios válidos en derecho". Como término antónimo a la residencia habitual define el art. 3.j) LDV la "residencia secundaria" como "toda aquella vivienda que se utiliza por su propietario para estancias temporales o intermitentes, y que no constituye su residencia habitual".

---

26 "El elemento interno, hace referencia a la normativa técnica que las diferentes administraciones públicas fijan en cada momento, como estándares mínimos que debe cumplir una vivienda para considerarse como tal".

27 "El elemento externo requiere que la vivienda sea adecuada, lo que supone la necesidad de que las políticas públicas tengan en cuenta las circunstancias personales, familiares, sociales y económicas de los beneficiarios de las mismas".

28 Ampliamente sobre el "factor X" en Caballé Fabra (2020, pp. 191-197).

29 Así lo define Nasarre Aznar (2022, p. 13), criticando que no se haya tomado en consideración al incluir las definiciones en la LDV.

Varios son los problemas que derivan tanto de la propia elección de la terminología, el uso indistinto de los tres términos mencionados (vivienda, residencia y domicilio) en la LDV[30] y la propia definición.

Empezando por la elección de la terminología, lo cierto es que, en nuestro ordenamiento jurídico, la mayoría de las normas se refieren a la "vivienda habitual" [31], mientras que la residencia y domicilio suelen utilizarse en ámbitos concretos. Por ejemplo, el art. 40 CC, que se refiere a la residencia habitual y al domicilio, cuando dice: "Para el ejercicio de los derechos y

---

30 Esta confusión terminológica no se limita a la LDV, ya que el legislador al modificar por medio de sus disposiciones otras normas del ordenamiento, emplea los términos "vivienda habitual", en vez de "residencia habitual". Véase en este sentido la reforma efectuada por la Disposición Final Primera al art. 10 LAU. Se añade una referencia a los "contratos de arrendamiento de vivienda *habitual*" y en el art. 17 LAU, al crear un nuevo punto 6.º, también se menciona el "contrato de arrendamiento de vivienda *habitual*". Otros ejemplos son las reformas que, por medio de la Disposición Final Quinta, se hacen a la LEC. Referencias a la "vivienda *habitual* del ejecutado" (art. 655 bis LEC) o a la "vivienda *habitual* del deudor" (art. 685 LEC), que previamente no se hacían en la LEC.

31 Valga como ejemplo la Ley 5/2019, de 15 de marzo, reguladora de los contratos de crédito inmobiliario; el Real Decreto-ley 19/2022, de 22 de noviembre, por el que se establece un Código de Buenas Prácticas para aliviar la subida de los tipos de interés en préstamos hipotecarios sobre vivienda habitual, se modifica el Real Decreto-ley 6/2012, de 9 de marzo, de medidas urgentes de protección de deudores hipotecarios sin recursos, y se adoptan otras medidas estructurales para la mejora del mercado de préstamos hipotecarios, que en el propio título hace referencia a la "vivienda habitual" y posteriormente la menciona èn otras 19 ocasiones; o la Ley 1/2013, de 14 de mayo, de medidas para reforzar la protección a los deudores hipotecarios, reestructuración de deuda y alquiler social, que se refiere a la "vivienda habitual" en 34 ocasiones.

el cumplimiento de las obligaciones civiles, el domicilio de las personas naturales es el lugar de su residencia habitual, y, en su caso, el que determine la Ley de Enjuiciamiento Civil".

En cuanto al uso indistinto de los tres términos "residencia", "vivienda" y "domicilio", debemos admitir que no es un problema exclusivo de la LDV, ya que una lectura detallada de otras normas de nuestro ordenamiento jurídico demuestra que se utilizan indistintamente los términos "vivienda" "residencia" o "domicilio", añadiéndoles bien el término "habitual" o incluso "habitual y permanente"[32]. Pero, precisamente por eso, el legislador podría haber aprovechado la oportunidad para eliminar este caos terminológico, estableciendo una base común.

Una crítica adicional a la definición se centra en la asociación que hace el legislador entre la habitualidad y el carácter permanente, ya que ello colisiona y difícilmente casa con otras normas de nuestro ordenamiento jurídico en las que no se exige ese carácter permanente[33].

---

32 A modo de ejemplo, se puede citar el Real Decreto 42/2022, de 18 de enero, por el que se regula el Bono Alquiler Joven y el Plan Estatal para el acceso a la vivienda 2022-2025, que como buena representación de este caos terminológico se refiere indistintamente a la "vivienda habitual" (33 menciones), a la "vivienda habitual y permanente" y a la "residencia habitual" (9 menciones).

33 En efecto, recuerda el CGPJ en el número 72 de su Informe Jurídico al Anteproyecto LDV (pp. 28 y 29) que según la Dirección General de Seguridad Jurídica y Fe Pública (DGSJFP) no es necesaria la permanencia en la vivienda habitual para recibir la protección registral, puesto que "habitual" y "permanente" no son términos equivalentes. A lo que añade Nasarre Aznar (2022, p. 13) que el legislador "distorsiona el concepto de 'residencia habitual' (art. 40 CC) al que exige que sea 'permanente' al contrario de lo que sucede en el art. 2.3 Ley suelo (+183 días al año) o el art. 7 LAU".

3. Sinhogarismo

Precisamente por constituir una de las manifestaciones más crueles dentro del problema de la vivienda en la sociedad, el "sinhogarismo" recibe un estudio y una dedicación muy intensa entre los legisladores y estudiosos no solo del Derecho, sino también de otras disciplinas como la psicología, la política, etc.

Es por ello por lo que, naturalmente, se han logrado mayores avances y puntos de entendimiento a nivel internacional sobre lo que deba entenderse por "sinhogarismo". Tanto es así, que la ONU, el Observatorio Europeo del sinhogarismo y FEANTSA (European Federation of National Associations Working with the Homeless), han definido, utilizando el método de la categorización, las distintas situaciones que reciben la calificación de "sinhogarismo".

En concreto, el Observatorio Europeo creó la denominada "tabla ETHOS" (*European Typology on Homelessness and housing exclusion*), que divide en cuatro las situaciones de sinhogarismo: sin techo (roofless), sin vivienda (houseless), la vivienda insegura (insecure housing) y la vivienda inadecuada (inadecuate housing)[34].

Esta tabla no se incluye en la LDV, lo cual llama *nuevamente* la atención, sobre todo a la vista de los antecedentes legislativos. Para aclarar la anterior afirmación, basta con consultar la Proposición de ley 122/000166 de garantía del

---

34 Caballé Fabra (2020, p. 219) explica que estas cuatro categorías ayudan a entender el mencionado factor X, en el sentido de que el sinhogarismo no implica únicamente que la persona afectada no tenga vivienda, sino también los casos en los que esta vivienda no representa para esa persona su hogar. La vivienda implica "algo más", dice la autora, y "precisamente este 'algo más' es lo que hace que una vivienda pase a tener la consideración de 'hogar'".

derecho a la vivienda digna y adecuada, publicada el 8 de octubre de 2021, en el Boletín oficial de las Cortes Generales, n.º 192-1, que dedicaba su art. 4 a definir un número bastante superior de términos y, al menos, en cuanto al "sinhogarismo" de forma más adecuada para la finalidad perseguida (*o que debería haber perseguido el legislador*). No solo se refería dicho artículo a la Tabla Ethos para definir a la "persona sin hogar o en situación de sinhogarismo", sino que enumeraba las distintas situaciones que recibirían esa calificación[35].

---

[35] En concreto, decía el art. 4.h): "Persona sin hogar o en situación de sinhogarismo: Todas aquellas personas que se encuentren, de acuerdo a la tipología ETHOS (*European Typology of Homelessness and Housing Exclusion*), en alguna de las siguientes situaciones: Personas que viven a la intemperie, personas que viven en las calles o un espacio público o exterior, sin albergue que pueda ser definido como vivienda; Personas en alojamientos de emergencia, personas sin lugar habitual de residencia que hacen uso nocturno de albergues; Personas en alojamientos destinados a personas sin hogar; Personas alojadas en refugios por cortos intervalos debido a experiencias de violencia doméstica o de género; Personas residentes en Centros de Internamiento para Extranjeros; Personas alojadas temporalmente en residencias para trabajadores temporeros. Personas dependientes de instituciones penitenciarias, sanitarias o tuteladas que carecen de vivienda estable; Personas residentes en centros para personas mayores sin hogar; Personas residentes en alojamientos inseguros, temporalmente, sin derechos legales o en condiciones de ocupación; Personas con requerimiento de desahucio de vivienda; Personas que viven bajo amenaza de violencia proveniente de sus convivientes; Personas que viven en alojamientos móviles, viviendas no convencionales o estructuras semi-temporales como chabolas o cabañas; Personas que viven en alojamientos sin cédula de habitabilidad; Personas que residen en viviendas hacinadas o sobreocupadas. Se entenderá que la vivienda se encuentra sobreocupada cuando superan el estándar nacional de ocupación de personas (número

En cambio, La letra l) del art. 3 LDV define el sinhogarismo como: "circunstancia vital que afecta a una persona, familia o unidad de convivencia que no puede acceder de manera sostenida a una vivienda digna y adecuada en un entorno comunitario y aboca a las personas, familias o unidades de convivencia que lo sufren a residir en la vía pública u otros espacios públicos inadecuados, o utilizar alternativas de alojamiento colectivo institucionalizado de las diferentes administraciones públicas o de entidades sin ánimo de lucro, o residir en una vivienda inadecuada, temporal o no, inapropiada o masificada, en una vivienda insegura, sin título legal, o con notificación de abandono de la misma, o viviendo bajo amenaza de violencia. Podrá calificarse como sinhogarismo cronificado, cuando la situación de sinhogarismo continúe o se produzca a lo largo de un periodo de tiempo igual o superior a un año".

Esta es, sin duda, una de las definiciones más criticadas de la ley[36]. Entre otras razones, cabe pensar que la categorización que establece la tabla ETHOS obedece a la necesidad de diferenciar entre situaciones distintas, con entidad propia, que requieren medidas adecuadas a sus particularidades. Sin embargo, la LDV no diferencia estas situaciones, siendo por tanto las medidas iguales con independencia de las circunstancias que concurran en el caso concreto.

---

máximo de ocupantes por vivienda en función del número de habitaciones, o espacios de uso común, y de la superficie de estas)".

36 Entre otros, Argelich Comelles (2023, p. 88) y Nasarre Aznar (2022, p. 13).

4. Hogar

Después de establecer una definición para el "sinhogarismo", ¿no resulta igualmente importante definir qué significa "hogar"? Ciertamente, no es tarea sencilla definir en el ámbito jurídico el "hogar". El legislador utiliza este término en la LDV, sin definirlo previamente, como sinónimo de "unidad de convivencia" o "unidad familiar"[37], lo que le resta a este concepto el alcance que debería tener, al menos, en estándar internacional. Es respecto del hogar donde el "factor X" tiene su máxima expresión. No sin razón concluye Caballé Fabra (2020, p. 265) que el hogar que exista en una vivienda digna y adecuada es (...) la máxima expresión del concepto de vivienda.

Estimamos innecesario hacer referencia, en este apartado, a las demás definiciones que recoge el art. 3 LDV, pues con las definiciones mencionadas ha quedado demostrada la conclusión y respondida la pregunta que encabeza este epígrafe: el citado precepto no sirve para unificar las definiciones recogidas en otras leyes o normas incluyendo los avances y esfuerzos realizados en materia de vivienda y ello aun a pesar de las advertencias recibidas durante la tramitación parlamentaria.

Es más, el legislador no es que no logre unificar, estableciendo unas bases comunes[38], las definiciones existentes, facili-

---

37 Argelich Comelles (2023, p. 87) propone sustituir todas las referencias al "hogar" por la de "unidad familiar", para equipararlo al concepto estándar en materia de subvenciones.

38 Nasarre Aznar (2022, p. 12) opina que una de las razones fundamentales que justifica la existencia de una ley estatal de vivienda es, precisamente, la posibilidad de establecer unos conceptos básicos y comunes que puedan a su vez ser adaptados en otras normas, como las autonómicas.

tando la comprensión de los términos empleados en la ley, sino que, por el contrario, complica sobremanera la comprensión de la ley y su alcance.

Lo más sorprendente es que tuviera a su disposición sobradas soluciones y propuestas contrastadas y bien fundamentadas, conocidas, además, por los intervinientes y expertos del sector que, desde luego, trataron de advertir al legislador de su existencia.

IV. ¿SIRVEN LAS DEFINICIONES DE VIVIENDA PARA APORTAR COHERENCIA INTERNA Y FACILITAR LA APLICACIÓN DE LAS MEDIDAS RECOGIDAS EN LA LEY?

Descartadas, pues, las dos finalidades anteriores, esto es, delimitar el ámbito de aplicación de la ley o unificar conceptos incluyendo los avances y propuestas doctrinales y legislativas nacionales e internacionales, queda una última pregunta que contestar, una última finalidad que podría haber perseguido el legislador al introducir el artículo 3.º LDV: aportar coherencia interna y facilitar la aplicación de las medidas recogidas en la ley.

Si la finalidad de introducir un listado de definiciones en la LDV es la de lograr coherencia interna y facilitar la aplicabilidad de las medidas que regula, entonces un estudio sistemático de la ley debería arrojar luz sobre el cumplimiento o no de este objetivo.

No es posible exponer todas las incoherencias o dificultades que se aprecian tras dicho estudio sistemático de la ley, por lo que pondremos el foco sobre tres ejemplos que permiten contestar negativamente a la pregunta que encabeza este apartado.

Para el primer ejemplo, partiremos de la definición de "vivienda digna y adecuada" (art. 3.c) LDV), en relación con lo dispuesto en el apartado primero del art. 8 LDV y en la Exposición de Motivos de la ley. En concreto, queremos llamar la atención sobre el hecho de que en la Exposición de Motivos se introduzca la idea del disfrute de la vivienda a través de "otras tipologías de tenencia" y en el art. 8 LDV se especifique que todos los ciudadanos tienen derecho a disfrutar de la vivienda digna y adecuada no solo en régimen de propiedad y arrendamiento, sino de "cualquier otro régimen legal de tenencia".

Pues bien, estas "otras tipologías" o regímenes de tenencia, no se definen en la ley, y quizás sea esa la razón por la que el legislador olvida extender a aquellas la aplicación de las medidas que regula.

Pongamos un ejemplo: el art. 31 LDV, encargado de regular el derecho de información, se circunscribe a la compraventa o al arrendamiento de una vivienda, sin hacer mención alguna a las "otras tipologías" o a otros regímenes de tenencia.

Pero, si una propiedad compartida o temporal (regulada en el ordenamiento catalán) tiene por objeto una "vivienda digna y adecuada", y el art. 8 LDV nos dice que, efectivamente, el derecho a una vivienda digna y adecuada se puede disfrutar no solo en régimen de propiedad o arrendamiento, sino también en "cualquier otro régimen legal de tenencia", ¿qué justificación tiene que la medida del art. 31 LDV no se extienda a este supuesto?

Es solo uno de los ejemplos de una obligación que impone la LDV, sin que quede delimitado su ámbito de aplicación.

Enlazando el problema expuesto con otra definición, en concreto la de "condiciones accesibles conforme al esfuerzo financiero", que recoge el art. 3.d) LDV, en ese precepto se circunscribe la aplicabilidad de cualquier medida a los negocios de compraventa y arrendamiento, al mencionar el "precio de venta o alquiler" en la definición.

Una restricción en cuanto a sus efectos difícilmente compatible con las exigencias internacionales. En efecto, de una lectura detallada de los textos autonómicos[39], nacionales e internacionales[40], de la doctrina e incluso de la propia LDV, resulta patente la relevancia del criterio de la "asequibilidad"[41], lo que hubiera justificado un esfuerzo adecuado para definirlo

---

39 La mayoría de las CCAA directamente se refieren a la "vivienda digna, adecuada y asequible" o incluso a la "vivienda digna y asequible", o incluyen la referencia a la "asequibilidad" en la propia definición de la vivienda adecuada.

40 Sobre la asequibilidad dice el Relator Especial de las Naciones Unidas en su Informe sobre una vivienda adecuada como elemento integrante del derecho a un nivel de vida adecuado de 13 de febrero de 2008 (A/HRC/7/16, p. 8): "La asequibilidad, uno de los elementos del derecho a una vivienda adecuada, constituye uno de los principales factores de la carencia de vivienda", lo que probablemente ha provocado que el elemento de la asequibilidad haya ido generando mayor preocupación y ganando más protagonismo. El Comité Económico y Social Europeo, en su Dictamen "Acceso universal a una vivienda digna, sostenible y asequible a largo plazo" (TEN/707-EESC-2020), hace referencia a la "crisis de la vivienda asequible", cuyo momento más crítico se dio como consecuencia de la Pandemia Covid-19. Como consecuencia de dicha *crisis,* el Comité, aun asumiendo la competencia de los Estados miembros en materia de vivienda, propone ciertas medidas encaminadas a garantizar el derecho de los ciudadanos a una vivienda asequible. Destacables son las propuestas de crear un "fondo europeo de inversión en viviendas *asequibles,* dignas y adecuadas" y de organizar, por parte de la Comisión, una cumbre europea de la vivienda asequible.

41 Dice Busto Lago (2023, p. 3) que el informe "The estate of housing in the UE 2017", elaborado por Housing Europe "destaca como el carácter asequible de la vivienda se ha convertido en una cuestión clave, impactando en la vida de millones de ciudadanos en Europa".

correctamente y con la suficiente amplitud, de manera que no se excluyan otros regímenes de tenencia o acceso a la vivienda, distintos de la compraventa o el arrendamiento.

El segundo ejemplo que habíamos anunciado está íntimamente relacionado con el mencionado arrendamiento de vivienda. Recordemos que el art. 3.a) LDV define la "vivienda" y que el art. 3.i) LDV define la "residencia habitual". Pues bien, la Disposición adicional 1.ª[42], que se refiere a la elaboración de una base de datos sobre arrendamientos de

---

[42] Según la Disposición Adicional 1.ª: "1. Para el desarrollo de lo previsto en esta ley, se conformará una base de datos de contratos de arrendamiento de vivienda, a partir de la información contenida en los actuales registros autonómicos de fianzas de las comunidades autónomas, en el Registro de la Propiedad y otras fuentes de información de ámbito estatal, autonómico o local, con el objeto de incrementar la información disponible para el desarrollo del Sistema de índices de referencia del precio del alquiler de vivienda establecido en la disposición adicional segunda del Real Decreto-ley 7/2019, de 1 de marzo, de medidas urgentes en materia de vivienda y alquiler.
2. Se promoverán los mecanismos de colaboración con las comunidades autónomas y otros organismos e instituciones, para disponer de la información veraz sobre los contratos de arrendamiento de vivienda vigentes, a través de los datos recogidos en los distintos registros autonómicos y estatales, con el objeto de realizar un adecuado seguimiento del conjunto de medidas incluidas en esta ley y determinar el progreso en el cumplimiento de los objetivos de incrementar la oferta de vivienda en alquiler a precios asequibles.
3. A la entrada en vigor de la ley se iniciará un proceso específico de colaboración entre el Departamento Ministerial competente en materia de vivienda y las comunidades autónomas que hayan desarrollado sistemas de referencia del precio del alquiler en sus respectivos ámbitos territoriales para asegurar la colaboración entre sistemas, la atención a las especificidades territoriales que deban te-

vivienda para el desarrollo de la ley, no especifica a qué tipo de vivienda se refiere.

Así que obliga a plantearse si se refiere a arrendamientos que tengan por objeto una vivienda, según la definición contenida en el art. 3.a) LDV, es decir, a cualquier arrendamiento que recaiga sobre un "edificio o parte de un edificio de carácter privativo y con destino a residencia y habitación de las personas, que reúne las condiciones mínimas de habitabilidad exigidas legalmente, pudiendo disponer de acceso a espacios y servicios comunes del edificio en el que se ubica, todo ello de conformidad con la legislación aplicable y con la ordenación urbanística y territorial".

Si es este el caso, entonces la base de datos englobaría arrendamientos de vivienda habitual, de temporada y de vivienda turística.

En caso contrario, esto es, si la intención del legislador es la de crear una base de datos que únicamente recopile información de los arrendamientos de vivienda habitual, conforme a la definición que recoge el art. 2 LAU en relación con el art. 7 LAU, ¿por qué no se refiere a la "residencia" –según terminología del art. 3.i) LDV– o al menos a la "vivienda" habitual?

Otro ejemplo, por tanto, de una obligación impuesta por la LDV, sin aclarar su ámbito de aplicación[43].

---

nerse en cuenta, así como el establecimiento de plazos para agilizar su aplicación efectiva".

43 La única pista que nos da el legislador es la remisión a la Disposición adicional 2.ª del RD-ley 7/2019, de 1 de marzo, de medidas urgentes en materia de vivienda y alquiler. Aunque esta disposición adicional también se refiere al "arrendamiento de vivienda" a los efectos de crear un índice de referencia de

Para finalizar, el tercer ejemplo que permite contestar negativamente a la pregunta que encabeza este apartado es que existe un artículo específicamente destinado a definir términos de relevancia para la aplicación de las medidas reguladas en la LDV, pero aun así hay definiciones que en vez de aparecer en el listado, se incluyen en otro punto de la ley.

Por ejemplo, la definición de la vivienda vacía o desocupada[44], que no se incluye en el art. 3 LDV, sino en la Disposición final tercera, a los efectos de modificar la redacción del apartado 4.º del art. 72 del Texto Refundido de la Ley reguladora de las Haciendas Locales, aprobado por Real Decreto Legislativo 2/2004, de 5 de marzo: "A estos efectos tendrá la consideración de inmueble desocupado con carácter permanente aquel que permanezca desocupado, de forma continuada y sin causa justificada, por un plazo superior a dos años, conforme a los requisitos, medios de prueba y procedimiento que establezca la ordenanza fiscal, y pertenezcan a titulares de cuatro o más inmuebles de uso residencial".

---

precios de alquiler, lo cierto es que en esta ley sí tiene sentido esa referencia, por cuanto que su ámbito de aplicación es dual: arrendamiento de vivienda y para uso distinto de vivienda, siendo así que el arrendamiento de temporada se integra dentro del segundo grupo, mientras que el arrendamiento turístico, siempre y cuando esté sometido a normativa sectorial turística, queda excluido del ámbito de aplicación de la ley.

44 Esta omisión en el art. 3 LDV sorprende, sobre todo, a la vista de los recursos de inconstitucionalidad que se presentaron ante el TC respecto de las medidas autonómicas adoptadas por algunas CCAA en relación con las viviendas vacías. Ampliamente sobre este tema: Gálvez Criado (2023, pp. 342-353).

V. CONCLUSIONES Y PROPUESTAS DE MEJORA

Primera. Consideramos demostrado que las definiciones contenidas en el art. 3 LDV no han servido para ninguna de las tres finalidades que podría haber cumplido en beneficio de la claridad, coherencia, aplicabilidad y utilidad de la LDV.

No sirve para delimitar el ámbito de aplicación de la LDV, ni para coordinar las normas existentes y que deben coexistir con la LDV, ni tampoco para unificar los conceptos en materia de vivienda, incluyendo los avances legislativos, jurisprudenciales y doctrinales de ámbito internacional e interno. Tampoco sirve para aportar coherencia al texto de la ley, ni de hilo conductor entre los términos que emplea la ley y las medidas que regula, ni para facilitar la aplicación de la LDV.

En definitiva, no sirve más que para crear mayor confusión en un tema tan complejo como este, sobre todo, a la vista del caos terminológico existente en la actualidad. La técnica legislativa, a su vez, complica sobremanera la aplicabilidad de esta ley. Tanto es así que aun habiendo establecido ciertas definiciones en el art. 3 LDV, el propio legislador las ignora en el texto, utilizando indistintamente otros términos para referirse, aparentemente, a lo mismo.

Segunda. Dos son las propuestas principales que haremos para mejorar este artículo y en general su relación con el resto del cuerpo legal. En primer lugar, y aunque se puede intuir claramente de la estructura del trabajo, proponemos que se cumplan y se dé respuesta a las tres preguntas planteadas.

Para empezar, debería introducir el legislador, sin duda, una disposición general destinada a aclarar el ámbito de aplicación de esta ley. Como refuerzo de esta delimitación, en el art. 3 LDV se pueden introducir ciertas definiciones o aclaraciones, si se quiere, de términos que se emplearán a lo largo del texto.

Una lectura correctora detallada del texto llevará a eliminar todas aquellas expresiones o términos que se empleen como sinónimos de los términos definidos.

Centrándonos en las definiciones en sí, se utilizará esta magnífica oportunidad para introducir en un texto legal de alcance estatal, cuando las haya, las novedades en cuanto a la terminología y sus conceptos cuando hayan sido acogidas por la mayoría de la doctrina y demás agentes especializados en materia de vivienda.

Finalmente, y a la vista de que las definiciones deben servir para aclarar el propio texto y facilitar la aplicación en la práctica de las medidas reguladas en la ley, se revisará la propia coherencia interna de la ley, esto es, que el objeto de cada medida quede claro y que el art. 3 LDV sirva para que, al nombrar dicho objeto, por ejemplo: vivienda adecuada u hogar, sea fácilmente localizable, interna -en el texto- y externamente -en la práctica-.

Contestamos así a las tres preguntas.

Tercera. La segunda propuesta se refiere a los propios términos elegidos y al contenido de las definiciones y su conexión con las medidas.

En concreto, entendemos que una forma útil hubiera sido utilizar los criterios de la OG n.º 4, definirlos siguiendo dicha OG, y estructurar la ley entorno a estos criterios.

Así, la primera definición sin duda debe ser la de la "vivienda adecuada", que debe reunir estos criterios: la seguridad jurídica de la tenencia, la disponibilidad de servicios, materiales, facilidades e infraestructura, los gastos soportables en el uso de la vivienda, la habitabilidad de la vivienda, su asequibilidad en el momento del acceso, la ubicación, la adecuación cultural y, finalmente, el factor espiritual o factor X, que nos llevará a otros dos términos a definir: el hogar y la vivienda familiar.

Una vez establecidos dichos criterios, decíamos, sirven para estructurar la propia ley de vivienda. ¿Cómo? Pues estableciendo las medidas propias y apropiadas para lograr que los ciudadanos puedan acceder y permanecer en una vivienda que según estándar internacional y -ya- nacional sea *adecuada.*

Y, como previamente se ha establecido el ámbito de aplicación de la LDV y se ha aclarado que, según el reparto competencial, cada uno puede regular hasta el límite contenido en dicha asunción competencial, esas medidas deberán ser en unos casos directamente aplicables y en otros una base a considerar por las CCAA en su propia normativa.

Para ejemplificar la propuesta anterior, se incluirían en el capítulo correspondiente a la "habitabilidad", los estándares básicos a cumplir para que una vivienda pueda considerarse adecuada; y en el capítulo correspondiente a la "asequibilidad", las medidas que, dentro de su competencia, pueda elaborar el legislador estatal para facilitar el acceso de *todos* los ciudadanos a una vivienda adecuada y, especialmente, de aquellos ciudadanos en una situación especial de vulnerabilidad.

La pequeña aclaración que nos hemos permitido en cuanto a "todos" los ciudadanos, no ha sido casual, sino intencionada y mucho, por cuanto que, a veces, al legislador se le olvida que en la actualidad los estándares de vulnerabilidad regulados no se corresponden con la realidad, dada la magnitud del problema de acceso a la vivienda. Lo cierto es que la mayoría de las medidas de apoyo al acceso a la vivienda se refieren a un colectivo muy concreto que realmente no representa al grueso de la sociedad que sufre las consecuencias de la dificultad de acceso a la vivienda, por lo que las medidas en cuanto a su alcance subjetivo debe ser ampliado y prever todos esos sectores de la sociedad que, aun aparentemente en una buena situación económica, en las circunstancias actuales no se pueden permitir ninguna de las tenencias actualmente previstas en nuestro ordenamiento.

Volviendo a la estructura de la ley, supondría que la suma de todas las medidas, la suma de todos los capítulos de la ley, llevarían a un resultado: la vivienda adecuada. El acceso y permanencia en la vivienda adecuada.

Cuarta. A nuestro parecer, esta estructura es lógica, se anunciaría desde el principio y se reflejaría en todo el texto legal: primero, en la disposición general que delimite el ámbito de aplicación de la ley; en segundo lugar, en la disposición general destinada a definir y conceptualizar determinados términos posteriormente utilizados en la ley; y en tercer lugar, en la estructura de la ley, en la que previsiblemente se recogerán por capítulos todas aquellas medidas encaminadas a responder a los criterios que llevarán a los ciudadanos a alcanzar el derecho a la vivienda adecuada, el objetivo final e indiscutible que debería perseguir esta ley.

VI. REFERENCIAS BIBLIOGRÁFICAS

Argelich Comelles, C. (2023). *Ley por el derecho a la vivienda.* Aranzadi.

Busto Lago, J. M. (2023). "La Ley por el derecho a la vivienda y el derecho subjetivo a la vivienda". *Cuadernos de Derecho Privado,* (6), 2-9.

Caballé Fabra, G. (2020). *La intermediación inmobiliaria ante los nuevos retos de la vivienda.* Tirant lo Blanch.

Gálvez Criado, A. (2023). "Los límites impuestos a las facultades de uso y disfrute de los propietarios para luchar contra la escasez de vivienda disponible en alquiler en España". *Actualidad Jurídica Iberoamericana,* (19), 334-383.

López y López, A.M. (2022). "Propiedad privada y disciplina del mercando en el Proyecto de Ley por el derecho a la vivienda". *Cuadernos de Derecho Local,* (59), 14-34.

Molina Roig, E. (2018). *Una nueva regulación para los arrendamientos de vivienda en un contexto europeo.* Tirant lo Blanch.

Nasarre Aznar, S. (2022). "El Proyecto de Ley de vivienda 2022". *Fedea* (2022/11), 1-25.

Pérez Daudi, V. (2022). "El derecho a la vivienda digna y la tutela judicial efectiva en los Tratados Internacionales". En: Cervilla Garzón, M. D. y Zurita Martín, I. (dirs.) *Vivienda y colectivos vulnerables*. Aranzadi, 329-350.

Prats Albentosa, L. (2022). "Derecho a gozar de una vivienda digna y adecuada y derecho a la propiedad de vivienda (A propósito del Proyecto de Ley por el derecho a la vivienda)". *Revista Jurídica del Notariado*, (115), 131-174.

Pisarello, G. (2009). "El derecho a la vivienda como derecho social: implicaciones constitucionales". *Revista catalana de dret públic*, (38), 1-13.

Sainz-Cantero Caparrós, M. B. (2023). "La redefinición del derecho de propiedad sobre la vivienda en la era de los bienes comunes. A propósito de la Ley 12/2023 del derecho a la vivienda". *Actualidad Jurídica Iberoamericana*, (19), 630-663.

Capítulo II

El principio de igualdad y no discriminación en las relaciones entre particulares en torno a la vivienda

M.ª DOLORES CERVILLA GARZÓN
ISABEL ZURITA MARTÍN
Catedráticas de Derecho civil
Universidad de Cádiz

I. APUNTE INTRODUCTORIO

La Ley 12/2023, de 24 de mayo, por el derecho a la vivienda, dedica un precepto específico de su Título preliminar al principio de igualdad y no discriminación, sobre el que giran las reflexiones que exponemos en este capítulo. El legislador, a pesar de la publicación anterior de la Ley 15/2022, de 12 de julio, integral para la igualdad de trato y no discriminación, decidió reflejar este principio de forma expresa en su artículo 6 en el marco de una ley especial, aunque también de carácter general, dedicada a la vivienda.

Es evidente que este precepto no recoge novedad alguna en la materia que abordamos, pues tan solo viene a completar, en el sector concernido, la normativa integrante del Derecho antidiscriminatorio propio del ordenamiento jurídico español, que ya contaba como referente fundamental con la mencionada Ley integral.

Nuestra intención en las páginas que siguen no puede ser otra más que reflexionar sobre la trascendencia de esta nueva norma, tratando de ofrecer una visión práctica de las consecuencias de su aplicación en el tráfico jurídico inmobiliario destinado al acceso a la vivienda. A estos efectos, prestaremos primeramente atención a su contextualización dentro del marco del Derecho antidiscriminatorio, comenzando por apuntar con brevedad el concepto y orígenes de este, así como los principios fundamentales afectados, para hacer con posterioridad un repaso igualmente conciso a su previo reflejo legal en las normas comunitarias y españolas que anteceden a la Ley especial sobre el derecho a la vivienda que aquí comentamos.

II. EL PRINCIPIO DE AUTONOMÍA PRIVADA Y LA EFICACIA HORIZONTAL DE LOS DERECHOS FUNDAMENTALES

Iniciar este trabajo afirmando la trascendencia del principio de autonomía privada en materia contractual no parece sino una obviedad. Es claro que dicho principio constituye un pilar básico de nuestro ordenamiento jurídico patrimonial, permitiendo a las partes negociar libremente el contenido de una relación contractual o, de entrada, si van a llevar a término la misma. La autonomía de la voluntad se ha defendido tradicionalmente frente a cualquier agresión de tercero, ya sea este otro particular o los poderes públicos, sobre la base de la libertad del sujeto de autorregular sus propios intereses; todo ello con los únicos límites dispuestos en el artículo 1255 CC, esto es, que no se manifieste contradicción alguna con la ley, la moral y el orden público.

Este preludio no resulta, sin embargo, desatinado si observamos la debilitación que dicho principio ha ido experimentando en su convergencia con derechos fundamentales de la persona. Ciertamente, debemos partir en este estudio de la autoridad del principio de autonomía de la voluntad en materia contractual sin desviar, al mismo tiempo, nuestra mirada de la supremacía de derechos básicos constitucional e internacionalmente reconocidos.

Como es sabido, en su sentido tradicional el principio de autonomía privada no solo concede a los contratantes la libertad de contratar o no, sino también la de elegir a su contraparte y, llegado el caso, la de establecer cuantos pactos, cláusulas y condiciones tengan por conveniente. Bien es cierto que, tal cual se ha indicado, este principio no es absoluto, en tanto sometido a los límites legalmente establecidos. Pero tales límites se han contemplado arraigadamente como excepcionales, además de extraños, respecto de la propia relación contractual entre las partes[1]. Esta percepción, sin embargo, ha ido evolucionando, de forma parsimoniosa pero constante, en las últimas décadas del siglo pasado y primeras del presente, tanto en el ámbito doctrinal como en el jurisprudencial y legal, hasta alcanzar un estadio que ha hecho modificar la concepción de un principio que se nos aparecía como sagrado en el Derecho contractual recogido por los decimonónicos códigos europeos.

No podemos dejar de referirnos a la controversia acerca de la eficacia horizontal o entre particulares de los derechos fundamentales (*Drittwirkung der Grundrechte),* sobre la que no se manifiesta expresamente nuestra Constitución, y cuyo análisis, obviamente, desborda el objeto de nuestro estudio. Debemos meramente partir de la doctrina del Tribunal Constitucional, que de modo reiterado se ha decantado por el reconocimiento de la eficacia directa de tales derechos en el ámbito de las relaciones privadas[2], sin dejar de observar al

---

1 Para Jiménez Horwitz (2014, p. 494), los límites eran considerados por motivos extracontractuales, incidiendo de manera excepcional en las relaciones económicas de los particulares.

2 Entre otras muchas, puede observarse su evolución, en distintas esferas y derechos, en las SSTC 78/1982 de 20 de diciembre (TOL78.969);

mismo tiempo las peculiaridades que en dicha órbita suscita su aplicación. Llama la atención la doctrina sobre el hecho de que no se trata aquí simplemente de un problema de eficacia mediata o inmediata de los derechos fundamentales, sino de un supuesto de colisión de derechos, por cuanto el principio de autonomía de la voluntad ostenta una posición prevalente en el Derecho privado y cuenta asimismo con reconocimiento constitucional por virtud del artículo 10.1 de nuestra Carta Magna, como manifestación del derecho al libre desarrollo de la personalidad. En verdad, los derechos fundamentales no operan de igual forma en el ámbito de las relaciones privadas que en las existentes entre los ciudadanos y los poderes públicos, de ahí que la doctrina de la eficacia horizontal de estos derechos deba reservarse a los supuestos más graves de violación de valores esenciales, sobre todo aquellos que supongan un atentado contra el orden público[3]. Es preciso contar, pues, con criterios que midan la incidencia de los derechos fundamentales en las relaciones entre particulares, como pueda ser la existencia de una situación de desigualdad entre los sujetos o el hecho de verse comprometida de forma directa la propia dignidad de la persona[4].

Y en este escenario, es palmario que, entre los derechos fundamentales que reconoce nuestra Constitución, es el derecho a la igualdad y no discriminación el protagonista de la mencionada evolución. En este terreno, es obligado partir de la

---

22/1984 de 17 de febrero (TOL79.311); 114/1984 de 29 de diciembre (TOL79.403): 19/1985 de 13 de febrero (TOL79.434); 177/1988 de 10 de octubre (TOL80.025); 99/1994 de 11 de abril (TOL82.507); o 140/2014 de 11 de septiembre (TOL4.517.082).

3 Así lo explica Llorente San Segundo (2018, pp. 31-32).

4 Véanse Díaz Revoiro (2015, pp. 115 y ss.) y Alfaro Águila-Real (1993, pp. 108-112).

consideración de la autonomía privada como un principio que aparece, en nuestro actual Derecho, decididamente mediatizado por la defensa de la igualdad de trato a la que toda persona tiene un derecho constitucionalmente reconocido.

III. EL DERECHO ANTIDISCRIMINATORIO

1. La prohibición de discriminación como principio normativo

El Derecho antidiscriminatorio –considerado por la doctrina como una rama emergente del Derecho[5]–, puede ser entendido como todo un conjunto de normas y principios en cuya virtud "el Estado se compromete a garantizar la misma dignidad social a todas las personas, tratando de remover todos los obstáculos de carácter económico o social que impidan la realización de la igualdad sustancial"[6]. En el seno de esta nueva disciplina se entremezclan principios constitucionales, tratados internacionales y todo un entresijo de normas que conciernen de modo directo tanto a los poderes públicos como a los ciudadanos en particular, proyectándose, desde luego, sobre las relaciones privadas entre los individuos.

La progresión en el reconocimiento de los efectos del principio de igualdad en las relaciones entre particulares encuentra, no obstante, sus raíces en postulados ajenos al Derecho de nuestro continente, trayendo causa del Derecho antidiscriminatorio surgido de distintas enmiendas a la Constitución estadounidense. Es lugar común afirmar que el origen del Derecho antidiscriminatorio se encuentra en Estados Unidos,

---

5 Rey Martínez (2020).

6 Así lo define Barba (2023, p. 16).

reflejándose inicialmente en las Enmiendas Decimotercera –que supuso la abolición de la esclavitud en este país– y Decimocuarta, prohibitiva de la negación a los ciudadanos de la igual protección de las leyes. Su traslado a las relaciones entre particulares se fue materializando, a mediados del siglo pasado, a través de la publicación sucesiva de una serie de normas en distintos ámbitos, primero en el laboral, extendiéndose con posterioridad a otros hasta alcanzar carácter general[7]. Resulta de especial relevancia para nuestro objeto de estudio la *Fair Housing Act* de 1968, que proscribe la discriminación en el acceso a la vivienda, prohibiendo cualquier tipo de prácticas en la venta, alquiler y financiación de viviendas que sean discriminatorias por razón de raza, religión, sexo o nacionalidad; esta norma, aprobada en la Cámara de Representantes en los días posteriores al asesinato de Martin Luther King, es considerada como el último gran logro legislativo de la era de los derechos civiles en Estados Unidos.

En nuestro Derecho, el punto de referencia al que debemos dirigir nuestra mirada al objeto de este análisis no es otro que el artículo 14 de la Constitución Española, que postula que "los españoles son iguales ante la ley, sin que pueda prevalecer discriminación alguna por razón de nacimiento, raza, sexo, religión, opinión o cualquier otra condición o circunstancia personal o social"[8]. Este básico precepto no es solo el faro del principio de igualdad en nuestro ordenamiento, sino que alberga asimismo

---

7 Llorente San Segundo (2018, p. 15). Señalan Barrère Unzueta y Morondo Taramundi (2011, p. 4) que, aunque con importantes matices, el Derecho antidiscriminatorio europeo se ha edificado sobre el norteamericano.

8 Art. 9 CE: "…Corresponde a los poderes públicos promover las condiciones para que la libertad y la igualdad del individuo y de los grupos en que se integra sean reales y efectivas…".

una expresa prohibición de discriminación por razones diversas y concretas, que en modo alguno deben considerarse exhaustivas. Bajo el paraguas de "cualquier otra condición o circunstancia personal o social" pueden cobijarse otros posibles factores de discriminación que ha ido reconociendo el propio Tribunal Constitucional bajo determinadas circunstancias, como la edad, la discapacidad, la enfermedad o la orientación sexual[9].

Y es que los denominados factores de riesgo –aquellas condiciones o características, reales o presuntas, que definen la identidad o los rasgos de una persona o colectivo exponiéndolos al trance de ser sujetos de discriminación– se encuentran en constante evolución, constituyendo, como ha descrito la doctrina, una realidad líquida[10] que, en un momento concreto, es determinante del trato menos favorable e injustificado recibido por una persona o personas respecto de otra u otras.

La igualdad de trato conlleva observar la prohibición de discriminación desde dos perspectivas: la directa y la indirecta. La discriminación directa, también llamada "de trato", tendría lugar cuando una persona es tratada de forma menos favorable que otra en una situación análoga por razón de algún rasgo, característica o circunstancia personal, tales como género, orientación sexual, origen étnico, discapacidad, edad, enfermedad, etc. En este marco, el acoso discriminatorio es considerado una modalidad de discriminación directa, en tanto se identifica con toda conducta realizada por razón de alguna de las causas de discriminación antedichas, con el objetivo o la consecuencia de atentar contra la dignidad de una persona o grupo en que se integra y de crear un entorno intimidatorio, hostil, degradante,

---

9 Sobre la evolución de la jurisprudencia constitucional en la materia, véase Rey Martínez (2017, pp. 138-140).

10 Así los califica Barba (2023, p. 35).

humillante u ofensivo. Por su parte, la discriminación indirecta se produce cuando una disposición, criterio o práctica en apariencia neutrales crea una desventaja concreta a una persona por causa de algún factor de riesgo en comparación con otras, salvo que aquellas actuaciones puedan justificarse por una finalidad legítima y los medios para conseguirla sean apropiados o necesarios. Este concepto nació en la jurisprudencia norteamericana con la finalidad de evitar que un empresario pudiera evadir la prohibición de discriminación racial a través de prácticas aparentemente neutrales, pero dirigidas a alcanzar el objetivo discriminador[11]. En España es acogido por primera vez por el Tribunal Constitucional en su sentencia 145/1991, referente a unas limpiadoras de hospital cuyo salario era inferior al de los peones, cuando ambas categorías laborales realizaban el mismo trabajo. En el ámbito de la vivienda, una excepción por razón de origen ético en una oferta de arrendamiento sería un supuesto de discriminación directa; una limitación que atendiera al modo de vestir que solo fuera propio de determinados colectivos, significaría una discriminación indirecta.

Estos conceptos y postulados se han ido recogiendo progresivamente en la legislación, primero en normas comunitarias y, posteriormente, en el Derecho español, afectando a la globalidad del ordenamiento jurídico. El deber de no discriminación,

---

11 Como especifica Rey Martínez (2017, p. 141), la discriminación indirecta es una creación del Tribunal Supremo Federal de EEUU, por medio de la sentencia Griggs v. Duke Power Company, de 8 de marzo de 1971, que resuelve una supuesto de discriminación respecto de los empleados negros frente a los blancos en una empresa. Explica también el citado autor que la discriminación indirecta también es llamada "de impacto", frente a la discriminación directa o "de trato", porque supone una comparación del distinto impacto que una diferencia jurídica de trato, en principio neutral, produce sobre los miembros del grupo a proteger respecto de los de la mayoría.

como se ha señalado doctrinalmente, es un principio-valor de todos los ordenamientos jurídicos contemporáneos, trascendente tanto en las relaciones verticales como en las horizontales[12]. En realidad, con independencia de las normas específicas o sectoriales que lo alberguen, el principio de no discriminación se configura como un principio normativo, que bebe de la fuente del ordenamiento jurídico en su conjunto, confirmando así su aplicación directa incluso sin la preexistencia de una norma reguladora expresa[13]. Y justamente por configurarse como tal principio normativo, la interdicción de la discriminación, como norma imperativa, se aplica a las relaciones jurídicas entre particulares y atañe a todas las fases por las que puede devenir un contrato[14]. A la aplicación del principio de no discriminación en cada una de las distintas fases contractuales nos referiremos al analizar su incidencia en el acceso a la vivienda.

2. *Evolución normativa*

2.1. El Derecho comunitario

El papel del Derecho comunitario en la integración del Derecho antidiscriminatorio en el ordenamiento español es particularmente reseñable, en virtud de la transposición de diferentes Directivas europeas referentes a la prohibición de la discriminación en distintos ámbitos jurídicos. Si bien las primeras disposiciones se refieren a la esfera laboral –por motivos competenciales–, deben resaltarse especialmente tres Directivas emitidas a partir de la modificación en 1997 del Tratado

---

[12] Barba (2023, p. 26).

[13] En este sentido, García Rubio (2002, p. 304).

[14] Barba (2023, p. 40).

Constitutivo de la Comunidad Europea: Directiva 2000/43/CE, relativa a la aplicación del principio de igualdad de trato de las personas independientemente de su origen racial o étnico; Directiva 2000/78/CE, relativa al establecimiento de un marco general para la igualdad de trato en el empleo y en la ocupación; y Directiva 2004/113/CE, por la que se aplica el principio de igualdad de trato entre hombres y mujeres al acceso de los bienes y servicios y su suministro. Estas normas comunitarias responden a causas distintas de discriminación, por lo que debe realizarse una labor de generalización entre todas ellas. En su conjunto, parten del concepto del principio de igualdad de trato como "la ausencia de toda discriminación, tanto directa como indirecta" y distinguen entre discriminación directa e indirecta, dando entrada no solo a actuaciones reales ya producidas sino a situaciones hipotéticas de discriminación, al referirse expresamente a la situación en que una persona haya sido o pudiera ser tratada de manera menos favorable.

Es la Directiva 2000/43/CE la primera referencia en la materia que nos ocupa, máxime cuando en su artículo 3 especifica que "... se aplicará a todas las personas, por lo que respecta tanto al sector público como al privado, incluidos los organismos públicos, en relación con: .... h) el acceso a bienes y servicios disponibles para el público y la oferta de los mismos, incluida la vivienda". Posteriormente, la Directiva 2004/113/CE se referirá a su ámbito de aplicación de forma más general, disponiendo también en su artículo 3 que "... se aplicará a todas las personas que suministren bienes y servicios disponibles para el público, con independencia de la persona de que se trate, tanto en lo relativo al sector público como al privado, incluidos los organismos públicos, y que se ofrezcan fuera del ámbito de la vida privada y familiar, y a las transacciones que se efectúen en ese contexto". Queda patente así que la prohibición de discriminación afecta tanto a las relaciones de carácter público como a las de naturaleza privada.

En cualquier caso, el punto del que debemos partir es el destinatario al que se ofertan los bienes y servicios, que nos lleva a la distinción entre dos esferas subjetivas distintas de contratación. Efectivamente, ambas directivas contemplan la prohibición de discriminación en la esfera de los servicios ofertados al público, limitando la libertad contractual a los efectos de garantizar el acceso a estos servicios a todas las personas en condiciones de igualdad. Quedaría al margen, pues, el ámbito puramente privado de relaciones contractuales *inter partes*, como sería el caso de un arrendamiento de vivienda de vacaciones o de una habitación a una persona del círculo familiar o íntimo del arrendatario[15].

En definitiva, cuando se oferte el servicio al público en general debe garantizarse el acceso al mismo a todas las personas en condiciones de igualdad, sin que pueda excluirse a algún grupo o colectivo por razón de circunstancias de las que resulte su discriminación, ya se encuentren estas relacionadas con el sexo o con el origen racial o étnico de aquel.

No puede hablarse, sin embargo, de una limitación de la autonomía privada de carácter absoluto, en tanto se han de tener en cuenta ciertas precisiones que realiza la normativa comunitaria referentes a la libertad de elección de la otra parte contratante y a las diferencias de trato justificadas. Así, el artículo 3.2 de la Directiva 2004/113 especifica que la misma "no afectará a la libertad de la persona a la hora de elegir a la otra parte contratante, siempre y cuando dicha elección no se base en el sexo de la persona contratante". Seguidamente, el artículo 4.5 no prohíbe las diferencias de trato en tanto las mismas "estén justificadas por un propósito legítimo y los medios para lograr ese propósito sean adecuados y necesarios". La adecuación de esos propósitos

---

15 Así lo ejemplifica García Rubio (2007, p. 145).

habrá de apreciarse atendiendo a los criterios jurisprudenciales del Tribunal de Justicia de la Unión Europea, a los que dirige el considerando 16 de la Directiva 2004/113, junto a la ejemplificación de supuestos de diferenciación legítimos: "Sólo podrán admitirse diferencias de trato cuando estén justificadas por un propósito legítimo. Pueden constituir propósitos legítimos, por ejemplo, la protección de las víctimas de la violencia por razón de sexo (en supuestos como la creación de refugios para personas de un solo sexo), las razones de intimidad y decencia (en supuestos como la oferta de alojamiento hecha por una persona en una parte de su vivienda), la promoción de la igualdad de género o de los intereses de los hombres o de las mujeres (por ejemplo, organizaciones de voluntarios del mismo sexo), la libertad de asociación (en los casos de afiliación a clubes privados para un solo sexo), la organización de actividades deportivas (por ejemplo, acontecimientos deportivos para uno de los sexos). No obstante, toda limitación deberá ser adecuada y necesaria a tenor de los criterios emanados de la jurisprudencia del Tribunal de Justicia de las Comunidades Europeas".

Por medio de estas directivas, el legislador comunitario insta a los Estados miembros a incorporar a sus ordenamientos normas que regulen la discriminación entre particulares, con la finalidad última de que el principio de no discriminación se eleve a la categoría de principio general del Derecho contractual[16]. Entre otras cuestiones, para la aplicación efectiva de este

---

16 Para Llorente San Segundo (2018, pp. 20-21), la intención del legislador comunitario es que el principio de no discriminación sea considerado como un principio general del derecho contractual. La autora llama la atención sobre la evolución que ha experimentado la concepción misma del principio de igualdad en el Derecho comunitario, desde su configuración como una obligación negativa (prohibición de discriminación), hasta convertirse en una obligación positiva de promoción de la igualdad.

principio se introduce una inversión de la carga de la prueba, estableciéndose que los Estados deberán adaptar su legislación para que esta recaiga en la parte demandada cuando se aporten indicios de discriminación. A los mismos efectos, se ordena a los Estados miembros prever sanciones efectivas, proporcionadas y disuasivas aplicables en caso de incumplimiento de las obligaciones impuestas por ambas directivas.

No debe perderse de vista en todo caso que la norma comunitaria establece requisitos mínimos, reconociendo a los Estados miembros la facultad de adoptar o mantener disposiciones más favorables para la protección del principio de igualdad, sin que en ningún caso se pueda reducir, con motivo de la aplicación de estas directivas, el nivel de protección contra la discriminación ya garantizado por los ordenamientos nacionales.

2.2. El Derecho español

La transposición de las directrices comunitarias al ordenamiento español se realizó por medio de distintas normas que vinieron a afectar a diferentes ramas del Derecho, tanto en el marco del Derecho público como del Derecho privado. Las dos normas que derechamente derivan de las mencionadas directivas, ocupándose de acoger en el Derecho interno el principio de no discriminación, son la Ley 62/2003, de 30 de diciembre, de medidas fiscales, administrativas y del orden fiscal, y, con posterioridad, la Ley Orgánica 3/2007, de 22 de marzo, para la igualdad efectiva de mujeres y hombres.

La Ley 62/2003 dedica un capítulo específico dentro del título II al establecimiento de diversas medidas para la aplicación del principio de igualdad de trato, que vienen a reforzar y complementar –tal como manifiesta su Preámbulo– las numerosas normas que ya conforman nuestro ordenamiento jurídico en materia de no discriminación por todas las causas amparadas por el artículo 14 de la Constitución. Mediante estas medidas se adecúa la legislación española a lo previsto no solo en la Directiva 2000/43/CE, relativa a la aplicación del principio de igualdad de trato de las personas independientemente de su origen racial o étnico, sino también en la Directiva 2000/78/CE del Consejo, de 27 de noviembre de 2000, relativa al establecimiento de un marco general para la igualdad de trato en el empleo y la ocupación, que pretende luchar contra las discriminaciones basadas en la religión o convicciones, la discapacidad, la edad y la orientación sexual. Con esta norma se establece un marco legal general para combatir la discriminación por el origen racial o étnico de las personas en todos los ámbitos, abordándose la definición legal de la discriminación, directa e indirecta, en parecidos términos que lo previsto por las directivas que transpone, incluyéndose, por tanto, la conducta meramente hipotética a la que anteriormente se hizo referencia[17].

El artículo 27 de la Ley 62/2003 da entrada al capítulo III del título II, dedicado a las medidas para la aplicación del principio de igualdad de trato, delimitando su objeto y ámbito de aplicación: "Este capítulo tiene por objeto establecer medidas para la aplicación real y efectiva del principio de igualdad de trato y no discriminación, en particular por razón de origen racial o étnico, religión o convicciones, discapacidad, edad u orientación sexual, en los términos que en cada una de sus secciones se establecen". A partir de esta disposición común, se recogen normas en secciones distintas para diferenciar la

---

17 Para Aguilera Rull (2007, p. 6), el legislador español, siguiendo estas definiciones del Derecho comunitario, estaría sancionando una conducta meramente hipotética, que no ha tenido lugar, lo que podría conducir al demandado a tener que responder de un comportamiento inexistente. Indemnizando un daño que no se ha producido.

no discriminación por el origen racial o étnico de las personas (sección 2.ª, arts. 29-33) de la relativa a la igualdad de trato y no discriminación en el trabajo (sección 3.ª, arts. 34-43). Las primeras atienden a las esferas de la educación, la sanidad, las prestaciones y servicios sociales, la vivienda y, en general, la oferta y acceso a cualesquiera bienes y servicios. Las segundas, en cambio, dirigidas al ámbito laboral, se extienden no solo a la discriminación racial o étnica, sino a la producida por motivos de religión o convicciones, discapacidad, edad u orientación sexual de una persona. Como puede fácilmente colegirse, la determinación de los ámbitos de aplicación de las dos referidas secciones resulta incongruente, además de confusa, sin poder explicarse a ciencia cierta la distinción entre los distintos colectivos a que cada cual afecta. ¿Es que debemos entender que en la esfera de la prestación de cualesquiera servicios –entre ellos el acceso a la vivienda– no es posible la discriminación por motivos distintos al origen racial o étnico de las personas, o que los que se produzcan no se encuentran protegidos por las medidas previstas en la ley? Es evidente que ninguna de estas cuestiones parece factible, por lo que debe entenderse que se trata simplemente de una disfunción normativa. En cualquier caso, como ha hecho notar la doctrina, las aplicaciones normativas referidas únicamente a determinados factores de riesgo solo deben considerarse emblemáticas o útiles en la medida en que ofrecen simplificaciones de regulación o medios para probar la discriminación...[18]. Como ha quedado señalado con anterioridad, debemos atender al carácter normativo del deber de no discriminación, a la naturaleza imperativa de un principio que deriva del conjunto del ordenamiento y que informa el mismo, con independencia de la existencia de una norma especial o sectorial en la materia.

---

18 Barba (2023, p. 36).

Ambas secciones se titulan de forma más que amplia, al referirse a las "medidas para la aplicación del principio de igualdad", si bien el contenido de la sección 2.ª se ciñe a referencias a la justificación de medidas de acción positiva, a la legitimación de las personas jurídicas en defensa de estos colectivos para actuar en procesos judiciales, a la carga de la prueba, o a la creación del Consejo para la eliminación de la discriminación racial o étnica. La sección 3.ª introduce modificaciones en diversas normas, entre otras, el Estatuto de los Trabajadores o la Ley de Integración Social de los Minusválidos. En general, la doctrina concurre en que el legislador español se limitó, al transponer las directivas, a incorporar a su ordenamiento la prohibición de discriminación en el acceso a bienes y servicios, sin incluir disposiciones de desarrollo más específicas y aclaratorias[19].

Más concreta, al tiempo que ambiciosa, es la Ley Orgánica 3/2007, de 22 de marzo, para la igualdad efectiva de mujeres y hombres. En su artículo primero esta ley establece que su objeto no es otro que hacer efectivo el derecho de igualdad de trato y de oportunidades entre mujeres y hombres "en cualesquiera ámbitos de la vida y, singularmente, en las esferas política, civil, laboral, económica, social y cultural". El artículo 4 dispone expresamente que "la igualdad de trato y de oportunidades entre hombres y mujeres es un principio informador del ordenamiento jurídico", especificando de formada redundante que, "como tal, se integrará y observará en la interpretación y aplicación de las normas".

La LO 3/2007 acoge, desde luego, algunas de las prescripciones anteriormente mencionadas sobre los conceptos de discriminación directa e indirecta, legitimación o prueba, si bien, a los efectos de nuestro estudio, resultan particularmente trascendentes los artículos más relacionados con los efectos de la

---

19 Llorente San Segundo (2018, p. 24).

discriminación en el marco de las relaciones contractuales entre particulares; en concreto, los artículos 10, 69 y 72.

Es el artículo 69 de la LO 3/2007 el que se centra en la igualdad de trato en el acceso a bienes y servicios en los términos en los que lo hace la mencionada directiva comunitaria. Siguiendo su apartado primero, "todas las personas físicas o jurídicas que, en el sector público o en el privado, suministren bienes o servicios disponibles para el público, ofrecidos fuera del ámbito de la vida privada y familiar, estarán obligadas, en sus actividades y en las transacciones consiguientes, al cumplimiento del principio de igualdad de trato entre mujeres y hombres, evitando discriminaciones, directas o indirectas, por razón de sexo". El apartado segundo, no obstante, aclara que ello no afecta a la libertad de contratación, "incluida la libertad de la persona de elegir a la otra parte contratante, siempre y cuando dicha elección no venga determinada por su sexo". La redacción de la norma en este punto resulta, cuando menos, peculiar. Por una parte, es insostenible afirmar que la norma antidiscriminatoria deja incólume la libertad de elección de la contraparte, pues es patente que no lo hace; por otra, resulta de complicada interpretación la referencia a la elección determinada por el sexo. En este sentido, el apartado tercero del artículo 69 contempla la excepción a la restricción de la libertad de elección de la contraparte, al establecer que las diferencias de trato serán admitidas, no obstante, "cuando estén justificadas por un propósito legítimo y los medios para lograrlo sean adecuados y necesarios". Como ya se indicó, la adecuación y necesidad habrán de ser concretadas por los tribunales en cada caso, debiendo atenderse a la jurisprudencia española[20] y comunitaria a estos efectos.

---

20 La STS 811/2001 de 8 de febrero (TOL4.964.639) declaró contrario al principio de igualdad la inadmisión de cinco mujeres en una comunidad de pescadores de El Palmar, al entenderse que esta prohibición no puede ampararse en la autonomía de la voluntad de asociaciones

Identificada la actuación discriminatoria, debemos atender a las consecuencias jurídicas que se prevé frente a tal conducta, a las que se refiere el artículo 10 de la LO 3/2007, estableciendo que "los actos y las cláusulas de los negocios jurídicos que constituyan o causen discriminación por razón de sexo se considerarán nulos y sin efecto, y darán lugar a responsabilidad a través de un sistema de reparaciones o indemnizaciones que sean reales, efectivas y proporcionadas al perjuicio sufrido, así como, en su caso, a través de un sistema eficaz y disuasorio de sanciones que prevenga la realización de conductas discriminatorias". De un lado, el ámbito de aplicación de la norma parece bastante amplio, al poder entenderse extensible a cualquier clase de actuación discriminatoria, en la esfera pública o privada, con incidencia en el acceso a bienes o servicios, superando así el tenor mínimo de la Directiva 2004/113 (art. 13). De otro, el precepto realiza una enunciación general de las posibles consecuencias jurídicas de tales actuaciones, mez-

---

privadas cuando esta comunidad ocupa una posición privilegiada que concede una explotación en exclusiva de un dominio público. Por su parte, el Tribunal Supremo ya se manifestó desde 2022 sobre ciertas fiestas populares, considerando nulos determinados actos administrativos que impedían la participación de mujeres en los mismos, como el Alarde de Irún (STS 19 septiembre 2002, ECLI: ES:TS:2002:5976) o el de Hondarribia (STS 13 septiembre 2002, ECLI:ES:TS:2002:5853). No obstante, una vez se desvincularon las asociaciones organizadoras de estos Alardes de la financiación y colaboración de las administraciones públicas, ya no se les consideró con el carácter de asociación privilegiada o dominante y, en consecuencia, se negó la infracción de los derechos de igualdad y no discriminación (STS 15 enero 2007, ECLI:ES:TS:2007:279; y 28 mayo 2008, ECLI:ES:TS:2008:3410). Esta doctrina fue reiterada recientemente por la STS 23 diciembre 2021 (ECLI:ES:TS:2021:4855), en relación a la prohibición de la participación de mujeres en una cofradía de Tenerife, resolución comentada por Fernández-Rivera González (2023, pp. 645-681).

clando efectos propiamente civiles –nulidad y reparación del daño– con un sistema de sanciones preventivo, escasamente concretando la Directiva 2004/113.

A esta declaración general debe sumarse lo preceptuado por el artículo 72, dentro del título dedicado al acceso a bienes y servicios y su suministro, que restringe las consecuencias de las conductas discriminatorias en el ámbito de aplicación del artículo 69 a la indemnización por los daños y perjuicios sufridos. Lo hace, no obstante, "sin perjuicio de otras acciones y derechos contemplados en la legislación civil y mercantil".

Quizás la redacción y encaje de ambos preceptos no resulten afortunados en cuanto a la enunciación de las posibles consecuencias jurídicas del atentado contra la prohibición de discriminación. Pero debemos entender esta normativa desde una mirada de conjunto del ordenamiento, posibilitando a la víctima la utilización de cualquier recurso jurídico dentro del amplio abanico que contempla nuestro Derecho contractual que la legitime a exigir al infractor el cumplimiento de la obligación de no discriminar y a la indemnización que, en su caso, corresponda por los daños sufridos. Entre estos recursos, la doctrina aboga por la posibilidad de ejercitar por analogía la acción de cesación que prevé la LO 1/1982, de 5 de mayo, de protección civil del derecho al honor, a la intimidad personal y familiar y a la propia imagen[21].

Con todo, la atención escrupulosa a lo recogido en el artículo 10 LO 3/2007 no siempre responderá a la protección más eficaz de la persona discriminada. Piénsese que la nulidad del contrato puede no satisfacer los intereses de la víctima, que recibirá en determinados casos mayor protección si se le permitiera la modificación de las condiciones que le son adversas.

---

21 García Rubio (2007, p. 155).

Esta posibilidad la contempla específicamente el artículo 72.2 LO 3/2007 para el contrato de seguro, otorgando al contratante perjudicado el derecho a reclamar la asimilación de sus primas y prestaciones a las del sexo más beneficiado; pero no se prevé con carácter general en la citada Ley ni en el Código Civil. Es posible, en todo caso, solo la declaración de nulidad de las cláusulas discriminatorias del contrato siempre que las subsistentes permitan que este perviva; así como la reclamación de la resolución cuando la actuación discriminatoria lleve al incumplimiento contractual[22].

Cuestión igualmente trascendente en este marco es la de determinar, ante el silencio de la normativa, si es viable la imposición del contrato a la persona que llevó a cabo la conducta discriminadora. Como hace ver la doctrina, habrá de atenderse aquí a la ponderación de los dos intereses en juego –la libertad de elegir a la otra parte contratante y el derecho a la igualdad de trato– para llegar a la solución de permitir la imposición contractual en aquellas situaciones en las que solo de esa forma se puede alcanzar el derecho efectivo a obtener la prestación (bienes o servicios en monopolio o situaciones de necesidad). En caso contrario, esto es, cuando la víctima de la discriminación puede acceder a los bienes o servicios por otros cauces, el remedio indemnizatorio será la respuesta más acorde con dicha ponderación[23] y, probablemente, la más satisfactoria para los intereses de la persona que sufrió la discriminación, que no desee ya contratar con el autor de la misma.

Como ya vimos, el derecho a la indemnización por los daños sufridos lo recoge la LO 3/2007 de forma general en el artículo 10 y, de forma concreta y sin perjuicio de otras acciones o

---

22 En este sentido, García Rubio (2007, p. 159).

23 García Rubio (2007, p. 158).

derechos contemplados en la legislación civil y mercantil, en el artículo 72.1. Esta referencia expresa es lo que lleva a pensar que se trata de una acción específica, al margen de las previstas por el artículo 1101 y 1902 del Código Civil, en sede de responsabilidad contractual y extracontractual respectivamente. Estamos ante una acción de responsabilidad contractual de carácter objetivo, que ha de responder a la reparación de la víctima por los daños efectivamente sufridos –aunque el art. 6.1 LO 3/2007 tipifique el riesgo de sufrir discriminación como supuesto de discriminación directa[24]–, ya sean de carácter material o moral, con un plazo de prescripción, a falta de indicación legal concreta, de cinco años, *ex* artículo 1964.2 CC.

Con posterioridad a estas leyes, la Ley 26/2011 de 1 de agosto, de 1 de agosto, de adaptación normativa a la Convención Internacional sobre los Derechos de las Personas con Discapacidad, introdujo varias modificaciones en distintas normas incluyendo precisiones sobre la prohibición de discriminación en el acceso a bienes y servicios para dar cumplimiento al principio de igualdad de oportunidades de las personas con discapacidad, evitando discriminaciones, directas o indirectas, por razón de la misma (entre otras, la del art. 10 bis de la Ley 51/2003, de 2 de diciembre, de igualdad de oportunidades, no discriminación y accesibilidad universal de las personas con discapacidad).

Como ha podido comprobarse, las respuestas del legislador español a las directrices comunitarias sobre la protección del

---

[24] Como explica García Rubio (2007, p. 162), en el caso de que exista riesgo de discriminación, pero no se haya materializado daño alguno, no habrá obligación de indemnizar, sin perjuicio de que el solo riesgo de ser discriminado haya llegado a producir daños, ya sean materiales (gastos destinados a evitar la discriminación) o morales (sufrimiento por el riesgo de ser discriminado).

principio de igualdad de trato fueron escalonadas, además de parciales o sectoriales, lo que propiciaba que, a pesar del reconocimiento generalizado de la necesidad de interdicción de la discriminación, no se contemplara el Derecho antidiscriminatorio con una visión o vocación de conjunto dentro de nuestro ordenamiento.

Ello se alcanzó finalmente con la Ley 15/2022, de 12 de julio, integral para la igualdad de trato y no discriminación, que se presenta, como expone su preámbulo, con la vocación de convertirse en el mínimo común normativo que contenga las definiciones fundamentales del Derecho antidiscriminatorio español y, al mismo tiempo, albergue sus garantías básicas.

Consecuente con su carácter integral, el ambicioso propósito de esta ley queda reflejado en las más exhaustivas definiciones de los fenómenos discriminatorios –discriminación directa e indirecta, por asociación y por error, múltiple e interseccional, acoso discriminatorio...–, y en la amplitud de su ámbito de aplicación, tanto subjetivo como objetivo. Así, en relación al primero, el artículo 2.1 extiende considerablemente los factores de riesgo respecto de las normas anteriores en la materia, declarando que nadie podrá ser discriminado por razón de nacimiento, origen racial o étnico, sexo, religión, convicción u opinión, edad, discapacidad, orientación o identidad sexual, expresión de género, enfermedad o condición de salud, estado serológico y/o predisposición genética a sufrir patologías y trastornos, lengua, situación socioeconómica, o cualquier otra condición o circunstancia personal o social; y aun no elimina la cláusula final abierta que pueda dar cabida a cualquier otra condición o circunstancia que la evolución social lleve a ser razón de discriminación.

En lo que respecta el ámbito objetivo, el artículo 3 lo describe de forma igualmente amplia, mencionándose la vivienda de forma expresa en el apartado 1. k): "Acceso, oferta y suministro de bienes y servicios a disposición del público, incluida la vivienda, que se ofrezcan fuera del ámbito de la vida privada y familiar". La Ley dedica a cada uno de esos ámbitos preceptos concretos, destinando al derecho a la igualdad de trato en el acceso a la vivienda el artículo 20, que constituye el marco regulador desde el que hay que dar virtualidad e interpretar el artículo 6 de la Ley 12/2023.

A la aplicación concreta de las reglas y principios hasta aquí comentados al ámbito del acceso a la vivienda dedicaremos el resto de este trabajo.

IV. DISCRIMINACIÓN CONTRACTUAL Y DERECHO A LA VIVIENDA

1. El derecho a la vivienda y el principio de no discriminación

A pesar de no encontrarse en el elenco de los derechos fundamentales, el derecho a una vivienda digna[25] se perfila como uno de los derechos-pilares del estado social que demanda una regulación acorde con las necesidades de una ciudadanía que, cada vez más, interioriza que esta es una obligación asistencial que debe asumir el Estado moderno[26].

---

25 Manero Salvador (2015, pp. 333-364) y Ferrando Nicolau (1992, pp. 305-322).

26 Artículo 47 CE: "Todos los españoles tienen derecho a disfrutar de una vivienda digna y adecuada. Los poderes públicos promoverán las condiciones necesarias y establecerán las normas pertinentes para hacer efectivo este derecho, regulando la utilización del suelo de acuerdo con el interés general para impedir la especulación. La comunidad participará en las plusvalías que genere la acción urbanística de los entes públicos."

El derecho a una vivienda digna supera, con mucho, el derecho a acceder a una vivienda digna, pero es evidente que este es el eje sobre el que el primero se sustenta. Basta con asomarse a las noticias de prensa para percatarse de que el ámbito del acceso a la vivienda materializa la mayor tasa de discriminación y que así es percibido por los medios de comunicación y por la sociedad[27]. En consecuencia, el tratamiento que del mismo se dispensa por parte de los poderes públicos, evitando que criterios discriminatorios ligados a los llamados "factores de riesgo" incidan en la determinación del sujeto que, por vía contractual, accede al derecho de usar y disfrutar de una vivienda con propósito habitacional, se convierte en un elemento esencial para atender al derecho que reconoce nuestra Constitución.

El hecho de que el derecho a la vivienda lleve aparejada la calificación de "digna" (es decir, que reúna requisitos de habitabilidad suficientes para cumplir el fin que persigue la norma) implica que, para su salvaguarda, este aspecto deba ser objeto de protección por las normas que regulan el ejercicio del derecho. En este sentido, el legislador debe establecer controles encaminados a asegurar la calidad y las condiciones de habitabilidad del inmueble.

Finalmente, el derecho a la vivienda debe proteger al ciudadano frente a medidas de "acoso inmobiliario" encaminadas a que el usuario abandone el inmueble que, hasta ese momento, había constituido su residencia habitual. Detrás del mismo suelen esconderse conductas discriminatorias hacia colectivos marginales o vulnerables, y no solo las de carácter económico ligadas a situaciones de arrendamientos antiguos con rentas muy bajas para la propiedad. Sea cual fuere el motivo, el ordenamiento

---

27 https://www.rtve.es/noticias/20230218/discriminacion-alquiler-viviendas-espana/2424447.shtml https://www.accem.es/la-vivienda-la-esfera-de-la-vida-con-mayor-tasa-de-discriminacion/

jurídico está obligado a proteger a los residentes frente a conductas ilícitas de este tipo por carecer de justificación alguna.

Identificados, pues, los supuestos en los que el principio de no discriminación[28] se proyecta en el derecho a la vivienda, debemos dedicar unas líneas a delimitar la viabilidad del principio en las relaciones contractuales cuando estas versan sobre alguno de los aspectos señalados. Es decir, si es factible menoscabar la autonomía negocial cuando el objeto del contrato afecte al derecho a la vivienda por observarse la vulneración del principio de no discriminación.

Tal y como hemos visto en epígrafes precedentes, el principio de no discriminación elevado a principio de derecho contractual europeo[29] no opera en el ámbito de las relaciones *intuitu personae*, caracterizadas, precisamente, por ser las cualidades del deudor las que son tenidas en cuenta por el contratante a la hora de concertar la relación jurídica. En este ámbito, la confianza que ese sujeto despierta en la otra parte del contrato es la que justifica su elección. Ciertamente, las cualidades del inquilino son determinantes a la hora de celebrar un contrato de arrendamiento, como también pudieran serlo, en menor medida, las de la persona a la que queremos vender nuestra vivienda. Ahora bien, dado el carácter de "bien esencial para la vida ordinaria de los ciudadanos"[30], las razones subjetivas que se encuentran tras la elección del inquilino o

---

28 No vamos a referirnos al principio de igualdad, pues cuando este se proyecta en la contratación se traduce en la cláusula de no discriminación; que es la sede en la que la vulneración de la igualdad se manifiesta. No se tiene derecho, en vía contractual, a ser tratado de forma igualitaria sino a no ser discriminado sin causa que lo justifique. Vid. Zubero Quintanilla (2018, pp. 312-313).

29 Carapezza Figlia (2019, pp. 287 y ss.) así lo considera.

30 Aguilera Rull (2009, p. 10).

comprador no pueden alinearse con los factores de riesgo que denotan, por sí mismos, la existencia de discriminación.

En otras palabras, si la elección, o la redacción del clausulado o una actuación en el marco del contrato, se fundamentara en la presencia en el sujeto afectado de un factor de riesgo del artículo 14 de la Constitución ("por razón de nacimiento, raza, sexo, religión, opinión o cualquier otra condición o circunstancia personal o social") o de los factores de riesgo enunciados en la lista abierta del art. 2.1 de la Ley 15/2022, se vulneraría el principio de no discriminación con las consecuencias jurídicas que ello acarrearía. Cualquier otro motivo subjetivo que concomitara en la celebración, redacción o actuación contractual (como la solvencia económica, el hecho de ser fumador o tener mascotas), no infringiría la prohibición de discriminación y, por consiguiente, gozaría de plena eficacia jurídica al estar respaldada por la autonomía negocial de los sujetos. Pues dicha autonomía nos permite y nos legitima para elegir a la persona que, por los motivos que sea (excepto los que hemos llamado "factores de riesgo") consideramos apropiada para realizar un contrato[31].

La conexión con la exclusión social que comporta impedir el acceso a la vivienda o su habitabilidad es, probablemente, la causa de que la vivienda haya estado siempre presente en los textos jurídicos (legales o jurisprudenciales[32]) que se han

---

[31] Jiménez Horwitz (2014, p. 500).

[32] La sentencia de la *Cour de Cassation, Chambre criminelle*, de 7 junio 2005, (https://www.legifrance.gouv.fr/juri/id/JURITEXT000007609194) que falla apreciando discriminación por razón de origen etnia y nacionalidad como consecuencia de las instrucciones que el propietario le había dirigido a la agencia inmobiliaria excluyendo al colectivo de extranjeros como posibles inquilinos.

ocupado[33] y se ocupan de limitar la autonomía de la voluntad en los contratos que a ella le afectan. Sin que se haya entrado a discutir ni debatir si, por ser un ámbito familiar o íntimo o la relación arrendaticia una relación *intuitu personae*, se encuentra excluida de la aplicación del principio de no discriminación. Pues el interés público que el derecho a la vivienda digna comprende, supera, con mucho, cualquier otro aspecto negocial y justifica que este no se imposibilite ni dificulte por motivos discriminatorios.

---

[33] Así, por ejemplo, en EE.UU. cuna del derecho antidiscriminatorio, tal y como hemos visto, se promulga el 11 de abril 1968 la ley para la vivienda justa (*Fair housing*) cuyo objetivo era facilitar la igualdad de los estadounidenses en el acceso a la vivienda. En este sentido, prohíbe la discriminación por motivos de raza color, religión, sexo (que puede incluir orientación sexual o identidad de género), discapacidad o situación familiar (tener hijos) y/o nacionalidad. Enumera una serie de acciones y prácticas ilegales cuando se considera que discriminan: negarse a vender o alquilar una propiedad a una persona por pertenecer a una clase protegida; discriminar en los términos, condiciones y/o privilegios de venta o alquiler por pertenecer a una clase protegida; discriminar en la publicidad, en particular, efectuar, imprimir, publicar o hacer que se efectúe, publique o imprima, cualquier aviso, declaración o anuncio que indique algún tipo de preferencia, limitación o discriminación por pertenecer a una clase protegida; distorsionar la disponibilidad de viviendas por el hecho de que la persona pertenece a una clase protegida; organizar acoso inmobiliario (*blockbusting*) o maniobras de redireccionamiento (*steering*); negarse a aceptar personas con discapacidades al no permitirles realizar adaptaciones razonables en la vivienda; discriminar a la hora de conceder préstamos para transacciones inmobiliarias, incluyendo la compra, construcción, mejora, reparación y/o mantenimiento de una vivienda; coaccionar, intimidar, amenazar o interferir con el ejercicio o goce del derecho a una vivienda justa de cualquier persona o de otros individuos que hayan colaborado o fomentado el ejercicio o goce de este derecho.

Sentada, pues, la viabilidad de la aplicación del principio de no discriminación en las relaciones jurídicas entre los particulares en torno a la vivienda cuando esta se destina a atender las necesidades habitacionales de los sujetos contratantes, procederemos a discernir sobre la regulación actual en la reciente Ley por el derecho a la vivienda.

2. El principio de no discriminación entre los particulares en el artículo 6 de la Ley por el derecho a la vivienda: identificación de los supuestos

2.1. Antecedente: el artículo 20 de la Ley 15/2022, de 12 de julio integral para la igualdad de trato y la no discriminación

La Ley 15/2022, no solo por ser anterior en el tiempo, sino, sobre todo, por diseñar el régimen jurídico integral para la igualdad de trato y no discriminación en nuestro ordenamiento, constituye el marco en el que se debe insertar el artículo 6 de la Ley por el derecho a la vivienda. Por ese motivo, antes de desmenuzar el precepto concreto que ocupa estas reflexiones, es del todo necesario mencionar el contenido del artículo 20 que se refiere, expresamente, a la aplicación del principio de igualdad y no discriminación en la vivienda.

A pesar de que el precepto se titula, textualmente, "Derecho a la igualdad de trato y no discriminación en el acceso a la vivienda", su ámbito objetivo se presenta más amplio de lo que pudiera desprenderse de estos términos. En primer lugar, porque la protección va más allá del mero "acceso a la vivienda", extendiéndose al periodo de vigencia del contrato en el que tiene causa el uso del inmueble. En segundo lugar, porque el párrafo tercero, con el que finaliza el redactado, amplía el régimen jurídico expresado en los dos párrafos anteriores al local de negocio[34] lo que impide conectar la norma con el derecho a la vivienda del art. 47 de la Constitución, que no lo comprende.

Se estructura la norma en tres párrafos, claramente diferenciados. El primero dirigido a la Administración pública (estatal, autonómica y municipal), encargándole tomar las medidas pertinentes y necesarias en su ámbito competencial para prevenir la "discriminación, incluida la segregación residencial, y cualquier otra forma de exclusión", con especial dedicación a las personas vulnerables.

El segundo de los párrafos[35] contempla, como destinatarios, a un grupo de sujetos muy variopintos ("prestadores de servicios de venta, arrendamiento, intermediación inmobiliaria, o cualquier otra persona física o jurídica") en relación a una concreta actividad: la oferta pública, para obligarlos, con carácter general, "a respetar en sus operaciones comerciales el derecho a la igualdad de trato y no discriminación". La redacción es confusa, con una deficiente técnica jurídica que puede generar problemas interpretativos. A fin de aclarar el sentido, podemos identificar a los "prestadores de servicios de venta, arrendamiento e intermediación inmobiliaria" con las agencias inmobiliarias; los "portales de anuncio", con las páginas webs en las que se publicitan los anuncios inmobiliarios y los terceros mencionados de forma genérica ("cualquier otra persona física o jurídica") con los titulares de derechos sobre inmuebles que le posibiliten para ceder, bien la titularidad, bien

---

34 "3. Lo previsto en los párrafos anteriores será de aplicación también a los locales de negocio".

35 "2. Los prestadores de servicios de venta, arrendamiento, intermediación inmobiliaria, portales de anuncios, o cualquier otra persona física o jurídica que haga una oferta disponible para el público, estarán igualmente obligados a respetar en sus operaciones comerciales el derecho a la igualdad de trato y no discriminación".

el uso, este último, por medio de un contrato de alquiler o cualquier otra modalidad contractual que así lo posibilite. La dicción del precepto, no obstante, nos invita a calificar como personas físicas o jurídicas, titulares de derechos y obligaciones y, en consecuencia, destinatarios naturales de las normas jurídicas, a los sujetos que denomina "prestadores" y "cualquier otra persona". Pero, ¿qué son los "portales de anuncios" desde un punto de vista jurídico? Un interpretación lógica y sistemática nos conduce a corregir el texto, considerando que los llamados a respetar el principio de igualdad de trato serían los titulares de dicho dominio web (persona física o jurídica) que se beneficia del mismo como medio para publicitar la oferta inmobiliaria.

Siguiendo con la interpretación de la norma, estas personas obligadas lo son en la medida que realicen una oferta pública, lo que justifica, la mención a los "portales de anuncio" pues en ellos solo tiene lugar la publicitación de la oferta. El concepto "oferta pública" debe ser interpretado en esta norma de forma amplia, abrazando a la "invitación a ofrecer"; técnica de frecuente recurso en los portales de anuncio.

La obligación atenerse a la igualdad de trato entre los interesados se extiende a todas las "operaciones comerciales" que, debemos inferir, tienen lugar a causa de la aceptación de la oferta o del contrato que se formalice tras las negociaciones establecidas como consecuencia de la respuesta a la invitación a ofrecer.

A continuación de este primer párrafo que pretende establecer la regla general se prohíben dos conductas en particular. La primera prohibición[36] solo es operativa si ha tenido

---

[36] "a) Rehusar una oferta de compra o arrendamiento, o rehusar el inicio de las negociaciones o de cualquier otra manera de impedir o denegar la compra o arrendamiento de una vivienda, por

lugar una oferta pública con relación a dos contratos que, si bien son las más usuales como medio de acceso a la vivienda, no son los únicos: compra y arrendamiento. En dicho reducido ámbito objetivo el legislador impide "rehusar la oferta", frase que debemos interpretar como que no puede rechazar la aceptación de la oferta, pues la oferta pública ya se ha realizado y, como hemos indicado, constituye el presupuesto de aplicación del precepto. También se dice que queda prohibido "rehusar el inicio de las negociaciones o de cualquier otra forma de impedir o denegar la compra o arrendamiento de una vivienda por razón de alguna de las causas de discriminación previstas en la ley"; texto de difícil interpretación ya que, en el ámbito de la oferta pública (interpretada en sentido estricto, excluyendo la invitación a ofrecer), no existen tratos preliminares ni tampoco cabe impedir o denegar la actividad ofertada si esta es aceptada por los destinatarios.

También se mencionan en la norma una serie de premisas que carecen de sentido alguno, ya que la mecánica oferta-aceptación-perfección del contrato así lo imposibilita. Conforme a dicha mecánica, formulada cualquier oferta pública (en sentido estricto), una vez que esta es aceptada, el contrato se ha perfeccionado, por lo que el oferente queda obligado y carece de legitimidad para "rehusar" o "impedir" nada (como, expresamente, indica el precepto). Pero, en caso de que "rehusara" o "impidiera", se generaría un incumplimiento contractual, con las consecuencias jurídicas que ello acarrearía y la protección

razón de alguna de las causas de discriminación previstas en la presente ley, cuando se hubiere realizado una oferta pública de venta o arrendamiento".

al contratante leal frente al incumplidor que dispensa nuestro Derecho privado[37].

La segunda prohibición[38] se incardina en la obligación enunciada en la regla general con la que comienza este segundo párrafo, dirigida a los sujetos que realicen una oferta pública, que quedan obligados en sus operaciones comerciales a respetar la igualdad y no discriminación (lo que, por cierto, es de todos sabido). Pues bien, en este escenario, el legislador, de forma redundante, reitera en la letra b) que está prohibido discriminar en los contratos de venta y arrendamiento con fundamento en las causas del artículo 2.1 de la Ley 15/2022. Es cierto que no se alude al requisito de la "oferta pública", lo que pudiera ampliar el ámbito de aplicación del precepto a cualquier contrato en el que se incluyan cláusulas discriminatorias a sujetos en los que concurra un factor de riesgo de los descritos. Pero también lo es que la estructura y redacción del texto dificulta mantener esta interpretación más extensa.

Finaliza el artículo 20.2 con una afirmación que, a modo de recordatorio, expande la obligación de no discriminar en el caso de contratos de larga duración (en particular, el arrendamiento) durante toda la vigencia del contrato[39]. La operatividad del principio, una vez que existe un contrato carente de condiciones o términos discriminatorios, es de difícil aprecia-

---

37 Sobre la imposibilidad de rechazar la aceptación de la oferta y su calificación como incumplimiento contractual en sede de aplicación del principio de no discriminación, Barba (2023, pp. 88-89).

38 "b) Discriminar a una persona en cuanto a los términos o condiciones de la venta o arrendamiento de una vivienda con fundamento en las referidas causas".

39 "La obligación de no discriminación se mantendrá durante todo el periodo posterior de uso de la vivienda, en el caso de los arrendamientos u otras situaciones asimilables".

ción y prueba, aunque pudiera detectarse en algunas cuestiones puntuales. Su implementación es de dudosa exigibilidad en todos los casos por poder afectar a la autonomía del sujeto que protege el artículo 10 de la Constitución. Así, ¿obligaría el principio de no discriminación al propietario a prorrogar el contrato al inquilino en el que concurre un factor de riesgo, por habérselo prorrogado a otro inquilino con un contrato igual o similar en el que no exista dicho factor, y no ser capaz de aportar una causa legítima, objetiva y suficiente que no sea su mera voluntad? Este tema, que merece una reflexión seria y pausada, será abordado con el detenimiento y precisión que merece más adelante en el epígrafe dedicado a las consecuencias jurídicas derivadas de la aplicación del principio de no discriminación en los contratos por medio de los cuales se accede y/o se garantiza el derecho a una vivienda digna. Permítanos el lector remitirnos a dicho epígrafe.

Concluyendo, el artículo 20.2 en el que se regula la aplicación del principio de no discriminación en la vivienda, no es, precisamente, afortunado. Contiene errores de redacción como los apuntados que dificultan su comprensión y enturbian su ámbito de aplicación. Pero, quizás sea la referencia a la previa existencia de una oferta pública, como presupuesto de su aplicación, el mayor obstáculo en orden a su implementación. Ello deja fuera de su cobertura todos los contratos en los que existe una oferta privada, incluso aquellos que se realizan por medio de condiciones generales.

2.2. La mejorable redacción del artículo 6 Ley por el derecho a la vivienda: un esfuerzo interpretativo

Al igual que su predecesor, el artículo 6 regula el principio de igualdad y no discriminación en relación a la vivienda de forma amplia, y no referido, exclusivamente, al ámbito de las relaciones entre los particulares. Pero, a diferencia del artículo

20 de la Ley 15/2022, que los trata de forma separada dedicando un párrafo al ámbito público y otro al privado, el mentado artículo 6 lo hace de forma conjunta. Ello nos obliga a desbrozar el precepto extrayendo de él las cuestiones relacionadas con nuestro objeto de estudio.

Tras una declaración general imbricada con el mandato constitucional, el legislador relaciona el derecho del uso y disfrute de una vivienda digna y adecuada con el "cumplimiento de los requerimientos legales y contractuales establecidos en la legislación y normativa vigente". Ciertamente, es difícil precisar a qué requerimientos se refiere, sobre todo cuando estos dice que proceden del contrato. A fin de llevar a cabo una interpretación ajustada a los conceptos mal utilizados por el legislador, podríamos entender que la desafortunada frase alude a los mandatos legales residenciados en normas imperativas ("¿requerimientos legales?") y al contenido obligatorio del concreto contrato que facilita o imposibilita el uso y disfrute de la vivienda, siempre que éste no infrinja norma imperativa alguna ("¿requerimientos contractuales?"). Asimismo, de la lectura del precepto puede colegirse (con cierta dificultad) que el principio de igualdad y no discriminación se superpone a los "requerimientos legales" (lo cual es obvio, pues la Constitución, sede del mismo en su art. 14 es preferente a la ley) y, al mismo tiempo, despliega su eficacia sobre el contrato en el que trae causa el derecho (o no derecho) de uso y disfrute de la vivienda[40].

---

40 "Artículo 6. *Principio de igualdad y no discriminación en la vivienda.* 1. En virtud del principio de igualdad y no discriminación en la vivienda, todas las personas tienen derecho al uso y disfrute de una vivienda digna y adecuada, cumpliendo con los requerimientos legales y contractuales establecidos en la legislación y normativa vigente, sin sufrir discriminación, exclusión, acoso o violencia de ningún tipo".

Como quiera que el principio de igualdad y no discriminación en relación al derecho de uso y disfrute de una vivienda digna es proyectado en dos ámbitos distintos (legal y contractual), puntualizamos que nuestro análisis se restringe, exclusivamente, al ámbito contractual, soslayando reflexiones sobre los límites constitucionales del legislador cuando regula la vivienda en cualquier faceta.

En este sentido, la norma se estructura sobre los espacios clásicos (discriminación directa e indirecta) en los que pudiera tener lugar situaciones discriminatorias en el ámbito contractual"[41].

Continúa el artículo comentado ocupándose de una cuestión ya fuera del ámbito contractual, como es el acoso inmobiliario[42],

---

41 "2. Las Administraciones competentes deberán garantizar el cumplimiento de lo previsto en el apartado 1, adoptando las medidas de protección necesarias para prevenir y hacer frente, de manera específica a las siguientes situaciones que afectan al uso y disfrute de la vivienda:
a) La discriminación directa, que se produce cuando una persona o grupo de personas recibe, en algún aspecto relacionado con la vivienda, un trato diferente del recibido por otra persona en una situación análoga, siempre que la diferencia de trato no tenga una causa legítima que la justifique objetiva y razonablemente, y los medios utilizados sean proporcionados, adecuados y necesarios.
b) La discriminación indirecta, que se produce cuando una disposición normativa, un plan, una cláusula convencional o contractual, un pacto individual, una decisión unilateral, un criterio o una práctica, aparentemente neutros, producen una desventaja particular para una persona o grupo de personas respecto de otras en el ejercicio del derecho a la vivienda. No existe discriminación indirecta si la actuación tiene una finalidad legítima que la justifica objetiva y razonablemente y los medios utilizados para alcanzar esta finalidad son proporcionados, adecuados y necesarios."

42 "c) El acoso inmobiliario, entendido como toda acción u omisión con abuso de derecho con el objetivo de perturbar a cualquier persona en el uso pacífico de su vivienda y crearle un entorno hostil, ya

que pudiera generar responsabilidad extracontractual al que lleva a cabo esta conducta. Es evidente que no es posible apreciar una conexión en cualquier supuesto, pues el objetivo del acoso no siempre ampara actitudes discriminatorias; pero sí lo es que dicha relación existe en algunos casos. Sea cual sea el fin al que se encuentren orientadas, el tratamiento jurídico es idéntico y las reglas que permiten obtener una satisfacción pecuniaria a la víctima se residencian en los artículos 1902 y ss. del CC. En la medida que este trabajo se circunscribe al ámbito contractual, no entraremos a estudiar y/o valorar esta situación.

Finalmente, la norma afronta un supuesto que, en verdad, nada tiene que ver con la eficacia del principio de no discriminación en las relaciones entre los particulares, expresando un mandato a la Administración como garante del cumplimiento del derecho a una vivienda digna. Dicho mandato consiste en estar vigilante y tomar las medidas oportunas, incluso, entendemos, sancionar cuando se percate de la existencia de "operaciones de venta, arrendamiento o cesión por cualquier título, completa o parcial, para residencia de una infravivienda, una vivienda sobreocupada y cualquier forma de alojamiento ilegal" (art. 6.2 d). La regla transcrita continúa diciendo: "o respecto a bienes sobre los que no se ostente un derecho legítimo que faculte al efecto o la representación del mismo". La frase merece ser calificada como de poco afortunada y de difícil comprensión en el marco de nuestro ordenamiento jurídico-privado, pues poco o nada tiene que hacer la Administración si el contrato se ha realizado por una persona que carece de legitimidad para ello (bien porque no es el titular del derecho o carece de legitimidad para disponer de él, sin que ello com-

---

sea en el aspecto material, personal o social, con la finalidad última de forzarla a adoptar una decisión no deseada sobre el derecho que le ampara de uso y disfrute de la vivienda".

porte, necesariamente, mala fe del sujeto). En cuyo caso, la respuesta al conflicto se solventaría por medio de la aplicación de la normativa general del contrato que predica su ineficacia. A no ser que el legislador le esté diciendo "entre líneas" a la Administración que sancione, también, a estos sujetos. Lo cual nos parece del todo improcedente, al no concurrir, a nuestro juicio, un interés digno de protección (como sí en el caso de infravivienda, vivienda sobreocupada y vivienda ilegal) que justifique sancionar al contratante sin derecho legítimo para llevar a cabo el contrato, simplemente, porque este verse sobre la vivienda y, como consecuencia de ello, pudiera ser declarado ineficaz por los Tribunales. Entrar en este campo es, llamémosle, "complicado" e innecesario.

2.3. Identificación de los supuestos discriminatorios en la contratación sobre la vivienda

Conforme acabamos de señalar, la prohibición de discriminación en el acceso y uso de la vivienda predicada en la norma se desglosa en el ámbito contractual en los ya conocidos espacios llamados discriminación directa y directa, sin que exista ninguna particularidad respecto del régimen jurídico general apuntado.

Nuestro discurso se circunscribirá, pues, a contextualizar la discriminación en el ámbito de la vivienda, identificando los supuestos contractuales más comunes en que estos pueden tener lugar. Este quizás sea el tema más espinoso e importante de los trabajos que versan sobre la aplicación del principio de no discriminación en cualquier ámbito, porque no toda diferencia de trato esconde una vulneración, siendo necesario que, además, carezca de justificación objetiva y, aun teniéndola, los medios empleados se califiquen como no adecuados y proporcionados.

En este sentido, y de los tres items (constatación de la diferencia, inexistencia de causa objetiva y/o desproporcionalidad de los medios y prueba) que deben converger para apreciar la infracción del deber de no discriminación, principio normativo que actúa como "parámetro de evaluación del contrato"[43], nos ocuparemos del segundo de los elementos reseñados. Ya que el primero carece de dificultad alguna y no precisa de análisis jurídico al referirse, simplemente, a hechos o circunstancias, y el tercero pertenece al ámbito de derecho procesal. Así, para detectar la diferencia es necesario contar con dos elementos para llevar a cabo la comparación, lo que exige tener siempre un referente (no siempre factible). Y para valorar la prueba aportada en el proceso, basta con aplicar la normativa general de la prueba ubicada en normas procesales, sin que esta materia demande especialidad alguna, que no sea las reglas relativas a la inversión de la carga de la prueba del artículo 30 de la Ley 15/2022 de 18 de julio, de la que nos ocuparemos más adelante.

Partimos de que el principio de no discriminación, como principio universal y general, se extiende no solo al clausulado del contrato examinado sino también a otras fases o momentos de ejecución del mismo, previos y posteriores al de su celebración.

2.3.1. Oferta discriminatoria: pública, privada y algorítmica

En el proceso de formación del contrato, el meritado principio limita la libertad del oferente, invalidando cualquier oferta, sea del tipo que sea, que impida concurrir a determi-

---

[43] Así lo califica Barba (2023, p. 29). Me remito a las pp. 28-32 en la que expone con claridad y precisión la fuerza normativa del aludido principio y su construcción dogmática.

nados sujetos o colectivos sin que dicha exclusión se encuentre justificada. En materia de vivienda esta restricción goza de predicamento, al imposibilitar *ab initio* el acceso a la vivienda a colectivos desfavorecidos y en los que se percibe, bien un factor de riesgo, bien una circunstancia subjetiva no protegible.

Dada la relevancia en el tráfico económico de las transacciones de inmuebles, es muy común que la oferta se publicite en distintos medios, con el fin de llegar al mayor número de destinatarios. Por ese motivo, la preocupación del legislador en relación a la discriminación en el acceso por medio de contratación privada al uso y disfrute de una vivienda digna se centra en la oferta pública, hasta tal punto que la ha elevado a eje vertebrador del régimen jurídico que expresa en el artículo 20 de la Ley 15/2022. Nos remitimos a los comentarios y apreciaciones que, sobre el particular, hemos vertido en el epígrafe precedente.

Aunque, técnicamente, existen diferencias palpables entre la llamada oferta pública, propiamente dicha, y la invitación a ofrecer[44], sobre todo en orden a las consecuencias jurídicas, utilizaremos el término "oferta pública" en este apartado en un sentido amplio, dado que el principio de no discriminación se proyecta con igual intensidad en ambos actos jurídicos. Solo distinguiremos cuando el discurso jurídico así lo demande.

En verdad, no es difícil detectar el posible contenido discriminatorio de una concreta oferta pública, lo que simplifica la tarea interpretativa previa al establecimiento de consecuencias jurídicas. Basta con que el destinatario no sea universal y que la restricción carezca de causa objetiva y legítima que la justifi-

---

44 No vamos a entrar a discernir sobre la diferencia entre "oferta pública" e "invitación a ofertar". Remito al lector que desee una mayor información a Zubero Quintanilla (2016, pp. 63-92; y 2017).

que. En este sentido, los anuncios públicos de venta o alquiler de vivienda en los que se inserta una limitación carente de justificación, evidentemente, vulneran el principio de no discriminación, que llevaría aparejada las consecuencias jurídicas que apuntaremos más adelante. Una oferta pública de alquiler de una vivienda para estudiantes que solo admita a chicas, podría ser un ejemplo de oferta discriminatoria por razón de género.

El ciber espacio se presenta como el escenario más usual en el que se desarrolla la oferta pública de vivienda, proliferando los portales de anuncios, a los que, expresamente, se refiere el artículo 20 de la Ley 15/2022 al enunciar en su párrafo segundo los "sujetos" destinatarios del mandato en dicha norma expresado. Con todo, este medio de difusión universal de la oferta pública de viviendas como tal no infiere en la determinación de su carácter discriminatorio (pero sí para la cuantificación del daño moral). En cambio, la incorporación de algoritmos para la selección de los sujetos que aceptan la oferta o la invitación a ofertar introduce un factor que dificulta la apreciación de las causas que excluyen a unos ciudadanos frente a otros, impidiendo analizar la legitimidad de estas.

La falta de transparencia[45] del algoritmo se erige como el motivo que entorpece el proceso de identificación de las órdenes que conducen al *software* a filtrar a los ciudadanos en función

---

[45] La falta de transparencia de los procesos que se llevan a cabo por medio de algoritmos es una preocupación/ocupación de los estudiosos del Derecho; sobre todo en el sector público. El contenido de la obligación de transparencia se encuentra determinada por la clasificación del sistema de inteligencia artificial en función del riesgo que comporta. Así figura desde el primer texto de la propuesta de Reglamento Europeo, actualmente en tramitación, y de forma específica en el acuerdo del Consejo y el del Parlamento de 8 de diciembre 2023 sobre el contenido de la ley europea que regulará de Inteligencia Artificial. Por todos, vid. Cotino Hueso (2023, pp. 17-63).

de los criterios que el humano haya preestablecido (sesgo algorítmico). La información para proceder a ello es la que se desprenda de los datos solicitados en los formularios que se cumplimenten. Basta el número de identificación para excluir a ciudadanos extranjeros, por ejemplo[46]. Esta oscuridad en el procedimiento de selección imposibilita acreditar la existencia de discriminación, pues el ciudadano preterido ni siquiera conoce el motivo de su exclusión.

El marco normativo general sobre el que articular nuestras reflexiones viene establecido en el Reglamento Europeo de 27 de abril de 2016, relativo a la protección de las personas físicas en lo que respecta al tratamiento de datos personales y a la libre circulación de estos datos. El concreto régimen jurídico aplicable a la protección del ciudadano ante la toma de decisiones automatizadas se localiza en dos preceptos: el artículo 15.1 y el artículo 22. En este sentido, el artículo 15. 1, g) faculta al ciudadano para solicitar información sobre: "la existencia de decisiones automatizadas, incluida la elaboración de perfiles, a que se refiere el artículo 22, apartados 1 y 4, y, al menos en tales casos, información significativa sobre la lógica aplicada, así como la importancia y las consecuencias previstas de dicho tratamiento para el interesado en el

---

46 La discriminación algorítmica se ha estudiado, sobre todo, en el ámbito de los recursos humanos relacionada con los procesos de selección de personal que excluyen, de forma injustificada, a determinados colectivos y de los sesgos que se implementan en el algoritmo a tal fin. Las conclusiones de estos trabajos sobre el particular son trasplantables al ámbito que nos ocupa. A modo de ejemplo, AA.VV. (2022).

artículo 22". Y el artículo 22[47]contiene una prohibición[48] de la toma de decisiones basadas únicamente en procesos automatizados, sin intervención humana alguna[49]

Conscientes de la facilidad con que se puede transgredir el principio de no discriminación, por medio del recurso a procesos automatizados en la toma de decisiones, la Ley 15/2022, en

---

47 "Decisiones individuales automatizadas, incluida la elaboración de perfiles 1. Todo interesado tendrá derecho a no ser objeto de una decisión basada únicamente en el tratamiento automatizado, incluida la elaboración de perfiles, que produzca efectos jurídicos en él o le afecte significativamente de modo similar. 2. El apartado 1 no se aplicará si la decisión: a) es necesaria para la celebración o la ejecución de un contrato entre el interesado y un responsable del tratamiento; b) está autorizada por el Derecho de la Unión o de los Estados miembros que se aplique al responsable del tratamiento y que establezca asimismo medidas adecuadas para salvaguardar los derechos y libertades y los intereses legítimos del interesado, o c) se basa en el consentimiento explícito del interesado. 3. En los casos a que se refiere el apartado 2, letras a) y c), el responsable del tratamiento adoptará las medidas adecuadas para salvaguardar los derechos y libertades y los intereses legítimos del interesado, como mínimo el derecho a obtener intervención humana por parte del responsable, a expresar su punto de vista y a impugnar la decisión. 4. Las decisiones a que se refiere el apartado 2 no se basarán en las categorías especiales de datos personales contempladas en el artículo 9, apartado 1, salvo que se aplique el artículo 9, apartado 2, letra a) o g), y se hayan tomado medidas adecuadas para salvaguardar los derechos y libertades y los intereses legítimos del interesado"

48 Comparto la tesis interpretativa más amplia de Sarra (2019, p. 45), frente a otra interpretación del precepto que considera que solo reconoce un derecho subjetivo al titular de los datos.

49 Para un mayor detalle del contenido y alcance del precepto, el comentario realizado por Sancho Villa (2021, pp. 1725-1745).

su artículo 25[50], expresa un mandato a la Administración de estar vigilante y, en el párrafo 3, encomienda a las empresas "promover el uso de una Inteligencia Artificial ética, confiable y respetuosa con los derechos fundamentales, siguiendo especialmente las recomendaciones de la Unión Europea en este sentido."

A pesar de la difícil prueba que comporta, el principio de no discriminación también se despliega en el ámbito de la oferta privada. Para su operatividad el perjudicado debe acreditar que el contenido tiene carácter discriminatorio al establecer cláusulas excluyentes o impositivas carentes de justificación. En consecuencia, si se le impone al sujeto alguna condición u obligación ligada a derechos de la personalidad o relacionada con los factores de riesgo, que de *facto* implique una discriminación

---

50 "Inteligencia Artificial y mecanismos de toma de decisión automatizados.1. En el marco de la Estrategia Nacional de Inteligencia Artificial, de la Carta de Derechos Digitales y de las iniciativas europeas en torno a la Inteligencia Artificial, las administraciones públicas favorecerán la puesta en marcha de mecanismos para que los algoritmos involucrados en la toma de decisiones que se utilicen en las administraciones públicas tengan en cuenta criterios de minimización de sesgos, transparencia y rendición de cuentas, siempre que sea factible técnicamente. En estos mecanismos se incluirán su diseño y datos de entrenamiento, y abordarán su potencial impacto discriminatorio. Para lograr este fin, se promoverá la realización de evaluaciones de impacto que determinen el posible sesgo discriminatorio. 2. Las administraciones públicas, en el marco de sus competencias en el ámbito de los algoritmos involucrados en procesos de toma de decisiones, priorizarán la transparencia en el diseño y la implementación y la capacidad de interpretación de las decisiones adoptadas por los mismos. 3. Las administraciones públicas y las empresas promoverán el uso de una Inteligencia Artificial ética, confiable y respetuosa con los derechos fundamentales, siguiendo especialmente las recomendaciones de la Unión Europea en este sentido. 4. Se promoverá un sello de calidad de los algoritmos".

al sujeto, la oferta se tildaría como tal. Por ejemplo, la oferta de arrendamiento de un inmueble a una pareja conviviente bajo la condición de que contrajera matrimonio, o a una mujer con la prohibición de que en el inmueble mantenga relaciones sexuales con su pareja femenina.

También podría calificarse como discriminatoria una oferta privada en relación a otra emitida por el mismo titular, sobre una vivienda de similares características, siendo más gravosa en un caso que en el que concurre un factor de riesgo. Siempre, claro está, que dicha diferencia de trato no se sustente en una justificación suficiente, en los términos establecidos en la ley.

2.3.2 La discriminación en los tratos preliminares

Si bien es cierto que no siempre es posible apreciar, previamente al lanzamiento de la oferta, la existencia de tratos preliminares, no es extraño que tengan lugar entre las personas interesadas en contratar en el ámbito de la oferta privada. Igualmente, cuando estamos frente a una invitación a ofrecer, tan usuales en el ámbito inmobiliario.

En este momento, en el que impera la libertad de las futuras partes contractuales para "elegirse" en función del contenido del contrato que negocian, dicha libertad también se encuentra mediatizada por el principio de no discriminación. Si la infracción del principio de buena fe da sustento jurídico a la indemnización de los daños y perjuicios ocasionados por la ruptura de los tratos preliminares, cuánto más si dicha ruptura se genera porque se prefiere a un contratante frente a otro por concurrir un factor de riesgo o por otra circunstancia carente de justificación. O, simplemente, se decide, por ese motivo, no seguir adelante con la negociación. Por ejemplo, se conculca el principio de no discriminación si se rompen los tratos preliminares porque el titular prefiere alquilar la vivienda a un nacional frente a un extranjero en idénticas condiciones a las que se estaban gestando con el segundo, simplemente por causa de su etnia. También si el motivo de la ruptura no es otro que descubrir que el sujeto es de una ideología distinta o su orientación sexual se dirige a las personas del mismo sexo.

Es importante apuntar que, para que se detecte la vulneración, estimamos como necesario que se hayan generado expectativas en el contratante discriminado, ya que el fundamento de la pretensión indemnizatoria se focaliza en la infracción de la buena fe contractual. Y ello no cambia porque, además, se sume otro motivo carente de fundamento jurídico[51]. Dicha circunstancia añadida tendrá su reflejo en las consecuencias jurídicas indemnizatorias, a las que nos dedicaremos más adelante.

Mantener que la ruptura de tratos preliminares no avanzados genera consecuencias indemnizatorias (¿a qué nos estamos refiriendo? ¿meras conversaciones? ¿) porque uno de los sujetos no desea contratar, sin haber creado expectativas, aun cuando el motivo interno se relacione con un factor de riesgo, es, a mi juicio, excesivo. De todos es sabido que la libertad de las personas también ampara discursos o posiciones ideológicas excluyentes

---

51 En contra, Barba (2023, pp. 69-72) considera que cuando la ruptura de los tratos preliminares infringe el principio de no discriminación no es preciso que las negociaciones estén avanzadas, ya que el sustento de la indemnización se ubica en la lesión a la dignidad del discriminado y no en la infracción del principio de buena fe. Entendemos que no es incompatible y que la doble lesión trae como consecuencia la “doble indemnización” (daños morales y patrimoniales) que no convergen cuando se lesiona la dignidad. Tampoco estoy de acuerdo con el profesor Barba no exigir un elemento de comparación para detectar la discriminación, tal y como hemos indicado en el texto principal.

que, lógicamente, marcan sus relaciones sociales[52]. El límite a esta libertad lo marca el artículo 16 de la Constitución en el orden público[53], así como la confluencia de otros derechos fundamentales que pudieran considerarse prioritarios, en un determinado momento, como los referidos en el artículo 14 de la Constitución. La tensión entre la libertad ideológica del artículo 16 y el derecho a autorregular sus intereses del artículo 10 que implica la liberta de decidir y escoger a los sujetos con los que le agrada relacionarse jurídicamente, con el principio de no discriminación del artículo 14, encuentra su punto de inflexión en el principio de buena fe. Si existe una infracción de este, como consecuencia de haber creado expectativas en el sujeto, surgirá obligación de reparar el daño causado. Si, además, el motivo de ruptura se conecta con un factor de riesgo, el daño generado se aumenta, pues se ha lesionado su dignidad como persona.

Fuera de este escenario, admitir la trasgresión del principio de no discriminación en la ruptura de tratos preliminares, amén de infringir los artículos 1.º y 16 de la Constitución, generaría inseguridad jurídica en el tráfico económico.

En relación a la prueba, la dificultad estriba, al igual que en otras fases del proceso de contratación, en acreditar la existencia de discriminación, a los efectos de que proceda la partida indem-

---

52 Tribunal Constitucional Auto 1227/1988: "la libertad ideológica que recoge el artículo 16.1 de la Constitución no constituye, como es obvio, una mera libertad interior sino que dentro de su contenido esencial se incluye la posibilidad de su manifestación externa que (...) no se circunscribe a la oral/escrita, sino que incluye también la adopción de actitudes y conductas".

53 Artículo 16 "Se garantiza la libertad ideológica, religiosa y de culto de los individuos y las comunidades sin más limitación, en sus manifestaciones, que la necesaria para el mantenimiento del orden público protegido por la ley. Nadie podrá ser obligado a declarar sobre su ideología, religión o creencias".

nizatoria correspondiente a esta lesión (ya que, los daños patrimoniales generados en la buena fe de que el negocio se formalizaría por lo avanzado de las negociaciones son fácilmente acreditables).

Por tanto, no basta con que confluya un factor de riesgo en alguno de los sujetos con los que el titular de la vivienda negocio su venta, alquiler o cesión del uso, para que de él se derive la discriminación si fuese, finalmente, preterido. Ni siquiera debe establecerse una presunción a su favor si concurre en uno de los que optan al contrato que se negocia. Pues para que la discriminación pudiera apreciarse en los tratos preliminares debe ir acompañada, en primer lugar, de las circunstancias objetivas referidas (negociación avanzada y similitud entre el contrato celebrado y el negociado en un espacio relativamente cercano en el tiempo entre ambos) y, en segundo lugar, que no converja un elemento de entidad que sustente la elección por otro contratante.

2.3.3. El contenido discriminatorio del contrato

Ya en fase de celebración del contrato propiamente dicha, y tal y como apunta el artículo 6, la discriminación puede aparecer bien de forma directa o indirecta y ambas modalidades se identifican, sin dificultad, en el ámbito de la vivienda. De forma semejante a como acaecía con la oferta contractual, el factor de publicidad se convierte en un instrumento esencial para detectar actitudes discriminatorias. En este sentido, si el contrato se lleva a cabo por medio de condiciones generales basta con analizar su contenido para enjuiciar si encierran una discriminación directa o indirecta para el contratante. No es preciso contar con otro elemento de comparación al estar dirigidas dichas condiciones, por su propia definición, a una colectividad.

En cambio, si el contrato por el que se accede a la vivienda es fruto de una negociación individual, la prueba de la discriminación demanda contar con otros de igual o similar índole y contenido. Porque la discriminación necesita la concurrencia de dos términos para la comparación, ya que se discrimina siempre "en relación a otro".

De ello se infiere que, si el titular del inmueble ha realizado varios contratos sobre inmuebles de similares características, siendo uno o alguno más gravoso en algún aspecto, la discriminación se desprende con facilidad de la propia realidad. Fuera de este supuesto, es muy difícil concluir que un contrato vulnera el principio de no discriminación solo porque sea más gravoso que otro, aun cuando exista una homogeneidad en las viviendas afectadas si dicha homogeneidad no se extiende al contenido del contrato y a las circunstancias que lo envuelven. Esta afirmación se conecta con la libertad de empresa en el marco de la economía de mercado (artículo 38 Constitución), que garantiza la libertad del titular de negociar las condiciones contractuales que, en ese momento, le resulten más beneficiosas (y que, no necesariamente, serán las objetivamente más favorables, porque la urgencia o la necesidad de liquidez justifica contratos por debajo del precio del mercado o la oportunidad por la urgencia del ocupante otros por encima de dicho precio).

Pero, como hemos indicado, la discriminación no solo exige una desigualdad de trato, sino que, además, no se perciba una *causa iuris* o "causa legítima" justificativa del trato diferente que pudiera evitar la aplicación del principio normativo de no discriminación.

Es difícil evitar la calificación de contenido discriminatorio del contrato, constatada la desigualdad de trato, cuando se encuentra presente un factor de riesgo en los hechos enjuiciados. "Nacimiento, raza, sexo, religión, opinión o cualquier otra condición o circunstancia personal o social", como prescribe el artículo 14 CE se erigen como elementos que muy difícilmente pueden orillarse. Si bien el meritado precepto no contiene un enunciado cerrado[54], podríamos recurrir al artículo 2.1 de la Ley 15/2022, por describir una enumeración, también abierta, pero más completa que la del texto constitucional, a la hora de considerar los factores de riesgo[55].

Aunque "nada hay imposible", y menos en el mundo de Derecho, acreditada la existencia de cualquiera de las circunstancias apuntadas encontrar una causa legítima que actúe como barrera a la aplicación del principio de no discriminación es complicado. Pues, como prescribe el referido art. 6, dicha causa debe ser "objetiva", lo que excluye las valoraciones subjetivas, y "razonable". Razonabilidad que debe medirse, obviamente, con parámetros también objetivos.

Distinto recorrido presenta la tarea de calificar como discriminatoria la diferencia de trato fuera del ámbito de los factores de riesgo mencionados en las normas (en concreto, en el artículo 14 de la Constitución y 2.1 de la Ley 15/2022). Para calibrar la "legitimidad" del motivo que la sustenta se han de conjugar los derechos en conflicto: por un lado, el derecho del titular del inmueble de negociar en el marco de la libertad de empresa y su facultad, si fuera propietario, de gozar y disponer del mismo "sin más limitaciones que las establecidas en las leyes" (art. 348 CC), por otro, el derecho a no ser discriminado del sujeto que quiere acceder a la vivienda. Todo ello en el marco del derecho a una vivienda digna que se protege en la Ley por el derecho a la Vivienda.

---

54 STC 75/1983 de 3 de agosto, primer pronunciamiento del Alto Tribunal que califica como lista abierta la del artículo 14 de la Constitución.

55 García Murcia (2022, pp. 15-16), concretamente sobre la valoración de los factores de riesgo del artículo 2.1.

En este escenario, la eficacia jurídica de la causa que se alega para motivar la diferencia de trato puede depender del análisis del marco normativo; pues, cuando existe una legislación que protege determinados intereses y que nos posiciona, como sociedad, frente a estos, no es defendible una causa que no se alinee con dichos intereses. El marco jurídico la dota de la objetividad y razonabilidad necesaria para que sea operativa e impida la aplicación del principio de no discriminación. Ello nos conduce, por ejemplo, a calificar como "causa legítima" aquella que excluye inquilinos fumadores, bien en fase de oferta o en el propio contrato; o les impone condiciones contractuales más gravosas con efectos disuasorios. También, las relacionadas con temas ambientales o de eficiencia energética, priorizando en la contratación, por medio de medidas directas o indirectas, a los ciudadanos que asuman un compromiso en este sentido como pudiera ser la instalación de paneles solares en la vivienda adquirida, la condición de reciclar la basura...

Me genera bastantes dudas[56] que la inclusión de medidas no justificadas que discriminen a los propietarios de mascotas impidiéndole el acceso y uso de la vivienda, tras la ley 17/2021, trasgreda el principio de no discriminación en las relaciones entre particulares. Con independencia del tratamiento de que se dota a los animales llamados "seres sintientes", pero calificados jurídicamente como "cosas"[57], no podemos obviar que

---

[56] Sobre todo después de leer la nota n.º 91 de la obra (Barba, 2023).

[57] La doctrina no se muestra pacífica en este punto. Uno de nuestros juristas que más se ha dedicado a estudiar la Ley y el régimen jurídico que instaura, el profesor Cerdeira Bravo de Mansilla, ha defendido con ahínco y contundencia que, el hecho de ser tenidos como seres sintientes no ha cambiado su calificación como "cosas". Pues, en verdad, en Derecho los entes físicos o son personas o son cosas, no existiendo un género intermedio como apuntan el profesor Torres Perea (2020), ferviente defensor de la tesis contraria conside-

se ha producido un cambio en cuanto a su consideración y, sobre todo, en la valoración y ponderación del bienestar animal, nuevo principio general[58], que debe ser tenido en cuenta por los tribunales a la hora de decidir sobre el destino de estos seres sintientes[59].

---

rando superado la inclusión en la categoría de "cosas". La brillantez del debate doctrinal hace difícil posicionarse o decantarse por una u otra tesis. Cierto es que la aplicación del régimen jurídico de las cosas le acerca más a esta categoría y permite, cuando ello sea necesario, trasplantar las normas que componen su régimen jurídico para solventar los conflictos generados en las crisis con mayor comodidad. Pero cierto es, también, que la alusión al adverbio "solo" del art. 333 (bis), pudiera parecer indicar que dicha aplicación es excepcional y se encuentra mediatizada a la compatibilidad con la "naturaleza", dice el precepto, lo que pone de manifiesto que los animales son algo distinto, por su naturaleza, a las cosas ("Solo les será aplicable el régimen jurídico de los bienes y de las cosas en la medida en que sea compatible con su naturaleza o con las disposiciones destinadas a su protección"). El "quid" de la cuestión, estriba en permitir el recurso al régimen jurídico de las cosas con mayor o menos holgura que, como hemos expresado, admite el CC. Así, por ejemplo, en caso de no existir consenso sobre el uso y disfrute de la mascota en situación de crisis, mantiene un sector de la doctrina que las mascotas deben ser consideradas una pertenencia de la vivienda familiar; ello conlleva que su uso correrá el destino del uso de la vivienda familiar como mantiene Cerdeira Bravo de Mansilla (2021, pp. 507-534).

58 Cerdeira Bravo de Mansilla (2022, pp. 111-126).

59 En primer lugar, los animales ya no son solo una cosa. Y ello conlleva una serie de consecuencias jurídicas que se cristalizan en el principio general de respeto del bienestar animal como inspirador de las decisiones jurisprudenciales o contractuales que sobre ellos se tomen. En segundo lugar, sea un tipo de cosa o un tercer género, lo cierto es que el régimen jurídico de las cosas y de los bienes es de aplicación en la medida en que sea compatible con su naturaleza o con las disposiciones destinadas a su protección.

En nuestra opinión, no se ha instaurado en nuestro ordenamiento jurídico una prohibición de discriminación en el acceso y uso de la vivienda a los tenedores de animales de compañía con los que convive. En primer lugar, porque, en relación a los arrendamientos, la LAU no lo contempla; en segundo lugar, si el uso del inmueble se cediera en precario, la gratuidad del mismo legitimaría al cedente para incluir condiciones (como la no convivencia con animales en el inmueble) que ni están prohibidas y que, en su caso, quedarían amparadas por la libertad contractual[60]. Finalmente, porque, con carácter general, no impera en nuestro ordenamiento jurídico un principio *favor animalis* en virtud del cual se justifique una presunción *iuris tantum* de discriminación, en los términos que hemos mantenido en líneas precedentes cuando se constata una desigualdad de trato como consecuencia de un factor de riesgo de los relacionados en las normas.

En consecuencia, la prohibición de convivencia con animales domésticos en el interior de la vivienda es legítima, objetiva y razonable lo que impide ser considerado como discriminatorio el hecho de que los tenedores de mascotas sean preteridos en la contratación sobre la vivienda. Siempre, claro está, no converjan otros intereses dignos de protección o factores discriminatorios no justificados. Por ejemplo, si la persona excluida debe convivir con su mascota por ser invidente y asumir la función de perro guía.

Por último, como sucede cuando estamos frente a conceptos jurídicos indeterminados, será el juez, quien, tras llevar a cabo la labor interpretativa y valorativa de los hechos, constatada la desigualdad en el trato, determinará si la causa alegada se considera legítima o los medios empleados gozan de la proporción adecuada. Valgan, pues, estas reflexiones como pautas de las que pudiera valerse a tal fin.

---

60 Vivas Tesón (2022, p. 13).

2.3.4. Ejecución discriminatoria del contrato

Cierto es que el artículo 6 la Ley por el derecho a la vivienda no menciona, expresamente, la discriminación con posterioridad a la celebración del contrato, ni tampoco podemos inferirlo de su redactado. Ahora bien, sí se refiere, expresamente, a ello el artículo 20.2 de la Ley 15/2022, marco normativo en materia de discriminación en nuestro ordenamiento jurídico. Así, dice: "La obligación de no discriminar se mantendrá durante todo el periodo posterior de uso de la vivienda, en el caso de los arrendamientos u otras situaciones asimilables".

Con bastante cautela, es posible apreciar una ejecución discriminatoria del contrato si las prestaciones que de él dimanan se exigen por parte del acreedor o se prestan de forma diferente, en función de las circunstancias del deudor. Por ejemplo, el contrato establece como domicilio de pago del arrendamiento el del acreedor, y cuando concurren los deudores el día del pago, el comportamiento del acreedor es diferente con los inquilinos extranjeros que con los nacionales. Con los primeros es más riguroso, no les deja entrar en su vivienda, no le permite hacer pagos por bizum, le recuerda el día antes el cobro.... Actitud que no muestra con los inquilinos españoles; o, a los inquilinos españoles les ha autorizado una obra para colocar aire acondicionado y a los extranjeros no, sin que haya causa alguna que fundamente esta decisión; o, a los inquilinos españoles les pinta anualmente la vivienda y atiende a todos sus peticiones, siendo más reticentes con los extranjeros ... En cualquier caso, para apreciar discriminación se debe aportar un elemento, al menos, que nos permita visibilizar la diferencia de trato.

Otra cosa distinta es alegar diferencia de trato en relación a cuestiones sobre las que el arrendador dispone de libertad, bien porque así se haya configurado en la LAU (prórroga del contrato a voluntad del arrendador en el espacio que la ley se lo permite), bien porque se trate de un acto de mera liberalidad que, como tal, entra dentro de la autonomía del sujeto (no elevar la renta o condonar o eximir el pago del alquiler durante un periodo). Dejamos aquí apuntadas estas cuestiones, que retomaremos con más detenimiento en el epígrafe siguiente dedicado a las consecuencias jurídicas derivadas de la inobservancia del principio de no discriminación en este ámbito concreto.

3. Las consecuencias jurídico-civiles y su valoración

Abordar la aplicación del principio de no discriminación en las relaciones jurídicas entre los particulares implica, no solo discernir sobre el ámbito en el que este opera, sino también, y sin duda lo más relevante, concluir sobre la reacción del ordenamiento jurídico en aras a restablecer la situación desactivando los efectos discriminatorios y/o compensar patrimonialmente los daños ocasionados.

Considerando que nuestro ordenamiento jurídico cuenta con una norma general e integral para la igualdad de trato y no discriminación, antes de analizar nuestras propuestas sobre las consecuencias jurídico civiles que pudieran tener lugar en los contratos que posibilitan el uso y disfrute de una vivienda digna, debemos detenernos a esbozar el marco que dicha norma posibilita, pues sobre dicho andamiaje incardinaremos las consecuencias jurídicas concretas, obviamente mediatizadas por las reglas generales.

3.1. El marco general previsto en la Ley 15/2022 de 12 de julio

La sanción civil que procede cuando tienen lugar las infracciones de mayor entidad del ordenamiento jurídico es la nulidad. En este sentido, la Ley 15/2022 en su artículo 26 así lo prescribe estableciendo su alcance tanto a las disposiciones legales o reglamentarias y a los actos administrativos como a las cláusulas de los negocios jurídicos que atenten al principio de no discriminación[61]. De forma semejante, pero con un entorno reducido a los contratos y negocios jurídicos, el art. 64[62] de la Ley 14/2023 de 28 de febrero para la igualdad real y efectiva de las personas trans y para la garantía de los derechos de las personas LGTBI, estableciendo que estas cláusulas que atenten a la no discriminación se tendrán por no puestas.

Pese a ello, no se limita el legislador a afirmar su ineficacia, sino que, también, menciona, muy acertadamente, la obligación de tomar medidas para detectar y prevenir situaciones o actos discriminatorios, y para que estos cesen (artículo 25)[63].

Antes de entrar en la nulidad del contrato, es conveniente apuntar que el contenido del artículo 25 alcanza a dos situaciones diferentes con consecuencias jurídicas no homogéneas. En primer lugar, el deber general de detectar y prevenir que se traduce, por obra del artículo 27.2, en una obligación jurídica de contenido indemnizatorio para las personas en él enunciadas (personas empleadoras o prestadoras de bienes y servicios), siempre que la discriminación

---

61 Artículo 26. Nulidad de pleno derecho. "Son nulos de pleno derecho las disposiciones, actos o cláusulas de los negocios jurídicos que constituyan o causen discriminación por razón de alguno de los motivos previstos en el apartado primero del artículo 2 de esta ley".

62 Artículo 64. Nulidad de los contratos y negocios jurídicos discriminatorios. "Las cláusulas de los contratos y negocios jurídicos que vulneren el derecho a la no discriminación por razón de orientación sexual, identidad sexual, expresión de género o características sexuales serán nulas y se tendrán por no puestas."

63 Artículo 25. Medidas de protección y reparación frente a la discriminación. "1. La protección frente a la discriminación obliga a la aplicación de métodos o instrumentos suficientes para su detección, la adopción de medidas preventivas, y la articulación de medidas adecuadas para el cese de las situaciones discriminatorias"

haya tenido lugar en su ámbito de organización y dirección, que genera, por consiguiente, responsabilidad patrimonial resarcitoria para la víctima a cargo de los sujetos mencionados[64]. Fuera de estos sujetos nos resulta difícil mantener una obligación para cualquier ciudadano, que no sea la obligación general de denunciar cuando se sospecha la comisión de un delito (si, efectivamente, así pudieran calificarse los actos discriminatorios).

En segundo lugar, y una vez se ha producido la discriminación, se alude a la necesidad de poner en marcha medidas encaminadas a que esta desaparezca. Dada la amplitud del concepto "medidas" y la heterogeneidad de los destinarios, no se conectan consecuencias jurídicas a la omisión de estas medidas de forma directa.

Por otro lado, el redactado del artículo 28[65], (en conexión con el 27.2), ya en sede procesal, contempla dos tipos de medidas: de carácter definitivo (cesar) y de carácter provisional

---

64 Artículo 27.2: 2. "Serán igualmente responsables del daño causado las personas empleadoras o prestadoras de bienes y servicios cuando la discriminación, incluido el acoso, se produzca en su ámbito de organización o dirección y no hayan cumplido las obligaciones previstas en el apartado 1 del artículo 25."

65 Artículo 28. "Tutela judicial del derecho a la igualdad de trato y no discriminación. La tutela judicial frente a las vulneraciones del derecho a la igualdad de trato y no discriminación comprenderá, en los términos establecidos por las leyes procesales, la adopción de todas las medidas necesarias para poner fin a la discriminación de que se trate y, en particular, las dirigidas al cese inmediato de la discriminación, pudiendo acordar la adopción de medidas cautelares dirigidas a la prevención de violaciones inminentes o ulteriores, la indemnización de los daños y perjuicios causados y el restablecimiento de la persona perjudicada en el pleno ejercicio de su derecho, con independencia de su nacionalidad, de si son mayores o menores de edad o de si disfrutan o no de residencia legal."

(cautelares) encaminadas a igual propósito, mostrando, así, una protección más completa que la dispensada en la Ley 3/2007[66].

Si bien la frase "y la articulación de medidas adecuadas para el cese de las situaciones discriminatorias" del artículo 25.1 nos conduce al contenido del artículo 28 que acabamos de mencionar, nos preguntamos si, dentro de estas medidas, podría incluirse la imposición de la obligación de contratar. Que no esté previsto, expresamente en la norma no es argumento suficiente para predicar su inoperancia. Es más, la finalidad reparatoria y resarcitoria y, en última instancia, disuasoria, nos legitima para plantearnos esta posibilidad. Además de establecer el segundo párrafo del citado artículo 25[67], que, apreciada la discriminación, las medidas de cese tendrán como objetivo la reparación del perjuicio (además de la indemnización). Nada más alineado con dicha finalidad que obligar a contratar al sujeto infractor con la víctima de la discriminación.

Para analizar la viabilidad de la imposición de una obligación como la descrita, sea por parte del legislador (que no lo ha hecho), sea por el juez, es necesario, en primer lugar, posicionarnos sobre su posible constitucionalidad. En este sentido, la libertad negocial, pilar del Derecho de contratos, como se ha reiterado, se encuentra protegida por el artículo 10 de la Constitución[68]; res-

---

66 García Rubio (2007, pp. 153-154).

67 "2. El incumplimiento de las obligaciones previstas en el apartado anterior dará lugar a responsabilidades administrativas, así como, en su caso, penales y civiles por los daños y perjuicios que puedan derivarse, y que podrán incluir tanto la restitución como la indemnización, hasta lograr la reparación plena y efectiva para las víctimas".

68 No opina así, García Rubio (2007, p. 156, nota al pie 64), que duda de la inconstitucionalidad de la imposición de contratación a persona discriminada, y considera que dicha posibilidad no estaría bajo el paraguas del art. 10, tal y como se ha afirmado en el texto principal. En esta línea, también Barba (2023, p. 139) y Grau Pineda (2014, p. 82).

paldada, asimismo, por "la libertad de empresa en el marco de la economía de mercado", que proclama el artículo 38.1 de la Carta Magna. Por lo que parece que, de entrada, estaría excluida tal posibilidad. Sin embargo, este supuesto versa, más que sobre la constitucionalidad de la medida, en sentido estricto, sobre el conflicto de dos derechos: la libertad negocial con el derecho a no ser discriminado del artículo 14. Ello nos invita, para su resolución, a establecer criterios dirigidos a priorizar uno u otro derecho.

En el recinto del Derecho privado, donde impera la protección de intereses particulares, la libertad del individuo ocupa un papel predominante. Por consiguiente, cuando tiene lugar la discriminación de un colectivo por un sujeto particular en el ámbito de relaciones contractuales privadas, la causa de la misma se enmarca dentro del espacio protegido por el artículo 16 de la Constitución, básicamente. Circunstancias ideológicas y religiosas son las que se encuentran detrás de decisiones excluyentes carentes de una causa objetiva y legítima que las justifique. Que dicha libertad, cuando se materializa en comportamientos discriminatorios, no esté amparada por la libertad del artículo 16 de la Constitución, se encuentra detrás de las consecuencias jurídicas establecidas en las normas españolas y extranjeras que regulan la operatividad del principio de no discriminación en las relaciones entre particulares. Es decir, en el mencionado conflicto se da preferencia al artículo 14, al establecer que la persona que discrimina en sus relaciones contractuales a otra, negándose a contratar con ellas, debe indemnizar los daños materiales y morales causados. Lo que impide esgrimir el derecho a la libertad ideológica como escudo protector ante dichas reclamaciones.

Sin que figuren otras circunstancias que introduzcan un interés público en la confrontación, el artículo 10 de la Constitución, pórtico de la enumeración de los derechos fundamentales y sede de los principios interpretadores de los mismos, impide que, además de las consecuencias jurídicas establecidas en las leyes, se le pueda imponer al ciudadano la obligación de contratar. Obligación que implica mantener relaciones jurídicas con un sujeto con el que este no desea hacerlo, fruto de un acto de su voluntad, de la libertad de todo ser humano. Al igual que el Derecho no puede ni debe imponernos relaciones de amistad o de afecto, tampoco está legitimado, con carácter general, a imponernos relaciones jurídico-patrimoniales.

Distinto sería que confluyeran otras circunstancias que justificaran, en aras de la protección de un interés superior, la obligación de contratar con determinados sujetos y en esa concreta coyuntura. Ello sucedería como consecuencia de que se aprecie la protección de otros derechos de carácter colectivo o de un valor superior que fuera prevalente. Así, por ejemplo, en una pandemia la imposición de vender mascarillas, o en una sequía la obligación de vender agua mineral. Fuera de estas circunstancias excepcionales, consideramos que la Constitución no otorga cobertura a la obligación de contratar.

Volviendo a las consecuencias jurídico-privadas, la nulidad como sanción general debe ser interpretada en un sentido amplio. Pues el principio de conservación del contrato y, en su caso, la protección del interés de la víctima demanda que, siempre que la cláusula discriminatoria no contenga un elemento esencial del contrato, se opte por la nulidad parcial del mismo. También tendría cabida, si los elementos concomitantes así lo posibilitaran, la llamada nulidad parcial coactiva con sustitución del contenido de la cláusula infractora por aquel fijado en la norma[69].

En verdad, solo se contempla en la ley la nulidad como consecuencia derivada del incumplimiento. Pero ello no debe interpretarse como exclusión de cualquier otra modalidad de ineficacia; ya que cualquier otra fórmula o vía de las previstas

---

69 Barba (2023, p. 165).

en nuestro ordenamiento jurídico que provea de una mayor satisfacción al perjudicado deberá ser admitida. En este sentido, la resolución del contrato sería factible si la víctima decidiera ejercitar esta acción en sustitución de la nulidad.

De acuerdo con lo establecido en el artículo 27, sea cual sea la modalidad de ineficacia que se ejercite, siempre estará abierta la indemnización por los daños y perjuicios ocasionados, que no exime de la obligación reparatoria (*in natura*), si fuera posible, del daño derivado de la vulneración. Dado que estos conceptos se corresponden con los daños patrimoniales, estos deben acreditarse y cuantificarse por la víctima en la oportuna reclamación, sobre la base de los criterios y reglas comunes a cualquier proceso de este tipo.

Expresamente se refiere la ley a la concomitancia de ambas modalidades de daño, consagrando la presunción *iuris et de iure* de existencia de daño moral si se acreditaba la infracción, además de apuntar los parámetros a tener en cuenta para su valoración[70]. La justificación jurídica del daño moral, en este caso, se residencia en la vulneración de un derecho fundamental que lesiona la dignidad como persona que no solo se aprecia en la responsabilidad extracontractual, si bien suela ser lo más frecuente. Aunque la responsabilidad contractual acostumbra

---

[70] Artículo 27. Atribución de responsabilidad patrimonial y reparación del daño. "1. La persona física o jurídica que cause discriminación por alguno de los motivos previstos en el apartado 1 del artículo 2 de esta ley reparará el daño causado proporcionando una indemnización y restituyendo a la víctima a la situación anterior al incidente discriminatorio, cuando sea posible. Acreditada la discriminación se presumirá la existencia de daño moral, que se valorará atendiendo a las circunstancias del caso, a la concurrencia o interacción de varias causas de discriminación previstas en la ley y a la gravedad de la lesión efectivamente producida, para lo que se tendrá en cuenta, en su caso, la difusión o audiencia del medio a través del que se haya producido".

a constreñirse al ámbito patrimonial, no excluye que, como consecuencia del incumplimiento, se infrinjan daños morales que, obviamente, deben ser indemnizados. En este caso, es la propia ley la que no deja espacio a la duda, al expresar esta presunción que, parece, no permite su omisión. Y que actúa como un elemento disuasorio al establecer una consecuencia patrimonial (a modo de sanción), siquiera por daño moral, que va a tener lugar siempre que se actúe contrariamente al principio de no discriminación.

Conscientes de la dificultad de la cuantificación del daño moral indemnizable, el precepto indica unos criterios, que en sí mismos no son novedosos, por ser los que, tradicionalmente se utilizan por los Tribunales para dicha valoración.

Por último, expresa la ley una presunción *iuris tantum* de existencia de discriminación y que, como tal, genera una inversión de la carga de la prueba, que se desprende de los indicios que debe aportar la persona que reclama y alega discriminación. Del redactado del artículo se colige que la presunción solo opera si existieran indicios fundados, de manera que la víctima debe acompañar a la demanda un principio de prueba que, a juicio del juez, se estime como suficiente para atisbar un comportamiento negocial discriminatorio. Por ejemplo, podría entenderse como indicio fundado a los efectos del artículo 30, la acreditación de los dos elementos de comparación necesarios para que tenga lugar la discriminación, que avalaría la aplicación de la presunción. Ello facilitaría el alcance de la reparación pecuniaria y, en su caso, *in natura*[71], al quedar liberada la víctima de probar la existencia

---

71 Artículo 30. Reglas relativas a la carga de la prueba. "1. De acuerdo con lo previsto en las leyes procesales y reguladoras de los procedimientos administrativos, cuando la parte actora o el interesado alegue discriminación y aporte indicios fundados sobre su existencia, corresponderá a la parte demandada o a quien se impute la

de discriminación y recaer en el demandado la carga probatoria de la concurrencia de causa legítima, objetiva y/o proporcionada que justifique la diferencia de trato.

3.2. Consecuencias específicas en la contratación sobre la vivienda

Partiendo de las líneas diseñadas en la normativa general de la Ley 15/2022 (nulidad como sanción, reparación *in natura* y/o indemnización por daños y perjuicios abarcando los daños morales que se presumen *iure et de iure*, obligación de establecimiento de medidas preventivas o de cese, definitivas o cautelares, e inversión de la carga de la prueba) y de las reglas que conforman la teoría general del contrato y de las obligaciones de nuestro Código Civil, procederemos a continuación a aplicarlas a la vivienda para fijar las consecuencias de las vulneraciones que pudieran acaecer en las distintas fases de la contratación.

3.2.1. Ineficacia de la oferta discriminatoria

Siguiendo el orden sistemático y cronológico del epígrafe 2.3, si la infracción tuviera lugar en la oferta, conforme a la nulidad como sanción establecida, esta se mantendría en todos sus extremos, excepto en aquello que atentara o generara discriminación (si ello fuera posible).

---

situación discriminatoria la aportación de una justificación objetiva y razonable, suficientemente probada, de las medidas adoptadas y de su proporcionalidad." En igual sentido el artículo 66.1 de la Ley 14/2023 de 28 de febrero para la igualdad real y efectiva de las personas trans y para la garantía de los derechos de las personas LGTBI.

El destinatario de la oferta, en sentido estricto, sea pública o privada, con contenido discriminatorio, puede optar entre impugnarla, antes de aceptarla, o aceptarla omitiendo *de facto* su preterición. En el primer caso, el juez valorará si la oferta pudiera mantenerse eliminando por ineficaz aquel contenido excluyente sin justificación o, por el contrario, debe retirarse cuando la supresión de aquello que fuera tachado como discriminatorio deja sin sentido la oferta en su conjunto. El oferente, no obstante, podría revocar la oferta si no existiera un plazo fijado en la propia oferta, dado que no se ha producido la aceptación[72]. Pues la reacción del ordenamiento jurídico no puede ir más allá, obligándole a mantener la oferta o, en su caso, a realizar un contrato por encima de su voluntad. Sin embargo, si el oferente retira la oferta al decidir que, al haberse eliminado el contenido discriminatorio, no está interesado en contratar con esa persona o colectivo injustamente preterido, dicha acción, por sí misma, genera una vulneración del derecho a la igualdad de trato indemnizable, incluyéndose en dicha cuantía, el daño moral.

En el segundo caso, el contrato se entiende celebrado como consecuencia de la mecánica normal en todo proceso de contratación (artículo 1262.I CC) y si el oferente se negara entraría en un supuesto de incumplimiento contractual con las consecuencias previstas, con carácter general, en el ordenamiento jurídico. Porque la mera aceptación del contratante ilegítimamente excluido, sin necesidad de hacer valer tal consideración, es suficiente para perfeccionar el contrato y generar efectos vinculantes para ambas partes.

Sentadas estas dos premisas, el lector avezado habrá comprobado la restricción a la "oferta en sentido estricto", en clara

---

72 Alventosa del Río (2018, p. 194).

referencia a la exclusión de la misma de la invitación a ofrecer; ya que en esta, al no contener los elementos del contrato proyectado, su aceptación no implica perfección de contrato alguno, lo que impide que proceda la segunda de las premisas contempladas.

Así, cuando la invitación a ofrecer contraviene el principio de no discriminación, al excluir a un determinado colectivo que no es llamado a esta, el colectivo o la persona lesionada podrá solicitar del juez la indemnización de los perjuicios causados, incluyendo los daños morales por lesión a su dignidad como persona. No estimamos, en cambio, procedente la obligación de mantener la invitación, ahora sin los términos discriminatorios, ya que el sujeto que "invita" está legitimado para retirar la invitación si no le resultara conveniente. De todas formas, aun cuando ello sucediere permitiendo al sujeto o sujetos discriminados realizar una oferta, al titular del inmueble le basta con no aceptarla sin tener que alegar causa alguna.

Es de observar que, cuando la oferta o la invitación a ofrecer se publicita en portales inmobiliarios y otros medios de difusión en internet (redes sociales), el criterio de la difusión del medio empleado del artículo 27.1 de la Ley 15/2022 otorga un amplio margen al juez para establecer importantes indemnizaciones, con un efecto disuasorio y de control del contenido de los publicado en los portales inmobiliarios y en las redes sociales.

Finalmente, en relación a la discriminación algorítmica, el hecho de que esta verse sobre la oferta o invitación a ofrecer en el marco de un proyectado contrato que posibilite el acceso a una vivienda, no añade ninguna especialidad al régimen general de ineficacia. La dificultad estriba en detectar dicha discriminación. La normativa general en esta materia (art. 15.1 y 22 del Reglamento de protección de datos, a la que hemos aludido en epígrafe anterior) despliega una función preventiva más que resarcitoria, aplicándose, si se detecta la discriminación, las medidas previstas en nuestro ordenamiento en los términos expuestos. Nos parece, muy interesante, en orden a evitar estas conductas indeseadas, la implementación de un sello de calidad de los algoritmos, que, como un *desiderátum*, lo menciona el art. 22.4 de la Ley 15/2022.

3.2.2. La indemnización por ruptura discriminatoria de los tratos preliminares

Es posible que, con carácter previo a la oferta privada o en la negociación que pudiera generarse ante la respuesta del sujeto a la invitación a ofrecer, tengan lugar tratos preliminares sobre el contrato que facilita la adquisición o el uso de la vivienda, y se hubiera producido una infracción del principio de no discriminación en los términos expuestos.

De acuerdo con la teoría general, se generaría responsabilidad civil patrimonial a cargo del que rompe los tratos de forma injustificada, que alcanza solo a los daños patrimoniales: es decir, los gastos producidos como consecuencia de la expectativa del sujeto en la realización del contrato, excluyendo las posibles ganancias que se pudieran haber derivado del contrato proyectado[73] . Como quiera que hemos considerado que, para que proceda la indemnización por vulneración del principio de no discriminación, es requisito que los tratos estuvieran avanzados y se hubiera generado expectativas negociales en el sujeto preterido, dichos daños patrimoniales siempre procederían, además de los que tengan causa en la existencia de discriminación.

---

73 Alventosa del Río (2018, p. 183).

A pesar de que la doctrina se muestra restrictiva sobre las partidas que comprenden el concepto "daño indemnizable" en esta fase precontractual, podríamos preguntarnos si, por haber tenido lugar un atentado a un derecho fundamental, sería posible sostener otra propuesta más extensa. En mi opinión, el tratamiento de los daños indemnizables debe ser idéntico en uno u otro supuesto, ya que el motivo de la preterición no debe influir en la cuantía patrimonial de los debido por el infractor. La vulneración de un derecho fundamental no es causa para incrementar la indemnización de otras partidas excluidas por la doctrina con argumentos jurídicos solventes, que no se desvanecen por la causa de la ruptura.

Sin embargo, no podemos obviar el hecho de que la causa aludida ha supuesto una lesión de la dignidad de la persona, que ha sido excluida por concurrir un factor de riesgo. La única forma de implementar el importe indemnizatorio abrazando dicha cuantía es incluyendo la cobertura del daño moral que se ha ocasionado. Ello lo permite la generosa redacción del artículo 27.1 Ley 15/2022 en el que se consagra la presunción *iure et de iure* de daño moral cuando se constata la infracción del principio de no discriminación. En consecuencia, la indemnización por ruptura de tratos preliminares con el objeto de discriminar a colectivos por causa de alguno de los factores de riesgo del artículo 2.1, de manera que se le imposibilite el acceso a una vivienda donde residir, obliga al sujeto activo a indemnizar, no solo por los daños patrimoniales, sino también por los daños morales.

De esta forma se penaliza al infractor que rompe los tratos preliminares vulnerando dos principios: el principio de buena fe contractual, que sustenta la indemnización de los daños patrimoniales, y el principio de no discriminación, fundamento de la indemnización por daños morales.

3.2.3. La ineficacia de la contratación discriminatoria: formas de proteger a la víctima

Una vez celebrado el contrato, ante cualquier cláusula o contenido discriminatorio, la sanción que se predica, con carácter general, es la nulidad, englobando, tal y como hemos ya referido, la nulidad parcial y la nulidad parcial coactiva. Esta premisa, del todo cierta e irrefutable, contextualizada al ámbito de la vivienda pudiera conducirnos a otras reflexiones sobre la actuación del juez integrando el contrato cuando ello lo demandara la protección de la víctima.

A nadie se le escapa que la sede fundamental de la discriminación, una vez celebrado el contrato, es el precio de la venta o del arrendamiento; o las condiciones más o menos gravosas de plazo o garantía, si bien en este ámbito, entendemos, que es más difícil probar la discriminación, ya que la libertad negocial ampara la diversidad de trato en atención a las circunstancias de las partes contratantes (así, por ejemplo, pactar mayores garantía porque las circunstancias económicas de la persona no nos ofrecen solvencia suficiente, a pesar de que en dicha persona concurra un factor de riesgo, se considera una causa legítima de la diversidad de trato). Circunstancias que, además, han sido aceptadas, fruto de una negociación individual (pues si se contrata bajo condiciones generales estas son idénticas en este aspecto).

Volviendo al precio de la venta o del arrendamiento[74] (contratos onerosos emblemáticos y comunes para acceder a una

---

[74] Las reflexiones que se vierten en el texto principal en orden a la integración del juez del precio, obviamente, no tienen lugar en el ámbito del artículo 17 de la LAU que establece una limitación del precio de los alquileres de las viviendas ubicadas en zonas tensionadas. En los supuestos descritos en los epígrafes 6 y 7, no se trata de alegar discriminación si el precio es superior al que resulte de

vivienda), si la víctima consigue acreditar la discriminación en los términos y por medio de los mecanismos apuntados, la nulidad pudiera ser del todo insuficiente, en la medida que se vería privado del inmueble, aun cuando fuera indemnizado por ello, comprendiendo en dicha cantidad una partida en concepto de daños morales por obra del art. 27.2 Ley 15/2022. A fin de servir al objetivo de las normas antidiscriminatorias (minimizar los daños, resarcir a la víctima y reparar el perjuicio causado) entendemos que sería viable otorgar a la víctima, junto a la acción de nulidad del contrato en su integridad (por ser el precio un elemento esencial del mismo), la acción de solicitar del juez la integración del precio, sustituyendo el más elevado y discriminatorio, por aquel que figura en el contrato o contratos alegados como elementos comparativos y referentes sobre los que se detecta la discriminación. La dualidad de acciones propuestas permitiría optar a la víctima por el remedio que mejor se acomodara a sus intereses familiares y personales (no olvidemos que la vivienda familiar goza de una importante protección en nuestro Derecho, fruto del relevante papel social que desempeña), que no siempre coincidirán con uno u otro. Obviamente, si opta por la conservación del contrato con el precio disminuido, el daño ocasionado sería de menor entidad y pudiera reducirse a los gastos derivados de la reclamación judicial y a la devolución de la parte del precio indebidamente cobrado. Aunque parece que el automatismo del artículo 27.2 de la Ley 15/2022 incluye el daño moral, si el sujeto discriminado puede continuar en la vivienda con un precio más ajustado nos resulta difícil apreciar, en este caso concreto, el daño moral al que se alude en dicho precepto.

---

la aplicación de las reglas para su fijación contenidas en la norma, sino, simplemente, nulidad de la cláusula por vulneración de norma imperativa y sustitución por la cantidad que impone la ley.

Como para que se genere la discriminación es imprescindible contar con referentes frente a los que esta se produce, la integración del juez es una tarea fácil y segura, pues basta acoplar el precio al de los demás contratos y, en caso de que entre ellos existan pequeñas variantes, al precio medio. Consideramos, asimismo, que dicha integración es factible y no vulneraría derecho alguno del titular de la vivienda, ya que el contrato está formalizado y la libertad de contratación ya ha sido ejercida y materializada. Tampoco es esta una técnica novedosa ni extraña, pues en materia de reclamación de precio de servicios profesionales impugnados como abusivos, el juez se encuentra legitimado para reducir el precio al medio del mercado, pudiendo valerse de las normas profesionales, si las hubiere, con carácter orientativo (pues si fueran imperativas el juez no integra, sino que impone la norma transgredida).

La nulidad parcial, en cambio, sí es un remedio adecuado cuando la discriminación tiene lugar por la imposición de condiciones en el precio o en cualquier otro aspecto más gravosas; su mera eliminación permite al contrato continuar desplegando sus efectos sin las trabas o gravámenes injustificados.

Si bien la sanción de nulidad prevista en las normas no excluye la posibilidad de ejercitar otras acciones como fuera la resolución del contrato, opinamos que, dado el régimen jurídico establecido que, expresamente, indica la indemnización de los daños y perjuicios, incluido los morales en los términos expuestos, como consecuencia de la infracción del principio, así como la interpretación amplia de la nulidad propuesta, resulta complicado en materia de vivienda encontrar un espacio para otras fórmulas de ineficacia contractual. Pues con el régimen jurídico diseñado se protege a los intereses de la víctima, permitiéndole jugar con sus preferencias en orden a alcanzar la solución jurídica más amoldada a su situación personal.

En este sentido, la opción por la nulidad total o parcial, si el contrato contiene cláusulas discriminatorias, sin excluir la indemnización según las consecuencias de una u otra fórmula, ubica a la víctima en una posición jurídica flexible que se amolda a sus intereses y a los de su familia. Asimismo, si la discriminación se detecta en el precio, más elevado o gravoso, el perjudicado puede optar por la nulidad del contrato (por ser el precio elemento esencial) o su reducción, conforme a los postulados expresados en líneas anteriores. Este es el "escenario" al que nos referíamos y que, a nuestro juicio, de hecho, no deja espacio para el ejercicio de otras acciones dirigidas a la reparación de la víctima.

3.2.4. Ejecución discriminatoria de un contrato válido: la tensión entre libertad e igualdad

Por último, el principio de no discriminación también despliega su eficacia en un tercer momento de la vida del contrato: su ejecución. Si bien este es una fase en la que es más difícil detectar, con carácter general, la operatividad de este principio, en materia de arrendamiento de vivienda, contrato de larga duración, puede generar algunas cuestiones de las que debemos ocuparnos.

En primer lugar, en relación a la aplicación de la cláusula de revisión del precio pactada o, en su caso (como ha sucedido como consecuencia del COVID) la limitación de la subida establecida en la norma. ¿Se calificaría como discriminatoria la decisión del arrendador de aplicar la subida en alguno de los contratos y en otros no? La cuestión estriba en dilucidar si el arrendatario, en el que se aprecia un factor de riesgo por su nacionalidad o etnia, por ejemplo, pudiera exigir ante los tribunales la no aplicación de la subida pactada o la legalmente establecida sobre la base de que no se ha exigido a otros inquilinos, de nacionalidad y etnia distinta, en una situación jurídica, con causa en contratos idénticos o muy similares, semejante a la suya. La respuesta, muy resbalosa, pone en pugna dos valores esenciales en nuestro ordenamiento jurídico: la igualdad y la libertad. Ciertamente, si el sujeto arrendador fuera una entidad pública, ni siquiera pudiéramos hablar de conflicto por no existir dos elementos a confluir, pues la libertad es un derecho conectado a la persona privada.

Si bien, aplicando los postulados del Derecho antidiscriminatorio que hemos ido desgranando a lo largo de este discurso, y siendo formalmente consecuentes con ellos, la respuesta debería ser afirmativa, vamos a invitar a lector a mirar el supuesto desde otra perspectiva con el objetivo de que nos acompañe en esa duda jurídica que nos hace cuestionar la respuesta anteriormente expresada.

En nuestro ordenamiento jurídico los sujetos se encuentran obligados, por supuesto, por las normas, y por los contratos que dice el Código tienen "fuerza de ley" entre las partes contratantes. Asimismo, de la causa de los contratos onerosos, como lo es el arrendamiento urbano, se deriva que el arrendador reciba una retribución económica que, a fin de mantener idéntica onerosidad, debe actualizarse periódicamente. Actualización que no es imperativa, porque las partes están legitimadas para excluirla en el contrato o, simplemente, no referirse a ella de lo que la LAU infiere su exclusión (art. 18.1 LAU). La libertad para determinar el importe de la renta y su actualización no es plena, y se encuentra limitada, que no eliminada, si la vivienda arrendada se protege por la LAU como consecuencia de ser destinada a uso habitacional, con limitaciones mayores si esta se sitúa en zona tensionada en los términos de los epígrafes 6 y 7 del artículo 17 LAU.

Continuando con el argumento, la libertad de contratación garantizada por el artículo 10 de la Constitución, pilar de la economía de mercado y de la libertad de empresa que lo caracteriza (artículo 38 de la Constitución), también comprende la capacidad del individuo de concertar actos jurídicos que, no necesariamente, le reporten un beneficio económico. Condonar deudas, prestar bienes muebles o inmuebles de forma gratuita, donar cosas...son prácticas usuales en el tráfico económico. En este ámbito, el sujeto puede decidir, sea por el motivo que sea, exigir la subida de la renta pactada, es decir el cumplimiento de un contrato o, en su caso, aplicar el índice de referencia sobre la renta acordada para elevarla en los términos que la normativa dispone, a un determinado inquilino y no hacerlo a otro al que, por mera liberalidad, simpatía o amistad, así lo considera conveniente u oportuno.

El concepto "discriminación" implica un tratamiento negativo o perjudicial sin causa justificada, que debe ser corregido eliminando el contenido discriminatorio, sustituyéndolo, en su caso, por el adecuado, e indemnizando los perjuicios derivados de dicha situación creada. El concepto "perjuicio" es material, e implica un menoscabo. Perjudicar no es equiparable a "no ser beneficiado" cuando no existe obligación legal alguna de atribuir dicho beneficio, y este no es más que una consecuencia de un acto generoso por parte del que lo concede.

Partiendo, pues, de la premisa anterior, y al igual que hemos defendido que, en nuestra opinión, no es jurídicamente defendible imponer la obligación de contratar alegando que no realizar el contrato discrimina a aquel que lo desea, ni siquiera invocando el derecho a una vivienda digna, tampoco lo sería obligar al arrendador a eximir de la elevación de la renta simplemente por haber decidido, en uso de su libertad, renunciar o no ejercitar este derecho ni ser requerido a justificar el motivo de dicha opción, pues ello entra dentro de su libertad como persona que le garantiza el artículo 10 de la Constitución.

Siguiendo en sede de arrendamiento de viviendas, la LAU otorga libertad al arrendador para excepcionar la prórroga del contrato, una vez transcurrido los plazos de duración mínima y prórroga obligatoria, otorgándole un amplio margen de decisión para optar si continúa con la relación jurídica o, por el contrario, sea por el motivo que sea, poner fin a la misma. De nuevo, podría plantearse la operatividad del principio de no discriminación en este ámbito, cuando el arrendador decide prorrogar el contrato de un inquilino o varios frente a otro u otros en los que concurre un factor de riesgo. ¿Está obligado el arrendador a alegar el motivo para evitar que se le obligue a prorrogar? O, lo que es peor, ¿es nula la prórroga por ser discriminatoria para el no prorrogado? La segunda cuestión, obviamente, es insostenible jurídicamente, porque la nulidad como sanción tiene un propósito resarcitorio de la víctima y en este caso no serviría a tal fin (amén de perjudicar a los no prorrogados privándoles de la que, hasta entonces, era su vivienda).

Así como en el supuesto de la no elevación de renta, renunciando al derecho legal o contractualmente reconocido, ha sido calificado como un acto de liberalidad que perjudica al que lo emite y beneficia al receptor, el hecho de no excepcionar la prórroga del contrato no es más que un acto de libertad, propio de la facultad de uso y disfrute que ostenta el arrendador y que lo legitima para asumir dicho rol. Conectado con este planteamiento, parece evidente que no es defendible exigir al arrendador una justificación sobre una decisión sobre la que tiene plena potestad; como tampoco lo es obligarle a realizar un acto jurídico, tal cual es prorrogar un contrato, solo porque carece de una justificación objetiva y proporcionada que no sea su mera voluntad, para optar prorrogar un contrato frente a otro y que, por ejemplo, el no prorrogado pertenezca a una etnia distinta del prorrogado y/o de la del arrendador.

V. REFERENCIAS BIBLIOGRÁFICAS

AA.VV. (2022). *Discriminación algorítmica en materia de Derecho laboral,* dirigida por Rivas Vallejo, P. Aranzadi Thomsons Reuters.

Aguilera Rull, A. (2007). "Discriminación directa e indirecta". *InDret,* (1), 1-17.

(2009). "Prohibición de discriminación y libertad de contratación". *InDret,* (1), 1-30.

Alfaro Águila-Real J. (1993). "Autonomía privada y derechos fundamentales". *Anuario de Derecho Civil,* 57-122.

Alventosa del Río, J. (2018). "La perfección del contrato en el Código Civil español". *Rev. Boliv. de Derecho* (26), julio, 176-217.

Barba, V. (2023). *Principio de no discriminación y contrato.* Colex.

Barrère Unzueta, M.A. y Morondo Taramundi, D. (2011). "Subordiscriminación y discriminación interseccional: elementos para una teoría del Derecho antidiscriminatorio". *Anales de la Cátedra Francisco Suárez* (45), 15-42.

Carapezza Figlia, G. (2019). "El principio de no discriminación en el Derecho contractual europeo". *Actualidad Jurídica Iberoamericana,* (10), 282-315.

Cerdeira Bravo de Mansilla, G. (2021). "Crisis de pareja y animales domésticos: ¿una "pertenencia" de la vivienda familiar?". *Familia y Derecho en la España del siglo XXI: libro homenaje al profesor Luis Humberto Clavería Gosálbez* (dir. Marín A. Velarde/ Cabezuelo Arenas, A. L./ Moreno Mozo, F.). Reus, 507-534.

(2022). "El bienestar animal como ser sintiente: un "nuevo" principio general para el derecho de animales". *Un nuevo Derecho civil sobre los animales.* Reus.

Cotino Hueso, L. (2023). "Qué concreta transparencia e información de algoritmos e inteligencia artificial es la debida". *Revista Española de la Transparencia,* (16), primer semestre, enero-junio, 17-63.

Díaz Revoiro, A.J. (2015). *Discriminación en las relaciones entre particulares.* Tirant Lo Blanch.

Ferrando Nicolau, E. (1992). "El derecho a una vivienda digna y adecuada". *Anuario de Filosofía del Derecho,* (IX), 305-322.

Fernández-Rivera González, P. (2023). "Perspectiva de género, asociaciones religiosas y Ley 15/2022, de 12 de julio, integral para la igualdad de trato y la no discriminación: reflexiones a propósito de la STS

925/2021, de 23 de diciembre de 2021 ". *Anuario de Derecho Eclesiástico del Estado,* vol. XXXIX, 645-681.

García Murcia, J. (2022). "La Ley 15/2022 integral para la igualdad de trato y la no discriminación". *Foro. Revista de Ciencias Jurídicas y sociales,* vol. 25, núm. 2, 11-22.

García Rubio, M.P. (2002). "La eficacia *inter privatos (Drittwirkung)* de los derechos fundamentales". *Libro Homenaje a Ildefonso Sánchez Mera,* vol. I, Consejo General del Notariado, Madrid, 297-314.

(2007). "Discriminación por razón de sexo y derecho contractual en la Ley Orgánica 3/2007, de 22 de marzo, para la igualdad efectiva de mujeres y hombres". *Derecho Privado y Constitución,* (21), enero-diciembre, 131-166.

Grau Pineda, M. C. (2014). *Mujeres, contratos y empresa desde la igualdad de género.* Tirant lo Blanch.

Jiménez Horwitz, M. (2014). "La protección contra la discriminación en las relaciones entre particulares: la evolución entre la responsabilidad extracontractual hasta los remedios por incumplimiento". *Anuario de Derecho Civil,* tomo LXVII, fasc. II, 475-515.

Llorente San Segundo, I. (2018). *La discriminación de personas con discapacidad y otros colectivos vulnerables en la contratación entre particulares.* Ed. Ramón Areces.

Manero Salvador, A. (2015). "El derecho a la vivienda desde la perspectiva del Derecho internacional. Análisis del caso español". *Los derechos sociales y su exigibilidad: libres de temor y miseria* (coord. Ribotta, S./ Rosetti, A.). Dykinson, 333-364.

Rey Martínez, F. (2020). *Derecho antidiscriminatorio,* Aranzadi.

(2017). "Igualdad y prohibición de discriminación: de 1978 a 2018". *Revista de Derecho Político,* UNED (100), 125-171.

Sancho Villa, D. (2021). *Comentario al Reglamento General de Protección de Datos y a la Ley Orgánica de Protección de Datos personales y Garantía de los Derechos Digitales.* Troncoso Reigada, A. (dir.) Vol. 1, 1725-1745.

Sarra, C. *(2019).* "Il diritto di contestazione delle decisioni automatizzate nel GDP". *Anuario Facultad de Derecho. Universidad de Alcalá XII,* 33-69.

Torres Perea, J. M. de (2020). *El nuevo estatuto jurídico de los animales en el Derecho civil: De su cosificación a su reconocimiento como seres sensibles.* Reus.

Vivas Tesón, I. (2022). "Si los animales son seres sintientes, ¿es posible prohibir la tenencia de un animal de compañía en una vivienda?". *Revista CESCO de Consumo,* (41), 1-20.

Zubero Quintanilla, S. (2018). "Límites a la autonomía de la voluntad en las asociaciones privadas". *Anuario de Derecho Civil,* Tomo LXXXI, fasc. II, 267-338.

- (2016). "La interpretación del valor del contenido de las declaraciones publicitarias en la jurisprudencia española y francesa". *Aranzadi civil-mercantil. Revista doctrinal,* (6), 63-92.

- (2017). *Las declaraciones publicitarias para contratar.* Tirant lo Blanch.

Capítulo III

Los derechos y deberes del ciudadano en relación con la vivienda

JOSÉ MANUEL DE TORRES PEREA
Profesor Titular de Derecho civil
Universidad de Málaga

I. INTRODUCCIÓN

La Ley 12/2023 regula el "Estatuto básico del ciudadano" en los artículos octavo y noveno que conforman el Capítulo I de su Título I. Se trata de una regulación que se inspira en el Capítulo I del Título I del RD Legislativo 7/2015, (TOL9.218.609) que aprobó el Texto Refundido de la Ley del Suelo y Rehabilitación Urbana, Capítulo que tiene la misma denominación, "Estatuto básico del ciudadano", y que se divide también en dos artículos: el primero (art. 5) sobre los derechos del ciudadano, y el segundo (art. 6) sobre los deberes del ciudadano. El paralelismo entre ambos es evidente.

En este capítulo, en primer lugar, vamos a analizar los distintos derechos y deberes previstos por el Estatuto básico del ciudadano de la Ley 12/2023, y en segundo lugar vamos a centrar la atención en el primero de ellos, es decir, el derecho a una vivienda digna y adecuada.

El art. 47 de la Constitución reconoce el derecho a una vivienda digna como un mero principio rector, sin reconocerlo como derecho fundamental. La doctrina, que no considera que constituya un verdadero derecho subjetivo, discute desde hace cuatro décadas sobre los efectos normativos que se derivan de dicho principio rector. Por tanto, el art. 47 CE requiere de una norma que lo desarrolle y lo configure, si procede, como un derecho subjetivo. Precisamente, esta es la función que, *prima facie*, podría pensarse que podría cumplir el artículo 8 de la Ley 12/2023 al desarrollar el citado art. 47 CE junto con la Ley del Suelo, los propios estatutos de autonomía y otras disposiciones.

Este análisis ha de hacerse conectando la norma con la realidad social a la que va dirigida. Las estadísticas muestran el progresivo deterioro de determinados sectores sociales en España para los que el acceso a una vivienda digna se ha convertido en un reto inasumible. En poco más de una década se ha pasado de ser un país de propietarios, en el que se estimaba innecesario el fomento de la vivienda pública, a un país en el que el acceso a la propiedad o al alquiler ya no está al alcance de todos. Se afirma que estamos ante un cambio de paradigma en la evolución social de la tenencia de vivienda que ha de reflejarse en un cambio de paradigma en su regulación.

II. DERECHOS DEL CIUDADANO

El art. 8 de la Ley tiene por título: "Los Derechos del ciudadano en relación con la vivienda". Comienza señalando que la norma es aplicable a todos los ciudadanos sin hacer distingos, por lo que reconoce la titularidad de estos derechos de forma universal, lo cual ha de ser considerado como un avance respecto al art. 47 CE[1] que excluye a los ciudadanos que no ten-

---

1 El art. 47 de la Constitución dice que "todos los españoles tienen derecho a disfrutar de una vivienda digna y adecuada", añadiendo que "los poderes públicos promoverán las condiciones necesarias y establecerán las normas pertinentes para hacer efectivo este derecho, regulando la utilización del suelo de acuerdo con el interés general para impedir la especulación".

gan la nacionalidad española. Por otro lado, hay que señalar que la LDV recoge conjuntamente derechos de naturaleza jurídico-privada, v.gr. derecho a "disfrutar" la vivienda, con otros de naturaleza jurídico-pública sin distinguir entre unos y otros.

Analizamos ahora los distintos derechos recogidos en el art. 8 LDV, si bien el derecho recogido en su primer apartado (derecho a disfrutar una vivienda digna y adecuada) se estudiará en la segunda parte de este capítulo.

1. Artículo 8 letra b): derecho a acceder a la información de que dispongan las Administraciones públicas

En la letra b) se recoge el derecho de naturaleza jurídico-pública a "acceder a la información de que dispongan las Administraciones públicas sobre los programas públicos de vivienda y a las condiciones de acceso", información que ha de incluir las correspondientes "prestaciones, ayudas y recursos públicos disponibles para garantizar el acceso a la vivienda por parte de las personas y familias en situación de vulnerabilidad". Finalmente, la norma específica que la información aportada por las Administraciones públicas ha de hacerse en formatos accesibles para personas con discapacidad.

Por tanto, el precepto recoge el derecho de acceso a la información pública del art. 105.b de la Constitución, que ha sido desarrollado por la Ley 19/2013 de Transparencia, acceso a la información pública y buen gobierno (TOL4.029.419)[2], para el

---

[2] Art. 37 Ley 19/2013: Del derecho de acceso a la información pública: Los ciudadanos tienen derecho a acceder a la información pública, archivos y registros en los términos y con las condiciones establecidas en la Constitución, en la Ley de transparencia, acceso a la información pública y buen gobierno y demás leyes que resulten de aplicación.

cual hoy se dispone de herramientas informáticas a las que puede acceder todo ciudadano, a través del portal de transparencia. La obligación de que se disponga de "formatos accesibles por personas con discapacidad" no deja de ser una redundancia, pues dicha obligación ya se recoge en el art. 5.5 de la Ley 19/2013[3].

La norma va dirigida a todos los ciudadanos, si bien incide especialmente en "las personas y familias en situación de vulnerabilidad". En este sentido, cabe preguntarse por los indicadores que nos puedan informar sobre cuando existe dicha "vulnerabilidad". Este concepto viene recogido en distintas disposiciones normativas si bien no siempre se ha acompañado de parámetros que puedan ayudar a su determinación[4]. De hecho, en la Exposición de Motivos de la Ley 12/2023 se recoge la necesidad de fijar parámetros concretos de vulnerabilidad económica basados en criterios de carácter objetivo[5]. Es en la DF 5.ª en la que la Ley 12/2023 ha introducido parámetros objetivos para cuantificar la

---

3 Art. 5.5 Ley 19/2013 que dice que toda la información será comprensible, de acceso fácil y gratuito y que estará a disposición de las personas con discapacidad en una modalidad suministrada por medios o en formatos adecuados de manera que resulten accesibles y comprensibles, añadiendo que se han de aplicar los principios de accesibilidad universal y diseño para todos.

4 Puede servirnos de referencia la situación de "riesgo de exclusión" a que se refiere el "Código de buenas prácticas" que el gobierno ha negociado con el sector bancario estableciendo unos mínimos a fin de proteger a los deudores bancarios fijando un umbral de pobreza que da lugar a la protección legal. Umbral que si bien es de mínimos a fin de conseguir la aceptación del sector bancario ha sido recientemente ampliado.

5 Si bien el artículo 14 tiene por rúbrica "situaciones de especial vulnerabilidad" lo cierto es que hace referencia a políticas en materia de vivienda para regenerar zonas urbanas con situaciones de infravivienda, así como otras medidas para combatir el "sinhogarismo" sin mayores especificaciones o criterios objetivos para determinar la "vulnerabilidad"

vulneración económica, con ocasión de la modificación del art. 441, 1 bis y 5 de la Ley 1/2000, de 7 de enero, de Enjuiciamiento Civil (TOL172.336) que pueden ser de ayuda[6].

Por otra parte, en este DF 5.ª se recogen determinados deberes de información a cargo de los tribunales, cuyo objetivo es dar a conocer a los demandados el riesgo de ser desahuciados; que existen administraciones públicas autonómicas y locales que pueden prestarle protección, con obligación de prestarles los datos exactos de identificación de las administraciones y el modo de tomar contacto con ellas, a fin de que puedan apreciar su situación de vulnerabilidad.

Volvemos a la coletilla final de la letra b) donde se dice que la información aportada por las Administraciones públicas ha de ofrecerse en formatos accesibles para personas con discapacidad. Respecto a las personas con discapacidad (mejor terminología que la recogida por esta ley), cabe cuestionarse si se hace referencia solo a las personas con discapacidad psíquica o se recoge un concepto más amplio conforme a la Convención de las Naciones Unidas de Nueva York, en tanto que el CC distingue entre uno y otro supuesto tras la Ley 8/2021 (TOL8.339.953), atribuyéndoles un distinto régimen jurídico[7].

---

6 El tribunal apreciará situación de vulnerabilidad económica del hecho que el importe de la renta (en el caso de juicio de desahucio por falta de pago), más el de los suministros de electricidad, gas, agua y telecomunicaciones suponga más del 30 por 100 de los ingresos de la unidad familiar y que el conjunto de dichos ingresos no alcance, con carácter general, el límite de 3 veces el Indicador Público de Renta de Efectos Múltiples mensual (IPREM).
Vid. Rodríguez Tirado (2024, pp. 7 y ss.) y Hernández Díaz-Ambrona (2015, p. 9).

7 La Ley 8/2021 modifica la DA 4.ª del CC señalando que la referencia a la discapacidad que se realiza en los artículos 96, 756 número 7.º, 782, 808, 822 y 1041, se entenderá hecha al concepto definido

Del hecho de que la Ley 12/2023 no especifique nada al respecto debe deducirse que opta por un concepto amplio conforme a la Convención y por tanto que se refiere a discapacidad física, mental o sensorial[8].

2. *Artículo 8 letra c): derecho a solicitar la inscripción en los registros de demandantes de vivienda protegida*

En la letra c) se refiere el derecho a solicitar la inscripción en los registros de demandantes de vivienda protegida constituidos al efecto por las Administraciones públicas competentes en la materia, y en los diferentes programas, prestaciones, ayudas y recursos públicos para el acceso a

---

en la Ley 41/2003, añadiendo que a los efectos de los demás preceptos de este Código toda referencia a la discapacidad habrá de ser entendida como aquella que haga precisa la provisión de medidas de apoyo para el ejercicio de la capacidad jurídica, en referencia a la situación de discapacidad mental o psíquica.

8 A mayor abundamiento, el marcado carácter administrativo de la ley que comentamos, determinado por su terminología en la que encontramos expresiones como "protección de operaciones de compra y arrendamiento de vivienda", nos informa sobre el enfoque con el que se ha redactado y, por tanto, que procede una interpretación administrativa de la expresión "persona discapacitada" conforme al RDL 1/2013. De la Fuente García Romero de Tejada (2016, pp. 81-99), considera que el concepto jurídico de persona con discapacidad que es utilizado por la Ley General de Discapacidad de 2013 se concibe como un concepto jurídico indeterminado que tiene su núcleo de certeza en la inexistencia de apoyos necesarios o ayudas para superar las barreras sociales que sufren las personas con deficiencias (físicas, mentales, intelectuales o sensoriales) previsiblemente permanentes. Se refiere al Real Decreto Legislativo 1/2013, de 29 de noviembre, que aprobó el Texto Refundido de la Ley General de derechos de las personas con discapacidad y de su inclusión social.

la vivienda, en función de su situación social y económica, así como de sus circunstancias personales y familiares. Esta inscripción se solicitará ante los registros municipales que existen para tal fin de forma presencial o telemática y se regula por lo dispuesto por la correspondiente normativa autonómica. Por ejemplo, en el caso de Andalucía rige el Reglamento Regulador de los Registros Públicos Municipales de Demandantes de Vivienda Protegida, aprobado por Decreto 1/2012, de 10 de enero (TOL2.392.233, BOJA n.º 19 de 30 de enero de 2012), modificado por Decreto 161/2018, de 28 de agosto (TOL6.753.858, BOJA n.º 172 de 5 de septiembre de 2018). Por tanto, en este punto la norma se limita a reiterar un derecho ya reconocido legalmente y de desarrollo autonómico.

3. Artículo 8 letra d): derecho a participar en los programas públicos de vivienda y acceder a las prestaciones, ayudas y recursos públicos en materia de vivienda

Por último, en la letra d) se refiere el derecho a participar en los programas públicos de vivienda y acceder a las prestaciones, ayudas y recursos públicos en materia de vivienda en los términos y condiciones establecidos en su normativa reguladora, que vuelve a ser un derecho de naturaleza jurídico-pública. Aquí el precepto completa el derecho a la información y el derecho al registro con el derecho a participar en los programas públicos de viviendas y de otras prestaciones y ayudas correspondientes. Se trata, por tanto, de una reiteración, que al igual que en las letras b) y c) poco o nada aporta a lo ya regulado. Respecto a la normativa aplicable nos remitimos a la Ley 19/2013 de Transparencia (TOL4.029.4199), las distintas regulaciones autonómicas en materia de Registros Públicos Municipales de Demandantes de Vivienda Protegida y regulación complementaria.

III. DEBERES DEL CIUDADANO

El art. 9 de la Ley regula los deberes del ciudadano en relación con la vivienda. Al igual que el art. 8 esta norma no discrimina entre españoles y extranjeros pues va dirigida a "todos los ciudadanos".

1. Artículo 9 letra a): deber de respetar y contribuir a preservar el parque de vivienda, evitando la realización de cualquier actividad molesta o insalubre

En primer lugar, en la letra a) se refiere el deber de respetar y contribuir a preservar el parque de vivienda, evitando la realización de cualquier actividad molesta o insalubre, que sea perturbadora del ejercicio del derecho de uso y disfrute señalado en el artículo anterior. En este punto la norma nos recuerda los arts. 5-6 de la Ley del Suelo (TOL257.543)[9]. De hecho, el primero de ellos (art. 5) recoge que la vivienda digna que constituya el domicilio de un ciudadano deberá estar "libre de ruido u otras inmisiones contaminantes de cualquier tipo que superen los límites máximos admitidos por la legislación aplicable"; y el segundo (art. 6), obliga a respetar y contribuir a preservar el paisaje urbano y de las infraestructuras y los servicios urbanos", añadiendo el deber de eliminar o reducir los efectos negativos causados por actividades insalubres, nocivas y peligrosas.

Estas conductas perjudiciales se encuadran dentro de la regulación de las relaciones de vecindad, que constituyen un límite al derecho de propiedad. Por tanto, nos encontramos ante un deber de naturaleza jurídico-privada derivada del derecho de

---

9 R.D. Legislativo 7/2015, de 30 de octubre, por el que se aprueba el texto refundido de la Ley del Suelo y Rehabilitación urbana.

propiedad como derecho subjetivo. Dicha regulación se cimienta sobre el principio de la buena fe y la prohibición del abuso de derecho consagrados en el art. 7 CC (TOL220.310), estando especialmente reflejada en la LPH (TOL230.715) y la LAU (TOL231.076). En especial el art. 7.2 LPH establece que ni el propietario, ni el inquilino de una vivienda urbana o local podrán realizar actividades en dicha vivienda que contravengan las disposiciones generales sobre actividades molestas, insalubres, nocivas, peligrosas o ilícitas, previendo que en su caso se podrá ejercitar la acción de cesación contra el infractor. Por su parte, el art. 27.e LAU señala que el arrendador podrá resolver de pleno derecho el contrato cuando en la vivienda tengan lugar actividades molestas, insalubres, nocivas, peligrosas o ilícitas. Por otro lado, en el derecho catalán se ha regulado la acción negatoria de inmisiones ilegítimas, que da lugar al cese de la actividad molesta y a la reposición de la finca a su estado anterior a la perturbación[10]. Por último, nos referimos a la jurisprudencia del TEDH que interpreta el art. 8 CEDH, que regula el derecho a la intimidad personal y familiar, que garantiza el derecho de toda persona a disfrutar con dignidad de su vivienda, protegido de inmisiones perturbadoras[11]. En este apartado la LDV se limita a recoger un enunciado general sin descender a concretar su contenido. En especial se echa de menos en la ley toda referencia a la legitimación para poder exigir el cumplimiento de este deber, así como un mínimo desarrollo de su contenido. Derecho que, en todo caso, de ser ejercitado por el propietario podría sustanciarse por medio de la llamada acción negatoria.

---

10 Art. 546-13 CCCat (TOL230.654), refiere un listado abierto de inmisiones en las que incluye las de humo, ruido, gases, vapores, olor, calor, temblor, ondas electromagnéticas, luz y demás similares. La perturbación ha de reunir las características de mediatividad, materialidad y continuidad.

11 STEDH 16798790, Caso López Ostra contra España, de 9 de diciembre de 1994, (TOL178.976).

2. Artículo 9 letra b): deber de realizar las actuaciones de conservación, reparación o mejora que correspondan

En segundo lugar, la norma recoge el deber de realizar las actuaciones de conservación, reparación o mejora que correspondan de acuerdo con el régimen legal de tenencia en virtud del cual se dispone de la misma. Esta norma puede conectarse con los arts. 389 y 1907 del Código Civil (TOL220.310) que obligan a reparar e indemnizar por los daños que el edificio pueda causar a terceros. Sin embargo, estos últimos se refieren a un caso extremo en el que se establece un deber de mantenimiento y reparación del edificio para evitar su ruina. Caso distinto sería el de aquella obra de mantenimiento "menor" en supuestos en los que no exista riesgo de ruina del inmueble, supuestos a los que no serían de aplicación las normas referidas, y respecto a los cuales, salvo que medie una regulación específica aplicable al supuesto que establezca otra cosa, en principio podría entenderse que cada propietario es libre de mantener la vivienda en la situación que desee y es a él a quien corresponde decidir si opta por realizar obras de mantenimiento o no, obras que en ocasiones pueden resultarle excesivamente onerosas. No obstante, debe señalarse que desde el Gobierno se ha apostado por un cambio de modelo en el sector de la construcción que apuesta por un urbanismo sostenible, basado en la rehabilitación, por lo que se busca en el fomento de la rehabilitación de inmuebles una salida sostenible al sector de la construcción. Esto implica que poco a poco el ámbito de la autonomía de voluntad del propietario va quedando limitado por una legislación garantista que impone deberes de mantenimiento/rehabilitación de la vivienda cada vez más exigentes[12].

---

12 En esta línea podemos citar el art. 16 de la Ley de Ordenación de la Edificación que señala que dispone la obligación del propietario de conservar en buen estado la edificación mediante un adecuado uso

3. Artículo 9 letra c): deber de respetar la pacífica tenencia de la vivienda ajena

En tercer lugar, se recoge una norma similar a la dispuesta en la letra a) pero en esta ocasión aplicable a la vivienda ajena, a disposición de otras personas, hogares o entidades públicas y privadas, estableciendo el deber de respetar la pacífica tenencia de la misma y abstenerse de la realización de cualquier tipo de actividad que la impida o dificulte. Aquí vuelve a echarse de menos toda referencia a quien pueda quedar legitimado para exigir el cumplimiento de este deber. Si bien el propietario/poseedor puede ejercitar las acciones correspondientes para la defensa de su posesión y propiedad, tal como se ha indicado anteriormente, cabe plantearse si la ley prevé también que pueda ser exigido por la propia Administración que, en todo

---

y mantenimiento. Normativa que prevé la obligación de entrega a los adquirentes de viviendas la documentación que integra el Libro del Edificio, entre la cual se han de incluir las instrucciones de uso y mantenimiento del edificio.

Igualmente, la normativa autonómica se hace eco de este deber. A título de ejemplo, la Ley 1/2010, de 8 de marzo, reguladora del Derecho a la vivienda en Andalucía (TOL1.791.221), en cuyo Título III se regula la conservación, mantenimiento y rehabilitación de las viviendas, estableciendo la obligación de los propietarios de velar por el mantenimiento a su costa de las viviendas en condiciones de calidad, dignas y adecuadas. Las medidas de fomento e intervención administrativa recogidas en la ley afectan muy especialmente a la rehabilitación de barrios y centros históricos. En la misma línea, entre otras, la Ley 8/2004 de 20 de octubre, de la Vivienda de la Comunidad Valenciana (TOL500.181), o la Ley 2/1999, de 17 de marzo de 1999, de medidas para la calidad de la edificación de la Comunidad Autónoma de Madrid (TOL75.967), que obligan a propietarios y usuarios a mantener las viviendas en buen estado de conservación, uso, mantenimiento y seguridad. Igualmente nos podemos referir a las obligaciones del propietario recogidas en el art. 9 LPH.

caso, quedará legitimada cuando se trate de vivienda ajena "a disposición de entidad pública".

Por otra parte, respecto a la defensa de la pacífica posesión de la vivienda nos remitimos a la regulación de los juicios sumarios que han venido a reemplazar los anteriores interdictos de recobrar y retener la posesión[13] y muy especialmente al derecho de todo poseedor[14] a ser respetado en su posesión ex arts. 441 y 446 CC y concordantes;[15] y con respecto a la defensa por parte del propietario a la acción negatoria.

4. Artículo 9 letra d): deber de cumplimiento de los deberes legalmente establecidos para el transmitente o intermediario

Por su parte, en la letra d) se señalan unos deberes de muy distinta naturaleza, lo cual evidencia la falta de sistemática que caracteriza a este artículo. Se trata de deberes en relación con las operaciones de compra o alquiler de vivienda, que exigen el cumplimiento de los deberes legalmente establecidos para el transmitente o intermediario definidos en el título IV de la Ley 12/2023, el cual tiene por título "sobre medidas de protección y transparencia en las operaciones de compra y arrendamiento de vivienda", y demás normativa aplicable en dichas

---

13 Lo cierto es que, aunque la LEC (TOL172.336) habla de acción posesoria de retener y recobrar, la jurisprudencia en ocasiones sigue denominándola como "interdicto", vid. STS 467/2016, 7 de julio de 2016.

14 El art. 230 LEC (TOL172.336) se refiere a la posesión o tenencia.

15 En la LEC 1/2000 (TOL172.336) se han suprimido los interdictos como procedimiento especial, pues han sido subsumidos en el ámbito del juicio verbal, que queda regulado en el Título III del Libro II, bajo la rúbrica "de los procesos declarativos ordinarios", en los cuales la sentencia sigue sin producir efectos de cosa juzgada.

operaciones. En el art. 30 de la Ley se refieren distintos derechos de la persona demandante, adquirente o arrendataria de vivienda, por lo que lo que en este artículo se configura como un derecho, en el art. 9 se recoge como un deber. Dentro de estos derechos/deberes se incluyen en el art. 31 los siguientes: aportar la información mínima en las operaciones de compra y arrendamiento de vivienda recogida, que incluyen información sobre la identidad del vendedor o arrendador, condiciones económicas de la venta o alquiler, características esenciales de la vivienda y edificio, la información jurídica del inmueble y en su caso el carácter protegido de la vivienda o del edificio con protección arquitectónica. Hay que señalar que el art. 30.1.a de la Ley atribuye a la persona demandante, adquirente o arrendataria de vivienda los derechos reconocidos en el texto refundido de la Ley General para la Defensa de los Consumidores y Usuarios aprobado por el Real Decreto Legislativo 1/2007, modificado por Ley 3/2014 de 27 de marzo (TOL4.147.248). Por tanto, a *sensu contrario*, deberá cumplir con los deberes que se derivan de la aplicación de esta normativa. Por lo demás, hay que prestar atención a lo dispuesto por la normativa estatal y autonómica aplicable a la compraventa y arrendamiento de vivienda y muy especialmente a la normativa autonómica que regula las viviendas de protección oficial, que a la fecha ha sido normada por las Comunidades de Andalucía, Aragón, Asturias, Canarias, Cantabria, Castilla-La Mancha, Castilla y León, Cataluña, Extremadura, Galicia, Madrid, Navarra y Valencia[16].

---

16 La regulación estatal de Viviendas de Protección Oficial se encuentra en el RDL 31/1978 (TOL955.556) y el RD 3148/1978 (TOL138.532), que rigen de forma supletoria en aquellas Comunidades Autónomas que no hayan asumido la competencia en materia de vivienda.

5. Artículo 9 letra e): deber de atender a la importancia del parque público de vivienda como instrumento de acción en favor del derecho a la vivienda y velar por su adecuado mantenimiento

Finalmente, en la letra e) en relación con el parque público de vivienda, media el deber de atender a su especial importancia como instrumento de acción en favor del derecho a la vivienda y velar por su adecuado mantenimiento y conservación, para que pueda servir a los hogares con mayores dificultades. Lo cual parece ser más una simple declaración de buenas intenciones que un deber jurídico. Podría considerarse en todo caso como un deber general dirigido a todo ciudadano, si bien podría pensarse que el primer obligado al cumplimiento de este deber debería ser la propia Administración. En todo caso, este propósito contrasta con el hecho de que España está a la cola de Europa con respecto a los porcentajes de parque público de vivienda respecto al total del parque inmobiliario, pues si bien en determinados países europeos se supera el 30%, en España la vivienda social solo alcanza al 2,5% del stock nacional de viviendas ocupadas[17]. Esto muestra un desajuste estructural que no puede solventarse a corto plazo y que en parte es causado por el hecho de que hasta hace poco no se consideraba que fuera necesario el fomento de la vivienda pública. Ha sido en la última década y a partir de la crisis de 2008 cuando el problema del acceso a la vivienda se ha convertido en un reto difícil de enfrentar por amplios sectores de la sociedad.

---

17 Porcentaje nimio si lo comparamos con el de otros países de la UE: el 30% en los Países Bajos, el 24% en Austria, el 17,6% en el Reino Unido y 16,8% en Francia. Housing Europe, "State of Housing within the EU, 2019", octubre de 2019, http://www.housingeurope.eu/resource-1323/the-state-of-housing-in-the-eu-2019, 85-87.

Una vez analizadas las distintas letras de los artículos 8 y 9 de la Ley podemos concluir que su función es la de recoger una serie de derechos y deberes ya legislados en otras normativas pero que el legislador ha estimado oportuno recoger conjuntamente en estas dos normas. Por tanto, estas disposiciones no aportan novedades importantes.

Sin embargo, hemos dejado pendiente el estudio del derecho contenido en la letra a) del art. 8, esto es el "derecho a la vivienda digna y adecuada", cuyo análisis procede ahora.

IV. EL DERECHO A DISFRUTAR DE UNA VIVIENDA DIGNA Y ADECUADA

El art. 8 letra a) señala que todos los ciudadanos tienen derecho a disfrutar de una vivienda digna y adecuada, en los términos dispuestos por esta ley, ya sea en régimen de propiedad, de arrendamiento, de cesión de uso, o de cualquier otro régimen legal de tenencia. Es este el derecho básico reconocido por la Ley 12/2023 y al que responde el título de esta.

El art. 8 de la LDV refiere el término "disfrutar" recogiendo la terminología del art. 47 CE. No obstante, no parece que el legislador tenga clara la distinción que existe a efectos civiles entre el uso y el disfrute de una cosa. Lo cierto es que el concepto jurídico de uso y el concepto jurídico de disfrute no tienen un mismo alcance. Si bien la facultad de disfrutar conlleva la de usar, no ocurre a la inversa. El *Ius fruendi* o derecho de goce de la cosa conlleva la facultad de hacerse con los frutos que ésta produce, pudiendo ser arrendada o subarrendada, facultades de las que se carece cuando se tiene un mero "*ius utendi.* Por tanto, en el art. 8 LDV se recoge una acepción del término "disfrutar" que parece asimilarse a la de "uso".

Tal y como hemos dicho, la primera cuestión que se plantea es si esta norma de desarrollo del art. 47 CE lo configura como derecho subjetivo exigible ante los tribunales conforme al art. 53 CE. Para dar respuesta a esta pregunta vamos a analizar en primer lugar cuál es la función que desde la doctrina se ha reconocido al art. 47 CE como principio rector y su desarrollo en la normativa estatal y autonómica. En segundo lugar, si del análisis conjunto e integrado del art. 8 con el resto de la Ley 12/2023 puede deducirse que el derecho a una vivienda digna y adecuada queda configurado como un derecho subjetivo o, incluso, como un derecho humano. Finalmente, conectamos el art. 8 con la realidad social a la que va dirigido a fin de concretar si la Ley 12/2023 ha sido suficiente o por el contrario no ha cumplido las expectativas que podrían haberse puesto en ella.

1. Análisis sobre el derecho a la vivienda contenido en el art. 47 CE y en la normativa estatal y autonómica, interpretación, función y ámbito de aplicación

Si bien el derecho a la vivienda ha sido reconocido en distintos tratados internacionales y constituciones como derecho humano o fundamental[18], la Constitución Española de 1978

---

[18] El art. 11 del Pacto Internacional de Derechos Económicos, Sociales y Culturales de 1966, conforme al art. 25 de la Declaración Universal de los Derechos Humanos de 1948 (TOL147.461), reconoce el derecho de la persona y la familia a un nivel de vida adecuado que incluye una vivienda digna. El cumplimiento de este pacto es vigilado por el PIDESC o sistema de garantías establecido por el Consejo Económico y social (ECOSOC), existiendo un Comité de Derechos Económicos, Sociales y Culturales. Por otro lado, la Carta Social Europea de 18 de octubre de 1961, revisada en 1996 (TOL207.990), recoge el compromiso de las partes a adoptar "medidas destinadas a favorecer el acceso a la vivienda de una calidad suficiente; a prevenir y paliar la situación de carencia de hogar con vistas a eliminar progresivamente dicha situación y a hacer asequible el precio de las

(CE) fue muy cauta al regularlo. En el art. 47 CE se afirma que "todos los españoles tienen derecho a disfrutar de una vivienda digna y adecuada", añadiendo que "Los poderes públicos promoverán las condiciones necesarias y establecerán las normas pertinentes para hacer efectivo este derecho (derecho a la vivienda), regulando la utilización del suelo de acuerdo con el interés general para impedir la especulación. La comunidad participará en las plusvalías que genere la acción urbanística de los entes públicos". No obstante, este precepto quedó recogido en el Capítulo III del Título I dedicado a los "Principios Rectores de la Política Económica y Social". Esto se ha interpretado en el sentido de que conforme al art. 53 CE[19], la

---

viviendas a las personas que no dispongan de recursos suficientes". Igualmente, el art. 34.3 de la Carta de los Derechos Fundamentales de la Unión Europea de 7 de diciembre de 2000 (TOL131.225) establece que, a fin de combatir la exclusión social y la pobreza, la Unión reconoce y respeta el derecho a una ayuda social y a una ayuda de vivienda para garantizar una existencia digna a todos aquellos que no dispongan de recursos suficientes.

Por otro lado, el TEDH ha interpretado el art. 8 de la CEDH (TOL164.153) en un sentido amplio de forma que dentro del derecho a la vida privada y familiar se incluye la protección de la vivienda como un bien jurídico que ha de ser adecuado y digno. Véase López Ostra v. España (TOL178.976) o Hatton y otros v. Reino Unido. También ha de destacarse la Carta de Niza (TOL230.652) que, en el ámbito de la UE, reconoce el derecho a una ayuda social y a una ayuda de vivienda que garanticen una existencia digna. Finalmente, la resolución del Parlamento Europeo de 21 de enero de 2021, sobre el acceso a una vivienda libre y asequible para todos considera que "el acceso a una vivienda adecuada constituye un derecho fundamental y debe considerarse una condición previa para ejercer y obtener el acceso a los demás derechos fundamentales y a una vida digna".

19 Art. 53 CE: Los principios rectores "sólo podrán ser alegados ante la jurisdicción ordinaria de acuerdo con lo que dispongan las Leyes que los desarrollen".

efectividad del derecho a la vivienda contenido en el art. 47 CE queda condicionada a su desarrollo mediante ley, no siendo posible, por tanto, su protección judicial por la vía de amparo sin dicho desarrollo legislativo.

Ahora bien: ¿cuál es la interpretación, función y ámbito de aplicación que ha de darse al art. 47 CE? Ya en 1979 Peces-Barba Martínez, padre de la Constitución, señaló que los principios rectores, entre ellos el que aquí nos ocupa, no dejaban de ser meros "derechos aparentes" que en ningún caso podían generar verdaderos derechos subjetivos o, en su caso, fundamentales[20]. Procede pues aclarar si además de contener un "mero derecho aparente" cabe interpretar que este artículo, aún sin desarrollo legislativo, pueda cumplir una función útil, que lo dote de cierta eficacia jurídica y, en ese caso, el alcance que ha de darse a la misma.

En este punto, desde un primer momento, una parte de la doctrina se ha decantado por considerar que los "Principios Rectores de la Política Económica y Social" no son sino meros enunciados constitucionales sin un mínimo ropaje que incluya un mandato y articule la correspondiente sanción, por lo que carecen de una mínima estructura lógica que permita considerarlos como "normas". En esta línea, para Garrido Falla estos preceptos no articulan verdaderas normas, por lo que carecen de consecuencia jurídica. De ahí podría deducirse que se trataría de una parte prescindible de la Constitución y que el art. 47 CE contendría un simple enunciado[21].

---

[20] En esta línea se han manifestado: Peces-Barba Martínez (1979, p. 41); Garrido Falla (1980, p. 581); Serrano Moreno (1987, p. 103); Prieto Sanchís (1990); Carmona Cuenca (1992, p. 106); García Macho/ Soler Vilar (1994, p. 16); López González (1997, p. 164).

[21] Garrido Falla (1980, pp. 579-580) y Cossío Díaz (1989, pp. 252-253), quien reconducía al absurdo el planteamiento de Garrido Falla,

Sin embargo, otra línea doctrinal considera que un análisis conjunto de la Constitución otorga a los principios rectores una importante función como orientadores de la acción del Estado y los fines determinados de su misión, posición defendida entre otros por García de Enterría[22]. Por tanto, y conforme a este criterio, los principios rectores son una herramienta para orientar la realización de los valores superiores recogidos por el art. 1 CE y dar forma al Estado social[23]. Igualmente, se ha afirmado el carácter obligatorio del mandato dirigido al legislador por la CE para que complete el texto constitucional con las adecuadas normas de desarrollo a fin de que los derechos que encarnan los principios rectores sean operativos[24]. Por tanto, no es aceptable afirmar que los artículos del Capítulo III del Título I de la Constitución sean prescindibles o meros enunciados. No obstante, esta importante función del art. 47 CE, encarnada en el mandato dirigido al legislador, queda desdibujada por el hecho de no preverse plazo alguno para su efectividad, por lo que, una vez transcurridas más de cuatro décadas desde la promulgación de la Constitución parece que el "mandato" constitucional ha quedado diluido en una mera opción que el legislador no ha considerado oportuno desarrollar. En todo caso, este hipotético "mandato" queda condicionado por un "desarrollo normativo" que a nivel estatal y hasta la fecha de entrada en vigor de la Ley 12/2023 no terminaba de verse con claridad. Por ello, la doctrina mayoritaria afirma que la única función que podría predicarse que ha cumplido

---

pues de él se derivaría la consideración de la Constitución como un simple ejercicio de retórica carente de sustancia normativa.

22 García de Enterría (2001, p. 68).

23 En esta línea véase Pérez Royo (2007, p. 202), cuando destaca el compromiso del Estado social con los sectores más vulnerables de la sociedad.

24 Fernández Rodríguez (1997, p. 464).

hasta la fecha el art. 47 CE es la función informadora, si bien hay que añadir que a nivel autonómico sí se ha regulado el derecho subjetivo a la vivienda.

Por otro lado, para Gálvez Montes[25] el texto constitucional se limita a compeler a los poderes públicos a promover las condiciones necesarias para que todo español pueda disfrutar de una vivienda digna y adecuada. Es decir, el objetivo de las actividades públicas es conseguir la efectividad de ese disfrute, pero su contenido no consiste en materializar el uso o disfrute y, por tanto, no se materializa en prestaciones concretas o específicas. Desde otra perspectiva se ha querido dotar al art. 47 CE de un cierto contenido pues se afirma que el reconocimiento constitucional que se hace en él del derecho a una vivienda digna debería tener ciertas consecuencias. Este contenido se encontraría en el reconocimiento constitucional articulado mediante un conjunto de normas que vendrían a constituir un "haz de derechos ordinarios, en el sentido de los derechos legales en los que se traduce el derecho a la vivienda"[26]. Este planteamiento en realidad solo tendría consistencia si existiera una norma de desarrollo estatal que dotase al derecho a la vivienda digna enunciado en la Constitución de una verdadera estructura de norma jurídica, norma que, repetimos, a nivel estatal *prima facie* parece difícil de establecer.

Algún autor llega todavía más lejos cuando proclama la vigencia de unos llamados "derechos sociales de provisión" que, en síntesis, permitirían a las personas exigir del Estado aquello que podrían adquirir de un particular si existiera una oferta razonable y suficiente en el mercado y se dispusiera de una solvencia capaz

---

25 Gálvez Montes (2001).

26 Espínola Orrego (2019, p. 15).

para realizar la adquisición[27]. Teoría que a primera vista parece más bien recoger un enunciado de corte iusnaturalista que una consecuencia del derecho positivo aplicable.

Otra línea doctrinal ha intentado encontrar en la legislación existente en materia de vivienda la norma de desarrollo esperada que permitiría hacer efectivo el derecho contenido en el art. 47 CE[28]. De hecho, muy lentamente, se empiezan a apreciar señales que muestran que éste es el camino, en especial en el ámbito autonómico.

También se ha considerado que estos principios programáticos servirían como canon de constitucionalidad de las leyes[29]. En todo caso la pasividad del legislador dificulta sobremanera la exigencia de un derecho formulado en abstracto cuyo contenido a la postre quedaría a voluntad del legislador. La jurisprudencia contencioso-administrativa no aporta mucho más que recordarnos el valor normativo del art. 47 CE.

Cierta doctrina ha querido ver una vía alternativa para dar amparo al derecho a la vivienda en el art. 10.2 CE, en tanto que establece que las normas relativas a los derechos fundamentales y a las libertades que la Constitución reconoce se interpretarán de conformidad con la Declaración de Derechos Humanos y los tratados internacionales sobre las mismas materias ratificados por España. Según estos autores esta norma sería aplicable a todo el Título I de la Constitución, por lo que

---

27 Alexi (1993, pp. 189-195).

28 García Macho (1982, pp. 165-207) y Beltrán Felipe (2000, pp. 165-166).

29 Díez-Picazo (2008, p. 68) afirma que si bien estos principios constitucionalizados no son directamente aplicables ello no significa que solo tengan un carácter meramente programático, pues el texto constitucional goza en su integridad de fuerza normativa, por lo que contendrían directrices que pueden operar como canon de constitucionalidad de las leyes.

debería considerarse los tratados internacionales en materia de derechos y libertades como parte integrante del bloque de constitucionalidad y los derechos fundamentales deberían ser colmados a través de los tratados internacionales. Esto permitiría justificar el reconocimiento del derecho a la vivienda como derecho subjetivo de su proclamación en los textos internacionales de derechos humanos. Se trata de una posición minoritaria y de muy difícil encaje[30].

Sin embargo, poco a poco se ha ido abriendo paso la idea de que el derecho a la vivienda debe ser considerado un derecho subjetivo entre cierto sector de la doctrina, comenzando por Muñoz Castillo[31] y Ponce Solé[32], quienes afirman la exigibilidad del derecho proclamado en el art. 47 CE. De esta forma, para Pisarello se trata de un auténtico derecho subjetivo materializado en la legislación de vivienda desarrollada[33].

Una posición avanzada que merece reflexión es la planteada por López Ramón cuando afirma que el artículo 53.3 CE debe ser interpretado distinguiendo entre principios y derechos. Conforme a ello esta norma solo descartaría el reconocimiento de los "principios" establecidos en el capítulo III del Título I de la Constitución como derechos subjetivos, pero sin que ello afectara a los derechos recogidos en los arts. 43, 44, 45 y 47 de dicho capítulo[34]. Ello derivaría del hecho de que el art. 53.3 CE solo se refiere a los "principios reconocidos en el capítulo tercero" sin mencionar los "derechos" reconocidos en dicho capítulo. Estos últimos quedarían protegidos por el

---

30 Recoge esta línea doctrinal Martínez Quiñones (2024, pp. 528-532).

31 Muñoz Castillo (2000, pp. 115 y 118).

32 Ponce Solé (2008, pp. 44-80).

33 Pisarello (2009, pp. 51-52).

34 López Ramón (2014, p. 58).

derecho fundamental a la tutela judicial efectiva de derechos e intereses legítimos recogidos por el art. 24 CE. Concluye López Ramón que la fuerza de la conciencia social sobre la necesidad de comprometerse con ciertos valores del texto constitucional, potenciada por la Declaración Universal de Derechos Humanos y el Derecho comparado justifican interpretar el término "derecho" del art. 47 CE como "derecho subjetivo". Un derecho subjetivo que sería constitucional, pero no fundamental, cuyo efecto sería proporcionar acción judicial a su titular.

Quizá la norma estatal de mayor trascendencia en la que se ha recogido y normado el derecho a una vivienda digna haya sido el Texto Refundido de la Ley de Suelo y Rehabilitación urbana (TOL5.534.447)[35], que dedica un capítulo a los derechos y obligaciones del ciudadano, incluyendo en el art. 5 como primer derecho, el derecho a una vivienda digna[36]. De hecho, en el art. 1b de la Ley se señala como su misión la de establecer las condiciones básicas para "...asegurar a los ciudadanos una adecuada calidad de vida y la efectividad de su derecho a disfrutar de una vivienda digna y adecuada"[37]. Si atendemos al art. 3 de la Ley cuando señala que los poderes públicos promoverán las condiciones para que los derechos y deberes de los ciudadanos establecidos en esta Ley sean reales y efectivos, podría pensarse

---

35 R.D. Legislativo 7/2015, de 30 de octubre, por el que se aprueba el texto refundido de la Ley del Suelo y Rehabilitación urbana.

36 El art. 5 del R.D. Legislativo 7/2015 regula los derechos del ciudadano, señalando que todos los ciudadanos tienen derecho a: "(a) Disfrutar de una vivienda digna, adecuada y accesible, concebida con arreglo al principio de diseño para todas las personas, que constituya su domicilio libre de ruido u otras inmisiones contaminantes de cualquier tipo que superen los límites máximos admitidos por la legislación aplicable y en un medio ambiente y un paisaje adecuados.

37 El texto refundido recoge el concepto "infravivienda" para cuando la vivienda no reúne las condiciones mínimas previstas por la Ley.

que estamos ante la esperada norma de desarrollo del art. 47 CE[38]. Sin embargo, se ha afirmado que la Ley del Suelo tampoco recoge un derecho subjetivo al acceso a la vivienda, pues

---

38 Por otro lado, la Ley del suelo de 2015 no ha servido para superar la carencia de vivienda estructural en España. Por un lado, al desdoblar los supuestos en los que ha de reservarse suelo para vivienda sujeta a régimen de protección pública se ha reducido el porcentaje del 30% al 20% para cuando se trate de actuaciones de reforma o renovación de urbanización. Y, por otro lado, prevé que el límite se reduzca optándose por una reserva inferior o que sea eximido en actuaciones de nueva urbanización en determinadas circunstancias. Con todo, lo cierto es que los resultados obtenidos son desalentadores. Art. 20.b: "Para hacer efectivos los principios y los derechos y deberes enunciados en el título preliminar y en el título I, respectivamente, las Administraciones Públicas y, en particular, las competentes en materia de ordenación territorial y urbanística, deberán: destinar suelo adecuado y suficiente para usos productivos y para uso residencial, con reserva en todo caso de una parte proporcionada a vivienda sujeta a un régimen de protección pública que, al menos, permita establecer su precio máximo en venta, alquiler u otras formas de acceso a la vivienda, como el derecho de superficie o la concesión administrativa. Esta reserva será determinada por la legislación sobre ordenación territorial y urbanística o, de conformidad con ella, por los instrumentos de ordenación, garantizará una distribución de su localización respetuosa con el principio de cohesión social y comprenderá, como mínimo, los terrenos necesarios para realizar el 40% de la edificabilidad residencial prevista por la ordenación urbanística en el suelo rural que vaya a ser incluido en actuaciones de nueva urbanización y el 20% en el suelo urbanizado que deba someterse a actuaciones de reforma o renovación de la urbanización. No obstante, dicha legislación podrá también fijar o permitir excepcionalmente una reserva inferior o eximirlas para determinados municipios o actuaciones, siempre que, cuando se trate de actuaciones de nueva urbanización, se garantice en el instrumento de ordenación el cumplimiento íntegro de la reserva dentro de su ámbito territorial de aplicación y una distribución de su localización respetuosa con el principio de cohesión social".

ni determina quién es el titular del derecho, ni quién el obligado a garantizarlo, ni cuál será el contenido de la prestación debida, ni las acciones necesarias para obtener el resultado[39]. Tornos Mas recuerda que un derecho subjetivo supone el reconocimiento en una norma de una acción para exigir de un tercero un comportamiento o una abstención, de forma que se puede obtener amparo judicial para imponer al tercero la acción u omisión que se reclama, lo cual no se reconoce en la Ley del Suelo. Igualmente, para Enériz Olaechea la Ley del Suelo[40] no aporta nada nuevo al derecho a la vivienda

Sin embargo, de contrario se ha postulado que tampoco puede encontrarse en la Ley del Suelo una limitación que impida considerar el derecho a una vivienda digna recogido en el art. 5 como un derecho subjetivo pues no lo condiciona a un posterior desarrollo normativo. Por tanto, conforme a esta doctrina ex art. 24.1 CE, cabría exigir la protección judicial de un interés legítimo a una vivienda digna y adecuada en la vía ordinaria y ello porque de la indeterminación de una norma no puede colegirse su ineficacia[41].

A nivel autonómico (estatutario, según cierta doctrina) encontramos que el derecho a la vivienda se ha recogido en los distintos estatutos de autonomía de una forma general o abstracta, pero sin vincular su exigibilidad a un posterior desarrollo normativo. En todo caso la STC 31/2010, de 28 de junio (TOL1.880.189)[42] distinguió entre derechos subjetivos *stricto sensu* y mandatos de actuación a los poderes públicos, ya es-

---

39 Tornos Mas (2014, p. 2855).

40 Enériz Olaechea (2008, p. 166), cuando comenta el art. 4.ª del Texto Refundido de la ley del Suelo de 2008 que ya recogía un derecho a una vivienda digna similar al de la vigente ley.

41 López Ramón (2014, p. 66).

42 STC 31/2010, de 28 de junio, BOE-A-2010-11409.

tén expresamente denominados como "principios rectores, ya enunciados literalmente como derechos" lo cual ha llevado a considerar que la referencia estatutaria al "derecho a la vivienda" no deja de ser un mero mandato de actuación. Esta jurisprudencia no ha estado exenta de crítica por quienes consideran que sí debería ser considerado como un derecho subjetivo, especialmente en casos como el del art. 38.2 del Estatuto de Cataluña (TOL962.780) cuando prevé la tutela judicial de los derechos en él proclamados, entre ellos el recogido en el art. 26 sobre el derecho a acceder a una vivienda digna por quienes no dispongan de recursos suficientes. Por ello se afirma que en dicho Estatuto sí queda sancionado un derecho subjetivo a la vivienda[43]. Es precisamente en el desarrollo autonómico de los distintos estatutos de autonomía donde podemos encontrar leyes que de una forma clara y explícita configuran un auténtico derecho subjetivo a la vivienda, estableciendo tanto sus titulares, como las condiciones de ejercicio, como los responsables de hacerlo efectivo. Se trata de una normativa que analizaremos más adelante[44].

Podemos concluir que a nivel estatal y hasta la aprobación de la Ley 12/2023 poco o nada se ha hecho por desarrollar el art. 47 CE a fin de dotarle de un contenido positivo que dé tutela al ciudadano para el ejercicio de dicho derecho.

---

43 López Ramón (2014, p. 64).

44 El legislador autonómico ha articulado diversas políticas conforme a dicho mandato, por ejemplo, para el fomento de la vivienda de protección oficial, iniciativa pública de promoción y gestión de vivienda, considerando a la vivienda como un servicio económico de interés general, conforme a lo dispuesto por la sentencia del TJUE de 8 de mayo de 2013, asunto C-197/11 (TOL3.659.986). De hecho, en algún supuesto ha llegado más lejos, tal como veremos más adelante.

2. Artículo 8 a) de la Ley 12/2023: ¿norma de desarrollo del art. 47 CE ex art. 53 CE?

Podría pensarse que la Ley 12/2023 tiene como propósito dar respuesta a la Resolución del Parlamento Europeo de 21 de enero de 2021, sobre el acceso a una vivienda libre y asequible para todos[45], en la que se pidió a la Comisión y a los Estados miembros que el derecho a una vivienda digna y adecuada fuera reconocido y ejecutado como un derecho humano fundamental mediante disposiciones legislativas europeas y nacionales aplicables.

Analizar si efectivamente el art. 8 a) ha cumplido o no con estas expectativas es vital para determinar si la Ley 12/2023 es suficiente de cara a configurar el derecho a una vivienda digna y adecuada como un derecho humano.

A tal fin nos proponemos ahora analizarlo para precisar su contenido y alcance. Son cuatro las cuestiones a dilucidar: ¿qué ha de entenderse por una vivienda digna y adecuada? ¿Contempla la Ley cualquier régimen legal de tenencia? ¿Cuáles son los "términos dispuestos por la ley" a los que hace referencia esta norma, y de los cuales puede depender que el derecho a la vivienda quede incardinado como un derecho subjetivo?¿Puede el derecho a la vivienda como derecho humano dimanar del art. 47 CE?

---

45 Resolución del Parlamento Europeo, de 21 de enero de 2021, sobre el acceso a una vivienda digna y asequible para todos (2019/2187(INI)).

2.1. ¿Qué ha de entenderse por una vivienda digna y adecuada?

El concepto "vivienda digna y adecuada" no solo se recoge en la Ley 12/2023, sino también en la Constitución española y en la Carta social europea. El hecho de que se trate de una vivienda digna y el salto de convertir el estado democrático de derecho en un estado social implican que no toda vivienda vale y que debe cumplir unos estándares mínimos de calidad[46].

En todo caso, la ley por el derecho a la vivienda se limita a señalar en su art. 2 que constituyen fines comunes de la acción de los poderes públicos en materia de vivienda asegurar la habitabilidad de las viviendas[47]; y en el art. 3 c que la vivienda ha de responder a las necesidades de residencia de la persona o unidad de convivencia en condiciones asequibles conforme al esfuerzo financiero, constituyendo su domicilio, morada u hogar en el que poder vivir dignamente[48]. Respecto al carácter

---

46 Si, por ejemplo, la preservación de un determinado entorno natural limitase la construcción o exigiese unos determinados estándares constructivos, habría que dirimir entre priorizar dar respuesta a una necesidad actual y acuciante o dar más peso al legado que vamos a dejar a nuevas generaciones.

47 Aporta una definición para este concepto: "entendida como el conjunto de los requisitos mínimos de calidad, funcionalidad y accesibilidad universal que, atendiendo a la normativa aplicable, deben cumplir las mismas para garantizar la dignidad y la salud de las personas, para satisfacer sus necesidades de habitación en las diferentes etapas de su vida. El art. 2 añade que se prestará especial atención a las necesidades de los menores, para quienes la vivienda constituye además un espacio fundamental de desarrollo, seguridad y cobijo, y como base para el efectivo ejercicio de derechos y libertades.

48 Añadiendo que la vivienda ha de salvaguardar la intimidad y garantizar el disfrute de las relaciones familiares o sociales, favoreciendo el pleno desarrollo y la inclusión social de las personas. En todo caso, las normativas que regulan la calidad de la vivienda,

"asequible"[49] de la vivienda el art. 3 d) señala que viene determinado por el esfuerzo financiero que suponga, que no debe superar con carácter general el 30% de los ingresos de la unidad de convivencia. Es esta una materia desarrollada en otro capítulo al cual nos remitimos.

2.2. ¿Contempla la Ley cualquier régimen legal de tenencia?

El art. 8 a) de la Ley señala que el derecho a una vivienda digna ha de serlo "en régimen de propiedad, de arrendamiento, de cesión de uso, o de cualquier otro régimen legal de tenencia". En este punto nos planteamos si "cualquier otro

---

aunque sea de forma parcial, son la Ley de Ordenación de la Edificación (TOL239.204) de 1999 y el Código Técnico de la Edificación aprobado por RD 314/2006 de 17 de marzo (TOL834.409) en tanto que buscan asegurar la seguridad, salubridad, comodidad y otras cualidades que han de reunir las nuevas construcciones. A la que hay que añadir la larga normativa que se ha desarrollado para garantizar la conservación de inmuebles, destacando la Ley 8/2013 (TOL3.785.131) que recogió el informe de evaluación de los edificios.

El concepto de infravivienda queda definido en el art. 3. c: "La edificación, o parte de ella, destinada a vivienda, que no reúne las condiciones mínimas exigidas de conformidad con la legislación aplicable. En todo caso, se entenderá que no reúnen dichas condiciones las viviendas que incumplan los requisitos de superficie, número, dimensión y características de las piezas habitables, las que presenten deficiencias graves en sus dotaciones e instalaciones básicas y las que no cumplan los requisitos mínimos de seguridad, accesibilidad universal y habitabilidad exigibles a la edificación".

49 Entre los fines que recoge el art. 2 destaca el de (ñ) garantizar la igualdad, desde la perspectiva de género, edad, capacidad y perspectiva territorial, en todas las políticas y acciones en materia de vivienda, a todos los niveles y en todas sus fases de planificación, ejecución y evaluación.

régimen de tenencia" sería siempre compatible con la calificación de la vivienda como "digna". La más avanzada normativa autonómica está apostando por dar preferencia a la tenencia de la vivienda en régimen de alquiler y solo por el tiempo necesario, es decir, en tanto que las circunstancias que dieron lugar a la adjudicación de la vivienda se mantengan. Finalmente, nos plateamos si la vivienda en régimen de ocupación sería también una vivienda digna que merece la protección de la LDV. Se trata de una cuestión que sobrepasa los propósitos de este trabajo, si bien parece que para el legislador de la LDV sí lo es.

2.3. ¿Cuáles son los "términos dispuestos por la ley" a los que hace referencia esta norma, y de los cuales puede depender que el derecho a la vivienda quede incardinado como un derecho subjetivo?

En principio, la letra a del artículo 8 parece contener un mero enunciado programático, que se limita a repetir lo establecido por el art. 47 CE y que, salvo que de "los términos dispuestos por la ley" se dedujera lo contrario, no contendría una estructura de derecho subjetivo por carecer de una consecuencia jurídica que habilitara al ciudadano para poder exigir la efectividad del derecho a la vivienda ante los tribunales. Procede, por tanto, concretar cuáles son esos "términos dispuestos en la ley 12/2023" que podrían haber convertido el derecho a la vivienda en un derecho subjetivo.

En primer lugar, debe señalarse que la realidad condiciona a la LDV, pues gran parte de las medidas adoptadas están dirigidas a que la propiedad privada contribuya a garantizar el acceso a la vivienda, lo cual podría ser consecuencia del escaso porcentaje del parque de vivienda existente en España que se destina a vivienda pública en comparación con otros países europeos, si bien puede que no sea la medida adecuada.

De hecho, la Ley incluye una serie de medidas, principalmente procesales y contenidas en los modificados artículos 439-440-441-675-685-704 LEC, para desincentivar la práctica de la tenencia de viviendas vacías, e incluye la figura del "gran tenedor" como medidas destinadas a ampliar el mercado inmobiliario disponible. Medidas que igualmente buscan limitar el precio del alquiler y que aunque buscan amparo en la función social de la propiedad han sido criticadas por entenderse que pueden afectar al contenido esencial de la propiedad constituyendo una intervención delimitadora, en cuyo caso serían auténticas medidas expropiatorias[50]. Además, se trataría de una expropiación sin compensación, en palabras de Nasarre Aznar, pues los propietarios de viviendas ven como sus facultades dominicales sobre las mismas les son sustraídas del régimen general del Código Civil y pasan a la tutela del Estado, vaciando de contenido civil la institución de la propiedad[51]. De hecho, cierta doctrina ha llegado a ver en esta normativa una categoría distinta a la de los bienes privados o públicos, calificándola de "bien común", por lo que se caracteriza por el hecho de que ha de facilitarse su acceso a quien tenga dificultades por resultar esencial para todo individuo, tal como lo sería el agua o el aire[52]. En todo caso, las medidas referidas no eximen a la Administración de su obligación de facilitar que la vivienda

---

[50] Medidas que serían constitucionales si respetaran los principios de racionalidad, proporcionalidad y adecuación, pero que López y López considera de aplicación indiscriminada por poder afectar a un modesto propietario para quien sea vital la renta del alquiler (López y López, 2022, pp.14-34).

[51] Nasarre Aznar (2022b, pp. 1-6), sobre las medidas concluye que el entonces proyecto de Ley recogía que "las que no resulten inconstitucionales resultarán contraproducentes para facilitar el acceso a la vivienda."

[52] Sainz-Cantero Caparrós (2023, pp. 630-663).

pública sea una opción real –y primaria– a la hora de paliar el alto déficit de vivienda existente en España.

En segundo lugar, la Ley 12/2023 modifica el texto refundido de la Ley de Suelo y Rehabilitación Urbana, aprobado por Real Decreto Legislativo 7/2015 (TOL5.534.477), de 30 de octubre, incrementando el porcentaje de reserva de suelo destinado a vivienda protegida del 30% al 40% de la edificabilidad residencial prevista por la ordenación urbanística en el suelo rural que vaya a ser incluido en actuaciones de nueva urbanización, y del 10% al 20% en el caso de suelo urbanizado que deba someterse a actuaciones de reforma o renovación de la urbanización. Si bien el hecho de que exista una obligación de reservar suelo para vivienda protegida no garantiza que dicha construcción llegue a hacerse efectiva. De las restantes medidas incluidas en la ley 12/2023 tampoco puede deducirse que garanticen que el derecho a la vivienda sea efectivo y reclamable ante las autoridades públicas.

Por tanto, de una primera lectura del articulado de la Ley se podría deducir que la misma ni eleva a rango constitucional el derecho del ciudadano a una vivienda digna, ni lo configura como un verdadero derecho subjetivo, y, por tanto, no garantiza que sea efectivo el acceso a la vivienda digna de las personas que se encuentran en situación de chabolismo o lo que la ley califica como "sinhogarismo" (posible traducción del término anglosajón *homeless*), –art. 2.n– y muy especialmente a la juventud. Sin embargo, se afirma que la Ley 12/2023 supone un auténtico cambio hacia un modelo que da prioridad al derecho a la vivienda[53]. Cabe preguntarse si esta Ley puede ser entendida e interpretada de forma diferente, o si hay que dar

---

[53] Méndez Juez (2023, p. 120), afirma que la Ley ha supuesto el cambio de un "modelo escalera" a un modelo "*housing first*". Sin embargo, creemos que en realidad queda mucho por andar para que tal afirmación pueda ser real.

la razón a Busto Lago, cuando dice que el derecho al disfrute de una vivienda digna no se concibe en la Ley 12/2023 como un derecho ejercitable frente a los poderes públicos, ni como un límite a la actuación de terceros para proteger el interés de su titular, "sino como un instrumento al servicio de las Administraciones Públicas con competencias en materia de vivienda para alcanzar los objetivos definidos en sus planes". A similar interpretación ha llegado el TC para salvar la constitucionalidad de algunas normativas autonómicas en relación con las viviendas vacías o desocupadas, al afirmar que no inciden en el derecho de propiedad, sino que se trata de instrumentos al servicio de políticas públicas en materia de vivienda[54].

En todo caso la parquedad y falta de precisión del art. 8 a) de la Ley 12/2023 ponen de manifiesto las grandes carencias que presenta para ser considerado un verdadero derecho subjetivo.

2.4. ¿Puede el derecho a la vivienda, como derecho humano, dimanar del art. 47 CE?

En la resolución del Parlamento Europeo de 21 de enero de 2021 sobre el acceso a una vivienda libre y asequible para todos[55] se pide a la Comisión y a los Estados miembros que el derecho a una vivienda adecuada sea reconocido y ejecutado como un derecho humano fundamental mediante disposiciones legislativas europeas y nacionales aplicables, pues afirma que "el acceso a una vivienda adecuada (...) debe considerarse

---

[54] Busto Lago (2023, p. 5).

[55] Resolución del Parlamento Europeo, de 21 de enero de 2021, sobre el acceso a una vivienda digna y asequible para todos (2019/2187(INI)).

una condición previa para ejercer y obtener el acceso a los demás derechos fundamentales y a una vida digna."[56]

La expresión "derecho humano" contenida en la Resolución ha de ser interpretada conforme a la jurisprudencia del TJUE, que en repetidas sentencias (vid. caso C-34/13 Monika Kusionova y Smart Capital TOL9.914.612) califica al derecho a la vivienda como un derecho fundamental o humano, incluyendo ambos adjetivos. Igualmente, en la terminología de Naciones Unidas se suelen identificar los derechos humanos con los derechos fundamentales[57].

No obstante, para poder configurar este derecho como derecho humano en el Ordenamiento español nos sirve de poco el art. 47 CE que no pasa de ser un principio programático, por tanto, el fundamento ha de encontrarse en otro lugar.

Nos puede guiar el preámbulo de la ley 12/2023, cuando comienza afirmando que el derecho al disfrute de una vivienda digna recogido en el art. 47 CE incide en el goce del contenido de otros derechos constitucionales, declarados incluso fundamentales, refiriéndose principalmente a la integridad física y moral –art. 15 CE– y a la intimidad personal y familiar –art. 18 CE–, si bien, también lo conecta con la protección de la salud –art. 43 CE– y con un medio ambiente adecuado -art. 45 CE–. Y concluye que todos estos derechos guardan una relación estrecha con los valores de la calidad de vida y del libre desarrollo de la personalidad en sociedad, consagrados por el

---

56 https://www.europarl.europa.eu/doceo/document/TA-9-2021-0020_ES.html

57 Vid. Informe sobre el derecho humano a una vivienda adecuada del relator especial sobre la vivienda adecuada de las Naciones Unidas. https://www.ohchr.org/es/special-procedures/sr-housing/human-right-adequate-housing

art. 10.1 CE. Además, en tanto que dicho artículo sitúa la dignidad humana como fundamento del orden público, también se proyecta sobre los llamados derechos sociales. Igualmente, en el preámbulo se señala que "Todo lo cual modula tanto el derecho de propiedad como la libertad de empresa, cuando operan en el sector de la vivienda, desde el doble punto de vista de la función social que deben cumplir y del interés general, respectivamente", citando los arts. 33.2, 38, 128.1 y 131.1 CE. Por ello, en el preámbulo se califica la vivienda como "bien esencial de rango constitucional" y "pilar central del bienestar social" en cuanto que es el lugar de desarrollo de la vida privada y familiar y centro de todas las políticas urbanas.

Por su parte, López Ramón considera que el disfrute de la vivienda está garantizado por la tutela judicial según se establece en el art. 42 CE, afirmando que es un derecho susceptible de ser exigido demandando ante los tribunales de justicia a quienes lo hayan desconocido total o parcialmente, y añadiendo que la efectividad de los derechos reconocidos en las leyes no está condicionada a la previsión presupuestaria del gasto[58].

De todo ello deducimos que el derecho a la vivienda como derecho humano no emana del art. 47 CE sino de una serie de preceptos constitucionales que inciden directamente en el mismo, en especial el derecho a la integridad física y moral y el derecho a la intimidad personal y familiar (arts. 15 y 18 CE y, correlativamente, art. 8 CEDH (TOL164.153)). Además, como hemos dicho anteriormente, el derecho a la vivienda como derecho humano queda consagrado como una manifestación de la dignidad humana por el art. 10 CE. Por tanto, es en estos artículos y no en el 47 CE en los que podemos encontrar la justificación de la configuración del derecho a la vivienda como un derecho humano.

---

58 López Ramón (2020, pp. 297-308 y pp. 300-301).

3. Derecho a la vivienda en la normativa internacional y europea

La Declaración Universal de Derechos Humanos de 1948 reconoció en su artículo 25 el derecho a una vivienda adecuada como parte del derecho a un nivel de vida adecuado. Derecho que igualmente se recogió en el art. 11.1 del Pacto Internacional de Derechos Económicos, Sociales y Culturales de 16 de diciembre de 1966 (TOL207.989). Este ha sido interpretado como un derecho a vivir en seguridad, paz y dignidad en un sitio determinado por el Comité de Derechos Económicos, Sociales y Culturales de las Naciones Unidas, en especial en las Observaciones 4.ª del Comité relativa al derecho a una vivienda adecuada de 1991 y la n.º 7 relativa a los desalojos forzosos de 1997. En las mismas se refieren las siguientes libertades que se derivan del derecho a una vivienda adecuada: la protección contra los desalojos forzosos y la destrucción y demolición arbitrarias del hogar, el derecho de ser libre de injerencias arbitrarias en el hogar, la privacidad y la familia, el derecho a la seguridad de la tenencia y el acceso no discriminatorio y en igualdad de condiciones a una vivienda adecuada. Además, señala que la vivienda no es adecuada si sus ocupantes no cuentan con ciertas medidas de seguridad de la tenencia que les garanticen protección jurídica contra el desalojo forzoso, el hostigamiento y otras amenazas.

En el ámbito europeo destaca en primer lugar el art. 34.3 de la Carta de los Derechos Fundamentales de la Unión Europea (TOL131.225)[59] que refiere que "con el fin de combatir la exclusión social y la pobreza, la Unión reconoce y respeta el derecho a una ayuda social y a una ayuda de vivienda para

---

[59] (2000/C 364/01) entró en vigor el 1 de diciembre de 2009, junto con el Tratado de Lisboa.

garantizar una existencia digna a todos aquellos que no dispongan de recursos suficientes, según las modalidades establecidas por el Derecho comunitario y las legislaciones y prácticas nacionales". Por tanto, la Carta se limita a programar "una ayuda" a la vivienda, sin llegar a reconocer un verdadero derecho subjetivo a la vivienda.

El siguiente paso lo ha dado el TJUE en el asunto C-34/13 Monika Kusionova y Smart Capital a.s. (TOL9.914.612), en el que se dictó sentencia el 10 de septiembre de 2014, considerando que "en el Derecho de la Unión, el derecho a la vivienda es un derecho fundamental garantizado por el art. 7 de la Carta que el Tribunal remitente debe tomar en consideración al aplicar la Directiva 93/13", es decir, lo incardina dentro del derecho a la vida privada y familiar, al domicilio y de sus comunicaciones. Por su parte, en 2020 en el Dictamen sobre "acceso universal a una vivienda digna, sostenible y asequible a largo plazo" del Comité Económico y Social Europeo se estableció que la UE debía velar por la consagración de un verdadero derecho universal a la vivienda, en particular mediante un Reglamento sectorial, con arreglo al procedimiento legislativo ordinario[60]. Finalmente, traemos a colación la Resolución del Parlamento europeo de 21 de enero de 2021 sobre el acceso a

---

60 Dictamen del Comité Económico y Social Europeo sobre acceso universal a una vivienda digna, sostenible y asequible a largo plazo, Dictamen de iniciativa EESC 2020/01076. Este señala que la política de vivienda no se puede limitar al único objetivo de ayudar a las personas que viven por debajo del umbral de riesgo de pobreza o en torno a él, va más lejos, pues considera que debe garantizarse una vivienda digna y asequible a largo plazo a todo ciudadano.

una vivienda libre y asequible para todos[61], ya comentada, que califica el derecho a la vivienda como un derecho humano[62].

4. *¿Cómo debería quedar configurado el derecho a la vivienda como un derecho subjetivo?*

Podemos concluir que con la Ley 12/2023 se ha perdido una oportunidad para consolidar una noción exhaustiva del derecho a la vivienda, tal como señalaba Ruiz-Rico respecto a la Ley del Suelo que de igual forma se limitó a recoger una noción abstracta de este derecho[63]. Es cierto que se trata de un reto mayor no exento de riesgos, pero un reto que debe ser afrontado[64].

Seguidamente analizamos cómo podría quedar configurado este derecho subjetivo, para lo cual partiremos de la normativa

---

61 Resolución del Parlamento Europeo, de 21 de enero de 2021, sobre el acceso a una vivienda digna y asequible para todos (2019/2187(INI)).

62 En este texto se conecta la asequibilidad de la vivienda con patrones de justicia distributiva debido al aumento excesivo de los costes de la vivienda, y se alerta de que las deficiencias en el mercado de la vivienda ponen en peligro la cohesión social en Europa por los niveles que alcanzan tanto la carencia de vivienda como la pobreza, situación que en última instancia mina la confianza en la democracia, lo cual obliga a las autoridades nacionales y locales a potenciar la vivienda social y asequible.

63 Ruiz-Rico Ruiz (2008, p. 103).

64 La experiencia francesa nos ha de servir de advertencia, pues aunque en Francia se ha intentado afrontar este reto, se ha hecho con una regulación – la llamada Ley "Dalo" *Droit au logement opposable* de 5 de marzo de 2007–que ha generado nuevos problemas y ha creado importantes lagunas, vid. Ruiz-Rico Ruiz (2008, pp. 15-22), y Inserguet-Brisset (2010, pp. 30-64).

vigente a nivel autonómico, a fin de determinar si podemos identificar un modelo a seguir. Puede afirmarse que a partir de 2011 se han ido promulgando una serie de Leyes de vivienda autonómicas de segunda generación en las que de forma explícita y detallada se regula este derecho como subjetivo[65].

Nos referimos, en primer lugar, a la derogada Ley 1/2011, de 10 de febrero, de Garantías en el Acceso a la Vivienda en Castilla-La Mancha (TOL8.147.989)[66] que fue la primera normativa autonómica que puso en marcha una normativa de este tipo[67], si bien nunca llegó a ser efectiva por su derogación justo un año después de su publicación.

---

65 Si bien con excepciones, pues no todas las regulaciones autonómicas que se han adoptado recientemente en materia de vivienda han optado por considerar que el derecho a la vivienda quede configurado como un derecho subjetivo de forma explícita, a título de ejemplo véase la Ley 11/2019, de 11 de abril, de promoción y acceso a la vivienda de Extremadura (TOL7.218.333).

66 Ley 1/2011, de 10 de febrero, de Garantías en el Acceso a la Vivienda en Castilla-La Mancha, BOE-A-2011-7706 (TOL8.147.989), derogada por la Ley 1/2012, de 21 de febrero de Medidas Complementarias para la Aplicación del Plan de Garantías de Servicios Sociales (BOE-A-2012-10756; TOL2.455.106).

67 La Ley daba respuesta al Pacto por la Vivienda en Castilla-La Mancha firmado en mayo de 2008 para facilitar a todas las personas castellanomanchegas el acceso a una vivienda digna en el que el ejecutivo autonómico se comprometía a pagar por la diferencia entre la renta de la vivienda libre alquilada y la cantidad que le hubiera correspondido pagar por la adquisición de una vivienda protegida adjudicada, para lo cual el solicitante debería tener cumplidos 22 años, llevar empadronado en Castilla-La Mancha al menos tres años, disponer de una fuente regular de ingresos no superior a 2,5 veces el IPREM y haber participado en los procesos de adjudicación de vivienda protegida. https://www.castillalamancha.es/node/167043

En segundo lugar, hacemos referencia a la Ley vasca de vivienda de 2015 (TOL5.181.949)[68] desarrollada por el Decreto 147/2023, de 10 de octubre, del derecho subjetivo de acceso a la ocupación de una vivienda (BOPV de 25 de octubre de 2023; TOL9.739.109))[69], que la adapta a la Ley 14/2022, de 22 de diciembre, del Sistema Vasco de Garantía de Ingresos y para la Inclusión (TOL9.519.495)[70]. El artículo segundo del Decreto 147/2023 establece que el derecho subjetivo de acceso se satisfará a través de dos modalidades: primera, mediante la adjudicación de una vivienda en régimen de alquiler o de un alojamiento dotacional con canon; segunda, mediante la concesión de la prestación económica de vivienda, añadiendo que esta última solo se concederá con carácter subsidiario, cuando no sea

---

68 Ley 3/2015 del País Vasco, de 18 de julio, de vivienda.

69 Decreto 147/2023, de 10 de octubre, del derecho subjetivo de acceso a la ocupación de una vivienda (BOPV de 25 de octubre de 2023), que entra en vigor el 2 de abril de 2024. https://laadministracionaldia.inap.es/noticia.asp?id=1238136

70 Cuya Disposición Transitoria Segunda garantiza el reconocimiento y percepción de la prestación complementaria de vivienda en tanto no se aprueben las disposiciones reglamentarias que desarrollen la Ley 3/2015, de 18 de junio, de vivienda, en relación con el derecho subjetivo de acceso a la ocupación de una vivienda y el establecimiento del sistema de prestaciones económicas que, con carácter subsidiario, satisfaga aquel derecho. Igualmente, el Decreto 147/2023 la adapta a los parámetros establecidos por la Ley 19/2021, de 20 de diciembre, respecto al ingreso mínimo vital, (TOL8.691.962) facilitándose el acceso al derecho subjetivo a las personas perceptoras de dicho ingreso y a las perceptoras de las ayudas de emergencia social, y por otro lado, subiendo los ingresos máximos para la calificación de riesgo de exclusión social de 15.000 euros para las unidades de convivencia de tres o más miembros, 12.000 para las de dos miembros y 9.000 para las de un solo miembro, a 19.000, 17.000 y 13.000 euros respectivamente (art. 5.1).

posible adjudicar una vivienda o un alojamiento dotacional[71]. Este derecho se mantendrá en el tiempo mientras subsistan las circunstancias que motivaron su reconocimiento (art. 10.1). El procedimiento para el reconocimiento del derecho subjetivo de acceso queda sujeto a la Ley 39/2015 (TOL5.494.102)[72], siendo tres meses el plazo máximo para resolver y notificar la resolución, dando lugar a estimación por silencio administrativo positivo si al vencimiento de dicho plazo máximo no se hubiese notificado la resolución. La adjudicación de las viviendas se realizará siempre en régimen de arrendamiento y en régimen de cesión de uso en el caso de los alojamientos dotacionales en las condiciones que se establezcan en el acto de adjudicación (art. 21.2) que darán lugar, respectivamente, a la obligación de pagar una renta o canon.

Por otro lado, la prestación económica de vivienda es una prestación periódica que contribuirá a que las personas titulares del derecho subjetivo de acceso sufraguen la renta correspondiente a un contrato de arrendamiento, subarriendo o alquiler de habitaciones o la contraprestación económica a que se refiere el artículo 27[73], mientras no se les adjudique una vivienda o alojamiento dotacional (art. 26). Para poder acceder a esta prestación económica de vivienda de carácter subsidiaria se exige que el solicitante esté empadronado y tenga su domicilio en una

---

71 En el art. 3 del Decreto 147/2023 se establecen los requisitos para ser titular del derecho subjetivo de acceso.

72 Ley 39/2015, de 1 de octubre, del Procedimiento Administrativo Común de las Administraciones Públicas.

73 Art. 27: "A los efectos de este Decreto, son alojamientos colectivos los servicios de alojamiento, centros residenciales, establecimientos de alojamiento u otros que sirvan a idéntico fin que sean considerados como domicilio según la normativa reguladora del Sistema Vasco de Garantía de Ingresos y para la Inclusión, siempre que exijan una contraprestación económica a las personas que residan en ellos".

vivienda libre, en régimen de arrendamiento, subarriendo o alquiler de habitaciones, o en un alojamiento colectivo cuya renta o contraprestación económica equivalente supere el 30% de sus ingresos anuales[74]. Finalmente, la cuantía máxima mensual de la prestación económica de vivienda será de 300 euros a lo largo de los 12 meses de cada año natural conforme al art. 29.

La tercera regulación autonómica en reconocer un derecho subjetivo a la vivienda fue la Ley 2/2017, de 3 de febrero por la función social de la vivienda de la Comunidad Valenciana (TOL5.9561.896), que se aprobó dos años después de la vasca[75]. En su artículo segundo, párrafo tercero, se establece la obligación jurídica de la administración autonómica valenciana y de las administraciones locales de poner a disposición del solicitante la ocupación estable de un alojamiento dotacional, de una vivienda protegida o de una vivienda libre si ello fuera necesario, que cumpla con todas las condiciones para garantizar su libre desarrollo de la personalidad, su derecho a la intimidad y el resto de derechos vinculados a la vivienda. Añade además que se entenderá que existe tal puesta a disposición cuando se proceda al pago de las ayudas al alquiler reguladas por la ley[76]. En todo

---

74 Además, se exige que el solicitante figure expresamente como parte arrendataria o subarrendataria en el contrato de arrendamiento, subarriendo o alquiler de habitaciones de la vivienda, o como cliente en un documento emitido por el alojamiento colectivo, y que abone la renta o contraprestación económica equivalente mediante transferencia bancaria (art. 29).

75 Ley 2/2017, de 3 de febrero, por la función social de la vivienda de la Comunidad Valenciana.

76 Por su parte, en el artículo sexto titulado "Acción pública y derecho subjetivo exigible" se establece que toda persona, en ejercicio de la acción pública en materia de vivienda, puede exigir ante los órganos administrativos y jurisdiccionales competentes el cumplimiento de la normativa en materia de vivienda.

caso, el titular del derecho exigible a un alojamiento asequible, digno y adecuado conforme al artículo 2 de esta ley podrá ejercerlo ante la *conselleria* competente en materia de vivienda, que en un plazo máximo de seis meses deberá resolver su solicitud, plazo que podrá ampliarse excepcionalmente[77].

El siguiente paso se ha dado en la Ley foral 28/2018, de 26 de diciembre, sobre el derecho subjetivo a la vivienda en Navarra (TOL6.976.009), que siguiendo las pautas de la Ley Vasca configura el derecho a la vivienda como un derecho subjetivo. La Ley establece que, ante la insuficiencia del parque público de vivienda para satisfacer la necesidad estimada de vivienda, se opta por establecer de forma subsidiaria una prestación económica, que se configura como un impuesto sobre la renta negativo mediante una deducción en la cuota diferencial. Por tanto, se trata de deducciones fiscales por arrendamiento que serán abonadas de forma anticipada y que serán directamente exigibles ante la Administración de la Comunidad Foral de Navarra. Esta prestación tiene dos modalidades: una vinculada a la necesidad de alcanzar la emancipación, dirigida a las personas jóvenes y con un carácter temporal, y otra vinculada a la insuficiencia de recursos económicos para financiar el acceso a una vivienda, con carácter permanente en el tiempo en tanto en cuanto se sigan cumpliendo los requisitos[78]. Se trata de una normativa desarrollada anualmente

---

77 Si no se hubiera dictado resolución en plazo, se entenderá otorgado el uso de un alojamiento por silencio administrativo.

78 La Ley Navarra establece un calendario gradual del reconocimiento del derecho, en función de la suficiente dotación de los recursos económicos, materiales y organizativos necesarios, si bien se recoge una implantación muy ambiciosa desde el primer momento desde la entrada en vigor de esta ley foral. De hecho, la implementación de una política legislativa que imponga al Estado asumir la responsabilidad de garantizar el derecho subjetivo a la vivienda viene condicionada por la capacidad económica del estado de que

por distintas órdenes forales que en la práctica y para el año 2024 dan derecho a recibir una prestación que en el caso de "alquiler joven", es decir entre 23 y 32 años y con unos ingresos máximos de 22.000€, salvo que se trate de una unidad familiar de más de un miembro en cuyo caso sería de 33.000€, que alcanza el 50% de la renta mensual, con un máximo de 280€ mensuales, siempre que la renta de la vivienda no supere los 700€ mensuales y que ha de renovarse anualmente para comprobar que se siguen cumpliendo los requisitos exigidos.

Esta normativa se ha visto completada con otra que pone el énfasis en el control del mercado de vivienda para protegerlo de la especulación inmobiliaria, destacando la normativa catalana y en especial la Ley 11/2020, de 18 de septiembre (TOL8.078.382)[79], que trata de hacer frente a la fuerte subida de los precios que afecta especialmente a la juventud y las personas con bajo nivel de ingresos que sufren situaciones de exclusión residencial. Para ello, siguiendo las pautas adoptadas en Alemania y Francia, establece sistemas de contención o delimitación de las rentas de alquiler en las zonas en las que el mercado se ha tensado mucho y se generan dificultades graves de acceso a la vivienda en el conjunto del parque residencial[80]. No obstante, creemos que, si bien el objetivo buscado por la Ley

---

se trate, tal como se ha estudiado desde el análisis económico del derecho, distinguiéndose entre diversas prioridades básicas. Vid. Ott y Shafer (1991).

79 Ley 11/2020, de 18 de septiembre, de medidas urgentes en materia de contención de rentas en los contratos de arrendamiento de vivienda y de modificación de la Ley 18/2007, de la Ley 24/2015 y de la Ley 4/2016, relativas a la protección del derecho a la vivienda (TOL8.078.282).

80 Se trata de una medida excepcional con una duración máxima de cinco años comprendida en un procedimiento administrativo en el que deben acreditarse las circunstancias que la justifican

11/2020 es correcto, cabe preguntarse si las medidas adoptadas han sido acertadas. De hecho, podría plantearse si estos sistemas de contención o delimitación de las rentas de alquiler alteran el contenido esencial del derecho de propiedad, lo cual ha sido defendido por parte de la doctrina que tilda estas medidas de expropiatorias. Sin embargo, creemos que la clave nos la da la STC 89/1994 de 17 de marzo de 1994 (TOL82.497)[81] que abrió un amplio campo de actuación en poder del legislador[82].

---

y precisar las medidas que adoptarán las administraciones para revertir o atenuar la situación del mercado tenso.

81 Respecto a las medidas que se han promulgado limitando el derecho de propiedad cabe señalar que resulta clave la STC Pleno 89/1994 de 17 de marzo de 1994 (TOL82.497), cuestiones de inconstitucionalidad 2.010/1989 y 969/1991 (acumuladas) en relación con el art. 57 de la Ley de Arrendamientos Urbanos, la primera, y con los arts. 57, 70, 71 y 73 de dicha Ley, pues al entender que no quedaba afectado el contenido esencial del derecho de propiedad, ni por la prórroga legal forzosa, ni por la congelación de renta, mostraba el amplio campo de actuación que dejaba en poder del legislador.

82 Debemos también hacer referencia a las medidas previstas para proteger los centros históricos de muchas ciudades españolas de las tensiones causadas por el mercado de la vivienda turística, que poco a poco va expulsando a los habitantes locales, fijando unos alquileres que resultan astronómicos para el ciudadano de a pie. Entre ellas resaltamos el índice de rentas del sistema nacional de precios de alquiler que busca enfriar el mercado inmobiliario limitando los alquileres máximos en zonas tensionadas, presentado por el Gobierno el 27 de febrero de 2024, conforme a lo previsto por la Ley 12/2023, que no obstante no afectará a todas las viviendas sitas en una "zona tensionada", si bien tiene en cuenta las características de cada vivienda. No obstante, la ejecución de esta medida necesita de la colaboración de las CCAA y hasta la fecha solo Cataluña se ha comprometido a aplicar el índice. De hecho, en 28 de abril de 2024, el Partido Popular instó al Gobierno a retirar el índice con mociones en CCAA y ayuntamientos. Vid.: https://www.mivau.gob.es/vivienda/alquila-bien-es-tu-derecho/alquiler

En todo caso, el impulso autonómico marca el camino a seguir para hacer efectivo el derecho subjetivo a una vivienda digna y adecuada. En especial son las legislaciones vasca y navarra las que recogen un modelo más depurado. Es un acierto que la ley prefiera optar por ceder la tenencia de la vivienda, ya sea por alquiler o mediante la cesión del uso, mientras subsistan las circunstancias que motivaron su reconocimiento, o incluso que se otorgue dicho uso con un carácter anual, solo prorrogable si subsiste la situación de necesidad. Es decir, se ha optado por una fórmula eficiente que vincula la ayuda al mantenimiento de la situación de necesidad, lo cual no sería posible si la vivienda se hubiera enajenado mediante compraventa.

Igualmente nos parece un acierto la forma en la que desde la normativa autonómica se afronta el reto de identificar a la persona en situación de vulnerabilidad que pueda tener derecho a una vivienda digna y adecuada. Es este uno de los grandes desafíos que conlleva la implantación de un derecho subjetivo a la vivienda, que creemos que ha sido tratado con fórmulas correctas. Por un lado, tal como hemos indicado anteriormente, en la normativa navarra se distingue entre dos modalidades: una vinculada a la necesidad de alcanzar la emancipación, dirigida a las personas jóvenes, con un carácter temporal; y otra vinculada a la insuficiencia de recursos económicos para financiar el acceso a una vivienda, con carácter permanente en el tiempo en tanto en cuanto se sigan cumpliendo los requisitos. Por otro lado, nos parece bien orientado el Decreto 147/2023 que adapta la Ley Vasca 3/2015, de 18 de junio, de Vivienda, a los parámetros establecidos por la Ley 19/2021, de 20 de diciembre (TOL9.739.109) respecto al ingreso mínimo vital, facilitándose el acceso al derecho a la vivienda o a prestaciones económicas sustitutorias a las personas perceptoras del ingreso mínimo vital y a las perceptoras de las ayudas de emergencia social, y por otro lado, subiendo los ingresos máximos para la calificación de riesgo de exclusión social de 15.000 euros para las unidades de convivencia de tres o más miembros, 12.000 para las de dos miembros y 9.000 para las de un solo miembro, a 19.000, 17.000 y 13.000 euros respectivamente (art. 5.1). En todo caso, la situación de vulnerabilidad será una situación personal predicable de una persona que cumpla con los requisitos legales establecidos y exclusivamente durante el tiempo en que se mantenga en dicha situación. Por tanto, no se puede atribuir colectivamente la situación de vulnerabilidad a un grupo social o por pertenencia a una etnia, a fin de beneficiarse de las ayudas sociales que configuran el derecho subjetivo a la vivienda.

Es cierto que se podrá alegar que las medidas dispuestas por estas regulaciones autonómicas pueden fomentar la renuncia del beneficiario a intentar toda mejora económica para evitar la pérdida del derecho a la vivienda una vez adquirido o incluso ocultar ingresos con el mismo fin. No obstante, desde el punto de vista de la eficiencia económica, estas medidas que con carácter subsidiario se materializan en ayudas para el alquiler son acertadas y garantizan no distraer el erario público con ayudas que a la larga se demuestren equivocadas por innecesarias, tal como resultaría en el caso de una compraventa a favor de una persona o unidad familiar que al poco de la adquisición resultase estar en una mejor posición económica, por lo que no cumpliría con los requisitos para la adjudicación.

Lo que sí es exigible es que estas medidas vengan acompañadas de un control exhaustivo que podría gestionarse por las entidades locales a fin de supervisar que durante el tiempo que el beneficiario recibe la ayuda su situación económica sigue bajo parámetros de vulnerabilidad o bien limitando su vigencia a un plazo anual, renovable solo si se acredita continuar cumpliendo con los requisitos. Esta vigilancia es vital a fin de que las medidas sean efectivas y ha de ir más allá de la mera comprobación documental del contenido de la declaración de la renta y manifestación de salario percibido por el beneficiario. Por tanto, deberían valorarse otros indicios de los que se pueda deducir que este está disfrutando de un nivel de vida no justificado, tal como, por ejemplo, suele hacerse en los procedimientos de separación y divorcio al fijarse el derecho de manutención a favor de los hijos o la pensión por desequilibrio económico.

Este análisis de la normativa autonómica pone de manifiesto el hecho de que es a las autonomías y no al Estado a quien corresponde adoptar medidas efectivas que permitan al ciudadano poder exigir el derecho a la vivienda a los poderes públicos. La razón de ello radica en que se trata de una competencia transferida por el Estado que por tanto corresponde exclusivamente a las Comunidades Autónomas. Como señala Prats Albentosa todas las CCAA tienen competencia para regular la actuación jurídico-pública y, por tanto, para permitir al ciudadano exigir a la Administración que cumpla con su deber de concederle el disfrute de una vivienda digna y adecuada, una vez que sea reconocido como derecho subjetivo[83].

No obstante, señala López Ramón que por imperativo constitucional y estatutario corresponde a las Comunidades Autónomas definir y desarrollar las políticas de vivienda, pero que el establecimiento de las condiciones básicas que garanticen la igualdad de los ciudadanos en el ejercicio del derecho constitucional a la vivienda es competencia irrenunciable del Estado conforme a lo dispuesto por el artículo 149.1.1.ª de la Constitución. Ante lo cual afirma que las bases de tan especial derecho para el Estado social han de ser comunes y fundarse en el acceso preferentemente por medio del mercado a una vivienda de calidad y conservación de la misma, centrando las actuaciones directas en las familias y personas socialmente marginadas para proporcionarles alojamientos temporales y definitivos o

---

83 Prats Albentosa (2022, p. 172).

viviendas familiares en régimen preferentemente de alquiler o de uso gratuito cuando fuera preciso[84]. En estas bases debería garantizarse tanto el derecho de acceso a la vivienda, como el derecho a la conservación de la vivienda, como el derecho a la calidad de la vivienda. Además, conforme al art. 148.1.3, estas bases comunes servirían para la planificación y coordinación económica del Estado, incluyendo las facultades de dirección general de economía que la STC 152/1988 entiende aplicable al sector de la vivienda[85].

Finalmente, se debe señalar la dificultad de formular medidas ponderadas para actuar con equilibrio huyendo de todo activismo. No puede ignorarse la existencia de una importante economía sumergida en nuestro país. De hecho, la picaresca ha sido durante largo tiempo una de las "cualidades" con la que se ha caracterizado a la sociedad española. Si bien es cierto que nos encontramos en una coyuntura social y económica difícil, con un sector de la población en situación muy vulnerable, también lo es que simultáneamente hay otro sector de la sociedad que hasta la fecha no ha tenido acceso a ayudas sociales por considerarse perteneciente a la llamada clase media, población que está viendo como su poder adquisitivo disminuye progresivamente hasta el punto de que el propio acceso a una vivienda se convierte en un reto mayor. Más aún en muchas ocasiones el acceso supone un sacrificio que supera con creces los porcentajes recomendados. Esta realidad convive con otra en la que la economía sumergida permite el acceso a las ayudas mediante el fraude. Por ello, para que las necesarias medidas de política legislativa que conviertan el derecho a la vivienda en un derecho subjetivo real sean bien entendidas por la sociedad han de venir de la mano de medidas preventivas y de control que eviten que

---

84 López Ramón (2014, p. 85).

85 Así lo señala Nasarre Aznar (2022a, pp. 1-25).

los fondos públicos se desvíen y terminen satisfaciendo intereses fraudulentos[86].

V. CONCLUSIONES

1. El artículo 8 de la LDV mezcla derechos de muy distinta naturaleza: unos derivados del derecho de propiedad y otros de naturaleza jurídico-pública.
2. Los párrafos contenidos en las letras b, c y d del art.8 LDV no aportan novedad alguna y han de ser considerados como meras normas de remisión a lo dispuesto por la Ley 19/2013 de Transparencia, acceso a la información pública y buen gobierno; a la normativa autonómica en materia de vivienda, en especial la reguladora de los Registros Públicos Municipales de Demandantes de Vivienda Protegida y a las normativas complementarias.
3. El artículo 9 LDV carece de toda sistemática y rigor. Por un lado, recoge deberes/obligaciones generales que atañen a todos los ciudadanos y que están regulados en el ordenamiento jurídico, incluso en la Constitución; y, por otro, deberes/obligaciones de naturaleza jurídico-privada que se derivan del derecho de propiedad como derecho subjetivo.

---

86 Se ha afirmado que en todo caso reconducir el mercado de la vivienda a un mercado intervenido no puede ser la solución, pues dejaría sin viviendas a las clases medias en aras de una objetividad que suele terminar en sorteos entre solicitantes, consiguiendo los ganadores una vivienda intervenida que da lugar a guetos. Una intensa intervención administrativa del mercado de viviendas solo quedaría justificada en situaciones de riesgo puntuales, como fue en el caso de la burbuja inmobiliaria que estalló en 2008: López Ramón (2014, p. 77).

4. Los deberes/obligaciones recogidos por el art. 9 LDV se contemplan de forma general o abstracta sin concretar su contenido, ni la legitimación activa para exigir su cumplimiento, ni las acciones a través de las cuales puedan ser ejercitados. En las letras a, b y c se recogen supuestos relacionados con la preservación, conservación y protección de tenencia de la vivienda: en la letra (a) debe entenderse que se realiza una remisión a la normativa reguladora de las relaciones de vecindad (art. 7 CC, ART.7 LPH, ART. 27 LAU, art. 18 CE, art 8 CEDH). En la letra (b) se obvia que, salvo normativa específica y no amenazando ruina, es al propietario a quien corresponde decidir el estado en que desee mantener su vivienda. En la letra (c) la remisión ha de entenderse a los juicios sumarios que han venido a reemplazar los interdictos de recobrar y retener la posesión (arts. 441 y 446 CC y concordantes), así como la acción negatoria.

En la letra (d) se incluye un deber de muy distinta naturaleza que difícilmente encaja con los anteriores pues se refiere a las obligaciones del transmitente o intermediario y que manifiesta la falta de sistemática de la que adolece este artículo. Finalmente, en la letra (e) al referirse al deber de atender a la importancia del parque público de vivienda y velar por su adecuado mantenimiento, se recoge un deber elemental de respeto a la cosa ajena que ya se encontraba en la *tria precepta* romana formulada por Ulpiano y que en este caso encuentra desarrollo en el derecho público. En todo caso podría pensarse que el primer obligado al cumplimiento de este deber sería la propia Administración.

Por tanto, el art. 9 recoge un conjunto de deberes de muy diversa naturaleza, escogidos aleatoriamente, que carecen de un criterio aglutinador que les pueda dotar de una mínima sistemática.

5. El derecho a una vivienda digna y adecuada recogido en el art. 8.a se formula para dar respuesta a una necesidad

social creciente como es la falta de acceso a la vivienda en propiedad o alquiler por amplios sectores de la población, ya sea por su edad o por su situación de vulnerabilidad. Sin embargo, no encontramos en la ley lugar alguno donde se fije su contenido y condiciones de ejercicio, por lo que no se configura ni como un derecho subjetivo, ni como un derecho fundamental o humano. La LDV se limita a adoptar medidas, mayoritariamente procesales, que pretenden ampliar el mercado inmobiliario disponible a expensas, principalmente, del sector privado, medidas que han sido muy criticadas por un sector importante de la doctrina.

6. El derecho fundamental o humano a la vivienda no puede colegirse del art. 47 CE sino de los arts. 15, 14 y 10 CE en tanto manifestación de la dignidad humana, del derecho a la integridad y a la intimidad.

7. Es necesario configurar el derecho a la vivienda como un derecho subjetivo por exigirlo el carácter social del estado español y para dar respuesta a la situación de vulnerabilidad de amplios sectores de la población. La nueva realidad social exige adaptar la ley al hecho de que el derecho a la vivienda sea un derecho humano o fundamental, tal como se ha puesto de manifiesto en numerosos foros internacionales y europeos.

8. Si bien la construcción efectiva del derecho a la vivienda como derecho subjetivo corresponde a las CCAA por razón de competencia, la Ley 12/2023 ha perdido la oportunidad de establecer unas bases comunes para la configuración del derecho a la vivienda como un derecho subjetivo al amparo del art. 149.1.1.ª. Estas bases comunes o condiciones básicas lo serían para que el Estado pudiera garantizar que en el ejercicio de las competencias autonómicas se garantiza la igualdad de todos los españoles en el ejercicio de sus derechos y el

cumplimiento de sus deberes constitucionales (SSTC 16/2018,TOL6.537.959 y 32/2018,TOL8.485.295).

9. A nivel autonómico se han dictado leyes de vivienda de segunda generación que configuran el derecho a la vivienda como subjetivo estableciendo, con carácter subsidiario, el derecho del beneficiario a percibir una prestación para garantizar el acceso a la vivienda mediante alquiler. Estas leyes merecen ser tenidas en cuenta como modelo a seguir.

10. El derecho subjetivo a una vivienda digna no debe ser confundido con un derecho a adquirir una vivienda en propiedad. Por el contrario, las ayudas sociales deben encaminarse a garantizar el acceso a la vivienda durante el estricto tiempo en que perdure la situación de vulnerabilidad. Por ello, la medida idónea sería la que garantizase el acceso a la vivienda en régimen de alquiler durante el tiempo preciso que se mantuviera la situación de vulnerabilidad.

11. Estas ayudas deben venir acompañadas de un exhaustivo control que podrían ejercer las entidades locales, a fin de comprobar que los beneficiarios mantienen una situación de vulnerabilidad durante su percepción. Este control es vital para lograr tanto su efectividad como su aceptación social debido al enorme esfuerzo presupuestario que puede suponer extenderlas a nivel nacional.

12. Ante el reto de determinar quién pueda beneficiarse de estas ayudas nos parece acertado el enfoque aportado por la normativa autonómica vasca y navarra, al distinguir entre medidas dirigidas a los jóvenes y a personas con insuficiencia de recursos económicos para financiar el acceso a una vivienda, conectando el derecho subjetivo con criterios objetivos como el del ingreso mínimo vital, el ser preceptor de ayuda de emergencia social o tener una situación calificada como riesgo de exclusión social.

VI. REFERENCIAS BIBLIOGRÁFICAS

Alexy, R. (1993). *Teoría de los derechos fundamentales.* Centro de Estudios Constitucionales. Madrid.

Beltrán de Felipe, M. (2000). *La intervención administrativa en la vivienda. Aspectos competenciales, de policía y de financiación de las viviendas de protección oficial.* Lex Nova. Valladolid.

Busto Lago, J. M. (2023). "La Ley por el derecho a la vivienda y el derecho subjetivo a la vivienda". *Cuadernos de Derecho Privado,* n.º 6, 2-9.

Carmona Cuenca, E. (1992). "Las normas constitucionales de contenido social: Delimitación y problemática de su eficacia jurídica". *Revista de Estudios Políticos,* n.º 76, 103-125.

Cossío Díaz, J. R. (1989). *Estado social y derechos de prestación.* Centro de Estudios Constitucionales. Madrid.

De la Fuente Romero de Tejada, C. (2016). "Sobre el concepto jurídico de persona con discapacidad y la noción de apoyos necesarios". *La Revista Española de Discapacidad,* vol. 4, n.º 2, 81-99.

Díez Picazo, L. M.ª (2008) *Sistema de derechos fundamentales.* Aranzadi Thomson-Civitas. Cizur Menor.

Enériz Olaechea, F. J. (2008). "Los derechos y deberes de los ciudadanos, propietarios y promotores en la nueva Ley". *Comentarios a la Ley del suelo: Ley 8/2007, de 28 de mayo* (dir. Enériz Olaechea/ Beltrán Aguirre). Thomson Reuters Aranzadi. Cizur Menor, 153-224.

Espínola Orrego, G. (2010). *El derecho a una vivienda digna y adecuada en el ordenamiento jurídico español.* Tesis doctoral (dir. Carmona Cuenca). Alcalá de Henares.

Fernández Rodríguez, T. R. (1997). "Las garantías de los derechos sociales". *Las estructuras del bienestar en Europa.* Civitas. Madrid.

Gálvez Montes, J. (2001). "Comentario al art. 47 CE". *Comentarios a la Constitución* (dir. F. Garrido Falla). Civitas. Madrid.

García de Enterría, E. (2001). *La Constitución como norma y el Tribunal constitucional.* Civitas. Madrid.

García Macho, R./ Soler Vilar, A. (1994). *Legislación sobre la Vivienda. Normativa Estatal y Autonómica.* Tecnos. Madrid.

García Macho, R. (1982). *Las aporías de los derechos fundamentales sociales y el derecho a una vivienda.* Instituto de la Administración Local. Madrid.

Garrido Falla, F. (2001). *Comentarios a la Constitución.* Civitas. Madrid.

González Pérez, J. (2015). "Comentario al art. 4". *Comentarios a la Ley del Suelo* (dir. J. González Pérez). Thomson-Reuters-Civitas. Madrid.

Hernández Díaz-Ambrona, M.ª D. (2015). *Consumidor vulnerable.* Reus. Madrid.

Inserguet-Brisset, V. (2010). "La evolución del Derecho francés en materia de vivienda". *Construyendo el derecho a la vivienda* (coord. López Ramón). Marcial Pons, Madrid.

López González, J. L. (1997). "Reflexiones sobre los Derechos sociales y su eficacia jurídica". *Revista General de Derecho,* n.º 628-629, 159-168.

López y López, A. (2022). "Propiedad privada y disciplina de mercado en el Proyecto de Ley por el derecho a la vivienda". *Cuadernos de Derecho Local,* QDL 59, 14-34.

López Ramón, F. (2014). "El derecho subjetivo a la vivienda". *Revista Española de Derecho Constitucional,* n.º 102, 49-91.

López Ramón, F. (2020). "El reconocimiento legal del derecho a disfrutar de una vivienda". *Revista de la Administración Pública,* 212, mayo-agosto, 297-308.

Martínez Quiñones, A. (2024). "Reflexiones en torno a la naturaleza y eficacia del derecho constitucional a una vivienda digna y adecuada: ¿tan sólo un principio rector?. *El derecho a la vivienda en tiempos de incertidumbre* (dirs. Mate Satué, L. C./ Hernández Sáinz, E./ Alonso Pérez, M.ª T.). Aranzadi. Madrid, 525-535.

Méndez Juez, M. (2023). "Sinhogarismo y derecho de acceso a la vivienda en España: Definición, problemática y respuesta parlamentaria". *Joaçaba,* vol. 24, n.º 1, 105-124.

Muñoz Castillo, J. (2000). *El derecho a una vivienda digna y adecuada. Eficacia y ordenación administrativa.* Colex. Madrid.

Nasarre Aznar, S. (2022a). "El proyecto de Ley de vivienda 2022", *Apuntes 2022/11,* FEDEA, 1-25.

Nasarre Aznar, S. (2022b). "Ley de vivienda: El Estado renuncia a cumplir con el art. 47 de la Constitución", *El Notario del Siglo XXI,* n.º 109, 1-6.

Peces-Barba Martínez, G. (1979). "Reflexiones sobre la teoría general de los derechos fundamentales en la constitución". *Revista de la Facultad de derecho de la Universidad Complutense de Madrid,* n.º 2, 39-50.

Pérez Royo, J. (2007). *Curso de Derecho Constitucional.* Marcial Pons. Madrid.

Pisarello, G. (2009). "El dret a l'habitatge com a dret social: implicacions constitucionals". *Revista Catalana de Dret Públic,* n.º 38.

Ponce Solé, J. (2008). "El derecho a la vivienda. Nuevos desarrollos normativos y doctrinales y su reflejo en la Ley catalana 18/2007, de 28 de diciembre, del derecho a la vivienda". *El derecho de la vivienda en el siglo XXI: sus relaciones con la ordenación del territorio y el urbanismo* (dir. Ponce Solé/ Domenec Sibina). Marcial Pons. Madrid, 44-80.

Prats Albentosa, L. (2022) "Derecho a gozar de una vivienda digna y adecuada y derecho a la propiedad de vivienda". *Revista Jurídica del Notariado,* julio-diciembre, n.º 115, 131-174.

Prieto Sanchís, L. (1990). *Estudios sobre derechos fundamentales.* Debate. Madrid.

Rodríguez Tirado, A. M.ª, (2024). "La Ley por el derecho a la vivienda y la vulnerabilidad económico-social en los procesos arrendaticios de desahucio de vivienda habitual por falta de pago. ODS y Reglas de Brasilia". *La Ley, Derecho de familia,* n.º 41, 1-29.

Sainz-Cantero Caparrós, M. B. (2023). "La redefinición del derecho de propiedad sobre la vivienda en la era de los bienes comunes. A propósito de la Ley 12/2023 del derecho a la vivienda", *Actualidad Jurídica Iberoamericana,* n.º 19, 630-663.

Serrano Moreno, J. L. (1987). "Algunas hipótesis sobre los principios rectores de la política social y económica". *Revista de Estudios Políticos,* n.º 56, pp. 95-120.

Tornos Mas, J. (2014). "El acceso a la vivienda y la nueva función social del derecho de propiedad". *R.V.A.P.,* n.º especial 99-100, 2853-2871.

Capítulo IV

La configuración del contenido de derecho de propiedad de la vivienda en la ley 12/2023

MARÍA TERESA ALONSO PÉREZ
Catedrática de Derecho civil
Universidad de Zaragoza

I. PLANTEAMIENTO

El tratamiento jurídico de cualquier aspecto del derecho de propiedad en términos generales es complicado, más cuando se trata de la propiedad inmobiliaria y, más aún, si cabe, cuando el inmueble objeto del derecho es la vivienda, por todas las connotaciones que la misma conlleva, no sólo desde el punto de vista jurídico, sino económico, social y político. La dificultad es tanta que, en cuanto se empieza a abordar la tarea, enseguida se percibe que consiste en poner puertas al campo o contener el océano. Como este estudio se enmarca en un análisis colectivo de diferentes aspectos regulados por la Ley estatal 12/2023, de 24 de mayo, por el derecho a la vivienda (LDV) esta norma va a servir de asidero en torno al cual ir discurriendo sobre cómo se conforma el contenido del derecho de propiedad de la vivienda.

Empezaré reseñando el aparato normativo básico y estructural en la materia y las relaciones existentes entre los mismos (apartado II); en dicho apartado se estudian los tres límites constitucionales al desarrollo normativo de la regulación del contenido del derecho de propiedad de la vivienda y que son los de índole competencial (epígrafe 1), la necesidad de que se regule en norma con rango de ley (epígrafe 2) y, finalmente, el debido respeto al contenido esencial del derecho que se conecta con la función social que el mismo debe desarrollar (epígrafe 3). Tras exponer, en el apartado III, unas ideas generales sobre cómo el artículo 1.2 LDV delimita el alcance de la Ley integrando en él la regulación del contenido del derecho a la propiedad de la vivienda, analizaré el artículo 10.1 LDV en el que se regulan las facultades del propietario de la vivienda –apartado IV–. En el apartado V, estudiaré el desconcertante artículo 10.2 que regula cuestiones de muy diversa índole y que no siempre guardan relación con el contenido del derecho de propiedad, sino que afectan a cómo las Administraciones públicas pueden fomentar el acceso a la vivienda de los ciudadanos. El apartado VI versa sobre los deberes que pesan sobre el propietario. Acabaré exponiendo unas conclusiones finales acerca de cómo entiendo que la LDV regula el contenido del derecho de propiedad sobre la vivienda.

II. APARATO NORMATIVO BÁSICO REFERIDO AL CONTENIDO DEL DERECHO DE PROPIEDAD DE LA VIVIENDA

El conjunto de normas que regula aspectos relativos a la vivienda es inmenso y está integrado por normas correspondientes a muy diferentes sectores jurídicos, sin embargo, en lo atinente a la propiedad de la misma puede identificarse un aparato normativo primario o básico conformado por las normas constitucionales que se refieren directamente a dicho concepto y por la regulación básica del derecho de propiedad contenida en el Código civil. A ese aparato primario se incorpora la Ley estatal 12/2023, de 24 de mayo, por el derecho a la vivienda.

Los artículos de la Constitución española (CE) que guardan relación directa con la vivienda y con la propiedad privada son el artículo 47 y el artículo 33 respectivamente, sin perjuicio de que otros preceptos constitucionales incidan de manera indirecta o colateral en la regulación de una y otra realidad[1].

El artículo 47 CE reconoce el derecho de todos los españoles a disfrutar de una vivienda digna y adecuada, instando a los poderes públicos a promover las condiciones necesarias para ello; además, dispone este precepto que se establecerán las normas pertinentes para hacer efectivo este derecho. Este artículo no prejuzga la titularidad jurídica por la que los ciudadanos pueden acceder a la vivienda, de modo que no puede derivarse del mismo ninguna directriz que nos permita definir o perfilar el contenido del derecho de los que detentan su vivienda en concepto de propietarios.

Por su parte, el artículo 33 CE, reconoce el derecho a la propiedad privada y a la herencia. Según el párrafo segundo de este precepto, el contenido del derecho de propiedad debe venir delimitado por la función social que, de acuerdo con las leyes, está llamado a cumplir. Por otro lado, se indica, en el párrafo tercero, que nadie podrá ser privado "de sus bienes y derechos sino por causa justificada de utilidad pública o interés social, mediante la correspondiente indemnización y de conformidad con lo dispuesto por las leyes".

Además del artículo 33 CE, debe tenerse en cuenta que el artículo 348 del Código civil proporciona el contenido básico del derecho de propiedad disponiendo que, entre las faculta-

---

1 Así, por ejemplo, el artículo 33 se relaciona con el principio de libertad de empresa recogido en el artículo 38 CE. Vid. en este sentido López y López (2022, p. 19), la propiedad privada y la libertad de empresa tienen una relación dialéctica.

des que integran este derecho subjetivo, se encuentra la facultad de goce, la cual incluye el uso y el disfrute, entendido éste como la posibilidad de uso y de explotación económica de la cosa por el propietario; además, este precepto reconoce al propietario la facultad de disposición, la cual debe reconocerse en el más alto grado posible y tanto a través de negocios jurídicos *inter vivos* como *mortis causa* –lo que conecta esta previsión normativa con el derecho a la herencia que reconoce el artículo 33 de la Constitución como extensión del derecho a la propiedad privada–. Además, el artículo 348 del Código civil otorga al propietario la facultad para reivindicar la cosa, lo que supone el reconocimiento de la facultad de defensa de su derecho[2].

Estos preceptos representan la estructura jurídica básica que debe ser atendida primariamente a la hora de configurar el contenido del derecho de propiedad de manera genérica con independencia de cuál sea su objeto. A esta estructura normativa esencial se sobrepone un conjunto de diferentes normas sectoriales que regulan determinados tipos de propiedad en función de su objeto, como la propiedad intelectual, la de minas, la de aguas, etcétera[3]. En el caso de la vivienda, a la normativa genérica referida y que se condensa en el artículo 33 CE

---

[2] Incluye la facultad de exclusión entre las facultades del dominio Castán Tobeñas (1992, p. 185): son las facultades "*que permiten al propietario impedir la intromisión o perturbación causada por personas extrañas, en el goce o utilización de la cosa*". En la p. 192 especifica que: "El derecho de reivindicar consiste en la facultad que tiene el propietario de reintegrarse en el poder que le corresponde sobre la cosa, por medio de la correspondiente acción judicial, cuando haya sido perturbada o desconocida su propiedad".

[3] Para Verdera Server (2023, pp. 868 y 869), "la pluralidad de estatutos dominicales tiene una indudable conexión con la función social de la propiedad". Guardaría relación, según este autor, con el diverso uso y tipo de explotación que se asocia a cada bien.

y en el artículo 348 Código civil, se superpone la regulación específica del contenido del derecho de propiedad de la vivienda que efectúa la Ley 12/2023, de 24 de mayo, por el derecho a la vivienda. La aprobación de esta ley supone un cambio sustancial en términos de propiedad privada que queda muy afectada cuando la misma recae sobre una vivienda, pero, aunque la regulación de alguno de sus aspectos, como el del conjunto de facultades y deberes del propietario, resultan sorprendentes, no creo que pueda afirmarse que se trate de un derecho más próximo al derecho público que al privado[4]. La propiedad privada sigue siendo una noción que, como su propio nombre indica es privada y, en consecuencia, civil.

Por otro lado, existe una importante doctrina contenida en las resoluciones del Tribunal Constitucional referidas al derecho de propiedad sobre la vivienda que debe tenerse en cuenta a la hora de interpretar las normas que la regulan. Esta doctrina se ha dictado en relación al derecho de propiedad tanto con carácter genérico como cuando su objeto es específicamente la vivienda.

1. La relación entre el artículo 33 CE y el artículo 348 del Código civil

Es habitual contraponer el concepto de derecho de propiedad que aparece recogido en el artículo 33 de la Constitución y el del artículo 348 del Código civil, como si cada uno de ellos reflejará concepciones diametralmente opuestas y distintas de este derecho. Dicha visión resulta poco fructífera en términos de delimitación del contenido del derecho de propiedad, cuando no perturbadora.

Cada uno de estos preceptos responde a un contexto histórico, económico y social determinado, pero el distinto momento histórico en que se promulgan estos preceptos no per-

4 Simón Moreno (2023, p. 144).

mite afirmar que el derecho de propiedad a que uno y otro precepto aluden sean realidades jurídicas contrapuestas o que se den la espalda.

En realidad, la completitud del Ordenamiento jurídico nos debe llevar a la idea de que el derecho de propiedad definido en el artículo 348 del Código civil es exactamente el mismo derecho de propiedad que aparece recogido en el artículo 33 de la Constitución española. Cada uno de estos preceptos regula una dimensión diferente del derecho de propiedad, concretamente la que a cada una de estas normas corresponde según su naturaleza y alcance. Por ello, no conviene, a la hora de delimitar el contenido del derecho de propiedad, enfrentar dichos preceptos, sino que es menester dar a cada uno su papel y afirmar su complementariedad.

De algunas resoluciones del TC parece deducirse que estos preceptos tensionan el concepto de derecho de propiedad que lo colocarían ante la disyuntiva que supone una visión eminentemente individual del mismo –la del Código civil– y una concepción social de dicho derecho subjetivo, que sería recogida por la Constitución. Cuando se leen algunas resoluciones del Tribunal Constitucional, el artículo 33 CE parecería desdecir o incluso anular, siquiera parcialmente, la visión individualista de la propiedad privada que se dice recoge el Código civil. Sin embargo, no es acertado enfrentar ambos preceptos como si se refirieran a conceptos del derecho de propiedad irreconciliables y ello porque, insisto, el derecho de propiedad del artículo 348 del Código civil y el del artículo 33 CE son exactamente el mismo derecho subjetivo. Más aún, el concepto de derecho de propiedad que recoge el artículo 348 del Código civil es constitucional, es decir, respeta y se ajusta a la Constitución, también a lo previsto sobre la propiedad privada en el artículo 33 CE.

Estas reflexiones nos llevan a plantearnos cuál es el papel de cada uno de estos preceptos en la conformación del contenido del derecho de propiedad. Si se atiende al tenor literal de estos preceptos nos encontramos con que el artículo 33 CE no dice nada acerca del contenido del derecho subjetivo que reconoce, si bien, en su párrafo segundo, viene a precisar que quedará delimitado por su función social, sin precisar en qué consiste la misma. Nada establece acerca del haz de facultades y deberes que se atribuyen al titular. Sin embargo, el artículo 348 del Código civil sí alude al contenido del derecho, señala el alcance de la potestad que sobre una cosa concreta corresponde a quien es su propietario. Si bien la delimitación del contenido del derecho de propiedad no puede derivarse exclusivamente del contenido del artículo 348 del Código civil[5] sino que debe atenderse a las normas que inciden en el mismo conformándolo y delimitándolo.

Por otro lado, el artículo 348 Código civil sólo refiere las facultades que detenta el propietario y nada precisa de los deberes que comporta tal titularidad; eso no significa que el Código civil no contemple tales obligaciones, pues se derivan de otras normas que con rango de ley lo conformen, como ocurre con las obligaciones fiscales o con las relaciones de vecindad, entre otras.

2. La relación entre el artículo 348 Código civil y la Ley 12/2023

Entre los objetivos de la Ley 12/2023 (LDV) se encuentra la delimitación del contenido del derecho de propiedad de la vivienda y, por razones de jerarquía normativa, debe respetar lo previsto en la CE. Sin embargo, esta ley, puede incidir y modificar la regulación que, sobre la propiedad

---

5 En este sentido, Aguilera Vaqués (2018, p. 706).

en general, se contiene en el Código civil al tratarse de normas del mismo rango.

Así ha ocurrido previamente con la propiedad que recae sobre otro tipo de objetos como las aguas, las minas, la propiedad intelectual, los inmuebles urbanos y rústicos. Se dice, por ello, que la propiedad de la vivienda ha pasado a ser una propiedad estatutaria similar a la que recae sobre los objetos referidos. Esto supone que el contenido del derecho de propiedad que se desprende de la regulación general de la propiedad tal y como aparece recogida en el Código civil va a ser delimitado más precisamente cuando el objeto de dicha titularidad sea una vivienda. Pero, al igual que el artículo 348 del Código civil respeta la CE, esta delimitación estatutaria del contenido de la propiedad sobre la vivienda también debe respetar la Carta Magna[6].

---

6 Se han interpuesto varios recursos de inconstitucionalidad contra diferentes preceptos de esta Ley:
1306/2024 (Ref. BOE-A-2024-7202). Gobierno gallego
1278/2024 (Ref. BOE-A-2024-7201). Gobierno catalán
1301/2024 (Ref. BOE-A-2024-5393). Gobierno vasco
5580/2023, (Ref. BOE-A-2023-20500). Parlamento de Cataluña
5518/2023 (Ref. BOE-A-2023-20499). Gobierno de la Comunidad de Madrid
5516/2023, (Ref. BOE-A-2023-20498). Consejo de Gobierno de las Islas Baleares
5514/2023, (Ref. BOE-A-2023-20497). Grupo Parlamentario Popular en el Congreso
5491/2023 (Ref. BOE-A-2023-20496). Gobierno de la Comunidad de Madrid

3. Los límites constitucionales al desarrollo de la regulación del contenido del derecho de propiedad de la vivienda

El ajuste de la LDV a la Constitución, en lo que nos afecta, se manifiesta en tres aspectos importantes: la reserva de ley, el respeto al contenido esencial del derecho de propiedad, así como el ajuste al reparto constitucional de competencias. Empezamos analizando este último.

3.1. Distribución competencial en relación al contenido del derecho de propiedad

La regulación del Código civil en materia de propiedad y la Ley 12/2023 sobre el derecho a la vivienda son legislación estatal, pero hay que tener en cuenta que el artículo 148.1.3 de la Constitución española atribuye la competencia en materia de vivienda a las Comunidades Autónomas y, en uso de la misma, han dictado leyes generales sobre vivienda la mayoría de Comunidades Autónomas[7]. El resto de CCAA, aunque no

---

[7] Legislación autonómica dictada en materia de vivienda, ordenada cronológicamente:
Canarias: Ley 2/2003, de 30 de enero, de Vivienda de Canarias. https://www.boe.es/buscar/act.php?id=BOE-A-2003-4607&p=20201228&tn=6
Valencia: Ley 8/2004, de 20 de octubre, de la Vivienda de la Comunidad Valenciana. https://www.boe.es/buscar/act.php?id=BOE-A-2004-19752
La Rioja: Ley 2/2007, de 1 de marzo, de Vivienda de la Comunidad Autónoma de La Rioja. https://www.boe.es/buscar/act.php?id=BOE-A-2007-6609
CA catalana: Ley 18/2007, de 28 de diciembre, del derecho a la vivienda. https://www.boe.es/buscar/act.php?id=BOE-A-2008-3657
Andalucía: Ley 1/2010, de 8 de marzo, Reguladora del Derecho a la Vivienda en Andalucía.
https://www.boe.es/buscar/doc.php?id=BOE-A-2010-5218

dispongan de regulación genérica sobre la vivienda, han regulado algunos aspectos de la misma, es el caso del Principado de Asturias, Cantabria, Aragón, Castilla-La Mancha y Madrid.

La cuestión competencial fue objeto de un vivo debate durante el proceso de elaboración de la LDV que se refleja claramente en la tramitación –un tanto controvertida– del Informe que debía elaborar y elaboró el Consejo General del Poder Judicial[8] y también en el debate parlamentario del Proyecto de Ley.

Navarra: Ley Foral 10/2010, de 10 de mayo, del Derecho a la Vivienda en Navarra. https://www.boe.es/buscar/act.php?id=BOE-A-2010-8618
Castilla y León: Ley 9/2010, de 30 de agosto, del derecho a la vivienda de la Comunidad de Castilla y León. https://www.boe.es/buscar/act.php?id=BOE-A-2010-14849
Galicia: Ley 8/2012, de 29 de junio, de vivienda de Galicia https://www.boe.es/buscar/act.php?id=BOE-A-2012-11415
Murcia: Ley 6/2015, de 24 de marzo, de la Vivienda y lucha contra la ocupación de la Región de Murcia https://www.boe.es/buscar/act.php?id=BOE-A-2015-4747&p=20220601&tn=2
CA vasca: Ley 3/2015, de 18 de junio, de vivienda. https://www.boe.es/buscar/act.php?id=BOE-A-2015-7802
Islas Baleares: Ley 5/2018, de 19 de junio, de la vivienda de las Illes Balears. https://www.boe.es/buscar/act.php?id=BOE-A-2018-9774
Extremadura: Ley 11/2019, de 11 de abril, de promoción y acceso a la vivienda de Extremadura. https://www.boe.es/buscar/act.php?id=BOE-A-2019-7224

8 El Informe se aprobó tras el rechazo del que primeramente se había elaborado y que fue rechazado en el Pleno. Las discrepancias versaban fundamentalmente sobre el alcance de las competencias estatales en la materia frente a las autonómicas. El Informe del Consejo General del Poder Judicial puede consultarse en el siguiente enlace: https://www.poderjudicial.es/cgpj/es/Poder-Judicial/En-Portada/El-Pleno-del-CGPJ-aprueba-el-informe-al-anteproyecto-de-Ley-por-el-Derecho-a-la-Vivienda

No obstante la competencia autonómica en materia de vivienda, no puede negarse que el Estado ha regulado muchos aspectos de la misma como, por ejemplo y sin ánimo de exhaustividad, la propiedad horizontal, la responsabilidad por defectos en la construcción o compraventa de vivienda o el arrendamiento de vivienda, entre otros. Por lo tanto, el Estado tiene competencia para legislar en materia de vivienda, debiendo precisarse hasta dónde puede llegar en el ejercicio de la misma para que no colisione con la de las CCAA[9].

La regulación del contenido del derecho de propiedad de la vivienda es competencia estatal[10] [11]. La Disposición final 7.ª LDV alude a los preceptos constitucionales en que se apoya la competencia estatal para dictar dicha ley, refiriendo como título competencial, el artículo 149.1.1 CE, en base al cual se entendería que el derecho de propiedad sobre la vivienda debe tener el mismo contenido en todo el territorio nacional para garantizar la efectividad del principio de igualdad. Otros títulos competenciales que guardan relación con esta cuestión

---

9 Calduch Alonso (2023, p. 1483).

10 Moreu Carbonell (2018, p. 42): "la innovación en el contenido del derecho de propiedad corresponde al Estado y debe hacerse con pleno respeto a la igualdad de todos los españoles (art. 149.1.1 CE) y sin violentar el núcleo esencial del derecho (art 53 CE)". En el mismo sentido: Arias Martínez (2019, p. 117): "nadie cuestiona que la propiedad privada es una institución civil cuyo tratamiento normativo se encuentra en el Código civil y que esta es una materia que, conforme a lo previsto en el artículo 149.1.8 CE corresponde en exclusiva al Estado".

11 Aunque nos encontramos ante una competencia estatal, muchas normas autonómicas aluden expresamente a cuestiones que guardan relación con el contenido del derecho de propiedad de la vivienda. Puede servir de ejemplo el artículo 1.2 de la Ley canaria 2/2003, de 30 de enero, de Vivienda, comunidad autónoma que no tiene competencia en legislación civil.

y mencionados en dicha D F 7.ª son el artículo 149.1.13 CE, relativo a la planificación general de la actividad económica, y, obviamente, el artículo 149.1.8 CE que atribuye al Estado la competencia en materia de legislación civil.

3.2. Reserva de ley

El artículo 53 de la Constitución española exige que el ejercicio de los derechos incluidos en el capítulo II del Título I, entre los que se encuentra el derecho a la propiedad privada recogido en el artículo 33, debe regularse por ley, es decir, por una norma con rango legal, la cual en todo caso deberá respetar su contenido esencial.

Sobre la reserva de ley en materia de propiedad privada que es predicable de la de la vivienda, existen varias resoluciones del Tribunal Constitucional que se han dictado en relación a varias normas autonómicas.

Por su parte, la STC 93/2015 ventiló un recurso de inconstitucionalidad contra el Decreto-ley andaluz 6/2013, de 9 de abril, de medidas para asegurar el cumplimiento de la Función Social de la Vivienda. Esta STC, en relación al punto de la configuración del deber de uso de la vivienda como contenido del derecho de propiedad sobre la misma, sin entrar en el fondo del asunto, viene a decir que la regulación del contenido del derecho de propiedad de la vivienda no es una cuestión que pueda regularse mediante Decreto-ley, siendo necesario hacerlo por ley.

Coherentemente con la anterior resolución, la STC 16/2018 declara constitucional la misma regulación del deber de uso de la vivienda en una ley navarra porque se contiene en una norma con rango de ley. Se resolvía un recurso de inconstitucionalidad referido a varias normas de la Ley Foral 24/2013, de 2 de julio que modificaba la Ley Foral 10/2010, de 10 de mayo, del Derecho a la Vivienda en Navarra. En este caso el TC no considera que haya invasión de las competencias estatales que le son atribuidas en función del artículo 149.1.1 y 149.1.8 CE.

3.3. Contenido esencial del derecho: el equilibrio entre la utilidad individual y la función social de la propiedad

El tercer límite constitucional que se establece para la regulación de la propiedad privada contemplada en el artículo 33 y, por tanto, también para la que recae sobre la vivienda es el respeto al *contenido esencial del derecho* tal y como se deriva del artículo 53 de la Constitución. Ahora bien, no es fácil precisar cuál es el contenido esencial del derecho de propiedad de la vivienda[12], ni cuales deben ser las consecuencias de su vulneración por la ley[13]. La valoración de si una determinada regulación del derecho respeta o no el contenido esencial del derecho debe

---

12 El voto particular de Álvaro Rodriguez Bereijo, al que se adhieren Pedro Cruz Villalón y José Gabaldón López- a la STC 89/1994 viene a señalar que: "En la sentencia de la que disiento viene a dejarse en manos del Legislador ordinario, en aras de la función social de la propiedad, la definición del contenido esencial de las "propiedades"....En definitiva, el contenido esencial del derecho de propiedad será lo que el propio legislador diga en cada caso y según el criterio dominante en cada momento histórico. Así, huérfano de todo referente ex Constitutione, la garantía institucional que para el derecho de propiedad representa el concepto de "contenido esencial" se desvanece y volatiliza".

13 Según López y López (2022, p. 27), cuando se sacrifica el contenido esencial del derecho de propiedad estamos ante una expropiación forzosa. Verdera Server (2023, p. 879): "El legislador al amparo de la función social, puede intervenir en la delimitación de la propiedad privada, pero si esa delimitación afecta al contenido esencial, se activa la garantía expropiatoria y la intervención sólo es posible si se indemniza".

hacerse a la vista de los principios de razonabilidad, proporcionalidad y adecuación.

El único elemento que proporciona el texto constitucional para ayudar en la concreción del contenido del derecho de propiedad privada es la mención a la función social que el artículo 33.2 de la CE incorpora, pero no se precisa qué debe entenderse por tal; tampoco la proyección que debe tener sobre el contenido del derecho se formula de manera clara y terminante[14].

Se suele blandir la función social como exponente de la evolución de la visión que se dice que recoge el Código civil sobre la propiedad y que la referencia que se hace a la misma en el artículo 33.2 CE supondría una nueva perspectiva que ubicaría a este derecho en un plano que, trascendiendo la dimensión meramente individual que atendería los intereses particulares del propietario, tendría en consideración los intereses generales.

La proyección de la función social como elemento delimitador del contenido del derecho de propiedad en general y de la propiedad de la vivienda en particular se suele entender que configura el aspecto pasivo de este derecho, imponiendo cargas y deberes al propietario. Sin embargo, acaso debiera explorarse la proyección de este elemento sobre el lado activo en cuanto que, desde dicho plano referido a las facultades del titular, la propiedad privada atiende intereses generales cubriendo la necesidad de vivienda de sus titulares, permitiéndoles generar crédito –mediante su hipoteca– y, consecuentemente, activar la economía, también sirve de mecanismo de pro-

---

14 Se entiende que es un elemento que habilita a los poderes públicos para delimitar el derecho de propiedad privada. Vid. Aguilera Vaqués (2018, p. 707). Según Verdera Server (2023) "la función social es la expresión legislativa de la finalidad socioeconómica que cumplen los bienes sobre los que recae el dominio".

tección social[15]. Esas prerrogativas del propietario, en cuanto individuo que forma parte del tejido social, permiten que la propiedad de la vivienda atienda al mismo tiempo necesidades particulares y generales o sociales.

La función social como elemento delimitador del contenido del derecho de propiedad de la vivienda no puede reducir la utilidad individual del derecho hasta anularla, debiendo establecerse un equilibrio entre los intereses generales que está llamada a atender y la utilidad individual que tal derecho debe reportar a su titular[16]. Es más, Verdera Server hace prevalecer la utilidad individual frente a la función social[17]. Por su parte, la STC 37/1987 dijo, de manera muy significativa en este sentido, que: "Utilidad individual y función social definen, por tanto, inescindiblemente el contenido del derecho de propiedad sobre cada categoría de bienes".

En la práctica, la función social referida en la CE sobre la propiedad ha servido para, fundamentalmente, marcar deberes al propietario, es decir, para recortar la potestad del propietario sobre el objeto. En coherencia con esta forma de entender las cosas, la LDV menciona la función social como delimitador del contenido de la propiedad en el artículo 11, referido a los deberes y cargas que la propiedad conlleva para el propietario.

---

15 Messía de la Cerda Ballesteros (2020, pp. 3451 y 3452). Este autor alude a la incidencia de la función social en el lado activo del derecho.

16 Para Aguilera Vaqués (2018, p. 709): utilidad individual y función social definen el contenido del derecho de propiedad sobre cada categoría de bienes. En sentido similar, Verdera Server (2023, p. 895), viene a decir que la función social se ha utilizado para vaciar de contenido la utilidad individual del derecho y lo que se pretendía era más bien un "equilibrio entre el interés particular y el interés general y no la prevalencia de este frente a aquel".

17 Verdera Server (2023, p. 920).

La invocación a la función social y la imposición de deberes y cargas al propietario en atención a la misma no puede legitimar un recorte de la potestad que haga al derecho irreconocible porque debe respetarse su contenido esencial, tal y como prescribe el artículo 53 de la Constitución. Y la clave para discernir entre las delimitaciones amparadas por la función social y, por tanto, respetuosas con la Constitución, y las que no entran en dicho paraguas y, en consecuencia, podrían ser tildadas de inconstitucionales, radica en hallar qué deberes impuestos al propietario recortan su derecho de manera que lo hacen irreconocible.

Se ha entendido que la afectación del contenido esencial del derecho por una norma amparada en la función social origina un supuesto expropiatorio, lo que se solventaría indemnizando al propietario por la pérdida de su titularidad[18]. Ahora bien, cabe hablar de expropiación cuando se priva de la titularidad al propietario, no cuando, manteniéndosele en su derecho, se recortan tanto sus facultades que ya no puede considerarse que el derecho del que sigue siendo titular pueda identificarse con el derecho de propiedad, aunque se le denomine así. En este segundo caso, no nos encontraríamos ante un supuesto expropiatorio de los contemplados en el artículo 33.3 de la Constitución, sino ante una norma reguladora del derecho

---

18 Verdera Server (2023, p. 879): "el legislador, al amparo de la función social, puede intervenir en la delimitación de la propiedad privada, pero sí esa delimitación afecta al contenido esencial, se activa la garantía expropiatoria y la intervención solo es posible si se indemniza". No obstante, en la p. 921 matiza que la ablación o privación del derecho debe ser completa para que haya expropiación. Vid. también López y López (2022, p. 21). Según este autor si estamos ante limitaciones procede una indemnización por estar ante una expropiación y si estamos ante límites entendidos como una intervención delimitadora general no se genera derecho a indemnización alguna.

de propiedad que debe tacharse de inconstitucional y que, en consecuencia, se tutela a través del recurso de inconstitucionalidad del artículo 161.1.a) CE, como indica el artículo 53 CE.

El derecho de propiedad debe proporcionar una utilidad privada, individual y ello forma parte de su contenido esencial; tan es así que no precisa formulación explícita en el artículo 33 CE porque basta con mencionar el carácter de "privada" de la propiedad. Sin embargo, ha sido necesario delimitar el contenido de la propiedad mediante un mecanismo interno con base en la función social que está llamado a desempeñar. Como dice Verdera, si la función social es la que, según el artículo 33 CE, delimita el derecho, el elemento central es la utilidad individual[19], hasta el punto de que considera este autor que una regulación que elimine toda utilidad individual afecta al contenido esencial y no puede justificarse por la vía de la función social.

III. EL ARTÍCULO 1.2 LDV Y SU REFERENCIA AL CONTENIDO DEL DERECHO DE PROPIEDAD SOBRE LA VIVIENDA

La Ley 12/2023 regula el derecho a la vivienda en sentido amplio, es decir, el derecho a la vivienda contemplado en el artículo 47 de la Constitución española. En la medida en que la adquisición de su propiedad es una vía de acceso a la vivienda, parece tener sentido que esta ley regule la conformación del derecho de la propiedad de la vivienda. Y, en efecto, este es uno de los objetivos de la ley que aparece expresamente formulado en el párrafo segundo del artículo 1. Además de este precepto, se refieren de manera específica a este asunto los artículos 10 y 11.

---

19 Verdera Server (2023, pp. 913 y 914).

El artículo 10 regula las facultades del propietario y el artículo 11 se refiere a sus deberes y cargas.

El artículo 1.2 de la LDV establece, entre los objetivos de la ley, la regulación del contenido del derecho de propiedad *en relación con su función social*. En esta primera referencia que hace la ley al contenido de la propiedad de la vivienda, no se alude, llamativamente, al aspecto positivo o activo del derecho subjetivo en cuestión; no se enumeran aquí las actuaciones que puede realizar el propietario en calidad de tal. Sería esperable y hubiera sido adecuado que se indicara, del mismo modo que se hace con los deberes, el poder de actuación que otorga a su titular el derecho subjetivo en cuestión y se señalara, además, que los poderes públicos garantizarán su efectivo disfrute. Sin embargo, el precepto alude únicamente al aspecto pasivo de la titularidad, enumerándose una serie de deberes del propietario y estableciendo que los poderes públicos van a asegurar –¿vigilar?– el cumplimiento de tales deberes, aplicando las medidas que legalmente procedan.

Es decir, en el frontispicio de la ley que, sin duda, es el artículo 1 LDV, en relación al contenido del derecho de propiedad de la vivienda viene a decirse que el propietario tiene una serie de deberes cuyo cumplimiento va a ser controlado por los poderes públicos. Es cierto que, más adelante, concretamente en el artículo 10 LDV, se contiene la delimitación positiva del derecho y se enumeran las facultades del titular, pero, dado que, en este primer artículo, se alude a los deberes, hubiera sido más correcto incluir también las facultades del propietario.

La formulación del artículo 1.2 LDV responde a la concepción del derecho de propiedad como aquel que otorga a su titular el más amplio poder que el Ordenamiento reconoce sobre un objeto, de modo que parece que lo único que puede hacerse por la ley para delimitar su contenido es "recortar" ese poder, reducirlo. Por eso sólo se alude al aspecto pasivo refiriendo algunos de los deberes del propietario y aludiendo al control o vigilancia de su cumplimiento por parte de las Administraciones.

IV. LAS FACULTADES QUE INTEGRAN EL CONTENIDO DEL DERECHO DE PROPIEDAD SOBRE LA VIVIENDA: ARTÍCULO 10.1 LDV

El artículo 1.2 LDV no hace referencia a las facultades del propietario de vivienda, las cuales se regulan en el artículo 10 LDV que perfila el margen de actuación que a su titular otorga el derecho de propiedad sobre la vivienda.

En primer lugar, demostrando que no se puede hacer abstracción del Código civil porque es el cuerpo normativo que configura el contenido del derecho de propiedad en su vertiente positiva, este artículo se refiere a las facultades que, según dicho precepto, integran tal derecho con carácter genérico: mientras el artículo 10.1. a) de la LDV alude a las facultades de *uso, disfrute y disposición,* el artículo 348 del Código civil menciona literalmente las facultades de *gozar y disponer.*

A la facultad de uso nos referiremos en el apartado VI-1. En lo que respecta a la *facultad de disfrute,* que incluye la explotación económica de la propiedad de la vivienda, ha sido *recortada* por esta Ley en cuanto la Disposición final 1.ª LDV establece un sistema de contención de precios en el alquiler de vivienda y la prórroga forzosa[20].

---

20 La STC 89/1994 en relación a la prórroga forzosa que se impuso para los arrendamientos urbanos decía que "una disposición que supusiera el vaciamiento del contenido económico de la renta acordada podría representar la vulneración del derecho reconocido en el artículo 33 CE".

En relación a esta cuestión varias resoluciones del TC se dictaron al hilo de una serie de normas autonómicas dictadas con posterioridad a la crisis de 2008 y que fueron declaradas inconstitucionales. Por ejemplo, la STC 37/2022 en relación a la Ley catalana 11/2020. En esta resolución, el TC considera que la fijación del precio-renta de los alquileres forma parte de las bases de las obligaciones contractuales que están reservadas en exclusiva a la legislación estatal por el artículo 149.1.8 CE. Más allá de esta resolución y de otras similares, la cuestión es si el Estado puede fijar un sistema de contención de rentas como el que ha establecido en la LDV sin afectar el contenido esencial del derecho de propiedad. Y sobre eso no hay respuesta del Tribunal Constitucional; habrá que esperar a los recursos de constitucionalidad que se han interpuesto contra la DF primera de la Ley[21].

Incide también en el derecho de uso y disfrute, la suspensión de los lanzamientos de vivienda habitual establecida en el Decreto-ley 1/2021, de 19 de enero, de protección de los consumidores y usuarios frente a situaciones de vulnerabilidad económica y social. La STC 9/2023 dice que dicha suspensión no supone una afectación del contenido esencial del derecho de propiedad y desestima el recurso de inconstitucionalidad, al igual que la STC 15/2023[22].

---

21 Recursos contra diferentes apartados de la Disposición Final primera de la LDV: n.º 5491-2023 promovido por la Comunidad de Madrid; número 5514-2023 promovido por el grupo parlamentario popular del Congreso; número 5516-2023, promovido por el Consejo de Gobierno de las Illes Balears; y, finalmente, el número 1301-2024, promovido por el Gobierno Vasco.

22 Existe un voto particular a las dos SSTC referidas del magistrado Arnaldo Alcubilla.

En cuanto a la *facultad de disposición*, la STC 28/2012 se pronuncia en relación a Ley canaria 5/1999 que modificaba la legislación turística y constituía un derecho de retracto a favor de particulares, considerando inconstitucional este precepto porque Canarias no tiene competencia en materia de legislación civil (art 149.1.8 CE) [23].

El artículo 10 LDV no ha incluido una de las facultades que se recogen en el artículo 348 del Código civil y que consiste en la *posibilidad del propietario de defender su titularidad, reivindicándola.* Esta ausencia me parece muy significativa y entiendo que debería incorporarse expresamente como facultad del propietario en la LDV. Me refiero a esta cuestión en el apartado siguiente.

Podría decirse que esa primera letra a) del artículo 10.1 LDV es la única que realmente conforma el contenido del derecho de propiedad de la vivienda, pues, en cuanto a las facultades recogidas en las otras letras: la recogida en la letra b) no es propiamente contenido del derecho de propiedad y la letra c) alude a facultades que son meras manifestaciones o desarrollos de las recogidas en la letra a) como explico a continuación.

En cuanto al "derecho de consulta de las Administraciones públicas sobre la situación urbanística de la vivienda" o del edificio en que se encuentra –artículo 10.1. b) LDV–, se deriva de la obligación de transparencia que compete a las Administraciones públicas. La facultad de "consulta" no sólo la tiene el propietario de una vivienda en cuanto tal, sino cualquier

---

[23] Se anuncia la aprobación de un nuevo Decreto-ley por parte de la Generalidad catalana en el que se establece un derecho preferente de compra de la misma para adquirir las viviendas cuya propiedad corresponde a grandes tenedores. Noticia publicada (30.04/2024) https://www.eleconomista.es/vivienda-inmobiliario/noticias/12792987/04/24/la-generalitat-se-lanza-a-comprar-pisos-de-grandes-tenedores-sin-saber-cuantos-existen.html

ciudadano, sea o no propietario de la misma. El artículo 5 de Real Decreto Legislativo 7/2015, de 30 de octubre, por el que se aprueba el Texto Refundido de la Ley de Suelo y Rehabilitación Urbana (TRLSRU) recoge entre los derechos de los ciudadanos el derecho "a ser informados por la Administración competente, de forma completa, por escrito y en plazo razonable, del régimen y las condiciones urbanísticas aplicables a una finca determinada, en los términos dispuestos por su legislación reguladora". De tal manera que no puede entenderse que este derecho a la consulta de la situación urbanística de la finca sea contenido propio del derecho de propiedad sobre la vivienda, sino que es una facultad que el propietario tiene en su cualidad de ciudadano y, además, el propietario de la vivienda no puede impedir, en virtud de su titularidad, el acceso a dicha información por parte de otras personas.

En la letra c) de este artículo 10.1 LDV se reconoce al propietario de la vivienda la "facultad de realización de las obras de conservación, rehabilitación, accesibilidad universal, ampliación o mejora, de conformidad con las condiciones establecidas por la Administración competente". Este conjunto de actuaciones que el propietario puede realizar en la vivienda se deriva de sus facultades de uso, disfrute y disposición; no son, las contenidas en la letra c), facultades con una entidad propia y diferenciada de aquéllas otras. Adelanto que esta facultad, de manera contradictoria a lo preceptuado en este artículo 10 LDV, se establece como deber del propietario en el artículo 11 LDV, lo que nos pone sobre la pista de la intención del legislador en relación a la rehabilitación de viviendas, a la que más adelante haré referencia.

Lo dicho viene a reflejar que el precepto de referencia a la hora de precisar el contenido del derecho de propiedad de una vivienda sigue siendo el artículo 348 del Código civil, cuya centralidad en la materia queda avalada de manera contundente por el contenido de este artículo 10 LDV.

V. LA FACULTAD DE DEFENSA DE LA PROPIEDAD Y LA ADMINISTRACIÓN PÚBLICA COMO GARANTE DEL PLENO DISFRUTE DEL DERECHO DE PROPIEDAD DE LA VIVIENDA POR SU TITULAR: ARTÍCULO 10.2 LDV

Traigo de nuevo a colación el hecho de que el artículo 10.1 LDV no recoja expresamente la facultad de defensa de su titularidad que compete al propietario, que sí aparece recogida en el artículo 348.2 Código civil cuando menciona la posibilidad de reivindicación. Todo derecho conlleva una acción para su ejercicio y su defensa y, en este caso, el propietario, pese al silencio de la LDV sobre este punto, la detenta tanto en el orden jurídico-público o penal[24] como en el ámbito jurídico-privado o civil.

En el plano puramente civil, se dispone de la acción reivindicatoria como mecanismo de defensa del derecho de propiedad sobre la vivienda (artículo 348.2 Código civil). Además, el propietario dispone de las acciones encaminadas a declarar la titularidad del derecho de propiedad sobre una vivienda, de la acción negatoria y todas aquellas que pueden servir de instrumento de defensa del derecho de propiedad de la vivienda y de su integridad.

---

24 La facultad de defensa debe tomar en consideración las previsiones del Código penal que regula los delitos contra el patrimonio en los artículos 234 y siguientes y que, en lo atinente a la propiedad de los inmuebles como la vivienda, se concreta en los artículos 245 a 247 reguladores del delito de usurpación. Por otro lado, el delito de allanamiento de vivienda contemplado en el artículo 202 del Código penal también posibilita defender la vivienda de la invasión de un tercero, aunque, en este caso, no se protege exclusivamente su propiedad, sino el hecho que un inmueble constituya la vivienda de un sujeto, al margen del concepto en que la detente.

Quizás para suplir la falta de referencia expresa a dicha facultad del propietario de la vivienda, el párrafo segundo del artículo 10.2 LDV recoge el deber de la Administración pública de velar por el pleno ejercicio del derecho de propiedad sobre la vivienda. De este modo se refuerza el reconocimiento de las facultades del propietario que se hace en el párrafo primero colocando a la Administración como garante del disfrute pleno de la propiedad por su titular. Lo cual no obsta ni impide ni es incompatible con la defensa del derecho por parte de su titular.

Pero, realmente, la lectura del precepto revela que no parece ir destinado a regular la cuestión referida sino que, al delimitar la actuación de la Administración garantizando que los propietarios puedan disfrutar plenamente de su propiedad sobre las viviendas, el precepto obliga a los poderes públicos a actuar promoviendo las acciones pertinentes para favorecer el acceso a la vivienda, lo que no guarda relación con garantizar a sus titulares el disfrute del derecho de propiedad de la vivienda.

De lo antedicho cabe deducir que, en este párrafo 1 del artículo 10.1 LDV, se mezclan dos cuestiones diversas y distantes: por un lado, se convierte a la Administración en garante del pleno –y pacífico, añadiría yo– disfrute de la vivienda por su propietario y, por otro lado, se impulsa a la Administración pública a promover el acceso a la vivienda. Es más, como este inciso sobre la promoción del acceso a la vivienda por parte de los poderes públicos se ubica en un precepto relativo al derecho de propiedad, puede interpretarse en el sentido de que la Administración queda obligada a favorecer el acceso a la vivienda mediante la adquisición *de su propiedad.*

No obstante, coherentemente con la interpretación que hay que dar al precepto, se listan una serie de medidas susceptibles de ser adoptadas con el objetivo de promover el acceso a la propiedad de la vivienda. Algunas de las actuaciones recogidas en la enumeración se manifiestan como especialmente propicias como la relativa a las ayudas y subvenciones públicas o la que menciona los incentivos fiscales.

Sin embargo, las otras actuaciones recogidas en la lista carecen de conexión con el conjunto del precepto en que se ubican y están diseñadas para regular cómo la Administración debe favorecer el acceso a la vivienda en términos generales y no propiamente el acceso a su propiedad. Es decir, serían actuaciones que tendrían sentido en un precepto dirigido a regular el derecho a la vivienda del artículo 47 CE y carecen del mismo si se refieren a la propiedad de la vivienda. Así, se mencionan las siguientes medidas: gestión de los parques públicos de vivienda, colaboración con entidades del tercer sector para facilitar la gestión de los parques públicos de vivienda, fomentar la cesión de las viviendas por parte de los particulares para favorecer el incremento de oferta en alquiler social o a precio asequible.

En el último inciso del precepto se refiere también a una medida que, si bien guarda relación de manera indirecta con el ejercicio del derecho de propiedad, se enfoca a propiciar el acceso a la vivienda en términos generales y no a través del derecho de propiedad. Se dice en este precepto que la Administración va a "fomentar la intermediación en el mercado del arrendamiento de viviendas para propiciar su efectiva ocupación". El tenor del inciso no pretende amparar en modo alguno el ejercicio de la facultad de explotar la propiedad arrendando la vivienda por parte de su propietario; es más, se señala claramente cuál es el objetivo de la medida concreta que contempla: la *efectiva ocupación* de la vivienda. La Administración va a fomentar el alquiler, lo cual puede suponer el establecimiento de incentivos para animar a que el propietario de una vivienda en desuso la alquile voluntariamente o puede suponer el establecimiento de ciertos mecanismos de presión al propietario para que lo haga. En cualquier caso, no parece que sea ésta una medida que guarde relación con la defensa de la propiedad de la vivienda ni con el fomento del acceso a la misma. Al contrario, parece que esta medida puede afectar negativamente la plenitud de las facultades del propietario porque el objetivo no es la protección de su titularidad y de su ejercicio, sino la plena ocupación, quizás la consecución de ese objetivo se quiera conseguir a pesar del ámbito de actuación que, en principio, debería reportar la titularidad al propietario.

VI. LOS DEBERES Y CARGAS DEL PROPIETARIO DE VIVIENDA: ARTÍCULO 11 LDV

El artículo 11 LDV regula los deberes y cargas del propietario de una vivienda estableciendo un listado de los mismos. Cabe destacar, de entrada, dos aspectos: el primero es que se alude a la función social de la propiedad mencionada en el artículo 33.2 CE y, el segundo aspecto reseñable es que el precepto, al enumerar los deberes del propietario de vivienda, incluye algunos que, en el artículo 10 LDV, habían sido configurados previamente como facultades. Cabe plantear si es factible jurídicamente que una posibilidad de actuación del propietario sobre la vivienda pueda configurarse, a un tiempo, como deber y como facultad. A continuación, analizamos los deberes que este precepto pone a cargo del propietario de vivienda.

1. El deber de uso y disfrute propios y efectivos de la vivienda

Como hemos dicho anteriormente, el artículo 10 LDV conforma el uso y disfrute de la vivienda como una facultad del propietario que integra el contenido del dominio. Como toda facultad, se supone ínsita en ella la posibilidad del titular de ejercerla, pero también la de *no* ejercitarla. En relación a esta cuestión, puede traerse a colación la afirmación de Montés Penadés que se pronuncia en el sentido de que el no uso debe ser "considerado como goce y, por tanto, ejercicio del derecho"[25].

Tal consideración se admite sin problemas, sobre todo, si se predica del resto de facultades que integran el dominio: así, en relación a la facultad de reivindicar una cosa, el propietario que pierde la cosa o teme perderla puede reivindicarla –esto es, defender su titularidad– voluntariamente y puede, voluntariamente también, no ejercerla; la consecuencia de su no ejercicio es que, por efecto de la posesión, la cosa podrá llegar a ser propiedad de otro sujeto por intermedio de la prescripción adquisitiva si se dan el resto de requisitos. Lo mismo ocurre con la facultad de disposición: el titular de la propiedad decide voluntariamente si la ejercita, transmitiendo la propiedad del objeto del derecho, o si no la ejercita, reteniendo su titularidad.

Sin embargo, contrariamente a tales consideraciones que cabe hacer desde una perspectiva dogmática jurídico-privada, el artículo 11 LDV convierte al uso y disfrute en un deber del propietario. En este punto la ley estatal sigue el paso marcado con antelación por algunas normas autonómicas, como la Ley canaria 2/2003, de 30 de enero, de Vivienda, cuyo artículo 1.2, de manera muy similar a como lo regula la ley estatal, estipula que forma parte del contenido del derecho de propiedad privada de la vivienda "el deber de destinar de forma efectiva el bien al uso habitacional previsto por el ordenamiento jurídico en coherencia con la función social que debe cumplir ". Otras normas autonómicas sobre vivienda, como la catalana, la balear o la andaluza, aunque

25 Montés Penadés (1991, p. 960).

con diversa formulación a la recogida en la ley estatal, conectan el cumplimiento de la función social de la vivienda con el hecho de que la misma deba estar ocupada y disponen que se incumple dicha función social cuando la vivienda esta deshabitada de forma permanente e injustificada[26].

La consecuencia de esta doble configuración del uso y disfrute es que, en este precepto, se está privando al propietario de la facultad de *no usar* su vivienda, de no disfrutarla, de no explotarla económicamente. En este punto, algunas normas autonómicas introdujeron una regulación diversa en función de si el propietario era persona física o jurídica que es criticada por Simón Moreno que considera que ese diferente trato no está justificado ni por el artículo 33 de la Constitución ni por la interpretación que el TC ha hecho del mismo[27].

La razón de esta configuración del uso y disfrute de la vivienda se encuentra en su conexión con el cumplimiento de la función social que compete a la propiedad, como ya habían puesto de manifiesto algunas de las normas autonómicas antes referidas y como hace el artículo 11.1. a) LDV que también alude a esta conexión al incluir entre los deberes del propietario el del "uso y disfrute propios y efectivos de la vivienda conforme a su calificación, estado y características objetivas, de acuerdo con la legislación en materia de vivienda y la demás que resulte de aplicación, garantizando en todo caso la

---

[26] Puede confrontarse el art 5.b) de la Ley catalana 18/2007, de 28 de diciembre, del derecho a la vivienda, el artículo 4.2.b) de la Ley vasca 3/2015, de 18 de junio, de vivienda, el artículo 5 de la Ley balear 5/2018, de 19 de junio, de la vivienda. Presenta un matiz diferente el artículo 1.3 Ley 1/2010, de 8 de marzo, Reguladora del Derecho a la Vivienda en Andalucía porque no impone el uso de la vivienda expresamente al propietario.

[27] Simón Moreno (2023, p. 165).

función social de la propiedad". Es precisamente, en el último inciso donde se alude a la razón por la que la ley impone al propietario la obligación de hacer un uso y disfrute *propios y efectivos* de la vivienda: porque debe ejercer su titularidad *garantizando en todo caso la función social de la propiedad.* Se quiere remachar, de este modo, que la configuración como deber de lo que siempre ha sido y es una facultad del propietario a usar y disfrutar su vivienda (y que incluye la posibilidad de no ejercitarla), es debida a la proyección de la función social sobre el contenido del derecho.

En esta misma línea, que considera que la función social de la propiedad de la vivienda justifica que se obligue al propietario al uso y disfrute de la vivienda, se ha pronunciado reiteradamente el Tribunal Constitucional, que ha tenido ocasión de pronunciarse al hilo de las normas autonómicas ya referidas y que habían sido impugnadas por inconstitucionales. Las resoluciones dictadas por el TC sobre este asunto son las siguientes:

La STC 43/2018 (BOE 29 de mayo ECLI:ES:TC: 2018:43) resuelve un recurso de inconstitucionalidad contra la Ley 2/2014, de 20 de junio, de modificación de la Ley 2/2003, de 30 de enero, de vivienda de Canarias. El Tribunal Constitucional se muestra favorable a la constitucionalidad de la norma autonómica que regula el deber de usar la vivienda como contenido esencial del derecho de propiedad privada de la vivienda.

La STC 80/2018 declara constitucional el artículo 5.3.b de la Ley valenciana 2/2017 por la función social de la vivienda, que establecía que no se cumple esa función social si la vivienda esta deshabitada de forma permanente e injustificada.

La STC 106/2018 declara constitucional el art 1.3 Ley de vivienda de Extremadura: que incluye en el contenido del derecho de propiedad de la vivienda "el deber de destinar de forma efectiva el bien al uso habitacional previsto por el ordenamiento jurídico".

Por otro lado, la STC 8/2019, que resolvía un recurso de inconstitucionalidad contra determinados preceptos de la Ley catalana 4/2016, de 23 de diciembre, de medidas de protección del derecho a la vivienda de las personas en riesgo de exclusión residencial, conduce por desistimiento del Abogado del Estado a la constitucionalidad de la expropiación en caso de viviendas vacías que se había contemplado en el artículo 15 de dicha ley.

De modo que la doctrina contenida en las resoluciones del Tribunal Constitucional viene a decir que resulta ajustado a la Constitución la imposición al propietario del deber de uso y disfrute de la vivienda en aras al cumplimiento de su función social. Se considera que dicho deber no incide en el contenido esencial de la propiedad, sino que se relaciona con las actividades de fomento y movilización de vivienda deshabitada.

Pese a todo, que el uso y el disfrute de la vivienda por su titular sea facultad y deber es conceptualmente inadecuado como lo sería obligar al propietario a ejercitar la facultad de disposición o la de defensa[28]. El propietario debe tener la facultad de no usar ni disfrutar su derecho de propiedad. Mejor que la imposición que supone el deber de usar y disfrutar la vivienda, hubiera sido que la ley incentivara el uso y disfrute por el propietario, por ejemplo, mediante medidas fiscales que promovieran el alquiler de las viviendas no usadas ni explotadas por sus propietarios.

---

[28] En relación a las normas autonómicas que configuraban como deber la posibilidad de uso de la vivienda, Simón Moreno (2023, p. 156) considera que se ha excedido el marco constitucional de los artículos 33 y 149.1.8 de la Constitución. Coherentemente con esta apreciación, podría entenderse excedido dicho marco por la Ley estatal en este punto, de manera que este autor considera (p. 173) que son inconstitucionales los artículos 10.1.a) y 11.1.a) de la LDV.

Debe tenerse en cuenta que algunas normas autonómicas establecen la expropiación del uso como consecuencia del incumplimiento del deber de usar la vivienda por sus propietarios (artículo 42 Ley catalana 18/2007, art 52 Ley foral de Navarra 10/2010, art. 72 Ley vasca 3/2015). Esta consecuencia del incumplimiento del deber de uso de la vivienda ha sido declarada constitucional en las resoluciones del TC referidas más arriba[29].

2. *Obligación de mantenimiento, conservación y, en su caso, rehabilitación de la vivienda*

El segundo deber que el artículo 11.1.b) impone al propietario consiste en la obligación de mantener, conservar y, en su caso, rehabilitar la vivienda, según se establezca en esta ley, en la normativa urbanística y de vivienda y de los instrumentos dictados a su amparo. Al igual que en el caso anterior, dicha obligación se ha configurado en el artículo 10.1.c) LDV como facultad del propietario.

También la mayor parte de las normas autonómicas de vivienda contemplaban el mantenimiento de la vivienda como una obligación, aunque, en este caso, más prudentemente que la ley estatal, no lo habían configurado previamente como facultad[30].

---

29 Simón Moreno (2023, p. 160) entiende que entiende que nos encontramos ante una reducción desproporcionada y, por tanto, inconstitucional, del derecho de propiedad.

30 Puede comprobarse esta forma de operar en las normas autonómicas siguientes: Artículo 1.2 de la Ley canaria 2/2003, artículo 28 Ley valenciana 8/2004, artículo 17.2 Ley andaluza 1/2010, artículo 12 de la Ley gallega 8/2012, artículo 9 de la Ley murciana 6/2015, en otros se establece como un deber que recae sobre los poderes públicos (art 15 de la Ley Foral navarra 10/2010) y en otros casos se decreta que incumple su función social siempre que el estado de la

De nuevo nos topamos ante la configuración de una posibilidad de actuación del propietario en relación a la vivienda como facultad y como deber con todos los inconvenientes que ello comporta y que ya hemos puesto de relieve respecto al derecho de uso.

Actualmente los poderes públicos favorecen la rehabilitación de edificios, lo que incluye a las viviendas, y su conformación como deber es una manifestación más de una política decidida en esta dirección. De hecho, entre las facultades del propietario del suelo urbano recogidas en el artículo 14 TRLSRU se menciona la de "participar en la ejecución de actuaciones de reforma o renovación de la urbanización", lo que sería algo paralelo a la conservación y rehabilitación de viviendas que aquí estamos tratando; entre los deberes que contempla el artículo 15 del TRLSRU habla del deber de conservar al igual que este artículo 11.1.b) LDV.

Sin embargo, convertir la conveniencia de rehabilitar las viviendas en una obligación puede ser inconveniente. Es mejor impulsar dichas actuaciones con medidas de fomento y apoyo económico a los propietarios para que procedan a efectuarlas como la que supone el Real Decreto 853/2021, de 5 de octubre por el que se regulan los programas de ayuda en materia de rehabilitación residencial y vivienda social del Plan de Recuperación, Transformación y Resiliencia, entre cuyos objetivos, según su artículo 1.2 se encuentra el de "contribuir al cumplimiento de los objetivos del Plan de Recuperación, Transformación y Resiliencia en los referidos ámbitos de la rehabilitación residencial ". Este Real Decreto supone un impulso decidido a la rehabilitación de edificios y viviendas.

---

vivienda suponga un riesgo para las personas y se hayan dado ayudas públicas suficientes (artículo 5.2 Ley catalana 18/2007, artículo 4 de la Ley vasca 3/2015 y artículo 5.2 a) Ley balear 5/2018).

3. Evitar la sobreocupación o el arrendamiento para usos y actividades que incumplan los requisitos y condiciones de habitabilidad legalmente exigidos

En la letra c) del artículo 11.1 LDV se incorporan una serie de deberes a cargo del propietario de la vivienda formulados negativamente.

Así, se insta al propietario a "evitar la sobreocupación de la vivienda", lo cual puede plantear problemas si referimos este deber a la propia situación del propietario y su familia, por ejemplo, a aquellos casos en que el número de miembros que integran la unidad familiar sea elevado o cuando se incorpora a la vivienda un miembro, que vivía independientemente, con respecto al cual el propietario tiene que cumplir una obligación de alimentos. En estas situaciones y, seguramente otras que pueden darse como el nacimiento de un hijo o un nieto del propietario, es posible que este no pueda evitar la sobreocupación de la vivienda o que el cumplimiento de este deber determine el incumplimiento de otros deberes prioritarios como el de alimentos.

Ahora bien, no son estos supuestos de sobreocupación referidos a la situación familiar del propietario o propietarios de la vivienda[31] los que se contemplan en este precepto, sino el deber del propietario de evitar el arrendamiento de la vivienda a un

---

31 No sé si cabría esta interpretación en relación al artículo 19 de la Ley 11/2019, de 11 de abril, de promoción y acceso a la vivienda de Extremadura. Este precepto hace responsable de la sobreocupación: al propietario, al arrendador o al subarrendador, de modo que hay que entender que cuando la vivienda se encuentra sobreocupada y es el propietario quien habita con su familia también puede incurrirse en la infracción de hacinamiento contemplada en el artículo 93.i) de la Ley extremeña; esta infracción es calificada de grave y se imputa al propietario, arrendador o subarrendador.

número excesivo de inquilinos–pisos patera-, siendo la falta de rigor en la redacción del precepto la que ocasiona este problema de interpretación que puede salvarse, al menos parcialmente, si se conecta el deber de impedir la sobreocupación con el de no arrendar para usos y actividades que incumplan los requisitos y condiciones de habitabilidad legalmente exigidos. Según esta interpretación, el propietario no podría arrendar la vivienda de manera que la misma resulte sobreocupada ni tampoco para actividades que incumplan los requisitos y condiciones de habitabilidad. Lo que se quiere evitar es que el propietario arriende la vivienda a un número excesivamente elevado de personas o que se alquilen inmuebles que carecen de las condiciones de habitabilidad necesarias para ser usadas como viviendas.

Esa creo que es la forma correcta de entender esta letra c) del artículo 11.1 LDV, el cual plantea otros interrogantes que analizamos a continuación.

La LDV no proporciona ningún parámetro al que atenerse para saber cuántas personas determinan la sobreocupación de una vivienda. Sí los encontramos en algunas normas autonómicas como, por ejemplo y sin ánimo de exhaustividad, en el artículo 19 de la Ley extremeña de vivienda que regula también las sanciones correspondientes o en el artículo 62 de la Ley de vivienda del País vasco que reenvían al desarrollo reglamentario para el establecimiento de las consecuencias del incumplimiento de este deber.

Del contraste de la norma estatal con las autonómicas que hemos mencionado sobre esta cuestión resulta que la Ley del País Vasco, al igual que la estatal, no conecta esta situación de sobreocupación con el cauce jurídico por el que quienes ocupan la vivienda se encuentran en ella; sin embargo, el artículo 19 de la Ley extremeña hace responsable de la sobreocupación: al propietario, al arrendador o al subarrendador, de modo que hay que entender que cuando la vivienda se encuentra sobreocupada y es el propietario quien habita en ella con su familia también puede incurrirse en la infracción de hacinamiento contemplada en el artículo 93.i) de la Ley extremeña, la cual es calificada de infracción grave.

La norma extremeña nos pone sobre la pista de otro problema de interpretación del artículo 11.1 c) LDV, el cual parece entender que sólo el propietario puede arrendar, desconociendo que otros titulares de derechos sobre la vivienda pueden hacerlo, como el usufructuario o el titular de un derecho de superficie sobre el inmueble, aunque en este segundo caso puede entenderse mencionado al superficiario cuando se alude al propietario porque lo es. Incluso pueden arrendar sujetos que no detentan siquiera una titularidad jurídico-real sobre la vivienda, como un arrendatario a quien se le permite subarrendar en virtud del artículo 8 LAU. En lo atinente a este punto, la Ley extremeña de vivienda es más precisa, pues menciona al propietario, al arrendador y al subarrendador. La ley estatal no incluye ni al usufructuario ni al arrendatario con facultad de subarrendar –artículo 8 LAU–.

En realidad, el problema viene dado porque lo que se quiere regular en realidad es el arrendamiento, pero, al incorporar este precepto a un artículo que regula el contenido del derecho de propiedad sobre la vivienda, su alcance queda reducido al arrendamiento por parte del propietario, cuando en realidad también usufructuarios o arrendatarios con facultad de subarriendo pueden alquilar.

Tanto si se entiende que se está regulando el contenido del derecho de propiedad de la vivienda, como si se considera que se está regulando el arrendamiento de vivienda, puede afirmarse que estamos ante una materia de competencia estatal, pudiendo suscitarse un problema con el alcance de la competencia de las CCAA en esta materia que, como hemos visto, la han regulado.

4. Obligaciones de información precontractual en la compraventa y arrendamiento de vivienda

La letra d) del artículo 11.1 LDV establece a cargo del propietario –y de otros sujetos, entre los que, por ejemplo, se encuentran los intermediarios– una serie de obligaciones de información precontractual en caso de venta, arrendamiento de la vivienda o cualquier otro negocio jurídico de tenencia o disfrute de la vivienda, remitiendo al título IV de la propia LDV reguladora de la transparencia en contratos de venta y arrendamiento de vivienda; dicha remisión se concreta en los artículos 30 y 31 LDV, pues el resto de preceptos de dicho capítulo IV versan sobre obligaciones de información y transparencia a cargo de las Administraciones públicas.

El marco de este estudio impide un análisis exhaustivo y minucioso de esta cuestión, por lo que me voy a centrar en precisar a qué tipo de relaciones son aplicables estas obligaciones[32].

El artículo 30 LDV considera que las obligaciones de información precontractual son aplicables en caso de venta, arrendamiento de la vivienda o cualquier negocio jurídico del que resulte otro régimen jurídico de tenencia o disfrute. Podríamos incluir, entonces, la constitución por enajenación de un usufructo o de un derecho de superficie sobre la vivienda, aunque los artículos 30 y 31 LDV a los que se remite el artículo sólo se refieren a la compraventa y al arrendamiento. No se aplicarían, sin embargo, en caso de constitución por reserva de estos derechos al enajenarse la nuda-propiedad o el suelo en caso de un derecho de superficie. Probablemente estas obligaciones de información tampoco serán exigibles al propietario cuando ceda gratuitamente la tenencia de la vivienda.

---

32 Para un análisis completo me remito a mi libro sobre esta cuestión: Alonso Pérez (2010).

En otro orden de cosas debe estudiarse la operatividad de estas obligaciones de información precontractual en función de la cualidad de los sujetos que conciertan la venta, el arrendamiento o los otros negocios referidos. El artículo 30.1 LDV dispone que quienes quieren comprar o arrendar una vivienda tienen los derechos que se derivan del TRLGDCU; se entiende –aunque no lo explicita la ley– que se refiere a aquellos derechos relacionados con la información y transparencia en materia de compraventa y arrendamiento de vivienda y también cabe entender que dicha obligación sólo opera en el caso de que la relación pueda ser calificada como relación de consumo, lo que viene a ser confirmado por el párrafo segundo de este mismo artículo.

Por lo dicho, además de las normas de la LDV, a las relaciones de consumo serán de aplicación las normas previstas en el TRLGDCU con alcance general –fundamentalmente el artículo 60–, la normativa aplicable en materia de publicidad, así como el Real Decreto 515/1989, de 21 de abril, sobre protección de los consumidores en cuanto a la información a suministrar en la compraventa y arrendamiento de viviendas. Por último, deben atenderse las normas que en esta materia han dictado las comunidades autónomas que disponen de competencia en materia de vivienda y de consumo.

Cuando los contratos de venta y de arrendamiento de vivienda no merezcan la calificación de relación de consumo, sino que sean puramente civiles entre particulares, se aplicarán únicamente las obligaciones precontractuales de información que aparecen recogidas en los artículos 30 y 31 LDV. Y, aunque, probablemente los datos de que haya que informar al potencial adquirente o al futuro arrendatario de la vivienda sean los mismos con independencia de que la relación pueda ser calificada de consumo o no, las consecuencias del incumplimiento de estas obligaciones se diversificarán.

Si la relación es de consumo, la información suministrada en la fase precontractual, incluida la proporcionada en la publicidad, integra el contenido del contrato por aplicación del artículo 61 TRLGDCU, debiendo prevalecer sobre lo expresamente pactado si es más beneficioso para el consumidor. De modo que si, por ejemplo, se incumple el reclamo de "vistas al mar" en caso de que aparezca en los folletos publicitarios, puede entenderse concurrente un incumplimiento esencial del contrato y ser viable la resolución contractual por aplicación del artículo 1124 del Código civil. Cuando la relación es puramente civil, en tal supuesto, no podría instarse la acción del artículo 1124 del Código civil si en el contrato escrito o en la escritura pública no aparece recogido que tiene "vistas al mar", pero podría intentarse una anulación del contrato por vicios del consentimiento en base a la existencia de error o dolo, según el caso.

Si la relación es de consumo, en caso de falta de transparencia consistente en omisión de información, puede instarse la integración del contrato por aplicación del artículo 65 TRLGDCU[33]. Habría que ver en qué medida puede resultar de aplicación el artículo 83 TRLGDCU que determina la nulidad de una cláusula no transparente en el contrato.

Si la relación es puramente civil, para proteger al adquirente o arrendatario en caso de incumplimiento de los deberes precontractuales se podría instar la anulación del contrato por la existencia de vicios del consentimiento. El efecto que tiene el que estas obligaciones se hayan formulado normativamente es que la carga de la prueba de su cumplimiento recaería sobre el vendedor o el arrendador, de modo que el comprador o arrendatario partirían con ventaja en el proceso[34]. No obstante, el

---

33 En sentido similar, Mate Satué (2024).

34 Sobre esta cuestión, vid. mi libro, Alonso Pérez (2010, pp. 238 y ss.).

remedio que conlleva la declaración de nulidad del contrato puede no convenir al adquirente o arrendatario que pueden preferir mantenerse en la vivienda con una compensación.

5. *La carga de colaboración con los Ayuntamientos en zonas de mercado residencial tensionado*

Para cerrar la lista de deberes que integran el contenido del derecho de propiedad sobre la vivienda, el artículo 11.1, en la letra e) alude a una carga derivada de la existencia de zonas declaradas de mercado residencial tensionado que aparece regulada en los artículos 18 y 19 LDV. Dicha obligación de colaboración sólo se contempla para los grandes tenedores de vivienda en el artículo 19 LDV.

VII. CONCLUSIONES

1. Las normas básicas que regulan el contenido del derecho de propiedad de la vivienda son el artículo 33 de la Constitución, el artículo 348 del Código civil y los artículos 1.2, 10 y 11 de la LDV. Todas estas normas recogen el mismo concepto de derecho de propiedad, regulando cada una un aspecto del derecho subjetivo en función del diferente alcance y naturaleza de la norma en que se contienen.

2. La regulación del contenido del derecho de propiedad de la vivienda debe respetar tres límites establecidos por la Constitución. El primero de ellos es el reparto constitucional de competencias entre el Estado y las Comunidades Autónomas. Se entiende que es una materia cuya regulación corresponde al Estado en base al artículo 149.1.1, 149.1.3 y 149.1.8 de la Constitución. Del artículo 53 de la Constitución se derivan los otros dos límites: la regulación de esta materia debe hacerse en una norma

con rango de ley y, además, la regulación que se haga de la misma debe respetar el contenido esencial del derecho. Este último debe reflejar un equilibrio entre la utilidad individual que debe reportar el derecho a su titular y el desarrollo de la función social que le es propia, tal y como establece el artículo 33 de la Constitución española.

3. La Ley 12/2023, de 24 de mayo, regula el contenido del derecho de propiedad de la vivienda en el artículo 1.2 y, fundamentalmente, en el artículo 10, que establece las facultades que este derecho proporciona al propietario, y en el artículo 11, referido a los deberes. Esta regulación es defectuosa desde el punto de vista técnico, lo cual plantea muchos problemas de interpretación. Así puede considerarse un desajuste conceptual el hecho que varias posibilidades de actuación sobre la vivienda por parte de su propietario se configuren, a un tiempo, como facultades y como deberes, como con el uso y disfrute de la vivienda o con la conservación, mantenimiento y rehabilitación de la vivienda.

4. La enumeración de las facultades del propietario que se hace en el artículo 10 LDV recurre al artículo 348 del Código civil para incluir el uso y disfrute de la vivienda y la facultad de disposición, reconociendo, de este modo, el papel central que corresponde en esta materia a dicho Código. La LDV no va mucho más allá de lo que establece el artículo 348 Código civil en este punto porque el resto de facultades contempladas por el precepto no tienen entidad propia: así la facultad de conservar y rehabilitar la vivienda es desarrollo de las de uso, disfrute y disposición; por otro lado, la facultad consistente en la posibilidad de consultar el estado urbanístico de la vivienda o del edificio en que se encuentra no se desprende del derecho de propiedad sobre una vivienda sino que la detenta cualquier ciudadano en virtud del artículo 5 TRLSRU.

5. El propietario de una vivienda, aunque no se contemple expresamente entre las facultades del propietario de una vivienda recogidas en el artículo 10 LDV, también ostenta la facultad de defensa de su titularidad mediante su reivindicación –artículo 348.2 del Código civil-. Se trata de la *acción* que conlleva todo derecho subjetivo y que permite a su titular ejercitarlo y defenderlo ante los Tribunales. En esta línea, cabe destacar que el párrafo segundo del artículo 10 LDV convierte a las Administraciones en garantes de la posición del propietario de vivienda debiendo asegurarle el pleno ejercicio de su titularidad.

7. Por otro lado y regulando una cuestión que no guarda relación con las facultades del propietario contempladas en el párrafo primero, el párrafo segundo de este artículo 10 LDV insta a las Administraciones públicas a fomentar el acceso a la vivienda. Concretamente a favorecer el acceso a la propiedad de la vivienda mediante el establecimiento de incentivos fiscales. También contempla otras medidas encaminadas a favorecer otras vías de acceso a la vivienda como el alquiler.

8. El artículo 11 LDV alude a la función social que debe cumplir la propiedad de la vivienda para delimitar los deberes que incumben a su titular. En este precepto se impone, como deber, el uso propio y efectivo de la vivienda por su propietario que, en el artículo 10, había sido configurado como facultad. Lo mismo ocurre con la obligación de conservar y rehabilitar la vivienda –que el artículo 10.1 LDV establecía como facultad– en lo que supone un nuevo impulso a las políticas de rehabilitación, entre cuyos fines se encuentra el incorporar inmuebles en desuso al mercado inmobiliario de la venta y del alquiler.

9. Otro deber que se establece a cargo del propietario es el de evitar el arrendamiento a un número elevado de personas

para impedir la sobreocupación, así como el alquiler como vivienda de inmuebles que no cumplen con los requisitos de habitabilidad. La cuestión es que se está regulando el arrendamiento en un precepto referido al contenido del derecho de propiedad de la vivienda y, en consecuencia, esa limitación no se establece para titulares de otros derechos distintos a la propiedad que también pueden alquilar la vivienda, como el usufructuario o el arrendatario con facultad de subarrendar en base al artículo 8 LAU.

Como colofón final puede decirse que la regulación del contenido del derecho de propiedad de la vivienda en la LDV se caracteriza por las deficiencias técnicas como, por ejemplo, la inclusión de aspectos que no guardan relación con esta cuestión en los artículos que la regulan; también la configuración de ciertas actuaciones que pueden llevarse a cabo sobre la vivienda como facultades y como deberes resulta controvertida y difícil de entender desde un punto de vista conceptual. Por otro lado, el artículo 1.2 LDV, al precisar su alcance e indicar que es objeto de la misma la regulación del contenido del derecho de propiedad sobre la vivienda sólo alude a los deberes del propietario y a la obligación de la Administración de controlar su cumplimiento, prescindiendo de cualquier alusión al lado activo de la titularidad –las facultades– lo que, a mi modo de ver, evidencia que la perspectiva desde la que se formula esta regulación es sesgada, utilizándose el elemento de la función social de la propiedad a la que alude el artículo 33 de la Constitución sin tener en cuenta, entre otras cosas, que la misma debería proyectarse también sobre el plano activo del derecho.

VIII. REFERENCIAS BIBLIOGRÁFICAS

Aguilera Vaqués, M. (2018). "Comentario al artículo 33 de la Constitución española". *Comentario a la Constitución española. 40 aniversario 1978-2018. Homenaje a Luis López Guerra.* Volumen I. (dir. Pérez Tremps, P./ Saiz Arnaiz, A.). Tirant biblioteca virtual, 705-713.

Alonso Pérez, M. T. (2010). *Las obligaciones legales de información precontractual en la compraventa de vivienda: a través del laberinto normativo, estatal y autonómico, en materia de vivienda y de consumo.* Civitas.

Arias Martínez, M. A. (2019). "Las competencias autonómicas en materia de vivienda frente a las competencias estatales de carácter transversal en la reciente jurisprudencia constitucional". *Revista de Estudios de la Administración Local y Autonómica.* Nueva Época, n.º 11, abril-septiembre 2019, 106-121.

Calduch Alonso, M. (2023). "La distribución de competencias en materia de vivienda entre el Estado y las Comunidades Autónomas". *Revista Crítica de Derecho Inmobiliario,* n.º 797, 1447-1483.

Castán Tobeñas, J. (1992). *Derecho civil español, común y foral,* Tomo segundo, *Derecho de cosas,* Volumen primero, *Los derechos reales en general. El dominio. La posesión.* Reus.

Garrido Mayol, V. (2022). "Hacia una nueva configuración del derecho constitucional a la vivienda". *Vivienda y Colectivos Vulnerables,* (dir. Cervilla Garzón, M.ª D./ Zurita Martín, I.). Aranzadi, 267-296.

López y López, A. M.ª (2022). "Propiedad privada y disciplina del mercado en el Proyecto de Ley por el derecho a la vivienda". *Cuadernos de Derecho Local, 59,* 14-34.

Mate Satué, L. C. (2024). "La información precontractual en el arrendamiento de vivienda: ¿un mecanismo eficaz para la protección del arrendatario: especial atención al vulnerable?". *El derecho a la vivienda en tiempos de incertidumbre.* Aranzadi.

Messía de la Cerda Ballesteros, J. A. (2020). "La función social de la propiedad de la vivienda y las normativas autonómicas". *Revista Crítica de Derecho Inmobiliario,* n.º 782, 3417-3465.

Montés Penadés, V. (1991). "Comentario al artículo 348 del Código civil". *Comentario del Código civil.* Tomo I. Ministerio de Justicia.

Moreu Carbonell, E. (2018). "Viviendas vacías". *Nuevas vías jurídicas de acceso a la vivienda,* dir. por Alonso Pérez, M. Teresa. Aranzadi, 27-72.

Simón Moreno, H. (2023). "La evolución constitucional de la función social de la propiedad y el nuevo régimen del derecho de propiedad sobre una vivienda en la Ley por el derecho a la vivienda". *Derecho privado y Constitución*, n.º 42, 139-177.

Verdera Server, R. (2023). "*Pro proprietate.* Notas sobre la configuración constitucional de la propiedad privada". *Anuario de Derecho civil*, Tomo LXXXVI, 2023-III, pp. 859-988.

IX. RESOLUCIONES DEL TRIBUNAL CONSTITUCIONAL

STC 37/1987, de 26 de marzo, (BOE núm. 89, de 14/04/1987). ECLI:ES:TC:1987:37

STC 89/1994, de 17 de marzo, BOE núm. 89, 14/04/1994. ECLI:ES:TC:1994:89

STC 93/2015, de 14 de mayo, (BOE núm. 146, 19/06/2015) ECLI:ES:TC:2015:93

STC 16/2018, de 22 de febrero, (BOE núm. 72, 23/03/2018), ECLI:ES:TC:2018:16

STC 43/2018, de 26 de abril, (BOE núm. 130, 29/05/ 2018) ECLI:ES:TC: 2018:43

STC 80/2018, de 5 de julio, (BOE núm. 189, 06/08/ 2018), ECLI:ES:TC:2018:80

STC 106/2018, de 4 de octubre (BOE núm. 264, 01/11/ 2018), ECLI:ES:TC: 2018:106

STC8/2019,de17deenero,(BOEnúm.39,14/02/2019)ECLI:ES:TC:2019:8

STC 37/2022, de 8 de abril, (BOE núm. 84, 08/04/2022), ECLI:ES:TC:2022:37

STC 9/2023, de 22 de febrero, BOE núm. 77, 31/03/ 2023), ECLI:ES:TC:2023:9

STC 15/2023, de 7 de marzo (BOE núm. 89, 14/04/2023), ECLI:ES:TC:2023:15

Capítulo V

La publicidad y el deber de información en la ley 12/2023, de 24 de mayo, por el derecho a la vivienda

PAULA CASTAÑOS CASTRO
Profesora Titular de Derecho civil
Universidad de Málaga

I. CUESTIONES PREVIAS

Es bien sabido que el derecho a la vivienda es un derecho constitucional, concretamente el art. 47 CE establece que "Todos los españoles tienen derecho a disfrutar de una vivienda digna y adecuada. Los poderes públicos promoverán las condiciones necesarias y establecerán las normas pertinentes para hacer efectivo este derecho, regulando la utilización del suelo de acuerdo con el interés general para impedir la especulación (...)"[1]. Se trata de un deber que compete, sin excepción, a to-

---

1 No resulta en absoluto sorprendente que la CE haga referencia al derecho a la vivienda si tenemos en cuenta, –y en esto estamos de acuerdo con el preámbulo de la Ley– que la vivienda constituye, ante todo, un pilar central del bienestar social en cuanto lugar de desarrollo de la vida privada y familiar, y centro de todas las políticas urbanas. Es por ello que el derecho a la vivienda se recoge, no solamente ya en nuestra Carta Magna, sino en importantes declaraciones internaciones y en destacados textos europeos, como la reciente Resolución del Parlamento Europeo, de 21 de enero de 2021, sobre

dos los poderes públicos, que están obligados a cumplir con él en el marco de sus respectivas esferas de competencia.

Es justamente en el contexto del cumplimiento por parte del Estado de la obligación que le incumbe en la protección del derecho a acceder a una vivienda digna y adecuada, en el que debe entenderse –y así lo expone expresamente el Preámbulo de la Ley– la nueva LDV (Ley 12/2023, de 24 de mayo, por el derecho a la vivienda)[2].

Lo cierto es que, en la actualidad, las dificultades que existen para tener acceso a una vivienda representan un problema de envergadura, que sin duda merece ser abordado desde diferentes ámbitos, también en una ley estatal[3]. Por tanto, se considera del

---

el acceso a una vivienda digna y asequible para todos, en la que se pide a la Comisión y a los Estados miembros que se aseguren de que el derecho a una vivienda adecuada sea reconocido y ejecutable como un derecho humano fundamental mediante disposiciones legislativas europeas y estatales aplicables, y que garanticen la igualdad de acceso para todos a una vivienda digna.

2 Se trata de la primera ley estatal reguladora del derecho a la vivienda desde la aprobación de la Constitución.

3 Que el art. 148.3 de la Constitución señale que todas las Comunidades Autónomas tienen asumida en sus Estatutos de Autonomía, sin excepción, la competencia plena en materia de vivienda no significa que el Estado no pueda abordar esta materia, siempre que se trate de un marco normativo coherente que complemente las normas autonómicas. Al Estado, dicha competencia le viene atribuida, tal y como se señala en el preámbulo de la ley, por diversos preceptos constitucionales –arts. 149.1.1 o 149.1.13, entre otros–. La jurisprudencia del Tribunal Constitucional también se ha mostrado favorable a considerar competente al Estado para legislar en materia de vivienda. Así, por ejemplo, la STC 80/2018, de 5 de julio, señala que “No habiendo el legislador estatal ejercido la habilitación que el art. 149.1. 1.ª CE le otorga, resulta necesario afirmar que el legislador autonómico en materia de vivienda, en el momento en el

todo oportuno y necesario que el legislador nacional se ocupe de regular el derecho a acceder a una vivienda digna y adecuada y al disfrute de la misma en condiciones asequibles[4] –aunque una cosa es estar de acuerdo con la necesidad de aprobar una ley estatal de vivienda que complemente la legislación autonómica, y otra muy distinta estar de acuerdo con el contenido de *esta* LDV[5]–.

---

que realizamos este enjuiciamiento, no encuentra límites desde esta perspectiva constitucional". Esta misma doctrina jurisprudencial se reitera en otras sentencias actuales como la STC 32/2018, de 10 de abril, y la STC 43/2018, de 26 de abril.

4 La mayor parte de los problemas que giran en torno a este derecho son consecuencia del desequilibrio entre la nueva construcción y la actividad de rehabilitación, del desequilibrio entre la tenencia en propiedad y la tenencia en alquiler o del desequilibrio existente entre el ámbito rural y el urbano. A esto se añade que los parques públicos de vivienda en alquiler son del todo insuficientes para atender las necesidades de personas con escasos recursos económicos.

5 Se trata de una ley ambiciosa, con numerosos objetivos –aunque pocos de ellos cumplidos–, entre los que destacan los siguientes: establecer una regulación básica de los derechos y deberes de los ciudadanos en relación con la vivienda; facilitar el acceso a una vivienda digna y adecuada a las personas que tienen dificultades para acceder a una vivienda en condiciones de mercado; dotar de instrumentos efectivos para asegurar la funcionalidad, la seguridad, la accesibilidad universal y la habitabilidad de las viviendas; definir los aspectos fundamentales de la planificación y programación estatales en materia de vivienda; regular el régimen jurídico básico de los parques públicos de vivienda, o favorecer el desarrollo de tipologías de vivienda adecuadas a las diferentes formas de convivencia y de habitación. Además, se presta especial atención a colectivos vulnerables, en particular a los jóvenes, cuya situación es especialmente preocupante. Según los datos de la última encuesta anual de estructura salarial del INE, la ganancia media anual en 2019 de los trabajadores entre 20 y 24 años se situó en 12.640,65 euros, de los trabajadores entre 25 y 29 años en 17.772,31 euros y de los trabajadores entre 30 y 34 años en 20.969,47 euros. Por su parte, en cuanto

Es objeto de la Ley, según su artículo primero, regular el derecho a acceder a una vivienda digna, y regular asimismo el contenido básico del derecho de propiedad de la vivienda en relación con su función social, pero también, tal y como

---

a los precios de venta, el precio medio se sitúa en 1.649,2 euros el metro cuadrado en el conjunto de España, tal y como se señala en el Real Decreto 42/2022, de 18 de enero, por el que se regula el Bono Alquiler Joven y el Plan Estatal para el acceso a la vivienda 2022-2025. En relación con el nivel de sobreesfuerzo para el pago del alquiler de una vivienda, los hogares jóvenes se ven sobreexpuestos a la sobrecarga financiera que caracteriza este régimen de tenencia en nuestro país. Según los últimos datos de Eurostat, en España el 35,9 % de la población que reside en una vivienda en alquiler a precios de mercado destina más del 40 % de sus ingresos al pago del alquiler, lo que representa una cifra que está situada más de diez puntos porcentuales por encima del porcentaje medio de sobrecarga al pago del alquiler a precios de mercado del conjunto de la Unión Europea, que se encuentra en el 25,2 %. La relación entre los ingresos de los jóvenes y los precios de venta, así como el sobreesfuerzo que han de hacer para el pago del alquiler de las viviendas contribuye a que actualmente se constate un notable retraso en la edad de emancipación. En particular, según los últimos datos de la encuesta continua de hogares del INE, el 55,0 % de los jóvenes de entre veinticinco y veintinueve años vivía con sus padres o con alguno de ellos en 2020, tratándose de un porcentaje que ha crecido 6,5 puntos en los siete últimos años. Asimismo, el porcentaje de personas de treinta a treinta y cuatro años que vivían con sus padres o con alguno de ellos se situó en el 25,6 % en el mismo año 2020, registrando un aumento de 5,1 puntos desde 2013. Si se compara la edad media de emancipación con los principales países de nuestro entorno, se observa que, según los últimos datos de Eurostat correspondientes al año 2020, la edad media de emancipación en España se situó en los 29,8 años, incrementándose ligeramente con respecto al año anterior y situándose 3,4 años por encima de la media de la Unión Europea (26,4). En los últimos años se ha producido en nuestro país un incremento muy significativo de la edad media de emancipación, desde la cifra de 28,5 años registrada en el año 2011.

dispone el párrafo tercero del citado precepto "reforzar la protección del acceso a información completa, objetiva, veraz, clara, comprensible y accesible, en las operaciones de compra y arrendamiento de vivienda", de lo que se ocupa el Título IV de la norma. Ahora bien, ¿por qué el legislador ha decidido regular este tema en la LDV? ¿Por qué este asunto forma parte del objeto de la norma? Si tenemos en cuenta que, como veremos a continuación, antes de la entrada en vigor de la Ley ya existía -y sigue existiendo al no haber quedado derogada- una regulación de los deberes de información en la compraventa y arrendamiento de vivienda, ¿qué es exactamente lo que persigue el legislador introduciendo de nuevo esta cuestión? En el Preámbulo de la Ley encontramos la respuesta, pues en él se dice que el Título IV tiene como principal objetivo "reforzar la protección en las operaciones de compra o alquiler de vivienda, estableciendo una serie de garantías y obligaciones de información a la que tienen derecho las personas o entidades adquirentes o arrendatarias de vivienda, y una serie de responsabilidades derivadas de su incumplimiento, ya se trate del vendedor o del intermediario en la operación inmobiliaria"; no obstante, ¿verdaderamente es el refuerzo de la protección lo que busca el legislador o, más bien, dotar de contenido una ley que se presentaba como un logro de la legislatura? ¿Consigue la Ley su objetivo, o los artículos que la norma destina a regular tales cuestiones están vacíos de contenido, tratándose de normas superfluas e innecesarias? Desde un primer momento puede ya adelantarse, aunque se analizará con más detalle en apartados posteriores, que los preceptos que la Ley dedica a la cuestión que nos ocupa no solamente no ayudan a intensificar la protección que la legislación anterior a la LDV ya otorga, sino que complican el panorama significativamente, confundiendo conceptos tales como el de información precontractual y publicidad, utilizando una sistemática caótica para regular cuestiones, en su mayoría, ya reguladas en normas previas y creando problemas interpretativos que también tendremos oportunidad de tratar en las páginas siguientes.

En definitiva, si bien es cierto que si se quiere garantizar adecuadamente el derecho a acceder a una vivienda –ya sea en régimen de propiedad o de arrendamiento– es necesario regular pormenorizadamente la información que debe proporcionarse al propietario o arrendatario en cada una de las fases contractuales[6], no menos cierto es que esta Ley poco aporta a la regulación ya existente, ni siquiera desde un punto de vista formal, con lo necesario que hubiera sido, al menos, aportar cierto orden y sistemática en la regulación de un tema que, para garantizar el acceso a la vivienda, es primordial.

Dicho esto, el objetivo principal de esta investigación es triple: en primer lugar, comprobar si los preceptos de la ley encargados de regular tal cuestión (arts. 30 y 31) conviven con otras normas sobre la materia que también permanecen vigentes; en segundo lugar, señalar las principales aportaciones y las principales deficiencias de la actual LDV en materia de información precontractual y publicidad. Por último, realizar propuestas de mejora, encaminadas todas ellas a sistematizar y enriquecer el contenido de la LDV para el caso que nos ocupa.

---

6 Una óptima garantía del derecho a la vivienda implica que el interesado conozca –o al menos pueda conocer–, antes de la formalización del contrato, información que le resultará útil para formar su voluntad contractual. La información precontractual se erige así como instrumento fundamental para proteger al interesado en acceder a una vivienda, ya sea como propietario, como arrendatario, o en cualquier otro régimen jurídico de tenencia y disfrute.

II. NORMATIVA APLICABLE EN MATERIA DE PUBLICIDAD E INFORMACIÓN EN LOS CONTRATOS DE COMPRAVENTA Y ARRENDAMIENTO DE VIVIENDAS

La reciente Ley por el derecho a la vivienda no es la única ley que, actualmente, regula la publicidad y la información en los contratos de compraventa y arrendamiento de viviendas. Esta norma convive con otras –estatales y autonómicas– que también hacen referencia a tal cuestión. De este modo, el panorama actual en cuanto a normativa aplicable es el que sigue:

- ***Leyes especiales:***
  - Ley 12/2023, de 24 de mayo, por el derecho a la vivienda. Como ya se ha señalado anteriormente, es objetivo principal de esta norma reforzar la protección del acceso a información completa, objetiva, veraz, clara, comprensible y accesible, en los contratos de compra y arrendamiento de vivienda. La norma dedica el Título IV, dividido a su vez en dos capítulos, a desarrollar este objetivo. En dicho Título IV se establecen, como indica el Preámbulo de la Ley, garantías y obligaciones de información a la que tienen derecho las personas o entidades adquirentes o arrendatarias de inmuebles, se recoge la información que puede requerir el interesado en la compra o arrendamiento de una vivienda, antes de que se formalice la operación (información precontractual) y, por último, contiene una serie de medidas cuyo objetivo principal es contribuir a la mejora de la información y el compromiso con la transparencia en materia de vivienda. Ahora bien, es objeto de esta investigación analizar, únicamente, los aspectos civiles y procesales de la LDV. Al no ser aspectos jurídico-privados los deberes de información y transparencia que se

imponen al Estado en los arts. 32 y ss., dichos preceptos no serán aquí examinados. Por la razón expuesta, exclusivamente se estudiará el Capítulo I del Título IV, conformado por los arts. 30 y 31. En lo que respecta al art. 30, en él se establecen cuáles son los derechos de las personas demandantes, adquirentes o arrendatarias de vivienda: los reconocidos en la legislación de consumo, los que figuran en la legislación autonómica aplicable y el de recibir información, incluida la suministrada por la publicidad, completa, objetiva, veraz, clara, comprensible y accesible, sobre las características de las viviendas, sus servicios e instalaciones y las condiciones jurídicas y económicas de su adquisición, arrendamiento, cesión o uso. El párrafo segundo del citado precepto señala los sujetos que ostentan ese deber de información, haciendo alusión a todos los agentes que, operando en el sector de la edificación y rehabilitación de viviendas y la prestación de servicios inmobiliarios, estén facultados para transmitir, arrendar o ceder una vivienda en nombre propio o por cuenta ajena, esto es, promotores, personas propietarias y otras titulares de derechos reales, agentes inmobiliarios y administradores de fincas. Por último, el tercer y último párrafo del precepto define qué se entiende por información y publicidad –atribuyendo a ambas un mismo concepto–[7], y qué se entiende por información incompleta, insuficiente o deficiente[8].

---

7 Señala el precepto que se entiende por información y publicidad "toda forma de comunicación dirigida a demandantes de vivienda, usuarios o al público en general con el fin de promover de forma directa o indirecta la transmisión, el arrendamiento y cualquier otra forma de cesión de viviendas".

8 Será información incompleta, insuficiente o deficiente "la información que omita datos esenciales o los contenga en términos capaces

En lo que se refiere al art. 31, contiene tres apartados diferenciados. En el primero de ellos se menciona la información que puede ser requerida por la persona interesada en la compra o arrendamiento de una vivienda, antes de la formalización de la operación y de la entrega de cualquier cantidad a cuenta[9]. En el segundo párrafo

---

de inducir a error a los destinatarios o producir repercusiones económicas o jurídicas que no resulten admisibles, por perturbar el pacífico disfrute de la vivienda en las habituales condiciones de uso".

9 Esta información es la siguiente: a) Identificación del vendedor o arrendador y, en su caso, de la persona física o jurídica que intervenga, en el marco de una actividad profesional o empresarial, para la intermediación en la operación; b) Condiciones económicas de la operación: precio total y conceptos en éste incluidos, así como las condiciones de financiación o pago que, en su caso, pudieran establecerse; c) Características esenciales de la vivienda y del edificio, entre ellas: 1.º Certificado o cédula de habitabilidad; 2.º Acreditación de la superficie útil y construida de la vivienda, diferenciando en caso de división horizontal la superficie privativa de las comunes, y sin que pueda en ningún caso computarse a estos efectos las superficies de la vivienda con altura inferior a la exigida en la normativa reguladora; 3.º Antigüedad del edificio y, en su caso, de las principales reformas o actuaciones realizadas sobre el mismo; 4.º Servicios e instalaciones de que dispone la vivienda, tanto individuales como comunes; 5.º Certificado de eficiencia energética de la vivienda; 6.º Condiciones de accesibilidad de la vivienda y del edificio; 7.º Estado de ocupación o disponibilidad de la vivienda; d) Información jurídica del inmueble: la identificación registral de la finca, con la referencia de las cargas, gravámenes y afecciones de cualquier naturaleza, y la cuota de participación fijada en el título de propiedad; e) En el caso de tratarse de vivienda protegida, indicación expresa de tal circunstancia y de la sujeción al régimen legal de protección que le sea aplicable; f) En caso de edificios que cuenten oficialmente con protección arquitectónica por ser parte de un entorno declarado o en razón de su particular valor arquitectónico o histórico, se aportará información sobre el grado de protección y las condiciones y limitaciones para las intervenciones

se señala de forma expresa el derecho que tiene la persona interesada en la compra o arrendamiento de una vivienda de solicitar información acerca de la detección de amianto u otras sustancias peligrosas o nocivas para la salud. Finalmente, en el tercer y último apartado, se establecen una serie de obligaciones para el propietario –o persona que intervenga en la intermediación de la operación– que solamente serán de aplicación cuando la vivienda que vaya a ser objeto de arrendamiento se encuentre ubicada en una zona de mercado residencial tensionado y, además, vaya a tratarse de la vivienda habitual del arrendatario. En tales casos, el propietario deberá indicar tal circunstancia –vivienda ubicada en zona de mercado residencial tensionado– e informar, con anterioridad a la formalización del arrendamiento, y en todo caso en el documento del contrato, de la cuantía de la última renta del contrato de arrendamiento de vivienda habitual que hubiese estado vigente en los últimos cinco años en la misma vivienda, así como del valor que le pueda corresponder atendiendo al índice de referencia de precios de alquiler de viviendas que resulte de aplicación.

- RD 515/1989, de 21 de abril, sobre protección de los consumidores en cuanto a la información a suministrar en la compraventa y arrendamiento de viviendas. Señala el art. 1 que el presente Real Decreto es de aplicación "a la oferta, promoción y publicidad que se efectúe en el marco de una actividad empresarial o profesional, siempre que

---

de reforma o rehabilitación; g) Cualquier otra información que pueda ser relevante para la persona interesada en la compra o arrendamiento de la vivienda, incluyendo los aspectos de carácter territorial, urbanístico, físico-técnico, de protección patrimonial, o administrativo relacionados con la misma.

aquellos actos vayan dirigidos a consumidores...". Por tanto, se trata de una norma que se aplica exclusivamente a los contratos de compraventa y arrendamiento celebrados entre consumidor y empresario, excluyéndose su aplicación en los contratos entre particulares. Además, se trata de una norma estatal que reconoce expresamente la posibilidad de que existan normas autonómicas que regulen esta misma cuestión[10]. Así, la Disposición Adicional segunda señala al respecto lo siguiente: "Lo establecido en este Real Decreto será de aplicación supletoria respecto de las Comunidades Autónomas que estatutariamente hayan asumido la competencia plena sobre la defensa de los consumidores y usuarios, excepto los artículos 3.°, apartado 2, y 10, que tendrán vigencia en todo el Estado, en virtud de lo dispuesto en la regla 8.° del artículo 149.1,

---

[10] En Andalucía se dictó la Ley 13/2003, de 17 de diciembre, de defensa y protección de los consumidores y usuarios, en la que se dispone que las Administraciones Públicas de Andalucía adoptarán medidas eficaces dirigidas, entre otras cosas, a: "Facilitar a los consumidores toda clase de información sobre materias o aspectos que les afecten o interesen directamente y, de modo particular, sobre la construcción de viviendas" (art. 16.2). Precisamente para dar cumplimiento al mandato contenido en este precepto se dicta el Decreto 218/2005, de 11 de octubre, por el que se aprueba el Reglamento de información al consumidor en la compraventa y arrendamiento de viviendas en Andalucía. Una de las principales novedades que introduce esta regulación es la necesidad de que los sujetos que ofrezcan la venta o arrendamiento de viviendas deberán disponer de un Documento Informativo Abreviado. El contenido de este documento será distinto según se trate de venta o arrendamiento, y en caso de venta, según se trate de vivienda en proyecto, en construcción o ya construida. Otra de las cuestiones que debe destacarse de esta normativa es la relativa a la regulación de la información que debe facilitarse en segundas o ulteriores transmisiones de la vivienda, reconociendo así la importancia del mercado inmobiliario de segunda mano.

de la Constitución Española". Estamos ante una norma que consta tan solo de once preceptos que hacen referencia a diversas cuestiones: la necesidad de que tanto en la publicidad como en la información precontractual conste si la vivienda se encuentra en construcción o si la edificación ha concluido –en el caso de que la vivienda no se encuentre totalmente terminada se deberá tener a disposición del público y de las autoridades competentes copia del documento o documentos en los que se formalizan las garantías entregadas a cuenta–; la exigencia de ajustar la oferta, promoción y publicidad dirigida a la venta o arrendamiento de viviendas a las verdaderas características, condiciones y utilidad de la vivienda, así como también la exigencia de que dicha oferta, promoción y publicidad no induzca a error a sus destinatarios; la norma también señala, de igual modo que lo hace la legislación de consumo, que la publicidad vincula al empresario y será exigible aun cuando no figure en el contrato celebrado; se hace mención a cuál debe ser el contenido de la información que esté a disposición del público o, en su caso, de las autoridades competentes (plano de la vivienda, descripción de la vivienda, materiales empleados, precio...), información esta que se amplía considerablemente cuando la vivienda se promociona exclusivamente para la venta (art. 5), dedicando un artículo específico a la información relativa al precio de venta (art. 6); se establece asimismo la obligación del vendedor de entregar al adquirente de la vivienda, si este lo solicita, copia de todos los documentos a los que hace referencia este Real Decreto; se dispone la prohibición de incluir determinadas cláusulas en el contrato y, por último, se indican las consecuencias de incumplir lo dispuesto en la norma, aunque no desde la perspectiva del derecho privado, sino imponiendo una serie de sanciones administrativas según la infracción sea leve, grave o muy grave.

- Ley 38/1999, de 5 de noviembre, de Ordenación de la Edificación (LOE). Dispone el art. 7 de la LOE que el Libro del Edificio –en el que se contiene el proyecto, las modificaciones debidamente aprobadas, el acta de recepción, la relación identificativa de los agentes que han intervenido durante el proceso de edificación, las instrucciones de uso y mantenimiento del edificio y sus instalaciones–, debe ser entregado a los usuarios finales del edificio. Se trata, por tanto, de una obligación de carácter general, que no queda limitada a las relaciones de consumo. También cabe señalar al respecto que esta obligación solamente cobra verdadero sentido si la vivienda ya ha sido construida y que la obligación de información no es precontractual, sino contractual[11], de modo que, "cuando el bien que es la vivienda se entregue al adquirente en cumplimiento del contrato, debe incorporar el Libro del Edificio" (Alonso Pérez, 2010, p. 79). No obstante, algunas comunidades autónomas, como es el caso de Andalucía, exigen que el Libro del Edificio se ponga a disposición del interesado en la adquisición de una vivienda en el periodo precontractual. En este sentido, el art. 7. f) del Decreto 218/2005, de 11 de octubre, por el que se aprueba el Reglamento de información al consumidor en la compraventa y arrendamiento de viviendas en Andalucía, exige que se ponga

---

11 Esta idea queda reforzada por la propia redacción del precepto, en el que se señala que el Libro del Edificio debe ser entregado a los usuarios finales. En la práctica, como acertadamente señala Alonso Pérez (2010, p. 81), puede ocurrir que la entrega se lleve a cabo en un momento muy posterior a la perfección del contrato, puesto que, cuando se trata de venta de viviendas por construir, hasta que no se produce la entrega de la obra al propietario o promotor, no hay obligación de entregar el Libro del Edificio.

a disposición del consumidor el Libro del Edificio en el periodo precontractual y siempre que se trate de viviendas que ya han sido construidas[12].

- ***Leyes generales:***

  - Real Decreto Legislativo 1/2007, de 16 de noviembre, por el que se aprueba el Texto refundido de la Ley General para la defensa de consumidores y usuarios y otras leyes complementarias. En aquellos casos en los que el propietario o arrendatario ostenten la condición de consumidor –en el sentido del art 3–, será asimismo de aplicación la legislación de consumo. Concretamente, el art. 60 del citado texto legal, que se centra en regular el contenido, las características y los términos en que debe suministrarse la información previa al contrato. Asimismo, será igualmente de aplicación el art. 61, que regula la integración de la oferta, promoción y publicidad en el contrato, y en el que se dispone textualmente, en su apartado segundo, que "el contenido de la oferta, promoción o publicidad, las prestaciones propias de cada bien o servicio, las condiciones jurídicas o económicas y garantías ofrecidas serán exigibles por los consumidores y usuarios, aun cuando no figuren expresamente en el contrato celebrado o en el documento o comprobante recibido y deberán tenerse en cuenta en la determinación del principio de conformidad con el contrato"[13]. En este sentido, y como bien señala

---

12 Como se aprecia, el Decreto andaluz sí limita su aplicación a las relaciones de consumo. Sobre este particular *vid.* Alonso Pérez (2010, p. 80).

13 En definitiva, la publicidad vincula al empresario. La mayor parte de la doctrina se mantiene unida al sostener que aquello que el legislador trata de salvaguardar es la confianza negocial fundadamente generada en el ánimo de los consumidores y usuarios destinatarios de

tanto la doctrina como la jurisprudencia, la información contenida en medios publicitarios que contenga una referencia específica a las características del bien o servicio, "no sólo genera una legítima expectativa en el adquirente, sino que, y sobre todo, constituye parte de la oferta de quien realiza la actividad de promoción (...); por ello no es extraño que la jurisprudencia primero y la legislación después obligasen al empresario anunciador en los términos de lo ofrecido en su publicidad" (Picatoste Bobillo, 2011, p. 404). De igual modo se pronuncia la Propuesta de Anteproyecto de Ley de Modernización del Derecho de obligaciones y contratos[14] al declarar en su art. 1276 que "quedarán insertadas en el contrato y tendrán valor vinculante las afirmaciones o declaraciones efectuadas por un profesional en la publicidad o en actividades de promoción de un producto o servicio, salvo que pruebe que la otra parte conoció o debió haber conocido que tal declaración o afirmación era incorrecta". Por tanto, aunque en fase de publicidad las partes aún no se encuentran estrictamente en fase de negociaciones precontractuales, esto no es óbice para que la información aportada por aquélla

---

los actos publicitarios o promocionales. Sin embargo, resulta dudoso que la norma solamente haya sido confeccionada al servicio de los intereses de los consumidores; por ello estamos de acuerdo en afirmar que "también es una norma de ordenación del sistema jurídico de contratación con los consumidores y usuarios que preordena el *iter* o proceso de formación del contrato con consumidores y usuarios" (Font Galán, 2011, p. 157). Continúa diciendo este autor que lo realmente novedoso del art. 61 es que este precepto da pie para apreciar que en los contratos establecidos con consumidores el *iter* formativo del contrato se retrotrae "al punto de arranque que supone la puesta en práctica de estas operaciones o prácticas comerciales de promoción del contrato" (Font Galán, 2011, p. 167).

14 Boletín de información. Año LXIII, Madrid, enero de 2009.

deba ser real y vinculante[15]: así se deduce no solamente de preceptos generales del Código –arts. 7 y 1258[16]–, sino también del ya mencionado art. 61 RD 1/2007.

---

15 Además del carácter vinculante de la publicidad, es menester hacer referencia a determinadas cuestiones que pueden suscitar dudas en cuanto a la mencionada integración de la publicidad. En primer lugar, debe señalarse que la publicidad vincula con independencia de su procedencia, al igual que ocurre con la información precontractual. Así, por ejemplo, si la publicidad no proviene del vendedor o del arrendador, sino de una entidad financiera vinculada a cualquiera de ellos, igualmente vincula; en segundo lugar, la publicidad vincula sin importar si es dolosa o no, aunque los remedios que se activen a favor del perjudicado varíen según sea o no imputable el incumplimiento. En este sentido "no se considera necesario que hubiese culpa o dolo de parte del emisor de la publicidad: que de buena fe creyese en la exactitud de lo anunciado no es razón para considerarle no vinculado por ello, ya que ha suscitado la legítima confianza de quienes recibieron el mensaje" (Gómez Calle, 1994, p. 74). Se entiende que el art. 61 establece un régimen de responsabilidad objetiva (Morales Moreno, 1992, p. 685).

16 En la idea de la buena fe hace también especial hincapié Díaz Moreno (2006, pp. 323 y 233), para quien sería incompatible con la buena fe contractual negar las prestaciones expresadas en la publicidad cuando esta se realiza, precisamente, para incitar a contratar. Por último, el carácter vinculante de la publicidad ¿solamente se activa en aquellos casos en los que el contrato ha llegado efectivamente a celebrarse o, por el contrario, el consumidor puede exigir que el contrato se celebre en las condiciones anunciadas en la publicidad? La literalidad del art. 61 TR parece indicar que el precepto está pensando en un consumidor que ha celebrado el contrato. A idéntica conclusión nos lleva la interpretación del art. 1258 C.c., del que se desprende que los efectos de la buena fe se despliegan siempre que el contrato se haya perfeccionado (García Vicente, 2009, p. 784). Sin embargo, pese a la opinión de quienes defienden que antes de la celebración del contrato, no asiste al consumidor un derecho de integración

- Por último, habrá que tener en consideración otras normas que, como la Ley 34/1988, de 11 de noviembre, General de Publicidad, pueden resultar de aplicación. Además, el Código civil siempre regirá de forma supletoria.

Todas las normas que se acaban de mencionar son normas aplicables al caso que nos ocupa, esto es, la publicidad y el deber de información en los contratos de compraventa y arrendamiento de vivienda. Aunque la actual LDV contiene una disposición derogatoria según la cual "quedan derogadas a la entrada en vigor de esta ley cuantas disposiciones de igual o inferior rango se opongan a lo dispuesto en esta ley", lo cierto es que, en materia de publicidad e información precontractual, las normas que conviven con la nueva LDV no contradicen lo dispuesto en ella, por lo que se entiende que continúan vigentes.

---

publicitaria quedando, pues, desasistido de protección precontractual en lo concerniente a exigir que el contenido de las ofertas promocionales o publicitarias se incorpore al contenido de la oferta contractual, creemos que, a no ser que con anterioridad a la perfección del contrato el empresario haya revocado correctamente su oferta –y siempre que sea ésta revocable–, no hay razones jurídicas para que no queden integradas las ofertas publicitarias en el contenido prestacional. Es más, si el consumidor tiene derecho a exigir, una vez celebrado el contrato, que las ofertas publicitarias integren el contenido del mismo, no puede esto significar, sino que el comerciante tiene "la previa y antecedente obligación legal de integrar su oferta contractual con los contenidos negociales de sus ofertas comerciales, publicitarias o promocionales" (Font Galán, 2011, pp. 173 y 174).

III. APORTACIONES, DEFICIENCIAS Y PROBLEMAS INTERPRETATIVOS DE LA ACTUAL LEY DE VIVIENDA EN MATERIA DE INFORMACIÓN PRECONTRACTUAL Y PUBLICIDAD

En este apartado se analizarán las aportaciones y las deficiencias más significativas de la LDV, en particular en lo que se refiere a la información precontractual y a la publicidad (arts. 30 y 31). No obstante, antes de profundizar sobre ello, es necesario hacer una breve reflexión sobre una cuestión que impregna toda la Ley, también los preceptos que son aquí objeto de estudio. Se está haciendo referencia a la falta de rigor terminológico del legislador. En repetidas ocasiones se emplea una terminología inadecuada que revela, entre otras cosas, la ignorancia de los redactores de la ley en cuestiones relacionadas con el Derecho privado. Un claro ejemplo de que el legislador desconoce los términos que suelen emplearse en Derecho contractual, es precisamente cuando hace mención a *las operaciones de compra y arrendamiento de vivienda*[17]. Lo correcto y apropiado hubiera sido aludir a contratos de compraventa y arrendamiento.

Dicho esto, pasan a enumerarse las aportaciones y las deficiencias más relevantes de los arts. 30 y 31.

- El art. 30.1.a) de la Ley plantea un problema interpretativo importante difícil de resolver. Según este precepto: "Son derechos de las personas demandantes, adquirentes o arrendatarias de vivienda, o en cualquier otro régimen jurídico de tenencia y disfrute: a) Los reconocidos en el texto refundido de la Ley

---

17 El mismo Título IV de la Ley se titula "Medidas de protección y transparencia en las operaciones de compra y arrendamiento de vivienda". Dentro de este Título IV, el art. 31 regula la información mínima en las operaciones de compra y arrendamiento de vivienda.

General para la Defensa de los Consumidores y Usuarios y otras leyes complementarias, aprobado por Real Decreto Legislativo 1/2007, de 16 de noviembre, y en la legislación autonómica aplicable". Nos hacemos pues la siguiente pregunta: ¿establece este artículo la aplicación de la legislación de consumo a todos los sujetos o solamente a los que tengan la condición de consumidor? ¿Modifica este precepto el ámbito de aplicación de la legislación de consumo? Lo cierto es que existen argumentos en ambos sentidos. Veamos por qué. Para interpretar correctamente la norma, se acude al art. 3.1 CC, según el cual "Las normas se interpretarán según el sentido propio de sus palabras, en relación con el contexto, los antecedentes históricos y legislativos, y la realidad social del tiempo en que han de ser aplicadas, atendiendo fundamentalmente al espíritu y finalidad de aquellas". Pues bien, señala al respecto Díez-Picazo que "la interpretación según el contexto aconseja poner en conexión todos los preceptos legislativos que tratan de una determinada cuestión, por presuponerse que entre ellos hay una coherencia y una interdependencia"[18]. Es precisamente al poner en conexión el art. 30.1.a) de la Ley con el párrafo segundo de ese mismo precepto, donde surgen las dudas. El art. 30.2 dispone que "todos los agentes que (...) estén facultados para la transmisión, el arrendamiento y la cesión de viviendas (...) deben cumplir en su actividad el deber de información completa, objetiva, veraz, clara, comprensible y accesible conforme lo previsto en esta ley, así como en la legislación de defensa de consumidores y usuarios *cuando*

---

[18] Díez-Picazo y Gullón Ballesteros (2005, p. 169).

> *se trate de relaciones entre consumidores o usuarios y empresarios*". Este precepto demuestra que, cuando quiere matizar, el legislador lo hace. Así, el artículo indica que los promotores y demás agentes deberán cumplir con el deber de información impuesto en la legislación de consumo solamente en aquellos casos en los que el contrato de compraventa o arrendamiento se haya celebrado con un consumidor, esto es, con una persona que actúa al margen de su actividad empresarial o profesional. De lo contrario, bastará con que cumplan con lo establecido en la LDV. Por tanto, siguiendo este hilo argumental, si el legislador en el art. 30.1.a) no especifica y no matiza, como si lo hace en el art. 30.2., será porque quiere que *todas* las personas demandantes, adquirentes o arrendatarias de vivienda –sean o no consumidoras–, disfruten de los derechos reconocidos en la legislación de consumo. También puede argumentarse justamente lo contrario: si el legislador en el art. 30.2. limita la aplicación de la legislación de consumo a contratos celebrados con consumidores, este límite debe estar presente, para así establecer un criterio coherente y unitario, en toda la Ley, a no ser que la propia Ley estableciera expresamente algo distinto.

Si se sigue este segundo hilo argumental, que creemos más correcto que el primero, debe interpretarse el precepto restrictivamente: solamente se aplicará la legislación de consumo en contratos celebrados entre consumidor y empresario.

¿Es esto lo más lógico? ¿O hubiera sido esta una buena oportunidad para ampliar el ámbito de aplicación y extender los deberes de información a cualquier vendedor o arrendador? Dada la importancia del bien objeto del contrato, no es del todo descabellado afirmar que la normativa protectora se podría aplicar a cualquier comprador o arrendatario de vivienda, con independencia de quién sea el otro sujeto. ¿También en aquellos casos en los que el vendedor o el arrendador es un particular? En mi opinión, y aunque no parece haber sido esta la postura del legislador, entiendo que debieran atribuirse a cualquier vendedor o arrendador cargas informativas importantes. No tanto las propias de la legislación de consumo, al no estar el TRLGDCU pensando exclusivamente en viviendas, sino, más bien, las de un texto legal como el RD 515/1989, de 21 de abril, sobre protección de los consumidores en cuanto a la información a suministrar en la compraventa y arrendamiento de viviendas.

Volviendo a la norma que se analiza, es decir, a lo que la LDV dice, y no a lo que, a nuestro juicio, debiera decir, ha de concluirse que este problema interpretativo debe resolverse a favor de considerar que son derechos de las personas demandantes, adquirentes o arrendatarias de vivienda los reconocidos en la legislación de consumo, solo y exclusivamente cuando tales sujetos tengan la condición de consumidor, por actuar al margen de su actividad profesional o empresarial. Ahora bien, para llegar a esta conclusión no hacía falta ningún precepto de la LDV que lo indicara, puesto que esto mismo ya se establece en la legislación de consumo, siendo por tanto el artículo que nos ocupa una norma innecesaria y superflua.

- El art. 30. 1. b) introduce un matiz que consideramos acertado e innovador: Si el demandante, adquirente o arrendatario de vivienda es una persona con discapacidad o dificultades de comprensión, debe proporcionársele la información, incluida la suministrada por medios publicitarios, en un formato accesible[19].

---

19 Se trata de un precepto que está en consonancia con la Convención Internacional sobre los Derechos de las Personas con Discapacidad, en la que se reconoce, tal y como se señala en el Preámbulo de la

En última instancia, al mismo resultado se llegaría aplicando la buena fe contractual –art. 7 CC–[20].

- El art. 30.2, como ya se ha adelantado anteriormente, establece una diferencia según el contrato de compraventa o arrendamiento sea o no de consumo. Si lo es, los agentes que lleven a cabo la venta o el arrendamiento están sujetos a lo dispuesto tanto en la LDV como en la legislación de consumo –con respecto al deber de información–. Por el contrario, si el contrato no se celebra con un consumidor, solamente deberán cumplir los deberes de información impuestos por la LDV. Además, el precepto especifica quiénes pueden tener la consideración de *agentes*: promotores, personas propietarias y otras titulares de derechos reales –como puede ser, por ejemplo, el usufructo–, agentes inmobiliarios y administradores de fincas. Todos ellos deberán cumplir en su actividad el deber de información completa, objetiva, veraz, clara, comprensible y accesible conforme a la LDV. Ahora bien, como ya se ha dicho, solamente si el adquirente es consumidor, se aplicará, también, la legislación de consumo. En la práctica, esto se traduce en que en aquellos casos en los que el vendedor o arrendador sea un particular, con independencia de quien sea el adquirente o arrendatario, nunca se aplicará el

---

Ley, la necesidad de garantizar la accesibilidad como presupuesto fundamental para el disfrute de una vivienda adecuada.

20 Para Mate Satué se trata de una previsión insuficiente. Hubiera sido oportuno aprovechar la ocasión para añadir las especialidades que dentro del deber de información corresponderían a los consumidores vulnerables que celebran estos contratos. Es cierto que una persona con discapacidad puede tener la consideración de consumidor vulnerable, pero existen otras causas de vulnerabilidad que no resultan atendidas en esta norma (2024, p. 324).

TRLGDCU. ¿Es lo justo? ¿Querer garantizar el acceso a la vivienda no puede implicar atribuir a particulares deberes de información significativos? Sobre este particular se volverá más adelante, cuando se enumeren ciertas propuestas de mejora.

- El párrafo tercero del art. 30 es otro de los grandes despropósitos de la Ley. En este precepto se demuestra nuevamente la falta de rigor técnico del legislador, que no distingue conceptos tales como el de publicidad, información precontractual e información contractual. Este artículo establece un concepto único para la información y para la publicidad, a las que define como "toda forma de comunicación dirigida a demandantes de vivienda, usuarios o al público en general con el fin de promover de forma directa o indirecta la transmisión, el arrendamiento y cualquier otra forma de cesión de viviendas"[21]. El desacierto del legislador a la hora de redactar estos preceptos es evidente y, cuanto menos, sorprendente, máxime si tenemos en cuenta que es el propio legislador el que en el art. 1.3. de la Ley señala que la norma tiene por objeto, entre otras cosas, "reforzar la protección del acceso a información completa, objetiva, veraz, clara, comprensible y accesible, en las operaciones de compra y arrendamiento de vivienda".

Si se quiere garantizar el acceso a la vivienda a través de, entre otras muchas vías, una adecuada regulación de la publicidad y la información precontractual, qué menos que empezar por

---

21 Llama también la atención que el precepto establece un concepto unitario solamente "a los efectos de los apartados anteriores", ¿esto significa que a partir del art. 31 el concepto de información y publicidad ya no es el mismo?, es decir, ¿el concepto único solamente rige para el art. 30.1. y para el art. 30.2?

diferenciar ambos conceptos. Según el art. 2 de la ley 34/1988, de 11 de noviembre, General de Publicidad, se entiende por publicidad "toda forma de comunicación realizada por una persona física o jurídica, pública o privada, en el ejercicio de una actividad comercial, industrial, artesanal o profesional, con el fin de promover de forma directa o indirecta la contratación de bienes muebles o inmuebles, servicios, derechos y obligaciones". Por tanto, además de informar, el objetivo principal de la publicidad es promover la contratación de los productos y servicios que ofrece. Por su parte, la información precontractual se constituye como el derecho clave del que goza el sujeto interesado en comprar o arrendar una vivienda en el periodo precontractual. De modo muy general, puede decirse que se trata de un acto comunicativo mediante el cual se pretende dar a conocer el alcance y las características del servicio o del bien que va a ser contratado. De este modo, puede definirse la información como "un elemento de conocimiento suministrado obligatoriamente por una de las partes contratantes (deudor de la obligación de información) a la otra parte (acreedor de la obligación de información), teniendo como objeto principal la adecuada formación del consentimiento contractual de este último, tanto en lo referente a los aspectos jurídicos como materiales del negocio" (Llobet Aguado, 1996, p. 33).

Es cierto que a través de la publicidad se dan a conocer aspectos del bien o servicio ofertado, pero las diferencias entre ambas son notables[22]. Entre otras cosas, mientras que la publicidad tiene

22 También existen similitudes entre ambas: por ejemplo, la publicidad, al igual que ocurrirá con la información precontractual, vincula con independencia de su procedencia. Así, por ejemplo, si la publicidad no proviene del vendedor o del arrendador, sino de una entidad financiera vinculada a cualquiera de ellos, igualmente vincula. Además, la publicidad vincula sin importar si es dolosa o no, lo mismo que sucede con la información precontractual. Refiriéndose

carácter voluntario, la información precontractual no[23]. Además,

---

a la publicidad, señala Gómez Calle (1994, p. 74) lo que sigue: "no se considera necesario que hubiese culpa o dolo de parte del emisor de la publicidad: que de buena fe creyese en la exactitud de lo anunciado no es razón para considerarle no vinculado por ello, ya que ha suscitado la legítima confianza de quienes recibieron el mensaje".

23 Aunque la publicidad tenga carácter voluntario, en el momento en el que el empresario decide incluirla, ésta debe ajustarse a los requisitos impuestos por la ley. Así, por ejemplo, no solamente será vinculante, sino que, además, deberá cumplir el contenido mínimo previsto en el art. 20 de la legislación de consumidores. En cuanto al carácter vinculante de la publicidad podemos preguntarnos si solamente se activa en aquellos casos en los que el contrato ha llegado efectivamente a celebrarse o, por el contrario, el consumidor puede exigir que el contrato se celebre en las condiciones anunciadas en la publicidad. La literalidad del art. 61 TR parece indicar que el precepto está pensando en un consumidor que ha celebrado el contrato. Así, se refiere al derecho que tiene el consumidor a exigir el contenido anunciado en la oferta, promoción o publicidad, aun cuando aquellas condiciones o características publicitadas no figuren expresamente en el *contrato celebrado.* A idéntica conclusión nos lleva la interpretación del art. 1258 C.c., del que se desprende que los efectos de la buena fe se despliegan siempre que el contrato se haya perfeccionado. Sobre este particular *vid.* Picatoste Bobillo (2011, p. 409), Schulze (2006, p. 40) y García Vicente (2009, p. 784), según el cual "el efecto de la integración o fuerza vinculante sólo se produce en el caso de que el contrato se celebre". Sin embargo, pese a la opinión de quienes defienden que antes de la celebración del contrato, no asiste al consumidor un derecho de integración publicitaria, creemos que, a no ser que con anterioridad a la perfección del contrato el empresario haya revocado correctamente su oferta –y siempre que sea ésta revocable–, no hay razones jurídicas para que no queden integradas las ofertas publicitarias en el contenido prestacional. Es más, si el consumidor tiene derecho a exigir, una vez celebrado el contrato, que las ofertas publicitarias integren el contenido del mismo, no puede esto significar, sino que el comerciante "tiene la previa y antecedente obligación legal de integrar su oferta contractual con los contenidos negociales

como resulta lógico, los aspectos sobre los que se informa en la oferta comercial suelen ser bastante más reducidos que los que deben figurar en la información precontractual[24].

- Por lo que respecta al art. 31.1, en él se regula la "información mínima en las operaciones de compra y arrendamiento de vivienda", no dedicándose ningún precepto a regular, como hubiera sido lo deseable, el contenido de la información básica que debe figurar en la publicidad. La información a la que hace referencia este precepto debe completarse con lo dispuesto en los arts. 4 y ss. del Real Decreto 515/1989, así como también con lo establecido en la legislación de consumo, siempre que el adquirente o arrendatario tengan la condición de consumidor[25]. El precepto no establece un plazo dentro del cual debe suministrarse

---

de sus ofertas comerciales, publicitarias o promocionales" (Font Galán, 2011, pp. 173 y 174). Por tanto, se entiende que existe el derecho a exigir los contenidos publicitarios prometidos en las ofertas comerciales, siempre que, claro está, estas ofertas no hayan sido revocadas correctamente por el empresario que las realizó, o siempre que el contenido de dicha publicidad no haya sido expresamente excluido en la fase de tratos preliminares.

24 Esto ocurre en varias leyes, como, por ejemplo, en la propia legislación de consumo, donde se puede comprobar que el art. 60 (información previa al contrato) es mucho más extenso que el art. 20 (información necesaria en la oferta comercial). Del mismo modo sucede en la Ley 5/2019, de 15 de marzo, reguladora de los contratos de crédito inmobiliario. En el art. 6 de la misma se regula la información básica que debe figurar en la publicidad, mientras que el art. 10 es el encargado de regular lo referente a la información precontractual.

25 Debe apuntarse que alguna de esta información mínima a la que hace referencia este precepto es poco concreta y en ocasiones difícil de obtener. Así, por ejemplo, ocurre con la superficie útil –que frecuentemente no consta ni en el Registro de la Propiedad ni en el catastro–.

la información, ni tampoco hace referencia al plazo de validez de la información suministrada.

Llama una vez más la atención la pésima redacción del precepto, en el que se dice que "la persona interesada en la compra o arrendamiento de una vivienda (...) *podrá requerir*, antes de la formalización de la operación y de la entrega de cualquier cantidad a cuenta, la siguiente información (...)". Lo mismo ocurre en el art. 31.2, según el cual "la persona interesada (...) *podrá requerir* información acerca de la detección de amianto u otras sustancias peligrosas o nocivas para la salud". Parece obvio que, con independencia de la existencia de un artículo que lo reconozca, toda persona tiene derecho a pedir información sobre tales cuestiones. Hubiera sido mucho más exacto establecer que el sujeto en cuestión –promotor, propietario...– *debe facilitar* o, si se quiere, *está obligado a proporcionar* la información. Por tanto, la información a la que hace referencia el precepto debe ser proporcionada en todo caso, con independencia de que sea o no requerida por el interesado[26]. Si interpretamos el precepto según el tenor literal de sus palabras, se llega a una situación ilógica y absurda y, por supuesto, contraria al espíritu de la ley: solamente si el sujeto interesado lo solicita, tendrá derecho a ser informado sobre las cuestiones que enumera el art. 31.

- Por último, estos preceptos no dejan claro cuál es su verdadero ámbito de aplicación. ¿Se aplican a los arrendamientos de temporada? ¿A los arrendamientos de estudiantes realizados por particulares o por inmobiliarias? ¿A los arrendamientos en residencias de estudiantes? Además, mientras el art. 30 LDV establece que es apli-

---

26 Así se deduce también de la redacción del art. 30.2. de la Ley, según el cual "todos los agentes (...) *deben cumplir* en su actividad el deber de información (...)".

cable a "cualquier otro régimen jurídico de tenencia y disfrute", el art. 31 se refiere exclusivamente a la compraventa y al arrendamiento. Surgen muchos interrogantes: ¿se aplica lo establecido en estos preceptos al arrendamiento de habitaciones y al subarriendo? ¿Se aplica a las formas de tenencia intermedia de Cataluña (propiedad compartida y propiedad temporal)? ¿Se aplica a las compraventas de nudas propiedades que están aumentando en el mercado considerablemente? ¿Se aplica a los alquileres sociales que las administraciones públicas ponen a disposición de personas vulnerables, o estos se rigen por su normativa administrativa propia? En definitiva, ¿a qué contratos sobre qué "vivienda" se aplican estos artículos?

IV. PROPUESTAS DE MEJORA

Una vez expuestas las principales deficiencias de los artículos 30 y 31 de la LDV, se tratará ahora de proponer ciertos cambios que contribuyan a mejorar la regulación de la información precontractual en la ley que nos ocupa. Ya se ha adelantado en páginas anteriores que es conveniente que una ley cuyo propósito principal es garantizar el derecho a una vivienda digna y adecuada, regule una cuestión tan importante como es el deber de información precontractual; no obstante, creemos que se ha desaprovechado una excelente oportunidad, y que el legislador no ha prestado demasiada atención a este deber, pese a lo dispuesto en el art. 1.3 de la Ley. Es por ello que pasamos a enumerar algunas propuestas de mejora, cuyo fin último es enriquecer el contenido actual de la norma.

1. Los artículos 30 y 31 de la Ley conforman el Capítulo I del Título IV. No se entiende que en este mismo Título IV, concretamente en su Capítulo II, se regulen deberes de información y transparencia que se imponen al Esta-

do. Hubiese sido más oportuno que las cuestiones que regula el Capítulo II no hubiesen formado parte de un Título que se titula "Medidas de protección y transparencia en las operaciones de compra y arrendamiento de vivienda". En el Título IV se mezclan cuestiones que nada tienen que ver para formar parte todas ellas de un mismo Título. Una cosa es el deber de información precontractual en los contratos de compraventa y arrendamiento de vivienda –materia propia del derecho contractual–, y otra muy distinta lo regulado en los arts. 32 y ss. de la Ley. En definitiva, hubiese sido más exacto que el Título IV únicamente hubiese estado conformado por preceptos que hagan referencia a los aspectos tratados en los arts. 30 y 31, así como a aquellos que guarden relación con estos.

2. En segundo lugar, otra propuesta de mejora es la que tiene que ver con la terminología empleada por el legislador, y a la que ya se ha hecho referencia. No es habitual, entre los estudiosos del derecho privado, referirse al contrato de compraventa y arrendamiento de vivienda como "operaciones de compra y arrendamiento", lo que denota la ignorancia del legislador en cuestiones relacionadas con el derecho privado. Por tanto, se propone que la Ley utilice una terminología rigurosa y precisa, introduciendo los términos propios del derecho contractual.

3. En tercer lugar, la redacción de algunos preceptos también es deficiente, e induce a error. La redacción del art. 31.1 es buena muestra de ello, al establecer que es la persona interesada en la compra o en el arrendamiento la que puede requerir cierta información. Es evidente que el deber de información precontractual debe ejercitarse por parte del sujeto facultado para la transmisión o el arrendamiento en todo caso, con independencia de la actuación del interesado. Así, se propone sustituir la expre-

sión utilizada por los arts. 31.1 y 31.2 –*la persona interesada podrá requerir información*–, por otra que no induzca a error como, por ejemplo, "el vendedor o arrendador (...) está obligado a proporcionar la siguiente información (...)".

Estas tres primeras propuestas de mejora tratan cuestiones puramente formales, pero no por ello menos importantes. Sirven para intuir lo que se confirma cuando se estudian con detalle los artículos 30 y 31 de la Ley: que el legislador es un desconocedor del Derecho privado y que el deber de información precontractual, lejos de ser una prioridad, solamente sirve para dotar de contenido la Ley.

4. En cuarto lugar, se propone que la Ley sea categórica en cuanto a su ámbito de aplicación. Una cuestión tan importante no debe ser objeto de interpretaciones. Por las razones anteriormente expuestas se cree que el legislador ha optado por aplicar la legislación de consumo solamente a aquellos casos en los que el contrato haya sido celebrado entre consumidor y empresario, pero no cuando el contrato se celebre entre particulares. Se propone, por el contrario, dada la importancia que tiene el bien objeto del contrato –la vivienda–, atribuir a cualquier vendedor o arrendador cargas informativas considerables, no tanto las impuestas en el TRLGDCU –que no están pensadas específicamente para los contratos que nos ocupan–, sino más bien las que se atribuyen en el RD 515/1989, de 21 de abril, sobre protección de los consumidores en cuanto a la información a suministrar en la compraventa y arrendamiento de viviendas –o similares–. En definitiva, el comprador o arrendatario, con independencia de quien sea el vendedor o arrendador, merecen una información detallada de la vivienda que va a ser objeto del contrato. Esta información, obviamente, debiera ser distinta según se esté ante un contrato de arrendamiento o de compraventa y, en este último caso, según se esté ante una vivienda nueva o usada.

5. En quinto lugar, se propone cambiar la redacción del art. 30.2, no siendo necesario referirse a todos los posibles autores de la información. Habría bastado decir que la información tiene que ser veraz, completa, comprensible, clara y accesible, sin aludir a los sujetos.
6. En sexto lugar, se propone distinguir, cosa que no hace la norma, entre información precontractual y publicidad. Como ya se ha dicho, se trata de conceptos diferentes, y merecen ser regulados por separado.

Es cierto que tanto la publicidad como la información precontractual persiguen informar a la persona adquirente o arrendataria del inmueble; ahora bien, el propósito que cumple el deber de informar será distinto en cada etapa. Dicho de otro modo, la información proporcionada por la publicidad no persigue los mismos fines que la información precontractual. Concretamente, ¿qué se persigue cuando una ley regula la información publicitaria? Normalmente, proteger frente a prácticas publicitarias desleales o engañosas y también que se puedan comparar anuncios. Elaborar una lista de los datos que deben figurar en la publicidad es imprescindible para que se puedan contrastar distintas ofertas. Además, la información proporcionada a través de la publicidad sirve para que el interesado, al conocer los datos publicitarios, decida si está interesado o no en seguir informándose acerca del producto o servicio publicitado. Debe asimismo puntualizarse que dar publicidad al producto es un acto voluntario, no estando el empresario en absoluto obligado a ello; ahora bien, si decide publicitarlo, como será lo habitual en la práctica, serán de aplicación una serie de preceptos, como por ejemplo el art. 20 de la legislación de consumo. Por tanto, lo ideal hubiese sido que esta ley, como también hacen otras leyes –es el caso de la Ley 5/2019, de 15 de marzo, reguladora de los contratos de crédito inmobiliario–, hubiera destinado un precepto, o varios, a regular tal cuestión.

Dicho esto, nos hacemos la misma pregunta en sede de información precontractual, ¿qué persigue el legislador al regularla? Puede afirmarse que este tipo de información, que es obligatoria y no voluntaria, tiene como principal objetivo formar el consentimiento contractual del que la recibe. Es bastante más detallada que la información proporcionada por la publicidad y a través de ella se pretende dar a conocer el alcance y las características del bien o del servicio en cuestión. Sin duda, debiera haber sido tratada por separado, incluyendo la ley los preceptos oportunos para su regulación.

En definitiva, se propone que la nueva LDV regule la publicidad y la información precontractual en preceptos distintos e independientes, como se viene haciendo en otras leyes aprobadas recientemente. ¿Y cuál debiera ser el contenido de los preceptos encargados de regular la publicidad? ¿Y cuál el contenido de los preceptos destinados a regular la información precontractual?

En lo que a la publicidad respecta, se propone la inclusión de un único artículo, cuyo título podría ser "Información básica que deberá figurar en la publicidad de los contratos de compra y arrendamiento de vivienda". El contenido del precepto parece obvio: enumerar todas aquellas cuestiones que tienen que aparecer en la publicidad en aquellos casos en los que el promotor, –o cualquier otro sujeto– decidan dar publicidad a la venta o al arrendamiento. Tales cuestiones podrían ser las siguientes[27]: a) Si la vivienda está ya construida o se encuentra en construcción; b) Ubicación de la edificación; c) Datos que identifiquen al promotor de la edificación; d) Número exacto de viviendas; e) Si se menciona el número de dormitorios, así como el resto de piezas

---

27 Pese a enumerar las cuestiones que deberían constar en el precepto, nuestra intención no es tanto elaborar una lista detallada de tales cuestiones –que podrían ser objeto de discusión–, sino dejar constancia de la importancia de regular la publicidad.

de la vivienda y los anejos vinculados, o de cualquier otra forma se efectúa una descripción general de la vivienda, debe indicarse necesariamente su superficie útil; f) En caso de mencionar el precio de venta de la vivienda, se deberán indicar también los tributos y otros gastos que deben pagar el consumidor; g) Si se menciona la renta, ésta deberá incluir los gastos de comunidad si estos corren a cargo del consumidor; h) Cuando esté prevista la entrega de dinero antes de iniciar la construcción o durante la misma, se mencionará expresamente que las cantidades anticipadas se garantizarán conforme a ley, el nombre de la entidad garante, número de la póliza correspondiente, la existencia de una cuenta bancaria especial y exclusiva para los ingresos y su Código Cuenta Cliente. Deberá indicarse si el dinero se entregará como señal, como parte del precio o como parte del precio y señal. Esta enumeración está inspirada en el art 5 del Decreto 218/2005, de 11 de octubre, por el que se aprueba el Reglamento de información al consumidor en la compraventa y arrendamiento de viviendas en Andalucía. Ha de matizarse que muchas cosas, como se aprecia, solamente tienen sentido respecto de la vivienda nueva, y no tanto cuando se trata de una vivienda de segunda mano. Para tal fin, también podría resultar de inspiración la LCCI, que distingue entre publicidad, información general e información personalizada.

Además, el precepto debería contener una referencia, aunque sea mínima, a las características que debe reunir la información que se proporciona a través de la publicidad, señalándose, al menos, que debiera ser veraz, comprensible, accesible, suficiente y actualizada.

En cuanto a la información precontractual, se propone igualmente incluir preceptos que se encarguen de su regulación. Establece el art. 31 de la actual LDV en su apartado primero que "Sin perjuicio de los principios y requerimientos contenidos en la normativa autonómica de aplicación y con carácter mínimo, la persona interesada en la compra o arrendamiento de una vivienda que se encuentre en oferta podrá requerir, antes de la formalización de la operación y de la entrega de cualquier cantidad a cuenta, la siguiente información, en formato accesible y en soporte duradero, acerca de las condiciones de la operación y de las características de la referida vivienda y del edificio en el que se encuentra: a) Identificación del vendedor o arrendador y, en su caso, de la persona física o jurídica que intervenga, en el marco de una actividad profesional o empresarial, para la intermediación en la operación (...)". Ahora bien, el hecho de que el precepto aluda a la aplicación de la normativa autonómica, ¿justifica su pobre redacción y su falta de detalle? El precepto, como se aprecia, enumera una serie de cuestiones que parecen ser las mismas en todos los casos. No distingue entre contrato de arrendamiento de vivienda y contrato de compraventa. Tampoco entre viviendas en construcción y ya construidas. En caso de viviendas ya construidas, tampoco se diferencia la primera transmisión de ulteriores transmisiones. Aunque es cierto que la legislación autonómica -al menos en Andalucía-[28] si tiene en cuenta todas estas cuestiones, eso

---

[28] En Andalucía rige el Decreto 218/2005, de 11 de octubre, por el que se aprueba el Reglamento de información al consumidor en la compraventa y arrendamiento de viviendas en Andalucía. Esta norma dedica un precepto, concretamente el art. 6, a regular lo que la misma norma denomina "documento informativo abreviado en la venta de viviendas en proyecto o en construcción". Al respecto, señala este precepto que "Quien ofrezca, aun a título de simple intermediario, la venta de viviendas en proyecto o en construcción, deberá entregar gratuitamente un Documento Informativo Abreviado, conforme al modelo que se adjunta en el Anexo II, a cualquier consumidor que solicite información sobre dichas viviendas". Además, se especifica cuál debe ser el contenido exacto de dicho documento. El art. 7 es el encargado de regular este mismo Documento Informativo Abreviado en la venta de viviendas construidas en su primera transmisión y el art. 11 regula este Documento para el caso de arrendamiento de vivienda. Por otra parte, el art. 8 incide en la obligación de poner a disposición de los consumidores una nota ex-

no es óbice para que la norma estatal hubiera sido más rigurosa y precisa, unificando criterios.

En definitiva, se propone que la actual LDV dedique preceptos concretos a la regulación de la información precontractual, entendiendo que el contenido de la misma puede variar según estemos ante un contrato de compraventa o de arrendamiento, según estemos ante una vivienda ya construida o no, o según se trate de la primera transmisión de la vivienda o de otra posterior. Estas cuestiones debieran haber sido tratadas en la Ley, lo que sin duda hubiera supuesto un mayor esfuerzo por parte del legislador quien, como ya se ha apuntado, ha legislado, al menos en cuanto a la información precontractual se refiere, con cierta ligereza[29].

7. Se propone especificar cuál es el ámbito de aplicación objetivo de los artículos 30 y 31 LDV, resolviendo la contradicción que existe entre el actual art. 30 LDV –aplicable a cualquier otro régimen jurídico de tenencia y disfrute distinto de la compraventa y el arrendamiento– y el art. 31 LDV –que hace referencia exclusivamente a la compraventa y al arrendamiento–.

---

plicativa sobre el precio de la venta y las formas de pago. Por último, los arts. 9 y 10 de la norma regulan la información y documentación previa a la firma del contrato de compraventa en primera transmisión o en segundas o ulteriores transmisiones de la vivienda.

29 El art. 31.1 g) es buena muestra de ello. En él se dice que parte del contenido de la información precontractual debe ser "Cualquier otra información que pueda ser relevante para la persona interesada en la compra o arrendamiento de la vivienda, incluyendo los aspectos de carácter territorial, urbanístico, físico-técnico, de protección patrimonial, o administrativo relacionados con la misma". En definitiva, se introduce un cajón de sastre que de ningún modo contribuye a la seguridad jurídica.

8. El Preámbulo de la Ley señala específicamente que el Título IV regula las responsabilidades derivadas del incumplimiento, pero no aparece nada de esto en la letra de la Ley, por lo que se propone introducir un precepto que haga referencia a tal cuestión[30].

Es usual, y así ocurre también en la legislación de consumo, que cuando una norma regula la información precontractual, no contiene preceptos que señalen cuales son las consecuencias de incumplir este deber, aunque sería lo deseable.

Ciñéndonos a la LDV, no cabe duda de que, si se imponen una serie de deberes de información, el incumplimiento de tales deberes tiene consecuencias jurídicas[31]. Dicha Ley no solamente no establece, ni tan siquiera, un régimen de sanciones[32], sino que, desde la perspectiva del Derecho privado, tampoco hace mención a posibles actuaciones que puede llevar a

---

30 Sobre cuáles debieran ser los remedios frente al incumplimiento de la información precontractual en el arrendamiento *Vid.* Mate Satué (2024, p. 325).

31 En la práctica, al menos en el contrato de compraventa, se consigue que el deber de información se cumpla correctamente a través de la figura del Notario.

32 Por ejemplo, el artículo 81 de la Ley 7/2014, de protección de las personas consumidoras y usuarias de las Islas Baleares tipifica como hecho sancionable en materia de consumo "Poner a disposición de los consumidores productos, bienes o servicios sin la información mínima y/o relevante, veraz, correcta, suficiente y transparente sobre sus características esenciales, así como sobre sus condiciones de utilización". Se considera una infracción grave, con sanción de 4.500,01 € a 24.000 €. Ello será sin perjuicio, cuando corresponda, de la aplicación de otras sanciones (como las establecidas en el art. 47 TRLGDCU, si tuvieran cabida), sin incurrir nunca en duplicidad sancionadora.

cabo el sujeto al que se le ha proporcionado una información errónea o incompleta.

Así las cosas, se propone que la ley contenga un precepto en tal sentido; ahora bien, ¿cuál debiera ser su contenido?

Para dar respuesta a esta pregunta, debemos contestar otro interrogante: desde la óptica del Derecho privado, y ciñéndonos al caso que nos ocupa, ¿qué posibilidades de actuación tiene el sujeto interesado en la compra o arrendamiento de vivienda si no se cumple lo dispuesto en la publicidad o en la información precontractual? Antes de nada, ha de puntualizarse que la publicidad con valor informativo pasa, al igual que la información precontractual, a integrar el contenido del contrato, por lo que resulta conveniente tratar los incumplimientos de manera conjunta[33]. Dicho esto, se pasa seguidamente a exponer cuáles podrían ser las consecuencias jurídicas de incumplir lo señalado en la publicidad o en la información precontractual:

a. Integración del contrato: según el art. 65 TRLGDCU, los contratos celebrados con consumidores se integrarán, en beneficio del consumidor, en los supuestos de omisión de información precontractual relevante. Así, por ejemplo, si se omite información respecto de un

[33] ¿Qué se entiende por incumplimiento? ¿Qué comportamientos debe realizar el empresario para considerar incumplido su deber de información precontractual? Pues bien, dicha infracción puede tener lugar tanto mediante un comportamiento activo (comunicar una información falsa o engañosa), como mediante un comportamiento omisivo (no dar una información que se está obligado a dar). Por tanto, "el deber de informar implica, de un lado, que ha de transmitirse la información exigible y, de otro, que ha de omitirse cualquier información errónea" (Gómez Calle, 1994, p. 81). Ocurre lo mismo en el ámbito de la publicidad, siempre y cuando se haya optado por publicitar la venta o el arrendamiento.

derecho que tiene el consumidor, ese derecho pasa igualmente a ser contenido contractual.

b. La resolución contractual: el ejercicio de este remedio tiene sentido si, una vez celebrado el contrato de compraventa o de arrendamiento de vivienda, el comprador o arrendatario -sean o no consumidores-comprueban que se incumplió de forma grave y sustancial el deber precontractual de información, o bien lo publicitado no se corresponde en absoluto con la realidad. Ahora bien, la resolución no es un efecto, al menos directo, del incumplimiento del deber precontractual de información, sino más bien del incumplimiento de las obligaciones contractuales. El incumplimiento, pese a ser la información un deber precontractual, tiende a materializarse en un momento posterior a la celebración del contrato; por tanto, lo que se incumple ya no es un deber precontractual, sino auténticas obligaciones contractuales. A esto se añade que dicha resolución podrá ir acompañada de la correspondiente indemnización por daños y perjuicios, siempre que se den los presupuestos para ello.

c. La anulación del contrato por error: por último, siempre que el incumplimiento del deber precontractual de información, o la publicidad por falsa o engañosa, hayan causado un error sustancial y excusable en el comprador o en el arrendatario -con independencia de que sean o no consumidores-, podrá estos proceder a anular el contrato, alegando un consentimiento viciado por error, en aplicación de lo dispuesto en el art. 1266 y concordantes del Código Civil

d. Podría asimismo proponerse una solución alternativa que permitiera al comprador o arrendatario permanecer con el contrato (no se consigue ni con la resolución ni con la anulación), pero adaptando su contenido a las legítimas expectativas del comprador o arrendatario. Esto es especialmente importante en el caso del arrendamiento, dadas las enormes dificultades del acceso a la vivienda. En tal sentido, podría ser viable una solución basada en el saneamiento por vicios ocultos cuando la ocultación de información o su suministro de forma incorrecta pueda ser considerada un vicio oculto. Por otra parte, también podría ser posible el ejercicio de la acción "quanti minoris", esto es, rebaja del precio del alquiler (o del precio de la compraventa) en los términos de los arts. 1486 y 1553 CC.

V. CONCLUSIONES

1. La actual LDV tiene como principal objetivo garantizar el derecho a acceder a una vivienda digna y adecuada -ya sea en régimen de propiedad o de arrendamiento-. Para conseguir tal propósito es necesario regular la información que debe proporcionarse al propietario o arrendatario en la etapa precontractual, cuyo primordial fin es formar correctamente la voluntad contractual del adquirente. Una óptima garantía del derecho a la vivienda implica que el interesado conozca –o al menos pueda conocer–, antes de la formalización del contrato, información que le resultará útil para formar dicha voluntad contractual. La información precontractual se erige así como instrumento fundamental para proteger al interesado en acceder a una vivienda, ya sea como propietario, como arrendatario, o en cualquier otro régimen jurídico de tenencia y disfrute. Por tanto, es imprescindible que una ley estatal de vivienda regule la información en los contratos de compraventa y arrendamiento.

2. Tal importancia le otorga el legislador al deber de información, que el propio artículo 1.3 de la actual LDV

dispone que es objeto de la norma "reforzar la protección del acceso a información completa, objetiva, veraz, clara, comprensible y accesible, en las operaciones de compra y arrendamiento de vivienda".

3. Concretamente, la actual LDV dedica los artículos 30 y 31 a tal fin, ubicados ambos en el Capítulo I del Título IV. Estos preceptos conviven con otras normas que también permanecen vigentes, tales como el RD 515/1989, de 21 de abril, sobre protección de los consumidores en cuanto a la información a suministrar en la compraventa y arrendamiento de viviendas y los preceptos de la legislación de consumo que sean aplicables al caso que nos ocupa.

4. No se entiende que el Título IV de la Ley esté conformado, además de por el Capítulo I, por el Capítulo II. Este segundo Capítulo trata cuestiones que no guardan relación con las reguladas en el Capítulo primero, y que no debieran haber sido reguladas en este Título –deberes de información y transparencia que se imponen al Estado (arts. 32 y ss.)–.

5. Los artículos 30 y 31 de la LDV presentan notables deficiencias. Algunas de ellas de carácter puramente formal, otras de carácter material. Algunas de estas deficiencias están presentes a lo largo de todo el texto normativo, como es el caso de la terminología empleada por el legislador, que revela su ignorancia en lo que atañe al Derecho privado, más concretamente en lo que respecta al Derecho de contratos. Un claro ejemplo de ello es referirse a los contratos de compraventa y arrendamiento como *las operaciones de compra y arrendamiento de vivienda.*

6. La redacción de los preceptos mencionados induce en muchas ocasiones a error. Así, por ejemplo, el art. 31.1 dispone que "la persona interesada en la compra o

arrendamiento de una vivienda (...) *podrá requerir*, antes de la formalización de la operación y de la entrega de cualquier cantidad a cuenta, la siguiente información (...)". Lo mismo sucede en el art. 31.2, según el cual "la persona interesada (...) *podrá requerir* información acerca de la detección de amianto u otras sustancias peligrosas o nocivas para la salud". Es evidente que el deber de información precontractual debe ejercitarse por parte del sujeto facultado para la transmisión o el arrendamiento en todo caso, con independencia de la actuación del interesado. Por tanto, se propone sustituir la expresión utilizada por los arts. 31.1 y 31.2 –*la persona interesada podrá requerir información*–, por otra que no induzca a confusión como, por ejemplo, "*el vendedor o arrendador (...) está obligado a proporcionar la siguiente información (...)*".

7. Por su parte, el art. 30. 1 a) plantea un problema interpretativo. No queda claro si este precepto establece la aplicación de la legislación de consumo a todos los sujetos o, únicamente, a los que tengan la condición de consumidor. A nuestro entender, por las razones expuestas anteriormente en el texto principal y que no conviene aquí reiterar, se debe interpretar la norma vigente a favor de considerar que la legislación de consumo no será de aplicación cuando el contrato no se celebre entre consumidor y empresario. Se propone justo lo contrario, esto es, extender los deberes de información a todos los vendedores o arrendadores, de forma que, si el contrato se celebra entre particulares, el vendedor esté obligado a proporcionar al comprador –o el arrendador al arrendatario–, una información detallada y exhaustiva –por ejemplo, la propia del RD 515/1989, de 21 de abril–. Esta conclusión encuentra su justificación en la importancia que tiene el bien objeto del contrato.

8. El párrafo tercero del art. 30 es otra muestra de la falta de rigor técnico del legislador, que no distingue entre publicidad e información precontractual, estableciéndose un concepto único para ambas. Se entiende que la información proporcionada por la publicidad no persigue los mismos fines que la información precontractual, por lo que debieran ser reguladas por separado, en preceptos distintos e independientes. Así, por ejemplo, se hace en la LCCI, que puede resultar de inspiración.

9. Aunque el Preámbulo de la Ley dice expresamente que el Título IV regula las responsabilidades derivadas del incumplimiento, lo cierto es que no existe ningún precepto destinado a tal fin, por lo que se propone incluirlo. Se trataría de un artículo en el que se enumeraran las posibilidades de actuación de quien recibe una información incompleta o errónea. Además, podría haberse incluido en la norma un precepto en el que se estableciera un régimen de sanciones derivadas de infringir lo dispuesto en ella.

VI. REFLEXIÓN FINAL

Expuesto todo lo anterior, se concluye que los artículos 30 y 31 de la actual LDV –encargados de regular la información precontractual en el contrato de compraventa y arrendamiento de vivienda– presentan deficiencias muy significativas: en ellos no se emplea una terminología adecuada, su redacción induce a error, plantean problemas interpretativos, confunden conceptos –como el de publicidad e información precontractual– y no regulan cuestiones de interés, entre otras cosas. De su estudio se deduce que el propósito del legislador no ha sido tanto reforzar la protección del acceso a la información –como se señala en el art. 1.3 de la Ley–, sino, más bien, dotar de contenido una ley que se presentaba como un logro de la legislatura. Se ha desaprovechado una magnífica oportunidad para regular con precisión una cuestión decisiva para lograr el objetivo último de la norma: garantizar el acceso a una vivienda digna y adecuada, ya sea en régimen de propiedad o de arrendamiento.

A nuestro entender, lo ideal hubiera sido que el legislador hubiera regulado esta cuestión con mucho más detalle y rigor, introduciendo todas las propuestas de mejora que se han planteado. Habría sido más razonable, útil y lógico que se hubiese aprovechado este texto legal para replantearse todas las normas aplicables a la información en los contratos de compraventa y arrendamiento de vivienda, derogar otras normas en su caso, y acabar con una dispersión legislativa que solamente genera incertidumbres y que complica la tarea de los aplicadores del derecho.

VII. REFERENCIAS BIBLIOGRÁFICAS

Alonso Pérez, M. T. (2010). *Las obligaciones legales de información precontractual en la compraventa de vivienda.* Civitas.

Argelich Comelles, C. (2023). *Ley por el derecho a la vivienda.* Aranzadi.

Castilla Barea, M. (2000). *La imposibilidad de cumplir los contratos.* Madrid. Dykinson.

Díaz Moreno, A. (2006). "La protección del consumidor: una perspectiva global". *Estudios de Derecho Judicial,* (103), 209-278.

Díez-Picazo, L. y Gullón Ballesteros, A. (2005) *Sistema de Derecho civil,* Tecnos.

Font Galán, J. I. (2011). "Publicidad comercial y contrato con consumidores. Conexiones funcionales y normativas: sustantivación obligacional e integración contractual de las ofertas promocionales y publicitarias". *Revista de Derecho Patrimonial,* (26), 147-179.

García Vicente, J.R. (2009). "Comentario al artículo 61". *Comentario del Texto Refundido de la Ley General para la Defensa de los Consumidores y Usuarios y otras leyes complementarias.* Aranzadi.

Gómez Calle, E. (1998). *El contrato de viaje combinado.* Civitas.

Gómez Calle, E. (1994). *Los deberes precontractuales de información.* La Ley.

Llobet i Aguado, J. (1996). *El deber de información en la formación de los contratos.* Marcial Pons.

Martín Oviedo, J. M. (2010). *Memento práctico: Consumo 2010-2011.* Ediciones Francis Lefebvre.

Martínez de Aguirre Aldaz, C. (2008). *Curso de Derecho Civil (II). Derecho de Obligaciones.* Colex.

Mate Satué, L. (2024). "La información precontractual en el arrendamiento de vivienda: ¿un mecanismo eficaz para la protección del arrendatario: especial atención al vulnerable?". *El derecho a la vivienda en tiempos de incertidumbre.* Aranzadi.

Morales Moreno, A. M. (1992). "Información publicitaria y protección del consumidor (Reflexiones sobre el art. 8 de la LGCU)". En: *Homenaje a Juan Vallet de Goytisolo,* vol. VIII, Consejo General del Notariado.

Parra Lucán, M. A. (2015) "Riesgo imprevisible y modificación de los contratos". *Indret,* (4).

Picatoste Bobillo, J. (2011). El derecho de información en la contratación con consumidores". *Revista Jurídica de Doctrina y Jurisprudencia,* n.º 4.

Poillot, E. (2006). *Droit européen de la consommation et uniformisation du droit des contrats.* Librairie générale de droit et de jurisprudence.

Riesenhuber, K. (2003). *Europäisches Vertragsrecht.*

Schulze, R. (2006). "Deberes precontractuales y conclusión del contrato en el Derecho Contractual Europeo. *Anuario de Derecho Civil,* (1), 29-58.

Capítulo VI

La figura del administrador de fincas tras la ley 12/2023, de 24 de mayo, por el derecho a la vivienda

JESÚS MARTÍN FUSTER
Profesor Ayudante Doctor de Derecho civil
Universidad de Málaga

I. LA FIGURA DEL ADMINISTRADOR Y SU REGULACIÓN ACTUAL

1. Introducción: La figura del administrador

Nos centraremos en este capítulo en tratar la figura del administrador de las comunidades de propietarios, a raíz de la reciente Ley 12/2023, de 24 de mayo, por el derecho a la vivienda (LDV; TOL9.568.821).

Dichos administradores de las comunidades de propietarios tienen una función esencial en el ejercicio y desarrollo de dichas comunidades[1], siendo de interés su estudio actual,

---

[1] Loscertales Fuertes (2020, p. 139). Considera dicho autor que el administrador "cuando es un profesional, es el auténtico motor y asesor de la Comunidad, cuidando y vigilando el funcionamiento de los servicios comunes, así como la preparación y la presentación de

teniendo en cuenta que el impacto de dicha LDV en la figura del administrador apenas ha sido tratado por la doctrina.

La referida Ley comienza indicando en su preámbulo, respecto de los administradores de fincas, que es un colectivo profesional de gran importancia para asegurar la garantía y protección de los derechos de los consumidores, señalando que la disposición adicional sexta regula determinados aspectos de su actividad, por su particular relevancia en el ámbito de la vivienda.

Sin embargo, como se verá, ni se especifica ni se brinda una especial protección respecto a esa señalada importancia para los consumidores, ni se regulan realmente aspectos concretos de su actividad. De este modo, poca regulación de esta figura se realiza en dicha normativa, quedando prácticamente en su integridad en igual situación, con algunos matices que se señalarán.

Se analizará la regulación actual existente y los problemas que se plantean respecto de la figura de los administradores de las comunidades de propietarios, recogida como un órgano necesario de la comunidad en la Ley 49/1960, de 21 de julio, sobre propiedad horizontal (LPH; TOL230.715). Como se verá en este trabajo, esta figura necesaria del administrador no tiene que ir referido necesariamente al administrador de fincas que se regula en la disposición adicional sexta de la LDV (TOL9.568.821).

Dichos administradores son figuras esenciales en las comunidades de propietarios, que se han ido profesionalizando debido a que dicho cargo, como recogen los códigos deontológicos de administradores de fincas, puede requerir en muchas ocasiones amplios conocimientos en diversas áreas como el derecho, la fiscalidad, los seguros, la economía, la contabilidad, la informática, etc., que exigen cierta preparación, así

---

las cuentas u de los presupuestos, ejecutando los mismos, cobrando y pagando, así como otras cuestiones de índole económica".

como una continua dedicación y puesta al día. Por ello, suele ser habitual que, en aquellas comunidades formadas por varios vecinos, y muy especialmente cuando son numerosos, se contrate a un profesional externo para que lleven estas gestiones. Profesional que, no hay que olvidar, está prestando servicios a consumidores y usuarios como son la propia comunidad y sus vecinos, con la consiguiente protección que estos merecen según la normativa tuitiva existente[2].

Esta figura del administrador se suele asociar con el administrador de fincas profesional[3], lo que conlleva una de las principales problemáticas que será analizada, sobre la necesidad o no de colegiación por parte de aquel administrador externo a la comunidad que quiere ejercer dicho cargo. Debate antiguo que parece no tener fin, sin que la LDV (TOL9.568.821) ayude a aclarar esta cuestión, o incluso añada algunas dudas, como se examinará.

Además, y relacionado en cierta medida con el punto anterior, es interesante determinar la posibilidad de que una persona jurídica ejerza de administrador de fincas, ya que la nueva

---

2 La protección de los consumidores ha sido reconocida también por la reciente sentencia del TJUE de 27 octubre 2022. C-485/21, Caso S. V. OOD contra E. Ts. D. En ella se reconoce, entre otras cuestiones, que el hecho de que se haya celebrado por la Junta general el contrato con el administrador no impide que la persona física concreta sea considerada consumidor en la medida en que quepa calificarla de "parte" en ese contrato.

3 De hecho, algunos de los textos que se mencionarán en este trabajo, para referirse al administrador colegiado, suelen recogerlo textualmente usando la mayúscula en la primera letra, denominándolo "Administrador de Fincas", o "Administrador de finca". Posiblemente se haga así para darle cierto nombre propio y distinción del resto de administradores no colegiados. No obstante, en este texto usaremos el nombre siempre en minúscula, siguiendo la regla general de que las profesiones no llevan mayúsculas.

LDV (TOL9.568.821) sólo habla de personas físicas, resultando así de interés el estudio de esta cuestión.

Como se puede comprobar, algunas de estas cuestiones han sido largamente debatidas, tanto por la doctrina como por la jurisprudencia, por lo que resulta necesaria la debida intervención del legislador para ofrecer seguridad jurídica y acabar con los problemas que se plantean continuamente con estas cuestiones referentes a la figura del administrador. Así, es de destacar que el artículo 13 LPH (TOL230.715), que recoge en su apartado 6.º las referencias al órgano de administrador, así como en el 7.º respecto a su duración, no han sido objeto de modificación alguna, a pesar de que pudiera ser conveniente su actualización para poner fin a los debates que se plantean en la práctica.

A falta de esta legislación que aclare y arroje luz sobre estos asuntos, comenzaremos indicando la regulación y situación actual de la figura del administrador, para entrar después a analizar los requisitos que se exigen para dicha figura, examinando finalmente las "novedades" de la LDV (TOL9.568.821).

2. La regulación del administrador

El artículo 13.5 LPH (TOL230.715) señala que las funciones del administrador son asumidas por el presidente de la comunidad, salvo que los estatutos o por acuerdo mayoritario de la Junta se decida su ejercicio separado; el 13.6 indica que se puede acumular el cargo de secretario y administrador o nombrarse independientemente, para a continuación establecer los requisitos necesarios para ser administrador, que veremos a continuación. Y el art. 13.7 establece una duración general de un año –salvo acuerdo contrario en estatutos–, con posibilidad de remoción anticipada por la Junta.

Además de estas menciones, aparecen recogidas en el art. 20 las funciones que se le encomiendan, expresando que le corresponderá:

"a) Velar por el buen régimen de la casa, sus instalaciones y servicios, y hacer a estos efectos las oportunas advertencias y apercibimientos a los titulares.

b) Preparar con la debida antelación y someter a la Junta el plan de gastos previsibles, proponiendo los medios necesarios para hacer frente a los mismos.

c) Atender a la conservación y entretenimiento de la casa, disponiendo las reparaciones y medidas que resulten urgentes, dando inmediata cuenta de ellas al presidente o, en su caso, a los propietarios.

d) Ejecutar los acuerdos adoptados en materia de obras y efectuar los pagos y realizar los cobros que sean procedentes.

e) Actuar, en su caso, como secretario de la Junta y custodiar a disposición de los titulares la documentación de la comunidad.

f) Todas las demás atribuciones que se confieran por la Junta".

Además de ello, el art. 21 –reformado por la Ley 10/2022, de 14 de junio de 2022 (TOL9.002.109)– faculta al "secretario administrador profesional", si así lo acordare la junta de propietarios, para exigir judicialmente el pago a los propietarios morosos a través del proceso monitorio.

Junto a estos artículos, que son los que establecen la regulación sustantiva de dicha figura, debemos mencionar que existen otra serie de disposiciones referidas al "administrador de fincas" profesional colegiado, las cuales también regulan en cierta medida la profesión y su ejercicio. Entre ellas hay que destacar:

- **El Decreto 693/1968, de 1 de abril, por el que se crea el Colegio Nacional Sindical de Administradores de Fincas.**

Como se recoge en su preámbulo, se pretende canalizar orgánicamente esta actividad profesional mediante la creación de un Colegio Nacional Sindical que incorpore a quienes profesionalmente, es decir, con un carácter regular y habitual, y no en razón de mandato de confianza o familiar, se dediquen a administrar fincas rústicas o urbanas. Se pretende que sirva tanto como órgano representativo profesional, que se ocupe de velar por los derechos e intereses legítimos de los colegiados, como de Entidad de la Organización Sindical, así como de verdadero colaborador del Estado en todo lo que afecta a la regulación del ejercicio profesional de los encuadrados en él.

En este Decreto aparece la cualificación necesaria para ejercer la profesión de administrador de fincas (art. 5), así como la necesidad de colegiarse para "ejercer legalmente dicha profesión" (art. 2), cuestión esta de importancia como veremos a lo largo de este trabajo.

- **Estatutos del Colegio Nacional Sindical de Administradores de Fincas.**

La versión vigente de estos Estatutos fue aprobada por la Resolución de la Delegación Nacional de Sindicatos, de 28 de enero de 1969. En ellos se desarrolla la organización y funcionamiento colegial de esta institución, incorporando los deberes y derechos de los colegiados. Asimismo, establece un régimen disciplinario al que se someten dichos colegiados. Se debe destacar que se recoge igualmente la cualificación necesaria para ejercer de administrador de fincas, si bien desarrolla algo más este punto respecto del anterior Decreto, sin que se requiera actualmente de forma necesaria una titulación universitaria[4].

---

[4] Los requisitos respecto a la formación se recogen en el Artículo 12. D), requiriéndose "Estar comprendido en uno de los siguientes casos: -Primero. Acreditar la posesión de uno de los títulos siguientes: Licenciado en Derecho, en Ciencias Políticas, Económicas y Comerciales; profesor Mercantil; Procurador de los Tribunales de Justicia; Ingeniero Agrónomo e Ingeniero de Montes; Veterinario;

Se ha intentado actualizar y renovar estos Estatutos generales, sin éxito hasta el día de hoy. Se presentaron unos Estatutos Generales de los Colegios Profesionales de Administradores de Fincas y de su Consejo General, siendo aprobados en Pleno Extraordinario del Consejo General el 25 de marzo de 2010, iniciándose su tramitación parlamentaria mediante el Proyecto de Real Decreto de los Estatutos Generales de los Colegios Profesionales de Administradores de Fincas y su Consejo Superior, presentado en el año 2015. No obstante, estos Estatutos Generales no han llegado a ser aprobados.

Además de estos Estatutos Generales, también existen los Estatutos Colegiales de los distintos Colegios Territoriales de Administradores de Fincas. En estos se suele desarrollar y concretar la actividad colegial en el respectivo territorio, incorporándose una serie de infracciones relativas a la profesión, así como su correspondiente régimen sancionador.

- **Código deontológico europeo para profesionales inmobiliarios.**

Este Código, aprobado en marzo de 2006, está referido a los profesionales inmobiliarios, entre los que se incluyen

---

Ingeniero Técnico Agrícola e Ingeniero Técnico Forestal; Perito Agrícola y Ayudante de Montes.
-Segundo. Acreditando la posesión del título de Bachiller Superior, tanto general como técnico; Técnico de Grado Medio; Maestro de Enseñanza Primaria o Graduado social, será necesario además superar las pruebas de carácter técnico y especializado, con arreglo al programa vigente en el momento de su petición de ingreso a que hace referencia el artículo siguiente de estos Estatutos.
-Tercero. Quienes obtengan el título académico de Administrador de fincas, que será expedido por el Organismo competente, una vez aprobados los cursos de formación en la Escuela Oficial de Administradores de fincas o superados los exámenes que la misma proponga".

a los administradores de Fincas y a los agentes inmobiliarios, estableciendo que quedarán sujetos a él estos profesionales en la medida en que no entre en conflicto con la normativa nacional.

Se define al administrador de fincas como aquella persona que, de forma habitual, como actividad principal o secundaria, ejerce, por cuenta de terceros, en calidad de mandatario o intermediario, una actividad de gestión de propiedades o derechos inmobiliarios, o de administración de propiedades inmobiliarias en copropiedad.

Se desarrollan los deberes deontológicos que deben seguir estos profesionales, detallando sus relaciones con los clientes, así como con los demás compañeros. Se fija la necesidad de establecer una garantía financiera y un seguro de responsabilidad civil que cubran la responsabilidad en que puedan incurrir por los fondos depositados o bajo su confianza, así como la responsabilidad en el ejercicio de su profesión.

Se detalla además un código de conducta en materia de comercio electrónico, especificándose las normas de actuación y de información precontractual.

Asimismo, se establece un Consejo Europeo de Arbitraje, Mediación y Disciplina de los Profesionales Inmobiliarios que, además de tratar los conflictos profesionales que puedan surgir entre profesionales inmobiliarios de diferentes países de la Unión Europea o del espacio europeo afiliados, tienen la función de Consejo Disciplinario, pudiendo imponer sanciones por el incumplimiento de lo previsto en dicho Código Deontológico.

- **Código Deontológico del Administrador de Fincas.**

Con posterioridad al Código Deontológico Europeo, y basándose en gran parte en este, se crea en 2017 este Código Deontológico general a nivel estatal por el Consejo General de Colegios de Administradores de Fincas de España.

En él se expresa la necesidad de plantear el cumplimiento por parte de los administradores de fincas de unos criterios deontológicos y de buenas prácticas en el ejercicio de su profesión, "que han de servir de guía fundamental para el desarrollo de su actividad, sin perjuicio y más allá de la normativa profesional, que piensa en obligaciones con consecuencias disciplinarias", como se refleja en su exposición de motivos.

Se expresa el carácter de norma básica para la ordenación de la profesión en todo el territorio nacional, estando todos los administradores de fincas colegiados obligados a atender a las normas establecidas, sin perjuicio del deber de cumplimiento de las normas deontológicas aprobadas por el correspondiente Colegio Territorial al que estén adscritos.

Entre las obligaciones, basadas en el Código deontológico europeo, se recoge la necesidad de permanecer informado regularmente de las legislaciones vigentes en cada momento, actuando siempre conforme a la *lex artis*; conocer las condiciones de los mercados; no aceptar ningún encargo que sobrepase su área de experiencia; e informarse de todos los hechos esenciales relativos a cada una de las propiedades inmobiliarias que gestione. Se añaden asimismo directrices respecto a su relación con los clientes, basadas en criterios de transparencia y honestidad en su profesión, que eviten posibles conflictos de intereses; e igualmente se recogen respecto de las relaciones con los restantes administradores de fincas, velando por la lealtad en la competencia y respeto; así como las relaciones con el Colegio, prestando su colaboración y ayuda.

Se recoge igualmente la necesidad de disponer de un seguro de responsabilidad civil que garantice, por un lado, la responsabilidad en la que pueda incurrir, en general, por el error, mala praxis o incumplimiento de sus deberes en el ejercicio de su profesión y de cuya conducta deriven perjuicios patrimoniales a su cliente; así como la responsabilidad en la que pueda incurrir con respecto a los fondos, bienes y derechos que les hayan sido depositados o confiados en el ejercicio de su actividad profesional.

También se debe destacar que se recogen las particularidades sobre el uso de las nuevas tecnologías, siendo aplicable a todos los medios tecnológicos que se utilicen, incidiendo en la debida información precontractual y contractual, señalando la necesidad de distinguir entre la fase de oferta, y la fase de la firma, debiendo el cliente recibir una copia del contrato con carácter previo.

Igualmente, se establece un régimen disciplinario detallando las distintas infracciones según su gravedad, así como el régimen sancionador existente.

II. NATURALEZA Y REQUISITOS DEL ADMINISTRADOR

1. La necesidad o no de colegiación para ser administrador

Un primer problema que se ha planteado y debatido largamente por la doctrina es la naturaleza o requisitos que debe reunir una persona para poder ejercer como administrador, como puede ser su cualificación profesional o la necesidad o no de estar colegiado en algún colegio de administradores de fincas.

Sobre estos requisitos del administrador nada refiere la nueva LDV (TOL9.568.821), que se limita en todo momento a hablar de "administradores de fincas" en el sentido que luego veremos. Tenemos así que acudir a lo dispuesto por la LPH (TOL230.715), según la cual en todas las comunidades sujetas a esta Ley tiene que existir un administrador. Los estatutos no pueden excluir válidamente dicha figura[5]. Si bien, en el apartado quinto del art. 13 LPH (TOL230.715) se señala que dicho cargo, así como el de secretario, será ejercido por el presidente de la comunidad, salvo que los estatutos o la Junta de propietarios acuerden que dichos cargos se ejerzan de manera separada.

Por su parte, el apartado sexto señala que dichos cargos de secretario y administrador pueden acumularse en una misma persona o nombrarse de modo independiente, como ya indicamos. Y, lo que aquí más interesa ahora, recoge que tanto el cargo de administrador como el de secretario-administrador podrá ser ejercido por cualquier propietario (sin requerir nada más), "o por personas físicas con cualificación profesional suficiente y legalmente reconocida para ejercer dichas funciones. También podrá recaer en corporaciones y otras personas jurídicas en los términos establecidos en el ordenamiento jurídico".

Como hemos referido, el hecho de que se adjudique dicho cargo de administrador a un propietario parece que no plantea problemas, ya que no se exige nada más en cuanto a cualificación profesional, ni ningún otro requisito.

No obstante, en la práctica es común la designación de un administrador externo independiente de los propietarios de la comunidad, debido a la complejidad que puede conllevar en ocasiones

---

5 Alonso Sánchez (1994, p. 76): "Por las razones expuestas entiendo que no cabría otorgar validez alguna a la cláusula estatutaria que ordenase prescindir del Administrador como cargo de esa Comunidad pues, conforme a lo establecido en el párr. 2.º del art. 12 LPH (TOL230.715), simplemente viene permitido a los Estatutos exigir que el desempeño del cargo de Administrador recaiga en persona distinta del Presidente. La misma solución cabe mantenerse respecto de los acuerdos que fueran adoptados en Junta de Propietarios proponiendo idéntica exclusión del órgano en cuestión".

la gestión de dicha comunidad y la serie de tareas que dicho cargo lleva aparejado. Es entonces cuando se plantea el problema de qué requisito es necesario, especialmente si es exigible la colegiación como miembro de algún Colegio de Administradores de Fincas.

La Ley 2/1974, de 13 de febrero, sobre Colegios Profesionales (TOL5.003), en su artículo 3.2, reformado por la conocida como Ley ómnibus (Ley 25/2009, de 22 de diciembre, de modificación de diversas leyes para su adaptación a la Ley sobre el libre acceso a las actividades de servicios y su ejercicio (TOL1.742.735), recoge que la incorporación a un Colegio Profesional es requisito indispensable para ejercer una determinada profesión cuando así lo establezca una ley estatal.

Además, dicha Ley ómnibus recoge en su Disposición Transitoria 4.ª que "En el plazo máximo de doce meses desde la entrada en vigor de esta Ley, el Gobierno, previa consulta a las Comunidades Autónomas, remitirá a las Cortes Generales un Proyecto de Ley que determine las profesiones para cuyo ejercicio es obligatoria la colegiación". A fecha de hoy sigue sin haberse dictado tal normativa.

A su vez, dicha Ley 25/2009 acota los supuestos donde se debería seguir exigiendo la colegiación en la futura ley. Así, en su DT 4.ª señala que el Proyecto de Ley referido debe prever la continuidad de la obligación de colegiación "en aquellos casos y supuestos de ejercicio en que se fundamente como instrumento eficiente de control del ejercicio profesional para la mejor defensa de los destinatarios de los servicios y en aquellas actividades en que puedan verse afectadas, de manera grave y directa, materias de especial interés público, como pueden ser la protección de la salud y de la integridad física o de la seguridad personal o jurídica de las personas físicas".

Como hemos referido, y dejando a un lado la Ley 12/2023, de 24 de mayo, por el derecho a la vivienda, de la que después hablaremos, en España no hemos tenido una norma con rango de ley estatal que regule estas profesiones que necesitan colegiación. Actualmente, en el ámbito de los administradores de fincas, dicha necesidad de colegiación aparece recogida en el Decreto 693/1968, de 1 de abril, por el que se crea el Colegio Nacional Sindical de Administradores de Fincas, en donde se recoge que "Para ejercer legalmente la profesión de administrador de fincas rústicas y urbanas será requisito indispensable estar colegiado en la corporación profesional que se crea por el presente decreto".

Por ello, y como dispone la DT 4.ª de la ley ómnibus, hasta la entrada en vigor de la futura ley que determine la necesidad de colegiación, se mantendrán las obligaciones de colegiación vigentes.

Por su parte, la LPH (TOL230.715) exige no solo cualificación profesional suficiente, sino también que esté "legalmente reconocida para ejercer dichas funciones". Algunos autores, siguiendo una interpretación estricta de estas disposiciones, han entendido que es necesaria la colegiación para poder ser administrador de una comunidad[6].

Pero frente a ello, la mayoría de la doctrina y de los tribunales se inclinan por pensar que realmente tal requisito de colegiación no es exigible para ser administrador, ya que la ley no lo exige expresamente[7]. Este argumento lo podemos encontrar,

---

6 Así, por ejemplo, Rubiras Valenzuela (2021, p. 17): "En cuanto a la cualificación profesional suficiente, aunque la Ley nada diga al respecto, entendemos que se refiere a aquellas personas que están colegiadas en los Colegios de Administradores, con una cobertura legal reconocida".

7 No obstante, esta opinión no es unánime. Interesante resulta el debate planteado por el Magistrado Vicente Magro Servet acerca de la posibilidad de que un administrador colegiado atacase el nombramiento de otro sin colegiación ni demostrada cualificación profe-

entre otras resoluciones, en el Auto de la AP de Madrid núm. 7/2018 de 3 enero (TOL6.541.365), donde recogiendo lo dictado por la AP de Barcelona en su sentencia de 13 de junio de 2012 (TOL2.617.764), con citas de otras sentencias de Audiencias Provinciales, determina que el art. 13 LPH (TOL230.715) no exige que el cargo de secretario o administrador sea ostentado por persona que se halle de alta en el Colegio de administradores de fincas. Entiende así que la ley se limita a señalar la necesidad de "cualificación profesional suficiente y legalmente reconocida", "pero no indica el alcance mínimo o la titulación, ni quien ha de apreciar la suficiencia. Destaca la doctrina que no existe una titulación académica que acredite una idoneidad objetiva y previa para la administración de fincas urbanas".

Y en el mismo sentido se ha pronunciado el TS en su Sentencia de 11 de noviembre de 2008, Rec. 1602/2006 (TOL1.401.565), que recogemos por su interés:

> "[...] La modificación legal producida por la entrada en vigor de la Ley 8/1999, de 6 de abril, que reformó la 49/1960, de 21 de julio, de Propiedad Horizontal, despeja las dudas y vacilaciones que hasta ese momento pudieran derivarse de la normativa anterior y de las resoluciones judiciales dictadas en los

---

sional (Magro Servet, 2017, pp. 1-8). La mayoría de los magistrados encuestados entienden que la colegiación no es un requisito imprescindible, no obstante, se encuentran algunas opiniones como la de María Félix Tena Aragón, Presidenta de la AP de Cáceres, quien defiende que "Toda persona que ejerza las funciones de administrador en una propiedad horizontal debe pertenecer al correspondiente colegio profesional. Ello se recoge expresamente en el art. 13.6 LPH (TOL230.715) y está avalado por la jurisprudencia del TS en sentencias como la citada de 8 de noviembre de 2016", entendiendo que si se elige a alguien sin dicha colegiación "se trataría de una designación contraria a la ley, y por lo tanto impugnable por cualquier perjudicado, en este caso, cualquier otra persona que haya optado a ese cargo cumpliendo el requisito de colegiación correspondiente".

> distintos órdenes jurisdiccionales, en relación con la exigencia de titulación y colegiación para el ejercicio de la profesión de Administrador de Fincas, al menos en lo que respecta al campo de la propiedad horizontal.
>
> Si de la legislación anterior, representada sustancialmente por el Decreto 693/1968, de 1 de abril , y por el Real Decreto 1464/1988 , podía derivarse, no sin ciertas dificultades, que para el ejercicio de dicho cargo de administrador de fincas era necesario ostentar el correspondiente título y estar colegiado en el Colegio Profesional, y así lo entendieron las sentencias que en defensa de su pretensión ha aducido el recurrente –incluso la sentencia que cita de la Sala de lo Civil de 14 de octubre de 2002 se está refiriendo a un caso anterior a la Ley 8/99–, sin embargo, conforme a la nueva redacción dada al artículo 13 de la Ley de Propiedad Horizontal, ha desaparecido el carácter de exclusividad que pudiera haberse atribuido al Administrador de Fincas".

Argumento recogido en sentencias posteriores, como la STS de 28 de marzo de 2011, Rec. 4595/2008 (TOL2.083.619). En dicho caso se enjuiciaba un anuncio del Colegio Oficial de Administradores de Fincas de Galicia, donde indicaban que "eran los únicos que contaban con la autorización legal necesaria para ejercer su profesión con total acierto". Consideraba así que este anuncio evitaba o restringía la competencia en el mercado de los servicios de administración de fincas, abierto a otros operadores económicos y no susceptible de monopolizarse por los colegiados que agrupaba el Colegio Oficial de Administradores de Fincas de Galicia.

Así lo reconoce también la Comisión Nacional de Mercados y de la Competencia en un informe de 2010 sobre sobre los Colegios Profesionales tras la transposición de la Directiva de Servicios, donde señala que hasta ahora no solo no se exige estar colegiado para ejercer la actividad profesional de administración y gestión de fincas, sino que tampoco existe una titulación académica que configure una idoneidad objetiva para la administración de fincas urbanas. No obstante, reconoce que para poder denominarse "administrador de fincas" se debe estar colegiado, lo que en cualquier caso constituye una exclusividad, aunque de distinta índole. Aun así, considera dicho organismo que este tipo de restricciones, aunque referidas a la nomenclatura, pueden tener un efecto equivalente a la colegiación obligatoria, porque el profesional debe colegiarse para no tener una desventaja competitiva significativa respecto a los profesionales que pueden utilizar una denominación profesional adecuada[8].

De este modo, se puede comprobar cómo la situación es compleja, siendo difícil encontrar una solución a dicha problemática. Por un lado, tenemos vigente la necesidad de colegiación; por otro lado, se admite que no es necesario estar colegiado para ejercer de administrador. Para conjugar tales criterios, y con el fin de ofrecer una posible solución a tal debate, se podría considerar como necesario hacer una distinción. Habría que distinguir entre la figura del "administrador" de la comuni-

---

8 Muy crítica se suele mostrar la CNMC sobre esta necesidad de colegiación. Recoge en su informe del año 2016 sobre el Proyecto de Real Decreto por el que se aprueban los estatutos generales de los colegios profesionales de administradores de fincas y de su consejo general: "Las actividades habituales de los administradores de fincas no son servicios exclusivos de éstos, ni deberían serlo. No existe justificación en restringir la competencia a una profesión concreta en la gestión económica de una comunidad de vecinos, la elaboración y seguimiento de sus presupuestos, la atención y control de proveedores, la atención a la reparación de las averías o la contratación y seguimiento de las pólizas de seguros. Una mayor competencia es fundamental para una prestación más eficiente de estos servicios (...) Los beneficios de la competencia se deben traducir en el ofrecimiento de servicios en términos de competitividad, precio, calidad e innovación que sean preferidos por los vecinos, los demandantes en el mercado (...) No elementos externos a la libertad de contratación, introducidos en la normativa y, en este caso, faltos de justificación".

dad, que sería aquella persona que, con cualificación profesional suficiente, se encarga de las funciones atribuidas a dicho órgano de la comunidad. Y, por otro lado, la figura del "administrador de fincas" estrictamente hablando como cargo profesional, que sería la persona cuyo ejercicio profesional consiste en realizar de manera habitual y principal en dicha actividad, siendo en estos últimos casos cuando se exige la debida colegiación, por pertenecer a un determinado colectivo profesional cuya colegiación es necesaria según la normativa acabada de exponer.

De este modo, el Decreto 693/1968, de 1 de abril, expone que "A dicho efecto, se entenderá que ejercen profesionalmente dicha actividad, las personas naturales que, de forma habitual y constante con despacho abierto al efecto y preparación adecuada destinan la totalidad o parte de su trabajo a administrar fincas rústicas o urbanas de terceros, en beneficio de éstos, con sujeción a las leyes, velando por el interés común y recibiendo un estipendio". Se puede comprobar que sería el ejercicio habitual y constante de esta función lo que marcaría la diferencia.

A diferencia de estos profesionales, se puede entender que pueden existir otras profesiones que podrían estar legitimadas para ejercer de administradores sin dedicarse principalmente a ello, como podría ser el caso de los abogados. En todo caso, lo que sí parece claro es que no podrán usar la denominación de "administrador de fincas" si no están colegiados en el correspondiente colegio de administradores de fincas, ya que se estaría asumiendo una determinada categoría profesional que no corresponde, teniendo su respectiva regulación, control y garantías, principalmente a través de dichos colegios profesionales.

Esto último es lo que se planteó en la STS 2393/2016 de 8 de noviembre (TOL5.877.209), donde por parte del Consejo General de Colegios de Administradores de Fincas se presentó un recurso en contra de la inscripción de la marca "APAF, ASOCIACIÓN PROFESIONAL DE ADMINISTRADORES DE

FINCAS", siendo dicha asociación ajena al Colegio de administradores de fincas. Estima el Tribunal Supremo el recurso e impide la inscripción de dicha marca, estableciendo que la actividad del "administrador de fincas" se encuentra sometida a la colegiación según el Decreto del año 1968 antes referido, de modo que "el usuario medio puede percibir que los servicios que presta una empresa cuya marca se denomina "APAF ASOCIACIÓN PROFESIONAL DE ADMINISTRADORES DE FINCAS" se corresponde con los propios de los colegios profesionales oficiales de administradores de fincas y no con servicios de una empresa privada que presta servicios que, en principio, no están relacionados con esta actividad profesional".

Nótese igualmente que el Tribunal Supremo lo que examina en este caso es la inadecuación de la nomenclatura utilizada porque puede causar error o confusión con la actividad reglada de los "administradores de fincas", pero en ningún momento refiere que todos los administradores de comunidades deban estar colegiados[9] ni pone en duda la legalidad de que cualquier empresa determinada pueda llevar a cabo funciones de administración.

2. La posibilidad de que una persona jurídica ejerza de administrador

Otra de las cuestiones que se ha planteado igualmente es si una persona jurídica puede ostentar el cargo de administrador de una comunidad de propietarios. Aquí, nuevamente, el art. 13 de la LPH (TOL230.715) recoge este supuesto sin ser demasiado específico.

Así, dicho artículo, tras expresar lo mencionado anteriormente de que puede ser cualquier persona física con cualifi-

---

9 Como parece referir la magistrada María Félix Tena Aragón conforme a lo expuesto en nuestra nota n.º 7.

cación profesional suficiente y legalmente reconocida para ejercer dichas funciones, recoge que "También podrá recaer en corporaciones y otras personas jurídicas en los términos establecidos en el ordenamiento jurídico".

Es frecuente en la práctica que los administradores de fincas desarrollen su actividad a través de formas societarias, generalmente a través de Sociedades Limitadas. Sin embargo, en este caso no podrán constituirse como sociedad profesional (SLP)[10], ni podrán acceder, como tal persona jurídica, a la colegiación.

Por ello, la AP de Guipúzcoa en su sentencia núm. 386/2018 de 16 julio (TOL6.924.824), considera que, aunque la LPH (TOL230.715) admite que el cargo de administrador pueda recaer en una sociedad en los términos establecidos en el ordenamiento jurídico, "la imposibilidad de que se pueda acceder a la forma de sociedad profesional por la falta de titulación específica (el art. 1.1 de la Ley 2/2007 conceptúa como actividad profesional aquella para cuyo desempeño se requiere titulación universitaria oficial, o titulación profesional para cuyo ejercicio sea necesario acreditar una titulación universitaria oficial), determina que no se pueda considerar a una persona

---

[10] El Proyecto de 2015 de Real Decreto de los Estatutos Generales de los Colegios Profesionales de Administradores de Fincas y su Consejo Superior sí que recoge esta posibilidad de funcionar como una sociedad profesional, considerándolo una profesión regulada y exigiendo para ello una titulación universitaria. Este punto es criticado igualmente por la CMNC en su informe, quien considera que supone una restricción considerable a la libertad de empresa y al libre acceso y ejercicio de la profesión, no estando ésta justificada en cuanto a su necesidad, no debiendo considerarse la actividad de administración de fincas como profesión regulada, y siendo indebido asimismo la exigencia de la colegiación obligatoria.

jurídica como administradora de fincas[11], sin perjuicio de que administradores de fincas constituyan una sociedad con dicho objeto, que desarrollan ellos como personas físicas".

Por su parte, la AP de Alicante en su Sentencia núm. 169/2008 de 12 marzo (TOL4.213.994), recogiendo lo dictado por una anterior de 7 de octubre de 1999, comienza indicando que "el nuevo artículo 13. 6 LPH permite que el cargo de Administrador pueda recaer en «corporaciones y otras personas jurídicas en los términos establecidos en el ordenamiento jurídico», lo que viene a significar que para el legislador actual no existe inconveniente alguno para que ese cargo lo pueda ostentar una persona jurídica como es el caso de la actora, quedando así difuminadas las sombras de ilicitud que atribuye la apelante por el hecho de que no pueda estar colegiada".

No obstante esta referencia, sigue indicando que el administrador profesional o conjunto de administradores profesionales, siendo personas físicas, pueden ejercer su profesión bajo un mismo nombre comercial en forma societaria, siendo ello perfectamente legal. "En tal sentido, es factible que una comunidad de propietarios pueda contratar los servicios de una sociedad cuyo objeto social sea la administración de fincas, aunque cohabite con el ejercicio de otras actividades siempre que se module adecuadamente la responsabilidad del profesional y de la propia sociedad, en la medida en que su propia responsabilidad personal frente a la comunidad no podrá ser obviada por el administrador, o administradores, en ningún momento".

---

11 Y ello a pesar de la dicción literal de la Ley 3/2015, de 18 de junio, de vivienda del País Vasco, donde en su DA 1.ª dice expresamente que "También podrán considerarse administradoras de fincas las corporaciones y otras personas jurídicas en las condiciones y en los términos establecidos en el ordenamiento jurídico".

Parece así inclinarse finalmente por la posibilidad de ejercer las funciones a través de una sociedad, pero siendo realmente la persona física la responsable a efectos directos como "administrador de fincas" respecto a la comunidad. Concluye así la sentencia que una sociedad no puede actualmente "colegiarse", lo que no impide la realización de dicha actividad profesional de administración por un administrador colegiado, pero a través de sociedades mercantiles instrumentales.

Igualmente, los Estatutos de diversos Colegios de administradores de fincas, como los de Málaga o Almería recogen expresamente que "3. Los Administradores de Fincas podrán ejercer la profesión bajo cualquiera de las formas societarias legalmente establecidas. En ningún caso, el Colegio podrá por sí mismo o a través de sus Estatutos o del resto de la normativa colegial, establecer restricciones al ejercicio profesional en forma societaria".

A la vista de todo ello se podría concluir que igualmente el ejercicio de administrador de una comunidad puede ser llevado a cabo por una sociedad o persona jurídica, si bien la misma como tal no puede ser considerada como "administrador de fincas", debido a que no puede ser colegiada y tener la categoría de sociedad profesional. Pero en estos casos lo que ocurrirá es que estará integrada por una o varias personas físicas que tengan la consideración de administradores de fincas colegiados, quienes serán en última instancia quienes asuman la responsabilidad del buen funcionamiento de la comunidad.

No obstante, y enlazándolo con el punto anterior sobre la necesidad de colegiación, y debido a la parquedad del texto de la normativa, también cabría la posibilidad de que una persona jurídica ejerza de administradora sin tener a ningún miembro colegiado. Este podría ser el caso visto anteriormente de despachos profesionales de abogados con forma societaria, integrados por abogados (sin colegiarse como administradores de fincas) que también se ocupen de estas gestiones de administración de comunidades (así lo admite la SAP Valencia núm. 91/2011 de 22 febrero (TOL2.532.966)).

III. REGULACIÓN DEL ADMINISTRADOR DE FINCAS EN LA LEY 12/2023, DE 24 DE MAYO, POR EL DERECHO A LA VIVIENDA

1. Disposición adicional sexta de La Ley 12/2023

En dicha disposición adicional es donde se recoge principalmente cierta regulación específica para el cargo de los administradores de fincas en el texto de la LDV (TOL9.568.821). La misma se denomina "los administradores de fincas", y contiene cuatro apartados, que veremos a continuación. Curiosamente, dicha regulación contenida en esta Disposición Adicional es prácticamente un calco de lo previsto en el art. 54 de la Ley 18/2007, de 28 de diciembre, del derecho a la vivienda[12], de Cataluña (TOL1.222.100), en sus apartados 1, 2, 5 y 6.

---

[12] Se adjunta por su interés los apartados coincidentes con la versión de la Ley catalana, donde las variaciones son mínimas:
"Artículo 54. Los administradores de fincas.
1. A efectos de la presente ley y de las actividades que regula, son administradores de fincas las personas físicas que se dedican de forma habitual y retribuida a prestar servicios de administración y asesoramiento a los titulares de bienes inmuebles y a las comunidades de propietarios de viviendas.
2. Los administradores de fincas, para el ejercicio de su actividad, deben tener la capacitación profesional requerida y deben cumplir las condiciones legales y reglamentarias que les sean exigibles, te-

Ello se debe a que dicha disposición adicional, que no figuraba en el original Proyecto de Ley, fue introducida a propuesta de la enmienda presentada por la diputada de *Junts per Catalunya* Mariona Illamola Dausà (Grupo Parlamentario Plural) (enmienda n.º 425) con fecha de 22 de mayo de 2022, la cual propuso la misma redacción que la prevista en el ordenamiento catalán, si bien el texto finalmente aprobado descartó ciertos aspectos como veremos. Dicha enmienda tenía como justificación la "Mejora técnica para aportar mayor seguridad a los consumidores"[13].

---

niendo en cuenta que la pertenencia al correspondiente colegio profesional les habilita para el ejercicio de la profesión.
5. Los administradores de fincas, en el desarrollo de su actividad profesional, deben actuar con eficacia, diligencia, responsabilidad e independencia profesionales, con sujeción a la legalidad vigente y a los códigos éticos establecidos en el sector, con especial consideración hacia la protección de los derechos de los consumidores establecidos por el presente título
6. Para garantizar los derechos de los consumidores, los administradores de fincas deben suscribir un seguro de responsabilidad civil, que pueden constituir por medio del correspondiente colegio o asociación profesional".

13 Igualmente resulta llamativo que se incorporase esta regulación de los administradores de fincas, y no se incorporase la de los agentes inmobiliarios, que suelen tratarse conjuntamente en los derechos forales que tratan estas cuestiones, a pesar de ser solicitado por diversas enmiendas (enmiendas 281, 576 y 722 presentadas en el Congreso con fecha de 27 de mayo de 2022). Asimismo, sobre esta ausencia de los agentes inmobiliarios se pronunció el senador Josep Lluís Cleries i Gonzàlez a través de la enmienda n.º 37 de 9 de mayo de 2023, así como en su intervención en el Senado de 17 de mayo de 2023.

El primero de los apartados de la referida disposición adicional sexta dice así:

> "1. A los efectos de la presente ley y de las actividades que regula, son administradores de fincas las personas físicas que se dedican de forma habitual y retribuida a prestar servicios de administración y asesoramiento a los titulares de bienes inmuebles y a las comunidades de propietarios de viviendas".

Se dedica así a definir la profesión de administrador de fincas. Nótese una vez más que no se está refiriendo al cargo de "administrador" en general de una comunidad de propietarios, sino al administrador de fincas profesional, como hemos referido antes, requiriéndose así la habitualidad en este ejercicio, de forma similar a como se viene definiendo en el Decreto del año 68 y en las distintas disposiciones colegiales –estatutos y códigos deontológicos–.

Sí llama la atención el hecho de que se recojan sólo a las personas físicas y no las corporaciones o personas jurídicas –como sí hace la Ley 3/2015, de 18 de junio, de vivienda del País Vasco (TOL5.181.949)–. Aunque la explicación posiblemente se encuentre en lo expuesto anteriormente, queriéndose en este caso referir a la persona física que es la que tiene la profesionalización del cargo y es la única susceptible de colegiarse. No obstante, debido a que en la práctica se contrata directamente con estas personas jurídicas, sería recomendable una regulación que reflejase o aclarase estos extremos.

Respecto del segundo apartado, se recoge:

> "2. Los administradores de fincas, para ejercer su actividad, deben tener la capacitación profesional requerida y deben cumplir las condiciones legales y reglamentarias que les sean exigibles".

Este apartado, que visto *a priori* pudiera parecer que simplemente hace una alusión genérica a que los administradores de fincas deben cumplir aquellos requisitos legales o reglamentarios que marque el ordenamiento –alusión que parece innecesaria por su obviedad–, sí que puede plantear algunas dudas o controversias.

Y esto se debe a que, como se indicó anteriormente, los requisitos de necesidad de colegiación tras la Ley ómnibus deben establecerse por ley, y hay quien ha visto aquí la imposición de la obligatoriedad de colegiación, no ya de forma transitoria como hasta ahora, sino definitiva, al incorporarse a una norma con rango de ley.

Algunos autores parecen apuntar a esta idea[14], siendo por tanto necesaria la colegiación de todo "profesional" que ejerza la administración. Hay quien entiende incluso que esta nueva regulación en cierta medida responde y complementa la referencia que se hace en la LPH (TOL230.715) en su art. 21.2 cuando habla de "secretario administrador profesional"[15], entendiendo que al aludir el precepto al carácter de "profesional" implica su colegiación.

---

14 Así parece entenderlo, aunque sin confirmarlo de modo contundente, Nasarre Aznar (2023, p. 4): "A diferencia de lo que parece prever para los administradores de fincas (DA 6.ª), para los agentes inmobiliarios se consagra su liberalización (establecida desde 2000), lo que es un grave error, pues no garantiza ni su supervisión, ni suficiente formación ni colegiación obligatoria (art. 30.4 Proyecto)".

15 Loscertales Fuertes (2023, p. 1): "El mero concepto de "profesional", como se indicaba en los comentarios de la 14.ª Edición del Libro sobre Propiedad Horizontal, podía dar lugar a diversas interpretaciones. Conforme la redacción del art. 21, no había problemas legales si cualquier persona o empresa se daba de alta como tal, con independencia de que tuviera o no el título de Administrador de Fincas colegiado, pues ahora no es que se exija expresamente que deba tener este último requisito, pero hay que llegar a esa conclusión a tenor del apartado 2 de lo dispuesto en la Ley del Derecho a la Vivienda, pues en otro caso la nueva normativa no tendría ningún sentido".

No obstante, acerca del establecimiento definitivo de la necesidad de colegiación de los administradores de fincas por norma con rango de ley, no está exenta de dudas dicha cuestión. Llama poderosamente la atención que en este "traslado" de la normativa catalana, que se copia literalmente, precisamente se obviase la referencia al inciso final del artículo original (54.2 Ley 18/2007) donde termina recogiendo expresamente: "teniendo en cuenta que la pertenencia al correspondiente colegio profesional les habilita para el ejercicio de la profesión"[16].

Quizás el legislador, conocedor de tal polémica actualmente, y en espera de una futura ley sobre colegiación de actividades profesionales, haya preferido no hacer mención expresa a tal asunto. No obstante, aunque se haya omitido tal inciso, ya que alude a que se deben cumplir las condiciones legales y reglamentarias que sean exigibles, y estando vigente actualmente dicha necesidad de colegiación, como hemos visto, para ejercer la profesión de "administrador de fincas", sí que consideramos viable esa interpretación en el sentido de exigir como obligatoria la colegiación.

Ahora bien, dicho esto, no compartimos la opinión sobre el hecho de que con esta nueva Ley se extienda la necesidad de colegiación a todo administrador, ya que reiteramos que esta DA va referida únicamente al "administrador de fincas" en el sentido profesional definido anteriormente, y no a todo administrador, cuya regulación sigue igual en el art. 13 LPH (TOL230.715). De esta manera, se puede seguir entendiendo que cuando se hace referencia al "secretario administrador profesional" en el art. 21.2 LPH (TOL230.715) se quiere aludir al administrador ajeno al copropietario que requiere cualificación profesional suficiente, sin

---

16 Inciso que sí estaba incluida en la propuesta original de enmienda que introdujo dicha DA 6.ª, y pedida nuevamente por la enmienda en el Senado n.º 36 de Josep Lluís Cleries i Gonzàlez (GPN), con fecha de 9 de mayo de 2023.

que tenga que ser necesariamente colegiado, como se expuso con anterioridad. De hecho, si se examina el apartado 3.º de dicho art. 21, menciona al "secretario-administrador con cualificación profesional necesaria y legalmente reconocida que no vaya a intervenir profesionalmente en la reclamación judicial de la deuda" –utilizando aquí la misma descripción que menciona el art. 13–, por lo que entendemos que se está refiriendo siempre al mismo sujeto, sin que se quiera establecer una distinción.

Es decir, la Disposición Adicional en su primer apartado se limita a definir al administrador de fincas en sentido similar a como lo hacía el Decreto 693/1968, y en el segundo, suponiendo que admitiese la necesidad de colegiación, realmente la situación sería similar a la existente antes de la aprobación de la LDV. Como se ha expuesto, esta necesidad de colegiación se exigía –y se sigue exigiendo– por dicho Decreto, entendiéndose vigente dicha necesidad por la DT de la Ley ómnibus.

Ningún otro añadido o matiz se añade a este asunto, sin que se extienda dicha obligación al cargo de administrador de comunidades regulado en la LPH (TOL230.715), ya que, para ello, hubiera sido necesario, o al menos aconsejable, hacer una reforma del referido artículo 13 LPH (TOL230.715), donde se especificase la necesidad de colegiación del administrador profesional ajeno al copropietario. Sin embargo, nada de esto se dice.

En este sentido, hubiera sido aconsejable aprovechar la LDV para zanjar de forma definitiva el debate sobre la necesidad o no de colegiación de los administradores profesionales de las comunidades de propietarios, proporcionando la certeza jurídica de la que se carece en la actualidad[17]. Ha sido otra oportunidad perdida.

---

17 Considera Caballé Fabra (2024, p. 353) la necesidad de establecer una colegiación obligatoria, lo que proporciona un mecanismo ideal para garantizar que todos los profesionales cuenten con una

En cuanto al tercer apartado de la disposición adicional, se expone:

> "3. Los administradores de fincas, en el desarrollo de su actividad profesional, deben actuar con eficacia, diligencia, responsabilidad e independencia profesionales, con sujeción a la legalidad vigente y a los códigos éticos establecidos en el sector, con especial consideración hacia la protección de los derechos de los consumidores establecidos por las comunidades autónomas y en esta ley".

En este párrafo se puede observar una tendencia que parece que cada vez es más habitual en los legisladores actuales, de escribir textos legales con párrafos "políticamente correctos" pero que realmente nada aportan. No es necesario que exista ningún artículo en cada profesión que diga que esos profesionales deben actuar con eficacia, diligencia o responsabilidad, ya que presupone que todo profesional –y todo contratante– debe actuar conforme a los principios de la buena fe y ser diligente en su desempeño, conllevando responsabilidad en caso contrario, como no puede ser de otro modo y así se recoge en el Código Civil (TOL220.310) y demás normativa de aplicación.

De hecho, muchos de estos términos sí que suelen recogerse en los estatutos y códigos deontológicos de los administradores de fincas, donde tienen mayor cabida, además de que en dichos textos se enumeran una serie de derechos y deberes que plasman y concretan estas obligaciones genéricas, siendo de obligado cumplimiento por parte de todos los integrantes, y estando sometido dichos administradores de fincas al régimen disciplinario correspondiente.

---

misma formación, así como también supervisar su actividad y la recaudación de los datos.

Igualmente innecesaria es la referencia a que deben someterse a la legalidad vigente, adaptarse a sus códigos deontológicos, y respetar a los consumidores; texto superfluo que nada aporta al no establecerse determinados deberes o derechos para los sujetos implicados.

En lo relativo al apartado 4.º de la disposición adicional, se recoge:

> "4. Para garantizar los derechos de los consumidores, los administradores de fincas deben suscribir un seguro de responsabilidad civil, pudiendo hacerlo directa o colectivamente".

Este último párrafo viene a reiterar lo dictado por el Código Deontológico del Administrador de Fincas del Consejo general de Administradores de Fincas de España, acerca de la necesidad del seguro. Si bien eleva a rango legal la necesidad de dicho seguro, no supone en la práctica ninguna novedad, recogiéndose además en dicho Código Deontológico de manera más detallada[18].

---

[18] Artículo 6. Seguros de responsabilidad civil profesional y de caución. Sin perjuicio de la regulación legal vigente en cada momento, los Administradores de Fincas colegiados deberán disponer de unos seguros que garanticen:
a) La responsabilidad en la que pueda incurrir, en general, por el error, mala praxis o incumplimiento de sus deberes en el ejercicio de su profesión y de cuya conducta deriven perjuicios patrimoniales a su cliente.
b) la responsabilidad en la que pueda incurrir con respecto a los fondos, bienes y derechos que les hayan sido depositados o confiados en el ejercicio de su actividad profesional.
A solicitud del interesado, el Administrador de Fincas colegiado habrá de exhibir el correspondiente certificado que acredite la vigencia de dichas garantías que, en todo caso, cubrirán, como mínimo, los límites de responsabilidad establecidos por la normativa colegial aplicable en cada momento.

Como se puede observar de los párrafos comentados de la disposición adicional sexta, nada aporta realmente esta "nueva" regulación sobre los administradores de fincas[19], ya que repite lo que ya existe, o recoge texto superfluo como hemos visto, y ello a pesar de que en la Exposición de Motivos se decía que en dicha DA se regulan determinados aspectos de su actividad al ser un colectivo profesional de gran importancia.

Y como se ha expuesto, dicha regulación se hace extrayendo el texto de diversos apartados de la normativa catalana, pero en dicha normativa regional sí que tiene incluso un mayor peso, ya que en el referido artículo 54 de la Ley 18/2007 se recoge además en sus apartados 3.° y 4.° cuándo dichos administradores de fincas tienen que cumplir con obligaciones adicionales impuestas a los agentes inmobiliarios. Así, se refiere en dicha normativa que en la medida en que estos administradores de fincas compatibilicen su actividad con la "prestación de servicios de transacción inmobiliaria de operaciones de compraventa, permuta o cesión de bienes inmuebles distintos del traspaso o arrendamiento de los bienes que administran" tienen la condición de agentes inmobiliarios, y deben cumplir con los requisitos que se fijan en dicha normativa. Por el contrario, si dichos administradores de fincas sólo prestan "servicios de mediación en operaciones de arrendamiento respecto a los bienes sobre los que tienen encomendada la administración" no están sujetos al cumplimiento de dichos requisitos establecidos[20].

---

19 De modo similar considera Caballé Fabra (2024, p. 344) que la Ley 12/2023 no ha incorporado ninguna novedad en la regulación de los agentes inmobiliarios ni en la de los administradores de fincas.

20 Igualmente, dichos párrafos sí estaban incluidos en la propuesta original de enmienda que introdujo dicha DA 6.ª, y pedida nuevamente por la enmienda en el Senado n.° 36 de Josep Lluís Cleries i Gonzàlez (GPN), con fecha de 9 de mayo de 2023. Sin embargo, no fue aceptado su incorporación.

En conclusión, se puede constatar, como hemos argumentado, que se ha desaprovechado la oportunidad en dicha regulación estatal de regular o aclarar algunos de los aspectos controvertidos que hemos mencionado como el hecho del ejercicio del cargo de administrador por quien no está colegiado, o la existencia de personas jurídicas que ocupen dicho cargo.

2. Otras normas de la LDV relativas a los administradores de fincas

Además de lo dispuesto principalmente por la DA 6.ª mencionada, la LDV (TOL9.568.821) hace un par de alusiones a los administradores de fincas.

La primera de ellas se limita a encuadrarlos dentro de los "agentes que intervienen en el mercado inmobiliario", donde en el art. 2 r) LDV (TOL9.568.821) se establece como fin común de la actividad de los poderes públicos el "Fomentar la transparencia y garantizar la participación en el desarrollo de las políticas públicas de vivienda" de estos agentes, entre los que se incluye a los administradores de fincas.

Esta primera referencia tiene su origen en unas enmiendas presentadas en el Congreso al Proyecto de Ley con fecha de 27 de mayo de 2022, por el Grupo Parlamentario Confederal de Unidas Podemos-En Comú Podem-Galicia en Común (enmienda n.º 86), y más concretamente la del Grupo Parlamentario Euskal Herria Bildu (enmienda n.º 204), cuyo fin era "recoger y reforzar los mecanismos de dialogo social en el marco del derecho a la vivienda". No obstante, ninguna otra especificación se hace al respecto, ni se menciona cómo estos administradores de fincas pueden ser afectados o participar en las referidas políticas públicas de vivienda.

Además de esta referencia, se mencionan a los administradores de fincas en el artículo 30.2 LDV (TOL9.568.821), en relación con el art. 31, mención que sí puede tener algo más de peso en la medida en que establecen ciertas obligaciones, aunque no son de gran calado como veremos.

Como estos preceptos van referidos a los derechos de información que impone la LDV, no se tratará aquí en profundidad al ser analizados por otro de los capítulos de esta obra. No obstante, conviene señalar algunos aspectos en lo que respecta a los administradores de fincas.

En primer lugar, es relevante establecer a qué administradores de fincas se les atribuye este deber de información. Así, en dicho art. 30.2 se recoge que todos los agentes –entre los que incluye a los administradores de fincas– que "operando en el sector de la edificación y rehabilitación de viviendas y la prestación de servicios inmobiliarios, estén facultados para la transmisión, el arrendamiento y la cesión de las viviendas en nombre propio o por cuenta ajena (...) deben cumplir en su actividad el deber de información completa, objetiva, veraz, clara, comprensible y accesible conforme lo previsto en esta ley, así como en la legislación de defensa de consumidores y usuarios cuando se trate de relaciones entre consumidores o usuarios y empresarios, quedando sujeta la publicidad que realicen a la legislación general que la regula, con prohibición, en particular, de cualesquiera actos publicitarios con información insuficiente, deficiente o engañosa".

Al incluir a los administradores de fincas, la LDV posiblemente pensando en aquellos supuestos, entre otros, que suelen ser frecuentes en la práctica, donde se le encarga al administrador la gestión del arrendamiento de alguna vivienda al que tiene encargada la administración, siendo la comunidad en general la que se beneficia de las rentas de dicho arrendamiento.

De este modo, en estos casos se establece que se debe informar según se indica en dicha Ley –que es lo que se recoge en el art. 31–. El resto de la norma que menciona que se debe dar cumplimiento a la legislación de consumo y publicidad es, una vez más, totalmente innecesaria por ser obvio que toda actuación se somete a la legalidad vigente.

Sí es de destacar que, debido al alcance amplio de la definición de agentes que intervienen según el art. 30.2, en estos casos sí que podrían estar incluidos todos aquellos administradores no colegiados, en la medida en que estos realicen las actividades enunciadas de transmitir, arrendar o ceder viviendas. Es decir, si un abogado –sin estar colegiado como administrador de fincas– ejerce las funciones de administrador de una comunidad de propietarios en virtud del art. 13.6 LPH (TOL230.715), debe cumplir con estos deberes de información si tiene facultades para enajenar, arrendar, o ceder viviendas.

Y respecto a esta obligación de información que se impone a dichos administradores, sin entrar en detalle sobre la misma, tenemos que destacar dos aspectos que son relevantes:

El primero de ellos es que realmente no impone un deber de actuación activo por parte de los administradores, sino que, como se recoge el art. 31, toda la información que se detalla en dicho artículo es la que "podrá requerir" la persona interesada en la compra o arrendamiento de una vivienda.

Y, en segundo lugar, si se analiza el contenido de dicha información, se observa que se trata principalmente de las características esenciales de la operación y de la vivienda, no suponiendo ningún cambio de gran trascendencia dicha regulación. De hecho, no existiendo esta regulación o fuera del ámbito de la misma, todo comprador o arrendatario estaría legitimado para solicitar esa información, en la medida en que es información necesaria o útil acerca de la operación u objeto del contrato[21].

---

21 La única salvedad parece encontrarse en el apartado tercero, donde en los casos en donde se vaya a arrendar una vivienda que se encuentre en una zona de mercado residencial tensionado, al menos literal-

Como se puede observar de todo el análisis de la LDV, este deber de información es básicamente el único nuevo deber que incorporaría la Ley a los administradores –aunque como hemos visto mayormente no supongan una gran novedad–, y sólo cuando ejerzan estas funciones de transmisión, arrendamiento y cesión de la vivienda. Pero a pesar de ello, y de que el preámbulo de la LDV señalaba que se regulaba la actividad de estos administradores por su importancia en la garantía y protección de los consumidores, llama la atención que ninguna consecuencia se ha establecido en dicha normativa para el caso de incumplimiento de estos deberes de información.

Desde luego, ninguna consecuencia directa contractual puede extraerse de ello –más allá de los remedios generales como el dolo, error-vicio, o supuestos de responsabilidad extracontractual que el incumplimiento del deber de información puede conllevar–. Pero podría haberse dispuesto, como suele ocurrir en estos casos, que el incumplimiento de dicha normativa supone una infracción administrativa, y más concretamente, en materia de consumo. No obstante, se podría defender que su incumplimiento sí puede ser considerado una infracción administrativa en dicha materia y tener la correspondiente sanción administrativa en la medida en que se considere subsumible en los supuestos genéricos tipificados ya

---

mente sí parece imponerse una obligación, al referir que el propietario o intermediario "deberá" indicar tal circunstancia e informar "con anterioridad a la formalización del arrendamiento, y en todo caso en el documento del contrato, de la cuantía de la última renta del contrato de arrendamiento de vivienda habitual que hubiese estado vigente en los últimos cinco años en la misma vivienda, así como del valor que le pueda corresponder atendiendo al índice de referencia de precios de alquiler de viviendas que resulte de aplicación".

sea en la normativa estatal –arts. 47.1 g) y p) del TRLGDCU[22] (TOL1.175.543) –, o de alguna norma autonómica –como se recoge de modo más concreto en el art.71.2.1.ª de la Ley 13/2003, de 17 de diciembre, de Defensa y Protección de los Consumidores y Usuarios de Andalucía[23] (TOL330.321) –.

IV. CONCLUSIONES

1. La Ley 12/2023, de 24 de mayo, por el derecho a la vivienda no parece suponer un avance o novedad respecto a los administradores de fincas.

A pesar de las intenciones anunciadas, exponiendo en el preámbulo que se regulan determinados aspectos de su actividad, y señalando su importancia para asegurar la garantía y protección de los derechos de los consumidores, la realidad está lejos de tales pronunciamientos.

---

22 "Art. 74. 1. Son infracciones en materia de defensa de los consumidores y usuarios las siguientes: g) El incumplimiento de las normas relativas a registro, normalización o denominación de productos, etiquetado, envasado y publicidad de bienes y servicios, incluidas las relativas a la información previa a la contratación; p) La obstrucción o negativa a suministrar las condiciones generales de la contratación que establece el artículo 81.1 de esta ley o cualquier otra información requerida por la Administración competente en el ejercicio de sus competencias de acuerdo con esta ley".

23 "Art. 71.2. Serán infracciones por incumplimiento de requisitos y condiciones de elaboración y comercialización de bienes o por incumplimiento de las condiciones técnicas de la instalación o de la prestación del servicio. 1.ª Elaborar, distribuir, suministrar u ofertar bienes o servicios sin cumplir correctamente los deberes de información que impongan o regulen las leyes y los reglamentos en relación con cualquiera de los datos o menciones obligatorios o voluntarios y por cualquiera de los medios previstos para tal información".

2. Es necesario distinguir entre los administradores de fincas colegiados y administradores no colegiados.

Es importante tener claro que, según entiende la mayoría de la doctrina y la jurisprudencia, el art. 13 LPH (TOL230.715) no exige que el cargo de administrador sea desempeñado necesariamente por un "administrador de fincas" profesional que esté colegiado, sino que permite que dicho cargo sea ejercido por cualquier profesional que tenga competencia suficiente, aunque no esté colegiado como administrador de fincas, como puede ser el caso de que dicho puesto lo ocupe un abogado.

Hay que destacar que, según entendemos, la regulación de la DA 6.ª de la LDV va realmente dirigida a dichos profesionales que desempeñan el cargo de administradores de fincas de forma habitual, esto es, a los administradores colegiados, sin que abarque a todos los administradores mencionados en el art. 13 de la LPH (TOL230.715).

3. Se puede entender subsistente la necesidad de colegiación del "administrador de fincas" profesional.

No se ha dictado ninguna ley estatal que regule las profesiones que requieren colegiación, por lo que debemos entender vigente la necesidad de colegiación del Decreto 693/1968, según dispone la Ley ómnibus.

Con la LDV tenemos una ley estatal que refiere que el administrador de fincas debe cumplir con "las condiciones legales y reglamentarias que les sean exigibles", lo que algunos autores han considerado que establece, ahora sí por rango de ley, la necesidad colegiación. No obstante, como suele insistir la CNMC, no se especifican las justificaciones que, respetando los principios de no discriminación y proporcionalidad, exigirían tal colegiación por razones de necesidad e interés general.

A pesar de todo ello, aun entendiendo que esta nueva Ley confirma esta necesidad de colegiación, la situación no cambia en la práctica, puesto que, en primer lugar, dicha necesidad de colegiación estaba presente como se ha expuesto; y, en segundo lugar, no se ha modificado el art. 13 LPH (TOL230.715), por lo que, al igual que hasta ahora, se permitiría que todo profesional con competencia suficiente, aunque no esté colegiado, pueda ejercer dicho cargo de administrador. No existe previsión alguna que indique lo contrario.

4. El legislador debe dar cumplimiento en el tiempo más breve posible a su obligación de establecer las profesiones que necesitan colegiación.

Como se señala por la Comisión Nacional del Mercado y de la Competencia en su informe de 2016[24], la regulación liberalizadora de restricciones de acceso y ejercicio de servicios profesionales se viene reclamando insistentemente desde diversas instancias internacionales, de manera sistemática por el FMI o la OCDE, y muy especialmente por la propia UE en las recomendaciones específicas a los Programas Nacionales de Reforma. Como se ha examinado, es un tema pendiente de regulación desde el año 2009 con la mencionada Ley ómnibus, que todavía el legislador no se ha atrevido a realizar.

Esta situación de indeterminación no es nada aconsejable, siendo conveniente la oportuna regulación para ofrecer la necesaria seguridad jurídica, no sólo en este ámbito, sino respecto de todas las profesiones que se puedan ver afectadas.

5. El término "administrador de fincas" está reservado a los administradores colegiados.

---

[24] Informe del año 2016 sobre el Proyecto de Real Decreto por el que se aprueban los estatutos generales de los colegios profesionales de administradores de fincas y de su consejo general.

Debido a esta distinción de conceptos entre administradores colegiados y no colegiados, el Tribunal Supremo ha dispuesto que el término "administrador de fincas" debe estar reservado para aquellos administradores colegiados.

Entiende que dicha nomenclatura va asociada a estos profesionales colegiados, con la mayor protección y garantía que supone este cuerpo específico, por lo que no permite su uso por cualquier profesional que ejerza el cargo de administrador de una comunidad.

6. Apenas se aporta nueva protección a los consumidores, estableciéndose unos deberes de información poco relevantes.

Respecto de los administradores de fincas, y a pesar de lo anunciado en el preámbulo de la LDV, no hay en la práctica un reforzamiento de la protección de los consumidores[25], ya que apenas existe regulación al respecto.

La imposición del seguro de responsabilidad civil para los administradores de fincas profesionales es algo que ya existía como obligatorio según los códigos deontológicos, regulándose en estos de manera más completa y detallada.

---

[25] Si se quiere mejorar la protección a los consumidores a nivel legislativo, la práctica cotidiana en relación con los administradores presenta otros problemas que no se han tratado aquí por exceder de la temática objeto de estudio, pero que podrían ser de interés para el legislador si se quiere mejorar dicha protección y el ejercicio de los administradores de comunidad de propietarios. Ejemplo de ello sería lo relativo a los problemas que se suscitan con la duración del cargo, la facultad de remoción fijada en el art. 17 LPH (TOL230.715), así como la indemnización procedente, que son los temas que suelen generar más controversias con las comunidades de propietarios, existiendo además muchas incertidumbres por las posiciones contradictorias de las diferentes Audiencias provinciales.

Y respecto los nuevos deberes de información, se imponen en supuestos limitados y de forma no imperativa, ya que "pueden" ser solicitados por los interesados, siendo además información ordinaria de la operación, y sin que ni siquiera la normativa establezca sanción alguna a dicho incumplimiento.

7. La LDV supone una oportunidad desaprovechada.

El legislador podría haber aprovechado la ocasión para dar respuestas y ofrecer la necesaria seguridad jurídica en muchas de estas cuestiones examinadas. Lejos de ofrecer claridad, en ocasiones produce más dudas si cabe, sin aportar ninguna mejora significativa para los que desempeñan este cargo o para los consumidores, ni para la sociedad en general. El tiempo dirá cómo se acaban interpretando tales disposiciones y el impacto final de las mismas, pero parece claro que esta normativa no satisface la necesidad de resolver las cuestiones aquí planteadas.

V. REFERENCIAS BIBLIOGRÁFICAS

Alonso Sánchez, B. (1994). "El administrador de fincas en el régimen de la propiedad horizontal". *Revista Doctrinal Aranzadi Civil–Mercantil* vol.III. BIB 1994\10.

Caballé Fabra, G. (2024). "Los profesionales inmobiliarios en los arrendamientos". *Derecho a la vivienda en tiempos de incertidumbre* (dirs. Mate Satué, L. C./ Hernández Sainz, E./ Alonso Pérez, M.ª T.). Aranzadi.

Carreras Maraña, JM. (2006). "La responsabilidad civil del Administrador de Fincas" En *Boletín de Propiedad Horizontal y Derechos Reales.* El Derecho, 30 de mayo de 2006, n.º 64.

Flores Rodríguez, J. (2008). "El Administrador de Fincas en Europa. ¿Es posible un modelo uniforme de Administrador de Fincas Europeo?". *ElDerecho.com.* EDC 2008/34937.

López Navarro, JF. (Coord.) (2024). *GPS Propiedad Horizontal. Guía Íntegra para la Administración de Fincas.* 10.ª Edición. Tirant. (TOL7.177.754)

Loscertales Fuertes, D. (2020). Propiedad Horizontal: Legislación y Comentarios. Dykinson.

Loscertales Fuertes, D. (2023). "La nueva ley de Vivienda y los administradores de fincas" *www.abogacía.es.* Actualidad. Opinión y Análisis. https://www.abogacia.es/actualidad/opinion—y—analisis/la—nueva—ley—de—vivienda—y—los—administradores—de—fincas/ Recuperado el 21 de enero de 2024.

Magro Servet, V. (2017). "¿Puede un administrador de fincas impugnar el nombramiento de otro en junta alegando no cualificación profesional?" *ElDerecho.com* EDC 2017/513300.

Nasarre Aznar, S. (2023). "Ley de Vivienda: el Estado renuncia a cumplir con el art. 47 de la Constitución". *El Notario del Siglo XXI,* n.º 109.

Pérez Miralles, JA. (2017). "Hacia una nueva Ley de Propiedad Horizontal. El proyecto del consejo general de colegios de administradores de fincas de España desde la prespectiva de la mejora de la protección de los consumidores y usuarios". *Revista Consumo y Empresa,* n.º 4, Febrero 2017.

Rosat Aced, Ignacio (2015). *Dossier PH nº 8: El administrador: características y nombramiento.* Tirant. TOL5.584.923.

Rubiras Valenzuela, V. (2021). "Régimen de Propiedad Horizontal". En: Moreno Ubrich, R. *Manual de Derecho Inmobiliario.* Grandes Tratados. Aranzadi.

Capítulo VII

La modificación de la ley 29/1994, de 24 de noviembre, de arrendamientos urbanos por la ley por el derecho a la vivienda

JOSÉ MANUEL MARTÍN FUSTER
Profesor Ayudante Doctor de Derecho civil
Universidad de Málaga

I. INTRODUCCIÓN

En este capítulo nos centraremos en examinar las principales novedades introducidas por la Ley 12/2023, por el Derecho a la Vivienda (LDV) en la Ley 29/1994, de arrendamientos urbanos (LAU; TOL231.076). En concreto, en este trabajo nos centraremos en analizar las modificaciones de la LAU contenidas en la Disposición Final Primera de la Ley 12/2023, y que afectarán al arrendamiento de vivienda.

En primer lugar, debemos destacar la finalidad de las reformas introducidas. En este sentido, la intencionalidad es clara y pertinente: se pretende mejorar la accesibilidad a la vivienda y evitar los incrementos de renta desproporcionados.

Actualmente nos encontramos con un mercado de alquiler que algunos autores califican de "injusto", donde resulta en muchas ocasiones más caro pagar una renta de alquiler que hacer frente a la cuota de un préstamo hipotecario para la compra de esa misma vivienda[1]. Por ello, parece conveniente establecer algunas medidas para evitar encontrarnos con un mercado tensionado y con unas rentas cada vez menos asequibles. Cuestión distinta es cuáles son las medidas correctas que puedan mejorar o solucionar la situación; siendo una cuestión más compleja, ya que depende de factores económicos, sociales y culturales estructurales, que están provocando una alta demanda de vivienda en nuestro país, tanto en propiedad como en alquiler, y donde encontramos una escasez de oferta en ambos tipos de tenencia.

Como se reconoce en el preámbulo de la LDV, la Constitución española (TOL173.304) reconoce, en su artículo 47, el derecho al disfrute de una vivienda digna y adecuada, imponiendo a los poderes públicos el deber de promover las condiciones necesarias que garanticen la igualdad en el ejercicio de los derechos y el cumplimiento de los deberes constitucionales y de establecer las normas pertinentes para hacer efectivo el referido derecho. Además, se recuerda que la vivienda es un bien esencial de rango constitucional, que conecta con otros derechos que además son derechos fundamentales, como la integridad física y moral, o el derecho a la intimidad personal y familiar. Por ello, la vivienda constituye un pilar central del bienestar social en cuanto lugar de desarrollo de la vida privada y familiar.

Además, se señala que todo ello permite modular tanto el derecho de propiedad como la libertad de empresa, cuando operan en el sector de la vivienda, desde el doble punto de

---

1 Así lo refleja, Miguel Hernández (2023, p. 1). Señala el autor además que la vivienda es un bien verdaderamente de "primera necesidad" y aunque los datos varían según unas u otras fuentes (OCDE, sindicatos, estudio Infojobs-Fotocasa, etc.), se observa que los españoles dedicamos entre un 40 % y 60 % del salario al pago de la renta de alquiler, y hasta un 90 % en el caso de los jóvenes.

vista de la función social que deben cumplir y del interés general, respectivamente.

En nuestro ordenamiento, el arrendamiento de vivienda, que constituye uno de los instrumentos fundamentales en materia de política de vivienda, se regula por la Ley de Arrendamientos Urbanos de 1994. Esta ley ha sido objeto de varias reformas, por lo que el régimen actual convive con el de los contratos celebrados bajo la vigencia de las leyes de arrendamientos anteriores.

Entre las últimas reformas, encontramos las medidas de carácter social adoptadas durante el estado de alarma decretado a causa del COVID-19, como el Real Decreto-ley 11/2020, de 31 de marzo, por el que se adoptan medidas urgentes complementarias en el ámbito social y económico para hacer frente al COVID-19 (TOL7.854.128), o el Real Decreto-ley 6/2022, de 29 de marzo, por el que se adoptan medidas urgentes en el marco del Plan Nacional de respuesta a las consecuencias económicas y sociales de la guerra en Ucrania (TOL8.886.633). Algunas de las medidas introducidas en estas leyes, como son la prórroga extraordinaria o la limitación en la actualización de la renta, tienen protagonismo en las novedades que analizaremos a continuación.

En este marco, las principales novedades que afectan a la regulación del arrendamiento se recogen en la Disposición Final Primera de la ley, titulada "medidas de contención de precios en la regulación de los contratos de arrendamiento de vivienda", que constituirá el objeto de estudio de este capítulo[2].

---

2 Como señala Molina Roig (2023, p. 2), durante la tramitación del Proyecto de Ley se introdujeron algunas modificaciones en el Congreso de los Diputados, pero permaneció idéntico tras su paso por el Senado, al no aprobarse las enmiendas presentadas en dicho momento. Entre las modificaciones en su tramitación parlamentaria, la citada autora destaca las siguientes: se han relajado los criterios para

Como se verá a continuación, se trata de una serie de medidas que vienen a retocar el sistema actual, con algunos añadidos puntuales a la LAU, pero sin llegar a provocar cambios profundos en el sistema actual, que quizá hubiera sido lo conveniente[3].

Una de las cuestiones que llama la atención, es que en el Preámbulo se declaren como "mecanismo de carácter excepcional y acotado en el tiempo" las novedades que introduce, justificado con el fin de intervenir en el mercado para amortiguar las situaciones de tensión y "conceder a las administraciones competentes el tiempo necesario para poder compensar en su caso el déficit de oferta o corregir con otras políticas de vivienda las carencias de las zonas declaradas de mercado residencial tensionado". Y llama la atención porque esta ley y estas medidas tienen carácter general y de permanencia, y su aplicación no está limitada en el tiempo, como se manifiesta en el Preámbulo de la propia

---

que las Comunidades Autónomas puedan declarar las zonas de mercado residencial tensionado; se establece la obligación de informar sobre las rentas y los índices de precios aplicables a los contratos de alquiler; se introducen modificaciones en la definición de gran tenedor de vivienda, y al sistema para frenar las subidas abusivas de las rentas de alquiler, tanto respecto a las rentas iniciales como a la actualización de las rentas vigentes; se crea una nueva prórroga extraordinaria del contrato; se prohíbe la posibilidad de cobrar al arrendatario los gastos de gestión inmobiliaria y formalización del contrato de alquiler, o de obligar a realizar el pago de la renta en metálico. Además, se sustituye la creación a nivel estatal de un Registro de contratos de alquiler por una simple base de datos.

3 A este respecto, conviene destacar la obra de Martens Jiménez (2023). En dicho trabajo se realiza un estudio profundo del actual sistema de arrendamientos, y se hace una propuesta de reforma del sistema actual muy interesante, que va más allá de meros retoques, y se propone un nuevo sistema más acorde a la realidad social en que vivimos.

ley, que señala que uno de los objetivos es "establecer una regulación básica de los derechos y deberes de los ciudadanos en relación con la vivienda". Es cierto que algunos de los elementos introducidos, como las zonas de mercado residencial tensionado, tienen un plazo, pero dicho plazo es prorrogable indefinidamente, y como la propia ley tiene carácter permanente, esto puede conllevar a alargar las medidas en el tiempo.

Hechas estas consideraciones previas, a continuación, pasaremos a enumerar las novedades introducidas en la Ley de Arrendamientos Urbanos, y que posteriormente serán analizadas con detalle en este capítulo:

I. Como primera novedad, se establece una prórroga extraordinaria de un año duración del contrato de arrendamiento de vivienda, que requiere dos requisitos: por un lado, que el arrendatario demuestre que está en situación de vulnerabilidad económica y social mediante un informe de los servicios sociales, y por otro, que el arrendador tenga la condición de gran tenedor de vivienda.

II. Además, se introduce otra prórroga extraordinaria, por plazos anuales hasta un máximo de tres años de duración, para los casos en que el inmueble se encuentre en una zona de mercado residencial tensionado. Dicha declaración de mercado residencial tensionado es competencia de la administración autonómica, y, por tanto, esta prórroga solo será aplicable en el caso de que dicha declaración tenga lugar. Asimismo, conviene recordar que estas dos prórrogas son obligatorias para el arrendador y potestativa para el arrendatario.

III. Por lo que se refiere a una de las medidas más controvertidas, la limitación de la renta, la Ley mantiene la regla de la libertad de pactos, pero establece excepciones cuando se den determinados supuestos:

1.º Si la vivienda se encuentra en una zona de mercado residencial tensionado la renta no podrá exceder de la del último contrato vigente una vez actualizada según lo pactado, salvo algunas excepciones justificadas;

2.º Por otro lado, en los casos donde concurra la doble circunstancia de que la vivienda se encuentre en una zona de mercado residencial tensionado y el arrendador sea gran tenedor, la renta no podrá exceder del límite máximo del precio aplicable conforme al sistema de índices de precios de referencia. Se trata de un nuevo índice que se ha aprobado en marzo de 2024.

3.º Esta misma limitación se aplicará a los contratos de arrendamiento de vivienda en los que el inmueble se ubique en una zona de mercado residencial tensionado, y sobre el que no hubiese estado vigente ningún contrato de arrendamiento de vivienda en los últimos cinco años.

A este respecto, conviene señalar que la ley, para evitar eludir esta limitación de la renta, prohíbe repercutir al nuevo inquilino cuotas y gastos no contemplados en el contrato anterior.

Todas estas limitaciones que acabamos de señalar solo son aplicables a los contratos celebrados con posterioridad a la entrada en vigor de la Ley.

IV. Respecto a los contratos anteriores a la entrada en vigor de la ley, debemos señalar que sí les afectará otra limitación; en este caso, la referente a la actualización de la renta. Así, en los contratos en los que hubiera correspondido la actualización de la renta entre el 31 de marzo de 2022 y el 31 de diciembre de 2023, el límite máximo para la actualización de la renta sería del 2% (variación anual del IGC). Y, para las rentas que deban ser actualizadas en 2024, se establece un límite máximo de un 3%. Cabe señalar que estos límites se aplicarán salvo pacto con el arrendador que no sea gran tenedor.

V. Por otro lado, se han introducido otras modificaciones destacables, como son una ampliación de los deberes de información del arrendador, el establecimiento del pago por medios electrónicos como método predeterminado, o la imposición de los gastos de formalización y de gestión inmobiliaria al arrendador.

Estas son las novedades que analizaremos a continuación. Como vemos, el legislador ha sido poco ambicioso. Parece que considera suficiente con introducir algunas reformas puntuales en el arredramiento de vivienda, con medidas cuya eficacia se discute, y que además difícilmente beneficiará a la generalidad de arrendatarios, ya que tienen un alcance temporal y subjetivo limitado.

Además, señalaremos las posibles cuestiones controvertidas que puedan plantearse en cada una de estas medidas[4]. Entre estas cuestiones, se señalan algunos problemas que pueden plantearse con esta nueva normativa, como cuestiones de competencia, la limitación del derecho de propiedad y a la libertad de precios, la excesiva remisión a la legislación autonómica, o su aplicación efectiva[5].

---

[4] Algunos autores, como Nasarre Aznar (2022, p. 23), se muestran críticos con las medidas de la nueva ley, señalando que "las medidas excepcionales propuestas para viviendas situadas en zonas tensionadas no solo tendrán previsiblemente un alcance limitado, sino que, además, si superan su constitucionalidad, muchas de ellas se han demostrado ineficientes e, incluso, contraproducentes tanto históricamente en toda España, como recientemente en Cataluña y a nivel comparado (Alemania y Estados Unidos, entre ellas)".

[5] En este sentido, Flores González (2024, p. 226) señala que "la nueva Ley se encuentra con importantes obstáculos para su aplicación

II. PRÓRROGAS EXTRAORDINARIAS

La primera novedad introducida por la Disposición Final Primera es la modificación del art. 10 de la LAU, referente a las prórrogas del contrato.

En cuanto a prórrogas legales, cabe decir que las previstas en el artículo 9 (prórrogas hasta 5 años o 7 en caso de arrendador persona jurídica) y el art. 10.1 de la Ley de Arrendamientos Urbanos (prórroga de hasta 3 años más, transcurridas la prevista en el artículo 9) se mantiene exactamente con la misma regulación que les dio el Real Decreto-ley 7/2019, y se añaden dos nuevas prórrogas legales extraordinarias que entran en juego cuando se agotan las anteriores y se dan determinadas circunstancias previstas en el artículo 10:

1.ª El artículo 10 apartado 2 establece que, agotada la prórroga obligatoria del art. 9 o la tácita del art. 10.1, se da la posibilidad de una prórroga extraordinaria de un año de duración, que será obligatoria para el arrendador que sea gran tenedor y potestativa para el arrendatario, y que opera cuando el arrendatario se encuentra en situación de vulnerabilidad social y económica. Por lo tanto, son dos los requisitos para esta prórroga extraordinaria: la situación de vulnerabilidad del arrendatario, y la condición de gran tenedor del arrendador.

Respecto a la vulnerabilidad, es competencia de la administración determinar si el arrendatario es o no vulnerable. Para ello, el arrendatario deberá presentar un informe o certificado

---

práctica. No solo por la cuestión competencial. Hay comunidades autónomas que han manifestado su reticencia a la aplicación de la Ley, cuando su cooperación fundamental pues son las administraciones competentes en materia de vivienda. Por otro lado, los estudiosos de esta norma han denunciado que no establece mecanismos jurídicos adecuados para dar cumplimiento a la misma".

emitido en el último año por los servicios sociales de ámbito municipal o autonómico. Sin embargo, la LDV no suministra el concepto de vulnerabilidad social y económica a los efectos previstos, por lo que deja libertad a la administración para que determine si el arrendatario es o no vulnerable, no quedando claro, por tanto, el criterio a aplicar[6].

Y en cuanto a lo que es un gran tenedor, se define en el art. 3.k de la LDV, y lo comentaremos en un apartado a continuación.

2.ª El apartado 3 del artículo 10 establece que cuando el inmueble se ubique en una "zona de mercado residencial tensionado" podrá prorrogarse de manera extraordinaria el contrato de arrendamiento por plazos anuales, por un período máximo de tres años, salvo que las partes hayan pactado un nuevo arrendamiento o hayan acordado otros términos o condiciones. Respecto al concepto de zona de mercado de residencial tensionado, se contempla en el art. 18 de la LDV. Así, las administraciones competentes en materia de vivienda podrán declarar zonas de mercado residencial tensionado a aquellos ámbitos territoriales en que exista un especial riesgo de oferta insuficiente de vivienda para la población en condiciones que la hagan asequible para su acceso en el mercado. La vigencia de la declaración de un ámbito territorial como zona de mercado residencial tensionado será de tres años, pudiendo prorrogarse anualmente.

Por su parte, el arrendador podrá eludir esta segunda modalidad de prórroga siempre que el arrendador haya comu-

---

[6] Así lo indica Flores González (2024, p. 240). Además, se podría pensar si sería aplicable los criterios de vulnerabilidad de la disposición final 5.ª, en concreto, los del actual art. 441.7 LEC, pero parece que el legislador no lo ha querido así, o al menos, no lo ha previsto expresamente, pues sólo se hace referencia a un informe o certificado de los servicios sociales.

nicado en los plazos y condiciones establecidos en el artículo 9.3 de la LAU, la necesidad de ocupar la vivienda arrendada para destinarla a vivienda permanente para sí o sus familiares en primer grado de consanguinidad o por adopción o para su cónyuge en los supuestos de sentencia firme de separación, divorcio o nulidad matrimonial.

Cabe recordar que estas dos prórrogas extraordinarias sólo podrán aplicarse a los nuevos contratos de alquiler que se hayan formalizado a partir de la entrada en vigor de la LDV, ya que los contratos vigentes con anterioridad siguen vinculados al régimen legal en el momento de su celebración.

Otra cuestión interesante aquí es qué ocurre en los casos que pueden considerarse como novación del contrato, o en supuestos de tácita reconducción, ya que estas cuestiones no se abordan en la LDV.

En la Ley, como decimos, no se regula qué se entiende por nuevo contrato de arrendamiento, lo que podrá dar lugar a diferentes planteamientos sobre qué sucede cuando el contrato se encuentra en tácita reconducción y sobre lo que se entiende por novación contractual. En este sentido, se puede plantear que los contratos de arrendamiento de vivienda que se renueven por tácita reconducción (arts. 1566 y 1581 CC) a partir de la entrada en vigor de la LDV, deban quedar sometidos a sus normas, entendiendo que desde ese momento nace un nuevo contrato, siguiendo la jurisprudencia del Tribunal Supremo[7]. Aunque esta solución no parece del todo satisfactoria. Téngase

---

7 Así lo indica Molina Roig (2023, p. 8), que señala que, con esta situación de tácita reconducción, "de acuerdo a la reiterada jurisprudencia del Tribunal Supremo (SSTS de 26 de septiembre de 2018 (CENDOJ 28079110012018100527), 14 de junio de 1984, de 21 de febrero de 1985, de 9 de abril de 1985, y de 15 de octubre de 1996.) nace un nuevo contrato".

en cuenta que la tácita reconducción supone, en el frecuente caso de pago mensual de la renta, que el contrato tenga una duración de 1 mes. Y si aplicamos la nueva ley, tendría derecho el arrendatario a la nueva prórroga extraordinaria, que, de aplicarse, haría que el contrato pasara a tener una duración de 1 año o incluso 3. Parece más correcto plantear que la LDV está pensando en contratos de arrendamiento de duración anual (5 o 7 años de prórroga obligatoria y hasta 3 de prórroga legal tácita).

Lo mismo ocurriría cuando se prorroga el contrato existente por un pacto entre las partes que comporta un considerable incremento de renta, al entender que en estos casos existe *animus novandi* debido a un cambio sustancial del objeto del contrato, que sería incompatible con la obligación inicial, lo que da lugar a la extinción del contrato anterior y al nacimiento de un nuevo contrato, al que debe aplicarse la nueva normativa[8].

1. Incorporación del concepto de gran tenedor

Conviene ahora hacer una breve mención a un concepto que se añade en la LDV y que tiene relevancia en la LAU. Vemos en este artículo 10 que acabamos de analizar el concepto de

---

8 A ello hace referencia Molina Roig (2023, p. 8), que entiende que sería conveniente que se incluyera esta precisión en la norma, para aclarar la aplicación de la misma y para evitar conflictos entre las partes, "como realizaba la Disposición transitoria de la Ley 11/2020 en Cataluña, al establecer que «en caso de novación del contrato con posterioridad a la declaración del área como área con mercado de vivienda tenso, siempre que suponga una ampliación de la duración del contrato o una modificación de la renta, se aplicará lo establecido en la presente ley», haciendo además, mención expresa a los efectos de la tácita reconducción".

"gran tenedor". Hay que tener en cuenta que nos encontramos ante un concepto que se incorpora por primera vez a la LAU.

Una de las particularidades de la legislación española referente a las nuevas medidas introducidas, como las obligaciones de los arrendadores en el ámbito de la contención de rentas y los procesos de desahucio, es la definición de gran tenedor, un concepto que no está presente en ninguna de las legislaciones de los países que nos rodean[9].

La definición de carácter general del gran tenedor, como señala la citada autora, no es una novedad de la LDV, ya que anteriormente había sido recogida por el legislador autonómico en el marco de la crisis derivada de las ejecuciones hipotecarias por parte de las entidades bancarias, y también por el propio legislador estatal en la normativa proteccionista relacionada con la pandemia sanitaria, en el ya citado Real Decreto-ley 11/2020, de 31 de marzo, por el que se adoptan medidas urgentes complementarias en el ámbito social y económico para hacer frente al COVID-19[10].

Así, es gran tenedor la persona física o jurídica titular de más de diez inmuebles urbanos de uso residencial, excluyendo garajes y trasteros, o una superficie construida de más de 1.500 m2, con la posibilidad de particularizar ese número (cinco o

---

9 Así lo señala Gifreu Font (2023, p. 71).

10 Se indica además que esta condición de gran tenedor es utilizada para imponer unos parámetros diferentes, según sea el arrendador, "en aras de evitar el crecimiento de los precios de alquiler o de mejorar las condiciones del lanzamiento del ocupante de la vivienda, como el establecimiento de un procedimiento de conciliación o intermediación (DF quinta), la imposición de un deber de colaboración y suministro de información en ZMRT (art. 19) o la limitación de rentas en ZMRT (DF primera, apartado tres)" (Gifreu Font, 2023, p. 72).

más inmuebles urbanos de uso residencial) cuando estén ubicados en zonas tensionadas (art. 3.k de la LDV). Esta definición de gran tenedor, al igual que el resto de las definiciones recogidas en la LDV, se establecen a los efectos de lo dispuesto en la misma, respetando por supuesto las definiciones adoptadas en el marco de las legislaciones de las comunidades autónomas sobre vivienda.

Se critica en este punto que nos encontramos en un país donde la mayoría de los casos se trata de pequeños propietarios, siendo evidente que los grandes tenedores están en clara minoría. Se señala que, hasta 2021, los grandes tenedores son propietarios de 98.639 viviendas, que representan un 4,4% del total de viviendas arrendadas sobre un total de más de 2,4 millones. Por lo que el 95,6% de las viviendas arrendadas son propiedad de particulares[11].

Por este motivo, se alega que asignarles una posición dominante en el mercado de la vivienda o considerar que tienen la capacidad económica suficiente para reemplazar al Estado en su función asistencial es claramente un error. Pero ello no impide que legítimamente se les pueda exigir un mayor compromiso social que a los pequeños propietarios[12].

Por otro lado, hay autores que critican la consideración cuantitativa de gran tenedor, o la precisión del concepto, indicando además que para evitar situaciones desiguales, las medidas relativas a los grandes tenedores deberían referirse a grandes tenedores que sean una persona jurídica y cuya actividad empresarial se refiera al sector inmobiliario, es decir, entidades financieras, filiales inmobiliarias y entidades de gestión de ac-

---

11 Argelich Comelles (2023, p. 3).

12 Gifreu Font (2023, p. 72).

tivos inmobiliarios, en el mismo sentido que se ha establecido en la legislación autonómica[13].

Tampoco queda claro a si se refiere al derecho de propiedad plena o también afectará a los titulares de otros derechos. En relación con esto, en los “Criterios orientadores aprobados por los Jueces de Primera Instancia de Barcelona, con motivo de la entrada en vigor de la Ley 12/2023, de 24 de mayo, por el Derecho a la Vivienda” se ha manifestado que computarán no solo las situaciones de titularidad plena, sino también computarán las titularidades de derechos reales que conlleven por sí solas la facultad de uso y disfrute del inmueble, o la posibilidad de hacer actos de administración ordinaria sobre el mismo, como es el caso del usufructo[14].

---

13 Argelich Comelles (2023, p. 4). Señala la autora que la consideración cuantitativa de gran tenedor necesita replantearse cualitativamente: “Es desacertada la rebaja de la consideración de gran tenedor de vivienda de diez a cinco viviendas, si las Comunidades Autónomas acreditan zonas de mercado residencial tensionado, tras la tramitación parlamentaria de la norma, pues no cabe convertir ex lege en gran tenedor a quien no lo es”. Igualmente crítica con el concepto es Flores González (2024), que señala que aquí entra en juego la discrecionalidad de las administraciones autonómicas, lo que supone que se podrá llegar a soluciones diversas en distintas partes del territorio español, con la inseguridad jurídica que esto supone.

14 Así lo indica Flores González (2024, p. 240), que señala además que, siguiendo estos criterios, “en el caso de copropiedad o cotitularidad de derechos reales los jueces de Barcelona proponen que se incluyan la titularidad de cuotas superiores al 50%; en el caso de titularidades compartidas, para alcanzar el cómputo de los 1.500 metros cuadrados de uso residencial, se deberán tener en cuenta las partes alícuotas, incluyendo todas las fincas sobre las que tenga alguna cuota de participación. La razón de esta interpretación descansa en el argumento de que la Ley emplea el término de titularidad, que no es sinónimo de plena titularidad del inmueble. Por último, se precisa que para hacer el cómputo general de titularidad de «*más de diez inmuebles*» se tendrán en cuenta todos los ubicados en España”.

2. Zona de mercado residencial tensionado

Otro elemento nuevo en la LAU es el concepto de “mercado residencial tensionado”, que define la LDV en su art. 18. Sin entrar en detalle, sólo haremos una breve referencia a este concepto, ya que va a tener relevancia de cara a los arrendamientos de vivienda.

Dicho artículo de la Ley regula la posibilidad para las Comunidades Autónomas de realizar la declaración de zona de mercado tensionado, ya que son las que ostentan la competencia exclusiva en materia de vivienda según el artículo 148.1.13 CE.

La declaración de zona de mercado residencial tensionado podrá realizarse en aquellos ámbitos territoriales en los que exista un especial riesgo de oferta insuficiente de vivienda para la población, que conlleve que el acceso a la vivienda sea inasequible.

En dicho artículo encontramos una ampliación, respecto de su redacción inicial, de los criterios para la declaración de zona de mercado residencial tensionado por parte de las Comunidades Autónomas: procederá dicha declaración si la carga media del coste mensual del arrendamiento o préstamo hipotecario, más los gastos y suministros básicos, superan el 30% de la renta media de la unidad familiar, o si el precio de la compraventa o arrendamiento de la vivienda se ha incrementado en tres puntos por encima del IPC en los cinco años anteriores.

Cabe señalar que inicialmente el Proyecto de Ley requería que se cumplieran ambos requisitos y que el crecimiento acumulado del IPC fuera de al menos cinco puntos, por lo que se han flexibilizados los requisitos[15].

---

15 Señala Molina Roig (2023, p. 5) que esta modificación era necesaria para que la medida fuera proporcional con la realidad que ha experimentado el mercado y el incremento del IPC en el último

Por otro lado, se ha añadido un deber de información específico en caso de viviendas que se encuentren en una zona de mercado residencial tensionado. En estos casos, se debe informar en las ofertas y en el contrato de arrendamiento sobre la cuantía de la última renta del contrato que hubiese estado vigente en los últimos cinco años en la misma vivienda, así como del valor que le pueda corresponder atendiendo al índice de referencia de precios de alquiler de viviendas.

Por último, conviene resaltar que actualmente, Cataluña ha sido la única comunidad autónoma en solicitar la declaración de zonas de mercado tensionado, que se ha concretado en la Resolución de 14 de marzo de 2024, de la Secretaría de Estado de Vivienda y Agenda Urbana, por la que se publica la relación de zonas de mercado residencial tensionado[16]. Además, la Generalitat de Cataluña ha comenzado los trámites para ampliar de 140 a 271 el número de municipios catalanes dentro de la zona de mercado tensionado. Como consecuencia, se señala que el control de precios en estas zonas ha hecho caer la oferta de viviendas en alquiler un 13%, y, además, en el conjunto de Cataluña, el precio se incrementó un 4,3%[17].

---

año, "pues de otra manera quedarían fuera casi todos los municipios fuertemente tensionados por la oferta insuficiente de vivienda en alquiler y las elevadas rentas de los existentes".

16 Resolución de 14 de marzo de 2024, de la Secretaría de Estado de Vivienda y Agenda Urbana, por la que se publica la relación de zonas de mercado residencial tensionado que han sido declaradas en virtud del procedimiento establecido en el artículo 18 de la Ley 12/2023, de 24 de mayo, por el derecho a la vivienda, en el primer trimestre de 2024. BOE núm. 66, de 15 de marzo de 2024.

17 Estudio realizado por idealista. Disponible en: https://www.idealista.com/news/inmobiliario/vivienda/2024/04/15/816612-el-stock-de-alquileres-permanentes-en-cataluna-cae-un-13-tras-la-entrada-en-vigor (última consulta, 9/05/2024).

III. LIMITACIÓN DE LA RENTA

1. Nuevas rentas

Posiblemente una de las medidas más polémicas introducidas por la LDV sea la limitación de la renta del alquiler[18]. Esto se encuentra en dos nuevos apartados, los números 6 y 7 del artículo 17, y que van a afectar a aquellas viviendas que se encuentren en zonas de mercado residencial tensionado, con una especial mención en caso de que el arrendador tenga la condición de gran tenedor.

Pues bien, podemos comenzar señalando que, aunque en el art. 17.1 de la Ley de Arrendamientos Urbanos se consagra la libertad de pactos cuando se dispone que la renta será la que libremente fijen las partes, la nueva ley, como decimos, viene a limitar

---

[18] Como indica Martín (2023, p. 1), el legislador ha llevado a cabo una serie de polémicas modificaciones tanto en la Ley 1/2000 de Enjuiciamiento Civil como en la Ley 29/1994 de Arrendamientos Urbanos. Siendo una de las medidas más controvertidas la limitación añadida a la LAU en relación con la facultad de las partes para establecer libremente el precio de la renta en los nuevos contratos de arrendamiento que se celebren sobre viviendas dentro de las zonas calificadas de "mercado tensionado". Señala el citado autor que dicha medida ha provocado gran revuelo entre los arrendadores al considerar que supone una verdadera injerencia por parte del Estado en la libertad de pacto que preside nuestro Código Civil. Además, el Tribunal Constitucional ha admitido a trámite los recursos de inconstitucionalidad contra esta Ley que han sido promovidos por el Grupo Parlamentario Popular en el Congreso de los Diputados, el Parlamento de Cataluña y por los Consejos de Gobiernos de las CCAA de Andalucía, Islas Baleares y Madrid, considerando estos últimos que esta norma invade sus ámbitos competenciales en materia de ordenación del territorio, urbanismo y vivienda.

esta libertad en los apartados 6 y 7 del artículo, señalando restricciones a la autonomía de la voluntad en los siguientes supuestos:

1.º Se modifica el art. 17.6 LAU: Cuando el arrendamiento de vivienda se ubique en una zona de mercado de residencial tensionado, la renta pactada al inicio del nuevo contrato no podrá exceder de la última renta de contrato de arrendamiento de vivienda habitual que hubiese estado vigente en los últimos cinco años en un contrato anterior en la misma vivienda. Y ello, una vez aplicada la cláusula de actualización anual de la renta del contrato anterior.

En estos casos, se puede plantear una duda. ¿Qué ocurre en los casos donde la renta no se ha actualizado durante varios años, aun constando el método de actualización en el contrato? Esto puede ocurrir por cualquier motivo: por ejemplo, es posible que, debido a situaciones de crisis nacional, como durante el COVID, o atendiendo a situaciones personales del arrendatario, el arrendador haya decidido no actualizar la renta. En estos casos, no está claro si la nueva norma permitiría al arrendador aplicar la renta actualizada al nuevo contrato. Pues bien, en estas situaciones, además de por el tenor literal de la norma, que señala "una vez aplicada la cláusula de actualización anual de la renta del contrato anterior", entendemos que es posible actualizar la renta del contrato anterior de cara al nuevo contrato, aunque no se haya aplicado previamente, ya que, si no se hizo, fue debido a un trato de favor del arrendador que ahora, y por una nueva ley, no le puede afectar negativamente.

Por tanto, en las zonas de mercado residencial tensionado, la renta del último contrato debidamente actualizada se convierte en la renta del nuevo contrato. Y para evitar la elusión o repercusión de este límite por otros métodos, se establece que no se podrán fijar nuevas condiciones que establezcan la repercusión al arrendatario de cuotas o gastos que no estuviesen recogidos en el contrato anterior.

Establecida dicha limitación, se señala una excepción: sólo podrá excederse dicho límite hasta un máximo de un 10 por ciento sobre la última renta actualizada del anterior arrendamiento de vivienda en los siguientes supuestos:

a) Cuando la vivienda haya sido objeto de rehabilitación en los dos años anteriores en los términos previstos en el apartado 1 del artículo 41 del Reglamento del Impuesto sobre la Renta de las Personas Físicas[19].

b) Cuando se hayan realizado en la vivienda actuaciones de rehabilitación o mejora que supongan un ahorro de energía primaria no renovable del 30 por ciento.

c) Cuando en los dos años anteriores a la fecha de la celebración del nuevo contrato de arrendamiento se hubieran finalizado actuaciones de mejora de la accesibilidad, debidamente acreditadas.

---

19 El citado artículo menciona que tendrán la consideración de rehabilitación las obras en la vivienda "que cumplan cualquiera de los siguientes requisitos:
a) Que se trate de actuaciones subvencionadas en materia de rehabilitación de viviendas en los términos previstos en el Real Decreto 233/2013, de 5 de abril, por el que se regula el Plan Estatal de fomento del alquiler de viviendas, la rehabilitación edificatoria, y la regeneración y renovación urbanas, 2013-2016.
b) Que tengan por objeto principal la reconstrucción de la vivienda mediante la consolidación y el tratamiento de las estructuras, fachadas o cubiertas y otras análogas siempre que el coste global de las operaciones de rehabilitación exceda del 25 por ciento del precio de adquisición si se hubiese efectuado ésta durante los dos años inmediatamente anteriores al inicio de las obras de rehabilitación o, en otro caso, del valor de mercado que tuviera la vivienda en el momento de dicho inicio. A estos efectos, se descontará del precio de adquisición o del valor de mercado de la vivienda la parte proporcional correspondiente al suelo".

d) Cuando la duración del arrendamiento de vivienda sea igual o superior a diez años. O cuando se incluya en el contrato una prórroga potestativa para el arrendatario de diez o más años, respetando las mismas condiciones.

2.º Se modifica el art. 17.7 LAU: Cuando se dé la doble circunstancia de que el arrendador sea un gran tenedor de vivienda y el inmueble se sitúe en una zona de mercado residencial tensionado, la renta pactada al inicio del nuevo contrato no podrá exceder del límite máximo del precio aplicable conforme al "sistema de índices de precios de referencia", atendiendo a las condiciones y características de la vivienda arrendada y del edificio en que se ubique.

3.º Igualmente, se aplicará el sistema de índices de referencia para la renta cuando la vivienda se sitúe en zonas de mercado residencial tensionado y no hubiese estado vigente ningún contrato de arrendamiento de vivienda en los últimos cinco años. En este caso, es independiente que el arrendador tenga la condición o no de gran tenedor. Y precisamente esta equiparación puede ser criticable, ya que los límites impuestos al derecho de la propiedad y libertad de empresa no deberían equipararse para ambos, y más teniendo en cuenta que la gran mayoría de arrendadores son pequeños propietarios.

Aquí podemos mencionar que la LDV añade una Disposición transitoria séptima, referida a la aplicación de las medidas en zonas tensionadas. Dicha disposición establece que la regulación señalada en el art. 17.7 se aplicará a los contratos que se formalicen desde la entrada en vigor de la Ley 12/2023, de 24 de mayo, por el derecho a la vivienda, y una vez se encuentre aprobado el referido sistema de índices de precios de referencia. Además, se indica que la resolución del Departamento ministerial competente en materia de vivienda que apruebe el referido sistema de índices de precios de referencia se realizará por ámbitos territoriales, considerando las bases de datos, sistemas y metodologías desarrolladas por las distintas comunidades autónomas y asegurando en todo caso la coordinación técnica.

Pues bien, sobre este índice, podemos decir que el Ministerio de Vivienda y Agenda Urbana ya ha aprobado el Índice de Precios de Referencia, concretamente, se ha hecho en la Resolución de 14 de marzo de 2024, de la Secretaría de Estado de Vivienda y Agenda Urbana, por la que se determina el sistema de índices de precios de referencia a los efectos de lo establecido en el artículo 17.7 de la Ley 29/1994, de 24 de noviembre, de Arrendamientos Urbanos[20]. En dicha resolución, se señala en su art. 2 que la fijación del límite máximo de la renta de los nuevos contratos de arrendamiento de vivienda en los supuestos recogidos en el referido artículo 17.7 LAU, "se realizará de acuerdo con la aplicación en línea a la que se accede a través del portal de Internet del Ministerio de Vivienda y Agenda Urbana, tomando el valor superior del rango de valores individualizados que resultan de la localización y las diferentes características de la vivienda objeto de arrendamiento".

Este índice ofrece un rango de valores de precios de alquiler, atendiendo a la localización, superficie y demás características, para viviendas alquiladas situadas en edificios de tipología residencial colectiva de más de cinco años de antigüedad, y cuya superficie construida no sea inferior a 30 m2 ni superior

---

20 También se puede acceder al Sistema Estatal de Referencia del Precio del Alquiler de Vivienda, disponible en: https://www.mivau.gob.es/vivienda/alquila-bien-es-tu-derecho/serpavi Recuperado el 12 de abril de 2024. Por otro lado, hay que decir que el Gobierno catalán ha interpuesto un recurso contencioso-administrativo contra dicha resolución, al entender que no se incluye ninguna aplicación territorial ni tiene en cuenta los índices de referencia que ya pudieran existir en alguna comunidad autónoma ni ningún mecanismo de colaboración entre los diversos índices.

a 150 m2. Para estimar la renta se ha tomado el dato obtenido de las declaraciones de IRPF por parte de arrendadores de vivienda cuyo uso sea exclusivamente de vivienda habitual. Otra cuestión clave de dicho índice es si realmente se ajusta o no a la realidad actual[21].

No obstante, hay que señalar que la aplicación de estas medidas mencionadas no es inmediata ni generalizada, puesto que tiene como presupuesto necesario la declaración de zonas de mercado residencial tensionado, que depende de las comunidades autónomas. Como precedente, el legislador catalán fue el primero en España en regular el control de rentas, aunque dicho recorrido terminó debido a las SSTC 37/2022 (TOL8.889.749) y 57/2022 (TOL8.916.712), las cuales consideraron que la Ley catalana 11/2020 infringía una norma básica en materia de arrendamientos urbanos, afectando a la autonomía de la voluntad a la hora de fijar la renta[22].

---

21 Como declara la OCU, el problema de este índice es que "la última información disponible en 2024 proviene de datos de 2022, y se explotan datos desde el año 2015, con lo que existe un decalaje temporal importante con la realidad actual. Por lo tanto, los precios resultantes del índice tienen poco que ver con los precios que un aspirante a inquilino se encuentra cuando va al mercado a buscar una casa en alquiler". Disponible en: https://www.ocu.org/fincas-y-casas/gestion/alquiler/analisis-gratis/2024/02/indice-precios-alquiler-ministerio (última consulta, 11/05/2024).

22 Según el TC, las reglas para la determinación de la renta derivadas del contrato de arrendamiento de vivienda del art. 6 de la Ley 11/2020 vulneran el principio de libre determinación de la renta de alquiler establecido en el art. 17.1. LAU, en conexión con el principio de libertad de pactos que emana del art. 1255 del Código Civil. Este principio de libre estipulación de la renta en los arrendamientos urbanos constituye una base de las obligaciones contractuales, por cuanto define uno de los elementos estructurales del contrato. Como ha señalado el TC:

De hecho, este precedente tiene relevancia en la actual normativa, ya que el índice de referencia de la LDV tiene su influencia en el índice de precios que existe en Cataluña, y este nuevo índice afecta a la mayor parte de las medidas sobre limitación de la renta del alquiler establecidas por el legislador estatal, incluida la obligación de los arrendadores de incluir en sus ofertas el precio de referencia, de la que se hablará después. Pero también es interesante resaltar que incluía otras medidas que el legislador español no ha incluido, como el derecho de los arrendatarios de obtener la restitución de las cantidades abonadas en exceso e incrementadas con el interés legal de dinero más tres puntos.

Y todo esto es lo más controvertido respecto a las medidas aplicadas, puesto que limitan la autonomía de la voluntad en la fijación de la renta, afectando a la libertad de pactos entre las partes. Como señala Gifreu Font, la regulación de la propiedad no es pura, en términos jurídicos. Se trata de un ámbito regulatorio que combina normas de derecho público y de derecho privado, por lo que el legislador debe encontrar un equilibrio entre ambos: de una parte, la defensa del interés individual del titular del derecho y, de otra, la incorporación de finalidades

---

"El Estado resulta así ser el competente para fijar legalmente el principio de libertad de pactos en el establecimiento de la renta del contrato de alquiler de vivienda, así como para, en su caso, determinar sus eventuales modificaciones o modulaciones en atención al designio del legislador estatal respecto a la necesidad de ajustar o no el funcionamiento del mercado inmobiliario en atención a lo dispuesto en el art. 47 CE o en otros preceptos constitucionales. [...] La reserva al legislador estatal de la competencia para la regulación tanto de la regla general como de sus posibles excepciones o modulaciones excluye, por tanto, que cualquier legislador autonómico pueda condicionar o limitar la libertad de las partes para determinar la renta inicial del alquiler de viviendas" (FJ 4 de la STC 37/2022).

sociales que satisfagan intereses públicos a los que se vincula la propiedad privada. Entre dichos intereses, tenemos la garantía a los ciudadanos del disfrute de una vivienda digna, incidiendo en la delimitación del derecho de propiedad, y concretamente en la definición de su función social[23]. El problema aquí es que esta medida comporta un coste patrimonial que no es asumido por las Administraciones Públicas, sino que se traslada a los propietarios. Además, se plantea la posible inseguridad jurídica que pueda producir debido a las diferencias que pueda haber entre comunidades autónomas[24].

Igualmente se critica que la combinación de plazos tan extensos de duración del arrendamiento de vivienda, con regímenes de control de rentas, ocasiona una disminución de la oferta de vivienda en arrendamiento, como ya sucedió con la prórroga forzosa y la congelación de la renta en las Leyes de arrendamientos urbanos de 1946, 1956 y 1964[25].

---

23 Gifreu Font (2023, p. 62).

24 Así lo critica Flores González (2024, p. 241), señalando que estas limitaciones alteran "los cimientos del régimen jurídico de los arrendamientos urbanos. Pero lo único inquietante no es que se restrinja la libertad de las partes a la hora de acordar cual debe ser el «alquiler» a pagar; por añadidura estos límites no están fijados de forma clara y precisa por la Ley".

25 En este sentido, Argelich Comelles (2023, p. 7). Señala la autora que "El control de rentas previsto en la Ley por el derecho a la vivienda es un legal transplant de lo dispuesto en septiembre de 2018 en el § 556 BGB como Mietpreisbremse, para los nuevos contratos de arrendamiento de vivienda, tras su aplicación en Berlín, así como a nivel global en otras ciudades como Nueva York, San Francisco, París, Copenhague o Estocolmo, entre otros". Y manifiesta que este mecanismo ha fracasado donde se ha aplicado remitiéndose a unos estudios estadísticos comparados del Prof. Dr. Kholodilin.

Así, como efectos negativos del control de rentas, se suelen señalar los siguientes: la reducción de la oferta, lo que además conlleva el incremento del plazo de espera para los arrendatarios y aumento de la renta; la disminución de la movilidad residencial, habitualmente por razones laborales; el aumento del coste oculto o no declarado sobre la renta de la vivienda; y, finalmente, la falta de ejecución de las obras de conservación necesarias y el deterioro de las viviendas[26].

Por otro lado, también hay quien defiende estas medidas, alegando varias razones para justificar su pertinencia y su constitucionalidad[27]:

a) Es una medida que persigue un interés general, ya que tiene como fin destensar el mercado del alquiler en las zonas en las que se detecta un riesgo especial de falta de vivienda a precios asequibles para la población;

b) Se alega asimismo el carácter excepcional y limitado temporal de las zonas de mercado tensionado;

c) Cuenta con excepciones, considerando situaciones particularizadas, como una duración del contrato superior a diez años o la inversión realizada por el propietario en la vivienda, permitiendo un incremento de renta del 10% a las viviendas que han sido rehabilitadas;

d) Se alega objetividad en el procedimiento, con la previsión de unos trámites administrativos mediante los cuales se identifica, con datos objetivos, la situación del mercado inmobiliario de los municipios susceptibles de ser declarados zona de mercado residencial tensionado, lo que descarta la posible toma de decisiones discrecionales y arbitrarias;

---

26 Así lo recoge Argelich Comelles (2023, p. 8).

27 Gifreu Font (2023, p. 97).

e) Se trata de una medida que, aunque reduce la renta obtenida por el propietario, afectando al contenido de su derecho y pudiendo afectar a la economía de mercado, no le impide la consecución de un beneficio económico ni le priva de la libre disposición sobre la vivienda;

Por otro lado, cabe decir que, en el ámbito comparado, esta no es una medida novedosa, ya que son muchos los países del entorno europeo e internacional que han apostado por esta fórmula para intentar solventar los problemas de falta de oferta de vivienda en alquiler. Así, señala Gifreu Font que, siguiendo las estela de esas experiencias comparadas – sobre todo, Alemania[28] y Francia[29], entre otras (Escocia, Irlanda, Estados

---

28 En Alemania, Gifreu Font (2023, p. 65), destaca la Ley Mietpreisbremse de 2015, que permitía su aplicación facultativa en las ciudades tensionadas que lo solicitaran y cuya vigencia se ha ampliado hasta 2025, y respeta el nivel de rentas previo y establece que la renta de los nuevos arrendamientos solo puede incrementarse un 10% respecto a las aplicadas en viviendas equivalentes de la misma zona (precio comparativo medio de la zona), excepto que se produzcan mejoras en la vivienda. Y en 2020 se aprobó una Ley que fijaba un tope legal de los precios de los alquileres, así como la congelación de la actualización de esas rentas de acuerdo con los niveles de junio de 2019 (medidas popularmente conocidas con los términos «Mietendeckel» y «Mietenstopp»). No obstante, en marzo de 2021 se declaró inconstitucional esa Ley por invasión de competencias federales.

29 En Francia, se puede destacar la ley ALUR (Loi n.º 2014-366 pour l'accès au logement et un urbanisme rénové, de 24 de marzo de 2014, vigente hasta 2017), que permitió declarar como "tensionadas" determinadas "zonas de urbanización continua" y fijar un techo en los precios de los alquileres, que no podían superar el 20% de la media. Y en 2018 se aprobó la Ley ELAN (Loi n.º 2018-1021 portant évolution du logement, de l'aménagement et du numérique, de 23 de noviembre de 2018), que recupera la posibilidad (carácter facultativo) de techar las rentas del alquiler, mediante la fijación de un precio de referencia, teniendo en cuenta criterios como

Unidos, Austria, Suiza, Nueva Zelanda...)–, se introduce ahora un cambio en el patrón intervencionista que, yendo más allá de la congelación o minoración de los incrementos anuales de renta, gira ahora hacia un control directo del precio a través de un índice de referencia aplicable en determinados ámbitos territoriales previamente declarados.

Por último, es de resaltar que se han analizado estudios empíricos sobre el control de rentas, pero la conclusión, aunque no es muy positiva, tampoco parece definitiva.

Por un lado, se ha estudiado el control de rentas que tuvo lugar en Cataluña. Y la conclusión de dicho estudio es que el control introducido en 2020 dio lugar a un notable descenso en los precios, pero también a una moderada reducción de la oferta. Y, cuando se eliminó dicho control, se constató un gran incremento del precio de la renta. Por lo tanto, se concluyó que el efecto total del control de renta es un poco ambiguo y complejo, ya que se obtuvo una reducción de la renta, a cambio de una reducción de la oferta de viviendas, por lo que no es un sistema que sea totalmente recomendable para solucionar el problema de la vivienda[30].

Por otro lado, a nivel general, se ha hecho un análisis de diferentes estudios empíricos sobre el control de alquileres publicados en revistas especializadas entre 1972 y 2022. Y se indica que, aunque el control de alquileres parece ser eficaz para lograr su objetivo principal, esto es, que haya alquileres más bajos, está dando lugar a varios efectos no deseados. Por lo tanto, parece que a corto plazo se consigue el objetivo, pero es

---

la localización, el número de habitaciones o el año de construcción de la vivienda (Gifreu Font, 2023, p. 65).

30 Esto se recoge en Kholodilin/ López/ Rey Blanco/ González Arbués (2022, p. 40).

posible que surjan inconvenientes como los mencionados anteriormente. Por todo ello, se concluye que no está claro cuál es el impacto general de la política de control de alquileres en la sociedad. Además, se indica que el análisis se complica aún más por el hecho de que esta política no se adopta en el vacío, ya que simultáneamente con el control de alquileres, se efectúan otras políticas de vivienda, como la protección de los inquilinos contra el desalojo, las ayudas para la vivienda y la estimulación de la construcción de edificaciones de viviendas. Incluso otras políticas, como las políticas bancarias y fiscales también pueden modificar los resultados de las regulaciones de control de alquileres[31].

2. *Actualización de la renta*

Respecto a las medidas limitadoras, cabe destacar también las referentes a la actualización de la renta.

En esta materia, la Disposición final sexta de la LDV señala cuál es la "limitación extraordinaria de la actualización anual de la renta de los contratos de arrendamiento de vivienda". Y para ello, remite al Real Decreto-ley 6/2022, de 29 de marzo, por el que se adoptan medidas urgentes en el marco del Plan Nacional de respuesta a las consecuencias económicas y sociales de la guerra en Ucrania, que lo modifica y queda en los siguientes términos.

1.º La primera limitación está prevista para los contratos en los que corresponda la actualización anual de la renta

---

31 Todo esto se refleja en el estudio llevado a cabo por Kholodilin (2022). En este trabajo se analiza una amplia gama de estudios empíricos sobre el control de alquileres publicados en revistas especializadas entre 1972 y 2022.

entre la entrada en vigor del decreto (el 31 de marzo de 2022) y el 31 de diciembre de 2023. Aquí se distinguen dos supuestos: Si el arrendador es un gran tenedor, se regirá el incremento de la renta por el pacto entre las partes, sin que éste pueda superar la variación anual del Índice de Garantía de Competitividad (IGC), que como máximo será del 2%. Sin embargo, si el arrendador no es gran tenedor, primero habrá libertad total de pacto, sin límite, y en su defecto, habrá el límite del 2% citado. Por lo tanto, este límite únicamente no será operativo cuando el arrendador no sea un gran tenedor y haya un nuevo acuerdo entre las partes en relación con la actualización de la renta.

2.º Para los casos en que la renta deba ser actualizada dentro del período comprendido entre el 1 de enero de 2024 y el 31 de diciembre de 2024, el criterio es igual al anterior mencionado, pero el límite máximo cambia al 3%. Es decir, sea o no gran tenedor el arrendador, si no hay un nuevo pacto entre las partes, la revisión de la renta no podrá exceder del 3%. Por el contrario, si el arrendador no es gran tenedor, primará el pacto sobre el límite del 3%, que solo operará en ausencia de acuerdo.

Además, en el apartado cinco de la disposición final primera de la Ley por el Derecho a la Vivienda se añade una disposición adicional undécima a la LAU de 1994, redactada de la siguiente forma: el Instituto Nacional de Estadística creará, antes del 31 de diciembre de 2024, un índice de referencia para la actualización anual de los contratos de arrendamiento de vivienda, que se fijará como límite de referencia a los efectos del artículo 18 de la ley. Y el propósito es evitar incrementos desproporcionados en la renta de los contratos de arrendamiento. En caso de que a partir de diciembre de 2024 no se defina este índice, las actualizaciones de renta se harán de acuerdo con la Ley de Arrendamientos Urbanos y, por lo tanto, el índice que en ningún caso podrá superarse será el Índice de Precios al Consumo (art. 18.1 LAU).

El hecho de crear un nuevo índice que requiere de una posterior aprobación para que sea aplicable es cuestionado por la doctrina, que señala que, en caso de no aprobarse, se corre el riesgo de que se vuelva al IPC, lo que implicará un aumento desproporcionado para los contratos de arrendamiento. Para evitar esto, quizá hubiera sido mejor acudir al IGC, utilizado para fijar las actualizaciones de la renta cuando no se pacta un índice determinado[32].

IV. OTRAS NOVEDADES: MÉTODO DE PAGO, GASTOS Y DEBERES DE INFORMACIÓN

Para terminar, podemos señalar otras novedades que se han introducido y que afectarán también a los arrendamientos de vivienda.

Así, se modifica el apartado 3 del artículo 17 de la LAU, indicando ahora que el pago se efectuará a través de medios electrónicos. Y sólo excepcionalmente, cuando alguna de las

---

32 Así lo manifiesta Molina Roig (2023, p. 10). Esta autora indica además que "El IGC establece una tasa de revisión de precios consistente con la recuperación de competitividad frente a la zona euro. Esa tasa será igual a la del Índice de Precios al Consumo Armonizado (IPCA) de la UEM menos una parte de la pérdida de competitividad acumulada por España desde 1999. Cuando la tasa de variación de este índice se sitúe por debajo de 0% por ciento, se tomará este valor como referencia, lo que equivaldría a la aplicación de la regla de no revisión. Cuando la tasa de variación de este índice supere el 2%, se tomará este valor como referencia. De esta forma, se asegura que los contratos a los que se aplique este nuevo índice contribuyan a garantizar el mantenimiento de la competitividad de la economía en el medio plazo".

partes no tenga cuenta bancaria o acceso a medios electrónicos de pago, podrá solicitar que el pago sea en metálico y en la vivienda arrendada.

La novedad aquí radica en que el método predeterminado pasa a ser el pago por medios electrónicos. Anteriormente, en defecto de pacto, el pago debía hacerse en metálico. De esta forma, el propósito de la norma es evitar el fraude fiscal. En este sentido, cabe mencionar el problema que tenemos actualmente en este ámbito, ya que, de acuerdo con los datos de Gestha, el 40,8% de los alquileres no se declaran[33].

Respecto a esta modificación, se resalta que esta medida resulta más bien como una declaración de intenciones, ya que, pese a que se indique que es obligatorio pagar por medios electrónicos, basta con que el arrendatario alegue que no tiene acceso a medios electrónicos de pago para que pueda pagar al contado en el propio hogar en que habita, para lo cual debería desplazarse el arrendador a cobrar físicamente al inmueble arrendado[34].

Por lo que respecta a los gastos, se ha modificado el apartado 1 del artículo 20 LAU. La mayor parte de la redacción sigue en los mismos términos, pero la novedad reside en que ahora se señala que los gastos de gestión inmobiliaria y los de formalización del contrato serán a cargo del "arrendador". Es decir, ahora es irrelevante si el arrendador es persona física o jurídica para tener que soportar estos gastos, ya que siempre estará obligado al pago de los gastos de gestión inmobiliaria y de formalización del contrato. Recordemos que, en la redac-

---

[33] GESTHA, Informe sobre el alquiler sumergido según los datos IRPF 2018, 2021.

[34] Magro Servet (2023, p. 22).

ción anterior del precepto, solamente estaba obligado a ello el arrendador que era persona jurídica.

Cabe mencionar que esta disposición fue añadida durante la tramitación parlamentaria, ya que esta modificación no se contemplaba en el texto inicial que aprobó el Consejo de ministros.

Por otro lado, no se precisa qué se entiende por gastos de formalización, ni está claro si este artículo se está refiriendo a la formalización en documento privado o en escritura pública. Recordemos que el art. 1553 CC. señala que los gastos de la escritura serán de cuenta del arrendatario. A este respecto, podemos entender que todos los gastos de formalización, independientemente de su naturaleza pública o privada, serán de cuenta del arrendador en estos casos, ya que la nueva redacción de la LAU no distingue entre supuestos, y se trata de una norma con aplicación especial para esta materia.

Respecto a esta medida, aunque se pueda criticar que se atribuyan los gastos siempre al arrendador, creemos que es una medida razonable, ya que, especialmente con los servicios inmobiliarios, es el arrendador quien los contrata para que ofrezcan su vivienda al mercado, por lo que debe ser asumido por ellos.

Sin embargo, se puede plantear la duda de qué sucede en los casos de las renovaciones de contratos vigentes ya que esta cuestión no se especifica en la norma. En estos casos, respecto a los gastos entendemos que debe aplicarse el mismo criterio y que deben ser abonados por el arrendador, teniendo en cuenta la finalidad de la norma. Se señala que, si se ha querido proteger al arrendatario, excluyéndole del pago de la formalización y gastos de gestión inmobiliaria, cuando es posible que haya alguna prestación de un servicio de búsqueda de vivienda y formalización, más debería protegérsele en las renovaciones en las que no existe dicha prestación de servicio y el trámite es más sencillo y menos laborioso para el arrendador[35].

En cuanto a los deberes de información, y sin profundizar en este capítulo en ello puesto que hay otro dedicado a esta materia, sí vamos a hacer una breve referencia a aquellos deberes de información que tienen relevancia de cara al arredramiento de vivienda.

Así, en primer lugar, podemos decir que se amplían las obligaciones de información en los arrendamientos de vivienda: El artículo 31 de la LDV establece que, en la oferta de compra o arrendamiento de vivienda, la persona interesada podrá requerir, antes de la formalización de la operación y de la entrega de cualquier cantidad a cuenta, las condiciones económicas de la operación, entre otra información relevante.

Por otro lado, como ya se mencionó anteriormente, en caso de viviendas que se encuentren en una zona de mercado residencial tensionado, se debe informar en las ofertas y en el contrato de arrendamiento sobre la cuantía de la última renta del contrato que hubiese estado vigente en los últimos cinco años en la misma vivienda, así como del valor que le pueda corresponder atendiendo al índice de referencia de precios de alquiler de viviendas.

Además, se establece el respeto a la normativa autonómica, debido a que hay Comunidades que ya regulan obligaciones de información, como es el caso de Cataluña (arts. 60 y ss. Ley 18/2007, de 28 de diciembre, del derecho a la vivienda).

Cabe tener en cuenta que la obligación de información se regula como una potestad del interesado y no como una obligación del oferente, de forma que si no se acredita que se ha

---

35 Así lo indica igualmente, Molina Roig (2023, p. 12).

pedido no existe incumplimiento de la obligación. Por otro lado, es de destacar que no se establece un régimen sancionador por el incumplimiento de estas obligaciones, lo que dificulta su aplicación efectiva.

V. CONCLUSIONES

1. Podemos concluir que el legislador ha realizado una serie de reformas a la LAU que resultan insuficientes y cuya eficacia parece discutible. En el preámbulo de la LDV se reconoce la necesidad de una reorientación de las políticas para reforzar el alquiler, ya que tradicionalmente se ha apostado por la propiedad como forma esencial de acceso a la vivienda[36]. Sin embargo, no se producen modificaciones con suficiente entidad para provocar un cambio de modelo.
2. El momento era oportuno para hacer una reforma del sistema de arrendamientos, pero finalmente el legislador ha sido poco ambicioso. En la actualidad, existe una preocupación social evidente por las dificultades que conlleva el acceso a la vivienda. Por ello, se precisaba la intervención legislativa, de ámbito estatal, para mejorar la situación actual y promover un marco legislativo estable, que facilite el acceso y disfrute a la vivienda en alquiler. Pero en vez de optar por una reforma profunda del sistema de arrendamientos de vivienda, se ha preferido hacer unos meros retoques a la actual ley, que además no beneficiarán a la generalidad de los arrendatarios.

---

36 Además, el propio preámbulo reconoce la necesidad de un aumento de la vivienda pública en alquiler. Necesidad que también se reclama en el informe de la OCU (2023). "Estudio sobre el consumidor vulnerable y barreras de acceso a la vivienda en alquiler".

Por otro lado, la técnica legislativa también parece mejorable, con numerosos conceptos indeterminados o una excesiva remisión al ámbito autonómico. Igualmente, la efectividad de la norma plantea ciertas incertidumbres. Más allá de las cuestiones o controversias competenciales, la Ley introduce definiciones como la de zona de mercado residencial tensionado, gran tenedor, o vulnerabilidad económica y social, cuya aplicación va a depender de las administraciones autonómicas, lo que puede conllevar desigualdades, y, además, muchas de ellas no se han mostrado muy favorables a su aplicación[37]. Esta falta de efectividad inmediata y la dependencia de las comunidades autónomas serán factores relevantes que pueden debilitar la efectividad de las nuevas medidas previstas.

Asimismo, cabe destacar que la Ley no regula un régimen de inspección y sanción para el incumplimiento de las obligaciones que establece. Por lo tanto, deberán ser las Comunidades Autónomas las que lo prevean dentro del ámbito de sus competencias.

Por otro lado, conviene advertir que nos encontramos ante una serie de medidas que han sido aprobadas sin un apoyo general de los principales partidos políticos, por lo que es probable que con un cambio de Gobierno se reforme esta ley y aparezca otro régimen jurídico más que añadir a los regímenes jurídicos ya vigentes en materia de arrendamientos urbanos, lo que redunda en inseguridad jurídica[38]. Por ello, se reclama un acuerdo general donde encontrar puntos comunes, para

---

37 De hecho, a día de hoy, mayo de 2024, Cataluña ha sido la única comunidad autónoma en solicitar la declaración de zonas de mercado tensionado, como ya se apuntó anteriormente.

38 Así lo advierte Flores González (2024, p. 233).

poder así solucionar, en un marco de seguridad jurídica, el problema del acceso a la vivienda[39].

3. No afectará a una gran parte de los arrendamientos. El legislador se ha contentado, para mejorar el arredramiento de vivienda, con adoptar medidas cuya eficacia se discute, de alcance temporal y subjetivo limitado, y que en ningún caso podrá beneficiar a la generalidad de los arrendatarios. Se limitan a incluir prórrogas extraordinarias cuando se den determinados requisitos, y unas limitaciones a la renta, aplicables en las zonas de mercado tensionado. Son medidas insuficientes para promover un cambio de modelo en el que el arrendamiento tenga mayor peso.

4. Nos encontramos con medidas claramente intervencionistas y cortoplacistas, que puede tener un efecto contrario al deseado: la reducción de la oferta y el aumento de la renta. Se crean nuevas obligaciones y cargas para los propietarios de viviendas, especialmente para una categoría nueva de propietarios, los grandes tenedores de vivienda. Se trata de medidas consistentes en trasladar las responsabilidades de la Administración Pública al propietario de la vivienda, lo que puede provocar un efecto no deseado, como es que estas viviendas se pongan a la venta o se desplacen hacia otras fórmulas más rentables y con mayor seguridad jurídica, como el alquiler de temporada, el alquiler de habitaciones, o para viviendas con fines turísticos.

---

[39] Además, por seguridad jurídica y transparencia, Moll de Alba Lacuve (2024, p. 406) expresa que debería existir un texto único que contuviera toda la normativa de arrendamientos de viviendas.

Además, se está poniendo el punto de mira en el alquiler de temporada y al alquiler de habitaciones[40]. Pero no parece que este sea el camino correcto, ya que más intervención pública conllevará menor oferta de alquiler residencial.

5. Las medidas que se consideran excepcionales y temporales no están correctamente delimitadas en el tiempo. Señala el legislador que la Administración "compra tiempo" para poder actuar y corregir con otras políticas de vivienda las carencias de las zonas declaradas de mercado residencial tensionado. Sin embargo, esto que parece algo temporal es posible que acabe alargándose en el tiempo, ya que la situación de la vivienda no parece que se pueda corregir en un corto plazo. Por ello, es probable que, al vencer el plazo de tres años de vigencia de las zonas de mercado tensionado, se renueven repetidamente, perpetuando así unas medidas con una pretendida intencionalidad temporal.

De entre las nuevas medidas, la limitación de la autonomía de la voluntad en el ámbito contractual, mediante la contención de rentas, es la más discutible, por no respetar la libertad de precios. Nos sumamos en este punto a la opinión de quienes, como Martens Jiménez, defienden que hubiese sido preferible, en vez de imponer dicho control, dar la opción de que sea una medida voluntaria. Así, en vez de imponer una limitación de la renta, se puede ofrecer al arrendador la posibilidad de optar por una renta libre, o acogerse a una renta referenciada en función de unos parámetros, a cambio de recibir una bonificación fiscal. Con ello, se incita a seguir unas

---

[40] De hecho, recientemente el Gobierno de Cataluña ha aprobado el Decreto ley 6/2024, de 24 de abril, de medidas urgentes en materia de vivienda, para aplicar las medidas de contención de rentas al alquiler de temporada y de habitaciones.

rentas acordes a los índices que se planteen, pero respetando la libertad de precios[41].

Asimismo, desde el punto de vista del arrendador, además del control de rentas, se enfrenta a otros problemas que no tienen una solución adecuada, como son los riesgos que asumen los propietarios frente al impago por el arrendatario, y los problemas que supone el desahucio. Esto además provoca una reducción de la oferta, y que los propietarios que accedan a ofrecer sus viviendas en arrendamiento eleven sus precios, exigiendo además una serie de requisitos al arrendatario que en muchas ocasiones es excesiva. Todo ello da lugar a una dificultad de acceso por parte de los arrendatarios. Por ello, una vía para aumentar la oferta de vivienda es aportar seguridad a los arrendadores particulares[42].

6. Se pierde la oportunidad de un Registro de contratos de alquiler. La Disposición Adicional Primera anteriormente establecía la obligación de crear un Registro de contratos de alquiler a nivel estatal, pero durante la tramitación parlamentaria de la LDV, se sustituyó la de dicho Registro por una simple base de datos. Con ello, compartimos la idea de autores como Molina Roig o Martens Jiménez, quienes entienden que se pierde la oportunidad de disponer de un organismo que, además de dar publicidad y tener una función estadística, se le podría atribuir competencias de control del cumplimiento de requisitos de los contratos de alquiler o de las malas prácticas arrendaticias[43].

---

41 Esta opción se defiende por Martens Jiménez (2023, p. 309).

42 Así lo entiende también Sáenz de Jubera Higuero (2024, p. 376). Considera que las medidas deberían centrarse en incentivar al propietario, ofreciéndoles además mayor seguridad jurídica.

43 Esta idea se puede comprobar en Molina Roig (2023, p. 15), así como en Martens Jiménez (2023, p. 307).

7. El planteamiento del legislador debió ser otro. El legislador en la LDV parece ocuparse exclusivamente de las dificultades económicas de los arrendatarios, y de la consideración de los grandes tenedores, no conllevando así ningún impacto global, que afecte a la mayoría de la población.

Debe tenerse en cuenta elementos clave como es la duración del alquiler. Y en este punto, compartimos la propuesta de Martens Jiménez, que indica que uno de los grandes inconvenientes que afrontan las personas que buscan acceder a una vivienda en alquiler es la inestabilidad en el tiempo: es común que una persona o familia haga de su vivienda un hogar, pero por el transcurso de unos pocos años se vea abocada a trasladar su vida a otro lugar, lo que conlleva el percibir el arrendamiento como solución transitoria, no estable, para cubrir una necesidad básica. Por ello, se puede ir más allá y plantear incluso una reforma profunda, procediendo incluso a repensar la tipología actual, e incorporando medidas novedosas en nuestra regulación como el arrendamiento indefinido, siguiendo otros modelos de países donde funciona[44].

Por todo lo expuesto, y teniendo en cuenta que la situación de la vivienda no es un tema sencillo, sino complejo donde

---

44 En este sentido, podemos remitirnos a la interesante propuesta de regulación de los arrendamientos de vivienda que propone Martens Jiménez. Señala la autora que "en países donde el arrendamiento es una verdadera alternativa a la propiedad, se puede apreciar la tendencia hacia los arrendamientos de vivienda indefinidos. Principalmente, Alemania, que solamente en casos tasados admite los arrendamientos de duración determinada, Suiza, que permite que los contratos de arrendamiento puedan ser de duración determinada o indefinida, o Austria, que establece una normativa proteccionista para los contratos de arrendamiento indefinido" (Martens Jiménez, 2023, p. 249).

influyen diferentes factores económicos, sociales y culturales, parece claro que unas reformas puntuales e intervencionistas como las de la LDV no serán suficiente para mejorar la situación actual de la vivienda en alquiler. Puede ser necesario un nuevo planteamiento del alquiler de vivienda, buscando un nuevo sistema acorde a la realidad actual.

VI. REFERENCIAS BIBLIOGRÁFICAS

Argelich Comelles, C. (2023). "Derecho sin utopías sobre la propiedad privada y legal transplants adecuados: las medidas y carencias de la Ley 12/2023, de 24 de mayo, por el derecho a la vivienda". *Aranzadi digital* núm. 1/2023 parte Estudios y comentarios

Flores González, B. (2024). "Alquilar una casa con la Ley por el Derecho a la Vivienda. (Otro régimen jurídico más en materia de arrendamientos urbanos)". *LA LEY Derecho de familia,* n.º 41, Sección A Fondo, Primer trimestre de 2024, LA LEY 3821/2024

Gifreu Font (2023). "Intervención pública en el mercado libre de vivienda en alquiler en municipios con áreas tensionadas. Un análisis desde la fallida experiencia catalana. *Revista española de Derecho Administrativo,* núm. 230/2023 parte Estudios. Pamplona

Kholodilin, K. (2022). "Rent Control Effects through the Lens of Empirical Research", *Deutsches Institut für Wirtschaftsforschung,* 139, disponible en: https://www.diw.de/documents/publikationen/73/diw_01.c.833177.de/diw_roundup_139_en.pdf. Recuperado el 12 de abril de 2024.

Kholodilin, K./ López, F./ Rey Blanco, D./ González Arbúes, P. (2022). "Lessons from an Aborted Second-Generation Rent Control in Catalonia". *DIW Berlin Discussion Paper,* No. 2008, disponible en SSRN: https://ssrn.com/abstract=4159469 or http://dx.doi.org/10.2139/ssrn.4159469 Recuperado el 12 de abril de 2024.

Magro Servet, V. (2023). "Análisis práctico y sistemático de los aspectos relevantes de la nueva Ley de vivienda 12/2023 de 24 de mayo". *Diario LA LEY,* n.º 10300, LA LEY 4622/2023

Martens Jiménez, I. L. (2023). *Los arrendamientos de vivienda. Tipos, problemas y propuestas de modernización.* Atelier.

Martín, R., (2023). "La Cara y la Cruz: Pros y contras de la Ley 12/2023", por el Derecho a la Vivienda". *Actualidad Jurídica Aranzadi*, núm. 1000/2023 parte Cara.

Miguel Hernández, L. (2023). "La Cara y la Cruz: Pros y contras de la Ley 12/2023, por el Derecho a la Vivienda". *Actualidad Jurídica Aranzadi* núm. 1000/2023, parte Cruz.

Molina Roig, E. (2023). "Principales novedades sobre arrendamientos urbanos en la Ley 12/2023 por el derecho a la vivienda". *Actualidad Civil*, n.º 5, Sección Derecho Inmobiliario. LA LEY 4424/2023

Moll de Alba Lacuve, C. (2024). "La ley 12/2023 por el derecho a la vivienda y los contratos de arrendamientos". *El derecho a la vivienda en tiempos de incertidumbre* (dirs. Mate Satué, L. C./ Hernández Sáinz, E./ Alonso Pérez, M.ª T.). Aranzadi.

Nasarre Aznar, S. (2022). "El Proyecto de Ley de vivienda, Informes y papeles del Grupo de Trabajo Mixto Covid-19". *Apuntes Fedea* 2022/11.

OCU (2023). "Estudio sobre el consumidor vulnerable y barreras de acceso a la vivienda en alquiler". Informe OCU.

Sáenz de Jubera Higuero, B. (2024). "Intervención, control y limitación del precio de los arrendamientos. ¿Solución eficaz ante el problema de alquiler residencial en España?", en *El derecho a la vivienda en tiempos de incertidumbre* (dirs. Mate Satué, L. C./ Hernández Sáinz, E./ Alonso Pérez, M.ª T.). Aranzadi.

Capítulo VIII

Viviendas indebidamente ocupadas

MARÍA LUISA MORENO-TORRES HERRERA
Catedrática de Derecho civil
Universidad de Málaga

I. EL FENÓMENO DE LA OCUPACIÓN ILEGAL EN LA LEY 12/2023, POR EL DERECHO A LA VIVIENDA. PANORAMA GENERAL

1. Precisiones conceptuales

Se puede hablar de viviendas indebida o ilegalmente ocupadas en un doble sentido, amplio y restringido. En sentido amplio una vivienda estará ilegalmente ocupada cuando esté siendo poseída por un sujeto que carece de título (vigente) para ello, aun cuando se introdujera en el inmueble de manera legítima (como arrendatario, usufructuario, comodatario, etc.). En sentido estricto, una vivienda está ilegalmente ocupada cuando uno o varios sujetos se han introducido en ella sin consentimiento de su actual poseedor. La expresión ocupación ilegal se utiliza en este caso como sinónimo de despojo. Por despojo entiende la doctrina civilista el acto de apropiación de una cosa que está en poder de otro sujeto, sin la autorización de este, tanto si es violento (con empleo de la fuerza) como si no[1], y con independencia de su carácter

---

[1] Entre otros muchos, Bendersky (1961, p. 30).

clandestino. También es característica del despojo el tratarse de una conducta no amparada por las leyes (García Valdecasas, 2006, p. 65) y el hecho de dejar privado al sujeto despojado del poder fáctico que anteriormente ostentaba sobre la cosa[2]. Este último requisito permite distinguir el despojo de la perturbación, que se da cuando se realizan actos que alteran o atentan al poder de hecho que otro sujeto ejerce sobre una cosa, pero que no suponen su privación[3]. Aunque son diferentes, perturbación y despojo tienen en común el ser perturbaciones de hecho y por eso sus autores van a ser considerados perturbadores de la posesión ajena o despojantes, incluso si son titulares de un derecho a poseer.

Conforme a lo expuesto, despojar no es exactamente lo mismo que *okupar*, al menos conforme a la definición del diccionario de la RAE, según el cual okupar es "tomar una vivienda o un local deshabitados e instalarse en ellos sin el consentimiento de su propietario", y okupa "Dicho de un movimiento radical: Que propugna la ocupación de viviendas o locales deshabitados". Parece razonable entender que un inmueble *deshabitado* es un inmueble en el que no vive nadie y no un inmueble en

---

2 Por eso dice Hernández Gil (1980, p. 705) que el despojo es una privación ilegal, si bien no determina la pérdida de la posesión para el despojado hasta que transcurra el plazo de un año, de conformidad con lo dispuesto en los arts. 444 y 460, 4.º CC.

3 Dice al respecto la STS 15 diciembre 2020 (TOL8.248.959) que "[...] hay que distinguir entre las nociones de despojo y de perturbación. La primera (despojo) se corresponde con aquellos hechos materiales que se concretan en la privación total o parcial del goce de la cosa poseída. La segunda (perturbación) se identifica con las conductas que, sin la voluntad del poseedor o en contra de ella, suponen una invasión o una amenaza de invasión de la esfera posesoria que, sin llegar a su privación, la pone en duda e impide o dificulta su libre ejercicio, tal y como venía realizándose antes de la inquietación".

el que sus habitantes no se encuentran en el momento en que un tercero toma posesión de él. Por otra parte, el dato definitorio del despojo no es la falta de consentimiento del *dueño,* sino del poseedor actual, sea cual sea su título posesorio.

Por su parte, la actual la actual doctrina civilista utiliza el término *okupación* para referirse a aquel sujeto que se introduce en el inmueble sin consentimiento. Así, Cuena Casas (2024, p. 230) afirma que "Técnicamente okupa es aquel que se instala en propiedad ajena (vivienda o local) sin el consentimiento del propietario o poseedor legítimo, frecuentemente adquirida por violencia", advirtiendo que no tiene tal consideración el inquilino que se niega a abandonar el inmueble cuando expira su contrato o deja de pagar la renta, ni tampoco el precarista.

2. La ocupación ilegal de viviendas en la Ley por el derecho a la vivienda

La LDV contempla el fenómeno de la ocupación ilegal de viviendas, entendida en el sentido estricto anteriormente indicado, casi exclusivamente desde la óptica de la persona que realiza la ocupación, al que es preferible llamar despojante y no ocupante, toda vez que esta última expresión también su utiliza para aludir a cualquiera que se encuentra en el uso del inmueble. La disposición final quinta de la Ley 12/2023 modifica la Ley 1/2000, de Enjuiciamiento Civil (TOL172.336), en bastantes aspectos, muchos de los cuales conciernen directamente, entre otros, a quienes han realizado actos de despojo de viviendas. Estas novedades legislativas están dirigidas en su mayor parte a tutelar a los ocupantes de la vivienda en riesgo de desalojo (arrendatarios, usufructuarios, precaristas..., y también despojantes), aunque contienen además algunas mejoras técnicas de normas anteriores cuya vigencia se mantiene.

De las muchas modificaciones realizadas en la LEC por la LDV –todas las cuales se contienen en la disposición final quinta– aquí interesan tan solo las que son aplicables en los procesos judiciales que se siguen contra quienes ocupan viviendas tomadas por la vía de los hechos, sin consentimiento ni expreso ni tácito del anterior poseedor[4].

Varios de los preceptos de la LEC a los que da nueva redacción la LDV han sido posteriormente modificados por el Real Decreto-ley 6/2023, de 19 de diciembre, mediante el que se aprueban medidas urgentes para la ejecución del plan de recuperación, transformación y resiliencia en materia de servicio público de justicia, función pública, régimen local y mecenazgo (TOL9.803.574). No obstante, la mayor parte de estas últimas modificaciones no inciden sobre los aspectos objeto de análisis en este capítulo.

Dicho de modo muy sucinto, las novedades legislativas que introduce en la Ley procesal la LDV son fundamentalmente las siguientes:

1.ª Obligación de notificar las resoluciones procesales que contengan fijación de fecha para el lanzamiento de quienes ocupan una vivienda "a las Administraciones públicas competentes en materia de vivienda, asistencia social, evaluación e información en situaciones de necesidad social y atención inmediata a personas en situación de riesgo o exclusión social, por si procediera su actuación" (art. 150.4 LEC).

---

4 Nos referimos a las acciones de tutela sumaria de la posesión que se siguen por los trámites del juicio verbal –antiguos interdictos– a los que alude el n.º 4 del art. 250.1 LEC. Según este "Se decidirán en juicio verbal, cualquiera que sea su cuantía, las demandas siguientes: 4.º- Las que pretendan la tutela sumaria de la tenencia o de la posesión de una cosa o derecho por quien haya sido despojado de ellas o perturbado en su disfrute".

Con anterioridad a la LDV, la notificación a las Administraciones públicas ya estaba prevista, pero requería el consentimiento de los interesados[5].

Este deber de notificación lo hace extensivo la disposición adicional séptima LEC, de nuevo cuño, a los procedimientos penales que se sigan por el delito de usurpación del art. 245.2 CP, en los términos que en ella se detallan[6].

2.ª En el caso de las acciones recuperatorias de la posesión que se siguen por los trámites del juicio verbal de los números 1.º, 2.º, 4.º y 7.º del art. 250.1 LEC, se ordena que no se admitan las demandas que pretendan la recuperación de una finca en la que no se especifiquen los siguientes extremos: a) si el inmueble constituye vivienda habitual de la persona ocupante[7]; b) si concurre en la parte demandan-

---

5 La redacción previa de esta norma –que introdujo la Ley 5/2018, de 11 de junio, de modificación de la Ley de Enjuiciamiento civil en relación a la ocupación ilegal de viviendas (TOL6.632.464)–, era la siguiente: "Cuando la notificación de la resolución contenga fijación de fecha para el lanzamiento de quienes ocupan una vivienda, se dará traslado a los servicios públicos competentes en materia de política social por si procediera su actuación, siempre que se hubiera otorgado el consentimiento por los interesados".

6 En concreto, "siempre que entre quienes ocupen la vivienda se encuentren personas dependientes de conformidad con lo dispuesto en el apartado 2 de la Ley 39/2006, de 14 de diciembre, de Promoción de la autonomía personal y Atención a las personas en situación de dependencia, víctimas de violencia sobre la mujer o personas menores de edad" (TOL1.796.120).

7 Pese a que también este requisito se establece literalmente para los casos de los números 1.º, 2.º, 4..º y 7º del apartado 1 del art. 250 y pese a que se habla de "vivienda habitual de la persona ocupante", y no del arrendatario, se ha afirmado que "no queda claro si se refiere al art. 250.1, 4.º LEC o se ha de entender extensible a los demás supuestos posesorios de recuperación de la posesión de los

te la condición de gran tenedora de vivienda, c) en el caso de que la parte demandante tenga la condición de gran tenedor, si la parte demandada se encuentra o no en situación de vulnerabilidad económica (art. 439, 6 LEC). Además, en aquellos casos en los que la actora tenga la condición de gran tenedora, el inmueble objeto de demanda constituya vivienda habitual y su ocupante se encuentre en situación de vulnerabilidad económica, "no se admitirán las demandas en las que no se acredite que la parte actora se ha sometido al procedimiento de conciliación o intermediación que a tal efecto establezcan las Administraciones públicas competentes [...]" (art. 439.7 LEC).

Por vivienda habitual hay que entender la residencia habitual, es decir, "la vivienda que constituye el domicilio permanente de la persona que la ocupa y que puede acreditarse a través de los datos obrantes en el padrón municipal u otros medios válidos en derecho" (art. 3 i) LDV).

Para determinar cuándo se reúne la condición de gran tenedora, la LEC se remite al art. 3 k) LDV, el cual dispone que es gran tenedor "la persona física o jurídica que sea titular de más de diez inmuebles urbanos de uso residencial o una superficie construida de más de 1500 m2 de uso residencial, excluyendo en todo caso garajes y trasteros".

Finalmente, la LDV establece cómo se debe acreditar la concurrencia o no de vulnerabilidad económica[8] y cómo se debe

---

números 1.°, 2.° y 7.° LEC" (Rodríguez Tirado, 2024, p. 19), algo que pensamos que está fuera de duda.

8 "Para acreditar la concurrencia o no de vulnerabilidad económica se deberá aportar documento acreditativo, de vigencia no superior a tres meses, emitido, previo consentimiento de la persona ocupante de la vivienda, por los servicios de las Administraciones autonómicas y locales competentes en materia de vivienda, asistencia social, evaluación e

acreditar que la parte actora se ha sometido al procedimiento de conciliación o intermediación establecido al efecto por las Administraciones públicas competentes[9].

3.ª Se introduce una norma que dispone que en todos los decretos o resoluciones judiciales que tengan como objeto el señalamiento del lanzamiento, independientemente de

---

información de situaciones de necesidad social y atención inmediata a personas en situación o riesgo de exclusión social que hayan sido específicamente designados conforme a la legislación y normativa autonómica en materia de vivienda". Se añade, no obstante, que el requisito del apartado c) del art. 439.6 LEC podrá también cumplirse mediante: 1.º La declaración responsable emitida por la parte actora de que ha acudido a los servicios indicados anteriormente, en un plazo máximo de cinco meses de antelación a la presentación de la demanda, sin que hubiera sido atendida o se hubieran indicado los trámites correspondientes en el plazo de dos meses desde que presentó su solicitud, junto con justificante acreditativo de la misma.
2.º El documento acreditativo de los servicios competentes que indiquen que la persona ocupante no consiente expresamente el estudio de su situación económica en los términos previstos en la legislación y normativa autonómica en materia de vivienda. Este documento no podrá tener una vigencia superior a tres meses".

9 De conformidad con lo dispuesto en el art. 439.7 LEC, el citado requisito "podrá acreditarse mediante alguna de las siguientes formas: 1.º La declaración responsable emitida por la parte actora de que ha acudido a los servicios indicados anteriormente, en un plazo máximo de cinco meses de antelación a la presentación de la demanda, sin que hubiera sido atendida o se hubieran iniciado los trámites correspondientes en el plazo de dos meses desde que presentó su solicitud, junto con justificante acreditativo de la misma.
2.º El documento acreditativo de los servicios competentes que indique el resultado del procedimiento de conciliación o intermediación, en el que se hará constar la identidad de las partes, el objeto de la controversia y si alguna de las partes ha rehusado participar en el procedimiento, en su caso. Este documento no podrá tener una vigencia superior a tres meses".

que este se haya intentado llevar a cabo con anterioridad, se deberá incluir el día y hora exacta en los que tendría lugar el mismo[10]. Ello se hace mediante la adicción de un nuevo apartado 5 al art. 440 LEC. Sin embargo, posteriormente, el ya citado Real Decreto-ley 6/2023, de 19 de diciembre, ha dado nueva redacción al art. 440 LEC, eliminando de este precepto la exigencia de inclusión del día y hora del lanzamiento, que se lleva ahora a otro lugar: al art. 438.6, pfo. 2.º. Consecuentemente, se suprime la remisión que el art. 549.3 LEC hacía al art. 440.5, el cual mantiene, no obstante, la referencia al día y hora exacta del lanzamiento.

4.ª Se ordena informar a la parte demandada, ya en el decreto de admisión a trámite de la demanda, en los casos de los números 1.º, 2.º, 4.º y 7.º del art. 250.1 LEC, "siempre que el inmueble objeto de la controversia constituya la vivienda habitual de la parte demandada", "de la posibilidad de acudir a las Administraciones Públicas autonómicas y locales competentes en materia de vivienda, asistencia social, evaluación e información de situaciones de necesidad social y atención inmediata a personas en situación o riesgo de exclusión social" (art. 441.5, pfo. 1.º LEC).

5.ª Se impone al Juzgado el deber de comunicar inmediatamente y de oficio la existencia del procedimiento a las administraciones anteriormente referidas, "a fin de que puedan verificar la situación de vulnerabilidad y, de existir esta, presentar al Juzgado propuesta de alternati-

---

10 Desaparece, pues, la posibilidad de lanzamientos sin fijación de día y fecha, algo que hasta ahora era una práctica frecuente, cuyo objetivo era evitar las movilizaciones que se producían para intentar impedirlos. Este aspecto no presenta mayor interés a los fines de este trabajo, por lo que no se volverá sobre esta norma.

va de vivienda digna en alquiler social [...] y propuestas de medidas de atención inmediata [...] así como de las posibles ayudas y subvenciones de las que pueda ser beneficiaria la parte demandada" (art. 441.5, pfo. 2.º LEC).

6.ª Se atribuye al tribunal que conozca del desahucio la decisión sobre la suspensión del proceso, a la vista de la información recibida de las Administraciones Públicas competentes y de las alegaciones de las partes. Esta suspensión tiene por finalidad hacer posible que se adopten las medidas propuestas por dichas administraciones y puede tener una duración máxima de dos meses si el demandante es una persona física, y de cuatro si se trata de una persona jurídica; transcurrido este plazo, continuará el procedimiento (art. 441. 6 LEC). La decisión sobre la suspensión no es automática, sino que la deberá tomar el tribunal "previa valoración ponderada y proporcional del caso concreto, apreciando las situaciones de vulnerabilidad que pudieran concurrir también en la parte actora y cualquier otra circunstancia acreditada en autos" (art. 441.7 LEC).

El ámbito de aplicación de las normas anteriormente referidas cambia de unos casos a otros, si bien ninguna de ellas es de aplicación exclusiva a los procesos previstos específicamente en la Ley procesal para la recuperación del bien objeto de despojo (art. 250, 4.º LEC), sino que contemplan, además, otros procesos judiciales distintos. En concreto: los procesos mediante los que se pretenda la recuperación de la finca rústica o urbana, dada en arrendamiento ordinario o financiero, o en aparcería (art. 250.1, 1.º LEC); los que pretendan la recuperación de la plena posesión de una finca, rústica o urbana, cedida en precario (art. 250.1, 2.º LEC); finalmente, los que demanden la efectividad de derechos reales inscritos en el Registro de la propiedad frente a quienes se opongan a ellos o perturben su ejercicio, sin disponer de título inscrito que legitime la oposición o la perturbación (art. 250.1, 7.º LEC).

No es necesario extenderse en más detalles sobre la reforma de la Ley procesal por la LDV para constatar que, en términos generales, está orientada a la protección de quienes en el ámbito de un proceso judicial van a ser desalojados del inmueble en donde tienen su vivienda habitual. Se tiene en cuenta su vulnerabilidad económica y también la llamada vulnerabilidad social. Este camino se había iniciado ya por normas anteriores, si bien es cierto que nacidas como soluciones de urgencia y de carácter extraordinario para dar respuesta a situaciones coyunturales, como las derivadas de la pandemia[11]. La Ley pone el foco en el sujeto que se encuentra en posesión de la vivienda y ha sido demandado en un proceso judicial que puede derivar en su lanzamiento, lo que no quiere decir que ignore absolutamente las circunstancias del demandante. De hecho, se establecen soluciones distintas dependiendo de que el demandante sea o no gran tenedor. Además, se hace una referencia expresa a la posible vulnerabilidad del demandante, aspecto este en el que se profundizará igualmente más adelante, al efectuar el análisis detallado de las distintas normas.

---

11 Es el caso del Real Decreto-ley 37/2020, de 22 de diciembre, de medidas urgentes para hacer frente a las situaciones de vulnerabilidad social y económica en el ámbito de la vivienda y en materia de transportes (TOL8. 241-535), que introdujo un nuevo art. 1 *bis* en el Real Decreto-ley 11/2020, de 31 de marzo, de medidas urgentes complementarias en el ámbito social y económico para hacer frente al COVID-19. Inicialmente se había previsto la posibilidad de suspensión de los desahucios de arrendatarios vulnerables y sin alternativa habitacional, pero luego se amplió a otros procesos, entre ellos el del n.º 4 del art. 250.1 LEC, sobre tutela sumaria de la posesión. El citado art. 1 *bis* tiene ahora la redacción que le ha dado el art. 87 del Real Decreto-ley 8/2023, de 27 de diciembre, mediante el cual se adoptan medidas para afrontar las consecuencias económicas y sociales derivadas de los conflictos en Ucrania y Oriente Próximo, así como para paliar los efectos de la sequía. (TOL9.841.182).

II. VALORACIÓN DE LAS NORMAS APLICABLES A LOS OCUPANTES ILEGALES DE VIVIENDA CONTENIDAS EN LA LDV

Tras presentar en el apartado anterior una imagen de conjunto sobre el tratamiento de la ocupación ilegal de viviendas en la LDV –que se circunscribe al ámbito procesal–, se abordan ahora con mayor detalle las distintas normas, al objeto de valorarlas, no con carácter general, sino en referencia, estrictamente, a los supuestos de ocupación ilegal.

A nuestro modo de ver, merecen distinta opinión las normas que contienen deberes de información al propio demandado, o de notificación a las Administraciones públicas, que aquellas otras que establecen requisitos para la admisibilidad de la demanda o permiten la suspensión del proceso, por más que todas posean una *ratio* común, que no es otra que evitar el sinhogarismo[12]. Dentro de la definición legal de sinhogarismo del apartado 1 del art. 3 LDV se incluyen a quienes residen en una vivienda "sin título legal o con notificación de abandono de la misma".[13] Las medidas contenidas en la disposición final quinta

---

[12] Sin embargo, según Nasarre Aznar (2024, p. 36) internacionalmente la *okupación* está considerada como una situación de sinhogarismo.

[13] El art. 3 l) LDV define el sinhogarismo como la "circunstancia vital que afecta a una persona, familia o unidad de convivencia que no puede acceder de manera sostenida a una vivienda digna y adecuada en un entorno comunitario y aboca a las personas, familias o unidades de convivencia que lo sufren a residir en la vía pública u otros espacios públicos inadecuados, o utilizar alternativas de alojamiento colectivo institucionalizado de las diferentes administraciones públicas o de entidades sin ánimo de lucro, o residir en una vivienda inadecuada, temporal o no, inapropiada o masificada, en una vivienda insegura, sin título legal, o con notificación de abandono de la misma, o viviendo bajo amenaza de violencia".

LDV están todas orientadas a facilitar y promover la protección social, por parte de la Administración, de personas vulnerables. También, al menos desde la óptica del legislador, por más que ello no sea fácil de comprender, el procedimiento de intermediación o conciliación previsto en el art. 439.7 LEC, respecto del cual declara el Preámbulo de la Ley que "facilitará a las Administraciones competentes dar adecuada atención a las personas y hogares afectados, ofreciendo respuesta a través de diferentes instrumentos de protección social y de los programas de política de vivienda".

1. Información al demandado y notificaciones a la Administración

En el caso de entablarse una acción posesoria frente al sujeto que ha ocupado ilegalmente una vivienda, y "siempre que el inmueble objeto de controversia constituya la vivienda habitual de la parte demandada", surgen a cargo del Juzgado, en virtud de lo ordenado en el art. 441.5 LEC, dos deberes distintos: el primero, informar al demandado, en el decreto de admisión a trámite de la demanda, de la posibilidad de acudir a las Administraciones públicas competentes en materia de vivienda, asistencia social, evaluación e información de situaciones de necesidad social y atención inmediata a personas en situación o riesgo de exclusión social; el segundo, el de comunicar inmediatamente y de oficio la existencia del procedimiento a las citadas Administraciones[14].

---

14 Surge la duda, no obstante, acerca de si esto último se ordena solo para el caso en que el inmueble objeto de la demanda sea vivienda habitual. Y surge porque el pfo. 2.º del art. 441.5, en el que se impone este deber, no menciona, como sí hace el anterior, este requisito. No obstante, la interpretación sistemática y teleológica conduce a esta conclusión. Además, no tendría sentido que la noti-

Lo primero que se piensa es que, a diferencia de lo que ocurre en los casos en que el poseedor de la vivienda sea un arrendatario (art. 250.1, 1.º LEC), un precarista (art. 250.1, 2.º), o incluso un poseedor o perturbador sin título inscrito (art. 250.1, 7.º), en el caso del despojante (art. 250,1, 4.º), difícilmente va a ocurrir que el demandado tenga su residencia habitual en el inmueble cuya posesión se está reclamando. O, quizás más exactamente, difícilmente va a ocurrir que el demandante haya manifestado que es esa la condición del inmueble reclamado, salvo en el improbable supuesto, por más que pueda estimarse teóricamente posible[15], de que el despojante se haya empadronado en la vivienda ocupada. Distinto es que pueda interpretarse –y quizás esto es lo que estaba en el ánimo del legislador– que quien ha tomado ilegalmente posesión de una vivienda tiene en ella su residencia habitual cuando carece de otra, pudiendo entenderse, únicamente a partir de este hecho, que la vivienda en cuestión constituye el domicilio permanente de quien la ocupa. Es cierto que el concepto de vivienda o residencia habitual es un dato fáctico, ajeno a los títulos jurídicos que respalden el uso, y que, consecuentemente, no es imposible que el ocupante ilegal tenga su vivienda habitual en el inmueble del que se haya apropiado.

Cuando esto ocurra, el Juzgado habrá de cumplir con los deberes de información que le impone el art. 441.5 LEC. Y es lógico que se hayan hecho extensivos al supuesto del que se está tratando, dada su finalidad. En el caso de la notificación a los propios demandados el objetivo es poner en su conocimiento que existen Administraciones a quienes pueden acudir en caso

---

ficación por el Juzgado a la Administración se exigiese con carácter general siempre que se iniciase alguno de los procedimientos de los números 1º, 2º, 4.º y 7.º del art. 250.1 LEC.

15 Así lo defiende Cuena Casas (2023, pp. 313-315).

de vulnerabilidad, y también facilitarles el acceso a ellas. De ahí que se disponga que “La información deberá comprender los datos exactos de identificación de dichas Administraciones y el modo de tomar contacto con ellas [...]”. En el caso de la notificación de oficio a las Administraciones se persigue que puedan estas verificar la situación de vulnerabilidad para, de constatarse así, presentar al Juzgado propuesta de alternativa de vivienda digna en alquiler social [...] y propuestas de medidas de atención inmediata [...] así como de las posibles ayudas y subvenciones de las que pueda ser beneficiaria la parte demandada” (art. 441.5 LEC).

Lo mismo cabe decir del traslado a las Administraciones públicas competentes de la notificación de las resoluciones procesales que contengan fijación de fecha para el lanzamiento de quienes ocupan una vivienda, la cual se ordena “por si procediera su actuación” (art. 150.4 *in fine* LEC). Hay que tener en cuenta que corresponde a las Administraciones competentes en materia de vivienda luchar contra el fenómeno del sinhogarismo, “promoviendo en su ámbito territorial el acceso a soluciones habitacionales de alojamiento en condiciones adecuadas [...]” (art. 14.3 LDV). Desde esta óptica, posee todo el sentido que les sean notificados, no solo los lanzamientos de arrendatarios, precaristas, usufructuarios y demás, sino también y, sobre todo, los de quienes ocupan viviendas sin título ni consentimiento para hacerlo. Es más, en el caso de los ocupantes ilegales la notificación cobra especial sentido, habida cuenta de que, como la doctrina pone de manifiesto, el fenómeno que se ha generalizado en los últimos tiempos, aunque no sea el único[16], es el de la denominada ocupación ilegal de necesi-

---

16 Autores como Mayor del Hoyo (2022, p. 2) hacen referencia también a las ocupaciones ilegales que se producen en el ámbito de la delincuencia organizada.

dad, practicada por personas que carecen de recursos para un acceso legal a la vivienda (Cuena Casas, 2023, p. 297).

Ahora bien, una cosa es el acierto del legislador al exigir las notificaciones referidas, y otra distinta que a causa de ellas se demore la recuperación de la posesión por el actor, que es justamente lo que va a ocurrir en aplicación de lo dispuesto en los párrafos 3.º a 5.º del art. 441.5 LEC. Aunque se han defendido otras interpretaciones, en nuestra opinión, se concede a la Administración competente un plazo de diez días para comprobar la vulnerabilidad económica y notificarla al órgano judicial, transcurrido el cual, y tanto si la Administración presenta su informe como si no, se dará traslado a las partes, "para que en el plazo de cinco días puedan instar lo que a su derecho convenga, procediendo a suspender la fecha prevista para la celebración de la vista o para el lanzamiento, de ser necesaria tal suspensión por la inmediatez de las fechas" (art. 441.5 *in fine* LEC).[17]

---

[17] En contra, Rodríguez Tirado (2023, p. 15), quien sostiene que no se prevé un plazo expreso para que las Administraciones públicas competentes emitan informe de la valoración de la vulnerabilidad económica y, en su caso, social del arrendatario, sino, simplemente, un plazo de notificación de dicho informe al juzgado. Hay que admitir que la redacción de la norma no es clara y que la interpretación literal avala la conclusión defendida por la autora. Sin embargo, y al margen del hecho de que cuesta creer que el legislador haya dejado sin determinar el plazo para que la Administración realice su tarea y pueda reanudarse el procedimiento, la interpretación sistemática de los párrafos 3.º y 5.º del art. 441.5 LEC y los precedentes de la norma conducen a la conclusión defendida en el texto. En efecto, en el Proyecto de ley por el derecho a la vivienda (BOCG, Congreso de los diputados, XIV legislatura, serie A, n.º 89-1, de 18 de febrero de 2022), tras establecerse en la redacción propuesta para el art. 441.5 LEC que la Administración debía confirmar y notificar la vulnerabilidad económica del afectado, se continuaba

No resulta fácil aceptar que el poseedor despojado tenga que soportar las dilaciones derivadas de una intervención administrativa[18], por más que esta esté dirigida a proteger a los vulnerables, pero no es menos cierto que la demora parece inevitable, por cuanto que se precisa un mínimo de tiempo para verificar la situación de vulnerabilidad y presentar en su

---

diciendo "De no recibirse respuesta en dicho plazo [...]", en alusión al plazo máximo de diez días. Es claro, pues, que en el proyecto el plazo era para la emisión y notificación del informe, y es ello lo que debe entenderse respecto del texto definitivo, pese a mantenerse en él una redacción que da a entender que el plazo es de notificación. Se dice, en efecto: "En caso de que estas Administraciones Públicas confirmasen que el hogar afectado se encuentra en situación de vulnerabilidad económica y, en su caso, social, se notificará al órgano judicial a la mayor brevedad y en todo caso en el plazo máximo de diez días". Puede todavía añadirse un último argumento a favor de la interpretación que defendemos: el Real Decreto-ley 37/2020, texto legislativo en el que con anterioridad a la LDV se regula (y con detalle) la suspensión de los juicios verbales de los arts. 250.1, 2.°, 4.° y 7.° LEC, muy tenido en cuenta por los redactores de la Ley 12/2013. En el apartado 4 del art. 1 bis del citado Real Decreto-ley se dice literalmente: "El Letrado de la Administración de Justicia deberá trasladar inmediatamente a los servicios sociales competentes toda la documentación y solicitará a dichos servicios informe, *que deberá ser emitido en el plazo máximo de quince días,* en el que se valore la situación de vulnerabilidad de la persona o personas que hayan fijado en el inmueble su vivienda, y se identifiquen las medidas a aplicar por la administración competente".

18 En este sentido, el Informe jurídico sobre el anteproyecto de Ley por el derecho a la vivienda emitido por el CGPJ el 27 de enero de 2022 advertía de que los numerosos plazos previstos para efectuar las oportunas comunicaciones a las Administraciones Públicas "determinarán sin duda amplias dilaciones adicionales, tiempos muertos procesales fatalmente derivados de la propia mecánica del funcionamiento de nuestros Tribunales" que pueden convertir los lanzamientos en trámites muy costosos (p. 63).

caso propuestas de alternativa de vivienda en alquiler social, de atención inmediata, o de ayudas y subvenciones económicas.

2. Los requisitos de procedibilidad

Dejando de lado otras valoraciones realizadas por la doctrina[19], de los requisitos de procedibilidad contenidos en el nuevo apartado 6 del art. 439 LEC se extraen dos ideas importantes: la primera, que el legislador entiende que no todos los ocupantes de vivienda que estén en riesgo de desalojo merecen ser protegidos, sino únicamente los vulnerables que tengan en el inmueble su vivienda habitual; la segunda, que el legislador

---

19 López Simó (2023, p. 8) cree, con Cortés Domínguez, que el nuevo apartado 6 del art. 439 LEC impide el normal ejercicio del derecho a la tutela judicial efectiva, "agravando la situación procesal del demandante con una inusual e irregular, desde el punto de vista constitucional, exoneración de la carga de la prueba del demandado, que hace quebrar el principio de la igualdad de partes en el proceso y el de la distribución de la carga de la prueba de acuerdo con el elemental principio de la facilidad probatoria (art. 217 LEC). Pueden existir, pues, serias dudas de la constitucionalidad de la reforma que estamos analizando". En parecido sentido, afirma Climent Esteve (2023, p. 3) que los nuevos requisitos de procedibilidad restringen considerablemente el inicio del proceso de desahucio y la actuación de los órganos judiciales. Por su parte, Gómez Linacero (2023, p. 3), observa que "nos encontramos ante un hito sin precedentes en la historia reciente de nuestro sistema procesal privado, por ser la primera vez [...] en que, para el ejercicio de acciones basadas en un derecho subjetivo privado, en el marco de una relación entre particulares regida por el Derecho civil y por los principios, en el orden procesal, dispositivo y a instancia de parte, se exige el cumplimiento de presupuestos de derecho público, concretamente de naturaleza administrativa, para la admisión de la demanda o acción ejercitada". Se suman a las críticas Álvarez Suárez (2024, p. 11) y Nasarre Aznar (2023, p. 22), entre otros muchos.

considera que esta protección se debe organizar de manera diferente en función que el demandado tenga o no la consideración de gran tenedor. Así, solo si el actor es un gran tenedor (y siempre y cuando se den el resto de circunstancias que recoge el art. 439.7 LEC) será preceptivo que las partes se sometan a un procedimiento de conciliación o intermediación previo al proceso judicial. Y solo "cuando la parte actora sea una gran tendedora de vivienda y hubiera presentado junto a la demanda documento acreditativo de la vulnerabilidad de la parte demandada" están obligadas las Administraciones públicas a efectuar directamente, dentro del plazo de diez días que establece el art. 441.5, pfo. 3.º LEC, "la propuesta de medidas de atención inmediata a adoptar, así como de las posibles ayudas económicas y subvenciones de las que pueda ser beneficiaria la parte demandada [...]. De entrada, no se comprende bien la norma, que suscita la pregunta sobre las razones que pueden explicar que únicamente se exija a la Administración adoptar medidas inmediatas de atención a favor del demandado vulnerable cuando el demandante sea gran tenedor. Sin embargo, la clave para comprender esta diferencia de trato se encuentra en el último párrafo del art. 441.5, en donde se ordena dar traslado a las partes durante cinco días, para instar lo que a su derecho convenga, y, lo que es más importante, "suspender la fecha prevista para la celebración de la vista o para el lanzamiento". Puede pues afirmarse que el legislador ha considerado que la dilación del procedimiento derivada de las actuaciones de las Administraciones públicas es más tolerable para el demandante que tenga la consideración de gran tenedor.

El dato importante es que, a la hora de diseñar el sistema protector, se realiza una valoración de la situación de los sujetos implicados. Aunque se trata de un planteamiento que tiene detractores, a nuestro modo de ver no es necesariamente criticable. Lo que ocurre –y este es el aspecto cuestionable– es que la LDV ha tomado en consideración, exclusivamente, criterios socio-económicos, olvidando totalmente los jurídicos. En concreto, ha equiparado al despojante –que es un sujeto que realiza un acto ilícito– con la persona que se introdujo en la vivienda al amparo de un título jurídico[20].

Sin entrar aquí en el análisis del acierto o desacierto del legislador a la hora de determinar la vulnerabilidad, o de configurar el concepto de "gran tenedor", y dejando para más adelante la importante cuestión sobre si el despojante, aun vulnerable, merece esta protección frente al despojado, incluso si es gran tenedor, lo que no resulta aceptable en ningún caso es que se imponga al sujeto que se ha visto privado contra su voluntad de la posesión de la vivienda, la carga de especificar si el inmueble constituye la vivienda habitual del demandado –lo que puede tener sentido en el caso de que el ocupante sea un arrendatario, por ejemplo, pero no respecto del despojante– y, lo que es peor aún, la carga de acreditar su vulnerabilidad[21].

Merece un juicio muy negativo el hecho de que se haga recaer sobre el demandante la especificación o prueba, según el caso, de dos extremos o circunstancias relativas al demandado y que deberían haber constituido instrumentos de defensa para este último, pero no requisitos para la admisibilidad de la de-

---

20 Se muestra muy crítica con la igualdad de trato que reciben sujetos como el precarista y el okupa Cuena Casas, quien afirma: "Los dos supuestos son distintos por más que en ambos el derecho del propietario padezca en la misma medida: se ve privado del uso y disfrute de su inmueble, está obligado a atender todos los gastos derivados de la propiedad del mismo e incluso puede verse obligado a responder civilmente de los daños causados a terceros por el ocupante de la vivienda (arts. 1902 y 1910 CC), y todo sin obtener compensación alguna por parte del poseedor del inmueble" (2023, p. 298).

21 *Vid.* art. 439.6 LEC.

manda[22]. Igualmente parece contrario a toda lógica exigir que las partes (que en este caso son despojante y despojado), se sometan a un procedimiento de conciliación o intermediación[23]. Este requisito de procedibilidad puede encontrar justificación cuando se persigue el desalojo de quien se introdujo legítimamente en el inmueble, pero no en el caso que es ahora objeto de análisis. No es aceptable que el despojado, ni siquiera si es gran tenedor, venga obligado a someterse a un procedimiento dirigido a alcanzar un acuerdo con quien le privó de la posesión. Distinto es que el objetivo de la conciliación o intermediación no sea lograr una solución consensuada, sino facilitar una respuesta de la Administración competente a personas afectadas –que es lo que deja entrever el Preámbulo de la Ley[24]–. Pero incluso en este caso la dilación que supone la medida no nos parece admisible, ni siquiera en referencia a la llamada ocupación de

---

22 En este sentido se manifiesta Malo Valenzuela (2023, p. 2), quien afirma que "Exigir a los ciudadanos a quienes se ha ocupado ilegalmente su vivienda la carga de especificar o acreditar tales extremos para que se admita su demanda no solo implica trasladarles problemas cuya solución, en su caso, compete a los poderes públicos, sino que dificulta el ejercicio del derecho a la tutela judicial efectiva en caso de ocupación y genera una grave inseguridad jurídica".

23 Opinión distinta es la de Simón Moreno (2021, p. 2195), quien proponía, antes de la publicación de la Ley, "impulsar que el propietario de la vivienda y sus ocupantes (siempre que se encontrasen en una situación acreditada de exclusión residencial) llegasen (voluntariamente) a una solución consensuada mediante un servicio de mediación/intermediación proactiva (que podría dar audiencia, también, a las comunidades de propietarios)".

24 En el Preámbulo se dice que "La aplicación de este procedimiento facilitará a las Administraciones competentes dar adecuada atención a las personas y hogares afectados, ofreciendo respuesta a través de diferentes instrumentos de protección social y de los programas de política de vivienda".

necesidad[25], porque contraviene el principio que prohíbe que cualquier sujeto se tome la justicia por su mano[26] .

3. La suspensión del procedimiento judicial

De todas las medidas dirigidas a la protección del demandado en riesgo de desalojo la que ha sido objeto de mayor atención y suscitado mayor rechazo[27] ha sido la suspensión del procedimiento judicial entablado para recuperar la posesión del inmueble, una medida que se presenta como instrumento de tutela ante situaciones de vulnerabilidad.

La suspensión de los procedimientos (o tan solo de los lanzamientos) en atención a la vulnerabilidad económica y social del poseedor de la vivienda no es una novedad de la Ley 12/2023. El legislador español la toma del Derecho catalán y la introduce por primera vez a favor de los despojantes, aunque como facultad judicial y sujeta a ciertos requisitos, en el Real Decreto-ley 37/2020, de 22 de diciembre, anteriormente citado, cuya disposición adicional segunda preveía, además, una compensación económica para ciertos propietarios[28].

---

25 Distingue Bastante Granell (2018, pp. 10-13) tres tipos de ocupantes sin título: el poseedor "okupa" con animus político, el poseedor "ocupa" con animus social y el poseedor "ocupa" con animus delictivo.

26 Véanse, no obstante, otras opiniones y un más profundo análisis del procedimiento de conciliación o intermediación en esta misma obra, en el capítulo "Conciliación e intermediación en materia de vivienda".

27 Entre otros muchos, Julià-Pijoan (2023, p.15), quien considera que "representa una quiebra para el derecho fundamental a un proceso público sin dilaciones indebidas" y que, en consecuencia, se debería proceder a su reformulación.

28 Mayores detalles sobre textos legales en los que se contempla la suspensión de los lanzamientos en atención a la vulnerabilidad,

Tras diversas prórrogas, continúa vigente el art. 1 bis del Real Decreto-ley 37/2020, en donde se regula la "Suspensión hasta el 31 de diciembre de 2024 del procedimiento de desahucio y de los lanzamientos para personas económicamente vulnerables sin alternativa habitacional en los supuestos de los apartados 2.º, 4.º y 7.º del artículo 250.1 de la Ley 1/2000, de 7 de enero, de Enjuiciamiento Civil, y en aquellos otros en los que el desahucio traiga causa de un procedimiento penal". Esto significa la coincidencia temporal de dos textos legales distintos en los que se contempla la suspensión: por una parte, el citado Real Decreto-ley, y por otra, la LEC, en la redacción dada por la LDV y cuya entrada en vigor se produjo el 26 de mayo de 2023 (disposición final novena LDV). Ahora bien, mientras que las medidas de suspensión previstas en el Real Decreto-ley "se establecen con carácter extraordinario y temporal" (art. 1 bis), no ocurre lo mismo con las recogidas en el art. 441 LEC[29]. Por consiguiente, hay que entender en principio que aquello que esté previsto en el Real Decreto-ley, pero no en la LEC, dejará de tener aplicación a partir del 1 de enero de 2025.

En la LDV, texto en el que se centra este análisis, la suspensión del procedimiento previsto en el art. 250. 4.º LEC, que es el procedimiento específico dirigido a la recuperación de la cosa objeto de despojo, está sujeta a importantes exigencias.

---

previos y posteriores al Real Decreto-ley citado en el texto, pueden verse en Cuena Casas (2023, pp. 319-324) y en Lafuente Torralba (2020-2021, pp. 144-154), quien realiza un buen análisis de las normas del Decreto-ley 37/2020 (llamado por los medios "decreto anti-desahucios") que regulan la citada suspensión.

29 La redacción actual del apartado 1 proviene del Real Decreto-ley 6/2023.

La determinación de cuáles son los requisitos que han de concurrir para decretar la suspensión no está exenta de dificultades, debido a la técnica legislativa empleada. Los nuevos apartados 5, 6 y 7 del art. 441 LEC presentan el gran inconveniente de que no ordenan ni distinguen bien las cuestiones, lo que exige un esfuerzo importante al intérprete y al aplicador del Derecho. Una cosa son los requisitos necesarios para poder suspender el procedimiento, otra los criterios que ha de tomar en consideración el juez para hacerlo, y otra diferente las normas puramente procedimentales. Sin embargo, todo ello aparece entremezclado y, además, regulado conjuntamente con los deberes que se hacen recaer sobre las administraciones competentes, originándose una confusión importante. Hay que lamentar que el legislador no se haya inspirado, mejorándola, en la más adecuada sistematización del Real Decreto-ley 37/20. Este último es, además, un texto considerablemente más completo, pues incluye, por ejemplo, unas causas de exclusión, lo que, como seguidamente se expondrá, no resulta superfluo.

Por lo que se refiere a los requisitos que para la suspensión del procedimiento establece la Ley 12/2023, el primero de ellos es que haya sido posible la identificación del receptor de la notificación de la demanda (cfr. art. 441.1 bis LEC, párrafo 3.°, *in fine*), algo que no siempre ocurrirá, toda vez que se permite que, en los casos en los que el objeto cuya posesión se pretende recuperar sea una vivienda o una parte de ella, la notificación se haga a quien se encuentre habitando en ella, pudiéndose hacer, además, a los ignorados ocupantes de la vivienda[30]. Se dispone, no obstante, que "A efectos de proceder a

---

30 Esta posibilidad se introdujo por la ya citada Ley 5/2018, de modificación de la Ley 1/2000, de Enjuiciamiento Civil, en relación a la ocupación ilegal de viviendas.

la identificación del receptor y demás ocupantes, quien realice el acto de comunicación podrá ir acompañado de los agentes de la autoridad" (art. 441.1 bis, pfo. 1.º)[31].

Los otros requisitos que han de concurrir para que pueda suspenderse el proceso, y que se deducen de los apartados 5, 6 y 7 del art. 441 LEC son: que el inmueble objeto de la controversia constituya la vivienda habitual de la parte demandada y que el Tribunal aprecie vulnerabilidad económica o social en el demandado. No es necesario, en cambio, que la actora tenga la condición de gran tenedora.

Por lo que se refiere a lo primero, y en línea con lo argumentado anteriormente, hay que entender que la manifestación sobre el carácter de vivienda habitual contenida en el escrito de demanda no es decisiva, y que para determinar este y otros extremos se tendrá en cuenta la información recibida de las Administraciones públicas competentes, así como las alegaciones de las partes (art. 441.6). Puede aventurarse que será altamente probable que los jueces concluyan que se trata de la vivienda habitual del demandado siempre que no conste que este posee un domicilio distinto.

En cuanto a la situación de vulnerabilidad, sin entrar en demasiados detalles, por cuanto que este aspecto es objeto de análisis detenido en otro capítulo de esta obra, y no ofrece peculiaridades en el caso de las viviendas indebidamente ocupadas, basta destacar que los criterios para su apreciación establecidos en la Ley son puramente orientativos y que el arbitrio judicial es amplio, tal y como se colige del art. 441. 6 LEC,

---

31 Un problema distinto, pero que también puede plantearse y demorar el lanzamiento, es el de la dificultad para notificar la sentencia que lo ordena. Sobre esta cuestión, *Vid.* Cruz Jiménez y García del Val (2024, pp. 285-287).

que ordena al tribunal resolver por auto sobre la suspensión, "a la vista de la información recibida de las Administraciones públicas competentes y de las alegaciones de las partes". Ahora bien, que el arbitrio judicial sea amplio no quiere decir que sea absoluto. Así, si la Administración ha verificado la situación de vulnerabilidad e informado al Juzgado de ella, este no la puede ignorar, lo que no quiere decir que tenga que decretar necesariamente la suspensión. En cambio, entendemos que el Juez puede apreciar la vulnerabilidad aun no habiéndolo hecho los servicios sociales. Hay que tener en cuenta que, como ha destacado Julià-Pijoan (2024, p. 9), la valoración de la situación de vulnerabilidad se realiza en dos ocasiones: primeramente, por los servicios sociales competentes, y, posteriormente, por el órgano judicial, que debe apreciar tanto la documentación de los servicios sociales como otra documentación que puedan eventualmente aportar las partes.

Finalmente, no es requisito de la suspensión la condición de gran tenedor del demandante, frente a lo exigido en el Real Decreto-ley 37/2020[32]. Distinto es que el plazo máximo de suspensión varíe en el caso de los grandes tenedores.

Interesa advertir que lo que la LDV hace es permitir al juez suspender el proceso, –suspensión que tendrá la duración máxima prevista en la ley–, y no limitar *sine die* los lanzamientos. Con esta advertencia no se quiere en absoluto restar importancia al perjuicio que la medida supondrá para el demandante, sino evitar el uso de un lenguaje que pudiera inducir a creer que la LDV consiente y propicia la permanencia ilimitada del despojante en el inmueble, lo que es inexacto.

---

32 El apartado 2 del art. 1 bis del Real Decreto-ley 37/2020 dice: "Será necesario para poder suspender el lanzamiento conforme al apartado anterior, que se trate de viviendas que pertenezcan a personas físicas titulares de más de diez viviendas [...]".

La finalidad de la suspensión es hacer posible que las Administraciones públicas adopten medidas de protección para el demandado, pero la suspensión no se mantiene hasta que esto ocurra[33], sino que está sujeta a un plazo, transcurrido el cual se alzará automáticamente y continuará el procedimiento. La duración máxima de la suspensión es de dos meses si el demandante es una persona física o de cuatro meses si se trata de una persona jurídica.

Se establecen reglas diferentes en función de quién sea el demandante, pero la norma presenta dificultades de coordinación con lo dispuesto en el art. 250.1, 4.° LEC. Todo induce a afirmar que la LDV está pensando en un demandante propietario y en el caso de la apropiación ilegal de una vivienda que se encontraba deshabitada, olvidando que el legitimado activo para interponer la acción de tutela sumaria de la posesión no es el dueño de la vivienda, sino el poseedor despojado, y que tanto si se trata de un poseedor de hecho (pfo. 1.° del art. 250.1, 4.° LEC) como de un poseedor con título (pfo. 2.° del art. 250.1, 4.° LEC) puede tener en el inmueble litigioso su vivienda habitual. Se trata esta de una circunstancia verdaderamente importante y que, sin embargo, la LDV ha ignorado a la hora de regular la suspensión. Entendemos que lo acertado habría sido excluir totalmente cualquier suspensión o dilación del procedimiento en aquellos supuestos en que el demandante (o incluso cualquier otro sujeto distinto) tenga su residencia habitual en la vivienda cuya posesión se reclama

---

33 Durante la tramitación parlamentaria de la Ley 5/2018, el Grupo Parlamentario Confederal de Unidos Podemos-En Comú Podem-En Marea presentó una enmienda mediante la que se pretendía garantizar que no se produjera el lanzamiento sin que el ocupante en riesgo de exclusión residencial tuviera garantizada una alternativa habitacional, pero esta enmienda fue rechazada (BOCG, Congreso de los Diputados, XII legislatura, serie B, núm. 78-4, de 29 de septiembre de 2017, p. 2).

y que ha sido objeto de despojo, pero no se ha previsto así. Es cierto, no obstante, que es lo razonable pensar que tal resultado nunca se producirá, en aplicación de lo dispuesto en el art. 441.7 LEC, a cuyo tenor el tribunal tomará la decisión relativa a la suspensión "previa valoración ponderada y proporcional del caso concreto, apreciando las situaciones de vulnerabilidad que pudieran concurrir también en la parte actora y cualquier otra circunstancia acreditada en autos". La tutela de derechos fundamentales del poseedor despojado, como el libre desarrollo de la personalidad (art. 10.1 CE), la integridad física y moral (art. 15 CE), la intimidad personal y familiar (art. 18.1 CE) y la inviolabilidad del domicilio (art. 18.2), todos los cuales aparecen mencionados en el Preámbulo de la LDV, (por su conexión con el derecho a disfrutar de una vivienda digna y adecuada), deberán llevar al juez a descartar la suspensión del procedimiento en perjuicio del demandante despojado que tuviera su vivienda habitual en el inmueble cuya restitución reclama. Piénsese, además, que uno de los objetivos de la LDV es "proteger la estabilidad y la seguridad jurídica en la propiedad, uso y disfrute de la vivienda [...]" (art. 2 e). El problema estriba en que se ha pensado exclusivamente en proteger al demandado, al poseedor sin título. De ahí que el precepto mencionado continúe diciendo "con especial atención a las personas y hogares en situación o riesgo de vulnerabilidad, y específicamente a familias, hogares y unidades de convivencia con menores a cargo, a través de medidas efectivas en materia de vivienda y asegurando la debida coordinación con medidas complementarias de atención social, formación, empleo y otras acciones de acompañamiento". Ahora bien, el art. 47 CE ordena a los poderes públicos proteger no solo el derecho de los ciudadanos a acceder a una vivienda digna y adecuada, sino que ordena también garantizarles su disfrute, lo que conlleva, obviamente, el establecimiento de medidas que les aseguren la continuidad en el disfrute[34] y, por lo tanto, adecuadas para combatir eficazmente las perturbaciones de hecho que provengan de otros sujetos.

Sin embargo, la LDV parece haber obviado este importante aspecto, al centrar toda su atención en el demandado-despojante en situación de vulnerabilidad, ignorando el derecho del demandante-despojado a continuar en la posesión de la vivienda. Es un olvido grave, que contradice declaraciones contenidas en el propio Preámbulo de la Ley y en textos internacionales que en él se mencionan[35]. La estabilidad y la permanencia adquieren una relevancia capital cuando la vivienda se concibe como uno de los espacios esenciales para el adecuado desarrollo de la personalidad (Anderson, 2018, p. 390), y también cuando se pone en relación con la inviolabilidad del domicilio y con el derecho a la intimidad personal y familiar. Estos son, de hecho, los bienes jurídicos protegidos por el delito de allanamiento de morada del art. 202.1 CP[36].

---

34 Afirma López Ramón (2014, p. 77) que la conservación de la vivienda es contenido mínimo del derecho a la vivienda establecido en el art. 47 CE, y que este derecho a la conservación de la vivienda "impone ante todo abstenciones por parte del poder público y de los demás sujetos, obligados todos a respetarlo".

35 Caso de la Declaración Universal de Derechos Humanos, adoptada y proclamada por la 183.ª Asamblea General de la Organización de las Naciones Unidas, de 10 de diciembre de 1948 (TOL147.461). Es contenido del derecho a la vivienda que consagra su art. 25 la seguridad en la tenencia (Pacto Internacional de Derechos Económicos, Sociales y Culturales de 16 de diciembre de 1966 (TOL163.591), y Dictámenes del Comité de Derechos Económicos y Culturales, que pueden verse con detalle en: Simón Moreno (2021, pp. 2174-2176).

36 Navarro Massip (2020, p. 17) y Muñoz de Dios Sáez (2020, p. 18), entre otros muchos. Como explica Ramos Martínez (2021, pp. 293-294), el concepto de morada es instrumental, es cualquier espacio de intimidad con cierta vocación de permanencia y que está destinado a actividades de la vida privada.

Aunque el marco legal ofrece al juez instrumentos para no ordenar la suspensión del procedimiento en los casos en los que el demandante u otro sujeto tuviera su vivienda habitual en el inmueble del que se pretende lanzar al demandado, en atención exclusivamente a esa circunstancia, no se pueden descartar otras interpretaciones. Podría entenderse que lo que el juez debe hacer en tales supuestos es valorar exclusivamente la vulnerabilidad económica y social de las partes y decantarse por la protección del sujeto más débil. Sin embargo, este sería un planteamiento incorrecto, porque, pese a lo que da a entender la letra de la ley, no se trata de un conflicto de vulnerabilidades, sino que lo que está en juego es el derecho del demandante a continuar en el disfrute de la vivienda, con independencia de cuál sea su situación patrimonial y sus circunstancias personales.

Habría sido deseable, pues, para evitar que los jueces tomen su decisión de suspender o no el procedimiento en base a una confrontación de las situaciones de vulnerabilidad económica y social de los litigantes, que se hubiese introducido en la LDV una norma del tipo de la contenida en el art. 1 bis, apartado 7, del Real Decreto-ley 37/2020, vigente hasta el 31 de diciembre de 2024. Según el citado precepto, en ningún caso procederá la suspensión del lanzamiento "Cuando se haya producido en un inmueble de propiedad de una persona física, si en dicho inmueble tiene su domicilio habitual o segunda residencia debidamente acreditada, sin perjuicio del número de viviendas de las que sea propietario". Esta exclusión del apartado a) se complementa con otra, contenida en el siguiente apartado, y que resulta igualmente muy oportuna por cuanto que en realidad lo relevante no es quién sea el propietario del inmueble, sino el hecho de que se haya despojado a cualquier sujeto de su vivienda, tanto si es habitual como si no. De ahí que se disponga que en ningún caso procederá la suspensión "Cuando se haya producido en un inmueble de propiedad de una persona física o jurídica que lo tenga cedido por cualquier título válido en derecho a una persona física que tuviere en él su domicilio habitual o segunda residencia debidamente acreditada". Aunque la técnica legislativa es mejorable y la redacción se puede simplificar, la solución es correcta. En términos generales cabe cualquier opinión, favorable o contraria, sobre la suspensión procesal como medida de protección. Unos la defenderán y otros no. Pero lo que parece estar fuera de duda es que no puede prevalecer el derecho a la vivienda del despojante sobre el derecho a la vivienda del despojado[37]. Téngase en cuenta, además, que tal y como se ha observado (Adán Domenech, 2023, p. 53), una orden judicial de desalojo no excluye que los poderes públicos deban atender las situaciones de exclusión residencial que pudieran producirse, en particular cuando afectaren a personas especialmente vulnerables.

Más criticable aún que el hecho de que el legislador no haya excluido expresamente la suspensión del procedimiento en aquellos casos en los que la vivienda cuya posesión se reclama estuviese siendo utilizada en el momento del despojo como vivienda habitual, es que el legislador permita con carácter general que tenga lugar la suspensión en el caso de ser la acción ejercitada la del art. 250. 4.º LEC.

Constituye este artículo la actual sede normativa de las acciones posesorias (antiguos interdictos), cuyo fundamento jurídico es la tutela de la paz jurídica. Si contamos con un instrumento procesal específico del que puede valerse cualquier poseedor, con o sin título, es para hacer efectivo el mandato contenido en el art. 441 CC, conforme al cual

---

[37] En parecido sentido, Mayor del Hoyo (2022, p. 33), quien señala que la ley defenderá el derecho a la vivienda del ocupante ilegal frente al derecho a la vivienda del titular legítimo.

"En ningún caso puede adquirirse violentamente[38] la posesión mientras exista un poseedor que se oponga a ello. El que se crea con acción o derecho para privar a otro de la tenencia de una cosa, siempre que el tenedor resista la entrega, deberá solicitar el auxilio de la autoridad competente". Nuestro sistema jurídico rechaza cualquier acto de desposesión no consentido, incluso si es realizado por quien tiene un derecho sobre la cosa frente a quien la posee sin título. Como dice el art. 446 CC "Todo poseedor tiene derecho a ser respetado en su posesión; y, si fuere inquietado en ella, deberá ser amparado o restituido en dicha posesión por los medios que las leyes de procedimiento establecen". Se legitima por ello a todo poseedor para interponer una acción sumaria dirigida exclusivamente a restaurar el estado de cosas anterior, a reprimir la perturbación o despojo, sea quien sea su autor. Es una protección provisional que no excluye un posterior juicio plenario en el que se debata y decida sobre el derecho a poseer. Este derecho a poseer fundado en un título jurídico (el *ius possidendi*) es irrelevante[39] para activar y conceder la tutela posesoria, cuyos presupuestos, además del plazo (art. 1968, 1.° CC), son solamente dos: la previa posesión del demandante y el despojo o perturbación del demandado. Si ambos resultan probados, el juez que conozca del caso deberá condenar al demandado a restituir la cosa al demandante, sin entrar en ninguna otra consideración. No obstante, esta sentencia no produce efecto de cosa juzgada (art. 447,2 LEC), quedando abierta la reclamación en el juicio ordinario correspondiente para debatir, no sobre el

---

[38] El TS considera que es adquisición violenta cualquier acto que no cuente con la voluntad del poseedor dirigida a la entrega de la cosa (STS 20 mayo 1946, TOL4.455.682).

[39] Sin perjuicio de lo que luego se matizará a propósito de la norma contenida en el párrafo segundo del art. 250. 4.° LEC.

hecho posesorio y el despojo, sino sobre los derechos que se ostenten sobre la cosa, que es, justamente, el aspecto que se habrá obviado en el juicio posesorio.

Lo que subyace en la tutela sumaria de la posesión no es otra cosa que la consideración de que toda perturbación y todo despojo deben ser reprimidos porque todos, independientemente de quien sea su autor y de quien la víctima, son injustos, puesto que alteran la paz jurídica y el orden público. Y, como decía Savigny[40], es contra esta injusticia contra la que se dirigen los interdictos posesorios.

Sobre la configuración de las acciones posesorias en el ordenamiento jurídico español y sobre su fundamento hay pleno acuerdo doctrinal y jurisprudencial. El planteamiento, que encuentra sus raíces en la *actio spolii*,[41] estaba ya presente en la antigua doctrina y llega hasta nuestros días[42], por más que algunos autores actuales parezcan luego olvidarlo a ciertos efectos. También en la jurisprudencia se ha mantenido en el tiempo la doctrina que considera que los procesos posesorios se centran en la posesión de hecho y que la tutela posesoria "se limita a mantener la posesión que tenía el demandante o recuperarla [...] sin alcanzar a la titularidad", cuestión que podrá dilucidarse en el juicio correspondiente (STS 25 noviembre 2008 (TOL1.408.456). Es muy clara al respecto la STS 15 diciembre 2020 (TOL8.248.959), que resuelve un caso en el que la acción posesoria (de retener la posesión) es interpuesta por la arrendataria de una vivienda

---

40 Cita tomada de la traducción de la obra de Ihering, *La posesión* (2022, p. 39).

41 *Vid.* Hernández Gil (1980, pp. 85-88).

42 Entre otros muchos, Hernández Gil (1980, pp. 42-43); García Valdecasas (2006, p. 64); Vázquez Barros (2005, p. 96); Cuena Casas (2011, p. 1799); Carrión Olmos (2023, p. 107).

contra la propietaria del edificio en el que aquella se ubicaba. El Juzgado de 1.ª instancia había desestimado la demanda con el argumento de que la tutela interdictal requiere que la perturbación llevada a cabo por la demanda sea ilícita, lo que no ocurría en este caso, en el que respondía a la ejecución de un acuerdo social. Por su parte, la Audiencia mantuvo el fallo, por considerar que faltaba el requisito del *animus spoliandi*, que definía como "la conciencia que el despojante tiene de que el acto que comete es fruto de un obrar arbitrario y sin título que lo autorice". El motivo del recurso de casación fue la infracción de los arts. 441 y 446 CC y de la doctrina jurisprudencial establecida en las SSTS de 7 de julio de 2016 (TOL5.775.269), 25 noviembre 2008, ya citada, y 21 abril 1979 (TOL1.740.800). El TS estima el recurso de casación, afirmando que "en la acción interdictal de retener o recobrar únicamente se reconoce legitimación activa a quien se encuentre en el disfrute de la cosa, y lo que pretenda sea una rápida protección para la continuación en el goce pacífico de la cosa, como situación de hecho, en la que haya sido perturbado o de la que haya resultado despojado". Y añade: "Señaló la sala en la sentencia de 21 de abril de 1979 que la protección sumaria interdictal halla su fundamento en la conveniencia de un logro acelerado y provisional de una paz jurídica inmediata, que, dando solución momentánea al conflicto suscitado, cumpla con unos fines pacificadores y de social armonía...viniéndose de este modo a prohibir aquellos actos de los particulares que unilateralmente, y por su propio poder, quieran imponer por propias vías de hecho, desentendiéndose de los instrumentos jurídicos y de los cauces jurisdiccionales que todo Estado de Derecho concibe y habilita, pues la apariencia posesoria debe ser absolutamente merecedora de respeto y toda destrucción de la misma ha de consumarse acudiendo a los medios jurídicos que el derecho proporciona". Concluye el Tribunal que "Como sucedía en el caso de los interdictos de retener y recobrar, los procesos derivados de las acciones de tutela sumaria de la posesión del art. 250.1, 4.º LEC, son procesos cautelares, conservativos y dirigidos a la tutela de la posesión como hecho, con objeto de evitar, por razones de orden público y paz social, la defensa privada, y en los que no se discute ni el derecho de propiedad ni cualquier otro que otorgue el mejor derecho a poseer, sino la realidad fáctica de la situación posesoria violentada".

Pues bien, si en nuestro ordenamiento jurídico no se permite que una situación posesoria sea atacada, no puede comprenderse que se permita ahora por la LDV la suspensión del procedimiento que se sigue contra el despojante y que está dirigido a restablecer el estado de cosas anterior. Si las leyes no amparan la actuación de quien, fuera de los procedimientos establecidos para ello, se toma la justicia por su mano y se apropia de una cosa que está poseyendo otro sujeto, ni siquiera cuando sea el despojante el que ostenta un derecho sobre la cosa, y no el despojado, es claro que no puede introducir –so riesgo de ser incoherente– ningún instrumento que suponga un respaldo a la conducta del despojante. Y ello ni siquiera con el argumento de tutelar a personas vulnerables. Desde esta óptica es criticable que se haya previsto la suspensión del procedimiento a favor de quien se haya apropiado indebidamente de un inmueble ajeno, sean cuales sean las circunstancias y los derechos de ocupante y ocupado. Se trata de una posibilidad que contraviene frontalmente el principio que fundamenta la tutela posesoria en nuestro ordenamiento jurídico y que cristalizó en el Derecho canónico en la regla *spoliatus ante omnia restituendus*. Este principio no cuenta, o no *contaba*, con excepción alguna, hasta el punto de que siempre se ha interpretado que alcanza incluso al dueño de la cosa que, tomándose la justicia por su mano, se la arrebata al ladrón, así como al "ocupa" (Bastante Granell, 2018, p. 6). Sin embargo, la LDV ignora absolutamente tan importante principio al contemplar la suspensión de los procedimientos verbales "que pretendan la tutela sumaria de la tenencia o de la posesión de una cosa o derecho por quien haya sido despojado de ellas o perturbado en su disfrute" (art. 250.1, 4.° LEC). Son perfectamente aplicables al tema que se está tratando las observaciones vertidas por el CGPJ en su Informe jurídico al Anteproyecto de Ley por el derecho a la vivienda cuando afirma que "el anteproyecto tiene una clara vocación jurídico-pública, lo que le lleva a obviar a menudo, cuando no contradecir, la regulación civil de las relaciones jurídicas de contenido patrimonial, tanto reales como obligacionales, alcanzadas por el anteproyecto. Es innegable que la atención a la dimensión social de la vivienda y la delimitación de la función social de la propiedad requieren de normas administrativas que traten de solventar, o al menos atenuar, los problemas de acceso y disfrute de la vivienda que se agudizan en determinados momentos. Ahora bien, esta regulación ha de articularse debidamente con el derecho codificado y, además, las normas de intervención de carácter excepcional y coyuntural no pueden desplazar a las generales y ordinarias con vocación de permanencia tal y como el anteproyecto hace" (pp. 5 y 6). Quizás lo que esté ocurriendo, como afirma Gutiérrez Sanz (2024, p. 253) es que, en el tema de la ocupación, impera lo ideológico, desdibujándose instituciones jurídicas hasta ahora aparentemente sólidas y socialmente reconocidas.

Todo lleva a pensar que la derogación por la LDV, a favor de los despojantes de viviendas, del principio que veta que los sujetos se tomen la justicia por su mano, no ha sido del todo consciente y que viene motivado, al menos en parte, por la ignorancia o el olvido por parte del legislador del hecho de que el legitimado activo para el ejercicio de las acciones posesorias no es el propietario, sino el poseedor, y que mediante estas acciones no se tutela derecho alguno sino, simplemente, el hecho previo de la posesión.

No es menos cierto que el propio legislador había ya contribuido previamente, de manera muy importante, a que se desdibujaran las tradicionales acciones posesorias. Dio un paso decisivo en este sentido la anteriormente citada Ley 5/2018, de modificación de la Ley de Enjuiciamiento civil, en relación a la ocupación ilegal de viviendas). Hasta esa fecha no existían especialidades en materia de tutela posesoria por el hecho de ser una vivienda el objeto del despojo, pero la alarma social que estaba generando la ocupación ilegal premeditada, y la idea de que ninguno de los cauces previstos en vía civil para procurar el desalojo resultaba plenamente satisfactorio, aparte de demorarse en exceso, determinó que se dictara la Ley 5/2018, dirigida, según su Preámbulo, a "articular los mecanismos legales ágiles en la vía civil que permitan la defensa de los derechos de los titulares legítimos que se ven privados ilegalmente y sin su consentimiento de la posesión de su vivienda, cuando se trata de personas físicas, entidades sin ánimo de lucro con derecho a poseerla o entidades públicas propietarias o poseedoras legítimas de vivienda social". El legislador invoca, además de las ya mencionadas, razones de interés social para justificar su intervención legislativa, al afirmar que "tampoco encuentra protección suficiente la función social que han de cumplir las viviendas que tienen en su haber las entidades sociales o instrumentales de las Administraciones públicas, para ser gestionadas en beneficio de personas y familias vulnerables, puesto que un porcentaje demasiado elevado del referido parque de viviendas se encuentra ocupado de forma ilegal [...]". Y manifiesta de modo expreso que "La ocupación ilegal, esto es, la ocupación no consentida ni tolerada, no es título de acceso a la posesión de una vivienda, ni encuentra amparo alguno en el derecho constitucional a disfrutar de una vivienda digna".

En atención a todo lo anterior se introduce una nueva acción en la LEC, distinta de la acción posesoria de recuperación de la posesión, la cual se mantiene sin modificaciones. Esta nueva acción, configurada en los arts. 250.1, pfo. 2.º[43], 437. 3 bis, párrafo 4.º[44] y 444.1 bis[45] LEC tiene como finalidad, ciertamente, la recuperación de la posesión de una vivienda (o de una parte) por quien se haya visto privado de ella sin su consentimiento. Sin embargo, no está legitimado para su ejercicio todo poseedor, sino, únicamente "la persona física que sea propietaria o poseedora legítima por otro título, las entidades sin ánimo de lucro con derecho a poseerla y las entidades públicas propietarias o poseedoras legítimas de vivienda social". Sin en-

---

43 "Podrán pedir la inmediata recuperación de la plena posesión de una vivienda o parte de ella, siempre que se hayan visto privados de ella sin su consentimiento, la persona física que sea propietaria o poseedora legítima por otro título, las entidades sin ánimo de lucro con derecho a poseerla y las entidades públicas propietarias o poseedoras legítimas de vivienda social". Esta es la redacción original de la norma, y no ha sido modificada posteriormente.

44 "Cuando se solicitase en la demanda la recuperación de la posesión de una vivienda o parte de ella a la que se refiere el párrafo segundo del numeral 4.º del apartado 1 del art. 250, aquélla podrá dirigirse genéricamente contra los desconocidos ocupantes de la misma, sin perjuicio de la notificación que de ella se realice a quien en concreto se encontrare en el inmueble al tiempo de llevar a cabo dicha notificación. A la demanda se deberá acompañar el título en que el actor funde su derecho a poseer." Esta es la redacción actual de la norma.

45 "Tratándose de un caso de recuperación de la posesión de una vivienda a que se refiere el párrafo segundo del numeral 4.º del apartado 1 del art. 250, si el demandado o demandados no contestaran a la demanda en el plazo legalmente previsto, se procederá de inmediato a dictar sentencia. La oposición del demandado podrá fundarse exclusivamente en la existencia de título suficiente frente al actor para poseer la vivienda o en la falta de título por parte del actor. La sentencia estimatoria de la pretensión permitirá su ejecución, previa solicitud del demandante, sin necesidad de que transcurra el plazo de veinte días previsto en el artículo 548." Esta es la redacción originaria de la norma, que se mantiene invariada.

trar en otras consideraciones, lo que ahora interesa destacar es que al demandante no le bastará probar la posesión previa y el despojo, sino que deberá acompañar a la demanda el título en que funde su derecho a poseer. Por su parte, el demandado podrá oponerse, exclusivamente, invocando la existencia de un título suficiente frente al actor para poseer la vivienda, o la falta de título por parte del actor.

Como resulta patente, los aspectos a probar, debatir y considerar no son puramente fácticos, –como ocurría en el interdicto de recobrar la posesión y sigue ocurriendo en el caso de la tutela sumaria de la posesión del art. 250.1. 4.°, párrafo primero–, sino que alcanzan a los títulos de los litigantes. Es por ello por lo que se trata de una acción nueva y distinta, por más que su fundamento último lo sitúe el Preámbulo de la Ley en el art. 441 CC. Es, ciertamente, una acción posesoria y otorga una tutela meramente provisional, que no excluye un posterior juicio plenario[46]; pero es un instrumento para la protección del *ius possidendi* o derecho a poseer o, quizás más exactamente, para la protección de su apariencia[47]. Por lo tanto, no se trata, frente a lo afirmado por el legislador, de una mera adecuación y actualización del tradicional interdicto de recobrar la posesión, sino de algo bien distinto, de la introducción de una acción diferente, con su propio régimen jurídico; acción que se suma, pero que no excluye, la acción posesoria

---

46 "En él no se discute o dilucida sobre el derecho, sino que se busca reintegrar la posesión de forma rápida con base en una apariencia de derecho a poseer fundada en un título" (Mayor del Hoyo, 2022, p. 21).

47 En este sentido, Mayor del Hoyo (2022, p. 15), quien observa con acierto que el legitimado activo es un sujeto que, dado que acredita un título, se entiende que es titular de un derecho a poseer.

del art. 240.1, 4.°, párrafo primero, mediante la que se protege el *ius possesionis.*[48]

En nuestra opinión, los cambios que introdujo la Ley 5/2018 han provocado que, por así decirlo, hayan caído en el olvido, en referencia a las viviendas, las verdaderas acciones posesorias, entendiendo por tales aquellas mediante las que se tutela la posesión como hecho (el *ius possesionis*) y que en la LEC de 1881 se conocían como interdictos. Y, lo que es más importante, parece haber caído en el olvido el fundamento y la razón de ser de la protección posesoria. La propia LDV no lo tiene en cuenta en su redacción de la disposición final quinta, pues de ser así no habría equiparado la posición jurídica del despojante y la de aquellos otros sujetos que iniciaron la posesión de una vivienda ajena con el consentimiento del dueño, del usufructuario o cualquier otra persona con derecho a poseer, algo que al parecer no ocurre en otros ordenamientos.[49]

En definitiva, desde la reforma de 2018, se piensa que la protección civil frente a la ocupación ilegal de viviendas es una protección reconocida al propietario[50]. Se considera que las normas mediante las que se regula el procedimiento que persigue el desalojo de quien ha ocupado ilegalmente una vivienda, resuelven (aunque sea provisionalmente) un conflicto entre el propietario y el ocupante de la vivienda. A ello contribuye, además, el hecho de que la mayor parte de las veces el actor es, en efecto, el propietario. Todos los autores que se han ocupado del tema razonan, sin excepción, en los mismos términos, entendiendo que en la

---

48 Así lo afirman Rodríguez Achútegui (2018, p. 3) y Arnau Moya (2021, pp. 320-321).

49 *Vid.* la información de Derecho comparado que recogen Simón Moreno (2021, pp. 2182-2185) y Malo Valenzuela (2023, pp. 3-4).

50 O, a lo sumo, al poseedor legítimo. Así, Pérez Daudí (2024, pp. 458 y ss.)

solución de este problema entra en juego el derecho de propiedad[51]. Lo mismo cabe decir del recurso de inconstitucionalidad resuelto por la STC de 16 de enero de 2024 (TOL9.863.735). El recurso cuestionaba la constitucionalidad de la norma del

---

51 Lafuente Torralba (2021, p. 115) escribe, a propósito de la ocupación ilegal: "Hallamos aquí a dos sectores enfrentados: por una parte, una masa creciente de desheredados sociales, castigados por el encadenamiento de crisis económicas, por el desempleo y la carestía de la vivienda, que se ven empujados a la ilegalidad o, al menos, seducidos por ella, para no verse a sí mismos y a los suyos arrojados más allá del umbral de exclusión; por otra parte, tenemos a los titulares de esas propiedades que están siendo usurpadas, que un buen día acuden a su piso y lo encuentran invadido por extraños que pretenden hacer política social a costa de ellos". La misma visión de la ocupación como un conflicto de intereses entre los propietarios y los ocupantes puede verse en Simón Moreno (2021, pp. 2161-2162). Mayor del Hoyo (2022, p. 31), en referencia a la figura de la suspensión, afirma que los propietarios afectados quedan en una situación de vulnerabilidad jurídica y que la suspensión "supone un obstáculo para la defensa de su derecho a poseer la vivienda ocupada ilegalmente, que afecta al propio derecho a la propiedad –art. 33 CE–, así como al derecho a la tutela judicial efectiva y a un proceso sin dilaciones indebidas –art. 24.1 y 2 CE–, dado que impide que pueda hacer efectivo por vía judicial el contenido de este derecho subjetivo sobre el inmueble". *Vid.*, también, p. 36. Adán Domenech (2023, p. 51), a propósito de la Ley 5/2018, dice que "pretende poner fin, o cuanto menos solucionar parcialmente, la confrontación entre el derecho que tienen todos los ciudadanos a disfrutar de una vivienda digna y adecuada –art. 47 CE– y el derecho a la propiedad privada –art. 33 CE–". Cuena Casas (2023, p. 334), quien se muestra muy crítica con el tratamiento de la ocupación ilegal por la LDV, considera que "La ocupación ilegal es un ataque frontal al derecho de propiedad que no puede justificarse con base en el derecho a una vivienda digna". También afirma que el derecho constitucional a la vivienda "es garantizado por el Estado a costa de limitar el derecho de propiedad de los particulares" (p. 331).

Real Decreto-ley 16/2021, de 3 de agosto, que atribuía al juez un margen de apreciación para acordar, en las circunstancias concurrentes de la pandemia, la suspensión del lanzamiento de la vivienda habitual habitada sin título para ello. Los recurrentes argumentaban que tal potestad judicial de suspensión afectaba al contenido esencial del derecho de propiedad y vulneraba el art. 86.1 CE en relación con el art. 33, lo que fue rechazado por el Pleno del TC, que desestimó el recurso.[52]

Es innegable que los propietarios de los inmuebles son víctimas de su ocupación ilegal. Sin embargo, lo que la LDV prevé es la suspensión de unas acciones recuperatorias de la posesión cuyo legitimado activo no es necesariamente un propietario, sino que basta que se trate de un poseedor. Es la tutela posesoria frente al despojo y su efectividad lo que está en juego, y no el derecho de propiedad. Y por ello debió razonarse, estrictamente, en clave posesoria. De haberse hecho así, no se habrían aplicado al caso del despojo las mismas normas procesales que a los arrendatarios o a los precaristas, cuya situación no es en absoluto comparable, por más que pueda afirmarse que también la posesión de estos últimos es indebida, una vez extinguida la relación contractual o el título en el que descansaba su disfrute[53].

---

52 La conclusión del TC fue que las normas impugnadas "prevén una medida limitada en cuanto a su ámbito de aplicación subjetivo, objetivo y temporal, que ni tiene por objeto una regulación directa y general del derecho de propiedad de la vivienda (art. 33 CE) ni afecta, por ello, a su contenido esencial. Medida que responde a una finalidad de interés social, que tiene una incidencia temporal y parcial sobre el citado derecho, y cuya efectiva adopción requiere la ponderación de las circunstancias del caso concreto por parte del órgano judicial competente".

53 Con este argumento, Arnau Moya (2021, p. 319) defiende que contra quienes ocuparon un inmueble sin consentimiento puede

La LDV parece haberse obcecado en su objetivo de proteger a los vulnerables. Los trata a todos por igual en los procesos verbales dirigidos a su desalojo, sin atribuir relevancia alguna al hecho de que hayan adquirido la posesión violentamente. Esto no es más que una muestra de lo que Muñoz Tabernero (2021, p. 143) ha descrito como una "visión buenista, demagógica y tolerante del fenómeno de la ocupación de bienes inmuebles" que según el autor, con el que coincidimos, afecta "a principios esenciales e indispensables para el mantenimiento de nuestro Estado de Derecho, tales como la seguridad jurídica, la confianza en nuestras instituciones y una respuesta indubitada y eficaz de nuestro Ordenamiento Jurídico ante la vulneración de los derechos o la mera injerencia en los mismos".

III. CONCLUSIONES

Primera: La LDV no contempla de una manera general el fenómeno de la ocupación ilegal de viviendas, pero tiene en cuenta la posición del despojante contra el que se siga la acción de tutela sumaria de la posesión regulada en el art. 250.1, 4.º LEC. Al despojante demandado (que la Ley no denomina así) se le va a aplicar la misma protección que a aquellos otros sujetos vulnerables que se hayan introducido en la vivienda de manera legítima, como arrendatarios y precaristas, y contra los que se haya iniciado un proceso verbal dirigido a su desalojo.

Segunda: La LDV, en línea con lo que habían hecho textos legales anteriores, contiene medidas tendentes a dificultar o

también utilizarse el desahucio por precario, una postura seguida también por numerosas sentencias de Audiencias provinciales, pero sobre la que el propio autor dice que el TS "ha pasado de puntillas" (p. 333).

retrasar desalojos forzosos de personas en situaciones de vulnerabilidad, con la finalidad de hacer posible la actuación de las Administraciones competentes en materia de protección social. Siendo este en términos generales un objetivo loable, la equiparación del despojante y de quienes iniciaron su posesión de modo legítimo, es decir, sobre la base de un título jurídico, no es aceptable, por cuanto que entra en contradicción con el fundamento al que responde en nuestro ordenamiento jurídico la tutela sumaria de la posesión. Ese fundamento no es otro que reprimir actos de perturbación de la posesión no consentidos, independientemente de quien sea el sujeto que los realice. Nadie puede tomarse la justicia por su mano. Tampoco quien esté necesitado de una vivienda.

Tercera: Por lo que se refiere al despojado, la LDV ha tomado el camino contrario al necesario. En lugar de garantizar su derecho a continuar en el disfrute del inmueble, mejorando la tutela sumaria de la posesión e intentando hacerla realmente eficaz y útil, ha incrementado las dificultades para que el despojado logre un rápido desalojo. Ello es especialmente grave en el caso en que el inmueble viniese ya siendo utilizado como vivienda, un supuesto que la Ley no ha contemplado. Queda, pues, confiar en el criterio judicial en lo que se refiere a la decisión sobre la suspensión del procedimiento. Sin embargo, otras normas procesales, cuya aplicación no depende del arbitrio judicial, provocarán irremediablemente una demora en el desalojo. La prevalencia, aunque sea temporal, del derecho a la vivienda del ocupante ilegal respecto del derecho a la vivienda de su anterior poseedor no encuentra justificación alguna y subvierte principios muy sólidos y arraigados en nuestro sistema, como la tutela de la paz jurídica.

Cuarta: El planteamiento unánime seguido por la doctrina a la hora de valorar el tratamiento que la LDV le da a la ocupación ilegal consiste en enfrentar el derecho a la vivienda del ocupante ilegal con el derecho de propiedad, siendo la opinión mayoritaria, al menos entre la doctrina civilista, la de que el derecho de propiedad no puede ceder ante el derecho a la vivienda del *okupa*. Muchos añaden que tampoco puede anteponerse al derecho de propiedad el derecho a la vivienda de quienes venían poseyendo sobre la base de un título jurídico, como arrendatarios o precaristas cuando ya su relación jurídica ha finalizado. Sin embargo, en el caso de la ocupación ilegal propiamente dicha (despojo), el planteamiento es incorrecto. No se trata de si el derecho de propiedad debe o no ser objeto de limitaciones en aras de la función social de la propiedad que proclama el art. 33 CE, ni tampoco de si el derecho a una vivienda digna y adecuada puede justificar o no esas limitaciones. Se trata de algo muy distinto: de amparar o no el acto de despojo. Este es el verdadero trasfondo y no debe quedar oscurecido por otras consideraciones como el derecho a la vivienda o la protección de los vulnerables. Todo despojo es contrario a las más elementales reglas de la convivencia y por lo tanto todo despojo debe ser reprimido y, por supuesto, de él no puede derivarse ventaja o beneficio alguno para su autor.

Quinta: La postura de la LDV respecto de la ocupación ilegal es una muestra de la tolerancia del legislador actual hacia el fenómeno y supone abrir una puerta a la autotutela, no admitida hasta ahora en el ámbito del Derecho privado. Las dilaciones que en los procesos posesorios pueden provocar las normas analizadas y, en particular, la suspensión del procedimiento, con la consiguiente permanencia en el inmueble de los autores del despojo, constituyen un incentivo a la ocupación ilegal. La situación actual es paradójica, por cuanto que mientras que el ocupante ilegal, si es considerado vulnerable, verá como en aplicación de las normas procesales se prolonga su posesión ilegítima, en cambio un hipotético propietario que despojase de su propia vivienda a un poseedor sin título resultaría condenado a reponer en su posesión al despojado, en aplicación igualmente de las normas procesales en vigor.

No hay, pues, más camino que replantearse el actual de estado de cosas y adoptar decisiones que hagan posible recuperar la coherencia: o se suprimen las normas procesales que prolongan la posesión en la vivienda del despojante vulnerable –que es lo que aquí se propugna–, o se elimina la tutela de la posesión como hecho –lo que supondría la derogación de un principio muy arraigado y básico en nuestro ordenamiento jurídico–.

IV. REFERENCIAS BIBLIOGRÁFICAS

Adán Domenech, F. (2023). "El juicio de desahucio de ocupas. Soluciones procesales al lanzamiento". *Actualidad Jurídica Iberoamericana*, n.º 19, 48-69.

Álvarez Suárez, L. (2024). "Los excesivos requisitos de procedibilidad y los menoscabos en las garantías procesales de los arrendadores a la luz de la nueva Ley sobre el derecho a la vivienda (1). *La Ley Derecho de familia*, n.º 41.

Anderson, M. (2018). "¿Un modelo distinto de propiedad para la vivienda?". *El derecho de propiedad en la construcción del Derecho privado europeo* (dir. Lauroba Lacasa, E.). Tirant Lo Blanch.

Arnau Moya, F. (2021). "Los procedimientos civiles para recuperar la posesión de inmuebles ocupados. *Revista Bolivariana de Derecho*, n.º 32, 314-365.

Bastante Granell, V. (2018). "La necesaria delimitación jurídico-social de los sujetos ocupa y ocupado". *Revista Aranzadi de Derecho Patrimonial*, n.º 47.

Bendersky, M.J. (1961). *Acciones posesorias y despojo*. Abeledo-Perrot.

Carrión Olmos, S. (2023). "La posesión". *Derecho civil III. Derechos reales* (coords. Verda y Beamonte/ Serra Rodríguez). Tirant Lo Blanch.

Climent Esteve, V. (2023). "Análisis de las principales modificaciones procesales introducidas por la Ley 12/2023, de 24 de mayo, por el derecho a la vivienda". *Diario La Ley*, n.º 10316, Sección Tribuna.

Cruz Jiménez, M.P. y García del Val, S. "Problemática en la recuperación judicial de la posesión". *El derecho a la vivienda en tiempos de incertidumbre* (dirs. Mate Satué/ Hernández Sainz/ Alonso Pérez). Aranzadi.

Cuena Casas, M. (2011). Comentario al artículo 441. *Código civil comentado* (dirs. Cañizares Laso y otros). Volumen I. Civitas.

Cuena Casas, M. (2023). "La ocupación ilegal de inmuebles: un necesario enfoque global". *Cuadernos de Derecho Transnacional,* Vol. 15, n.° 2, 293-336.

Cuena Casas, M. (2024). "¿La ocupación de inmuebles como instrumento para garantizar el derecho a la vivienda?". *El derecho a la vivienda en tiempos de incertidumbre.* (dirs. Mate Satué/ Hernández Sainz/ Alonso Pérez). Aranzadi.

López Ramón, F. (2014). "El derecho subjetivo a la vivienda". *Revista española de Derecho constitucional,* n.° 102, 49-91.

García Valdecasas, G. (2006). *La posesión.* Comares.

Gómez Linacero, A. (2023). "Preguntas y respuestas a la Ley por el derecho a la vivienda en clave procesal". *Diario La Ley,* n.° 10386, sección Tribuna.

Gutiérrez Sanz, M. R. (2024). "Entre la experiencia y la esperanza: ineficacia e ineficiencia de la tutela procesal penal frente a la ocupación ilícita". *El derecho a la vivienda en tiempos de incertidumbre* (dirs. Mate Satué/ Hernández Sainz/ Alonso Pérez). Aranzadi.

Hernández Gil, A. (1980). *La posesión.* Civitas.

Julià-Pijoan, M. (2023). "La suspensión de los procesos de desahucio por situaciones de vulnerabilidad: ¿constituye una dilación indebida? (1). *La Ley Derecho de familia,* n.° 41.

Lafuente Torralba, A.J. (2020-2021). "El *labyrinthus iudicorum* de la ocupación ilegal de viviendas: remedios en las vías penal y civil y análisis de su eficacia". *Revista de Derecho aragonés,* (XXVI-XXVII), 113-154.

López Simó, F. (2023). "Principales novedades procesales de la Ley 12/2023, de 24 de mayo, por el derecho a la vivienda. Breve análisis crítico (1)". *Diario La Ley,* n.° 10383, Sección Tribuna.

Malo Valenzuela, M.A. (2023). "La Ley por el derecho a la vivienda como forma de promover la ocupación ilegal de viviendas ajenas por vía legislativa". *El Notario del siglo XXI,* n.° 110, 40-45.

Mayor del Hoyo, M.V. (2022). "La propiedad de la vivienda ante la ocupación ilegal y las nuevas tendencias normativas: ¿una nueva vulnerabilidad?". *Vulnerabilidad patrimonial: retos jurídicos.* Aranzadi.

Muñoz de Dios Sáez, A. (2020). "La ocupación ilegal de inmuebles, una visión desde el Derecho penal". *El Notario del siglo XXI,* n.° 93,18-21.

Muñoz Tabernero, O.A: (2022). "Capacidades de prevención y lucha contra la usurpación (ocupación). Una visión multidisciplinar de la usurpación de inmuebles (especial referencia a la Comunidad Foral de Navarra". *Revista de Humanidades Cuadernos del Marqués de San Adrián*, n.º 14, UNED Tudela, 113-155.

Nasarre Aznar, S. (2023). "Ley de vivienda: el Estado renuncia a cumplir con el art. 47 de la Constitución". *El Notario del siglo XXI*, n.º 109, 18-22.

Nasarre Aznar, S. (2024). "Los retos de la vivienda tras quince años de crisis y la Ley 12/2023". *El derecho a la vivienda en tiempos de incertidumbre* (dirs. Mate Satué/ Hernández Sainz/ Alonso Pérez). Aranzadi.

Navarro Massip, J. (2020). "La insoportable levedad de la ocupación". *El Notario del siglo XXI*, n.º 93, 14-17.

Pérez Daudí, V. (2024). "Las opciones procesales frente a la ocupación y la protección de los colectivos vulnerables". *El derecho a la vivienda en tiempos de incertidumbre* (dirs. Mate Satué/ Hernández Sainz/ Alonso Pérez). Aranzadi.

Ramos Martínez, L.M. (2021). "Los derechos a la intimidad, a la propiedad y a la vivienda; una visión desde el delito de ocupación de bienes inmuebles". *Revista jurídica de la Universidad de León*, n.º 8, 287-296.

Rodríguez Achútegui, E. (2018). "Una visión judicial del interdicto de recobrar la posesión de la Ley 5/2028 frente a la ocupación ilegal de vivienda". *Revista Aranzadi Doctrinal*, n.º 10/2018.

Rodríguez Tirado, A.M. (2024). "La Ley por el derecho a la vivienda y la vulnerabilidad económico-social en los procesos arrendaticios de desahucio de vivienda habitual por falta de pago. ODS y Reglas de Brasilia". *Revista La Ley. Derecho de Familia*, n.º 41,1-29.

Simón Moreno, H. (2021). "La ocupación de viviendas sin título habilitante y los derechos fundamentales y humanos en conflicto". *Revista Crítica de Derecho Inmobiliario*, n.º 786, 2161-2212.

Vázquez Barros, S. (2005). *Los interdictos*. 2.ª edición. Bosch.

Von Ihering, R., traducción de Adolfo Posada (2022). *La posesión*. Ediciones Olejnik.

Capítulo IX

Vulnerabilidad y procesos arrendaticios de desahucio de vivienda habitual por falta de pago tras la ley por el derecho a la vivienda

ANA MARÍA RODRÍGUEZ TIRADO
Profesora Titular de Derecho procesal
Universidad de Cádiz

I. SOCIEDAD, VIVIENDA HABITUAL, VULNERABILIDAD Y JUICIO DE DESAHUCIO

1. Los procesos arrendaticios de vivienda habitual a través del juicio de desahucio por falta de pago en la Ley de Enjuiciamiento Civil: adaptación a la vulnerabilidad

En relación con el arrendamiento de viviendas, las reformas procesales desde el año 2011 hasta el año 2019[1] se diseñaron para atender la pronta recuperación de la posesión por el pro-

---

1 Cabe citar la modificación en materia de arrendamientos en la Ley de Enjuiciamiento Civil por la Ley 23/2003, de 10 de julio; Ley 19/2009, de 23 de noviembre; Ley 4/2011, de 24 de marzo; Ley 5/2012, de 6 de julio; Real Decreto-ley 7/2019, de 1.º de marzo; Ley 12/2023, de 24 de mayo, y el Real Decreto-ley 6/2023, de 19 de diciembre.

pietario o el titular legal en caso de incumplimiento contractual por parte del arrendatario, principalmente, por falta de pago o por expiración del plazo legal o contractual. El acierto de las reformas procesales, en particular, las de 2011 y 2013, ha sido dispar al dirigirse, básicamente, a la protección del arrendador con la finalidad de dinamizar el mercado del alquiler[2]. Así, quedaba relegada la protección procesal del arrendatario, sobre todo, del que se encuentre en situación de vulnerabilidad, inclusive la concurrencia de relaciones de consumo[3]. El legislador se había centrado, dada la alarma social generada, en la protección del deudor de los devastadores efectos de las ejecuciones hipotecarias y no tanto en el progresivo aumento de los desahucios de viviendas habituales arrendadas[4] vinculadas, asimismo, a situaciones de vulnerabilidad económica y social.

---

2 Según Perarnau Moya (2019, p. 2), en relación con el conato de reforma del derogado Real Decreto-Ley 21/2018 (TOL6.955.308) y con la reforma introducida por el Real Decreto-Ley 7/2019 (TOL.7.083.108), "la reforma pretende dar un giro social a los arrendamientos, ofreciendo una mayor seguridad a los inquilinos, incentivando la oferta de viviendas en alquiler a precios asequibles, incluso con estímulos fiscales, y apoyando a las familias más vulnerables, aliviando, por una parte, la carga financiera que el alquiler supone y, por otra, paliando, en lo posible, la dramática situación que el desahucio comporta. Un segundo grupo de medidas busca un incremento de la información y la transparencia en el sector de los arrendamientos de viviendas".

3 Sobre la nulidad de pleno derecho, nulidad relativa y anulabilidad de cláusulas abusivas, *vid.* Moreno García (2019, pp. 31 y ss.).

4 *Vid.* Rodríguez Tirado (2018, p. 16).
En el año 2022, se dictaron 18.687 sentencias dictadas en procesos arrendaticios urbanos de vivienda, de las cuales 14.449 relativas a la causa de la falta de pago de las rentas; por uso distintos del de vivienda, fueron 1.343 (https://www.poderjudicial.es/cgpj/es/Temas/Estadistica-Judicial/Estudios-e-Informes/Resoluciones-dictadas-en-procedimientos-sobre-Arrendamientos-Urbanos/) (Último acceso: 20/11/2023). Con respecto a las sentencias dictadas en caso

Hasta 2019, no existían mecanismos procesales a aplicar cuando se presentaba una demanda pretendiendo el desahucio por falta de pago de las rentas y cantidades debidas en atención a la situación económico-social del arrendatario, es decir, si la falta de pago obedecía a la concurrencia de un factor económico de vulnerabilidad, que dificultaba o impedía afrontar el pago de las rentas y cantidades adeudadas al arrendador. Tampoco se atendía a que dicho factor pudiera concurrir, al mismo tiempo, en el arrendador. Cabe reseñar la existencia de convenios de colaboración entre ayuntamientos, comunidades autónomas y el Consejo General del Poder Judicial, además del plan de vivienda estatal y de las normas autonómicas que inciden en la protección social y en la vivienda social. Por ejemplo, se suscribió en aquel momento un Convenio de colaboración entre el Consejo General del Poder Judicial, la Junta de Andalucía y la Federación Andaluza de Municipios y Provincias (FAMP), sobre la detección de supuestos de vulnerabilidad con ocasión del lanzamiento de vivienda familiar y medidas de carácter social y sobre cesión de datos en los procedimientos de desahucios y ejecución hipotecaria, hecho en Sevilla, a 1 de marzo de 2016[5].

---

de vivienda con causa en la falta de pago, no se indica el tipo de procedimiento seguido, pero parece encajar el juicio de desahucio del art. 440.3 LEC (TOL172.336) siendo el 86,9% sentencias totalmente estimatorias de la demanda y el 6,5%, totalmente desestimatorias. En cuanto a los decretos del LAJ, sólo podrán referirse al juicio de desahucio del art. 440.3 LEC como única causa la falta de pago para solicitar el desahucio (y, en su caso, acumular las rentas adeudadas), se han dictado 35.176 decretos para uso de vivienda (frente a las 4.290 por uso distinto del de vivienda), de los cuales 4.718 conllevaron la entrega de la posesión y 20.545 para la continuación para ejecución.

5 (Última prórroga vigente hasta 2018) (http://www.poderjudicial.es/stfls/CGPJ/RELACIONES%20INSTITUCIONALES/CONVENIOS/FICHERO/2016-9%20ANDALUC%C3%8DA%20DESAHUCIOS.pdf).

La reforma del ya citado Real Decreto-ley 7/2019, de 1.º de marzo[6], intentó atender, aunque tímidamente, al arrendatario vulnerable con la inserción del apartado 5 en el art. 441 LEC[7]. En la propia Exposición de Motivos, justifica los cambios legislativos en la urgencia de "adoptar medidas para corregir los efectos de la reforma practicada en la regulación de los contratos de arrendamiento en 2013, ampliando los plazos legales e introduciendo mecanismos que sirvan para responder a la grave situación que viven los hogares más vulnerables, estableciendo plazos y garantías en el proceso"[8]. Incidía en esta diferenciación en diversos artículos como, por ejemplo, en la duración del plazo de alquiler en uno y otro caso (cinco años –particular– o siete años –empresario–, respectivamente) o en la suspensión del proceso prevista en la anterior redacción del art. 441.5 LEC, que habría de referirse a vivienda habitual si se interpreta sistemáticamente con el art. 549.4.II LEC, tras la redacción conferida por la norma de 2019.

---

6 Convalidado por Resolución de 3 de abril de 2019, del Congreso de los Diputados, por la que se ordena la publicación del Acuerdo de Convalidación del Real Decreto-ley 7/2019, de 1 de marzo, de medidas urgentes en materia de vivienda y alquiler (BOE de 10 de abril de 2019, núm. 86).

7 Ni la Exposición de Motivos ni la redacción del apartado 5 del art. 441 LEC dejaba claro en qué momento procesal se ha de informar al demandado o efectuar la comunicación de oficio a los servicios sociales.

8 Refuerza la protección del arrendatario con la ampliación de la duración mínima del contrato y prórrogas obligatorias.
Para Molina Roig (2018, p. 239) "las vías utilizadas para ofrecer continuidad al arrendatario de una vivienda han sido dos": prórrogas obligatorias y acceso al Registro de la Propiedad para dar publicidad al arrendamiento.

Se constata, asimismo, un aumento de contratos de empresarios o profesionales del ramo inmobiliario como arrendadores[9]. Esta nueva realidad social reviste una mayor complejidad al coexistir con la relación de arrendador y arrendatarios particulares. El juicio de desahucio del art. 438.5 LEC no prevé *a priori* controlar judicialmente de oficio la existencia de cláusulas nulas en contratos de arrendamientos con consumidores en las que pudiera fundamentarse la pretensión como es la falta de pago o como causa expresa de oposición. No contiene mención alguna a la condición de consumidor del arrendatario, aunque no es objeto de este trabajo[10]. Se ha de tener en cuenta que la única causa tasada de oposición prevista es la de alegar y probar el pago o las circunstancias relativas a la procedencia de la enervación (art. 444.1 LEC).

Por su parte, la Ley 12/2023, de 24 de mayo, por el derecho a la vivienda ha revisado diversos preceptos de la Ley de Enjuiciamiento Civil a fin de afianzar no sólo la protección del arrendatario, sino también contemplar la del arrendador, en casos de concurrir en ellos factores de vulnerabilidad económico-social[11]. Al respecto se ha incorporado una suerte de incidente

---

9 Es acertado el análisis efectuado en esta línea por Molina Roig (2018, p. 71).

10 *Vid.* Pérez Daudí (2020) y Rodríguez Tirado (2019, pp. 93 y ss.).

11 No entramos en este trabajo, pero se suma un importante factor psicosocial en la falta de alternativa habitacional o ayudas para abonar las rentas debidas en los arrendamientos de vivienda. Se trató ya en "Los procesos de desahucio de vivienda arrendada por falta de pago y las situaciones de especial vulnerabilidad. Un acercamiento al análisis geográfico y estadístico del desahucio", p. 21.
Vid., v.gr., Bolívar Muñoz y otros (2016, p. 3).
En prensa, *v.gr.*, https://elpais.com/estilo-de-vida/2023-11-25/ansiedad-frustracion-e-incluso-claustrofobia-como-puede-afectar-a-la-salud-mental-vivir-de-alquiler.html

de suspensión del juicio de desahucio por falta de pago de vivienda habitual. Se sigue en la línea marcada por el legislador COVID19 que, a través del Real Decreto-ley 11/2020, de 31 de marzo[12], y sus posteriores actualizaciones, centró su atención en la creación de un incidente extraordinario de suspensión de los juicios de desahucio y de las ejecuciones de vivienda habitual en caso de vulnerabilidad económica vinculada a la pandemia COVID19, cuya aplicación se prorrogó hasta el 31 de diciembre de 2023 por mor del Real Decreto-ley 5/2023, de 28 de junio[13]. Se ha prorrogado de nuevo hasta el 31 de diciembre de 2024 por Real Decreto-ley 8/2023, de 27 de diciembre[14], por el que se adoptan medidas para afrontar las consecuencias económicas y sociales derivadas de los conflictos en Ucrania y Oriente Próximo, así como para paliar los efectos de la sequía.

Hay autores que consideran que, como consecuencia de la reforma de la Ley 12/2023 y la incorporación de los requisitos de admisibilidad (a los que aludimos como requisitos de procedibilidad), "los procedimientos de tutela posesoria pasan a serlo, también, de tutela habitacional"[15]. A nuestro juicio, sería

---

12 Real Decreto-ley 11/2020, de 31 de marzo, por el que se adoptan medidas urgentes complementarias en el ámbito social y económico para hacer frente al COVID-19 (TOL7.854.128).

13 Real Decreto-ley 5/2023, de 28 de junio, por el que se adoptan y prorrogan determinadas medidas de respuesta a las consecuencias económicas y sociales de la Guerra de Ucrania, de apoyo a la reconstrucción de la isla de La Palma y a otras situaciones de vulnerabilidad; de transposición de Directivas de la Unión Europea en materia de modificaciones estructurales de sociedades mercantiles y conciliación de la vida familiar y la vida profesional de los progenitores y los cuidadores; y de ejecución y cumplimiento del Derecho de la Unión Europea (TOL9.619.853).

14 Modifica los arts. 1 y 1 bis del Real Decreto-ley 11/2020.

15 La cita se completa como sigue: "[...] estableciendo un marco de acción de naturaleza procesal, e insertando en él algunas garantías

cuestionable hablar de tutela habitacional hacia el demandado en un proceso arrendaticio de desahucio por falta de pago, sino de la incorporación de medidas provisionales de protección social, como se analizará, ante la concurrencia de una situación de vulnerabilidad económica y, en su caso, social. No cambia la naturaleza de los procesos que se tramitan a través del juicio de desahucio del art. 438.5 LEC ni, por extensión, a los restantes de recuperación de la posesión a través del juicio de desahucio. La tutela privilegiada viene determina por su objeto concreto: la recuperación de la posesión y no por las medidas que "dilatan" su pronta tramitación en aras a la protección económica y social de una situación de vulnerabilidad. Para Montero Aroca (2016, p. 241) la tutela privilegiada puede consistir en crear un proceso especial declarativo, en otras –como es el caso– "en tramitar un asunto por el juicio verbal, sin referencia alguna a la cuantía (art. 250.1 LEC) y otras [...] consisten en conferir una tutela ejecutiva sin haber procedido antes a realizar un proceso de declaración". Ahora bien, este autor aclara que "no se trata de que el privilegio carezca siempre de razones objetivas que aconsejen su mantenimiento, sino de ser conscientes de que se trata realmente de un privilegio, aunque el mismo no sea siempre contrario al principio de igualdad ante la ley (art. 14 CE)". No en vano se ha venido reforzando, como se ha indicado, desde 2011 la posición del colectivo de propietarios o titulares legítimos para recuperar rápidamente la posesión inmediata para no perjudicar sus intereses económicos frente a quien no paga las rentas o cantidades asimiladas del arrendamiento o frente a quien no abandona el inmueble arrendado en caso de expiración del plazo del contrato. Desde 2019, pues, se ha matizado esta tutela exprés –sobre

técnicas para la protección especial de un derecho social (Ferrajoli, 1999, p. 64), como es el derecho al hogar del individuo vulnerable" (Rodríguez-Izquierdo Serrano, 2024, p. 53).

todo en caso de falta de pago de las rentas y cantidades asimiladas– cuando se advierte socialmente que hay arrendatarios que no pagan por concurrir en ellos una situación de vulnerabilidad económica que puede ir acompañada de vulnerabilidad social. La sensibilización del legislador se centra en que el inmueble arrendado tenga la condición de vivienda o vivienda habitual.

2. El concepto de vivienda habitual en los procesos arrendaticios por falta de pago. Incidencia en el juicio de desahucio

La LDV ha incluido definiciones que contribuyen positivamente en la determinación del uso de un inmueble arrendado de cara a la determinación de la vulnerabilidad del arrendatario y, en su caso, del arrendador. Ahora bien, son definiciones que serán de aplicación[16] "en tanto no entren en contracción con las reguladas por las administraciones competentes en materia de vivienda, en cuyo caso, [...] prevalecerán aquéllas" (art. 3.I LDV). Por ello, las utilizaremos a modo de referencia y sin perjuicio de un tratamiento más completo en otro capítulo de esta obra.

Así, *vivienda* será un "edificio o parte de un edificio de carácter privativo y con destino a residencia y habitación de las personas, que reúne las condiciones mínimas de habitabilidad exigidas legalmente, pudiendo disponer de acceso a espacios y servicios comunes del edificio en el que se ubica, todo ello de conformidad con la legislación aplicable y con la ordenación urbanística y territorial" (art. 3.a LDV)[17]. Por vivienda digna[18]

---

16 *Vid.* Nasarre Aznar (2023, p. 2).

17 Contra el art. 3 LDV se ha admitido recurso de inconstitucionalidad por el Tribunal Constitucional, núm. 1306-2024 (BOE núm. 90, 12 de abril de 2024), interpuesto por la Xunta de Galicia.

18 Sobre el derecho a la vivienda digna y el derecho a la tutela judicial efectiva, *vid.* Pérez Daudí (2020, pp. 17 y ss). Este autor llega a la

y adecuada, siguiendo la terminología constitucional, será "la vivienda que, por razón de su tamaño, ubicación, condiciones de habitabilidad, accesibilidad universal, eficiencia energética y utilización de energías renovables y demás características de la misma, y con acceso a las redes de suministros básicos, responde a las necesidades de residencia de la persona o unidad de convivencia en condiciones asequibles conforme al esfuerzo financiero, constituyendo su domicilio, morada u hogar en el que poder vivir dignamente, con salvaguarda de su intimidad, y disfrutar de las relaciones familiares o sociales, favoreciendo el pleno desarrollo y la inclusión social de las personas" (art. 3.c) LDV). Y *residencia habitual* es considerada como "la vivienda que constituye el domicilio permanente de la persona que la ocupa y que puede acreditarse a través de los datos obrantes en el padrón municipal u otros medios válidos en derecho" (art. 3.i) LDV)[19]. Enlazan los conceptos de vivienda y residencia habitual, corrigiendo así nuestra referencia a "vivienda habitual"[20], que es la que mantiene la Ley de Enjuiciamiento

---

conclusión, tras analizar la doctrina del TC y del TEDH, que, con respecto al derecho a la vivienda digna como principio rector de la política social y económica, "no se puede solicitar la aplicación judicial [...]. Sin embargo, lo que sí se puede exigir es que la interpretación que realicen los tribunales de las normas legales aplicables lo tomen en consideración y que lo reconozcan, respeten y protejan [...]". Y desde esta perspectiva se puede denunciar la vulneración de un principio rector de la política social y económica si se relaciona con la interpretación de un derecho fundamental" (*ibidem*, p. 19) como pudiera ser el derecho a la tutela judicial efectiva.

19 También se define la "residencia habitual" como la vivienda "que se utiliza por su propietario para estancias temporales o intermitentes, y que no constituye su residencia habitual" (art. 3.j) LDV).

20 Pallarés Neila (2017, pp. 158-161) diserta sobre la consolidación del concepto de vivienda habitual o de vivienda habitual de la familia. Considera que el carácter de la habitualidad es el dato más impor-

Civil al referirse a la vivienda habitual en casos de vulnerabilidad de los arts. 439.6 y 7, 441.5 y 549.4 LEC. Entendemos que serían equivalentes vivienda habitual y residencia habitual a efectos de su utilización en los procesos arrendaticios de vivienda "habitual" arrendada por falta de pago.

La regulación del juicio de desahucio del art. 438.5 LEC no alude al uso de la vivienda. Se limita a referirse a la pretensión de desahucio por falta de pago de rentas o cantidades debidas de inmueble sin especificar el uso al que esté destinado. El propio articulado del juicio verbal puntualiza, en ocasiones, con la mención a finca urbana (art. 437.3 LEC con respecto al compromiso de condonación); en otras ocasiones, engloba tanto a fincas urbanas como rústicas al no diferenciarlas como ocurre en el art. 438.5 LEC (anterior art. 440.3 LEC). En la redacción vigente de los arts. 439 y 441 LEC, se hace uso de la expresión "vivienda habitual" en casos de vulnerabilidad por falta de pago de inmueble arrendado además de extenderlo a otros procesos de recuperación posesoria.

Por su parte, la Ley 29/1994, de 24 de noviembre, de Arrendamiento Urbanos (TOL231.076) define el "arrendamiento de vivienda" como aquel arrendamiento que recae sobre una edificación habitable cuyo destino primordial sea satisfacer la necesidad permanente de vivienda del arrendatario. Esta nor-

---

tante que permite "distinguir entre la vivienda habitual y el mero alojamiento o la mera residencia [...], debe constituirse como sede jurídica de la persona física, el centro de sus intereses y este título sólo lo puede ostentar el lugar en el que la persona disfruta habitualmente de su intimidad personal y familiar, desde donde normalmente ejerce sus derechos y asume sus obligaciones". Quedarían excluidas las viviendas por temporada, las de recreo o las segundas viviendas. Señala que, a efectos tributarios y administrativos, difiere el concepto vinculado a una mínima permanencia temporal.

ma sigue haciendo uso de la expresión "vivienda habitual" a pesar de las definiciones de la Ley de 2023, por ejemplo, en los arts. 10 y 17 LAU.

No es desdeñable el dato de que el número de sentencias o decretos dictados se refieren porcentualmente a procesos por desahucio de finca arrendada por falta de pago con uso de vivienda[21]. Según la Estadística Judicial del CGPJ, la principal pretensión planteada en los procesos arrendaticios es la de desahucio por falta de pago dentro de los procesos arrendaticios de fincas urbanas destinadas, principalmente, al uso de vivienda.

Resulta relevante acercarse al concepto de "infravivienda", porque puede tener alguna repercusión en los contratos de arrendamiento suscritos y, en su caso, cuando se pretende la recuperación de la posesión por falta de pago siendo el uso para vivienda. Según el art. 3.b) LDV, por "infravivienda" se entiende "la edificación, o parte de ella, destinada a vivienda, que no reúne las condiciones mínimas exigidas de conformidad con la legislación aplicable. En todo caso, se entenderá que no reúnen dichas condiciones las viviendas que incumplan los requisitos de superficie, número, dimensión y características de las piezas habitables, las que presenten deficiencias graves en sus dotaciones e instalaciones básicas y las que no cumplan los requisitos mínimos de seguridad, accesibilidad universal y habitabilidad exigibles a la edificación".

---

21 *Vid.* Nota 9. La Estadística Judicial alude a "vivienda" o a "vivienda permanente" (https://www.poderjudicial.es/cgpj/es/Temas/Estadistica-Judicial/Estadistica-por-temas/Actividad-de-los-organos-judiciales/Estimacion-de-los-tiempos-medios-de-los-asuntos-terminados/) (Último acceso: 29/11/2023).

El INE[22] publica la información para el período 2004-2020 de la población que vive en hogares con determinadas deficiencias[23] en la vivienda (por comunidades autónomas y la Unión Europea y por características individuales). En el año 2020, la media en España fue del 19,7%. En la serie indicada, sólo en los años 2004 y 2010 ha estado por encima de dicho porcentaje.

En cualquier caso y como se ha indicado, la vulnerabilidad económica y, en su caso, social se atiende procesalmente cuando el arrendatario o, en general, el ocupante, estén en un inmueble con uso de vivienda o vivienda habitual ante la

---

22 El INE aclara que "desde la encuesta de 2021, se aplica el Reglamento (UE) 2019/1700 del Parlamento Europeo y del Consejo de 10 de octubre de 2019 por el que se establece un marco común para las estadísticas europeas relativas a las personas y los hogares. Con este nuevo reglamento las preguntas relativas a los problemas de la vivienda se dejan de recoger anualmente, pasando a incorporarse en un módulo dedicado a vivienda que tendrá una periodicidad de 3 años". (https://www.ine.es/ss/Satellite?L=es_ES&c=INESeccion_C&cid=1259948998406&p=1254735110672&pagename=ProductosYServicios%2FPYSLayout¶m1=PYSDetalleFichaIndicador¶m3=1259937499084) (Último acceso: 05/11/2023).

23 "La calidad de la vivienda recoge un conjunto amplio de aspectos, no sólo relacionados con la vivienda en sí misma, sino también con las condiciones del entorno en que esté situada (ruidos, contaminación, problemas medioambientales, delincuencia o vandalismo, etc.)." "Los problemas estructurales de la vivienda (como presentar problemas de goteras, humedades en paredes, suelos, techos, etc.), disponer del espacio suficiente o tener problemas de falta de espacio, sufrir determinadas deficiencias (falta de luz, no poder mantener una temperatura adecuada) constituyen elementos clave para determinar la calidad de la vivienda desde el punto de vista de condiciones materiales." (https://www.ine.es/ss/Satellite?L=es_ES&c=INESeccion_C&cid=1259948998406&p=1254735110672&pagename=ProductosYServicios%2FPYSLayout¶m1=PYSDetalleFichaIndicador¶m3=1259937499084) (Último acceso: 05/11/2023).

falta de uniformidad de las normas analizadas a pesar del catálogo de definiciones del art. 3 LDV.

3. Sociedad, desahucio y norma procesal

La vivienda es un tema recurrente dada la preocupación social manifestada por la ciudadanía[24] desde diverso prisma, que ha sido especialmente candente desde la crisis inmobiliaria de 2008. El nivel de preocupación entre la sociedad española sobre la vivienda ha variado según el CIS, pero sigue siendo uno de los problemas visibles. Así, en el baremo adelantado del mes de noviembre de 2023, ocupa el lugar noveno de los principales problemas de la sociedad española frente a enero de 2023, en el que se encontraba en el puesto vigésimo[25]. En marzo de

---

24 *Vid.*, *v.gr.*, https://elpais.com/economia/2023-05-16/dos-generaciones-atrapadas-por-los-precios-de-la-vivienda.html, https://elpais.com/economia/2023-04-24/alquilar-es-mas-dificil-que-nunca-la-renta-ya-se-come-el-43-del-salario.html, https://elpais.com/economia/2023-03-28/solo-el-2-de-los-hogares-en-riesgo-de-pobreza-reciben-ayudas-para-la-vivienda.html, https://www.elmundo.es/economia/ahorro-y-consumo/2023/05/19/6466667621efa0fd798b45d5.html, https://elpais.com/economia/2023-05-23/el-precio-de-las-casas-sube-al-ritmo-mas-rapido-desde-2006-con-el-alquiler-en-una-situacion-critica.html, https://elpais.com/economia/2024-04-01/el-precio-medio-del-alquiler-se-come-un-40-del-salario-minimo-en-espana.html (último acceso: 2/04/2024).

25 Https://www.cis.es/documents/d/cis/es3427mar_a (Último acceso: 21/03/2024). En el resultado adelantado del barómetro de julio de 2023, ocupa el puesto séptimo de los principales problemas de la sociedad española y el puesto sexto como problema que afecta personalmente (https://www.cis.es/documents/d/cis/es3413mar_a) (Último acceso: 20/11/2023). En el barómetro de enero de 2023, se sitúa en el puesto vigésimo dentro de los principales problemas de la sociedad española y el puesto noveno en cuanto a afectación perso-

2024, pasa al puesto décimo. Si se refiere a la afectación personal del problema de la vivienda en el encuestado, tiende a estabilizarse: se encuentra en el puesto octavo en noviembre de 2023 y noveno en enero de 2023. En marzo de 2024, se sitúa en el cuarto problema como afectación personal del encuestado[26]. Asimismo, se podría percibir el acceso a la vivienda o su permanencia en ella como problema, cuya intensidad varía según el territorio nacional en el que se encuentre ubicada y el nivel económico-social de las familias, todo ello en línea con los precios de compra o alquiler[27] y la existencia de áreas

---

nal (https://www.cis.es/documents/d/cis/es3390mar_a) (Último acceso: 22/03/2024).

En noviembre de 2022, la sociedad española, a través de las encuestas del CIS, colocó el problema de la vivienda en el puesto vigésimo sexto, mientras que se situó en el puesto decimoprimero como problema que afecta personalmente (https://www.cis.es/documents/d/cis/es3384mar_a) (Último acceso: 22/03/2024).

26 Https://www.cis.es/documents/d/cis/es3445marMT_a (Último acceso: 2/04/2024).

27 Es de interés la Resolución adoptada por el Defensor del Pueblo, el 1 de septiembre de 2023, con motivo de la solicitud de interposición de recurso de inconstitucionalidad contra la Ley 12/2023, de 4 de mayo, por el derecho a la vivienda. En esta Resolución se argumenta sobre la limitación de la renta arrendaticia en la Disposición final primera de la Ley 12/2023 en las zonas declaradas como de mercado residencial tensionado. Son varias las cuestiones planteadas sobre la inconstitucionalidad del artículo 3.k), la Disposición transitoria tercera, la Disposición final primera (apartados 1, 3, 5 y 6), la Disposición final quinta y la Disposición final sexta de la indicada Ley 12/2023. El Defensor del Pueblo resuelve no interponer el recurso de inconstitucionalidad solicitado. De momento, la primera sentencia dictada por el Tribunal Constitucional ante el recurso de inconstitucionalidad planteado por la Junta de Andalucía, la STC 79/2024, de 21 de mayo (ECLI:ES:TC:2024:79), publicada en el BOE de 24 de junio de 2024, núm. 152), declara inconstitucionales y nulos el art. 16, el inci-

tensionadas[28], y la respuesta estatal, autonómica y local que se confiera a través de políticas encaminadas a facilitar el acceso a la vivienda o la permanencia en ella a través de ayudas o subvenciones y, en su caso, con el establecimiento de alternativas habitacionales.

Directamente vinculado a la generación de situaciones de vulnerabilidad, emerge el factor económico, que determina la introducción de medidas de protección procesal en la norma procesal civil. Precisamente, el factor económico es relevante a la hora de acceder a una vivienda o mantenerse en ella, al que se puede sumar la vulnerabilidad social.

A modo de ejemplo y a fin de relacionar el factor económico en la sociedad actual con la vulnerabilidad económica, su afectación a la vivienda y la sensibilización del legislador procesal, el Estudio de Población Activa de 2022[29] permite conocer el decil de salarios del empleo principal en la sociedad española. Se obtiene así como resultado destacado que el salario mensual bruto fue de 2.128,4 euros, incluidas las pagas extraordinarias, y el salario medio fue de 1.814,0 euros[30]. Ahora bien, el 40% de los asalariados ganó entre 1.440.1 y menos de 2.373,7 euros en

---

so "que incluirá, con respecto a las viviendas de titularidad del gran tenedor en la zona de mercado residencial tensionado, al menos, los siguientes datos" del art. 19.3, el art 27.1 párrafo tercero, el art. 27.3 y la Disposición transitoria primera de la Ley 12/2023. No afectan a los analizados en la Resolución del Defensor del Pueblo.

28 *Vid., v.gr.,* Vives Miró; Rullán Salamanca; Albert Artigues i Bonet; Navarro Zurriaga (2023, pp. 1-26).

29 Https://www.ine.es/prensa/epa_2022_d.pdf (Último acceso: 2/04/2024, p. 1). Decil de salarios del empleo principal. Encuesta de Población Activa (EPA), 2022, publicado el 24 de noviembre de 2023 por el INE.

30 Es el valor que, "ordenando todos los individuos de menor a mayor salario, deja una mitad de los mismos por debajo de dicho valor y a la

2022 y el 30% ganó menos de 1.440,1 euros. Según este mismo estudio, "los trabajadores más jóvenes se concentran en los salarios más bajos, mientras que los de mayor edad tienen un peso relativo mayor en los más elevados"[31], de modo que, en 2022, "el salario medio bruto se incrementó con la edad, pasando de los 1.315,4 euros que, en promedio, percibieron los menores de 25 años, a los 2.381,2 euros de los asalariados de 55 y más años. El salario varía según la comunidad autónoma de que se trate, oscilando entre los 1.845,3 euros de Extremadura a los 2.545,8 del País Vasco[32]. Si relacionamos el precio del alquiler con el salario medio y la parte del territorio nacional en que se fije la residencia, el factor económico es relevante en la medida en que puede provocar dificultades reales de acceso a la vivienda o de mantenimiento en ella de una parte de la sociedad[33].

Ciertamente, las razones que provocan la escalada de los precios de alquiler o la falta de oferta de vivienda pública suficiente y las políticas sociales que se desarrollan o puedan desarrollar exceden con creces nuestro ámbito de estudio[34]. Tampoco pretendemos realizar un estudio sociológico sobre acceso a la vivienda.

---

otra mitad por encima" (https://www.ine.es/prensa/epa_2022_d.pdf (Último acceso: 2/04/2024), p. 1.

31 Https://www.ine.es/prensa/epa_2022_d.pdf (Último acceso: 2/04/2024), p. 2.

32 *Ibídem*, p. 7.

33 Https://www.fotocasa.es/fotocasa-life/alquiler/conoce-el-precio-de-la-vivienda-en-alquiler-en-tu-zona/ (Último acceso: 15/04/2024).

34 En desarrollo del art. 18 y de la Disposición final primera LDV, se ha aprobado el Decreto Ley 6/2024, de 24 de abril, de medidas urgentes en materia de vivienda de la Generalitat de Catalunya, para dar cabida a la limitación de renta en las zonas declaradas tensionadas (Diari Oficial de la Generalitat de Catalunya de 25 de abril de 2024, núm. 9150).

Si se habla de vivienda, la transversalidad está servida, puesto que se aborda desde distintas especialidades jurídicas y afecta a la delimitación del título competencial. En este sentido, el TC ha declarado que la vivienda "«no constituye un título competencial autónomo» [...], sino que puede recaer bajo distintos títulos competenciales estatales y autonómicos dependiendo de cuál sea el enfoque y cuáles los instrumentos regulatorios utilizados en cada caso por el legislador. Dicha complejidad competencial es consecuencia de las distintas dimensiones constitucionales que presenta la vivienda" como son la económica y la social[35].

El TC ha venido manifestando que, "al legislador estatal, en uso del amplio margen de apreciación del que dispone para adoptar disposiciones en materia social y económica (STC 32/2019, de 28 de febrero, fj 6, TOL7.108.515), le corresponde pronunciarse sobre la adecuación de las bases del derecho contractual a los principios rectores de la política económica y social y, en particular, al derecho de todos los españoles a disfrutar de una vivienda digna y adecuada consagrado por el art. 47 CE, en conexión con la función social de la propiedad privada reconocida por el art. 33.2 CE, introduciendo, en su caso, las modificaciones oportunas. Todo ello, sin perjuicio de las competencias autonómicas en la materia para desarrollar políticas de protección del derecho a la vivienda"[36].

---

35 STC núm. 37/2022, de 10 de marzo de 2022, fj 4 (TOL8.889.749).

36 STC núm. 37/2022, de 10 de marzo de 2022, fj 4: "La reserva al legislador estatal de la competencia para la regulación tanto de la regla general como de sus posibles excepciones o modulaciones excluye, por tanto, que cualquier legislador autonómico pueda condicionar o limitar la libertad de las partes para determinar la renta inicial del alquiler de viviendas. El establecimiento de dicha libertad o, en su caso, su modulación y, en general, la regulación de las reglas para la determinación de la renta, son aspectos esenciales del régimen

El art. 2 LDV fija los fines de las políticas públicas en materia de vivienda, buena parte de ellos han sido objeto de recurso de inconstitucionalidad ante el TC[37].

A nuestro juicio, el conocimiento de la realidad social repercute directamente en la respuesta procesal que el legislador proporcione, en particular y en lo que nos concierne, en el ámbito de la gestión y de la solución de conflictos jurídicos a través del Derecho Procesal.

Todo lo anterior repercute en que el legislador haya incluido medidas procesales en aras de proteger al vulnerable contra el que se dirija una acción de desahucio de la vivienda que ocupe con título o sin título legítimo. También tiene en cuenta otro factor adicional en quien alegue ser el titular de la vivienda (propietario, usufructuario…): el número de propiedades de que disponga y, en su caso, si puede ser vulnerable igualmente.

Se ha de tener en cuenta, además, que el legislador contemporáneo apostó por fomentar el mercado del alquiler[38] tras la crisis de 2008 frente a la preferencia de la sociedad española por la compra de vivienda habitual, si bien ya no queda tan claro que sea la tendencia. Se constata que sigue teniendo mayor implantación

---

obligacional de los contratos de arrendamiento de viviendas que, en su condición de bases de las obligaciones contractuales, deben ser comunes y, por tanto, establecidas por el Estado" (*ibidem*).

37 *Vid.* el recurso de inconstitucionalidad núm. 1278-2024, promovido por la Generalitat de Cataluña y admitido a trámite por el Tribunal Constitucional (BOE de 12 de abril de 2024, núm. 90).

38 Ley 37/2011, de 10 de octubre (TOL2.245.716) y Ley 4/2013, de 4 de junio (TOL3.954.706). El legislador de 2013 fija como objetivo "flexibilizar el mercado del alquiler para lograr la necesaria dinamización del mismo" (Parágrafo II de la Exposición de Motivos), motivado por la crisis financiera y el aumentado de ejecuciones hipotecarias de viviendas habituales.

en España la compra de vivienda frente al alquiler, si bien depende de la franja de edad de acceso a la vivienda[39].

El acceso a la vivienda o el mantenimiento en la vivienda puede generar conflictos jurídicos entre propietario y entidad bancaria con la que se haya constituido un contrato de préstamo hipotecario, entre quien ocupa un bien inmueble con el uso de vivienda y el propietario o el poseedor legítimo, entre arrendador y arrendatario o entre el precarista y el dueño, usufructuario o cualquier otra persona con derecho a poseer el bien inmueble. Al mismo tiempo, pueden concurrir factores que pueden situar a una persona en situación de vulnerabilidad, lo que puede producir efectos jurídicos en el ámbito procesal.

---

[39] Sobre las deficiencias del mercado de alquiler, sobre todo, en la menor oferta de viviendas de alquiler frente a la de vivienda en propiedad, la insuficiencia de las políticas sociales en materia de vivienda, *vid.*, *v.gr.*, Besser Valenzuela (2014, pp. 17 y ss.), y Molina Roig (2018, pp. 32, 113 y ss.).

Con datos actualizados, no queda tan clara la tendencia hacia la compra o el alquiler en términos absolutos. *Vid*, por ejemplo, https://www.ine.es/prensa/ecepov_2021_feb.pdf, https://www.fotocasa.es/fotocasa-life/sector/el-alquiler-vuelve-a-ganarle-terreno-a-la-compra-de-vivienda-el-46-quiere-alquilar-frente-al-44-que-busca-comprar/, https://www.bankinter.com/blog/finanzas-personales/porcentaje-espanoles-vivienda-propiedad, https://www.idealista.com/news/inmobiliario/vivienda/2023/10/17/808638-el-esfuerzo-familiar-para-alquilar-vivienda-en-espana-se-mantiene-en-el-31-de#:~:text=El%20porcentaje%20de%20los%20ingresos,del%20tercer%20trimestre%20de%202023, https://www.bde.es/f/webbde/SES/Secciones/Publicaciones/InformesBoletinesRevistas/BoletinEconomico/23/T2/Fich/be2302-art09.pdf, https://www.bde.es/webbe/es/estadisticas/compartido/datos/pdf/si_1_5.pdf, https://app.powerbi.com/view?r=eyJrIjoiMjE0ZWNjODctYzc4ZS00MjM4LWJkMDktNjRiNDM0ZDY4ZjRmIiwidCI6IjZhYjE1MmQ1LTllZDEtNDkwNi1iNWMyLWMwMjJhNzRhMzU2ZSIsImMiOjl9 (Último acceso: 2/04/2024).

Centrándonos en el mercado del alquiler y en la respuesta procesal ante la falta de pago de las rentas y cantidades asimiladas, que es la principal causa en los procesos arrendaticios judicializados, no entraremos a efectuar un análisis exhaustivo del juicio de desahucio del art. 438.5 LEC (antiguo art. 440.3 LEC) ni sobre su naturaleza jurídica, puesto que ya se ha trabajado con anterioridad[40]. El objetivo principal de este trabajo será analizar la vulnerabilidad económica como criterio introducido por legislador para establecer medidas de protección en situaciones de vulnerabilidad a través de la incorporación de tres principales novedades de la Ley 12/2023, de 24 de mayo, para la tramitación de los procesos de recuperación de la posesión de la vivienda arrendada por falta de pago: la objetivación de la vulnerabilidad económica del arrendatario, así como la valoración por el tribunal de la vulnerabilidad económica y social en arrendador y arrendatario, que se extiende a otros supuestos de recuperación de la posesión; el establecimiento de requisitos de procedibilidad para el demandante en caso de tratarse de gran tenedor, y la previsión de medidas de protección procesal como el incidente de suspensión del juicio de desahucio de la vivienda habitual en caso de situación de vulnerabilidad económica y, en su caso, social del arrendatario. Estas medidas de protección frente a la vulnerabilidad económica se circunscriben al inmueble arrendado que se destine al uso de vivienda habitual. Por ello, se ha dedicado especial atención al concepto de vivienda y vivienda habitual en la LEC, delimitándose también el concepto de vulnerable y vulnerabilidad antes de analizar los criterios objetivados en dicho texto legal.

Como inciso, es de destacar que el art. 118, setenta y cuatro, del Real Decreto-ley 6/2023, de 19 de diciembre, por el que se aprueban medidas urgentes para la ejecución del Plan

---

40 *Vid.* Rodríguez Tirado, (2018, pp. 1 y ss.).

de Recuperación, Transformación y Resiliencia en materia de servicio público de justicia, función pública, régimen local y mecenazgo (TOL9.841.185) traslada la regulación del que denominamos juicio de desahucio desde el art. 440.3 al art. 438.5 LEC, además de adaptar los apartados 4 y 5 del art. 440 como nuevo apartado 6 del art. 438 LEC. Así, fusiona en el apartado 6 del art. 438 LEC los antiguos apartados 4 y 5 del art. 440 LEC, cuyo apartado 5[41] había sido introducido por la LDV. Se prevé que la citación para recibir la notificación de la sentencia se efectuará de forma presencial o a través de la sede electrónica (art. 438.6 LEC).

No compartimos esta técnica legislativa. Este cambio sin justificación al hilo de una reforma con objetivos diferentes y sin una urgencia real, sobre todo si tenemos en cuenta que la LDV había abordado ya cambios en materia de vivienda en la LEC y, quizás, hubiera justificado en esta otra norma que legislativamente se abordara el cambio de ubicación. No obstante, a nuestro juicio y como hemos sostenido en otros trabajos, no es suficiente el cambio de ubicación de la regulación del juicio de desahucio por falta de pago del art. 440.3 al 438.5 LEC, porque, en realidad, incorpora la técnica monitoria, además de establecer una tutela privilegiada, con entidad suficiente para convertirse en una clase de proceso especial con su propio cauce procedimental más allá de las meras especialidades del juicio verbal[42].

---

41 La LDV adicionó el nuevo apartado 5 en el art. 440 LEC que disponía que, "en todos los casos de desahucio y en todos los decretos o resoluciones judiciales que tengan por objeto el señalamiento del lanzamiento, independientemente de que éste se haya intentado llevar a cabo con anterioridad, se deberá incluir el día y la hora exacta en los que tendrá lugar el mismo".

42 Rodríguez Tirado (2012).

II. PROTECCIÓN PROCESAL-SOCIAL EN LOS PROCESOS ARRENDATICIOS DE DESAHUCIO DE VIVIENDA HABITUAL A TRAVÉS DEL JUICIO DE DESAHUCIO POR FALTA DE PAGO

El juicio de desahucio de vivienda arrendada del art. 438.5 LEC (anterior art. 440.3 LEC) ha sufrido sucesivas adaptaciones por los vaivenes legislativos desde hace más de una década, siendo la reforma de 2019 la que, por primera vez, tiene en cuenta la vulnerabilidad del arrendatario como se ha referido con anterioridad. En la reciente de 2023[43], se objetivan los criterios que determinan la vulnerabilidad económica de arrendatarios básicamente (art. 441.5, 6 y 7 LEC), además de establecerse requisitos de procedibilidad (o de admisibilidad[44]) que condicionan la admisión de la demanda en caso de parte actora que sea considerada como gran tenedora (art. 439.7 LEC). Se acompañan de medidas de protección social mediante la suspensión temporal del juicio de desahucio que, en realidad, parece diseñarse como un incidente.

Se trata de medidas previstas para el ejercicio de la acción de desahucio para recuperación de la posesión del inmueble que constituya vivienda habitual de la parte demandada a través del juicio verbal (pretensiones de los números 1.°, 2.°, 4.° y 7.° del art. 250 LEC), que ceñiremos a los procesos arrendaticios de recuperación de la posesión de vivienda habitual por falta de pago de las rentas y cantidades asimiladas (art. 250.1.1.° LEC) mediante el juicio de desahucio del art. 438.5 LEC.

---

43 *Vid.* Torres López, (2023, pp. 7-12); Bonet Navarro (2021); Adán Doménech (2017).

44 En cualquier caso, la no concurrencia de dichos requisitos afecta a la admisión a trámite de la demanda. Como se apuntó, Rodríguez-Izquierdo Serrano (2024, p. 53), alude a criterios de admisibilidad.

La redacción del art. 438.5 LEC (hasta el 29 de marzo de 2024, ubicado en el art. 440.3 LEC) se ha mantenido en esencia desde la reforma incorporada por la ya citada Ley 4/2013, de 4 de junio. El Real Decreto-ley 7/2019, de 1 de marzo matizó las referencias a "fecha" por "día y hora señaladas" o "día y horas exactos" en la redacción del art. 440. 3 y 4 LEC e incorporó un nuevo apartado 5 en el art. 441 LEC[45], cuya redacción se ha modificado por la Ley 12/2023. Por primera vez, se insertaba en la norma procesal una medida de protección social con respecto al arrendatario vulnerable o que pudiera serlo por concurrir con factor de vulnerabilidad económico-social. En concreto, se preveía informar al demandado de la posibilidad –por parte del Letrado de la Administración Justicia en la cédula de emplazamiento con identificación de los servicios sociales– de que acudiese a los servicios sociales y, en su caso, de la posibilidad de autorizar la cesión de sus datos a los servicios sociales, a efectos de que pudieran apreciar la posible situación de vulnerabilidad. Por su parte, el juzgado informaría de oficio a los servicios sociales de la existencia del proceso arrendaticio iniciado. En principio, podría referirse a la pretensión de falta de pago como de expiración del plazo contractual o legal con o sin acumulación de la pretensión de reclamación de las rentas o cantidades debidas. Esta última posibilidad no se excluía al referirse a los casos del número 1.° del art. 250.1 LEC. Correspondía a los servicios sociales la confirmación de la situación de vulnerabilidad económico y/o social del hogar. No se contenían criterios objetivos para su determinación en la LEC.

---

[45] La norma de 2019 incorporó un nuevo párrafo en el art. 549.4 LEC como matización en caso de que se trate de vivienda habitual, con carácter previo al lanzamiento, se debía proceder conforme al art. 441.5 LEC, que preveía la comunicación con los servicios sociales. En mayo de 2023, se retoca sustancialmente la redacción del art. 441.5 LEC, como se analiza en este trabajo.

Cabe matizar que, durante la pandemia y prorrogadas en la post pandemia, se establecieron medidas de protección con elementos comunes que aún se siguen aplicando en los supuestos de vulnerabilidad para los que fueron establecidas, que no son coincidentes con los introducidos en la LEC por la LDV. Por ello, se abordará de forma separada, y como antecedente inmediato de la regulación general prevista en la LEC. En realidad, conviven dos regímenes en paralelo, sin que exista total homogeneidad de supuestos, requisitos y medidas establecidas, uno que es temporal y otro que se ha establecido como general.

1. Incidente de suspensión extraordinaria del proceso arrendaticio de desahucio de hogares vulnerables sin alternativa habitacional del art. 1.º del Real Decreto-ley 11/2020

Podemos encontrar como antecedente inmediato de las medidas incorporadas a través de la LDV en la Ley de Enjuiciamiento Civil, precisamente, la normativa COVID19 y, en concreto, el Real Decreto-ley 11/2020, de 31 de marzo, por el que se adoptan medidas urgentes complementarias en el ámbito social y económico para hacer frente al COVID-19 en sus sucesivas modificaciones[46]. Aunque la temporalidad se prometía más corta, aún sigue vigente y, en lo que concierne a la suspensión del proceso arrendaticio de desahucio o de la ejecución del lanzamiento en hogares vulnerables, su aplicabilidad se ha prorrogado hasta el 31 de diciembre de 2024.

---

46 Ha sido modificado por diversas normas posteriores desde el Real Decreto-ley 15/2020, de 21 de abril, al Real Decreto-ley 8/2023, de 27 de diciembre. Contra el Real Decreto-ley 11/2020, se ha planteado recurso de inconstitucionalidad resuelto por el Tribunal Constitucional en la STC 7/2024, de 16 de enero de 2024 (TOL9.863.735) fallando la pérdida sobrevenida del objeto y, en lo demás, la desestimación del recurso.

El Real Decreto-ley 8/2023, de 27 de diciembre, por el que se adoptan medidas para afrontar las consecuencias económicas y sociales derivadas de los conflictos en Ucrania y Oriente Próximo, así como para paliar los efectos de la sequía (TOL9.813.475) modifica el art. 1 del Real Decreto-ley 11/2020, de 31 de marzo, por el que se adoptan medidas urgentes complementarias en el ámbito social y económico para hacer frente al COVID-19. En concreto, prorroga el plazo para solicitar la suspensión extraordinaria del desahucio o del lanzamiento en procesos arrendaticios de desahucio por falta de pago o por expiración del plazo legal hasta el 31 de diciembre de 2024. Se refiere a una suspensión diferente a la prevista en el anterior art. 441.5 LEC (redacción de 2019)[47] y ahora en la vigente del art. 441.5, 6 y 7 LEC.

Ya son varias las prórrogas, por lo que no sería descartable una adicional[48] y, de esta manera, se continuaría contando con

---

47 "[...] se haya suspendido o no previamente en proceso en los términos establecidos en el apartado 5 del artículo 441 de la Ley 1/2000, de 7 de enero, de Enjuiciamiento Civil, la persona arrendataria podrá instar, de conformidad con lo previsto en este artículo, un incidente de suspensión extraordinaria del desahucio o lanzamiento ante el Juzgado por encontrarse en una situación de vulnerabilidad económica que le imposibilite encontrar una alternativa habitacional para sí y para las personas con las que conviva" (art. 1.1 Real Decreto-ley 11/2020).

48 En el caso de las medidas de suspensión de lanzamientos sobre la vivienda habitual para la protección de los colectivos vulnerables de deudores hipotecarios, se ha prorrogado la suspensión hasta 2028. Así, en el art.1.1.I de la Ley 1/2023, de 14 de mayo, de medidas para reforzar la protección a los deudores hipotecarios, reestructuración de deuda y alquiler social (TOL9.398.154), se modifica la duración de la medida de lanzamiento "hasta transcurridos quince años desde la entrada en vigor" de la indicada Ley. Se refiere tanto al lanzamiento derivado de un proceso judicial de ejecución hipotecaria como extrajudicial dentro del marco objetivo y subjetivo establecido en dicha norma.

un régimen general previsto en la Ley de Enjuiciamiento Civil y otro excepcional en la indicada norma de 2020 que, además, prevalece sobre el contenido en la Ley de Enjuiciamiento Civil. Así, el art. 1.1 Real Decreto-ley 11/2020 dispone que, referidos a juicios verbales de desahucio que pretendan la recuperación de la posesión (por falta de pago o expiración del plazo legal o contractual del contrato), "se haya suspendido o no previamente el proceso en los términos establecidos en el apartado 5 del artículo 441 de la Ley 1/2000, de 7 de enero, de Enjuiciamiento Civil, la persona arrendataria podrá instar, de conformidad con lo previsto en este artículo, un incidente de suspensión extraordinaria del desahucio o lanzamiento ante el Juzgado por encontrarse en una situación de vulnerabilidad económica que le imposibilite encontrar una alternativa habitacional para sí y para las personas con las que conviva".

El art. 1.1 Real Decreto-ley 11/2020 establece un incidente de suspensión extraordinaria del desahucio o del lanzamiento, que se abre dentro del juicio de desahucio que esté incoado, o de la ejecución, puesto que se refiere a la suspensión del desahucio o lanzamiento. Basta que la demanda de desahucio incluya la solicitud de ejecución para que se inicie directamente el proceso de ejecución una vez finalizado el declarativo (arts. 438.5 y 549.3 LEC).

El propio art. 1.1 Real Decreto-ley 11/2020 regula la tramitación específica para la resolución de dicho incidente. Si aún no hubiera fecha para el lanzamiento por no haber transcurrido los diez días del requerimiento efectuado al arrendatario o por no haberse celebrado la vista, se suspenderá aquel plazo o la celebración de la vista. Previamente, se habrá de acreditar la situación de vulnerabilidad económica por la parte arrendataria. Por ello, aunque no se indica en dicho precepto, se ha de entender que el juicio de desahucio quedará sujeto a lo que se resuelva sobre el incidente abierto para resolver sobre la suspensión extraordinaria solicitada, que podría considerarse de previo pronunciamiento a tenor del art. 390 LEC con tramitación especial y, por consiguiente, no se entrará a resolver sobre el fondo.

Los plazos de los apartados 2, 3 y 4 del art. 1 Real Decreto-ley 11/2020 no casan con los previstos para la resolución de incidente de suspensión previsto en el art. 441.5, 6 y 7 LEC.

Para ello, el incidente se inicia con la solicitud del arrendatario en la que deberá acreditar que se encuentra en alguna de las situaciones de vulnerabilidad económica de las letras a) y b) del art. 5.1 Real Decreto-ley 11/2020 con la aportación documental prevista en el art. 6.1 del mismo Real Decreto-ley: por ejemplo, acreditación de la situación legal de desempleo mediante certificado de la entidad gestora de las prestaciones o mediante certificado expedido por la Agencia Estatal de la Administración Tributaria o el órgano competente de la comunidad autónoma en caso de trabajadores por cuenta propia, la acreditación del número de personas que habitan en la vivienda habitual, nota simple del Registro de la Propiedad y la declaración responsable del deudor. La acreditación de vulnerabilidad presentada por el arrendatario será trasladada al arrendador por el Letrado de la Administración de Justicia para que, en el plazo de diez días, el arrendador acredite por los mismos medios si se encuentra en la situación de vulnerabilidad económica de la letra a) del mencionado art. 5.1 o en riesgo de situarse en ella en caso de que se adopte la medida de suspensión del lanzamiento.

El Letrado de la Administración de Justicia solicitará informe a los servicios sociales en el que valore la situación de vulnerabilidad del arrendatario y, en su caso, del arrendador, e identifique las medidas a aplicar por la administración competente. A tal efecto, trasladará inmediatamente a los servicios sociales competentes la documentación recibida de arrendatario y arrendador. El plazo para emitir informe será de diez días (art. 1.3 Real Decreto-ley 11/2020).

El incidente es resuelto por el juez mediante auto a la vista de la documentación presentada por las partes y del informe recibido de los servicios sociales. En el auto acordará la suspensión del lanzamiento si se considera acreditada la situación de vulnerabilidad económica y, en su caso, que no debe prevalecer la vulnerabilidad del arrendador. De no quedar acreditada la vulnerabilidad económica del arrendatario o deba prevalecer la del arrendador, el juez acordará la continuación del juicio de desahucio o de la ejecución del lanzamiento. El juez competente será el mismo que esté conocimiento del proceso principal.

El plazo de duración de la suspensión que se acuerde, como ya se ha indicado, se ha ido prorrogando desde la promulgación del citado Real Decreto-ley 11/2020. En la redacción vigente del art. 1.4 se determina que el auto que fije la suspensión señalará que el 31 de diciembre de 2024 se reanudará automáticamente el cómputo de los días a que se refiere el juicio de desahucio del art. 438.5 LEC o se señalará la fecha para la celebración de la vista (si hubiera oposición) y, en su caso, del lanzamiento, según el estado en que se encontrase al declararse la suspensión (art. 1.4 Real Decreto-ley 11/2020).

Durante el periodo de suspensión del juicio de desahucio o de la ejecución, las Administraciones públicas competentes deberán ejecutar las medidas establecidas en el informe de servicios sociales o adoptar aquellas otras que consideren adecuadas para satisfacer la necesidad habitacional de la persona en situación de vulnerabilidad que garanticen su acceso a una vivienda digna. Adoptadas estas medidas, se comunicará al juzgado para que, en el plazo de tres días, el Letrado de la Administración de Justicia acuerde el levantamiento de la suspensión del procedimiento mediante decreto.

Los requisitos para apreciar la situación de vulnerabilidad económica del arrendatario del art. 5.1 del Real Decreto-ley

11/2020 no son coincidentes con los establecidos en el art. 441.7 LEC. Aparte de que es diferente la tabla para obtener el cálculo del conjunto de ingresos económicos de la unidad familiar, se exige al arrendatario que pase a estar en situación de desempleo, Expediente Temporal de Regulación de Empleo (ERTE), o haya reducido su jornada por motivo de cuidados, en caso de ser empresario, u otras circunstancias similares que supongan una pérdida sustancial de ingresos, que no se exigen en la Ley de Enjuiciamiento Civil.

Ahora bien, se introduce un matiz relevante con respecto a la suspensión del art. 1 Real Decreto-ley 11/2020, de 31 de marzo. En el caso de que la demandante sea una gran tenedora de vivienda, se reanudará el juicio de desahucio o el lanzamiento a petición expresa de la parte demandante gran tenedora en caso de que acredite haberse sometido al procedimiento de conciliación o intermediación que establezcan las Administraciones Públicas y siempre que lo acredite conforme a lo previsto en la Disposición transitoria tercera de la LDV. La fecha a que se refiere es a partir del 31 de diciembre de 2024[49].

Y otro aspecto relevante que se prevé en la Disposición adicional segunda del Real Decreto-ley 37/2020, de 22 de diciembre, de medidas urgentes para hacer frente a las situaciones de vulnerabilidad social y económica en el ámbito de la vivienda y en materia de transportes (TOL8.241.535) es el derecho de los arrendadores y propietarios a recibir una compensación económica en caso de que se proceda a la suspensión extraordinaria del art. 1 Real Decreto-ley 11/2020 y de la que se hará cargo "la administración competente" en el caso

---

49 Disposición transitoria tercera modificada por el art. 88 Real Decreto-ley 8/2023, de 27 de diciembre.

de que esta no hubiera adoptado las medidas adecuadas para atender la situación de vulnerabilidad acreditada del arrendatario en los tres meses posteriores a la emisión del informe por dicha administración. Es decir, no se aplica en todos los supuestos, sino en caso de no adopción de dichas medidas en el indicado plazo.

Es interesante tener en cuenta la interpretación de la STC 15/2023, de 7 de marzo (TOL9.466.538)[50], que se pronuncia sobre un recurso de inconstitucionalidad contra la disposición final primera del Real Decreto-ley 1/2021, de 19 de enero, (TOL8.262.238) que modificó el art. 1 bis del Real Decreto-ley 11/2020. Si bien no cuestionó la medida de suspensión de desahucio de viviendas arrendadas, sino la de viviendas ocupadas-usurpadas de manera ilegal sin título habilitante en el marco de un proceso penal, se ha de tener en cuenta la argumentación relativa al derecho a la tutela judicial efectiva y la finalidad del interés social de la medida de suspensión. Aunque ya el contenido afectado se había derogado, el Tribunal Constitucional se pronuncia, en lo que se refiere a la medida de suspensión, declarando que "los incisos impugnados prevén una medida limitada en cuanto a su ámbito de aplicación subjetivo y objetivo, que ni tiene por objeto una regulación directa y general del derecho de propiedad de la vivienda (art. 33 CE), ni afecta a su contenido esencial. Una medida que responde a una finalidad de interés social, con una incidencia muy limitada y temporal

---

[50] Esta sentencia cuenta con tres particulares que disienten de la mayoría en relación a la falta de suficiente explicitación y de concurrencia del presupuesto habilitante de la extraordinaria y urgente necesidad, así como por la ausencia de conexión de sentido entre dicho presupuesto y la medida urgentemente adoptada, y la infracción de los límites materiales de los decretos-leyes.

sobre el citado derecho, y cuya efectiva adopción requiere la ponderación de las circunstancias del caso concreto por parte del órgano judicial competente" (fj 7)[51].

---

[51] La STC 9/2023, de 22 de febrero (TOL 9.440.831) resuelve un recurso de inconstitucionalidad sobre varios incisos de la Disposición final primera del Real Decreto-ley 1/2021, de 19 de enero, por el que modifica algunos preceptos del Real Decreto-ley 11/2020. Los recurrentes sostienen que "los incisos recurridos infringen [...] la prohibición que se desprende del art. 86.1 CE de afectar al derecho de propiedad, al ser este uno de los derechos regulados en el título I; incidiendo directamente en las vertientes subjetiva y objetiva que conforman el derecho. En la primera, al facultar al juez para suspender el desahucio en procesos penales respecto de personas "económicamente vulnerables y sin alternativa habitacional" que estén habitando viviendas sin ningún título habilitante para ello, se condiciona necesariamente el poder de disposición sobre dichas viviendas de sus legítimos propietarios. Y, en la vertiente objetiva, al hacer ceder el derecho de disposición de estos últimos en atención a una pretendida (aunque solo aparente e inadmisible constitucionalmente) función social".

Los recurrentes mantienen, asimismo, que "el derecho a la tutela judicial efectiva sin dilaciones indebidas (art. 24.1 CE) comprende [...] el derecho a la ejecución de las resoluciones judiciales, lo que incluye el derecho a que los fallos judiciales se cumplan "en sus propios términos" (STC 207/2003, por todas). Por ello, una regulación como la contenida en el Real Decreto-ley impugnado, de acuerdo con la cual el juez puede suspender el desahucio dispuesto en un procedimiento penal, incide directa y no tangencialmente en dicho derecho, razón por la cual ha de considerarse inconstitucional".

El Tribunal Constitucional fundamenta que, "en el presente caso, estamos ante la atribución al juez, por vía legal de urgencia, de una potestad de suspensión del lanzamiento de la vivienda habitual de aquellas personas que la estén habitando sin ningún título habilitante para ello, personas económicamente vulnerables y sin alternativa habitacional. Potestad de suspensión que el juez ha de adoptar previa valoración ponderada y proporcional de los intereses que están en juego en el concreto caso y de las circunstancias concurrentes. En otras palabras, la norma cuestionada se limita a otorgar un margen de apreciación al

Cabe concluir, pues, que el incidente de suspensión extraordinario previsto en el art. 1 Real Decreto-ley 11/2020 tiene un ámbito de aplicación más limitado para su adopción no siendo extensible a todos los procesos arrendaticios de desahucio por falta de pago en el que el arrendatario pudiera encontrarse en situación de vulnerabilidad del art. 438.5 LEC en relación con los arts. 441.7 LEC, sino a aquellos en los que, en el arrendatario, concurran concretos requisitos. En el caso de la norma COVID19, se vincula a unas determinadas situaciones laborales que no se prevén en la Ley de Enjuiciamiento Civil, que se ciñe a aplicar la tabla para determinar el conjunto de ingresos de la unidad familiar que, además, difiere de la contemplada en la norma de 2020.

Es de destacar que la regulación de la normativa COVID19 prevé la doble valoración de la vulnerabilidad de arrendatario y arrendador conforme a los mismos criterios objetivos del art. 5.1 Real Decreto 11/2020. No queda tan claro en el art. 441.7 LEC, salvo la acreditación que sí se prevé sin especificar concretos requisitos económicos.

---

juzgador para conceder o denegar, ponderadas las circunstancias del caso, la suspensión del lanzamiento de la vivienda habitual; decisión que, en cada caso concreto, será susceptible de revisión jurisdiccional. En definitiva, la ejecutividad de las sentencias no se discute ni se cuestiona por la norma impugnada, que solamente introduce la posibilidad de dilatarla en el tiempo y por un breve lapso temporal. Con este alcance limitado, la medida prevista no tiene por objeto una regulación directa y general del derecho a la ejecución de las resoluciones judiciales en sus propios términos, por lo que la queja relativa a los límites materiales del Real Decreto-ley previstos en el art. 86.1 CE debe ser desestimada en este punto (fj 5)". No obstante, el fallo se ciñó a declarar la pérdida de objeto del recurso de inconstitucionalidad.

2. Vulnerabilidad económico-social de arrendador y arrendatario de vivienda habitual en procesos de desahucio por falta de pago

Una definición general y transversal de vulnerable y de vulnerabilidad no es posible encontrarla en las normas procesales. Se articula como un concepto jurídico indeterminado[52] a concretar según el ámbito al que se aplique y con la concurrencia de determinadas circunstancias o factores endógenos o exógenos que sitúen a la persona con alta probabilidad de padecer un perjuicio o daños identificables con el riesgo de que no tenga capacidad o medios para protegerse a sí misma[53]. En este caso, se ha de circunscribir a los procesos arrendaticios de desahucio por falta de pago de vivienda, pudiendo colocarse en dicha situación tanto arrendatario como arrendador.

---

52 En este sentido, La Barbera (2019) pone de manifiesto que, "a pesar de su cada ver más explícito uso (vulnerabilidad), el contenido de la vulnerabilidad como categoría jurídica es aún ambiguo. Si bien es posible que el éxito de esta noción radique precisamente en su vaguedad e indeterminación, la utilización en decisiones judiciales de categorías ambiguas conduce a menudo a fallos contradictorios y efectos perversos en su implementación".

53 Sobre el concepto de vulnerabilidad -teniendo en cuenta que se centra en el ámbito de la salud pública y con las correspondientes prevenciones en su lectura-, *vid.* Zarowsky; Haddad; Nguyen (2013, pp. 95-97): "Nuestra idea de vulnerabilidad incluye tanto aspectos estáticos y más fácilmente medibles, como dimensiones dinámicas, que se centran más en la forma en que las personas reaccionan a los percances (shocks) recurrentes y se ven afectadas por ellos a lo largo del tiempo. De acuerdo con una extensa literatura al respecto, entendemos la vulnerabilidad a la vez como una situación y como un proceso: una situación de fragilidad extrema de una población o de un grupo concreto y un proceso que es potencialmente reversible o evitable si se aplican las medidas adecuadas" (*ibidem*, p. 95). Cabe añadir a población o grupo concreto.

Según la Real Academia de la Lengua, vulnerable significa, en una única acepción, "que puede ser herido o recibir lesión, física o moralmente"[54]. Para Trinidad Núñez, alguien correctamente protegido ya no será vulnerable[55]. Y vulnerabilidad, también en una única acepción, se representa como "cualidad de vulnerable"[56], es decir, calidad, condición o naturaleza de vulnerable. En este sentido, resultaría redundante hablar de "condición de vulnerabilidad", porque estaría refiriéndose a la naturaleza de vulnerable. Quizás y a fin de integrar ambos conceptos, podríamos aludir a personas o colectivos en los que concurran concretos factores endógenos o exógenos de vulnerabilidad[57] que los conviertan en vulnerables cuando esa causa se hace efectiva en ellos[58]. Por ejemplo, los menores son vulnerables por el hecho de ser menores, por un factor endógeno, y a los que nuestro ordenamiento, en general, brinda una especial protección de manera transversal. El arrendatario o el arrendador serán vulnerables si se cumplen las condiciones de vulnerabilidad al responder a factores exógenos, principalmente, económicos (aunque se sume la concurrencia de otros factores de vulnerabilidad como la convivencia con menores, con mayores o con personas con discapacidad necesitada de especial protección).

---

54 REAL ACADEMIA DE LA LENGUA, "Vulnerable", en *Diccionario de la lengua española*, https://dle.rae.es/vulnerable, s.f. (Último acceso: 22/3/2024).

55 Trinidad Núñez (2012, p. 129). Esta autora puntualiza que no ha de confundirse "debilidad" con "vulnerabilidad".

56 *Diccionario de la lengua española*, https://dle.rae.es/vulnerabilidad?m=form, s.f. (Último acceso: 22/3/2024).

57 Álvarez Alarcón (2023, pp. 24-25) analiza el concepto de persona en condición de vulnerabilidad y los elementos subjetivo y objetivo del mismo en relación con el acceso a la justicia.

58 *Vid.* Rodríguez Tirado (2023, pp. 94-97).

Hernández Díaz-Ambrona considera que, "por un lado, es difícil establecer normas jurídicas específicas que traten los casos de vulnerabilidad debido a la fluidez del propio concepto, y por otra parte, siempre será necesario investigar como surgen las situaciones de vulnerabilidad y que es lo que se puede hacer a fin de garantizar un menor número de casos en los que el consumidor resulte perjudicado precisamente por su propia vulnerabilidad"[59]. Se centra en el consumidor en general, aunque podría ser extrapolable al arrendatario consumidor[60].

---

59 Hernández Díaz-Ambrona (2015, p. 9). Esta autora continúa su reflexión sobre la dificultad de dar un concepto de vulnerabilidad. Distingue implícitamente entre consumidor vulnerable y consumidor en una situación de vulnerabilidad. En relación con el concepto utilizado de consumidor vulnerable, entiende que "se basa en la noción de vulnerabilidad endógena y hace referencia a un grupo heterogéneo de personas compuesto por aquellas consideradas de forma permanente como tales por razón de su discapacidad mental, física o psicológica, su edad, su credulidad o su género y además el concepto de consumidor vulnerable debe incluir asimismo a los consumidores en una situación de vulnerabilidad, es decir, el consumidor que se encuentra en una situación de impotencia temporal derivada de una brecha entre su estado y sus características individuales, por una parte y su entorno externo, teniendo en cuenta criterios tales como la educación, la situación social y financiera (por ejemplo el endeudamiento excesivo), o el acceso a las nuevas tecnologías)" (*ibidem*, p. 23). Considera que "la vulnerabilidad tiene en cuenta las características individuales del consumidor, así como su entorno, la educación recibida, su situación social y financiera, y el conocimiento de las nuevas tecnologías" (*ibidem*, p. 24).

60 Colomer Hernández (2022, p. 66) atiende a que "el elemento esencial para determinar el carácter vulnerable de una persona consumidora se encuentra en el hecho de no estar en condiciones de igualdad en el ejercicio de sus derechos como consumidores a consecuencia de encontrarse en una situación de subordinación, indefensión o desprotección".

El arrendador y el arrendatario en quienes concurra un factor de vulnerabilidad serán aquellas personas que presentan mayor posibilidad de ser dañadas patrimonialmente y moralmente. De ahí surge la necesidad de establecer medidas de protección para equilibrar su situación y, en lo que concierne a los procesos arrendaticios de desahucio por falta de pago, en su acceso a la justicia. No por el hecho en sí mismo de que el arrendatario no haya abonado las rentas o cantidades adeudadas, que si efectivamente no ha pagado según el contrato suscrito con el arrendador, debería hacerlo al ser causa de resolución del contrato, sino por el cumplimiento de un factor de vulnerabilidad que puede llevarlo al umbral de la exclusión social. Todo ello sin perder de vista el derecho del arrendador a recuperar la posesión inmediata de la vivienda y de las rentas que ha dejado de percibir. Por ello, abordamos la pobreza en este apartado, pues tiene una relación directa con la vulnerabilidad económico-social incluida por el legislador como causa que lleve a la suspensión del juicio de desahucio y que enlaza con el análisis económico-social del epígrafe I de este trabajo. Supone atender la concurrencia de derechos fundamentales y derechos sociales en el acceso a la justicia por ambas partes.

Con la modificación de la LEC por la LDV, se concreta, por primera vez, criterios objetivos la vulnerabilidad económica en el art. 441.7 LEC. Básicamente, se atiende al parámetro económico para determinar la vulnerabilidad económica que, a su vez, puede conllevar o ir acompañada de otros factores de vulnerabilidad social.

Según el Plan de Acción para la Implementación de la Agenda 2030 por España, en relación con el Objetivo 1 y su estado de situación, la tasa de riesgo de pobreza se corresponde con hogares formados por un adulto solo con uno o más hijos dependientes. De la serie 2008-2017, la tasa de riesgo de pobreza es del 40,6% (2017) en estos casos, lo que repercute, indudablemente, en el acceso a la vivienda digna y al pago de las rentas del arrendamiento o cuotas de préstamos hipotecarios[61].

Los Objetivos de Desarrollo Sostenible (ODS)[62] de Naciones Unidades incluyen en su Objetivo 16 el acceso a la justicia en condiciones de igualdad ("Paz, justicia e instituciones sólidas)[63]. Una de las acciones para lograr la meta 16.3, fijada en el Plan de Acción para la implementación de la Agenda 2030, aprobada por el Gobierno de España el 29 de junio de 2018, es la de Estado de Derecho y acceso a la justicia[64].

---

61 Gobierno de España, *Plan de Acción para la Implementación de la Agenda 2030. Hacia una Estrategia Española de Desarrollo Sostenible*, Agenda 2030, p. 16 (https://www.mdsocialesa2030.gob.es/agenda2030/documentos/plan-accion-implementacion-a2030.pdf) (Último acceso: 22/3/2024).
Atendido el nivel de educación, analizada la población mayor de 16 años, "los porcentajes más altos de la tasa de riesgo de pobreza en la serie de encuestas (2008-2017, base 2013) correspondieron a los niveles más bajo de educación primaria o inferior (26,5%) y secundaria 1.ª etapa (27,1%)". Si se atiende a la nacionalidad, esos niveles altos de riesgo de pobreza, se concentra en los de nacionalidad extranjera no de la UE alcanzando el 52,1% en 2017 (*ibidem*, pp. 17-19).

62 "Los objetivos de desarrollo sostenible (ODS) son una llamada a la acción a todos los países para erradicar la pobreza y proteger el planeta así como garantizar la paz y la prosperidad" (https://www.un.org/sustainabledevelopment/es/peace-justice/) (Último acceso: 22/3/2024).

63 En el Informe de Progreso 2023 de la Estrategia de Desarrollo Sostenible 2030, en el que se examina el cumplimiento de la Agenda 2030 en España (https://www.mdsocialesa2030.gob.es/agenda2030/documentos/IP23_EDS.pdf) (Último acceso: 22/3/2024). La Estrategia de Desarrollo Sostenible 2030 se aprobó en 2021 a través del despliegue de diversas políticas públicas. En España, se aprobó el Plan de acción para la implementación de la Agenda 2030 el 29 de junio de 2018.

64 Gobierno de España, *Plan de Acción para la Implementación de la Agenda 2030. Hacia una Estrategia Española de Desarrollo Sostenible*, Agenda 2030, *cit.*, p. 77.

Si se conecta con los Objetivos 1 –pobreza– y 11 –acceso a la vivienda– y con la existencia de falta de pago de rentas adeudadas o cantidades debidas en el arrendamiento de vivienda motivada por la concurrencia de circunstancias económico-sociales que sitúan al arrendatario en una situación de vulnerabilidad lo pueden convertir efectivamente en vulnerable en su acceso a la justicia[65]. Por ejemplo, la propia notificación

---

[65] Este mismo Plan, a partir del estudio realizado por la Fundación FOESSA, señala que el riesgo de los hogares españoles de quedarse atrás era elevado para un 20% de los hogares españoles; hogares que presentan las siguientes características: "aquellos en los que hay baja intensidad laboral, en los que el riesgo de exclusión social asciende al 55%, y en el caso de los hogares con desempleados, este riesgo afecta al 44%. En los hogares con personas de origen extracomunitario, el riesgo de exclusión social ascienda al 49%. Los hogares situados en un barrio degradado o marginal tienen un riesgo de exclusión social del 32% de los casos, asociado fundamentalmente a la inadecuación y la inseguridad de la vivienda y del barrio y al deficiente acceso a los servicios básicos". Según este mismo estudio, el riesgo de quedarse atrás se relaciona también con "características sociales y demográficas de las personas" como son la pertenencia a la etnia gitana, desempleados de larga duración o con empleos irregulares, inmigrantes de origen no comunitario y hogares con menores de 18 años (*ibidem*, pp. 19 y 20). La pobreza guarda relación con la exclusión social y los problemas relacionadas con la alimentación. Con respecto al Objetivo 16 y los retos a seguir en el Estado español, en el Plan de Acción, se sostiene que, "en el ámbito general de lucha contra la desigualdad ha de situarse la promoción de la igualdad de acceso a la justicia, como un elemento más de cohesión social, base de una sociedad democrática, porque la vulnerabilidad económica, social o educativa no puede ser nunca un obstáculo para obtener del Estado la protección jurídica que proporciona a los ciudadanos" (*ibidem*, p. 62). Entre las acciones comprometidas para alcanzar los ODS, se centran, entre otras, en la pobreza extrema y la pobreza relativa en todas sus dimensiones, el acceso a la vivienda y el Estado de Derecho y el acceso a la justicia (*ibidem*, pp. 73, 77 y 79).

del requerimiento de pago despachado por el Letrado de la Administración de Justicia puede generar dificultades por falta de entendimiento de lo contenido en el mismo. Se ha de tener en cuenta el plazo de diez días para adoptar algunas de las posiciones indicadas en el requerimiento (desalojar, pagar, enervar la acción u oponerse) y el exiguo plazo para reclamar la asistencia jurídica gratuita en caso de que pudiera ser beneficiario de ella (tres días siguientes a la práctica del requerimiento, art. 438.5.III LEC).

Según el Instituto Nacional de Estadística (INE) en relación con el cumplimiento de los Objetivos de Desarrollo Sostenible (ODS), en concreto, el Objetivo 11[66] y con el subindicador[67] y el 9,9% con gasto elevado en vivienda[68]. Junto al Objetivo 11, es relevante destacar el objetivo de sostenibilidad a través de la paz, la justicia y las instituciones sólidas, previsto en el Objetivo 16 de los ODS, a través de la meta 16.3 (Promover el estado de derecho en los planos nacional e internacional y garantizar la igualdad de acceso a la justicia para todos). La meta 11.1 del Objetivo 11 prevé, de aquí a 2030, "asegurar el acceso de todas las personas a viviendas y servicios básicos adecuados, seguros y asequibles y mejorar los barrios marginales". En atención

---

66 Objetivo 1: "Poner fin a la pobreza en todas sus formas en todo el mundo".
Objetivo 11: "Lograr que las ciudades sean más inclusivas, seguras, resilientes y sostenibles".

67 En el Plan de Acción para la Implementación de la Agenda 2030, en 2017, el porcentaje de población que vive por debajo del umbral nacional de la pobreza es de del 21,6% en España (*Plan de Acción para la Implementación de la Agenda 2030. Hacia una Estrategia Española de Desarrollo Sostenible*, Agenda 2030, *cit.*, p. 18).

68 https://ine.es/dyngs/ODS/es/indicador.htm?id=4909 (último acceso: 23/05/2023).

al *Plan de Acción para la Implementación de la Agenda 2030*, el umbral del riesgo de pobreza de los hogares de una persona se situaba en 8.255 euros al año y, en hogares compuestos por dos adultos y dos menores de 14 años, en 17.896 euros (datos base de 2017), lo que representaba el 21,6% de la población residente en España[69].

En esta línea, cabe citar el principio de política de vivienda en situaciones de especial vulnerabilidad desarrollado en el art. 14 LDV. Así, prevé que "las políticas en materia de vivienda tendrán especialmente en cuenta a las personas, familias y unidades de convivencia que viven en asentamientos y barrios altamente vulnerables y segregados, ya sea tanto en entornos urbanos como en zonas rurales, a las personas sin hogar, a las personas con discapacidad, a los menores en riesgo de pobreza o exclusión social, a los menores tutelados que dejen de serlo y a cualesquiera otras personas vulnerables que se definan en el momento de la actuación". Por su parte, el art. 15 LDV marca las líneas generales para garantizar el derecho de acceso a la vivienda estableciendo criterios básicos en el ámbito de la ordenación territorial y urbanística[70].

69 Gobierno de España, *Plan de Acción para la Implementación de la Agenda 2030. Hacia una Estrategia Española de Desarrollo Sostenible*, Agenda 2030, pp. 17-19.

70 Con la finalidad de ampliar la oferta de vivienda social o dotacional, con la finalidad de adaptar la vivienda a la demanda y facilitar el acceso a una vivienda digna y adecuada, con la finalidad de garantizar el derecho de acceso a la vivienda en los municipios en los que se hayan declarado uno o más ámbitos como zonas de mercado residencial tensionado, con la calificación de un suelo como de reserva para vivienda sujeta a un régimen de protección pública y su limitación de modificación, y con el porcentaje destinado a vivienda sujeta a algún régimen de protección pública de alquiler (art. 15.1 LDV).

La Ley por el derecho a la vivienda[71] alude a la vulnerabilidad social y económica al configurar los planes estatales en materia de vivienda y en materia de rehabilitación, regeneración y renovación urbana y rural. Como actuaciones que promoverán y apoyarán, se encuentran las actuaciones necesarias para la creación, ampliación y gestión de los parques públicos de vivienda, tanto derivadas de la nueva construcción como de la rehabilitación, orientados prioritariamente a atender a la necesidad de vivienda de las personas, familias y unidades de convivencia en situación de mayor vulnerabilidad social y económica. También establece aquellas encaminadas a favorecer el acceso a la vivienda por parte de jóvenes, así como a garantizar una vivienda digna y adecuada para personas en situación de mayor vulnerabilidad social, chabolismo, infravivienda o en emergencia habitacional (art. 24.2c) y f) LDV)[72]. Esta última puede venir derivada de un juicio de desahucio por falta de pago. Así, el art. 24.2 LDV dispone que el desarrollo de ambas actuaciones "podrán servir para atender a los trámites de intermediación y conciliación previos a la presentación de la demanda, así como a la atención a las personas y hogares sujetos a vulnerabilidad" en los procesos de recuperación de la posesión con las especiales del art. 439 LEC, en la subasta de bienes inmuebles del art. 655 bis y en la demanda de ejecución de viviendas hipotecadas con deudor en situación de vulnerabilidad económica del art. 685 LEC.

---

71 *Vid.* Anguita Ríos (2023).

72 El Real Decreto 42/2022, de 18 de enero, por el que se regula el Bono Alquiler Joven y el Plan Estatal para el acceso a la vivienda 2022-2025 (TOL8.741.877). Es previo a la aprobación de la Ley 12/2023, pero contiene programas de ayuda a personas objeto de desahucio de su vivienda habitual, así como a personas arrendatarias en situación de vulnerabilidad sobrevenida, por ejemplo, en los arts. 35 a 50. En esta línea, se redacta el art. 24.5 LDV.

El art. 16.4 LAU establece que no cabe pacto de renuncia al derecho de subrogación en caso de que quienes puedan ejercitarlo estén en situación de "especial vulnerabilidad" y afecte a menores, personas con discapacidad o personas mayores de sesenta y cinco años. Incluye la posibilidad de prórroga extraordinaria del contrato de arrendamiento en situación de vulnerabilidad económica y social en los términos del art. 10.2 LAU.

La vulnerabilidad económico-social para las pretensiones de desahucio de vivienda arrendada, en precario, recuperación de la posesión o la eliminación de la perturbación de la posesión, que afecte a algún derecho real inscrito por quien se oponga, se incorpora en los arts. 439.6 y 7, 441.5, 6 y 7 LEC a través del establecimiento de requisitos de procedibilidad y medidas de protección social a través del incidente de suspensión del proceso.

Si bien la confirmación de la vulnerabilidad económico-social se efectuará por los servicios sociales competentes, se fijan en el art. 441.7 LEC criterios objetivos para determinar la vulnerabilidad económica del arrendatario, aplicables exclusivamente al juicio de desahucio por falta de pago.

Así, el art. 441.7.I LEC dispone que el tribunal, para apreciar la vulnerabilidad económica, "podrá considerar el hecho de que el importe de la renta, si se trata de un juicio de desahucio por falta de pago, más el de los suministros de electricidad, gas, agua y telecomunicaciones suponga más del 30 por ciento de los ingresos de la unidad familiar y que el conjunto de dichos ingresos no alcance", con carácter general, el límite de 3 veces el Indicador Público de Renta de Efectos Múltiples mensual (IPREM)[73] (art. 441.7.I.a) LEC).

---

[73] En 2024, el IPREM mensual está fijado en 600 € (https://www.iprem.com.es). En este caso, el límite sería de 1.800 euros y el 30%, de 540 euros.

Este límite se incrementará según el art. 441.7.I.b) y c) LEC como sigue:

a. En 0,3 veces el IPREM por cada hijo a caso de la unidad familiar, que será de 0,35 veces por cada hijo en unidades familiares monoparentales o por cada hijo con discapacidad igual o superior al 33 por ciento[74].

b. En 0,2 veces el IPREM por cada persona mayor de 65 años miembro de la unidad familiar o personas en situación de dependencia a cargo[75].

En caso de que alguno de los miembros de la unidad familiar tenga declara discapacidad igual o superior al 33 por ciento, situación de dependencia o enfermedad que le incapacite acreditadamente de forma permanente para realizar una actividad laboral, el límite será de 5 veces el IPREM[76], sin perjuicio de los incrementos acumulados por hijo a cargo.

En cuanto al arrendador, como novedad en la norma procesal y como ya habíamos sostenido en anteriores trabajos, también considera que puede encontrarse en situación de vulnerabilidad económica. A tal efecto, el legislador confiere al tribunal, por lo que será el titular de la potestad jurisdiccional, excluyendo así al Letrado de la Administración de Justicia, que aprecie la vulnerabilidad "también" en el arrendador para tomar la decisión de suspensión del proceso. En concreto, el art. 441.7 LEC dispone que "el tribunal tomará la decisión [se refiere a la suspensión del proceso], previa valoración ponderada y proporcional del caso concreto, apreciando las situaciones

---

74 Con los datos de 2024, 0,3 veces equivale a 180 euros y 0,35 veces, a 210 euros.

75 Igualmente, con los datos de 2024, 0,2 veces supone 120 euros.

76 En tal caso, será de 3.000 euros con el IPREM de 2024.

de vulnerabilidad que pudieran concurrir también en la parte actora y cualquier otra circunstancia acreditada en autos".

Asimismo, el tribunal podrá considerar, para apreciar la vulnerabilidad social, el hecho de que en la vivienda se encuentren personas dependientes de acuerdo con el art. 2.2 de la Ley 39/2006, de 14 de diciembre, de Promoción de la Autonomía Personal y Atención a las personas en situación de dependencia, víctimas de violencia sobre la mujer o personas menores de edad (TOL1.796.120). En este art. 2.2, se define "dependencia" como "el estado de carácter permanente en que se encuentran las personas que, por razones derivadas de la edad, la enfermedad o la discapacidad, y ligadas a la falta o a la pérdida de autonomía física, mental, intelectual o sensorial, precisan de la atención de otra u otras personas o ayudas importantes para realizar actividades básicas de la vida diaria o, en el caso de las personas con discapacidad intelectual o enfermedad mental, de otros apoyos para su autonomía personal".

El arrendador, pues, podrá encontrarse en situación de vulnerabilidad económica, que deberá alegar en la demanda en el plazo de cinco días al que se refiere el art. 441.5.V LEC con la correspondiente acreditación mencionada. El plazo de cinco días es común para ambas partes, pudiendo instar ambas "lo que a su derecho convenga". El art. 441.7 LEC atribuye al juez que aprecie las situaciones de vulnerabilidad que pudieran concurrir también en la parte demandante, además de la posible vulnerabilidad del demandado. Sin embargo, el legislador no determina cuáles son los criterios de vulnerabilidad económica a alegar por el arrendador. La dicción textual del art. 441.7 LEC alude a que "el tribunal tomará la decisión previa valoración ponderada y proporcional del caso concreto, apreciando las situaciones de vulnerabilidad que pudieran concurrir también en la parte actora y cualquier otra circunstancia acreditada en autos". Los criterios del art.

441.7 LEC se centran en calcular la vulnerabilidad económica del arrendatario al partir de las rentas y otros gastos asumidos en relación con el IPREM visto.

Sería deseable, *de lege ferenda,* que se precisase si la vulnerabilidad económica y social es aplicable al resto de pretensiones de recuperación de la posesión del art. 250.1. 1.°, 2.°, 4.° y 7.° LEC. La determinación de criterios objetivos para determinar la vulnerabilidad económica sólo se ciñe al desahucio de vivienda arrendada por falta de pago y no a los restantes. Hubiera sido deseable que en los demás supuestos y en el del demandante vulnerable también se hubieran previsto. Es cierto que el art. 441.7 LEC alude a "acreditada" y a "valoración ponderada y proporcional" por el juez, pero la referencia iría en aras de garantizar la objetividad para ambas partes y en todos los supuestos de recuperación de la posesión como se hizo en la normativa COVID19.

3. Requisitos de procedibilidad: vivienda habitual, gran tenedor y situación de vulnerabilidad

La admisión de la demanda del arrendador vendrá condicionada al cumplimiento de los requisitos de procedibilidad[77] establecidos en el vigente art. 439.6 LEC, de aplicación a los procesos arrendaticios de vivienda por falta de pago, además de a otros procesos en los que se pretenda la recuperación de la posesión de los números 2.°, 4.° y 7.° del art. 250 LEC. En concreto, se ha de indicar en la demanda si el inmueble constituye vivienda habitual, si el demandante tiene la condición de

---

[77] Sobre requisitos de procedibilidad, *vid.* Álvarez Suárez (2024, pp. 79-102).

gran tenedor y, en tal caso, si el demandado se encuentra en situación de vulnerabilidad económica.

Así, la demanda ha de especificar si el inmueble constituye *vivienda habitual* de la "persona ocupante" (art. 439.6.a) LEC). De existir contrato de arrendamiento, debería constar si se trata de vivienda habitual o está destinada a otro uso en virtud del art. 2 LAU.

Además, la demanda ha de especificar si el demandante tiene la condición de *gran tenedor de vivienda*, que se conecta con el requisito de la letra b) del art. 439.6 LEC. El art. 3.k) LDV define como tal a los efectos de dicha Ley y, por consiguiente, del art. 439.6 y 7 LEC a la "persona física o jurídica que sea titular de más de diez inmuebles urbanos de uso residencial o una superficie construida de más de 1.500 m^2 de uso residencial". El legislador posibilita que se flexibilice dicha definición en atención al mercado residencial tensionado "hasta aquellos titulares de cinco o más inmuebles urbanos de uso residencial ubicados en dicho ámbito". Corresponde a la comunidad autónoma motivarlo en la correspondiente memoria justificativa, que habrá de ser pública.

Aun cuando sea evidente que se trata de una flexibilización en perjuicio del arrendador, siguiendo el tenor del art. 439.6 LEC, el demandante-arrendador tiene la carga de especificar si es gran tenedor, lo que implicaría conocer la indicada memoria justificativa publicada y accesible a la ciudadanía. Nada obstaría a que el arrendatario-demandado pudiera alegar esta circunstancia de gran tenedor del demandante como excepción si adopta la actitud de oponerse al requerimiento efectuado por vía del art. 438.5 LEC.

Si en la demanda se indica que el demandante no es gran tenedor, habrá de adjuntar certificación del Registro de la Propiedad en el que conste la relación de propiedades a su nombre (art. 439.6.b) LEC) de no ostentar dicha condición. Sin embargo, si el demandante alega ser gran tenedor, no será necesario adjuntar dicha certificación.

La demanda debe especificar si el demandado-arrendatario se encuentra o no en *situación de vulnerabilidad económica* en caso de que el demandante tenga la condición de gran tenedor. Es decir, este requisito se exigirá únicamente a los que cumplan dicha condición.

Igualmente, recae en el demandante-gran tenedor la carga de acreditar la concurrencia o no de vulnerabilidad económica del arrendatario-demandado o de haber intentado obtener el documento acreditativo. En otro caso, será motivo de inadmisión de la demanda.

Para obtener dicha acreditación o, al menos, haber intentado obtenerla, se podrá efectuar por alguna de estas vías del art. 439.6 LEC:

a. Aportación de documento acreditativo de los servicios de las Administraciones autonómicas y locales competentes en materia de vivienda, asistencia social, evaluación e información de situaciones de necesidad social y atención inmediata a personas en situación o riesgo de exclusión social asignados por la normativa autonómica correspondiente. La vigencia del documento acreditativo no podrá ser superior a tres meses y requiere el consentimiento previo de la persona *ocupante* de la vivienda para obtenerlo.

b. Declaración responsable del demandante-gran tenedor de haber acudido a los servicios sociales (o equivalentes) autonómicos o locales sin que hubiera sido atendida su solicitud en el plazo máximo de cinco meses de antelación a la presentación de la demanda o sin que se hubieran iniciado los trámites en el plazo de dos meses desde que presentó su solicitud. Se ha de adjuntar a la demanda el justificante acreditativo de su solicitud (art. 439.6.c) 1.° LEC).

c. Documento de los servicios sociales competentes que indiquen la negativa expresa del arrendatario de no someterse al estudio de su situación económica en los términos que prevea la normativa autonómica en materia de vivienda, siempre que dicho documento no tenga una vigencia superior a tres meses.

Sería deseable para facilitar la aportación de los requisitos indicados al proceso por parte del demandante la implementación de aplicaciones informáticas o plataformas que permitieran fácilmente obtener dichos datos directamente por el demandante o, incluso, directamente por el juzgado previa autorización de demandante y demandado. Bastaría con la declaración responsable de demandante con la autorización para obtener tales datos desde el juzgado mediante las herramientas tecnológicas existentes que podrían facilitar su desarrollo.

Se exige, asimismo, como requisito previo en el art. 439.7 LEC y, por ello, afectará a la admisión de la demanda que el demandante-gran tenedor acredite que se ha sometido a procedimiento de conciliación o de intermediación que establezcan las Administraciones Públicas competentes, con base en las circunstancias de ambas partes y de las posibles ayudas y subvenciones existentes en materia de vivienda según la normativa autónoma vigente en vivienda. Se ha de tratar de vivienda habitual y el arrendatario se ha de encontrar en situación de vulnerabilidad económica.

La acreditación de este requisito podrá realizarse por una de las dos formas siguientes:

a. Declaración responsable del demandante de que ha acudido a los servicios sociales (o equivalentes) autonómicos o locales sin que hubiera sido atendida su solicitud de conciliación o intermediación en el plazo máximo de cinco meses de antelación a la presentación de la demanda o justificante acreditativo de haber iniciado los trámites en el plazo de dos meses desde que se presentó su solicitud.

b. Documento acreditativo de los servicios sociales (o equivalentes) con el resultado de la conciliación o intermediación, que no podrá tener una vigencia superior a tres meses. Este documento habrá de contener la identidad de las partes, el objeto de la controversia y si alguna de las partes ha rehusado participar en dicho procedimiento.

En cuanto a la exigencia del previo sometimiento obligatorio al procedimiento de conciliación o intermediación por los servicios sociales competentes en caso de demandante-gran tenedor, resulta cuestionable dicha obligatoriedad y su viabilidad. A nuestro juicio, debería dejarse a la voluntad de las partes, previa información detallada a las partes o, incluso, la posibilidad de que las partes pudieran utilizar otros medios de solución de conflictos no jurisdiccionales. Quedará por ver cómo se desarrollan estos servicios de conciliación o intermediación en las respectivas comunidades autónomas, incluidos los plazos de resolución y la eficacia ejecutiva de dicho documento en caso de acuerdo.

4. Incidente de suspensión del juicio de desahucio como medida de protección social

El art. 441.5 LEC, en su redacción de 2023, prevé la suspensión[78] de la tramitación de los procesos arrendaticios del art. 250.1.1.° LEC y de otros posesorios, según lo ya analizado, como medida de protección social en favor del arrendatario y siempre que el arrendador no sea, igualmente, vulnerable. En realidad, esconde un incidente de suspensión de previo pronunciamiento.

En la redacción de 2019, sólo ser preveía para los procesos arrendaticios del art. 250.1.1° LEC. Se establecía, además, que la duración para que los servicios sociales pudieran adoptar

---

[78] Sobre suspensión, *Vid.* Julià Pijoan (2024, pp. 427-460).

medidas de ayudas o alternativas habitacionales era un mes a contar desde la recepción de la comunicación con la confirmación de los servicios sociales al órgano judicial (arrendador-demandante como persona física) o de tres meses en caso de que el arrendador-demandante fuera persona jurídica, con independencia de que existiera una relación de consumo. Cabe puntualizar que no se estableció un plazo para que los servicios sociales efectuaran la valoración de la vulnerabilidad y emitieran informe confirmando o no la concurrencia en el arrendatario o en el hogar del arrendatario. Por consiguiente, la pregunta sería si la tramitación del procedimiento seguía su curso, puesto que la suspensión procediese tras recibir la comunicación. No tenemos datos para conocer el tiempo de respuesta de los servicios sociales a los que se deriva por el juzgado competente con respecto a la evaluación de la vulnerabilidad. Tampoco los tenemos con respecto a la activación de las medidas sociales acordadas para el arrendatario. El legislador parece haber incorporado una cuestión incidental y el procedimiento para su resolución dentro del juicio verbal, ya que se extiende a otras pretensiones de recuperación de la posesión.

Contrasta la bonanza de los plazos del legislador con la realidad judicial. Si atendemos a la Estadística Judicial para duración media de los procesos arrendaticios tramitados por el juicio verbal –la mayoría a través del juicio de desahucio por falta de pago–, la duración media se ha incrementado desde 2016 de 4,7 meses a 7,7 en 2022[79]. También es cierto que las cifras más abultadas se

---

[79] En 2020, fue de 6,7 meses y, en 2021, de 7,3. En los años previos, la duración media estuvo en 5,1 en 2018 y 5,5 en 2019.
Desde la reforma de 2009, se ha producido una estabilización de la duración con oscilaciones leves al alza o a la baja. Si en 2009 la media nacional era de 4,8, en 2017 se situaba en 4,9. Es decir, el objetivo de acortar los plazos para lograr la recuperación no se logró (https://www.poderjudicial.es/cgpj/es/Temas/Estadistica-

centran en 2020, 2021 y 2022, afectados por las suspensiones de los juicios de desahucio por vulnerabilidad económico-social COVID19. Esta colaboración juzgados-servicios sociales ya se venía produciendo desde la crisis financiera de mediados del año dos mil como se ha indicado con anterioridad.

Una vez admitida la demanda, superados los requisitos de procedibilidad del art. 439.6 y 7 LEC y siempre que se trate de vivienda habitual del arrendatario, se informará a este mediante el decreto de admisión del Letrado de la Administración de Justicia sobre la posibilidad de acudir a las Administraciones Públicas autonómicas y locales competentes en materia de vivienda y que tengan servicios sociales o equivalentes para que evaluación la posible situación de vulnerabilidad del arrendatario. Esta información ha de redactarse en lenguaje claro y comprensible y en una lengua que comprenda el arrendatario, lo que repercutirá favorablemente en su acceso a la justicia. Además, se habrán de precisar los datos exactos de identificación de dichas Administraciones y el modo de tomar contacto con ellas.

Por su parte, el juzgado, que, como regla general, será el Letrado de la Administración de Justicia que haya dictado el decreto de admisión de la demanda junto con el requerimiento de pago del art. 438.5 LEC, comunicará de oficio de forma inmediata la existencia del proceso arrendaticio de vivienda por falta de pago a las indicadas Administraciones públicas para que verifiquen la situación de vulnerabilidad y que, en caso de existir, han de presentar al juzgado una propuesta de alternativa de vivienda digna en alquiler social a proporcionar por la Administración competente para ello, propuesta de medidas de atención inmediata a adoptar por

---

Judicial/Estadistica-por-temas/Actividad-de-los-organos-judiciales/Estimacion-de-los-tiempos-medios-de-los-asuntos-terminados/) (Último acceso: 29/11/2023).

la Administración competente y las posibles ayudas económicas y subvenciones de las que pudiera ser beneficiario el demandado (art. 441.5.II LEC).

Se prevé un plazo expreso para que las Administraciones públicas competentes emitan informe de la valoración de la vulnerabilidad económica y, en su caso, social del arrendatario, que se efectuará a la mayor brevedad y, en todo caso, en un plazo máximo de diez días (art. 441.5.III LEC). Por su parte, el Defensor del Pueblo entiende que este plazo engloba todas las actuaciones. Así, señala que "la reforma operada obligada a dar traslado, de oficio, desde el órgano judicial a las Administraciones públicas competentes a fin de que puedan comprobar la situación de vulnerabilidad de la parte demandada en los procedimientos judiciales de desahucio y, de existir esta, presentar al juzgado propuesta alternativa de vivienda digna en alquiler y de medidas de actuación inmediata, así como de las posibles ayudas económicas y subvenciones de las que pueda ser beneficiaria, todo en el plazo de diez días"[80]. Se vislumbra como un plazo exiguo, que abocará a su incumplimiento para emitir informe por parte de la administración competente y adoptar, asimismo, las medidas procedentes. Sería relevante la comunicación directa del juzgado con los servicios sociales o similares para agilizar el informe y la adopción, de proceder, de dichas medidas.

Si el demandante-gran tenedor hubiera presentado demanda con el documento acreditativo de la vulnerabilidad del arrendatario, en el oficio dirigido a la Administración pública se hará constar dicha circunstancia a fin de que efectúen directamente,

80 Resolución adoptada por el Defensor del Pueblo, el 1 de septiembre de 2023, con motivo de la solicitud de interposición del recurso de inconstitucionalidad contra la Ley 12/2023, de 24 de mayo, por el derecho a la vivienda.

en el mismo plazo –se entiende que los indicados diez días–, la propuesta de medidas de atención inmediata a adoptar, así como las posibles ayudas económicas y subvenciones de la que pueda ser beneficiaria la parte demandada y las causas, en caso de que proceda, que han impedido su aplicación con anterioridad (art. 441.5.IV LEC).

Recibido informe de los servicios sociales (o equivalentes) o transcurrido el plazo sin que lo hubiera emitido (por remisión al art. 439.6 y 7 LEC podría referirse al plazo de dos meses de la letra c) 1.° LEC), el Letrado de la Administración de Justicia dará traslado a las partes para que, en el plazo de cinco días, puedan instar lo que a su derecho convenga, es decir, se abre un incidente. El Letrado de la Administración de Justicia procederá a suspender la fecha prevista para la celebración de la vista o para el lanzamiento, de ser necesaria tal suspensión por la inmediatez de las fechas (art. 441.5.V LEC).

De presentar sus escritos las partes en el indicado plazo de cinco días o transcurrido el mismo, el juez resolverá por auto, a la vista de la información recibida de las Administraciones públicas competentes y de las alegaciones de las partes, si suspende el proceso para que se adopten las medidas de protección social hacia el arrendatario. La suspensión tendrá un plazo de dos meses si el demandante es una persona física o de cuatro meses si se trata de una persona jurídica (art. 441.6.I LEC).

La suspensión se alzará automáticamente por el transcurso de los dos o cuatro meses, respectivamente, o por la adopción efectiva de medidas por los servicios sociales (o equivalentes) (art. 441.6.II LEC).

Es relevante la incorporación *ex novo* por la LDV de la norma que establece que la decisión de la suspensión del proceso se efectuará por el tribunal previa valoración ponderada y proporcional del caso concreto, apreciando las situaciones de vulnerabilidad que pudieran concurrir también en la parte demandante y cualquier otra circunstancia acreditada en las actuaciones (art. 441.7.I LEC).

No se procederá al lanzamiento de vivienda habitual en caso de resolución de condena de desahucio por falta de pago (se extiende al desahucio por expiración legal o contractual del plazo) si previamente al lanzamiento no se ha procedido en los términos analizados del art. 441.5, 6 y 7 LEC (art. 549.4 LEC, modificado por la LDV).

Se ha de tener en cuenta que no se aplicará el plazo de espera legal del art. 548 LEC, es decir, el plazo de veinte días posteriores a aquel en que la resolución se hubiera dictado, cuando se trate de ejecución de resoluciones de condena de desahucio por falta de pago de rentas o cantidades debidas, o por expiración legal o contractual del plazo. En estos casos, se aplicará lo previsto en su regulación específica sobre la fijación de la fecha y horas exactas para proceder al lanzamiento (art. 438.5 y 6 LEC).

En este sentido, el art. 549.3 LEC dispone que se procederá a la ejecución directa, cuando en la demanda de desahucio se hubiera solicitado, "sin necesidad de ningún otro trámite para proceder al lanzamiento en el día y hora exacta señalados en la propia sentencia o en el día y hora exacta que se hubiera fijado al ordenar la realización del requerimiento al demandado (decreto que ponga fin al desahucio si no hubiera oposición al requerimiento).

El Defensor del Pueblo ha señalado que la incorporación de las medidas centradas en la vulnerabilidad económica incorporadas por la Disposición final quinta de la Ley 12/2023, en esencia, las analizadas en este apartado 2, entre ellas, la suspensión del proceso, "nada de todo ello modifica sustancialmente la regulación preexistente en la Ley de Enjuiciamiento Civil ni vacía de contenido el derecho de la propiedad de la vivienda”[81]. Si bien el establecimiento de compensaciones al demandante con las condiciones y requisitos que se prevean sería una solución adecuada mediante la suspensión del proceso o de la ejecución del lanzamiento, excede del ámbito procesal, más allá de acreditar dicha suspensión ante la administración que corresponda.

Queda aún pendiente de que el Tribunal Constitucional se pronuncie sobre los diversos recursos de inconstitucionalidad planteados contra la Ley 12/2023, algunos de los cuales afectan a la Disposición final quinta que modifica la regulación analizada de la Ley de Enjuiciamiento Civil.

III. CONCLUSIONES

Se ha consolidado la referencia al mercado de la vivienda en régimen de arrendamiento tenso o tensionado, de precios de alquileres inasumibles para una parte de las familias, del acceso a una vivienda digna y de políticas que permitan disponer de un parque de viviendas suficientes o de ayudas para las familias en riesgo de exclusión o que no disponen de renta suficiente para encontrar una vivienda que se ajuste a sus posibilidades económicas a tenor de las reglas del mercado inmobiliario, se habla de desahucio por falta de pago

---

81 “No se olvide en este punto que la función social forma parte del contenido esencial del derecho a la propiedad, tal y como se ha expuesto en los fundamentos jurídicos primero, segundo y tercero, y que la utilidad económica de la vivienda si bien se verá reducida a causa del tiempo que se demore su recuperación posesoria no desaparecerá” (Resolución adoptada por el Defensor del Pueblo, el 1 de septiembre de 2023, con motivo de la solicitud de interposición del recurso de inconstitucionalidad contra la Ley 12/2023, de 24 de mayo, por el derecho a la vivienda, fj sexto).

de la renta del arrendamiento y de familias en situación de vulnerabilidad debido a un factor económico, y también es posible hablar del arrendador en situación de vulnerabilidad al no percibir las rentas impagadas y del derecho a la propiedad privada, entre otras cuestiones. Es una materia que ha abierto un largo debate social, reclamando al Estado, a las autonomías y, en su caso, a las corporaciones locales, una respuesta que atienda las distintas sensibilidades y derechos en juego. También implica dar una respuesta procesal al problema del impago de rentas y no de forma lineal, sino atendiendo a las situaciones de vulnerabilidad que puedan concurrir y que el legislador es el que ha de delimitar. El juez queda sujeto al imperio de la ley, debiendo aplicar la norma al caso concreto.

Es un acierto que esta sensibilidad social también se traslade a la norma procesal en situaciones de vulnerabilidad económica y social, si bien el legislador deberá seguir puliendo cuestiones mejorables en la Ley de Enjuiciamiento Civil. Además de la reivindicada regulación independiente del juicio de desahucio por falta de pago, sería deseable que la suspensión del desahucio fuera regulada como incidente de previo pronunciamiento con regulación especial. También sería oportuno revisar lo relativo a la valoración de las situaciones de vulnerabilidad económica de demandante y los plazos de la suspensión, así como la valoración de la incorporación digital –aplicaciones o plataformas– para la comunicación entre las administraciones competentes y el juzgado para tramitar lo relativo a la calificación de vulnerabilidad económica y el resto de documentación a acreditar por el demandante gran tenedor.

IV. REFERENCIAS BIBLIOGRÁFICAS

Álvarez Suárez, L. (2024). "Los excesivos requisitos de procedibilidad y los menoscabos en las garantías procesales de los arrendadores a la luz de la nueva Ley sobre el Derecho de Vivienda". *La Ley Derecho de Familia: Revista jurídica sobre familia y menores,* n.º 41. 79-102.

Anguita Ríos, R. M. (2023). "La vivienda: entre el derecho humano y el activo financiero. A propósito de la Ley por el derecho a la vivienda". *Actualidad Civil,* n.º 11.

Adán Doménech, F. (2017). "El juicio verbal de desahucio por falta de pago de la renta y/o cantidad asimilada, con acción acumulada de reclamación de cantidad y juicio verbal de desahucio por expiración del plazo legal". *El juicio verbal de desahucio y el desalojo de viviendas okupadas* (coords. Izquierdo Blanco/ Picó i Junoy). Wolters Kluwer S.A. Barcelona.

Álvarez Alarcón, A. (2023). "Introducción. Concepto de acceso a la justicia y personas en condición de vulnerabilidad". *Acceso a la justicia de las personas vulnerables* (dirs. Álvarez Alarcón/ Ramírez Carvajal). Reus. Madrid.

Besser Valenzuela, G. (2014). *El proceso de desahucio por falta de pago.* Marcial Pons. Madrid.

Bolívar Muñoz, J., y otros (2016). "La salud de las personas adultas afectadas por un proceso de desahucio". *Gaceta Sanitaria,* vol. 30, n.º 1.

Bonet Navarro, J. (2021). *Los juicios por desahucio: especialidades procesales para la recuperación del inmueble arrendado, okupado o en precario.* Thomson Reuters Aranzadi. Cizur Menor.

Colomer Hernández, I. (2022). "Vulnerabilidad, datos personales y proceso civil". *Los vulnerables ante el proceso civil.* (dirs. Herrero Perezagua, J. F./ López Sánchez). Atelier. Barcelona.

Hernández Díaz-Ambrona, M.ª D. (2015). *Consumidor vulnerable.* Reus. Madrid.

Julià Pijoan, M. (2024). "La suspensión de los procesos de desahucio por situaciones de vulnerabilidad: ¿constituye una dilación indebida?". *La ley Derecho de Familia: Revista jurídica sobre familia y menores,* n.º 41, 427-460.

La Barbera, M. (2019). "La vulnerabilidad como categoría en construcción en la jurisprudencia del Tribunal Europeo de Derechos Humanos: límites y potencialidad". *Revista de Derecho Comunitario Europeo*, n.º 62 (doi: https//doi.org/10.18042/cepc/rdce.62.07), 235-257.

Lazo Jara, A. (2020). *Autonomía, vulnerabilidad, dependencia y derechos humanos*. Madrid.

Molina Roig, E. (2018). *Una nueva regulación para los arrendamientos de vivienda en un contexto europeo*. Tirant Lo Blanch. Valencia.

Montero Aroca, J. y otros (2016). *Derecho Jurisdiccional I*. Tirant Lo Blanch. Valencia.

Moreno García, L. (2019). *Las cláusulas abusivas. Tratamiento sustantivo y procesal*. Tirant Lo Blanch. Valencia.

Pallarés Neila, J. (2017). *La protección de la vivienda habitual como exigencia del derecho a vivir de forma independiente de las personas con discapacidad. Análisis teórico y realidad empírica*. Bosch. Barcelona.

Perarnau Moya, P. (2019). "Los arrendamientos de viviendas tras las reformas de los Decretos-Ley 21/2018 y 7/2019". *Diario La Ley*, n.º 9395, Sección Tribuna.

Pérez Daudí, V., (2020). *Tutela efectiva y derecho a la vivienda*. Atelier. Barcelona.

Rodríguez Tirado, A. M.ª (2023). "Acceso al sistema de justicia penal de las víctimas vulnerables menores de violencia sexual. La prohibición de la mediación penal". *Revista de la Asociación de Profesores de Derecho Procesal de las Universidades españolas*, n.º 7, 81-130.

Rodríguez Tirado, A. M.ª (2019). "La protección procesal del arrendatario-consumidor y la suficiencia del juicio de desahucio de vivienda arrendada por falta de pago del art. 440.3 LEC". *Revista Jurídica sobre Consumidores y Usuarios*.

Rodríguez Tirado, A. M.ª (2018), "Los procesos de desahucio de vivienda arrendada por falta de pago y las situaciones de especial vulnerabilidad. Un acercamiento al análisis geográfico y estadístico del desahucio", *Práctica de Tribunales*, núm. 135.

Rodríguez Tirado, A. M.ª (2013). "Reflexiones sobre los procesos de desahucio de finca arrendada por falta de pago y su enésima reforma: ¿un nuevo proceso civil especial y sumario?". *Proceso, eficacia y garantías en la sociedad global. Liber amicorum II*. Atelier. Barcelona.

Rodríguez Tirado, A. M.ª (2012). "El juicio «monitorio» de desahucio de finca arrendada por falta de pago del art. 440.3 LEC (modificado por Ley 37/2011)". *Revista Práctica de Tribunales*, n.º 96-97.

Rodríguez-Izquierdo Serrano, M. (2024). "La vulnerabilidad frente a los desalojos forzosos de vivienda como tendencia constitucional". *Revista Española de Derecho Constitucional*, n.º 130, 49-78.

Torres López, A. (2023). "Novedades en los juicios de desahucio tras la nueva Ley por el Derecho a la Vivienda". *Proceso civil: cuaderno jurídico*, n.º 152, 7-12.

Trinidad Núñez, Pilar (2012). "La evolución en la protección de la vulnerabilidad por el Derecho Internacional de los derechos humanos". *Revista Española de Relaciones Internacionales*, n.º 4, 125-168.

Vives Miró, S./ Rullán Salamanca, O./ Albert Artigues i Bonet, A./ Navarro Zurriaga, V. (2023). "Gentrificaciones financiarizadas, gentrificaciones de quinta oleada. Relaciones desahucios-clase en Palma (Mallorca-Islas Baleares)". *Scripta Nova: Revista electrónica de geografía y ciencias sociales*, vol. 27, n.º 2.

Zarowsky, C./ Haddad, S./ Nguyen, V.-K. (2013). "Ir más allá de los «grupos vulnerables: contextos y dinámica de la vulnerabilidad". *IUHPE-Global Health Promotion 1757-9759*, vol. 20, Sup. 1.

Capítulo X

Conciliación e intermediación en materia de vivienda[1]

VÍCTOR BASTANTE GRANELL
Profesor Titular de Derecho civil
Universidad de Almería

I. INTRODUCCIÓN

El 26 de mayo de 2023 entró en vigor la Ley 12/2023, de 24 de mayo, por el derecho a la vivienda (LDV). Dicha disposición legal –con sus adeptos y detractores– tiene por objeto garantizar el derecho a acceder a una vivienda digna y adecuada –así como su disfrute– y, para ello, también regula el contenido básico del derecho de propiedad en relación con su función social (art. 1 LDV). Se centra, pues, en conciliar –con mayor o menor acierto–, a través de distintas medidas, dos derechos recogidos en la Constitución española –el derecho a la vivienda y el derecho a la propiedad privada (arts. 47 y 33 CE, respectivamente)–. Sin duda alguna, en ocasiones resulta dificultoso conciliar tales intereses, que, de primeras, parecen absolutamente contrapuestos[2].

---

1 Dicho trabajo se desarrolla en el marco de la “Red temática para el estudio de los aspectos civiles y procesales de la Ley de vivienda” (D5-2022_03), financiada por el Plan Propio de investigación de la Universidad de Málaga.

2 Sobre este debate, véase Prats Albentosa (2022, pp. 131-174); Sáinz-Cantero Caparrós (2023, pp. 630-663).

Para ilustrar la compleja conciliación de tales intereses es suficiente poner la mirada en unos conflictos específicos en materia de vivienda, aquellas situaciones que nos acompañan –por su impacto social, jurídico y económico– desde la crisis económica de 2007. Me refiero a los problemas entre ocupantes y propietarios de vivienda, principalmente cuando aquel ostentaba la posición de deudor hipotecario o arrendatario, hallándose en situación de vulnerabilidad social y económica, sin haber podido sufragar la cuota del préstamo o de la renta; y, por consiguiente, con riesgo de sufrir un lanzamiento o desahucio. Hemos sido testigos de miles de ejecuciones hipotecarias y desahucios, de un drama social que ha marcado la vida de numerosas familias; pero también del advenimiento de medidas legales que han afectado a las facultades de múltiples propietarios de vivienda –sean o no grandes tenedores– (como la suspensión de lanzamientos o desahucios). Sin duda alguna, resulta complicado atisbar el equilibrio perfecto para proteger a personas en situación de vulnerabilidad, sin quebrantar los derechos de las personas propietarias de viviendas.

Aun así, La LDV es bastante clara cuando presenta como fin específico "proteger la estabilidad y la seguridad jurídica en la propiedad, uso y disfrute de la vivienda", y ello "con especial atención a las personas y hogares en situación o riesgo de vulnerabilidad" –específicamente a familias, hogares y unidades de convivencia con menores a cargo–. De este modo, la LDV emerge como una norma dirigida a amparar a personas con "vulnerabilidad habitacional", erigiéndose dicho propósito como fin básico de las políticas públicas de vivienda (art. 2 LDV).

Entre las medidas recogidas para asegurar su consecución, destaca el reconocimiento estatal y la inserción en la LEC de un nuevo trámite o procedimiento: la llamada "conciliación o intermediación" en materia de vivienda.

La LDV alude en numerosas ocasiones a la "conciliación o intermediación" o viceversa –intermediación o conciliación–. Ante ello, se plantean múltiples interrogantes: ¿qué tipo de procedimiento se presenta?, ¿cuál es su fundamento o finalidad?, ¿dispone de un marco regulatorio?, ¿quién es el organismo competente para su desarrollo?, ¿qué trámites deben seguirse para su inicio?, ¿qué normativa le resulta de aplicación, en defecto de regulación específica?, ¿constituye un nuevo requisito de procedibilidad?, ¿qué sujetos deben acudir a tal procedimiento?, etc. A lo largo del texto se intentará responder a tales cuestiones.

II. CONCILIACIÓN E INTERMEDIACIÓN EN MATERIA DE VIVIENDA: CONCEPTO, FUNDAMENTO Y REGULACIÓN LEGAL

Como indica el preámbulo de la LDV, a través de la disposición final quinta se introduce el llamado "procedimiento de conciliación o intermediación". En particular, se instaura como un nuevo "trámite previo", de carácter obligatorio y extrajudicial, dentro de distintos procedimientos de la Ley de Enjuiciamiento Civil: el juicio verbal (art. 439.7 LEC), la subasta de bienes inmuebles (art. 655 bis LEC) y, por último, la ejecución de bienes hipotecados (art. 685. 2 LEC).

Aunque la LDV presenta una "imprecisión conceptual" o "indefinición", como manifiesta Viola Demestre[3], podemos extraer su concepto. La conciliación e intermediación en materia de vivienda puede definirse como aquel procedimiento extrajudicial preceptivo en materia de vivienda, desarrollado a nivel autonómico –como veremos a continuación–, que tiene

---

3 Viola Demestre (2024, p. 478).

como finalidad, primeramente, gestionar conflictos entre propietarios –grandes tenedores– y ocupantes de vivienda –ante la posible pérdida de su vivienda habitual–; y, de no ser viable una solución, ofrecer soluciones o alternativas habitacionales mediante un sistema de protección social. Dicho con otras palabras, se dirige a ofrecer una labor de negociación y, de fracasar la misma, de protección, en beneficio de personas ocupantes con vulnerabilidad económica. No obstante, veremos si esa es la "esencia práctica" que irradia de tal procedimiento.

En principio, su finalidad es obligar a la parte actora –que, como veremos, debe ser un gran tenedor de vivienda– a acudir, antes de interponer la demanda pertinente, a un procedimiento extrajudicial para que las Administraciones competentes puedan ofrecer una respuesta a la persona afectada –ocupante de vivienda habitual en situación de vulnerabilidad económica, con riesgo de desahucio o lanzamiento– mediante instrumentos de protección social o programas de política de vivienda. Es decir, busca otorgar a las Administraciones una oportunidad para poder gestionar la controversia existente entre las partes implicadas (impago de rentas, impago de préstamo hipotecario, etc.), proponer soluciones y, de no ser posible, implementar medidas, ayudas o subvenciones, con la finalidad de garantizar a tales personas una vivienda digna y adecuada. En definitiva, pretende ofrecer soluciones –previas o posteriores– a la posible pérdida de la vivienda habitual[4].

---

4 "La nueva ley introduce un sistema alternativo de resolución de conflictos para favorecer un acuerdo de las partes en aras de evitar el lanzamiento, pensado para familias vulnerables. Asimismo, a falta de un acuerdo entre las partes, deberá concederse el tiempo necesario para que los servicios sociales del municipio donde radique la vivienda puedan ofrecer soluciones habitacionales y, en este sentido, se habilita al juez a establecer plazos superiores. Finalmente, se incluyen nuevas prorrogas en el lanzamiento y se estipula el acceso

Señalando esta misma finalidad, destaca lo manifestado por Viola Demestre[5] o López Simó:

> "Lo que persigue el legislador de la LDV con la introducción de esta nueva medida es —además, claro, de procurar que las partes en conflicto, arrendador e inquilino o propietario y ocupante ilegítimo, lleguen a un acuerdo amistoso que permita el mantenimiento del contrato de arrendamiento o de la situación de ocupación de un inmueble y, por tanto, se evite el inicio del correspondiente juicio de desahucio— facilitar a las Administraciones Públicas competentes, mientras se sustancia el procedimiento de conciliación o intermediación y por si éste finalizara sin acuerdo, la posibilidad de dar adecuada atención a las personas y hogares afectados por un eventual desalojo, ofreciendo respuesta a través de diferentes instrumentos de protección social y de los programas de política de vivienda (en estos términos se expresa al respecto el Preámbulo de la Ley, ap. III); es decir, se trata, en definitiva, de ganar tiempo en favor de la Administración (y a costa, obviamente, del propietario de la finca, que verá retrasado unos meses el inicio del juicio de desahucio), para que aquélla pueda, aprovechando la duración del procedimiento previo a la interposición de la demanda, atender adecuadamente a las personas ocupantes de la vivienda que, por su situación económica y/o social, puedan necesitarlo en el caso de que el futuro próximo proceso termine con una resolución de condena al desalojo"[6].

---

obligatorio a los mecanismos alternativos de resolución de conflictos para personal vulnerable, en la línea de la mediación arrendaticia extrajudicial, directamente remitida a las partes por el juez (DF quinta de la ley, que modifica el art. 441.5 y 6 de la LEC). Además, serán las propias comunidades autónomas las que articulen mecanismos propios de mediación, quienes podrán usar sus recursos habitacionales para dar solución de vivienda a las personas que han sufrido un desahucio" (Calvo San José, 2023, p. 153).

5 Viola Demestre (2024, p. 496).

6 López Simó (2023, LA LEY 10759/2023).

No obstante, como afirma cierta doctrina y atendiendo al tenor literal de la norma, parece que el legislador establece dicho trámite para brindar una ayuda social al ocupante, pero no para alcanzar un acuerdo entre las partes en conflicto, finalidad no retratada en la norma:

> "Llama poderosamente la atención que se hagan alusiones a los plazos máximos de antelación con que deben instarse declaraciones de las Administraciones Públicas y, muy destacadamente, que se institucionalice, no sólo la reclamación administrativa previa como en el orden contencioso, sino la mediación o conciliación, no para lograr un acuerdo entre las partes que ponga fin al litigio, sino para dar respuesta a una pretendida situación de vulnerabilidad del demandado o ejecutado, o eso parece deducirse de la reforma, que no indica, en ningún momento, cual es la finalidad de ese expediente preprocesal de conciliación o intermediación"[7].

De este modo, como señala Ruiz-Rico Arias, el legislador no parece mostrar mucha esperanza en el éxito de tal procedimiento extrajudicial:

> "Como es sabido, ese procedimiento no es de arbitraje, y por tanto no saldrá de él nada que no haya consentido la entidad financiera ejecutante, por lo que a la postre todo sigue en manos de esta última. Tampoco se especifica cuál será el objeto de esa conciliación o intermediación; sólo se alude vagamente al ... análisis de las circunstancias de ambas partes y de las posibles ayudas y subvenciones existentes en materia de vivienda..., lo cual, entendido así, resulta ser algo bastante modesto, que no servirá para solucionar el problema de vivienda generado en el ejecutado y su familia"[8].

Aun así, confío –aunque con escepticismo por parte de ciertas Administraciones públicas– en el papel que desempeñen

---

7 Gómez Linacero (2023, LA LEY 11025/2023).

8 Ruiz-Rico Arias (2023, LA LEY 10097/2023).

los servicios autonómicos, en aras a la consecución de soluciones que pretendan evitar el desahucio o el lanzamiento de la vivienda. Si bien, dicha labor será complicada dada la voluntariedad del procedimiento y su carácter no vinculante. Hasta aquí, el objetivo parece atractivo, dentro de un Estado de Bienestar Social. Ahora bien, ¿será un procedimiento funcional o, en su caso, un mero trámite vacío de contenido?

III. NUEVO REQUISITO DE PROCEDIBILIDAD

Posiblemente, en busca de la finalidad mencionada, el acogimiento al "procedimiento de conciliación e intermediación" se instaura como un nuevo requisito de procedibilidad. Nos encontramos ante un presupuesto o trámite preceptivo, de gran novedad en el sistema procesal civil[9]:

> "[...] se introducen en la LEC, en los supuestos que luego veremos, unos nuevos trámites, preceptivos, de intermediación y conciliación previos a la presentación de la demanda en los procedimientos de desahucio (en sentido amplio: procedimientos susceptibles de provocar el lanzamiento del demandado del inmueble) [...].Por lo tanto, a partir de ahora, en todos los casos indicados deberá hacerse o intentarse, antes de interponer la demanda de desahucio, una conciliación o intermediación; y si la parte actora no acredita que se ha sometido previamente a uno de estos procedimientos «no se admitirá la demanda», según establece el nuevo precepto mencionado en su párrafo primero"[10].

---

9 López Simó (2023, LA LEY 10759/2023).

10 *Id.* Sobre este asunto se ha pronunciado cierta jurisprudencia: "Del contingut d'aquestes normes resulta que la documentació que acrediti si la persona a la qual es vol desnonar és o no vulnerable i en aquest darrer supòsit si l'arrendador ha instat un procediment de conciliació amb ella, s'ha de presentar necessàriament amb la demanda i es constitueix com un requisit de procedibilitat o de viabilitat del procés. El

El legislador estatal no se encarga de regular el procedimiento de "conciliación e intermediación" en materia de vivienda –objeto de desarrollo normativo por las CCAA–, se centra en establecer su acogimiento como nuevo requisito de procedibilidad en la LEC. Lo anterior significa que estamos ante un requisito formal que debe cumplirse si la parte actora desea incoar alguno de los procedimientos judiciales afectados. En caso contrario, el juez deberá inadmitir la demanda presentada en el juicio o, en su caso, no iniciar la vía de apremio –si se trata de un proceso de subasta de bien inmueble con carácter de vivienda habitual–. Se configura así, el requisito de conciliación e intermediación en materia de vivienda como nueva exigencia de procedibilidad. No obstante, dicho requisito se aplica a unos ámbitos subjetivo y objetivo concretos.

1. Ámbito objetivo de aplicación

Como se adelantó previamente, se instaura como nuevo "requisito formal" dentro de diversos procedimientos de la Ley de Enjuiciamiento Civil. En particular, se aplica en el juicio verbal (art. 439.7 LEC), en la subasta de bienes inmuebles (art. 655 bis LEC) y, por último, en la ejecución de bienes hipotecados (art. 685. 2 LEC)[11].

---

que es tracta es de detectar, des de l'inici del procediment, si la part demandada és vulnerable i de si s'han posat en marxa els mecanismes institucionals per a la seva protecció que contempla la norma. La seva aportació és un requisit per admetre la demanda, sense que la seva omissió es pugi esmenar. Es configura, per tant, com una excepció a la regla general d'admissió de les demandes, precisament perquè legalment així s'ha establert" (SAP de Girona, n.º 305/2023, de 13 de diciembre de 2023, [ECLI:ES:APGI:2023:1287A]).

11 Señalar que tal requisito de procedibilidad también se aplica a los procedimientos suspendidos en virtud de los artículos 1 y 1

Respecto al juicio verbal, solamente afecta a las materias recogidas en los números 1.º, 2.º, 4.º y 7.º del apartado 1 del artículo 250 LEC. Es decir, se reserva a los juicios verbales que versen sobre a) impago de rentas (como las debidas por el arrendatario), 2) recuperación de la plena posesión de un bien inmueble cedido en precario, 3) tutela sumaria de la tenencia o de la posesión de un bien inmueble o 4) protección de la efectividad o ejercicio de derechos reales inscritos en el Registro de la Propiedad. Desde luego, el ocupante de una vivienda habitual puede verse afectado por las acciones que entablen sus legítimos propietarios u otros titulares (usufructuario, etc.). A modo de ilustración, podemos mencionar el juicio verbal iniciado por el arrendador ante el impago de rentas del arrendatario, en régimen de alquiler de una vivienda habitual. Lo mismo ocurre con un ocupante que ostente la condición de deudor hipotecario, pues ante el impago del

---

bis del Real Decreto-ley 11/2020, de 31 de marzo, por el que se adoptan medidas urgentes complementarias en el ámbito social y económico para hacer frente al COVID-19. Como dispone la disposición transitoria tercera de la LDV, "tras la entrada en vigor de esta ley, y a partir del 31 de diciembre de 2024, los procedimientos de desahucio y los lanzamientos indicados en los artículos 1 y 1 bis del Real Decreto-ley 11/2020, de 31 de marzo, por el que se adoptan medidas urgentes complementarias en el ámbito social y económico para hacer frente al COVID-19, que se encuentren suspendidos por aplicación de dichos preceptos, cuando la parte actora sea una gran tenedora de vivienda en los términos previstos por el artículo 3.k) de esta ley, sólo se reanudarán a petición expresa de la misma si la parte actora acredita que se ha sometido al procedimiento de conciliación o intermediación que a tal efecto establezcan las Administraciones Públicas, en base al análisis de las circunstancias de ambas partes y de las posibles ayudas y subvenciones existentes conforme a la legislación y normativa autonómica en materia de vivienda".

préstamo puede perder su vivienda habitual en sede ejecutiva o con la subasta del bien inmueble por la entidad bancaria. Por tal motivo, el legislador instaura la conciliación e intermediación como requisito de procedibilidad, para buscar soluciones en sede extrajudicial y, de no ser factible, permitir a las Administraciones competentes adoptar medidas de protección social.

2. *Ámbito subjetivo de aplicación*

Observado el ámbito objetivo de aplicación de tal requisito de procedibilidad, la cuestión ahora es la siguiente: ¿qué sujetos se ven afectados por el procedimiento de conciliación e intermediación? Para responder tal cuestión, basta acudir al preámbulo de la LDV:

> "La disposición final quinta también introduce un procedimiento de conciliación o intermediación en los supuestos en los que la parte actora tenga la condición de gran tenedor de vivienda, el inmueble objeto de demanda constituya vivienda habitual de la persona ocupante y la misma se encuentre en situación de vulnerabilidad económica. La aplicación de este procedimiento facilitará a las Administraciones competentes dar adecuada atención a las personas y hogares afectados, ofreciendo respuesta a través de diferentes instrumentos de protección social y de los programas de política de vivienda".

Dicha norma, al modificar la LEC, establece expresamente que únicamente debe cumplir, con tal requisito de procedibilidad, el denominado "gran tenedor de vivienda" –aunque el art. 655 bis también se refiere literalmente a "empresa de vivienda"–[12]. El art. 3, k) LDV define al "gran

---

12 "El procedimiento de conciliación o mediación se impone, pues, solo en los casos en que el ocupante se encuentra en situación

tenedor" como aquella "persona física o jurídica que sea titular de más de diez inmuebles urbanos de uso residencial o una superficie construida de más de 1.500 m2 de uso residencial, excluyendo en todo caso garajes y trasteros". No obstante, en la declaración de entornos de mercado residencial tensionado se pueden incluir aquellos titulares de cinco o más inmuebles urbanos de uso residencial ubicados en dicho ámbito, cuando así sea motivado por la comunidad autónoma en la correspondiente memoria justificativa. Por contraposición, se excluyen de tal concepto pequeños propietarios, sean personas físicas o jurídicas, si no cumplen las características anteriormente mencionadas.

Para acreditar el cumplimiento de tal requisito, la LDV fija dos posibles formas[13]. El "gran tenedor de vivienda" puede presentar una declaración responsable emitida por la parte actora de que ha acudido a los servicios de conciliación o intermediación, en un plazo máximo de cinco meses de antelación a la presentación de la demanda, sin que hubiera sido atendida o se hubieran iniciado los trámites correspondientes en el plazo de dos meses desde que presentó su solicitud, junto con justificante acreditativo de la misma. Otra opción es facilitar un documento acreditativo de los servicios competentes que indique el resultado del procedimiento extrajudicial, en el que se hará constar la identidad

---

de vulnerabilidad económica y la parte actora es gran tenedora de vivienda, por lo que sólo a los grandes tenedores de vivienda les es exigible la obligación de acreditar la concurrencia o no de vulnerabilidad económica, como expresamente se plasma en los arts. 439.6 y 655 bis de la LEC, por lo que entendemos que en dicho sentido debe interpretarse el art. 685.2, lo que conlleva la estimación del recurso de apelación" (SAP de Pontevedra, n.º 33/2024, de 16 de febrero de 2024, [TOL9.937.973]).

13 Véanse los arts. 439.7, 655 bis. 2 y 685.2 LEC.

de las partes, el objeto de la controversia y si alguna de las partes ha rehusado participar en el procedimiento, en su caso –dicho documento no puede tener una vigencia superior a tres meses–[14]. Con tales exigencias, los grandes tenedores deben presentar una solicitud ante la Administración competente para instar el inicio del procedimiento de conciliación o intermediación ante el ocupante de su inmueble –requiriendo un justificante de su presentación– y esperar un plazo de tiempo para poder interponer la demanda pertinente –entre dos o cinco meses, según se haya iniciado o no el procedimiento extrajudicial[15]–.

---

14 En el caso de que la empresa arrendadora sea una entidad pública de vivienda el requisito anterior se puede sustituir, en su caso, por la previa concurrencia de la acción de los servicios específicos de intermediación de la propia entidad.

15 Si la Administración competente no le comunica en dos meses que su solicitud está admitida y en trámite, podrá seguir con la vía judicial. En caso de haberse iniciado el procedimiento de conciliación o intermediación, debe transcurrir un plazo de cinco meses. No obstante, los jueces de primera instancia de Barcelona consideran que son plazos cumulativos, no independientes: "La declaración responsable emitida por el gran tenedor, en la que se diga que la Administración Pública no ha iniciado en el plazo de dos meses los trámites correspondientes (del procedimiento de conciliación o intermediación, o para la determinación de la vulnerabilidad de las personas ocupantes), sólo será eficaz a los efectos de considerar acreditados los requisitos procesales de la Disposición Transitoria Tercera de la Ley 12/2023, y de los arts. 439.6, 439.7, 655bis.1, 655bis.2 y 685.2 LEC, cuando se haya presentado ante el juzgado dentro de los cinco meses siguientes a la presentación de aquella solicitud ante aquella Administración Pública. Es decir, lo plazos de cinco meses y de dos meses que se contienen en esos párrafos no se aplican de manera independiente y para casos distintos, sino de manera conjunta y acumulada" (criterios orientadores aprobados por los Jueces de Primera Instancia

Atendiendo a tales requerimientos, cierta doctrina considera que los propietarios de inmuebles calificados como "grandes tenedores" se ven obligados a pasar por un trámite que en la mayoría de los casos acabará sin resultado alguno por la posibilidad del ocupante de prolongar su presencia en el inmueble más tiempo[16]. Como señala Viola Demestre, "desde la perspectiva procesal, se percibe como un impedimento al principio de tutela judicial efectiva, puesto que ralentiza el acceso a los tribunales"[17]. Sin embargo, aunque pueda ser así, permite actuar a las Administraciones públicas para buscar ayudas sociales o alternativas habitacionales. Asimismo, nada impide que se alcancen soluciones óptimas para impedir la pérdida de la vivienda habitual (reestructurar el préstamo hipotecario, carencias de pago, etc.).

Finalmente, hay que señalar que el requisito de conciliación extrajudicial es ineludible siempre que la persona demandada cumpla ciertas condiciones: por un lado, debe encontrarse en

---

de Barcelona, con motivo de la entrada en vigor de la Ley 12/2023, de 24 de mayo, por el Derecho a la Vivienda, 2023).

16 Magro Servet (2023, LA LEY 4622/2023). En este mismo sentido, Jiménez Muñoz (2024, LA LEY 3824/2024).

17 Viola Demestre (2024, p. 495).

situación de vulnerabilidad económica[18][19]; y, por otro, el bien

---

[18] "El tribunal tomará la decisión previa valoración ponderada y proporcional del caso concreto, apreciando las situaciones de vulnerabilidad que pudieran concurrir también en la parte actora y cualquier otra circunstancia acreditada en autos. A estos efectos, en particular, el tribunal para apreciar la situación de vulnerabilidad económica podrá considerar el hecho de que el importe de la renta, si se trata de un juicio de desahucio por falta de pago, más el de los suministros de electricidad, gas, agua y telecomunicaciones suponga más del 30 por 100 de los ingresos de la unidad familiar y que el conjunto de dichos ingresos no alcance: a) Con carácter general, el límite de 3 veces el Indicador Público de Renta de Efectos Múltiples mensual IPREM). b) Este límite se incrementará en 0,3 veces el IPREM por cada hijo a cargo en la unidad familiar. El incremento aplicable por hijo a cargo será de 0,35 veces el IPREM por cada hijo en el caso de unidad familiar monoparental o en el caso de cada hijo con discapacidad igual o superior al 33 por ciento. c) Este límite se incrementará en 0,2 veces el IPREM por cada persona mayor de 65 años miembro de la unidad familiar o personas en situación de dependencia a cargo. d) En caso de que alguno de los miembros de la unidad familiar tenga declarada discapacidad igual o superior al 33 por ciento, situación de dependencia o enfermedad que le incapacite acreditadamente de forma permanente para realizar una actividad laboral, el límite previsto en la letra a) será de 5 veces el IPREM, sin perjuicio de los incrementos acumulados por hijo a cargo. A estos mismos efectos, el tribunal para apreciar la vulnerabilidad social podrá considerar el hecho de que, entre quienes ocupen la vivienda, se encuentren personas dependientes de conformidad con lo dispuesto en el apartado 2 del artículo 2 de la Ley 39/2006, de 14 de diciembre, de Promoción de la Autonomía Personal y Atención a las personas en situación de dependencia, víctimas de violencia sobre la mujer o personas menores de edad" (art. 441.7 LEC). Por consiguiente, la situación de vulnerabilidad se determinará mediante una decisión ponderada y proporcional del tribunal. Acertadamente, como indica la doctrina, el art. 441.7 LEC fija criterios objetivos para determinar la vulnerabilidad económica (Rodríguez Tirado, 2024, LA LEY 3829/2024).

[19] "Para acreditar la concurrencia o no de vulnerabilidad económica de la parte ejecutada se deberá aportar documento acreditativo, de

vigencia no superior a tres meses, emitido, previo consentimiento de éste, por los servicios de las Administraciones autonómicas y locales competentes en materia de vivienda, asistencia social, evaluación e información de situaciones de necesidad social y atención inmediata a personas en situación o riesgo de exclusión social que hayan sido específicamente designados conforme a la legislación y normativa autonómica en materia de vivienda. Este requisito también podrá cumplirse mediante: 1.º La declaración responsable emitida por la parte actora de que ha acudido a los servicios indicados anteriormente, en un plazo máximo de cinco meses de antelación a la presentación de la solicitud de inicio de la vía de apremio, sin que hubiera sido atendida o se hubieran iniciado los trámites correspondientes en el plazo de dos meses desde que presentó su solicitud, junto con justificante acreditativo de la misma. En tal caso el Juzgado se dirigirá a las Administraciones competentes a fin de que confirmen, en el plazo máximo de diez días, si el hogar afectado se encuentra en situación de vulnerabilidad económica y, en su caso, social, así como las medidas previstas que se aplicarán de forma inmediata para que disponga de una vivienda. 2.º El documento acreditativo de los servicios competentes que indiquen que la parte ejecutada no consiente expresamente el estudio de su situación económica en los términos previstos por la legislación y normativa autonómica en materia de vivienda. Este documento no podrá tener una vigencia superior a tres meses" (arts. 439.6, 655 bis. 1 y 685. 2 LEC). Los trámites anteriores dilatan el tiempo para seguir con la vía judicial. Este procedimiento para acreditar la vulnerabilidad económica lo detalla de forma clara Magro Servet: "Pero ello exige que: primero haya pedido al ocupante que le firme el escrito y que este se niegue, y, a continuación, acudir a la Administración presentando una solicitud de que se averigüe si el ocupante es vulnerable y pedir justificante de que esta petición se le hace a la Administración. Más tarde esperar dos meses para comprobar si la Administración ha iniciado el trámite para determinar si es vulnerable, o no. Y, por último, esperar cinco meses sin ningún resultado para poder presentar la demanda de desahucio a la que acompañaría el escrito y justificante que presentó ante la Administración en los términos antes expuestos, la respuesta de la Administración en su caso, o la

inmueble que ocupa –como deudor hipotecario, arrendatario, etc.– debe ostentar la condición de vivienda habitual[20]. Repárese que la vulnerabilidad económica debe ser probada por la parte demandante –el gran tenedor de vivienda–, cuando lo coherente y oportuno sería que la justificase el ocupante.

3. Alabanzas y críticas del requisito de procedibilidad

Tras este breve comentario, al configurar la LDV la conciliación e intermediación como requisito de procedibilidad, resulta alabable que el legislador estatal fomente las alternativas de solución de conflictos de vivienda, antes de llegar a la vía judicial[21]. Al

---

constancia del silencio de la misma ante la petición que le formuló. Para el caso que el ocupante haya respondido al propietario a la primera petición formalmente de que no le autoriza a que se comprueben sus datos respecto a su posible vulnerabilidad se exigirá que se acompañe a la demanda de desahucio el documento acreditativo de los servicios competentes que indiquen que la persona ocupante no consiente expresamente el estudio de su situación económica en los términos previstos en la legislación y normativa autonómica en materia de vivienda con la particularidad de que en ese caso este documento no podrá tener una vigencia superior a tres meses, por lo que será preciso instar la demanda de inmediato. Para ello, el solicitante arrendador le habrá hecho constar en su escrito a la Administración que el ocupante se niega a autorizar que se investigue si es vulnerable" (Magro Servet, 2023, LA LEY 4622/2023).

20 Se considera "residencia habitual" aquella "vivienda que constituye el domicilio permanente de la persona que la ocupa y que puede acreditarse a través de los datos obrantes en el padrón municipal u otros medios válidos en derecho" (art. 3.i LDV).

21 En 2013, tras la instauración de la mediación o intermediación hipotecaria en numerosas ciudades españolas, reivindicaba su configuración como requisito previo, e incluso obligatorio, al procedimiento de ejecución hipotecaria. Véase Bastante Granell (2013, pp.

parecer, la cultura del diálogo comienza a adentrarse en la mente del legislador, con una finalidad eminentemente social –ayudar a la protección de personas ocupantes en situación de vulnerabilidad económica–. Si bien, digo "al parecer", por los siguientes motivos. Y es que la "fuerza ética" se desvanece al fijar la mirada en la falta de justicia y práctica social en ciertos territorios.

En primer lugar, llama la atención que se apueste únicamente por la conciliación e intermediación en materia de vivienda en ciertos procedimientos (juicio verbal, ejecución hipotecaria, etc.) sin incluir otros que pueden afectar negativamente al ocupante de una vivienda habitual. Ciertamente, son procedimientos más comunes, pero existen algunos que pueden derivar en la pérdida de tal bien inmueble. Piénsese, por ejemplo, en el proceso concursal –de persona física no empresaria–, procesos penales –iniciados por la ocupación de viviendas– o, a modo de ilustración, el cauce para realizar un desahucio administrativo –que puede aplicarse a viviendas sujetas a regímenes de protección pública–. Pues bien, ¿acaso las personas ocupantes de una vivienda, afectadas por tales procesos, no merecen que se exija previamente la tramitación de una conciliación e intermediación?, ¿se presume que no necesitan apoyo o asistencia social por las Administraciones públicas?

Para más inri, este requisito de procedibilidad solamente se aplica a los grandes tenedores de vivienda. Si la parte actora es una persona física o jurídica, que no reúne tal condición jurídica, no tendrá la obligación de acudir previamente a la conciliación o intermediación en materia de vivienda. Siendo así, si la finalidad es la búsqueda de soluciones y la protección social del

---

180-213). Más de diez años después, y tras la petición de cierta doctrina, el legislador estatal, con menor o mayor acierto, ha atendido a tales reivindicaciones al configurar la intermediación o conciliación como requisito de procedibilidad.

ocupante del bien inmueble (por ejemplo, un arrendatario), ¿por qué no se exige también a tales sujetos, aunque no sean grandes tenedores, acudir a la conciliación e intermediación en materia de vivienda?, ¿acaso la parte demandada, en posible situación de vulnerabilidad social y económica, no merece asistencia por la Administración competente? Si la finalidad es promover una "negociación social", no es comprensible despojar de tal opción a los ocupantes de pequeños tenedores. De ser así, podría pensarse que este requisito de procedibilidad tiene como finalidad exclusiva retrasar el desahucio o lanzamiento de grandes tenedores[22], sin ánimo de alcanzar soluciones a través de la intermediación de las Administraciones competentes. Estaríamos ante una aparente "medida social" que esconde como objetivo oculto dilatar el desahucio o lanzamiento iniciado por grandes tenedores de vivienda. Como opina Magro Servet, "resulta evidente que esta conciliación lo que introduce es un retraso en la recuperación de la posesión del inmueble obligando a los propietarios de inmuebles calificados como «grandes tenedores» a pasar por un trámite que en la mayoría de los casos será sin resultado alguno por la posibilidad del ocupante de prolongar su presencia en el inmueble más tiempo"[23]. No obstante, también es cierto que un "gran tenedor", por su condición, debería ser más "proclive a negociar" para alcanzar alguna solución habitacional –dado

---

22 Piénsese que realmente la LDV exige numerosos requisitos a los grandes tenedores de vivienda, entre otros, a) si el inmueble constituye vivienda habitual, b) si ostenta la condición de gran tenedor de viviendas, c) si la parte demanda está o no en situación de vulnerabilidad y d) si ha acudido a algún procedimiento de conciliación o intermediación autonómico. Con ello no se garantiza una recuperación inmediata de la vivienda. Al contrario, se quebranta la tutela de sus derechos, en "beneficio" de la protección social del ocupante.

23 Magro Servet (2023, LA LEY 4622/2023).

que tiene a su disposición un mayor número de bienes inmuebles, a diferencia de un pequeño propietario–.

Con tal regulación, existe una evidente discriminación jurídica para las personas ocupantes de vivienda, con vulnerabilidad económica y social, según la parte actora sea o no gran tenedor de vivienda. Lo coherente hubiera sido exigir tal requisito de procedibilidad, con independencia de la condición jurídica de la parte actora, con la finalidad de brindar a toda persona –ocupante y vulnerable– un cauce para la búsqueda de soluciones y la implementación de ayudas sociales a su favor. Sería una forma de potenciar la intermediación por las Administraciones competentes, con un propósito general de protección social. No obstante, nos encontramos con "ocupantes de primera" y "ocupantes de segunda", lo que, desde mi punto de vista, denota cierta discriminación jurídica.

IV. UN PROCEDIMIENTO "PENDIENTE" Y DEPENDIENTE DE LAS CCAA

A simple vista, podemos observar "buenas intenciones" –brindar apoyo social a ocupantes vulnerables, con la instauración de la intermediación y conciliación como requisito de procedibilidad–. No obstante, esas "bondades internas" no se extienden al gran tenedor de vivienda –cuya tutela del derecho a la propiedad se ve dilatada– o, bien, al ocupante vulnerable de un bien inmueble cuya propiedad no pertenezca a un gran tenedor de vivienda –parece ser que para tales sujetos no es necesario el trámite de intermediación y conciliación–.

Dejando de lado tales críticas, preocupa un aspecto esencial del mismo: su aplicación y funcionalidad social. En efecto, el legislador estatal presenta la intermediación y conciliación en la LDV y configura tal procedimiento extrajudicial como un requisito de procedibilidad. Ahora bien, no va más allá, estableciendo, por ejemplo, su regulación legal. Ello se debe a que ha considerado que tal procedimiento es competencia de las comunidades autónomas –aunque se refiere a las "Administraciones competentes" –. Por tal motivo, efectúa un reconocimiento a nivel estatal –en la LDV–, pero posteriormente indica que la intermediación y conciliación en materia de vivienda queda supeditada al desarrollo normativo que fijen las distintas Administraciones autonómicas[24], quienes serán las responsables de su diseño, gestión y desarrollo. Por consiguiente, la intermediación y conciliación en materia de vivienda queda supeditada al desarrollo normativo que fijen las distintas comunidades autónomas.

Efectivamente, desde hace años, ciertas comunidades ya disponen de procedimientos de "conciliación e intermediación" en materia de vivienda (como País Vasco o Cataluña). No obstante, algunas no están adaptadas a la nueva LDV. Otras, en cambio, no disponen de tal procedimiento. Ante tal situación, resulta criticable que el legislador no haya previsto alguna disposición transitoria para permitir a las comunidades autónomas adaptarse a la LDV y así poder regular y adaptar, si fuera necesario, tal procedimiento dentro de su cartera de servicios en materia de vivienda. Comprendo, en parte, que el legislador estatal traslade dicho procedimiento al ámbito autonómico, pero ha sido imprudente al no conceder un periodo de adaptación a las Administraciones autonómicas[25]. Tampoco

---

24 "[...] si la parte actora acredita que se ha sometido al procedimiento de conciliación o intermediación que a tal efecto establezcan las Administraciones Públicas, en base al análisis de las circunstancias de ambas partes y de las posibles ayudas y subvenciones existentes conforme a la legislación y normativa autonómica en materia de vivienda" (Disposición transitoria tercera LDV).

25 Tal proceder puede resultar comprensible –¿para qué se va a ocupar el Estado de gestionar un procedimiento extrajudicial si, en

recoge la LDV, por ejemplo, una fecha concreta para instaurar tal procedimiento. Siendo tal el panorama, nos encontramos ante una norma estatal que exige un requisito de procedibilidad, pero dicho procedimiento puede no estar operativo a nivel autonómico, afectando de forma notoria a la tutela judicial efectiva de los grandes tenedores de vivienda:

> "Serán, por tanto, las Administraciones Públicas con competencias en materia de vivienda (las Comunidades Autónomas, en virtud del art. 148.1.3.ª CE) [...] quienes tendrán que regular tales procedimientos; pero la LDV no dice cuándo, no hay —s.e.u.o. por mi parte— ninguna disposición de esta Ley que fije un plazo dentro del cual las CCAA deberán aprobar la normativa relativa a dichos procedimientos de intermediación y conciliación. Y es urgente que lo hagan, puesto que los aspectos procesales de la LDV, todos, ya están en vigor [...]; y sería completamente absurdo e inaceptable desde el punto de vista constitucional, por ser contrario al derecho fundamental a la tutela judicial efectiva en su dimensión de derecho de acceso a la jurisdicción, que un demandante no pudiera acreditar el cumplimiento de ese nuevo requisito de procedibilidad exigido por la LDV por no existir aún —porque la Comunidad Autónoma de que se trate todavía no los ha implementado— los nuevos trámites de intermediación y conciliación previos a la presentación de la demanda"[26].

Advertir que tal situación está causando actualmente numerosos problemas, como comprobaremos más adelante.

Como el legislador traslada este procedimiento extrajudicial a las comunidades autónomas, Ceuta y Melilla; las ha autorizado para que puedan utilizar los recursos de los planes

---

principio, ya se encuentra operativo en las CCAA?– Pero cuidado, dicho procedimiento no dispone de regulación en cada comunidad autónoma; y, además, su operatividad, principios y reglas pueden variar en cada territorio.

26 López Simó (2023, LA LEY 10759/2023).

estatales en materia de vivienda para cubrir los costes del proceso así como las compensaciones que puedan acordarse a solicitud de los propietarios de los inmuebles afectados (Disposición adicional cuarta de la LDV). Asimismo, el art. 24 LDV especifica que ciertas actuaciones de los Planes estatales en materia de vivienda y rehabilitación, regeneración y renovación urbana y rural podrán servir para atender a los trámites de intermediación y conciliación previos a la presentación de la demanda, así como a la atención a las personas y hogares sujetos a vulnerabilidad en los procedimientos recogidos en los arts. 439, 655 bis y 685 de la LEC. Aunque el legislador estatal se desentiende de la intermediación y conciliación en materia de vivienda, al menos parece brindar apoyo económico para su implementación y desarrollo autonómicos.

4. Desarrollo normativo y funcionamiento a nivel autonómico

Tras la entrada en vigor de la LDV, lo interesante es observar si tal procedimiento existe y funciona a nivel autonómico.

Para poder responder a tal interrogante, lo más adecuado era actuar desde la posición de un "gran tenedor de vivienda" e intentar buscar información para canalizar una posible solicitud de intermediación y conciliación. Era conveniente abandonar el plano teórico, para ahondar en la práctica y observar, así, como se inicia y desarrolla tal procedimiento en el ámbito autonómico. Por tal razón, decidí examinar la situación en varias comunidades autónomas –señalar que dicha búsqueda de información se realizó a fecha de 15 de marzo de 2024–. En concreto, he realizado gestiones (telefónicas, por correo electrónico, etc.) y búsqueda de información (en páginas de organismos oficiales, etc.) en Andalucía, Murcia, La Rioja, Valencia, Extremadura, Castilla La Mancha, Castilla y León, Aragón, País Vasco y Cataluña. En otras CCAA, como Madrid, la labor ha sido poco fructífera o nula –por falta de información en páginas oficiales, la complejidad para contactar con posibles responsables o el desconocimiento del personal de ciertos organismos–.

4.1. CCAA sin normativa, protocolo y/o procedimiento

Para comenzar, es oportuno ilustrar la situación en aquellas CCAA en las que el procedimiento no existe, bien porque no hay normativa específica o instrucciones y/o la intermediación no está operativa a través de algún servicio u organismo.

En Valencia o La Rioja todavía no se ha designado el organismo competente para desarrollar el procedimiento de intermediación o conciliación en materia de vivienda. En principio, en la primera CCAA, dicho procedimiento se desarrolla ante el "servicio de mediación en materia de vivienda", dependiente del Instituto de la Vivienda de La Rioja (IRVI)[27]. Si bien, todavía no se ha designado legalmente tal competencia por las autoridades competentes. Siendo así, la solicitud de inicio debe dirigirse a la Consejería competente en materia de vivienda. Lo mismo ocurre en Valencia, tras contactar con la Unidad de Ayuda ante Desahucios (UAD)[28], me informan que dicha solicitud de intermediación, como gran tenedor de vivienda, debe realizarse por registro de entrada a la administración competente, que es la Dirección General de Vivienda de Valencia. Situación similar ocurre en Extremadura. En dicha comunidad autónoma hay que presentar la solicitud a la Secretaría General de

---

27 https://www.irvi.es/mediacion_integral/index.html [Fecha de consulta: 15/03/2024].

28 https://habitatge.gva.es/es/web/vivienda-y-calidad-en-la-edificacion/unitat-ajuda-davant-desnonaments [Fecha de consulta: 15/03/2024].

Vivienda, Arquitectura y Regeneración Urbana (Consejería de Infraestructuras, Transporte y Vivienda). No obstante, no hay protocolo de actuación ni servicio de intermediación, aunque el expediente se traslada al servicio de adjudicación de viviendas sociales. Si acudimos a Aragón, el gran tenedor debe presentar la solicitud al Instituto Aragonés de Servicios Sociales (IASS) o, bien, a la Dirección General de Vivienda. No obstante, tras contactar con las personas responsables, me trasladan, por un lado, que no hay un protocolo de actuación –motivo por el cual la petición se puede presentar en ambos organismos–; y, por otro, que el servicio de mediación se encuentra inactivo, por lo que sería imposible realizar negociación alguna entre las partes. De este modo, podemos observar cómo en ciertas CCAA la actuación del gran tenedor se limitará a presentar la solicitud, guardar el justificante y esperar a que transcurra el tiempo fijado en la norma para iniciar la vía judicial.

Se comprueba cómo, ante la ausencia de una disposición transitoria, ciertas CCAA no han dispuesto de "tiempo suficiente" para organizarse y determinar los organismos competentes para desarrollar la intermediación con grandes tenedores –aunque puede que tampoco tengan intención de hacerlo–. De persistir tal situación, dicha solicitud seguramente terminará con el silencio de la Administración pertinente. Hay quien piensa que esto solamente perjudica a la parte actora, al gran tenedor, pero también afecta a la persona ocupante. Es cierto que el desahucio o lanzamiento se dilatará en el tiempo, ante la burocracia administrativa, pudiendo beneficiar al ocupante. Si bien, repárese que éste último no podrá contar con un adecuado sistema de apoyo social, al no poder disfrutar de un servicio capaz de alcanzar algún acuerdo habitacional con el gran tenedor o, en su caso, de buscar soluciones o ayudas, como un alquiler social. Por tal motivo, indica acertadamente la Plataforma de Afectados por la Hipoteca (PAH) lo siguiente:

> "Es clave que las Comunidades Autónomas dispongan de un buen proceso de intermediación o conciliación que garantice soluciones habitacionales para los desahucios de personas vulnerables de su vivienda habitual y, si hace falta, deberán desarrollarlo y/o regularlo, incluyendo medidas alternativas para garantizar el derecho a la vivienda"[29].

4.2. CCAA con normativa, protocolo y/o procedimiento

A pesar de esta situación de vacío normativo o de ausencia de instrucciones por ciertas Administraciones autonómicas, en otras Comunidades existe un procedimiento específico, se ha establecido un protocolo de actuación y/o se ha asignado un organismo competente para ofrecer el servicio de intermediación o conciliación a grandes tenedores.

Comenzando con Andalucía, en un principio pensé que el denominado "Sistema Andaluz de lucha contra la ocupación ilegal y de asesoramiento para la protección de la vivienda"[30] iba a ser el encargado de gestionar dicho procedimiento. Repárese que a través del mismo se encauza la intermediación hipotecaria y, además, ofrece protección tras la pérdida del domicilio habitual, como consecuencia de una ejecución hipotecaria o por morosidad del pago del alquiler con causas objetivas y justificadas. Si bien, finalmente no se ha designado

---

29 Plataforma de Afectados por la Hipoteca (PAH), Guía de documentos útiles para detener desahucios con la nueva Ley Vivienda (Ley 12/2023), p. 4.

30 Para más información, véase la página web de dicho servicio de la Junta de Andalucía: https://www.juntadeandalucia.es/organismos/fomentoarticulaciondelterritorioyvivienda/areas/vivienda-rehabilitacion/asesoramiento-ocupacion-ilegal.html [Fecha de consulta: 15/03/2024].

como organismo responsable. Lo coherente hubiera sido incluir la conciliación e intermediación en materia de vivienda como competencia de dicho servicio que, desde 2012, cuenta con experiencia sobre la temática. No habiendo optado por tal opción, Andalucía ha decidido trasladar la competencia de su gestión y desarrollo a los PIMA (Punto de Información para la Mediación en Andalucía), fijando un protocolo de solicitudes de información para la mediación en materia de desahucios por grandes tenedores conforme a la LDV[31].

---

[31] "1.- Solicitud de Sesión Informativa. El letrado, Procurador o la propia persona física o jurídica que sea considerada como gran tenedor solicitará por correo electrónico al PIMA (pima.sevilla.cjalfp@juntadeandalucia.es) una cita para la sesión informativa de mediación. En dicha solicitud se deberá recoger nombre de la persona solicitante (propietario/a), teléfono de contacto, una breve descripción del conflicto y su interés en participar en un proceso de mediación, nombre de la persona inquilina/ocupante, así como el domicilio de la vivienda que se pretende desahuciar. 2.- Citación a Sesión Informativa. Una vez recibida la solicitud, el PIMA procederá a citar a la parte actora mediante el mismo medio mencionado en el apartado 1. El PIMA tendrá un plazo máximo de dos meses para citar a las partes. 3.- Es necesario que la parte solicitante de la cita sea la que comunique su voluntad de asistir a la sesión informativa sobre mediación a la persona inquilina/ocupante a través de burofax u otro medio que deje constancia de la recepción, fecha y hora. El día de la cita deberá comparecer la parte actora o su representante (letrado, procurador, administrador…) con la documentación acreditativa de haber enviado la citación a la otra parte." 4.- Sesión Informativa de Mediación. Realizada la sesión informativa, se expedirá un informe, que contendrá los detalles de la sesión, la identidad de las partes, el objeto de la controversia, la confirmación de participación, así como su vigencia de tres meses. 5.- Derivación por órgano judicial tras la interposición de la demanda. Recibida la demanda en el juzgado y constatado por el Letrado de la Administración de Justicia (LAJs) que la demanda no cumple el requisito de procedibilidad sobre la conciliación exigida en la Ley de Enjuiciamiento Civil art. 439.7, el LAJ o el funcionario

Actualmente el gran tenedor debe solicitar por correo electrónico al PIMA una cita para la sesión informativa de mediación. Sin embargo, hay que señalar que la parte solicitante debe ser la que comunique su voluntad de asistir a la sesión informativa sobre mediación a la persona inquilina/ocupante a través de burofax u otro medio que deje constancia de la recepción, fecha y hora. Además, aunque no aparezca expresamente en el protocolo –al que pude acceder tras una petición por correo electrónico–, el gran tenedor debe presentar un informe de vulnerabilidad económica del ocupante. De no ser así, no tendría sentido iniciar el cauce de intermediación o conciliación en materia de vivienda. En este punto, indicar que algunos servicios sociales se están negando a ofrecer tal informe, al considerar que no es competencia su emisión. Siendo así, los grandes tenedores tienen problemas para su obtención y, consecuentemente, para poder solicitar e iniciar el procedimiento extrajudicial. De acudir al procedimiento judicial, sin abrir esta vía extrajudicial, el asunto sería derivado al PIMA por el órgano judicial.

Estas dudas sobre competencia se pueden observar en el "Informe sobre competencia municipal para la emisión de informe de situación de necesidad social y vulnerabilidad económica", emitido por la Diputación de Córdoba, n.º de expediente 199/2023-JADSC. Un Ayuntamiento solicitó informe para determinar su competencia para emitir un escrito a un banco sobre

---

designado para ello citará a las partes para sesión informativa a través del aplicativo @PIMA o por correo electrónico. En caso de hacerlo por este último medio se le responderá dando una fecha y una hora para la sesión informativa. El resultado de la sesión informativa se remitirá al juzgado (ya sea a través de la propia agenda o correo electrónico)" (protocolo de solicitudes de información para la mediación en materia de desahucios por grandes tenedores Ley 12/2023, de 24 de mayo, por el derecho a la vivienda).

situación de necesidad social o vulnerabilidad económica en relación con un procedimiento de ejecución hipotecaria. Se desconocía si la persona competente era la trabajadora social de dicha localidad o, en su caso, el Instituto Provincial de Bienestar Social (IPBS)[32]. No obstante, también se trataba la competencia del Ayuntamiento relativa al procedimiento de conciliación e intermediación. Respecto a la primera cuestión se establece lo siguiente:

> "Es por ello que, en función de lo anterior, no nos queda más que concluir que, efectivamente, la emisión de los informes de vulnerabilidad económica o situaciones de necesidades sociales corresponde emitirlos a los municipios como parte de las competencias propias que le son atribuidas por la Ley, ahora bien, conforme a dichas normas legales y reglamentarias, en la provincia de Córdoba, y para los municipios menores de 20.000 habitantes, dichas funciones son prestadas por los servicios del IPBS, como organismo autónomo local de naturaleza administrativa, creado por la Diputación Provincial de Córdoba".

En cuanto al segundo interrogante, especifica que "ninguna norma de rango legal o reglamentario regula o conduce a que el Ayuntamiento tenga que verse compelido a realizar tareas de intermediación y/o conciliación en términos extrajudiciales para materias civiles o mercantiles, por mor de que existan situaciones de necesidad social o vulnerabilidad económica derivadas o interconexas con procesos de ejecución hipotecaria". Siendo así, el informe se remite a los PIMA, como responsables de esta materia en Andalucía.

---

[32] Dicho informe aparece accesible en la siguiente página web: https://www.dipucordoba.es/wp-content/uploads/2017/06/Informe-sobre-competencia-municipal-para-emision-de-informe-de-situacion-de-necesidad-social.pdf [Fecha de consulta: 15/03/2024].

En este momento el panorama en Andalucía es poco alentador. En primer lugar, la competencia de tal procedimiento extrajudicial se traslada a los PIMA, cuando desde hace años existen oficinas de intermediación hipotecaria con experiencia en la materia, tanto a nivel autonómico como municipal. No hay una adecuada optimización de los recursos andaluces. Posiblemente el protocolo ante el PIMA se convierta en un mero trámite, sin funcionalidad social. En segundo lugar, las dudas sobre la competencia para la emisión del informe de vulnerabilidad perjudican a los grandes tenedores, perdidos ante la incertidumbre y la falta de información para conocer los organismos competentes de dicho escrito. Tal situación supone un obstáculo para iniciar la intermediación o conciliación en Andalucía. Por último, llama sumamente la atención que desde el PIMA no se especifica qué procedimiento va a desarrollarse (reglas, principios, mediadores, etc.). Deduzco que se gestionará como una "mediación civil" con mediadores que tengan conocimiento en materia de desahucios –al no tramitarse por oficinas de intermediación hipotecaria–. En consecuencia, aunque puede que me equivoque, el panorama andaluz no es muy halagüeño.

Pongamos el foco ahora en Murcia. La solicitud del procedimiento extrajudicial se debe realizar ante el Servicio de Mediación de la Vivienda de la Comunidad de Murcia[33]. Generalmente este servicio se ofrece a ciudadanos que se encuentren en situación de vulnerabilidad, originada por el impago de créditos hipotecarios o de arrendamiento. A través del mismo se ofrece información, conciliación o intermediación con entidades de crédito o grandes tenedores de vivienda. No obstante, con la LDV será el servicio responsable de gestionar

---

33 https://sede.carm.es/web/pagina?IDCONTENIDO=397&IDTIPO=240&RASTRO=c$m40288 [Fecha de consulta: 15/03/2024].

el procedimiento de intermediación con grandes tenedores. Tan es así que se informa expresamente en la página web de que los informes sobre vulnerabilidad social y económica, previstos en la LEC, son competencia de los servicios sociales municipales, por lo que los interesados (grandes tenedores) no deben usar este procedimiento para solicitar la emisión de dichos informes, sino presentar sus solicitudes directamente al Ayuntamiento que corresponda. De lo contrario, deberán ser clasificados y redistribuidos a los Ayuntamientos de toda la región, con el consiguiente retraso temporal en su tramitación. Cuando dispongan del informe, podrán solicitar la intermediación.

Debo señalar que, tras contactar telefónicamente con dicho servicio, la solicitud decaerá por silencio administrativo. No existe propósito por las autoridades competentes de resolver tales conflictos.

Poniendo el foco en el País Vasco, se dictó la Instrucción 2/2023, del Viceconsejero de Vivienda, sobre el procedimiento de conciliación o intermediación previsto en la LDV. Ello, por un lado, ante la falta de designación conforme a la legislación autonómica en materia de vivienda en relación a la Administración pública que debe prestar el servicio de conciliación o intermediación, previa declaración de vulnerabilidad económica de la parte demandada; y, por otro, por la premura ante la vigencia, sin disposición transitoria, de la LDV. Con buen acierto, ante la existencia de servicios que gestionan conflictos arrendaticios e hipotecarios, lo óptimo era canalizar la intermediación con grandes tenedores a través de los mismos. En concreto, nos referimos al servicio Bizilagun, dependiente del departamento de Planificación Territorial, Vivienda y Transportes; y el Servicio de Mediación Hipotecaria, gestionado por el departamento de Igualdad, Justicia y Políticas Sociales. El primero se encarga de tramitar los conflictos relacionados con el art.

439.6 y 7 de la LEC y los procedimientos suspendidos en virtud de los arts. 1 y 1 bis del RDL 11/2020. El segundo, por su parte, se centra en los procedimientos de los arts. 655 bis y 685 de la LEC.

Es importante señalar que dicha instrucción establece, respecto al servicio Bizilagum[34], el procedimiento a seguir, al atender a la Orden de 24 de octubre de 2007, del Consejero de Vivienda y Asuntos Sociales, por la que se regulan las funciones de mediación y conciliación en materia de propiedad horizontal y arrendamientos urbanos. De forma específica manifiesta: "A estos efectos, se considerará el procedimiento de conciliación o intermediación recogido en la Ley 12/2023 como el procedimiento de mediación establecido en la Orden previamente citada". Dicha norma define la mediación, fija unos principios y normas procedimentales. Además, la instrucción recuerda un objetivo: "la persona mediadora intentará que las partes implicadas en el conflicto lleguen por sí mismas a un acuerdo y podrá promover entre las partes un acuerdo o solución total o parcial del conflicto". La Instrucción también detalla el órgano competente para las solicitudes de grandes tenedores, documentación, datos necesarios o, bien, los trámites a seguir para la obtención del informe de vulnerabilidad económica. Es decir, el procedimiento no se constituye como un mero trámite prejudicial, sin negociación. Lo mismo sucede con el servicio de mediación hipotecaria, que cuenta con un procedimiento definido[35].

---

34 https://www.euskadi.eus/web01-a2bizila/es/contenidos/informacion/infor_eta_bitart/es_def/index.shtml [Fecha de consulta: 15/03/2024].

35 https://www.justizia.eus/servicio-de-mediacion-hipotecaria/webjus00-contentgen/es/#procedimiento [Fecha de consulta: 15/03/2024].

Respecto a Cataluña, en dicha comunidad autónoma se ha habilitado un procedimiento específico para atender a las solicitudes de intermediación o conciliación de los grandes tenedores de vivienda[36]. El organismo responsable es la Agencia de Vivienda de Cataluña (Agència de l'Habitatge de Catalunya)[37]. Como se señala en la página web, junto a la solicitud debe aportarse documento acreditativo de la vulnerabilidad económica del ocupante de vivienda. Por lo tanto, el gran tenedor debe tramitar previamente dicha solicitud y disponer del informe de vulnerabilidad[38]. En cuanto al procedimiento, observando la normativa aplicable, se desarrolla atendiendo a lo dispuesto en el Decreto 98/2014, de 8 de julio, sobre el procedimiento de mediación en las relaciones de consumo. Es decir, en Cataluña la intermediación o conciliación en materia de vivienda se sustancia a través de una mediación de consumo[39]. De esta forma,

---

36 https://habitatge.gencat.cat/ca/detalls/Tramits/23465_Intermediacio_compliment_requisits_Llei_12_2023 [Fecha de consulta: 15/03/2024].

37 Sin embargo, en Barcelona ciudad colabora con el Instituto Municipal de Vivienda de Barcelona (IMHAB).

38 No obstante, nada impide que, con el consentimiento expreso del ocupante, la Agencia de Vivienda de Cataluña, el Departamento de Derecho Sociales o los servicios sociales puedan consultar sus datos para realizar un estudio de su situación económica.

39 Hay que recordar que el art. 132-4 del Código de Consumo catalán señala que "los servicios públicos de consumo deben garantizar que, en los casos de ejecución hipotecaria de la vivienda habitual como consecuencia del incumplimiento del deudor, pueda llevarse a cabo un procedimiento de mediación destinado a la resolución extrajudicial de conflictos previo a cualquier otro procedimiento judicial o a la intervención notarial" (apartado 1). De hecho, en el segundo apartado se especifica que "el procedimiento de mediación debe tener por objeto buscar acuerdos entre las partes que hagan viable que la persona consumidora conserve la propiedad de la vivienda o, subsidiaria-

existe un cauce extrajudicial con sus reglas, principios y fases procedimentales. La Administración comunica el inicio de la mediación en dos meses, una vez obtenida la aceptación de la contraparte. Si en este plazo no hay acuerdo para el inicio de la mediación, se entenderá que no se llevará a término.

En Castilla y León, a través de la Ley 4/2024, de 9 de mayo, de medidas tributarias, financieras y administrativas, se modificó la Ley 9/2010, de 30 de agosto, del derecho a la vivienda de tal Comunidad, con la finalidad de introducir el servicio de mediación para grandes tenedores (disposición adicional quinta). Ciertamente, se establece una normativa específica que detalla, entre otros aspectos, el órgano competente de la mediación (servicio territorial de la Junta de Castilla y León competente en materia de vivienda de la provincia donde radique el inmueble) y del informe de vulnerabilidad económica (servicio competente en materia de asistencia social del Ayuntamiento, o en su defecto de la Diputación Provincial), así como el procedimiento (solicitud, fases, documentación duración y modos de terminación).

---

mente, la posibilidad de mantener su uso y disfrute. En el marco de este procedimiento, las partes o el órgano de resolución extrajudicial de conflictos pueden solicitar un informe de evaluación social con un análisis socioeconómico del deudor y las posibles vías de resolución del conflicto en los términos del artículo 133-6". Por consiguiente, antes de la LDV, en Cataluña se utiliza la mediación de consumo para resolver conflictos hipotecarios, salvaguardando la vivienda del deudor. De hecho, en su apartado 3 se quiso establecer la mediación como requisito de procedibilidad, pero tal aspecto fue declarado inconstitucional (STC 54/2018 de 24 de mayo [TOL6.635.716]). No obstante, este propósito finalmente ha sido recogido por el legislador estatal en la LDV, al establecer la intermediación o conciliación como presupuesto procesal, no solamente en ejecuciones hipotecarias sino también en materia de arrendamientos.

Finalmente, cabe mencionar a Castilla-La Mancha, que de forma reciente, a través de la Orden 95/2024, de 3 de junio, de la Consejería de Fomento, regula la competencia y el procedimiento de conciliación previa a los procesos judiciales de desahucio, recuperación de la posesión y ejecución hipotecaria de personas en situación de vulnerabilidad económica, previsto en la Ley 1/2000, de 7 de enero, de Enjuiciamiento Civil, en la redacción dada por la Ley 12/2023, de 24 de mayo, por el derecho a la vivienda. Dicha norma establece como órganos competentes del procedimiento a las oficinas de intermediación hipotecaria de las Delegaciones Provinciales de la Consejería de Fomento. El desarrollo normativo es más escueto, algo simple, en comparación con otras CCAA, como el País Vasco o Castilla y León.

Respecto a otras CCAA[40], he de reconocer que no he sido capaz de encontrar información al respecto. No he podido contactar con responsables de vivienda por vía telefónica o, bien, no he recibido respuesta de correos electrónicos en los que solicitaba información sobre dicho procedimiento extrajudicial. En ocasiones, ha sido un auténtico "galimatías" el proceso para obtener información. Lo anterior es una prueba más de la dificultad para los representantes de los grandes tenedores respecto a la gestión y tramitación de tal asunto. Si bien, como solución a estos silencios, lo más óptimo es formular la petición al órgano competente en materia de vivienda –junto con la solicitud de informe de vulnerabilidad o, bien, tras haberlo obtenido por servicios sociales–. En algunas Administraciones autonómicas se corre el riesgo de sufrir

---

[40] En las Islas Baleares también se ha creado un trámite específico para grandes tenedores: https://www.caib.es/seucaib/es/201/empreses/tramites/tramite/6107278/ [Fecha de consulta: 15/03/2024]

la inadmisión de la solicitud, si ésta no viene acompañada del informe de vulnerabilidad[41].

5. "Caos procedimental" e inseguridad jurídica: un perjuicio para grandes tenedores y ocupantes vulnerables

La LDV entró en vigor el 26 de mayo de 2023. Desde entonces, numerosas comunidades autónomas –por pasividad, desidia, razones políticas o falta de recursos humanos y económicos– no han sido capaces de crear un protocolo o una norma relativos al procedimiento de intermediación y conciliación en

---

[41] "En consecuencia, parece incongruente promover el procedimiento de conciliación o intermediación sin cumplir, previamente, la condición objetiva para su iniciación, que es la previa declaración de vulnerabilidad. Con ello, aunque razones de operatividad y pragmatismo pudieran sugerir instar ambas peticiones simultáneamente (petición para declaración o no de vulnerabilidad y solicitud para someterse a un procedimiento administrativo de mediación), lo cierto es que: a) las Administraciones podrían inadmitir de plano la solicitud de procedimiento de mediación sin acreditar la previa declaración de vulnerabilidad y b) para el probable caso de que la Administración desatienda la petición de declaración o no de vulnerabilidad la segunda instancia carecerá de todo objeto. Lo cierto es que, el legislador, podía haber introducido un automatismo para agilizar toda esta secuencia de trámites preprocesales, obligando a la Administración que declarase la situación de vulnerabilidad a convocar a la mayor brevedad, de oficio, un acto de conciliación o intermediación. En todo caso, tampoco parece descabellado o ilógico ni contraviene los principios del procedimiento administrativo consagrados por la Ley 2015, presentar una instancia o reclamación conjunta frente a la Administración con una doble petición, en relación de causalidad entre su solicitud final, a saber: a) declaración de vulnerabilidad o no del demandado-ejecutado y, en caso de emitirse tal declaración, b) iniciación de un procedimiento de intermediación o conciliación" (Gómez Linacero, 2023, LA LEY 11025/2023).

materia de vivienda para grandes tenedores. En otras, dicho procedimiento se ha instaurado transcurrido más de un año desde su entrada en vigor. Si bien, debo ir más allá, porque algunas ni siquiera ofrecen una información que resulte clara y completa.

Existe, pues, un "caos burocrático o procedimental" que genera una gran inseguridad jurídica a los grandes tenedores de vivienda en diversas CCAA[42]. Ante tal situación, y con el fin de cumplir con el requisito de procedibilidad, su actuación se centrará en formular la solicitud de inicio del procedimiento extrajudicial al organismo de vivienda pertinente, dejando transcurrir el plazo legal y, tras ello, en presentar la demanda junto con la declaración responsable de haber solicitado la conciliación e intermediación y el justificante acreditativo. Lo lamentable es que en ocasiones pedirán el inicio de un procedimiento extrajudicial que ni siquiera existe en alguna comunidad autónoma. Ciertamente tal situación perjudica al gran tenedor, que al verse obligado a realizar un trámite administrativo vacío de contenido[43], ve dilatado su derecho a tutelar

---

42 "El problema de carácter práctico está en que, a la fecha de entrada en vigor de esta normativa, está pendiente de regulación dicho procedimiento de intermediación, lo cual es inadmisible, pues genera una enorme inseguridad jurídica sobre cómo proceder en estos casos hasta que se regulen procedimientos ad hoc. Hasta que no exista una regulación al respecto, nuestra recomendación es que se formule una solicitud de conciliación o intermediación ante la correspondiente consejería de vivienda autonómica para tratar de cumplir con este requisito de procedibilidad" (Fuentes-Lojo Rius, 2023).

43 "Si bien toda acción destinada a evitar un juicio es deseable, el problema que a este respecto plantea la LDV es que adolece de cierta imprecisión, dicho esto en el sentido de que el legislador remite a la fórmula genérica «procedimiento de conciliación o intermediación que a tal efecto establezcan las Administraciones Públicas», pero no especifica con suficiente claridad cuál es el órgano competente para sustanciar las conciliaciones ni el procedimiento a seguir. Este

con prontitud su propiedad. Pero cuidado, porque si esa es la intención del legislador español o de la comunidad o ciudad autónoma, también se impide al ocupante acceder a un servicio que puede brindarle un apoyo social o la búsqueda de una solución habitacional. Para obtener este resultado, hubiera sido más oportuno suspender el desahucio o lanzamiento un periodo de tiempo, evitando generar cargas burocráticas o administrativas sin finalidad alguna.

Quiero realizar un breve paréntesis y comentar lo trasladado por los responsables de un organismo autonómico destinado oficialmente a canalizar la conciliación e intermediación de grandes tenedores –sin considerar apropiado mencionar la Administración autonómica–. Por vía telefónica me llegaron a afirmar que el servicio ofrecido era un mero trámite administrativo, sin propósito de llevar a cabo ningún tipo de conciliación o intermediación. En concreto, me trasladaron que desde sus dependencias, por orden de las personas responsables, no iban a destinar recursos humanos y económicos relacionados con la LDV. Ante tal comentario, le transmití que, siendo así, estaban impidiendo que los ocupantes de las viviendas pudieran contar con asesoramiento o apoyo social –por ejemplo, para conocer posibles alternativas habitacionales–. Tras ello, me comentó que si el gran tenedor pretende ofrecer alguna alternativa, se podría iniciar una intermediación. No obstante, como no suele ser el caso, la unidad responsable deja transcurrir el tiempo

---

vacío es especialmente oneroso en el caso de aquellas comunidades autónomas que, en la medida en que ni siquiera habían legislado sobre la materia, seguramente no disponen de una mínima infraestructura para llevar a cabo las conciliaciones. En este sentido, es claro el riesgo añadido de que tal requisito de procedibilidad quede vacío de contenido en caso de que no se habiliten los medios que permitan la implementación de los procedimientos de resolución extrajudicial de conflictos" (Climent Esteve, 2023).

fijado en la LDV para que el gran tenedor de vivienda pueda acudir a los tribunales. Es decir, optan por ofrecer un silencio administrativo. Tales manifestaciones y palabras me resultaron sorprendentes, porque quiera o no quiera el gran tenedor ofrecer alguna solución, este procedimiento extrajudicial debe tener como finalidad que la Administración autonómica pueda ofrecer al ocupante un adecuado asesoramiento o ayudas sociales en caso de pérdida de la vivienda habitual. Supone negar al ciudadano una oportunidad de arreglo e información.

Hecha esta salvedad, resulta criticable que desde la entrada en vigor de la LDV ciertas Administraciones autonómicas no hayan configurado su procedimiento de conciliación e intermediación –sobre todo, cuando en la mayoría de ellas existen oficinas de intermediación de deudas hipotecarias o de alquiler, creadas a nivel autonómico, local o provincial, tras la crisis económica de 2007–. Han tenido tiempo más que suficiente los responsables autonómicos.

Ahora bien, no podemos trasladar toda la responsabilidad a las comunidades autónomas. El legislador estatal se ha limitado a mencionar en la LDV y en la LEC un "procedimiento de conciliación e intermediación" dependiente de las Administraciones competentes o autonómicas, sin identificar la legislación aplicable. Comprendo que los asuntos de vivienda corresponden a las CCAA, pero ello no significa, ni mucho menos, que el legislador estatal no pueda establecer unas pautas mínimas. A modo de ilustración, hubiera sido conveniente señalar el organismo competente (justicia, vivienda, protección social, etc.), así como diseñar unas pautas comunes para dicho procedimiento extrajudicial (principios, reglas, finalidad[44], personas

---

[44] "Llama poderosamente la atención que se hagan alusiones a los plazos máximos de antelación con que deben instarse declaraciones de las Administraciones Públicas y, muy destacadamente, que se institucionalice,

negociadoras, etc.). Incluso sería aconsejable definir la naturaleza jurídica de dicha conciliación o intermediación –¿se trata de una mera labor de negociación entre las partes?, ¿se configura como una mediación civil?, ¿estamos ante una mediación de consumo?–. De sus respuestas puede depender el régimen normativo aplicable. Hay que facilitar la tarea a las CCAA, unificar aspectos del íter procesal. Ante tal silencio, se desprende la poca esperanza sobre la utilidad de tal procedimiento por parte del legislador estatal, que parece haber creado un simple trámite administrativo, sin finalidad social.

Junto a las CCAA que no han establecido protocolo o procedimiento extrajudicial alguno, se encuentran aquellas que han previsto pautas de actuación. Sin embargo, el protocolo a seguir en algunas no me parece muy oportuno. Es el caso de Andalucía. El procedimiento de conciliación o intermediación debe iniciarse ante los PIMA. En vez de contar con la ayuda y asesoramiento de las oficinas de intermediación hipotecaria –con una mayor experiencia sobre la materia, al abordar negociaciones con entidades bancarias y con capacidad de ofrecer apoyo social–, se ha decidido canalizar el procedimiento por los PIMA. Posiblemente me equivoque, pero si el asunto se deriva a mediadores no especializados, el PIMA se convertirá en un trámite administrativo para poder interponer la pertinente demanda judicial. Si la Junta de Andalucía hubiera deseado gestionar correctamente la búsqueda de soluciones o alternativas habitacionales a favor del ocupante de vivienda, hubiera

---

no sólo la reclamación administrativa previa como en el orden contencioso, sino la mediación o conciliación, no para lograr un acuerdo entre las partes que ponga fin al litigio, sino para dar respuesta a una pretendida situación de vulnerabilidad del demandado o ejecutado, o eso parece deducirse de la reforma, que no indica, en ningún momento, cual es la finalidad de ese expediente preprocesal de conciliación o intermediación" (Gómez Linacero, 2023, LA LEY 11025/2023).

designado como unidad competente al Sistema de Asesoramiento ante ocupación ilegal y protección de la vivienda, dependiente de la Consejería de Fomento, Articulación del Territorio y Vivienda, que lleva prestado servicio desde el año 2012. Desde entonces ofrece labores de intermediación por impago de créditos hipotecarios o alquiler. En mi opinión, la decisión adoptada no ha sido la más correcta.

V. CONCLUSIONES: DEFICIENCIAS Y OPORTUNIDADES PERDIDAS

Al realizar una primera lectura de la LDV, y tras observar el reconocimiento de la conciliación e intermediación en materia de vivienda, pensé –por fin el legislador estatal apuesta por la cultura del diálogo y la búsqueda de soluciones en materia de vivienda–. Sin bien, ese pensamiento –de emoción y sorpresa– se fue desvaneciendo de forma paulatina conforme fue examinándose el texto legal y, desde luego, su aplicación práctica en la mayoría de CCAA.

Es alabable que el legislador estatal haya establecido la conciliación e intermediación como requisito de procedibilidad. No obstante, no es posible, lamentablemente, efectuar más alabanzas.

El primer y principal problema se presenta ante la falta de una definición legal de dicho procedimiento, la fijación con precisión de su finalidad u objetivos y, desde luego, la ausencia de creación de unas reglas comunes (trámites procedimentales, organismos competentes, reglas, principios, etc.). Respecto al primer punto, ¿ante qué tipo de procedimiento nos encontramos?, ¿estamos ante un procedimiento administrativo?, ¿ante un mero trámite de conciliación o negociación?, ¿se trata de una mediación civil o de consumo? Sin definir la naturaleza, ésta se modula por las Administraciones públicas, encontrándonos con procedimientos extrajudiciales a los que resulta de aplicación una normativa distinta en cada comunidad o ciudad autónoma. Por mucho que las CCAA dispongan de competencias en materia de vivienda, era viable crear un marco común. En cuanto al segundo asunto, el legislador no hace hincapié en la finalidad de búsqueda de soluciones habitacionales con el gran tenedor de vivienda. Ese debería ser uno de los objetivos fundamentales de dicho procedimiento extrajudicial, con independencia de que la Administración pública, en caso de fracaso, ofrezca ayudas o protección social. Da la sensación de que el legislador estatal no sabía o desconocía la razón esencial para instaurar ese procedimiento extrajudicial como requisito de procedibilidad –o, bien, era consciente de que era un mero trámite administrativo, infructífero, para dilatar el tiempo de desahucios y lanzamientos–. Éste último propósito cobra mayor peso al excluir a ocupantes con propietarios que no sean grandes tenedores, ¿acaso no necesitan también un apoyo social por las Administraciones públicas?, ¿qué importa la condición del propietario? Se trata de proteger a personas ocupantes en situación de vulnerabilidad económica.

Atendiendo al fundamento ético de una "auténtica conciliación e intermediación en materia de vivienda", el legislador ha sido parco de contenido, incoherente y poco ambicioso en su reconocimiento y regulación. Poca fe –o ninguna– manifiesta por tal "novedad procedimental". Tan es así que, desde el principio, se desentiende, y "tira la piedra" en el tejado de las CCAA.

Con tal situación, a las Administraciones públicas les llega una obligación legal –ofrecer un procedimiento de conciliación e intermediación a grandes tenedores de vivienda–, sin un marco legal claro y preciso.

Tras este mandato, recogido en la LDV, comienza la cuenta atrás para sentar las bases de este procedimiento extrajudicial en cada comunidad o ciudad Autónoma. Desde la entrada en vigor de dicha norma, ciertas Administraciones públicas todavía no han regulado o establecido las instrucciones para desarrollar dicho procedimiento a favor de sus ciudadanos –bastante tiempo han tenido hasta este momento, una pasividad que no resulta justificable–. Hay, pues, CCAA sin procedimiento alguno. Ante ello, para que los grandes tenedores puedan cumplir con el requisito de procedibilidad, se ven abocados a solicitar el procedimiento a organismos competentes en vivienda, sin conciliación o intermediación alguna, con el único propósito de cumplir con el trámite extrajudicial obligatorio de la LEC. Otras CCAA han dado instrucciones para canalizar dicho procedimiento. Ahora bien, el cauce procedimental escogido –como ocurre en Andalucía–, salvo error en el pronóstico, supondrá también un proceso vacío de contenido, un trámite administrativo y procesal, sin finalidad de alcanzar acuerdos entre las partes u ofrecer protección social. Otras Administraciones Publicas, como las del País Vasco, han sabido –con gran acierto– canalizar dicho procedimiento a través de organismos con experiencia en la negociación de conflictos hipotecarios o arrendaticios.

Ahora pongámonos en la posición de los grandes tenedores. Operen a nivel provincial, autonómico o nacional, se encuentran en una situación de "confusión legal", sin conocer en numerosas ocasiones los trámites a seguir para iniciar y desarrollar la conciliación o intermediación. En ciertas CCAA no hay información disponible o resulta imposible contactar para solicitar información. Asimismo, en algunas la solicitud del informe de vulnerabilidad se puede gestionar con el mismo organismo y en otras se debe pedir previamente a la solicitud del procedimiento extrajudicial –ello sin tener en cuenta la incertidumbre o posibles dudas del organismo encargado para emitir el mencionado informe–. Pocas CCAA ofrecen una óptima información sobre los aspectos comentados. Como consecuencia del déficit de información o la ausencia de procedimiento, los grandes tenedores desconocen las reglas aplicables a la conciliación e intermediación en materia de vivienda (reglas, principios, normativa, etc.). En ocasiones, si realizan correctamente la solicitud de inicio, solamente deben adoptar un comportamiento pasivo a la espera de que transcurran los plazos pertinentes para acudir a la vía judicial. Incluso aunque se inicie el procedimiento, dada su voluntariedad, y los pocos incentivos, muchos optarán por no alcanzar un acuerdo. La intermediación o conciliación se convierte, así, en una especie de "suspensión indirecta" de desahucios y lanzamientos, provocada por una "marabunta de gestiones y trámites" sin propósito. Ahora bien, no cabe duda de que ciertas Administraciones efectuarán un procedimiento extrajudicial, con base a los fines que deben guiarlo.

Finalmente, hay que poner de relieve a otro perjudicado, el ocupante vulnerable de vivienda. Entiendo que, a priori, se piense que tal situación le beneficia, al prolongarse la fecha de desahucio o lanzamiento. No obstante, una "conciliación o intermediación funcional" le permitiría buscar soluciones con el gran tenedor de vivienda, a través de la asistencia de un tercero o, en caso de fracaso, contar con apoyo y asistencia social de personas o servicios especializados. Sin una oportunidad real de negociación y asistencia, se pierden opciones o soluciones habitacionales, y ello cierra la entrada de posibles alternativas. Si el legislador estatal –y algunos autonómicos– no creen o apuestan por este procedimiento, si ni siquiera lo hacen ciertas Administraciones públicas, no podemos exigir a los grandes tenedores de vivienda que interioricen una verdadera cultura del diálogo en materia de vivienda. Por tal motivo, me atrevo a señalar que no solamente hay una víctima de esta medida –el gran tenedor de vivienda y la tutela de su derecho de propiedad–, también puede considerarse víctima indirecta al ocupante de vivienda despojado de posibles soluciones. No obstante, tal perjuicio también aflora de forma directa hacia ocupantes de bienes inmuebles pertenecientes a propietarios que no reúnen la condición de gran tenedor, sin derecho alguno a una conciliación e intermediación obligatorias, previas al correspondiente proceso judicial. Ello denota, sin duda alguna, una situación de discriminación jurídica.

Ante una "conciliación e intermediación funcional", dicho procedimiento extrajudicial, como requisito de procedibilidad, encuentra plena justificación dentro de la LDV y la LEC. Sin embargo, ante una "conciliación o intermediación disfuncional" –que se presente como un mero trámite, sin negociación alguna o que concluya con un silencio administrativo–, el presupuesto se presenta como una "suspensión indirecta" de desahucios y lanzamientos, lesionando la tutela efectiva del derecho a la propiedad de grandes tenedores y, desde luego, la oportunidad de protección social de los ocupantes vulnerables. Estamos, en general, ante un procedimiento extrajudicial, vacío de espíritu, pero también de forma, sin que el legislador estatal –y algunos autonómicos– hayan sabido cultivar para erigirlo como un mecanismo óptimo de gestión de conflictos en materia de vivienda. La cuestión es: ¿seremos capaces de mejorar su funcionalidad y garantizar su esencia básica? El tiempo, las políticas legislativas y la implicación de las Administraciones, permitirán divisar tal interrogante[45].

---

[45] Dicho procedimiento puede verse afectado por el Recurso de inconstitucionalidad n.º 5514-2023, contra la disposición final quinta dos y disposición final quinta seis de la Ley 12/2023, de 24 de mayo, por el derecho a la vivienda, entre otros preceptos, admitido a trámite.

VI. REFERENCIAS BIBLIOGRÁFICAS

Bastante Granell, V. (2013). "Mediación hipotecaria: una solución al problema del sobreendeudamiento de los particulares". *Anales de Derecho,* (31), 180-213.

Calvo San José, M. J. (2023). "Ley 12/2023, de 24 de mayo, por el derecho a la vivienda [BOE-A-2023-12203]: Aspectos civiles la protección del derecho a acceder a una vivienda digna y adecuada". *Ars Iuris Salmanticensis: AIS: revista europea e iberoamericana de pensamiento y análisis de derecho, ciencia política y criminología,* (2), 145–154.

Climent Esteve, V. (2023). "Análisis de las principales modificaciones procesales introducidas por la Ley 12/2023, de 24 de mayo, por el derecho a la vivienda". *Diario La Ley,* (10316).

Fuentes-Lojo Rius, A. (2023). "Modificaciones en los procesos de desahucio y ejecuciones hipotecarias en la Ley por el derecho a la vivienda". Práctica de tribunales: revista de derecho procesal civil y mercantil, (163).

Gómez Linacero, A. (2023). "Preguntas y respuestas a la Ley por el derecho a la vivienda en clave procesal". *Diario La Ley,* (10386).

Jiménez Muñoz, F. J. (2024). "El régimen de los grandes tenedores de viviendas en la ley por el derecho a la vivienda". *LA LEY Derecho de familia,* (41).

López Simó, F. (2023) "Principales novedades procesales de la Ley 12/2023, de 24 de mayo, por el derecho a la vivienda. Breve análisis crítico". Diario LA LEY, (10383).

Magro Servet, V. (2023). "Análisis práctico y sistemático de los aspectos relevantes de la nueva Ley de vivienda 12/2023 de 24 de mayo". *Diario LA LEY,* (10295).

Prats Albentosa, L. (2022). "Derecho a gozar de una vivienda digna y adecuada y derecho a la propiedad de vivienda". *Revista Jurídica del Notariado,* (115), 131-174.

Rodríguez Tirado, A. M. (2024). "La Ley por el derecho a la vivienda y la vulnerabilidad económico-social en los procesos arrendaticios de desahucio de vivienda habitual por falta de pago. ODS y Reglas de Brasilia". *LA LEY Derecho de familia,* (41).

Ruiz-Rico Arias, M. D. (2023). "La propiedad de vivienda habitual y la ejecución hipotecaria tras la ley del derecho a la vivienda de 2023: insuficiente protección y necesidad de reforma legal". *Diario LA LEY,* (10373).

Sáinz-Cantero Caparrós, M. B. (2023). "La redefinición del derecho de propiedad sobre la vivienda en la era de los bienes comunes. A propósito de la ley 12/2023 del derecho a la vivienda". *Actualidad jurídica iberoamericana*, (19), 630-663.

Viola Demestre, I. (2024). "El procedimiento de conciliación o intermediación en la Ley por el derecho a la vivienda: retos y desafíos". *El derecho a la vivienda en tiempos de incertidumbre*. Aranzadi, 477-499.

Capítulo XI

Una valoración de conjunto sobre el derecho privado en la ley 12/2023 por el derecho a la vivienda

ANTONIO GÁLVEZ CRIADO
Profesor Titular de Derecho Civil
Universidad de Málaga

La Ley 12/2023, de 24 de mayo, por el derecho a la vivienda (LDV; TOL9.568.821), no ha dejado indiferente a nadie. Su irrupción en el mercado de la vivienda y en el ámbito jurídico a ella inherente ha provocado reacciones de todo signo como pocas leyes lo han hecho en nuestra democracia. Probablemente ha sido así porque era inevitable que una ley llamada a regular por primera vez en nuestro país el derecho a disfrutar de una vivienda digna y adecuada del art. 47 CE (TOL173.304) viniera acompañada de una importante carga ideológica[1]. Piénsese, por ejemplo, en lo tan recurrente como absurdo por parte de algunos de presentar el mercado de la vivienda en alquiler reducido a una lucha entre grandes propietarios (tenedores, nos dirá esta ley) y vulnerables inquilinos.

---

1 Por ejemplo, Noguera Fernández (2022, pp. 17 y ss.) presenta la función social de la propiedad como una lucha de clases y la concibe como "un instrumento para la reasignación, reapropiación y/o redistribución de bienes" (pp. 23 y ss.).

Sin embargo, era razonable esperar que el legislador, a quien pacientemente se ha esperado durante 45 años para ello, realizara una tarea rigurosa, coherente y depurada en el manejo de la técnica legislativa, los conceptos jurídicos o la sistemática de los preceptos introducidos o modificados, para implementar unas medidas claras, bien pensadas y estructuradas. Pero nada de esto se ha hecho. Muy al contrario, se ha aprobado una ley de forma apresurada –por no decir improvisada y atropellada–, confusa y que genera más dudas que certezas a todos los operadores de un mercado tan necesitado de seguridad jurídica.

Pasado ya un año desde la entrada en vigor de las medidas aprobadas, nos atrevemos a afirmar que, a día de hoy, no solo no ha contribuido a mejorar el ya de por sí difícil acceso de los ciudadanos a una vivienda digna y adecuada en la actualidad, sino que se ha convertido en un obstáculo más para ello. Por ejemplo, pueden constatarse unos efectos sobre el mercado del alquiler residencial diametralmente opuestos a los deseados por el legislador: se ha reducido la oferta, se ha provocado una huida hacia otras fórmulas arrendaticias (alquiler turístico, alquiler de temporada y alquiler por habitaciones[2]) y se ha expulsado del mercado a los más vulnerables.

Puede decirse que, lejos de contribuir a la solución, se ha convertido en un nuevo problema.

En los capítulos anteriores se ha realizado un análisis por separado de los distintos aspectos civiles y procesales de la LDV. En ellos se ofrece información concreta y precisa sobre cada uno de

---

2 Ya contamos con una primera respuesta legislativa frente a esta reacción del mercado arrendaticio hacia el alquiler de temporada y el alquiler por habitaciones: el nuevo art. 66 bis de la Ley 18/2007, de 28 de diciembre, del derecho a la vivienda de Cataluña (añadido por el art. 5 del reciente Decreto-ley 6/2024, de 24 de abril). Parece la política del "más madera".

los temas y, lo que es más importante, una valoración crítica de las decisiones adoptadas por la LDV y, muy especialmente, de sus insuficiencias. Nos ha parecido necesario completar este análisis por temas con una reflexión de conjunto de la Ley. Este último capítulo no es, sin embargo, una síntesis o resumen de lo anterior, sino el lugar donde se han querido plasmar ideas y opiniones sobre los distintos aspectos de la LDV, no necesariamente tratados en los capítulos previos, pero que descansan y precisan de un conocimiento exhaustivo del texto legal, así como de la jurisprudencia y de la doctrina especializada.

PRIMERA. La primera conclusión general que puede sentarse sobre la LDV es la ausencia en la misma de un enfoque o mirada del derecho a la vivienda desde el punto de vista del Derecho civil. La Ley gira alrededor de la propiedad de la vivienda, de su uso y posesión, de su cesión a terceros a través del contrato de arrendamiento, etc, pero no se encuentran referencias claras a las implicaciones civiles. Parece una ley redactada "a espaldas" del Derecho civil y de sus conceptos básicos, comenzando por la consideración del derecho de propiedad como un derecho subjetivo.

La razón de ello parece encontrarse en su propio preámbulo, cuando afirma (en relación a la Resolución del Parlamento Europeo, de 21 de enero de 2021, sobre el acceso a una vivienda digna y asequible para todos; cursiva nuestra):

> "Como consecuencia, *sus postulados demandan un cambio de paradigma en la consideración jurídica de la vivienda*, para reforzar su función como servicio social de interés general. Sólo de este modo será posible garantizar el derecho efectivo a una vivienda digna y asequible para todas las personas jóvenes, y especialmente a aquellos colectivos en riesgo de exclusión, reconocido y ejecutable como un derecho humano fundamental e incluyendo en el mismo el acceso a los servicios básicos definidos en la legislación urbanística y de ordenación del territorio, contribuyendo así a garantizar la vida digna en un entorno adecuado y a erradicar la pobreza en todas sus formas".

Este "cambio de paradigma en la consideración jurídica de la vivienda" no parece otra cosa que una definitiva confusión entre el Derecho público y el Derecho privado respecto a las viviendas; es como si el mercado libre de estos inmuebles no existiera y los institutos jurídicos que rigen su funcionamiento nada aportaran al derecho a una vivienda digna y adecuada. La falta de claridad y deslinde (incluso equiparación) entre la normativa administrativa y la reguladora de las relaciones jurídico-privadas la pone de manifiesto también el CGPJ en su informe sobre el entonces anteproyecto de LDV (Acuerdo adoptado por el Pleno en su reunión del día 27 de enero de 2022, punto 19).

Como decimos, es posible que la Resolución del Parlamento Europeo, de 21 de enero de 2021, sobre el acceso a una vivienda digna y asequible para todos (2019/2187(INI)) (2021/C 456/14), haya tenido una influencia decisiva en este enfoque. Parece ir en la línea del punto n.º 50 de la Resolución: "Pide que se incluya al sector de la vivienda *como un servicio social de interés general*, y no solo la vivienda social, ya que esto es fundamental para garantizar el derecho a una vivienda asequible y digna para todos". Así parece ser entendido por el legislador español, de manera que ha considerado todo el sector de la vivienda (público y privado) como "un servicio social de interés general", expresión que parece querer ir mucho más allá de la idea de servicio público.

Este "olvido" del legislador por el Derecho civil no es nuevo; ya lo reprochaba García Rubio en el propio Congreso de los Diputados[3]. Lo llamativo es que esta situación se haya producido también en una ley específica con la declarada finalidad de regular aspectos centrales del derecho de propiedad y uso

---

3 García Rubio (2022, pp. 233-245).

sobre una vivienda. Parece como si la vivienda fuera un asunto que solo incumbiera al Derecho constitucional, al Derecho administrativo o al Derecho procesal. La explicación parece clara: todas estas ramas han venido siendo clasificadas tradicionalmente dentro del Derecho público, que es la óptica única del legislador, en detrimento de las implicaciones del Derecho privado, y muy especialmente en este ámbito, del Derecho civil.

Nasarre Aznar llega al extremo de hablar de una "incivilización" al tiempo que de una "administrativización" del derecho de propiedad sobre la vivienda, un nuevo "rediseño" o "reconfiguración" jurídica que le lleva incluso a mantener la derogación del art. 348 CC (TOL220.310) en relación a la vivienda[4]. Es como si el Derecho administrativo hubiera colonizado al Derecho civil respecto al derecho de propiedad y todas sus implicaciones sobre la vivienda, lo cual ya había sido defendido en la doctrina[5].

Sin duda, el legislador parece haber encontrado en las herramientas del Derecho administrativo el cauce jurídico sobre el que cimentar el exacerbado intervencionismo público en materia de vivienda. Lo curioso del caso es que, por esta vía, se afecten directamente a instituciones centrales del Derecho civil y el legislador lo haga precisamente sobre su competencia exclusiva en esta materia.

---

4 Nasarre Aznar (2022, pp. 7-11).

5 Así, Aparicio Wilhelmi (2022, pp. 86-87) que, tras considerar que la necesidad social de limitar el derecho de propiedad y la libertad contractual (art. 1255 CC) es una cuestión de orden público derivada del contenido mismo del derecho de propiedad, cuestiona la propia naturaleza iusprivatista de la relación arrendaticia, poniendo como ejemplo la situación en Québec "donde la resolución de los conflictos en relaciones de arrendamiento de vivienda se sustancia a través de un tribunal administrativo".

En parte, es un camino que ya habían intentado recorrer diversas CCAA, que frecuentemente han visto rechazadas sus pretensiones por el Tribunal Constitucional[6]. Pero, en este caso, es el Estado quien lo hace, pues, como dice Nasarre Aznar, "sin embargo, esta vez los arts. 10 y 11 Proyecto sí que están, a nuestro juicio, reconfigurando y limitando la configuración del derecho de propiedad sobre viviendas conforme a las competencias del Estado en materia de Derecho civil"[7].

Tampoco debe ocultarse una apreciable pérdida de ese peso que tradicionalmente ha venido teniendo la doctrina civilista en materia de vivienda en nuestro país. Hoy en día parecen preponderantes los enfoques doctrinales constitucional, procesal, penal y sobre todo administrativo, sobre el problema de la vivienda en España. Los legisladores estatal y autonómico (también en el ámbito local) parecen haber encontrado en el Derecho público y en sus estudiosos las herramientas que mejor sirven a su decidido intervencionismo público. Haría bien la doctrina civilista en recuperar el enfoque y el discurso privatista sobre la vivienda, que resulta hoy en día más necesario que nunca, sin miedo a críticas o tachas de egoísmo en la defensa de derechos individuales.

SEGUNDA. Resulta sinceramente indigno de "la primera ley estatal reguladora del derecho a la vivienda desde la aprobación de la Constitución" que la misma presente tantas deficiencias técnicas y tan evidentes, sin detenernos ahora en graves errores sintácticos (*vid.* la redacción del art. 13.2, por ejemplo). Creemos poder mantener la tesis de que la mayor parte de estas deficiencias se deben al hecho de haber prescindido

---

6 Gálvez Criado (2023, pp. 342-353); Messía de la Cerda (2020, pp. 3420 y ss).

7 Nasarre Aznar (2022, p. 9).

de los conceptos y categorías propias del Derecho civil, como acabamos de decir. Podemos poner varios ejemplos claros[8]:

-La Ley habla hasta en once ocasiones de "operaciones de venta (compra) y arrendamiento de vivienda", cuando lo correcto hubiera sido hablar de contratos de compraventa y arrendamiento de vivienda, porque se trata de contratos y este es el concepto jurídico correcto, no el de "operaciones".

-En los supuestos de discriminación indirecta del art. 6.2 b) se incluyen, entre otros y como supuestos diversos, "una cláusula convencional o contractual, un pacto individual", sin que pueda saberse muy bien cuál es el criterio para distinguirlos. Quizás se quiera distinguir entre una condición general de la contratación y una cláusula negociada individualmente, aunque no lo creemos sinceramente.

-La Ley no distingue entre el uso y el disfrute de una vivienda. Puede admitirse que se utilice la palabra "disfrute" como sinónimo de goce o uso, cosa que la Ley hace con frecuencia, como en el art. 8 a), cuando refiere el derecho de todos los ciudadanos a "disfrutar de una vivienda digna y adecuada, en los términos dispuestos por esta ley, ya sea en régimen de propiedad, de arrendamiento, de cesión de uso, o de cualquier otro régimen legal de tenencia". Pero lo que no puede aceptarse es que se utilicen ambos términos (uso y disfrute) conjuntamente y como si fueran sinónimos (arts. 2 b, 2 e, 6.1, 6.2, 9 o 29), porque entonces se llega al absurdo de hablar de "uso y disfrute propios y efectivos de la vivienda" (art.

---

8 Los pone también de manifiesto Argelich Comelles (2023, p. 37, notas 48 y 49), respecto al entonces proyecto de ley.

11.1 a), parece que en contraposición a un "uso y disfrute ajenos" de la misma".

La consecuencia de este galimatías jurídico es que legislador desconoce en qué consisten las facultades de uso y disfrute de un propietario o titular de otro derecho real sobre una vivienda, por más que el art. 10 hable de facultades como contenido del derecho de propiedad de la vivienda.

-En el fondo, el legislador desconoce que el derecho de propiedad es, ante todo, un derecho subjetivo (absoluto), del que derivan una serie de derechos (facultades) y obligaciones para su titular y que es oponible *erga omnes*, aunque pueda admitirse que se trata de un "derecho subjetivo debilitado"[9].

Esto provoca que se confundan derechos y obligaciones, y lo que en un sitio aparece como un derecho en otro lo haga como una obligación, o que se confundan los derechos y obligaciones de los propietarios con los derechos y obligaciones de los ciudadanos en general. Por ejemplo, en el art. 1.2 se habla del deber de mantener, conservar y rehabilitar la vivienda, mientras que en el art. 10.1 c) ello se configura como un derecho del propietario. Lo mismo ocurre con las facultades de uso y disfrute: un derecho en el 10.1 a) y un deber en el art. 11.1 a), aunque en este caso puede tener alguna justificación.

-En fin, resulta de todo punto absurdo establecer como contenido del derecho de propiedad (como facultad del propietario), "el derecho de consulta a las Administraciones competentes, sobre la situación urbanística de la vivienda y del edificio en que se ubica" (art. 10.1 b).

---

9 Colina Garea (1997, pp. 208 y ss.), reproduciendo palabras de la STC 111/1983 de 2 de diciembre. En el mismo sentido, Simón Moreno (2023, pp. 148-149).

-No puede dejarse de señalar el uso generalizado de normas jurídicas incompletas, que enuncian supuestos de hecho, pero cuyas consecuencias jurídicas no son fáciles de adivinar. Así ocurre, por ejemplo, con los "deberes del ciudadano en relación con la vivienda" enumerados en su art. 9, donde su incumplimiento, o carece de consecuencias jurídicas concretas o estas ya están establecidas en normas anteriores a las que nada añade la propia Ley. En el caso de los deberes y cargas derivados del derecho de propiedad enumerados en el art. 11, la declaración de incumplimiento (y sus posibles consecuencias) parece dejarse a la "la legislación de ordenación territorial y urbanística y la de vivienda" desarrollada por las AAPP competentes en materia de vivienda (art. 11.2).

En cualquier caso, se desconocen las posibles consecuencias civiles (aparte de las administrativas u otras) derivadas del incumplimiento de todos estos deberes o de otros: la nulidad/anulabilidad de ciertos contratos que no cumplan las exigencias de la ley (así, en las situaciones descritas en el art. 6.2), la responsabilidad precontractual por defectos de información, otras acciones indemnizatorias en supuestos, por ejemplo, de discriminación, acción de cesación, resolución del contrato, etc.

Los puntos anteriores parecen pruebas claras de esa "administrativización" mal entendida de la Ley que en algunos aspectos debiera resultar sonrojante para el legislador (desde la Comisión de Transportes, Movilidad y Agencia Urbana, que fue donde comenzó su camino).

TERCERA. En la línea de la conclusión anterior, resulta confusa con frecuencia la noción de "vivienda" destinataria de las medidas públicas de fomento para el acceso a una vivienda digna y adecuada incluidas en la LDV, y por tanto, de su ámbito de aplicación. Debe partirse en cualquier caso de la definición de su art. 3 a):

"a) Vivienda: edificio o parte de un edificio de carácter privativo y con destino a residencia y habitación de las personas, que reúne las condiciones mínimas de habitabilidad exigidas legalmente, pudiendo disponer de acceso a espacios y servicios comunes del edificio en el que se ubica, todo ello de conformidad con la legislación aplicable y con la ordenación urbanística y territorial".

Pero también deben tenerse en cuenta las definiciones de residencia habitual y residencia secundaria:

"i) Residencia habitual: la vivienda que constituye el domicilio permanente de la persona que la ocupa y que puede acreditarse a través de los datos obrantes en el padrón municipal u otros medios válidos en derecho.

j) Residencia secundaria: toda aquella vivienda que se utiliza por su propietario para estancias temporales o intermitentes, y que no constituye su residencia habitual".

Como se ha puesto de manifiesto en esta obra, la Ley carece de una norma destinada a concretar su ámbito general de aplicación, en especial, una norma que especifique la noción clara de vivienda y las relaciones jurídicas en torno a ella a la que resultan aplicables cada una de las medidas aprobadas. Tampoco en los distintos títulos y capítulos de la Ley se concretan siempre y de forma clara sus específicos ámbitos de aplicación.

Esta ausencia podría haber sido subsanada conectando las definiciones establecidas en su art. 3 con la regulación específica establecida por el legislador en cada caso, pero esto tampoco se ha hecho de un modo satisfactorio. Así, se utilizan en el texto legal definiciones sobre vivienda (domicilio, morada u hogar[10]) que no se encuentran en el elenco del art. 3, mientras que en otros casos se han incluido definiciones que después

---

10 Argelich Comelles (2023, p. 89).

se hayan ausentes en el resto de la ley. Todo ello se añade al hecho de que las definiciones que nos ofrece el legislador no responden a su desarrollo moderno y conforme a los estándares internacionales aceptados en la actualidad, como puso de manifiesto la doctrina desde el principio[11].

Un buen ejemplo de lo que estamos diciendo es la misma noción de vivienda "digna y adecuada" que ocupa el lugar central de la LDV. Unas veces (así, el art. 7) se utiliza para referirse a la vivienda habitual y se relaciona con el art. 47 CE y el cumplimiento de la función social al tratarse de "un bien destinado a satisfacer las necesidades básicas de alojamiento de las personas". Otras veces quiere decirse vivienda con fines residenciales, y finalmente y en otras tantas ocasiones, se usan estos términos en el sentido de la definición legal del art. 3 a), pues es razonable que toda vivienda sea "digna y adecuada".

Esto nos obliga a realizar un acto continuo de averiguación del ámbito de aplicación de las diferentes normas, lo cual no garantiza que se logre. Así por ejemplo, determinar con exactitud los contratos a los que resultan de aplicación los deberes mínimos de información del art. 31 resulta una tarea muy complicada.

La consecuencia de todo lo anterior es que, en el fondo, desconocemos el alcance exacto de la labor legislativa del Estado en materia de vivienda; esto es, el alcance real del ejercicio por parte del Estado de sus (escasas) competencias en este ámbito. Es posible aventurar que esta situación de cierta inseguridad deje al TC un amplio margen que le permita pronunciar sentencias interpretativas que declaren conforme a la CE tanto las normas aprobadas por el Estado en la LDV como las que puedan aprobar en el futuro las CCAA como competentes plenos en esta

---

11 Nasarre Aznar (2002, pp. 12-13).

materia. Esto es lo ocurrido con la primera sentencia ya dictada por el TC, como se concretará después.

CUARTA. A partir de un planteamiento muy particular y para el logro del derecho a una vivienda digna y adecuada, el legislador de la LDV no se centra tanto en el derecho de propiedad en sí que recae sobre una vivienda, como en los usos, y muy particularmente, en el uso residencial, lo que ya venía haciendo la legislación de las CCAA en esta materia.

La lucha por una vivienda digna parece haber sido concebida como una "lucha por sus usos". Por eso, la actuación de las AAPP no parece encaminada a incidir en el derecho de propiedad privada en sí mismo, sino en los usos (y disfrute) de este derecho.

Por otra parte, este tipo de medidas o incidencias sobre los usos derivados del derecho de propiedad parecen evidenciar el fracaso hasta el momento de la vivienda como servicio público: los parques de vivienda pública o afectada por algún régimen de protección pública, la política de suelo, la política urbanística, toda la acción pública puesta en marcha con los recursos públicos ha fracasado en su finalidad de mejorar el acceso de los ciudadanos a una vivienda digna. Parece pensar el legislador estatal que ha llegado el momento de incorporar los recursos privados (las viviendas en régimen de propiedad privada libre) e incluirlos en las políticas públicas de vivienda al mismo nivel que la vivienda pública o afectada por algún régimen protector. El gobierno del Estado parece tener un plan al respecto y una de sus piezas básicas para comenzar a discurrir por este camino es la LDV, que ha de servir de base jurídica para su posterior desarrollo por las CCAA, que son las competentes plenas en la materia.

Esta conclusión ya había sido alcanzada por la doctrina respecto a las legislaciones autonómicas sobre vivienda. Así, Garrido Mayol escribe: "finalizo esta reflexión poniendo de manifiesto que la inoperancia de los poderes públicos les lleva a trasladar la solución del problema a los particulares: no a otro objetivo responden las medidas normativas relativas a las viviendas vacías, a los grandes tenedores de viviendas o a intervenir el mercado del alquiler. Tratando de crear un estado de opinión en tal sentido: antes de constatar la ineficacia política de nuestros gobernantes se prefiere criticar a los propietarios de viviendas desocupadas, cuando son aquellos y no éstos quienes están incumpliendo la Constitución"[12].

QUINTA. Esta intervención pública sobre los usos no solo afecta a estos como facultades derivadas del derecho de propiedad o de otros derechos sobre los inmuebles desde una perspectiva estática, sino que incide directamente sobre el aspecto dinámico o capacidad de los propietarios de celebrar contratos de cesión de uso a terceros (en concepto de arrendamiento, por ejemplo); en otras palabras, se coarta la autonomía de la voluntad de estos titulares como una de las manifestaciones principales de la libertad de empresa del art. 38 CE.

Esta idea enlaza directamente con las "bases de las obligaciones contractuales" del art. 149.1.8 CE, en la interpretación seguida por la STC 37/2022, de 10 de marzo (TOL8.889.749), que declaró contrario a la CE el sistema de limitación de precios al alquiler de vivienda permanente en zonas de mercado tensionado establecido por la Ley de Cataluña 11/2020, de 18 de septiembre, de medidas urgentes en materia de contención de rentas en los contratos de arrendamiento de vivienda (TOL8.078.382).

En efecto y siguiendo a la STC 132/2019, de 13 de noviembre (TOL7.606.723), entiende el TC que estas bases "se refieren al núcleo esencial de la estructura de los contratos y a los principios

---

12 Garrido Mayol (2022, p. 296).

que deben informar su regulación", de manera que entre ellas es indudable que se sitúa el principio de autonomía de la voluntad que en este caso se concretaba en la libre determinación de la renta (art. 17.1 LAU; TOL231.076). Aparte, quedaban afectados también el art. 10.1 CE en cuanto libertad individual derivada del principio de la dignidad de la persona y del libre desarrollo de la personalidad, el art. 1.1 CE o el 38 CE en sus aspectos económicos. La misma doctrina se reiteró poco después en la STC 57/2022, de 7 de abril (TOL8.916.712) y en la STC 118/2022 de 29 de septiembre (TOL9.256.995), ambas dictadas frente a varias normas que modificaban también preceptos de la Ley 11/2020, de 18 de septiembre, de Cataluña. Y lo mismo en la STC 150/2022 de 29 de noviembre (TOL9.331.452), aunque en este caso las limitaciones impuestas a la renta lo eran respecto a locales de negocio y durante la pandemia del Covid-19, en Cataluña también.

Esta cuestión, ligada al aspecto dinámico del derecho de propiedad u otros derechos, también es ignorada por la LDV. Por el contrario, López y López le da el protagonismo que merece a esta vertiente dinámica vinculada a la libertad de empresa, pues "la garantía constitucional de la propiedad y la de la libertad de empresa tienen una dimensión común, aunque existen matices entre ambas"[13].

SEXTA. La intervención pública diseñada por la LDV, a pesar de lo que parecen ser sus intenciones iniciales, no se orienta tanto hacia el derecho a acceder a una vivienda digna, adecuada y asequible, sino más bien hacia el "derecho a permanecer" en ella. En realidad, no parece que el legislador esté pensando tanto en la dificultad de los ciudadanos para acceder a la vivienda (en régimen de alquiler en este caso), sino en el derecho de los inquilinos u otros usuarios, con título o sin él,

---

13 López y López (2022, pp. 14 y 31-33).

a permanecer en la vivienda que vienen ocupando por encima de todo, siempre que tengan la consideración de "persona vulnerable", concepto cada vez más amplio.

En la LDV parece que el derecho a permanecer en la vivienda forma parte del derecho a acceder a una vivienda digna. El art. 2 e) habla de proteger la "estabilidad" en el uso y disfrute de la vivienda, para referirse básicamente al colectivo vulnerable y en la misma línea parecen situarse la disposición final primera en cuanto a las medidas de intervención sobre los contratos de arrendamiento vigentes y la quinta en cuanto a la suspensión de los procesos de desahucio o lanzamiento y otras medidas con la misma finalidad (sometimiento previo y obligatorio a un procedimiento de conciliación o intermediación). Respecto a estas medidas, la LDV es continuadora de esta política de intervención pública anterior.

Este derecho a permanecer en la vivienda, aunque sea de forma temporal, se intensifica especialmente cuando quien pretende la recuperación del inmueble tiene la consideración de gran tenedor y el demandado es una persona vulnerable. En este caso y por ejemplo, la LDV (mediante las modificaciones introducidas por su disposición final quinta en la LEC; TOL172.336) dispensa una protección especial al ocupante vulnerable, incluso en el supuesto del apartado 4.º del art. 250 LEC. Este apartado se refiere a la tutela sumaria de la posesión por quien ha sido despojado de ella (mediante la okupación de viviendas), de manera que se imponen también en este caso al demandante gran tenedor importantes requisitos de procedibilidad de la demanda cuya acreditación dilatará inevitablemente la recuperación de la finca[14].

---

14 Detenidamente sobre los efectos de las medidas de la LDV sobre los procesos para la recuperación de las fincas por parte de los grandes tenedores en Magro Servet (2023, pp. 18 y ss.).

Otro tanto ocurre con la suspensión de los procesos de desahucio o lanzamiento conforme a los nuevos apartados 6 y 7 del art. 441 LEC, aunque a estos efectos se distingue entre personas físicas (suspensión máxima de dos meses) y personas jurídicas (cuatro meses). Pero en ambos supuestos la protección incluye también al okupa vulnerable. Son medidas cuyo precedente evidente es el art. 1 bis del Real Decreto-ley 11/2020, de 31 de marzo[15] (TOL7.854.128), que sigue en vigor y convive con la regulación derivada de la LDV.

En definitiva, parece que el legislador es consciente de la dificultad de aquellos que gozan del uso de una vivienda en la actualidad de acceder en el mercado a otra de similares características, de manera que parece una de sus prioridades la permanencia o continuidad (aunque sea provisional) de estos en la vivienda que ya vienen ocupando (cualquiera que sea su régimen de tenencia) como forma de asegurar su derecho a acceder a una vivienda digna, adecuada y asequible. En otras palabras, "acceder" puede significar también "permanecer"[16].

---

15 Este art. 1 bis fue añadido por el Real Decreto-ley 37/2020 de 22 de diciembre.

16 Así se expresa el art. 2 e) de la Ley:

"Constituyen fines comunes de la acción de los poderes públicos en materia de vivienda, en el ámbito de sus respectivas competencias:

(...)

e) Proteger la estabilidad y la seguridad jurídica en la propiedad, uso y disfrute de la vivienda, con especial atención a las personas y hogares en situación o riesgo de vulnerabilidad, y específicamente a familias, hogares y unidades de convivencia con menores a cargo, a través de medidas efectivas en materia de vivienda y asegurando la debida coordinación con medidas complementarias de atención social, formación, empleo y otras acciones de acompañamiento".

SÉPTIMA. Con la LDV se consagra de forma definitiva en la legislación estatal la distinción entre quien tiene la consideración de “gran tenedor” y quien no la tiene, cuando “tenedor” es un término que tradicionalmente se ha asociado con la posesión (así, arts. 430, 432, 441 y 463 CC) y no con la propiedad u otra titularidad sobre bienes, como se hace ahora.

En el específico ámbito de la vivienda está clara la procedencia de esta “cubertería jurídica”[17], que parece haber querido evitar los términos “gran propietario” con la evidente finalidad de incluir a todo “titular” de una vivienda y no solo al que la ostenta en este concepto, como claramente se deriva de la definición de “gran tenedor” en el art. 3 k) LDV. No obstante, esta distinción sigue conviviendo con la que discrimina entre persona física y persona jurídica para determinados efectos, y en ocasiones, ambas resultan aplicables de forma combinada (como en los arts. 10.2 y 17.7 LAU, por ejemplo).

Con independencia de los concretos conflictos que las diferencias conceptuales —estatal y autonómica— de “gran tenedor” están planteando ya[18], lo cierto es que por esta vía el legislador está diferenciando de una forma muy intensa

---

[17] El origen del término se encuentra en el art. 5.9 la Ley 24/2015, de 29 de julio, de medidas urgentes para afrontar la emergencia en el ámbito de la vivienda y la pobreza energética de Cataluña, y fue acogido posteriormente por el Estado en los Reales Decretos-leyes 11/2020, de 31 de marzo, y 15/2020, de 21 de abril.

[18] *Vid.* al respecto Fuentes-Lojo Rius (2024). Así, este autor da cuenta de que el día 9 de febrero se dictó un Acuerdo de Unificación de Criterios de las Secciones Civiles de la Audiencia Provincial de Barcelona considerando que el Instituto Catalán del Suelo y la Agencia Catalana de Vivienda no tienen la condición de grandes tenedores de vivienda en aplicación del concepto de gran tenedor contenido en el art. 5.9 de la Ley catalana 24/2015, dando prevalencia a la regulación autonómica sobre la estatal.

los deberes derivados de la función social de la propiedad en claro perjuicio de los grandes tenedores. Como dice Verdera Server, se asiste a un sustancial cambio en el estatuto jurídico de los bienes en función de quién sea su titular, cuando el art. 33 CE conecta la función social con el uso o destino del bien, no con quién sea su titular[19].

En la doctrina se había reclamado una más tajante discriminación entre grandes y pequeños propietarios en la LDV (por ejemplo, respecto a las medidas específicas aplicables a los contratos de alquiler), como un instrumento en términos de razonabilidad y proporcionalidad que sirvieran de justificación a las medidas intervencionistas del legislador, con la clara finalidad de proteger a ese modesto propietario para quien es vital la renta del alquiler[20]. Resulta notorio en nuestro país la adquisición de inmuebles, con fines de arrendamiento, como instrumento de ahorro e inversión por parte de amplias capas de la clase media, lo que debería ser también muy tenido en cuenta por parte del legislador[21], pues este pequeño ahorrador logra por esta vía complementar con su sacrificio los bajos sueldos y las escasas pensiones que caracterizan nuestro mercado laboral y el sistema público de pensiones.

Esta idea parece haber calado en el legislador estatal y resulta, desde luego, mucho más razonable que la simple distinción entre personas físicas y personas jurídicas, que no se ha abandonado como decimos. No obstante y respecto a esta distinción, también se había señalado en la doctrina que la vertiente individual (patrimonial) del derecho de propiedad de una vivienda conecta de una forma muy intensa con las personas

19 Verdera Server (2023, pp. 911-912).

20 López y López (2023, p. 33).

21 Al respecto, Messía de la Cerda (2020, pp. 3451-3453).

jurídicas a través de la libertad de empresa del art. 38 CE[22], lo que también ocurre en el caso de los grandes tenedores.

La inversión privada en vivienda es de todo punto necesaria para el equilibrio de nuestro mercado y desincentivarla de esta manera no contribuirá a mejorar la situación. Invertir en vivienda y ser titular (no necesariamente propietario), según los casos, de cinco u once inmuebles urbanos de uso residencial (destinados al alquiler, por ejemplo) no parece que merezca el castigo divino del legislador.

OCTAVA. Resulta extraño que la LDV prescinda de la tradicional distinción entre "límites" y "limitaciones" del derecho de propiedad. La justificación de ello parece estar en la creencia por parte del legislador de que, mediante el texto normativo aprobado, está simplemente cumpliendo con su declarado objetivo de concretar el contenido básico (esencial) del derecho de propiedad sobre una vivienda en el cumplimiento de su función social (los "límites" intrínsecos de esta propiedad específica). Así parece derivarse del art. 1.2 al declarar uno de los objetos de esta Ley (cursiva nuestra) de la siguiente manera:

> "Con objeto de asegurar el ejercicio del derecho a la vivienda, *será asimismo objeto de esta ley la regulación del contenido básico del derecho de propiedad de la vivienda en relación con su función social*, que incluye el deber de destinar la misma al uso habitacional previsto por el ordenamiento jurídico, en el marco de los instrumentos de ordenación territorial y urbanística, así como de mantener, conservar y rehabilitar la vivienda, atribuyendo a los poderes públicos la función de asegurar su adecuado cumplimiento, en el ámbito de sus respectivas competencias, a través de la aplicación de las medidas que legalmente procedan".

---

22 Simón Moreno (2023, p. 165).

En este sentido López y López se extraña de que el entonces proyecto no mencionara ni en una sola ocasión al instituto de la expropiación forzosa, pues este va unido a la idea de limitaciones del derecho de propiedad, no de límites ("dicho en términos más concretos y, según creemos, más útiles para la interpretación, allí donde se sobrepase la delimitación del derecho de propiedad, comienza la expropiación"). En realidad, este autor advierte de esta ausencia desde el principio: "el Proyecto desconoce absolutamente la expropiación forzosa, por un lado como si diera por supuesto que todas sus intervenciones en la propiedad o la libertad de contratar fueran meramente delimitadoras, y por otro, renunciando a la utilización de la figura, bien como instrumento directo de políticas de vivienda, bien como sanción al incumplimiento de los deberes que impone el Proyecto mismo".

Es muy clara también la conclusión final del autor: "llamativo resulta que la expresión «expropiación forzosa» no aparezca ni una sola vez en el Proyecto, y no menos curioso es que no se prevean para el caso de desocupación reiterada de la vivienda más que penalizaciones fiscales (disposición final tercera), siendo como es un clásico de la función social la exigencia de que el bien se dedique a su destino típico. El Proyecto de Ley ni siquiera se plantea una técnica similar; extraño, cuando de tantos vuelos (se ve que solo retóricos) e ínfulas se reviste. O tal vez no tan extraño: tal vez, a pesar de las proclamas, la Ley tiene objetivos muy modestos, que ellas no logran ocultar"[23].

Y es que, en un contexto de escasez de vivienda disponible, la expropiación temporal del uso de viviendas vacías o desocupadas, mediando la correspondiente indemnización, podría ayudar en alguna medida a incrementar el número de vivienda

[23] López y López (2022, pp. 26, 15 y 31, respectivamente).

social disponible, como un instrumento más de las políticas públicas de vivienda[24].

En relación a esta cuestión, causa perplejidad que, en cumplimiento de la función social de la propiedad que recae sobre una vivienda, las facultades (deberes para la LDV) de uso y disfrute de los propietarios queden concretadas por remisión a la normativa urbanística y en materia de vivienda que resulte de aplicación en cada momento, esto es, "conforme a su calificación, estado y características objetivas, de acuerdo con la legislación en materia de vivienda y la demás que resulte de aplicación, garantizando en todo caso la función social de la propiedad" (art. 11.1 a). En otras palabras, esa normativa de remisión "podrá **condicionar las facultades dominicales clásicas** (art. 348 CC), es decir, podrá decidir cuándo, cómo, quien, en qué circunstancias, el titular de una vivienda podrá usarla, disfrutarla, gravarla o disponer de ella". El resultado es que "**la vivienda dejará de ser un bien inmueble cualquiera**, que pueda usarse, disfrutarse, gravarse y enajenarse como el resto, sino que solo se pueden llevar a cabo **dichas facultades si tienen como fin satisfacer las necesidades básicas de alojamiento de las personas**; si no es así, se entenderá que es un **uso o disposición ilegal del inmueble**, con las **consecuencias civiles** (...) y **administrativas**"[25].

No se está hablando de cuestiones futuribles inimaginables, sino que ya incluso antes de la aprobación de la LDV existían propuestas políticas en el sentido de limitar las facultades de disposición de los propietarios en aquellas zonas declaradas como mercado residencial tensionado, de manera que únicamente sea posible la transmisión de viviendas cuando el adquirente la destine a vivienda habitual, con el objetivo de limitar

---

24 Sáinz-Cantero Caparrós (2023, pp. 656-657).

25 Nasarre Aznar (2022, págs. 7-8); negrita del autor.

el número de viviendas de uso turístico en dichas zonas al 2%. Otro partido político se ha sumado a esta iniciativa en 2024 y se quiere llevar a la LDV con ocasión de la aprobación de los próximos presupuestos generales del Estado (suponemos que los correspondientes al año 2025). Y otro más ha propuesto una prohibición temporal (durante 3 años) de la venta de viviendas a particulares y empresas extranjeras para luchar contra la especulación, según se dice.

Puede aceptarse que el legislador hable de "función social de la vivienda" y no de "función social de la propiedad (sobre una vivienda)", que sería lo correcto conforme al art. 33 CE[26]. De hecho, Díez-Picazo entendía que más que hablar de una función social de la propiedad de forma genérica, "lo que habría que pensar es que existe una función social de los bienes, porque es en cada especie de bienes donde esa concitación de intereses [de los propietarios y de terceras personas] se produce y donde se produce, además, en forma más nítida. De nuevo se puede pensar en la propiedad urbana, en la agraria, o en otras parecidas"[27]. Viene siendo habitual, por tanto, hablar de pluralidad de propiedades o de estatutos de propiedad.

Sin embargo, lo que no resulta admisible es que la letra del art. 33.2 ("de acuerdo con las leyes") permita hacer una remisión general a lo que las leyes, básicamente autonómicas, establezcan en cada momento respecto al contenido del derecho de propiedad y las facultades del propietario. De nuevo y siguiendo a Díez-Picazo, "no puede entenderse, en mi opinión, que haya una entera remisión de cualquier eventual contenido del derecho de propiedad a la simple acción legislativa. El hecho de reconocer el derecho con un nombre y unos apellidos, que

---

26 Verdera Server (2023, p. 905).

27 Díez-Picazo (2006, p. 21).

remiten a una tradición jurídica preconstitucional, parece que debe significar algunas cosas, que, en algún sentido, coartan la libertad del legislador antes de que pueda entrar en juego la denominada «función social». Así, debemos entender que el derecho de propiedad privada es una forma de atribución de los bienes, que no solo los coloca en los que en términos económicos se puede denominar el sector privado, sino que, además, los coloca en manos de particulares y permite a estos la adopción de medidas y decisiones respecto del destino de tales bienes o de la forma de extraer de ellos sus rendimientos o sus beneficios, y ello porque solo en este sentido resulta comprensible una institución que se denomine propiedad privada"[28].

La lógica de la LDV se mueve en una línea diametralmente opuesta, queriendo que el contenido esencial del derecho de propiedad sobre una vivienda (*ius utendi, ius fruendi, ius disponendi*) venga determinado por las normas urbanísticas y en materia de vivienda aprobadas por el legislador en cada momento. Desde esta perspectiva, la propiedad privada sobre una vivienda se parecerá cada vez más a una concesión administrativa, como había señalado algún autor con anterioridad[29].

NOVENA. Resulta llamativo el silencio del legislador sobre algunos temas que han ocupado un lugar central en las regulaciones autonómicas en materia de vivienda que han precedido a la LDV. En concreto, sobre las medidas para hacer frente al fenómeno de las viviendas vacías, deshabitadas o desocupadas y fomentar así el acceso de los ciudadanos a una vivienda. Varias CCAA habían establecido el deber de los propietarios de

---

28 Díez-Picazo (2006, pp. 18-19).

29 Díez-Picazo (1991, p. 1265). O como dice Colina Garea (1997, pp. 215-216), la propiedad dejaría de ser considerada un derecho subjetivo y su titular se convertiría en un mero funcionario dirigido en todo momento por la Administración.

destinar la vivienda a uso habitacional en cumplimiento de su función social y su incumplimiento podía dar lugar a multas o sanciones, a la obligación de ofrecer la vivienda en alquiler a determinadas personas (alquiler obligatorio en caso de ejecución hipotecaria o desahucio, en algunos casos con opción de compra o en forma de alquiler social), e incluso podía derivar en la expropiación temporal del uso de la vivienda, que fue la medida que más controversia jurídica generó.

Como es sabido, muchas de estas normas fueron objeto de recurso de inconstitucionalidad y el TC terminó declarando inconstitucionales algunas de ellas. En el tema de las viviendas desocupadas, la medida consistente en la expropiación temporal del uso (o usufructo) de la vivienda fue la que, claramente, mayor rechazo ha encontrado en el Alto Tribunal, precisamente por invadir la competencia exclusiva del Estado del art. 149.1.13.ª CE ("bases y coordinación de la planificación general de la actividad económica").

En el ámbito de las viviendas desocupadas, las principales sentencias que han recaído hasta este momento han sido: la STC 93/2015 de 14 de mayo (TOL5.001.983) frente a la Ley 1/2010, de 8 de marzo, reguladora del derecho a la vivienda en Andalucía (TOL1.791.221); la STC 16/2018 de 22 de febrero (TOL6.537.959) frente la Ley Foral 10/2010, de 10 de mayo, del derecho a la vivienda de Navarra (TOL1.833.475); la STC 32/2018 de 12 de abril (TOL8.485.295) frente a la Ley 4/2013, de 1 de octubre, de medidas para asegurar el cumplimiento de la función social de la vivienda de Andalucía (TOL3.954.706); la STC 43/2018 de 26 de abril (TOL6.599.093) frente a la Ley 2/2003, de 30 de enero de vivienda de Canarias (TOL238.774); la STC 97/2018 de 19 de septiembre (TOL6.814.938) frente a la Ley 3/2015, de 18 de junio, de vivienda del País Vasco (TOL5.181.949); la STC 80/2018 de 5 de julio (TOL6.673.371) frente a la Ley 2/2017, de 3 de febrero, por la función social de la vivienda de la Comunidad Valenciana (TOL5.951.896);

la STC 106/2018 de 4 de octubre (TOL6.861.895) frente a la Ley 2/2017, de 17 de febrero, de emergencia social de la vivienda en Extremadura (TOL5.960.918); la STC 8/2019 de 17 de enero (TOL7.028.467) frente a la Ley 4/2016, de 23 de diciembre, de medidas de protección del derecho a la vivienda de las personas en riesgo de exclusión social residencial de Cataluña (TOL5.918.568); y finalmente, la STC 16/2021 de 28 de enero (TOL8.310.392) frente al Decreto-ley 17/2019, de 23 de diciembre, de medidas urgentes para mejorar el acceso a la vivienda de Cataluña (TOL7.646.629).

Este deber de destinar la vivienda a uso habitacional aparece ahora recogido en el art. 11 a) de la LV como un deber de "uso y disfrute propios y efectivos de la vivienda conforme a su calificación, estado y características objetivas", pero no aparece anudado su incumplimiento a unas consecuencias jurídicas concretas. En particular, no se hace referencia alguna a la posibilidad de expropiación temporal del uso de la vivienda, que es una medida vetada a las CCAA por parte del TC, que la ha reservado a la competencia del Estado.

Por eso, no es fácil de entender que, en una ley estatal reguladora del derecho a la vivienda, el Estado haya prescindido de esta medida como un complemento a las medidas autonómicas respecto a las viviendas vacías o desocupadas. Ni siquiera puede pensarse en razones presupuestarias para prescindir de ella, pues la previsión que pudo haberse aprobado no obligaba a su ejercicio sino en la medida en que la dotación presupuestaria lo permitiera.

No es que estemos a favor de esta medida como una medida eficaz para hacer frente a la escasez de vivienda, sino que resulta llamativa su ausencia en una ley apoyada por los mismos partidos políticos que la han planteado en el ámbito autonómico y que había sido rechazada por el TC por quedar reservada precisamente a la competencia estatal. En realidad, el único precepto que se refiere a las viviendas vacías o deshabitadas es el art. 34 y lo hace en el contexto de las obligaciones estatales de información y transparencia en materia de vivienda (aparte de la modulación del recargo a este tipo de viviendas por la disposición final tercera). El Estado a lo único que se compromete es a ofrecer información sobre las viviendas vacías que se hallen en su ámbito territorial, de manera que esta información sea publicada anualmente en su sede electrónica, nada más.

Quizás y en efecto, puede afirmarse que se trata de una ley con unos objetivos reales mucho más modestos que los publicitados en su preámbulo.

DÉCIMA. Se aprecia en la LDV un importante esfuerzo por parte del legislador en orden a justificar su acomodo dentro del marco competencial atribuido al Estado en esta materia, consciente de la asunción por las CCAA de la competencia plena en materia de "ordenación del territorio, urbanismo y vivienda" conforme al art. 148.1.3 CE.

Por ejemplo, en el propio preámbulo de la Ley se citan hasta tres sentencias del TC del año 2018 (la 16/2018, la 32/2018 y la 43/2018) que son utilizadas para reprochar al Estado la falta de ejercicio de la competencia a él atribuida por el art. 149.1.1.ª: "la regulación de las condiciones básicas que garanticen la igualdad de todos los españoles en el ejercicio de los derechos y en el cumplimiento de los deberes constitucionales", en relación a la vivienda. Viene a decirse que el TC encuentra dificultades para delimitar las competencias de las CCAA al no disponer de los límites que debieran ser establecidos por el Estado en el ejercicio de sus propias competencias, entre ellas, esta en concreto.

En realidad, todo el apartado II del preámbulo se dedica a exponer y justificar los distintos títulos competenciales (exclusivos) del art. 149.1 CE que permiten al Estado incidir en esta materia. Aparte de las ya mencionadas materias 1.ª y 13.ª, se citan las siguientes: legislación mercantil (6.ª), legislación civil (1.ª), bases de la ordenación de crédito, banca y seguros (11.ª), hacienda general y deuda del Estado (14.ª), legislación básica sobre protección del medio ambiente (23.ª) y bases de régimen minero y energético (25.ª).

Precisamente, la disposición adicional séptima residencia en las materias 1.ª y 13.ª (la regulación de las condiciones básicas que garanticen la igualdad de todos los españoles en el ejercicio de los derechos y en el cumplimiento de los deberes constitucionales y las bases y coordinación de la planificación general de la actividad económica, respectivamente) los principales títulos competenciales que sirven de base a esta Ley. Se exceptúan de lo anterior algunos preceptos concretos que se declaran justificados en la competencia sobre legislación civil, procesal y en materia de hacienda general.

Pero resulta muy dudoso el encaje general de la Ley en los dos títulos competenciales señalados como principales, de manera que se corre el riesgo de que algunos preceptos puedan ser declarados inconstitucionales, como había puesto de manifiesto Nasarre Aznar[30]. Primero, porque, en realidad y aunque se anuncia como el principal objeto de la Ley (art. 1.1), no existe en ella una regulación clara de los derechos y obligaciones que corresponden a todos los españoles y les permitan un acceso a la vivienda en unas condiciones básicas de igualdad; el legislador no regula aquello que dice que va a regular sobre la base de la competencia que le otorga el art. 149.1.1.ª.

Además, resulta también muy dudoso que puedan incluirse en el concepto de "condiciones básicas" aspectos tales como las reservas de suelo (art. 15) o la declaración de zonas tensionadas (art. 18). En el texto definitivamente aprobado sí se

---

30 Nasarre Aznar (2022, pp. 4-6). Extensamente sobre el problema competencial en la LDV, Prats Albentosa (2022, pp. 154 y ss).

ha salvado esta posible tacha de inconstitucionalidad respecto a las definiciones del art. 3, tal y como aparecían en el anteproyecto, pues ahora se ha explicitado que, en caso de contradicción con las definiciones de las regulaciones autonómicas, prevalecerán estas últimas. Esto es importante, por ejemplo, para la definición de "gran tenedor" (letra k).

Estas dudas habían sido ampliamente puestas de manifiesto por el CGPJ, que dedicó una buena parte de su informe a la cuestión competencial (puntos 25-66)[31]. En concreto, respecto a la competencia del art. 149.1.1.ª, el CGPJ había considerado que la intervención del Estado no podía agotar la regulación de la materia, sino que tenía que ser la mínima imprescindible, de manera que las CCAA pudieran desplegar sus políticas propias en la materia (punto 43). También sobre la competencia estatal residenciada en el art. 149.1.13.ª había realizado unas consideraciones semejantes a la luz de la jurisprudencia del TC (básicamente punto n.º 45).

Respecto a la regulación del contenido básico del derecho de propiedad de la vivienda en relación con su función social, como segundo objeto de la ley (art. 1.2), aquí el problema se plantea respecto a esa posible y pretendida por el legislador "mutación" del régimen civil del derecho de propiedad hacía su "administrativización", como una expropiación encubierta y sin la garantía expropiatoria. La base competencial de los arts. 10 y 11 es claramente la legislación civil, pero la posible inconstitucionalidad puede venir por esta razón: convertir a la vivienda en un servicio público más desvirtuando su contenido esencial sin la correspondiente compensación a los propietarios. No obstante y dada la remisión hecha por el art. 11 a la

---

31 Amplio análisis de la cuestión competencial y del contenido del informe del CGPJ y de sus votos particulares en este punto en Calduch Alonso (2023, pp. 1456 y ss.).

"legislación en materia de ordenación territorial y urbanística de aplicación" es posible interpretar que el juicio sobre la constitucionalidad de esta ley es una tarea que debe quedar aplazada hasta conocer esos futuros desarrollos de la misma[32].

Aparte de ello y en el ámbito civil, también debe llamarse la atención sobre los arts. 32 a 36 de la Ley, dentro del Capítulo II ("información y transparencia en materia de vivienda y suelo") del Título IV, que se dicen amparados en la competencia estatal sobre legislación civil (disposición final séptima 2.a). Esto resulta incomprensible, porque son normas que se refieren a obligaciones de información del Estado respecto a su parque público de viviendas, a la inversión que realiza en programas de política de vivienda, sobre viviendas vacías o deshabitadas de su titularidad, sobre la demanda de vivienda y sobre el suelo público estatal disponible para vivienda. Estas normas no pueden estar amparadas en la competencia estatal sobre la legislación civil; se trata, sin duda, de una errata, que tiene su origen en el hecho de que estos preceptos regulaban en su momento en el proyecto de ley la percepción de cantidades a cuenta durante la construcción de las viviendas, dentro de un capítulo que posteriormente fue desechado en el texto definitivamente aprobado.

En cualquier caso, se interpusieron varios recursos de inconstitucionalidad entre 2023 y 2024 contra la LDV por razones básicamente competenciales (Gobiernos andaluz, gallego, catalán, vasco, de la Comunidad de Madrid, Baleares, Parlamento de Cataluña y Grupo Parlamentario del Partido Popular) y a día de hoy ya conocemos el primero de los pronunciamientos del TC, consecuencia de un recurso de inconstitucionalidad presentado contra la LDV por el Presidente de la Junta de Andalucía. Fi-

---

32 López y López (2022, p. 32).

nalmente, la STC 79/2024 de 21 de mayo (TOL10.040.324), ha declarado inconstitucionales los arts. 16 (sobre vivienda protegida), el inciso segundo del art. 19.3 (información a suministrar por los grandes tenedores a las AAPP), el art. 27 (parques públicos de vivienda) y la DT primera (régimen jurídico de las viviendas de protección pública). Si esta postura se reitera por el TC en los sucesivos pronunciamientos que están por venir, podrá afirmarse que las normas Derecho privado de la LDV aprobadas por el legislador se mantienen en sus términos.

UNDÉCIMA. Finalmente, ¿qué aporta, realmente, la LDV?

Para tratarse de una ley con la declarada pretensión de, nada más y nada menos, que alcanzar los 7 objetivos enumerado en su preámbulo[33], lo cierto es que detrás de tantas procla-

---

[33] Estos son los objetivos declarados en su preámbulo, como decimos: "Dentro de este marco, los objetivos perseguidos por la ley son los siguientes:

-Establecer una regulación básica de los derechos y deberes de los ciudadanos en relación con la vivienda, así como de los asociados a la propiedad de vivienda, aplicable a todo el territorio nacional.

-Facilitar el acceso a una vivienda digna y adecuada a las personas que tienen dificultades para acceder a una vivienda en condiciones de mercado, prestando especial atención a jóvenes y colectivos vulnerables y favoreciendo la existencia de una oferta a precios asequibles y adaptada a las realidades de los ámbitos urbanos y rurales.

-Dotar de instrumentos efectivos para asegurar la funcionalidad, la seguridad, la accesibilidad universal y la habitabilidad de las viviendas, garantizando así la dignidad y la salud de las personas que las habitan.

-Definir los aspectos fundamentales de la planificación y programación estatales en materia de vivienda, con objeto de favorecer el ejercicio del derecho constitucional en todo el territorio.

-Regular el régimen jurídico básico de los parques públicos de vivienda, asegurando su desarrollo, protección y eficiencia para

mas y ensoñaciones "habitacionales", resulta finalmente una ley tan modesta como prácticamente prescindible. La espera nunca fue tan decepcionante.

En este sentido, las principales novedades y medidas de más calado que pueden señalarse no aparecen en su articulado, sino en disposiciones adicionales, finales y transitorias. Esto es muy significativo, a nuestro juicio, porque revela con claridad que las principales medidas podían haberse aprobado al margen de una rimbombante Ley "por el derecho a la vivienda", como se venía haciendo ya de hecho. Entre estas medidas de más calado situadas fuera del cuerpo central de la LDV, podemos citar las siguientes:

a) La futura base de datos de contratos de arrendamiento de vivienda (DA primera)

b) La autorización a las CCAA para que puedan aplicar los recursos de los planes estatales en materia de vivienda en los trámites de intermediación y conciliación, y lo que es más importante, en las compensaciones a los propietarios en caso de suspensión de los procesos de desahucio y lanzamiento (DA cuarta).

c) Las normas relativas a los administradores de fincas (DA sexta).

d) Las reglas transitorias aplicables a los procedimientos de desahucio y lanzamiento que se hallaren suspendi-

---

atender a aquellos sectores de la población con mayores dificultades de acceso.

-Favorecer el desarrollo de tipologías de vivienda adecuadas a las diferentes formas de convivencia y de habitación, favoreciendo la adaptación a las dinámicas y actuales exigencias de los hogares.

-Mejorar la protección en las operaciones de compra y arrendamiento de vivienda, introduciendo unos mínimos de información necesaria para dar seguridad y garantías en el proceso".

dos conforme a los arts. 1 y 1 bis del Real Decreto-ley 11/2020, de 31 de marzo (DT primera).

e) Las modificaciones introducidas en la LAU para la contención de los precios de los alquileres o en relación a la duración de los contratos (DF primera).

f) El recargo en el IBI a los inmuebles de uso residencial desocupados (DF tercera).

g) Los nuevos requisitos introducidos en la regulación de la LEC de los procesos de desahucio y lanzamiento que afecten a personas vulnerables (DF quinta).

h) La limitación (¿extraordinaria?) a la actualización de la renta en los contratos de arrendamiento de vivienda (DF sexta).

¿Era necesaria toda una LDV para aprobar estas medidas de mayor impacto? Sinceramente, creemos que no. Por otra parte, muchos de los preceptos que conforman el cuerpo central de la Ley referidos a otras cuestiones, o aportan poco al desarrollo legislativo ya alcanzado a través de otras normas situadas fuera de la LDV, o suponen un claro retroceso.

Por ejemplo, poco aportan los arts. 30 y 31 respecto a los derechos e información básica de los demandantes en las llamadas operaciones de compraventa y arrendamiento de vivienda, pues el legislador no hace sino intentar reproducir —de forma técnicamente muy deficiente y planteando nuevos problemas interpretativos— deberes precontractuales y contractuales básicamente ya existentes en la legislación de consumo o en normas generales, estatales o autonómicas. La norma era necesaria y podía haberse aprovechado la ocasión para reforzar estos deberes de información a los oferentes, pero delimitando claramente los contratos a los que se aplican, los deberes exigibles en cada momento del proceso de contratación y estableciendo unas consecuencias jurídicas a su incumplimiento.

Como otra oportunidad perdida puede tildarse también la regulación de los administradores de fincas establecida en la DA sexta, pues ni siquiera se aprovecha por el legislador para aclarar de una forma definitiva la exigencia o no de colegiación obligatoria a los administradores externos de las comunidades de propietarios mencionados en el art. 13 LPH, que viene siendo una cuestión discutida. La LDV solo menciona a los administradores de fincas profesionales, respecto a los que sí parece existir acuerdo sobre la obligación de estar colegiados.

El resto de apartados de esta DA sexta no hacen sino recoger obligaciones ya exigibles en otras normas. En este sentido, hubiera sido interesante recoger en la LDV unas previsiones específicas para los agentes de la propiedad inmobiliaria en términos de capacitación y colegiación, más allá de los deberes de información mínima impuestos en sus arts. 30 y 31.

Otro tanto ocurre con el art. 6 sobre igualdad y no discriminación en el acceso a la vivienda, que no supone novedad alguna respecto a la protección ya dispensada en general por la Ley 15/2022, de 12 de julio, integral para la igualdad de trato y no discriminación (TOL9.113.969), incluido su art. 20 dedicado al específico ámbito de la vivienda. Asimismo, la técnica jurídica de la que se vale el legislador de la LDV para hacer frente a las discriminaciones directas o indirectas que puedan producirse consiste en un mandato general a las Administraciones Pública para que adopten las medidas de protección necesarias (art. 20.2). Se olvida, nuevamente, de los importantes efectos del principio de igualdad y no discriminación en el ámbito contractual, que deberá ser completado por las previsiones de la Ley 15/2022 (o también por la Ley 4/2023, de 28 de febrero, para la igualdad real y efectiva de las personas trans y para la garantía de los derechos de las personas LGTBI; TOL9.421.382).

Además y aparte del acoso inmobiliario (que sí puede generar situaciones de discriminación), se incluye en el art. 6

la infravivienda, la vivienda sobreocupada y otros casos de alojamiento ilegal o de falta de titularidad que no termina de adivinarse qué relación presentan con el principio de igualdad y no discriminación en materia de vivienda.

La "guinda" la constituyen los arts. 8 y 9 sobre los derechos y deberes "del ciudadano" en relación a la vivienda. Se aglutinan una serie de derechos y deberes de diversa naturaleza (pública y privada), sin hilo conductor entre ellos y sin concretar el supuesto y de hecho y la consecuencia jurídica de cada supuesta norma. En definitiva y para hacerse una idea del alcance de cada uno de estos derechos y deberes, habrá de acudirse a la legislación sectorial reguladora de cada uno de ellos, como si la LDV no existiera a estos efectos.

No obstante, este galimatías jurídico no puede ocultar la realidad de la cuestión central: la LDV no parece que suponga avance alguno en el reconocimiento del derecho a acceder a una vivienda digna y adecuada del art. 47 CE como un derecho subjetivo exigible directamente por todos los ciudadanos con un concreto contenido y unas específicas condiciones de ejercicio. El reconocimiento de este derecho subjetivo, o el establecimiento de unas bases comunes para ello que posteriormente las CCAA podrían desarrollar, hubiera justificado la aprobación de una ley que hubiera merecido el nombre de "ley por el derecho a la vivienda"[34]. Tampoco esto puede ser apuntado en el haber del legislador[35].

---

34 Salvo que se entienda que tal derecho subjetivo directamente exigible ante los tribunales de justicia (art. 24.1 CE) ya existe en nuestro Derecho desde la Ley de Suelo de 2007 (art. 5-a del TR de la Ley de Suelo de 2015): López Ramón (2020, pp. 297-308). Expresamente en contra de otorgar tal alcance a esta norma, Souvirón Morenilla (2018, p. 188).

35 Nasarre Aznar (2022, p. 5); Busto Lago (2023, pp. 2-9); Prats Albentosa (2022, pp. 149-150).

REFERENCIAS BIBLIOGRÁFICAS

Aparicio Wilhelmi, M. (2022). "La vivienda como derecho y la limitación de la propiedad privada y del mercado: comentarios a propósito de la Ley catalana 11/2020 de contención de precios del alquiler". *Regular los alquileres. La lucha por el derecho a una vivienda digna en España* (coord. Noguera Fernández). Tirant lo Blanch, 57 y ss.

Argelich Comelles, C. (2023). *Ley por el derecho a la vivienda.* Aranzadi.

Busto Lago, J. M. (2023). "La Ley por el derecho a la vivienda y el derecho subjetivo a la vivienda". *Cuadernos de Derecho Privado,* n.º 6, 2-9.

Calduch Alonso, M. (2023). "La distribución de competencias en materia de vivienda entre el Estado y las comunidades autónomas (A propósito de la Ley por el derecho a la vivienda)". *Revista Crítica de Derecho Inmobiliario,* n.º 797, 1447 y ss.

Colina Garea, R. (1997). *La función social de la propiedad privada en la Constitución Española de 1978.* Bosch.

Díez-Picazo, L. (2006). "Propiedad y Constitución". *Propiedad y Derecho civil* (coord. Guilarte Gutiérrez). Colegio de Registradores de la Propiedad y Mercantiles de España, 13-22.

Díez-Picazo, L. (1991). "Algunas reflexiones sobre el derecho de propiedad privada en la Constitución". *Estudios sobre la Constitución Española. Homenaje al profesor Eduardo García de Enterría* (coord. Martín-Retortillo). Tomo II. Civitas, 1257-1270.

Fuentes-Lojo Rius, A. (2024). "Guía interpretativa sobre el concepto de gran tenedor de viviendas". *Diario La Ley,* n.º 10486, Sección Tribuna, 16 de abril de 2024, 1-10.

Gálvez Criado, A. (2023). "Los límites impuestos a las facultades de uso y disfrute de los propietarios para luchar contra la escasez de vivienda disponible en alquiler en España". *Actualidad Jurídica Iberoamericana,* n.º 19, 334 y ss.

García Rubio, M.ª P. (2022). "Reivindicando el valor del Derecho Civil. El tratamiento del Derecho civil por los poderes normativos". *Revista de Derecho Civil,* n.º 1, 233 y ss.

Garrido Mayol, V. (2022). "Hacia una nueva configuración del derecho constitucional a la vivienda". *Vivienda y colectivos vulnerables* (dirs. Cervilla Garzón/ Zurita Martín). Aranzadi, 267-296.

López Ramón, F. (2020). "El reconocimiento legal del derecho a disfrutar de una vivienda". *Revista de Administración Pública,* n.º 212, 297-308.

López y López, A. M.ª (2022). "Propiedad privada y disciplina del mercado en el Proyecto de Ley por el derecho a la vivienda". *Cuadernos de Derecho Local,* n.º 59, 14-34.

Magro Servet, V. (2023). "Análisis práctico y sistemático de los aspectos relevantes de la nueva Ley de vivienda 12/2023 de 24 de mayo". *Diario La Ley,* n.º 10295, 26 de mayo de 2023, 1-59.

Messía de la Cerda, J. A. (2020). "La función social de la propiedad de la vivienda y las normativas autonómicas". *Revista Crítica de Derecho Inmobiliario,* n.º 782, 3417 y ss.

Nasarre Aznar, S. (2022). "El proyecto de Ley de vivienda 2022". *Apuntes* 2022/11 (Fedea), 1-25.

Noguera Fernández, A. (2022). "Regular los alquileres en un escenario hostil: la función social de la propiedad y la lucha por el derecho a la vivienda en España". *Regular los alquileres. La lucha por el derecho a una vivienda digna en España* (coord. Noguera Fernández).Tirant lo Blanch, 15 y ss.

Prats Albentosa, L. (2022). "Derecho a gozar de una vivienda digna y adecuada y derecho a la propiedad de vivienda: a propósito del Proyecto de Ley por el derecho a la vivienda". *Revista Jurídica del Notariado,* n.º 115, 131-174.

Sáinz-Cantero Caparrós, M.ª B. (2023). "La redefinición del derecho de propiedad sobre la vivienda en la era de los bienes comunes. A propósito de la Ley 12/2023 del derecho a la vivienda". *Actualidad Jurídica Iberoamericana,* n.º 19, 630-663.

Simón Moreno, H. (2023). "La evolución constitucional de la función social de la propiedad y el nuevo régimen del derecho de propiedad sobre una vivienda en la Ley por el derecho a la vivienda". *Constitución y Derecho privado,* n.º 42, 139 y ss.

Souvirón Morenilla, J. M.ª (2018). "El derecho a la vivienda y su garantía pública: entre el servicio público y la protección social". *Servicios de interés general y protección de los usuarios: (educación, sanidad, servicios sociales, vivienda, energía, transportes y comunicaciones electrónicas* (coord. González Ríos). Dykinson,181-224.

Verdera Server, R. (2023). "*Pro proprietate.* Notas sobre la configuración constitucional de la propiedad privada". *Anuario de Derecho Civil,* Vol. III, 859 y ss.